Space Radiation

Astrophysical origins, radiobiological effects and implications for space travellers

Online at: https://doi.org/10.1088/978-0-7503-5444-8

About the Series

The series in Physics and Engineering in Medicine and Biology will allow the Institute of Physics and Engineering in Medicine (IPEM) to enhance its mission to 'advance physics and engineering applied to medicine and biology for the public good'.

It is focused on key areas including, but not limited to:
- clinical engineering
- diagnostic radiology
- informatics and computing
- magnetic resonance imaging
- nuclear medicine
- physiological measurement
- radiation protection
- radiotherapy
- rehabilitation engineering
- ultrasound and non-ionising radiation.

A number of IPEM–IOP titles are being published as part of the EUTEMPE Network Series for Medical Physics Experts.

A full list of titles published in this series can be found here: https://iopscience.iop.org/bookListInfo/physics-engineering-medicine-biology-series.

Space Radiation

Astrophysical origins, radiobiological effects and implications for space travellers

James S Welsh

Department of Radiation Oncology, Loyola University Stritch School of Medicine, Maywood, IL, USA

IOP Publishing, Bristol, UK

Permission to make use of IOP Publishing content other than as set out above may be sought at permissions@ioppublishing.org.

James S Welsh has asserted his right to be identified as the author of this work in accordance with sections 77 and 78 of the Copyright, Designs and Patents Act 1988.

ISBN   978-0-7503-5444-8 (ebook)
ISBN   978-0-7503-5440-0 (print)
ISBN   978-0-7503-5441-7 (myPrint)
ISBN   978-0-7503-5443-1 (mobi)

DOI   10.1088/978-0-7503-5444-8

Version: 20240501

IOP ebooks

British Library Cataloguing-in-Publication Data: A catalogue record for this book is available from the British Library.

Published by IOP Publishing, wholly owned by The Institute of Physics, London

IOP Publishing, No.2 The Distillery, Glassfields, Avon Street, Bristol, BS2 0GR, UK

US Office: IOP Publishing, Inc., 190 North Independence Mall West, Suite 601, Philadelphia, PA 19106, USA

This book is dedicated to my parents, Lee J Welsh and William F Welsh, Sr.

Contents

18 Planet Earth

Preface

Our Universe is imbued with radiation from all angles. And we receive radiation from the Universe from all angles. Some of this radiation has been around from the dawn of time and new sources are found daily. Space radiation reaches us and bathes us here on Earth's surface in the form of cosmic rays. Space radiation was locked into the rocks which formed our planet at its inception and now contributes to our natural background radiation. And of course, space radiation is of importance to future space travelers and colonizers. Although most science historians say that the 'space age' began with the launch of Sputnik 1 in October 1957, arguably, the first breakthrough of the space age was James Van Allen's 1958 discovery of the Earth's radiation belts through instruments he designed for Explorer 1. Besides the challenges posed by our Van Allen belts, beyond Earth's protective atmosphere and magnetic field, space travelers will have to contend with unmitigated galactic cosmic rays, the steady solar wind, and unpredictable solar particle events.

This textbook is appropriate for a wide range of readers, including mid- to upper-level undergraduate students in astronomy, physics, chemistry, and biology. Students considering medical physics, radiation oncology, and radiation therapy may also find much of this material to be relevant and appealing. It could also serve admirably as the primary text for a dedicated course on space radiation or for an introductory course on space weather.

To those at any level who are interested in radiobiology, ranging from undergraduate and graduate students to active researchers, you may find much of this information stimulating and useful. To those considering careers in health physics, you will find this material to be especially fascinating and relevant.

Educators and researchers in the field of astrobiology may find several of the chapters to be thought-provoking. Of course, general readers who want an introduction to the subject of space radiation may find this book alluring. Those curious about space travel will learn that microgravity, frigid temperatures and the absence of air, along with the anticipated hazards of astounding acceleration, supersonic speeds, and rough landings are not the only concerns for astronauts and future space colonizers.

The book begins with two chapters that provide a broad overview of space radiation in our immediate vicinity, our Solar System, the rest of the Milky Way galaxy, and the Universe at large. Radiation permeates our Universe from the small to the large, and from the near to the far. We cannot escape it since our Earth and Moon are radioactive—and for that matter, so are we, our pets, and our food. The final chapter discusses space radiation over the entirety of time—from the Big Bang to the end of time in the Dark Era.

Chapters 3 and 4 provide a review of relevant astronomy, and basic physics and chemistry that may be helpful to those who have not recently studied these topics.

Chapters 5 and 6 cover radioactivity and radiation physics in greater detail. These chapters set the stage for the more directed discussions on the various planetary, astronomical, and cosmological sources of radiation in the rest of the text. As some of the material in these two chapters builds on some basic concepts in quantum theory, students may wish to review the material in chapter 7 prior to delving into the detailed radiation science discussions in chapters 5 and 6.

Since ionizing radiation is composed of photons and subatomic particles, the subject naturally deals with the science of the very tiny; that is, it is in the realm of quantum mechanics. Chapter 7 provides a brief introduction to relevant topics in quantum theory and prepares readers for the specialized discussions in subsequent sections.

Because the subject of space radiation routinely alludes to leptons, hadrons, neutrinos, bosons and fermions, chapter 8 introduces the Standard Model of particle physics and the associated terminology.

Chapter 9 reviews relevant topics in radiation chemistry, including free radicals, reactive oxygen species and reactive nitrogen species with a particular emphasis on aqueous radiation chemistry.

Chapters 10 and 11 focus on radiation biochemistry and molecular radiation biology. These chapters provide a general introduction to radiobiology for students of physics and astronomy who may be unfamiliar with these important topics.

Chapter 12 covers the cellular biology of respiration with a general review at the start and an emphasis on radiation-related oddities of interest.

Chapter 14 covers radiation-resistant organisms. While most of these are prokaryotes, a handful are metazoans. To prepare readers for this discussion, chapter 13 was written to provide a brief overview of invertebrate zoology for anyone who might benefit from such a review. Chapter 14 also addresses radiophilic fungi and the theoretical concept of radiosynthesis.

Chapter 15 provides an overview of the relevant astrophysics behind the thermonuclear fusion which supports stars during their constant battle with gravity, as well as the fates of stars as their battles come to an end. Sometimes these battles end with enormous radiation-spewing supernova explosions. During these supernova explosions (and during neutron star collisions) heavy radioactive elements may be created through the various processes reviewed here.

Chapter 16 reviews the structure and physics of our Sun, with an emphasis on the radiation it emits today, in the distant past, and in the far future.

Chapter 17 covers cosmic rays, the primary form of ionizing radiation in outer space.

Chapter 18 provides a general geology review, including some coverage of radiogenic heating and natural nuclear reactors.

A major topic in space radiation is space weather. Chapter 19 introduces the essential ideas behind space weather and provides some additional details on the Van Allen belts, the ionosphere, and radiation in the International Space Station.

Each of the chapters on the various planets, moons, and asteroids (chapters 20–26) begins with a general introduction to the celestial body in question but then focuses on the scientific aspects of the radiation environment on these entities.

Chapter 27 briefly reviews aspects of our Van Allen belts such as the South Atlantic anomaly and covers the concept of radiation belts around exoplanets and ultracool dwarfs.

Chapter 28 is a fairly long chapter which reviews the history of life on Earth. Students and other readers who are already well-acquainted with paleontology may prefer to skip this review until the final section on the Laschamps geomagnetic excursion.

Chapter 29 is an aside on our climate over Earth's history and the possible short- and long-term influence of space radiation on climate.

Chapter 30 focuses on a few mass extinction events over the Phanerozoic Eon. The discussion includes the Ordovician extinction, which was probably caused by plate tectonics-related climate changes but may have also been provoked by a gamma ray burst. The subject of gamma ray bursts, which signify the birth of black holes, is then covered in greater detail in chapter 31. Various aspects of black holes, including their possible death through the emission of (still hypothetical) Hawking radiation are covered in chapter 34.

Chapters 32 and 33 deal with various sources of x-rays and gamma rays in outer space, including the recently discovered Fermi bubbles, along with x-ray pulsars, microquasars, and active galactic nuclei such as blazers and quasars. Special attention is directed to magnetars, which power soft gamma repeaters and giant gamma ray flares when they undergo starquakes.

The book ends with a review of the radiation-filled first few minutes of the Universe, the Quark, Hadron and Lepton Epochs, neutrino decoupling, and Big Bang nucleosynthesis. The chapter also covers the Radiation Epoch, recombination (which released the previously trapped photons that now roam the Universe as the cosmic microwave background radiation), and the various other chronological periods until the final phases of the Universe. And according to one model, the Universe finishes as an endless sea filled with nothing but radiation. An apt ending indeed!

Acknowledgments

The list of people who have assisted in this project is extensive. Some of them contributed directly, whereas other folks helped indirectly (perhaps without them even realizing how helpful they were). Among those I wish to formally thank are all my professional colleagues at the Loyola University Department of Radiation Oncology, especially Abhishek Solanki, Alec Block, Bill Small, Anil Sethi and John Roeske. Chief physicist Niranjan Bhandare and the members of the medical physics team at Edward Hines Jr VA Hospital including David Johnstone and Jaya Matthew should be thanked for their thought-provoking comments and their patience with me throughout this long project.

Others who have been influential in the generation of this work include several members of SARI (Scientists for Accurate Radiation Information), including but not limited to Bill Sacks, Greg Meyers, Rick Sanders, Wade Allison, Gary Hoe. I would also like to thank Brant Ulsh, Heath Foxlee and Gayle Woloschak for their stimulating and inspiring discussions. Special thanks also go out to Mark Palmer of Brookhaven National Laboratory and to my friends and colleagues at Fermilab including Mary Anne Cummings, Tom Kroc, Maureen Hix, Amanda Early, and Carol Johnstone. I also appreciate the helpful email input from Bobby Scott over the years. Brian Thomas has also been helpful through his advice on my public presentations at the Fermilab 'Ask a Scientist' program. I would also like to express my gratitude to astrobiologist and textbook author, Jonathan Lunine for the instructive discussions on various astrobiology and space radiation topics over the years. A particular thank you goes to author, scientist, medical physicist and inventor, Thomas 'Rock' Mackie for the encouragement and the countess conversations that have contributed to this project. I would also like to thank Krzysztof Fornalski, Marek Janiak and Michael Waligórski of the UNSCEAR delegation from Poland for their invaluable comments and scientific contributions. Additionally, I want to thank 'Big Professor Arnuwandas' (Christian Debranin) for the inspiration over the years.

Special thanks go to my laboratory colleague Noy Rithidech for her support and collaboration on this and several other exciting scientific research projects. Additionally, I want to thank my friends and colleagues, SMJ (Javad) Mortazavi, Joseph Bevelacqua, Lembit Sihver, and Alireza Mortazavi for the many successful and enjoyable collaborative efforts on various space radiation related projects. Similarly, I wish to acknowledge the many invaluable conversations and professional collaborations with Andy Karam and Robert Peter Gale.

I also wish to thank my colleagues at Saint James School of Medicine for their inspiring conversations, including the chancellor Jose Ramirez, the dean Claude Iliou, as well as Kaushik Guha, Raj Mitra, among others. From Northern Illinois University, I want to thank George Coutrakon, Linda Yasui, Nick Karonis, Kirk Duffin, Vishnu Zutshi and Larry Lurio for their support and ideas. Special thanks also go to Fritz and Ethan DeJongh, and Victor Rykalin of Proton VDA for their many interesting contributions. I should also specifically thank Reinhard Schulte for his encouragement and for introducing me to IOP Publishing.

Finally, very special thanks go out to my sister Anna and my brother (and stellar astronomer) Bill for their constructive criticism and advice and extremely helpful reviews of the drafts.

And perhaps most importantly, I wish to acknowledge my loving and supportive wife, Teri, for her tolerance and support during this effort, especially at times when it looked like there was no end in sight.

Author biography

James S Welsh

Dr James Welsh is presently the Chief of Service at the Edward Hines Jr VA Hospital and Professor of Radiation Oncology at the Loyola University Stritch School of Medicine. He is an advisor to UNSCEAR with the Polish delegation and was a past member of the ACMUI (Advisory Committee for the Medical Uses of Isotopes) under the United States Nuclear Regulatory Commission (NRC). He was a medical consultant for the NRC as well. Dr Welsh is a former faculty member at the Johns Hopkins School of Medicine and was a professor of Neurosurgery, Radiology and Radiation Oncology at LSU Shreveport. He was also a clinical professor of Medical Physics and Human Oncology at the University of Wisconsin. He was a fast neutron therapy specialist at the former NIU Institute for Neutron Therapy at Fermilab and is an adjunct professor of Physics at Northern Illinois University, where he taught an Honors Astrobiology Seminar. He is currently a Visiting Professor with Saint James School of Medicine. Dr Welsh is a member of ASTRO (the American Society of Radiation Oncology) and the AAPM (American Association of Physicists in Medicine) and is a Fellow of the American College of Radiology. He is also a Fellow of the American College of Radiation Oncology and a past president and Gold Medal recipient. He is the author of over 500 peer reviewed papers, abstracts, and book chapters as well as the popular book, *Sharks Get Cancer, Mole Rats Don't: How animals could hold the key to unlocking cancer immunity in humans.* He is the recipient of several teaching awards and is board certified in radiation oncology and neuro-oncology. Dr Welsh obtained his MD at the State University of New York at Stony Brook and his MS in Molecular Biophysics and Biophysics at Yale University. He did his residency training in radiation oncology at Johns Hopkins.

Disclaimer:

The contents of this work do not represent the views of the U.S. Department of Veterans Affairs or the United States Government.

IOP Publishing

Space Radiation
Astrophysical origins, radiobiological effects and implications for space travellers
James S Welsh

Chapter 1

An introduction to space radiation

Given that radiation infuses our entire Universe, the subject of 'space radiation' is dauntingly vast. In fact, space radiation is literally as vast as the Universe itself. For starters, the omnidirectional and omnipresent 2.73 K cosmic microwave background (CMB) radiation began its journey nearly 14 billion years ago and has been roaming through all of spacetime ever since. Its presence is taken as strong evidence for the Big Bang. Although we will focus principally on ionizing radiation in this text, in many cases, where there is low-energy electromagnetic radiation, there also is, or was, ionizing radiation.

The Universe is, and has been, expanding since its birth around 13.8 billion years ago. And as the Universe expands, all of space expands with it, stretching the wavelengths of various forms of electromagnetic radiation roaming through space, such as the CMB. Since the energy of individual photons is inversely proportional to their wavelengths, this implies that electromagnetic radiation freely flying through space diminishes in energy over time thanks to the expansion. Hence, the CMB photons that we detect today, permeating all of space, started their flights when they were born as high-energy gamma rays made by countless electron–positron annihilations. Amazingly, these blackbody polyenergetic present-day microwave photons were born as monoenergetic 511 keV gamma ray photons when all of the Universe's primordial antimatter electrons (positrons or antielectrons) met up with their matter electron counterparts and underwent mutual annihilation for the last time. The redshifted remnants of this mass carnage, the CMB, trace their ancestry back to a time when the Universe was about 4 billion K, around 10 s after the Big Bang.

Our own planet Earth, our Moon, the interplanetary space in the Solar System, and the Milky Way Galaxy at large are all suffused with ionizing radiation of various sorts and species. In the way of introduction, we will embark on a brief tour of the Universe, starting at our humble home planet Earth and venturing through the vast extents of our Milky Way Galaxy and beyond. All the while we will focus on the diverse doses and types of ionizing radiation encountered along the way.

doi:10.1088/978-0-7503-5444-8ch1   1-1

1.1 Our radioactive Earth

As will be discussed in subsequent chapters, Earth is a 'living' planet with an amazing array of geological phenomena in the form of earthquakes, volcanoes and plate tectonic activity. While some of the astonishing 44 TW powering this commotion is the residual heat left over from the accretion of planetesimals (the 'building blocks' of the early planets) crashing into one another and clumping together under the force of gravity, approximately 50% of today's geological driving force is due to heat released from radioactive decay deep under our feet. Simply put, our Earth is a radioactive planet. Primordial radioisotopes of uranium, thorium and potassium are presently releasing energy that drives the geological activity that keeps our planet's volcanoes firing and our plates moving. Radioactivity will be covered in depth elsewhere, but students might recall that radioisotopes decay exponentially over time. Given this knowledge, one will readily realize that Earth must have been far more radioactive in the distant past compared to today. The half-lives of primordial radionuclides present at the time of Earth's inception (around 4.567 billion years back) include neptunium-237 (2.14 million years), potassium-40 (1.25 billion years), uranium-238 (~4.5 billion years) and thorium-232 (14.1 billion years). Some of these radioisotopes, like neptunium are already extinct in nature, implying that they contributed all their radiation energy to an even more radioactive planet many years ago.

Just as an aside, if the Earth is loaded with natural radioactivity, it should come as no surprise that our bodies are radioactive as well. Although we do not normally have much uranium-238 or thorium-232 in our cells, potassium is of great physiological importance and abundance. Potassium is the main intracellular cation (positively charged ion) and is present within many cells at typical concentrations of 140–150 mM. This provides a total body load of about 140 grams of potassium. At a natural potassium-40 abundance of 0.012%, this means that a human body might contain 0.0168 g or 16.8 mg of potassium-40. Depending on the mass of the individual in question, this equates to an activity of around 4000 Bq (becquerels) and a 70-year lifetime absorbed radiation dose of nearly 10 mGy (milligray). Thus, humans are radioactive. And as we will discuss later, potassium-40 is not the only radionuclide naturally found in our bodies.

While smaller bodies in our Solar System such as the Moon, Mercury and Mars were also created from the same radioactive stuff, their surface area to volume ratios have allowed them to cool off to the point that their accretion heat and residual radioactivity are no longer sufficient to keep them 'alive' the way Earth is. It is unclear whether we, here on Earth, were more richly blessed with radio-isotopes than our fellow planetary companions at the moment of conception, but it remains possible. And our reservoir of internal heat, largely thanks to our rich endowment of primordial radionuclides, allows another important, radiation-related phenomenon—the existence of a geomagnetic field. And a planetary magnetic field is critically important in its own right when it comes to space radiation.

1.2 Our geomagnetic cocoon—Earth's radiation safety officer

The relationship between our geomagnetic field and space radiation stems from the fact that this magnetic field deflects a good fraction of the charged particulate radiation (cosmic rays) that would otherwise be hammering our atmosphere and blasting our biosphere. Earth's radioactivity-generated internal heat permits its outer core to remain molten (the inner core, although even hotter, is so dense that the iron-nickel metal is forced into the sold state). As explained in chapter 18, any spinning mass of hot, ionized liquid metal can generate a magnetic field, and this is the origin of our geomagnetic shield. But without our residual radioactivity, it is possible that Earth would now be too cool to maintain a molten core and a protective geomagnetic shield around us. The importance of a protective geomagnetic field will be discussed in greater depth in chapter subsequent chapters. But for now, it is worth underscoring the ironic reality that radiation (in the form of radioactive decay) facilitates a molten iron core, which generates a magnetic field that, in turn, shields us from incoming charged particulate radiation from outer space.

It is curious that our geomagnetic field is not perfectly constant. In fact, it waxes and wanes in strength, and on occasions, the north and south magnetic poles flip—north becomes south and vice versa. Over the last ten million years or so, such reversals averaged about once every 200 000–300 000 years. And since the last major geomagnetic reversal was recorded about 780 000 years ago, we are well past due. During the reversal process, the overall magnetic field strength sharply declines. Although Earth's thick atmosphere continues to block much of the incoming space radiation, the dose to the biosphere definitely increases during geomagnetic pole reversals and excursions (which are essentially periods of reduced field intensity accompanied by magnetic pole wandering rather than all-out reversals). The paleontological and radiobiological effects of such phenomena are currently being examined.

Additionally, the early loss of a planetary magnetic field around smaller planets such as Mars, has had devastating domino effects. Whether due to small size and an unfavorable area:volume ratio or an unfortunate 'congenital' deficiency of primordial radioisotopes, Mars' core is presently unable to generate a global planetary magnetic field to boast of. This early loss of its protective magnetic field allowed the solar wind (charged particles, mostly protons, incessantly radiating from the Sun) to erode the thin Martian atmosphere over time. If Mars were larger or had a greater initial endowment of primordial radionuclides, it might well still have a liquid metallic core and a protective Martian magnetic field—and a thicker Martian atmosphere. The fact that it has none of these illustrates the critical, but easily overlooked, role of radioactivity and radiation in planetary evolution.

1.3 Radiation from above

Besides the radiation emanating from the primordial radioactive isotopes beneath our feet, Earth is constantly being irradiated from above. Although attenuated by our atmosphere and partially deflected by our magnetic field, cosmic rays continue to contribute a sizeable fraction of the natural background radiation that humans (and all inhabitants of the surface) receive. In fact, the only regions that are substantially spared from this unremitting radiation exposure from above are underground habitats and the deep sea (which have their own biological challenges).

Cosmic rays are covered in detail in chapter 17, but for now we can mention that the Earth is bombarded by 'primary' and 'secondary' cosmic rays. Primary cosmic rays are mostly energetic protons (hydrogen nuclei), alpha particles (helium nuclei), high-energy electrons, with a smattering of heavier nuclei. The heavy nuclei, which range all the way up to uranium, are small in number but high in 'biological effectiveness' when it comes to radiation dose. Although the nomenclature is a bit inconsistent, for our purposes, we will say that primary cosmic rays come from two separate sources—solar and galactic. The first source is the Sun, in the form of the relatively steady solar wind and sporadic punctuations by 'solar particle events' (SPEs) derived from solar flares, solar prominences and coronal mass ejections. Because of their lack of warning and their occasionally high doses, these erratic SPEs can pose a real radiation hazard to astronauts and future colonizers of bodies in our Solar System. The second source of primary cosmic rays originates from far beyond our Solar System and are collectively called 'galactic cosmic rays'. It is this galactic cosmic ray pool that contains the occasional heavy charged particle (i.e. nuclei larger than helium). Some of these heavy primary cosmic rays are especially potent radiobiologically (that is, they have high relative biological effectiveness (RBE)) and are given the nickname of '**HZEs**' for high atomic number (Z), high-energy (E) particles.

Naturally, if there are 'primary' cosmic rays, there must be something called 'secondary' cosmic rays. Although the terminology is confusing (as clarified in chapter 17), in this text we will reserve the term secondary cosmic rays for those particles that are generated by collisions between incoming primary cosmic rays with molecules and atoms in our atmosphere. These collisions greatly magnify the total number of secondary particles and spread the radiation out over a far wider range than the single original primary cosmic ray. While primary cosmic rays rarely make it all the way to our surface before bashing into atmospheric molecules, secondary cosmic rays generated by such collisions do bathe us with radiation here on the surface. An analogy between primary and secondary cosmic rays may be the replacement of a rifle firing a single bullet (primary cosmic ray) with a shotgun firing a spray of pellets (secondary cosmic rays). The amount of natural background radiation we receive from these secondary cosmic rays is a function of both latitude and altitude as will be discussed later.

Incidentally, were it not for cosmic rays, we would not have carbon-14 for radiometric dating of organic materials. Carbon-14 has a half-life of only 5730 years, which means that any carbon-14 present at the birth of the Earth 4.567 billion years ago will have long since decayed away. In other words, carbon-14 is not a primordial radionuclide; it is a cosmogenic radionuclide whose supply must be constantly refreshed in order to keep the amount constant. The relatively constant amount of carbon-14 found in our atmosphere and biosphere is thanks to the creation of secondary cosmic rays (fast neutrons in particular) that then convert nitrogen-14 into carbon-14. The carbon-14 then reacts with molecular oxygen to form radioactive carbon dioxide. The radioactive carbon dioxide then enters the biosphere via photosynthesis by plants and microorganisms, and the carbon-14 works its way up the food chain into organisms like us. Thus, in addition to the primordial potassium-40 mentioned earlier, part of our natural radioactivity is thanks to the cosmogenic carbon-14 we contain. And it is worth stating the obvious —all (living) plants and animals are radioactive thanks to the carbon-14 made by cosmic rays in our upper atmosphere. Cosmogenic nuclides, including cosmogenic *radio*nuclides like carbon-14 and hydrogen-3 (tritium) are the natural products of cosmic ray interactions with our atmosphere and are considered in chapter 17.

The Earth's geomagnetic field may trap some incoming primary cosmic rays and store them in the form of the Van Allen belts or it may deflect them away from us. But given the shape of our planetary magnetic field (very roughly approximated as a bar magnet), the charged particles from space will be directed towards the polar regions. Thus, regions nearer the north and south poles receive more cosmic radiation than equatorial regions. And although our atmosphere is the actual source of secondary cosmic rays, the atmosphere also absorbs a large fraction of both primary and secondary cosmic rays before they can reach the biosphere. Such a safeguard from incoming radiation is not provided on the Moon, Mars and other bodies in our Solar System that are devoid of a thick, hospitable and protective atmosphere.

1.4 Our radiation-absorbing atmosphere

In addition to the geomagnetic field, which defends us by deflecting charged particles from space, we are blessed with a thick and stable atmosphere. Our atmosphere cocoons us from space radiation, both particulate and electromagnetic. And while our atmosphere, composed nearly entirely of diatomic nitrogen and oxygen molecules, is transparent to visible light, it is largely opaque to x-rays and gamma rays. Therefore, the ambient energetic electromagnetic radiation of outer space does not reach Earth's surface. This is not the case on the Moon and other places deprived of an ample atmosphere.

And as will be discussed in chapters 18 and 29, the small but important trace quantities of 'greenhouse gases' in our atmosphere are transparent to incoming sunlight but opaque to reflected infrared. This keeps Earth significantly warmer than the frigid temperatures (about −18 °C or 0 °F) we would experience with an atmospheric chemical composition devoid of such gases.

Just where certain gases reside is dependent on what part of the atmosphere we are talking about and chemical aspects of the gas in question. For instance, our familiar ozone layer is located in the stratosphere. Students may be familiar with the classical, layered nature of Earth's atmosphere with our troposphere, stratosphere, mesosphere, thermosphere and exosphere in ascending order. These layers are based on the trends of temperature as one rises in altitude. For example, the lowest layer, the troposphere grows cooler with height but the next layer up, the stratosphere grows warmer with ascension.

As explained in chapter 19, another way of categorizing the regions in our atmosphere is based on the degree of gaseous mixing. We reside in the lower level (the homosphere), where gases readily mix regardless of molecular weight. But above the boundary called the turbopause lies the heterosphere, an atmospheric region in which gases do not mix as well. In the heterosphere, gases stratify by molecular weight, leaving heavier molecular oxygen, nitrogen, argon and carbon dioxide at the bottom, with lighter helium and hydrogen in the upper heterosphere.

Any artificial boundary between 'Earth' and 'outer space' will be arbitrary, but the turbopause is often used as a convenient, atmospherically-defined dividing line. From a practical perspective, it does describe an altitude where the atmosphere becomes too thin for aeronautical flight and rocket-powered flight becomes necessary. The boundary between 'space' and Earth is sometimes called the Kármán line. Thus, for regulatory purposes, 'aircrafts' can fly below the Kármán line and fall under different jurisdiction from 'spacecrafts', which can fly above it. As might be predicted, there is disagreement on exactly where the Kármán line lies legally. But in general, it can be approximated by the turbopause, which also happens to lie near the dividing line between the mesosphere and the thermosphere—in other words, the mesopause.

Closer to home, our stratospheric 'ozone layer' shields us from a great deal of biologically-damaging ultraviolet radiation. Although a small amount of ultraviolet is important for cutaneous vitamin D production, without the generous ozone layer, far more solar ultraviolet would sneak in and present very serious challenges to terrestrial life. Ultraviolet radiation is quite capable of causing damage to the key molecule of life, DNA (which is found in only one copy ('haploid') in prokaryotic cells, such as bacteria, and only two copies ('diploid') in most eukaryotic cells, such as human cells). This low copy number implies vulnerability but also dictates that high-fidelity repair mechanisms must be in place.

Ultraviolet-induced DNA damage in the form of 'pyrimidine dimers' (cyclo-butane dimers) and 4,6 photoproducts will be reviewed in subsequent chapters. Fortunately, normal healthy cells have means of coping with ultraviolet-induced DNA damage (which would be expected given the central biological role of DNA and the fact that it is present in such low copy numbers within cells). But it is somewhat surprising that, while many organisms have redundancy in their ability to ward off the damaging effects of ultraviolet radiation, humans and other placental mammals are relatively limited. This limited capacity for ultraviolet-induced DNA damage repair makes our species vulnerable to rare autosomal recessive genetic conditions such as xeroderma pigmentosum and Cockayne syndrome. The fact that other organisms have developed alternative means of dealing with ultraviolet-induced DNA damage (such as direct DNA damage reversal) offers some hope in the form of gene therapy, as will be discussed in chapter 11.

But just as our atmosphere protects us from some forms of biologically-damaging radiation, massive doses of ionizing radiation can damage our atmosphere. It might be fun to speculate what might happen if a high-dose gamma ray burst were to zap us. We will go into some detail on what impact a large dose of gamma rays would have on our atmosphere in general and on our ozone layer in particular, in chapter 30. Just as ionizing radiation can cause radiobiological effects by generating free radicals, reactive oxygen species and reactive nitrogen species within our cells, radiation can analogously cause damage to our atmosphere by generating chemically active radiolytic products through heavy doses of gamma rays or x-rays. Reactive nitrogen species might be particularly hazardous to our convivial ozone layer. And some scientists have speculated that powerful gamma ray bursts have scorched us at least once in the distant past, perhaps contributing to the Ordovician mass extinction event around 444 million years ago.

1.5 Radiation within our atmosphere

NASA's Chandra X-ray Observatory detected x-rays in a portion of Earth's outermost extended atmosphere called the geocorona. The geocorona is the luminous part of Earth's exosphere (covered in chapter 19). Earth's geocorona is luminous in the far-ultraviolet and has been seen with ultraviolet cameras and spectrometers on the Galileo spacecraft, Apollo 16, the Astrid satellites and others. The geocorona contains an abundance of escaping neutral hydrogen atoms that Earth's gravity just cannot hold in place. Collisions between these hydrogen atoms and energetic incoming carbon, oxygen and neon ions in the solar wind produce x-rays. The x-rays are generated by a process called charge exchange. In charge exchange, electrons are transferred from the neutral, upper atmospheric hydrogen atoms to the positively charged solar wind ions. As discussed in chapter 6, when the newly transferred electrons cascade down electron energy levels to reach the stable ground state, x-rays are emitted. Charge exchange x-rays are not only found in our geocorona but also in the extended atmosphere of other bodies in our Solar System, such as comets. Comets have bright visible extended atmospheres (the coma) that also shine in x-rays. Because different elements exhibit distinct x-rays that reveal their identities (thus the name characteristic x-rays), the charge exchange process in the extensive atmospheres of comets can help identify the various elements in the solar wind besides the abundant hydrogen and helium.

Another radiation health hazard in our atmosphere is the ionosphere. As covered in chapter 19, the ionosphere is the ionized portion of our upper atmosphere and includes parts of the mesosphere, thermosphere and exosphere. Being partly ionized, the ionosphere technically is a plasma. It is ionized primarily by solar ultraviolet radiation and soft x-rays. It extends from around 48 km (or ~30 miles) up to 965 km (or ~600 miles) above the surface. The ionosphere constitutes the inner edge, or base, of Earth's magnetosphere. (And as will be explained elsewhere, the magnetosphere can be loosely defined as the area in which a planet's (or star's) magnetic field can influence charged particles in its vicinity.)

One important exploitation of the ionosphere is in skywave radio communication wherein radio waves are repeatedly reflected off the ionosphere and back to Earth, allowing long-range propagation. But when strong solar flares emit hard x-rays, these x-rays can disrupt the D-region of the ionosphere and release additional electrons that absorb rather than reflect radio waves. This can lead to radio blackouts or sudden ionospheric disturbances over minutes to hours, depending on the specifics of the responsible solar flare. Similar disruptions in radio communication can occur in the polar regions because of a phenomenon called polar cap absorption (PCA), which is caused by the release of a huge number of protons by a solar flare or coronal mass ejection. These protons arrive later than the solar x-rays, which travel at c, the speed of light in a vacuum. Upon interaction with the geomagnetic field, these protons spiral down magnetic lines of force and strike the upper atmosphere, including the ionosphere. The result is further ionization of the D and E layers, which hinders reflection of radio waves and thereby affects radio communication. PCAs may persist hours to days (averaging 24–36 h).

1.6 The Van Allen belts

As we grow more adventurous and leave the confines of our planet, with its protective atmosphere, space travelers must be aware of the extensive bands of energetic trapped charged particles surrounding Earth. These charged particles are mostly mundane electrons and protons, but there are some highly unusual particles up there as well. These doughnut- or bagel-shaped (i.e. torus- or toroidal-shaped) regions of radiation are the so-called radiation belts or Van Allen belts. Space travelers (and even unmanned spacecrafts harboring sensitive equipment) must be careful not to ignore the potentially significant doses possible from traversing the Van Allen radiation belts. This high-dose, high-energy, charged particulate radiation can be radiobiologically hazardous and electronically devastating. To provide an idea of how much radiation might be encountered, depending on exactly where the satellite is located, a satellite hanging around in the Van Allen belts might receive a dose upwards of 50 Gy (the gray is the unit of absorbed ionizing radiation dose and is abbreviated Gy; radiation units and a general physics review are included in chapters 4, 5, and 6). One can calculate that if a space vehicle were to fly right through the Van Allen belts, the absorbed radiation dose to the spacecraft (and perhaps to the astronauts inside, depending on shielding) could be:

$$(50 \text{ Gy}) \times (1 \text{ year}/365 \text{ days}) \times (1 \text{ day}/24 \text{ h}) = 0.006 \text{ Gy}.$$

As will be covered later, the absorbed radiation dose in gray can be converted into the biologically based 'equivalent dose' measured in sieverts (Sv). Thus, assuming that 1 Gy = 1 Sv in this case, 1 h spent in the Van Allen belt delivers an equivalent dose of 0.006 Sv or 6 mSv in this hypothetical example. (A more realistic estimate, based on Apollo data, would be around 15—16 cGy or rads for a one hour transit through the Van Allen belts; this would translate into 150–160 mSv, again depending on shielding within the spacecraft.) But as we will discuss later, this assumption of 1 Sv per gray may be a bit conservative; according to the United States Nuclear Regulatory Commission (10 CFR §20.1004), the 'quality factor' (or weighting factor W_R) for protons is 10, meaning that 1 Gy of protons delivers 10 Sv. Thus, an hour in the Van Allen belt might give a physical absorbed dose of 6 mGy but a biologically equivalent dose (or as covered in subsequent chapters, an 'effective dose') of 60 mSv. However—and underscoring the controversy surrounding this topic—the International Commission on Radiological Protection (ICRP) publication 33 listed the radiation weighting factor for protons as 5. Subsequently, the

ICRP reduced this figure to 2 in ICRP publication 103. The dependence of biological effects on proton energy is extremely relevant in space radiation, since some of the protons encountered in space can attain unbelievably high energies. And perhaps somewhat surprisingly, the biological effectiveness tends to be inversely proportional to proton energy. For reference, an average amount of natural background radiation on Earth's surface in the continental United States is about 3.1 mSv per year (with substantial geographical variation depending on altitude and the radioactivity of the rocks in the region—for instance living in Denver, Colorado, might provide an additional 1 mSv (or 100 millirem to use another commonly used radiation unit) per year). In contrast, living at sea level altitude in the United States might yield only 1.5 mSv or 150 millirem of natural background radiation annually. Note that 1 Sv = 1000 mSv = 100 rem = 100 000 mrem.

Thankfully, the Van Allen belts are torus-shaped, implying that there are 'escape routes' through which we can navigate our way out without being blasted. As will be described in chapter 19, the Van Allen belts are largely double-layered structures with an inner belt and an outer belt. A mnemonic for students might be to recall atomic structure, with the negatively charged electrons on the outside and the positively charged nucleus on the inside. Such are the Van Allen belts, with the outer belt largely populated by electrons and the inner belt made mostly of protons.

The Van Allen belts are the products of geomagnetic forces on charged particles from space. Without strong magnetic fields, Venus, Mars and the Moon do not display radiation belts the way we do. But the Earth is certainly not the only planet in the Solar System with radiation belts; all planets with strong magnetic fields will trap charged particle radiation and form their own versions of our Van Allen belts. As will be discussed later, Jupiter, with a magnetic field ten times stronger than Earth's, has enormous radiation belts. Data from the Voyager and Galileo missions suggest that these Jovian radiation belts will prove quite challenging for anyone planning to venture to any of the inner Galilean moons such as Io, Europa or Ganymede. The radiation intensity in Jupiter's radiation belts may be a million-fold more than in our already dangerous Van Allen belts.

Intriguingly, antimatter has been observed in the Van Allen belts in the form of minute numbers of antiprotons. Thus, during our adventure as astronauts, we did not have to travel very far into space to encounter some extremely exotic forms of radiation.

1.7 Solar particle events

As we venture further beyond the boundaries of our protective atmosphere and the extensive shield of our geomagnetic field, we are at the mercy of various forms of space radiation that we are not normally bombarded with here at home. Among such hazards are SPEs. SPEs are sporadic, meaning they do not recur regularly (as do sunspots for instance, with their roughly 11-year cycle, which coincides with a solar north-south magnetic pole flip). Highly energetic protons (tens to hundreds of mega-electron volts (MeV) a unit of energy which will be expounded upon in chapter 4) can affect the function of electronic equipment as well as have radiobiological consequences. In fact, one needs only to revisit the famous Carrington event of 1859 (which shocked telegraph operators, caused telegraph wires to overheat and malfunction and led to auroras visible as far south as Hawaii and Cuba) to imagine what the consequences of such a direct shot would be on a small spaceship beyond the protection of our atmosphere and geomagnetic field. Powerful solar flares and coronal mass ejections like the Carrington event are relatively rare and are fortunately spread out across the solar surface. Therefore, the odds of a direct hit are unlikely. Nevertheless, there is certainly an element of luck involved in any space mission, as the biological and electronic effects of solar energetic particles from a robust head-on SPE could be devastating indeed. An unlucky strike from a particularly powerful SPE would likely put our mission to an abrupt end before it barely got started. Therefore, even a relatively short jaunt to the Moon is not without some serious radiation risks. SPEs will be reviewed further in chapter 19.

1.8 Solar electromagnetic radiation

Solar particulate radiation is not the only form of ionizing radiation emanating from our Sun. We are all familiar with the ultraviolet radiation that results in a suntan, sunburn or even skin cancer. While some texts do not include ultraviolet when dealing with 'ionizing radiation', technically ultraviolet can cause ionization and can be biologically hazardous, so in this text, we will often be including ultraviolet when we talk about space radiation in upcoming chapters. It is worth remembering that although ultraviolet can cause DNA damage and, in excess, have negative health consequences, it is also biologically beneficial since sunlight is a source of vitamin D. While this vitamin can be obtained through dietary intake, a major source is the ultraviolet-driven cutaneous conversion of cholesterol derivatives into the vitamin, which is then further activated in the liver and kidneys into the fully active hormone. Specifically, 7-dehydrocholesterol is first converted in the skin (primarily in the stratum basale and stratum spinosum of the epidermis) into cholecalciferol by the action of UV-B radiation. The cholecalciferol (also known as vitamin D3) is then hydroxylated in the liver into 25-hydroxycholecalciferol (also called calcidiol or calcifediol); this is then hydroxylated again in the kidneys into the active product, 1,25-dihydroxycholecalciferol (or calcitriol). Technically, a vitamin is an organic compound that we must ingest because our bodies cannot synthesize it. So, since we can biosynthesize vitamin D, it is technically not a vitamin but rather could be considered a hormone. Nevertheless, the name 'vitamin D' is here to stay regardless of the pedantic details. And although we can obtain our vitamin D through ingestion of food or pills rather than direct sunlight, the plants and animals we eat to get our vitamin D (and even the industrial synthesis of vitamin D in a factory) all ultimately depend on ultraviolet for the chemical synthesis. In fact, when one thinks about it, vitamin D production in humans could be considered a form of 'photosynthesis'.

In addition to the familiar ultraviolet radiation stemming from the Sun, x-rays are also produced in abundance. Students of physics might initially find this perplexing since the Sun's surface is measured to be 5780 K, and while a heat source of this temperature yields plenty of visible light, it is far shy of what is needed for x-ray production. Of course, sunlight is not monochromatic but is emitted as a broad spectrum that can be described by a so-called Planck blackbody radiation curve. But even when one considers the blackbody spectrum of sunlight, the measured quantity of x-rays far exceeds what one would calculate from basic principles. (And all the basic principles of blackbody radiation, Planck curves and the correlations between temperature and wavelengths will be reviewed in chapter 7.) The answer lies in the fact that while the visible surface of the Sun (its photosphere) is only 5780 K, the surrounding corona (the outer part of the solar atmosphere) reaches searing temperatures measured in the millions. And at these temperatures, x-rays certainly can be produced. As will be discussed in chapter 16, the exact mechanisms by which the corona is heated to these extreme, x-ray-generating temperatures are still under debate, but data from the instruments onboard the European Space Agency's SOHO (the SOlar and Heliospheric Observatory) and NASA's TRACE (the Transition Region And Coronal Explorer) are shedding light on the question.

1.9 Radiation on the Moon?

While there is plenty of space radiation between here and the Moon to worry about, what about on the lunar surface itself? Well, without an atmosphere to speak of or a lunar magnetic field to shield us, during our adventure we will be at the mercy of much more radiation than we are here on Earth. Without an atmosphere that converts primary cosmic rays into secondary cosmic rays, on the Moon, we will be bombarded directly with high-energy cosmic primaries. Some of these primary cosmic rays include the notorious HZEs that have extremely high linear energy transfer (LET). LET, as explained in subsequent chapters, is a mathematical description of the deposition of energy along the path of a particle or photon (that is, dE/dx). High LET radiation tends to be biologically potent, and this is reflected by an accompanying high RBE. But when we get beyond our geomagnetic shield and venture to places like the Moon, the LET of some of the incoming primary cosmic rays can be truly staggering. Here on Earth, we typically talk about ionizing radiation in the kilo or mega electron volt energy range, but in space we will be exposed to some galactic cosmic rays with absolutely astonishing energies. As will be covered in chapter 17, some galactic cosmic rays with energies in the peta and exa electron volt energy range have been documented. An individual incoming proton with such energy might pack the wallop of a well-thrown baseball (figure 1.1)!

And just as we mentioned that the Earth is laden with radioactive minerals, so too is the Moon. While some regions might be more radioactive than others, lunar radioactivity is unlikely to be much of a health concern. (And for that matter, natural background radioactivity on planet Earth is probably not really very biologically concerning either). As will be discussed in chapter 19, there are areas here on Earth known as high natural background radiation regions in parts of Iran, India, Brazil, China and elsewhere that receive significantly more radiation than the world's average thanks to radioactivity in the soil and rocks in the regions. There do not appear to be any negative health consequences of such natural background

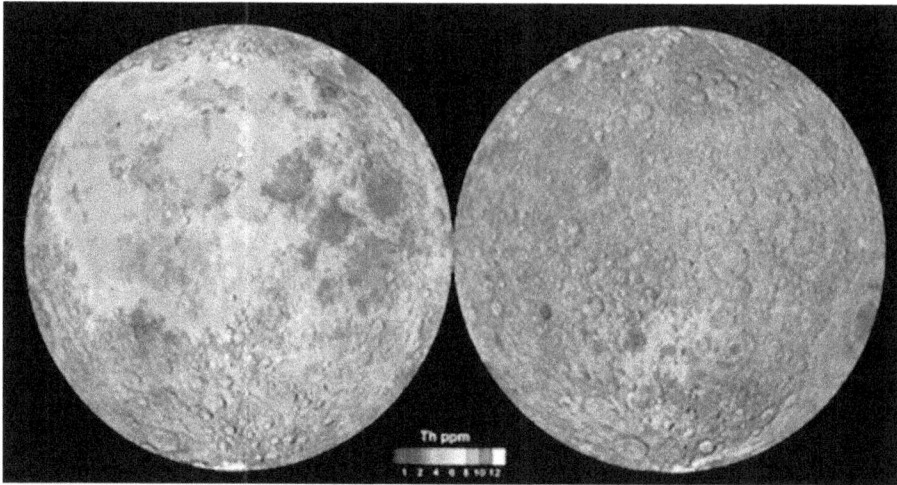

Figure 1.1. The nearside (left) and farside (right) of the Moon colored according to thorium concentration on the surface. The maps were generated from gamma ray data obtained from the Lunar Prospector mission. While most gamma rays on the Moon are products of galactic cosmic ray spallation, the 2.5 MeV gamma photons originating from decay of thorium-232 were used to create these maps. Regions of high thorium concentration indicate the presence of **KREEP** (potassium (K), rare-earth elements (REE) and phosphorus (P)). The KREEP-containing rocks on the nearside are primarily concentrated underneath the lunar mare called the **Oceanus Procellarum** and the crater known as the **Mare Imbrium**. Collectively, this high-KREEP lunar terrane is known as the **Procellarum KREEP Terrane**. On the far side, the concentration of radioactive thorium is primarily at the **Compton–Belkovich Thorium Anomaly**. Credit: NASA.

radiation, and in fact, there is a hint of the opposite—radiation hormesis, a health benefit from low-dose radiation, which will be discussed in upcoming chapters. But while low-dose radiation from radioactivity on the Earth or a brief trip to the Moon may be inconsequential from a health perspective, the overall dose of radiation from a prolonged stay on the Moon would no longer fall into the 'low dose' category.

Beyond the constant bombardment of cosmic rays and the very low-level radiation from natural radioactivity, another source of ionizing radiation on the Moon is x-ray fluorescence. When unabated space radiation strikes the lunar surface, inner shell electrons in atoms are ejected, leaving the atom in an excited, ionized state. Upon relaxation and rearrangement of orbital electrons back down to the ground state, these excited atoms emit characteristic fluorescent x-rays. Since the energies of these characteristic x-rays are unique to the element they originated from, analysis of these x-rays can serve as a sort of chemical fingerprint, revealing the atomic composition of the lunar minerals they radiate from (Figure 1.2).

1.10 Next stop, Mars

During this brief journey across the Universe, we will not be heading inwards towards the Sun and visiting searing hot Venus or Mercury. Rather, we will strictly be heading outward and will spend time on our nearest planetary neighbor, Mars. Unlike the Moon, Mars does host a thin atmosphere, but it is nothing like our own,

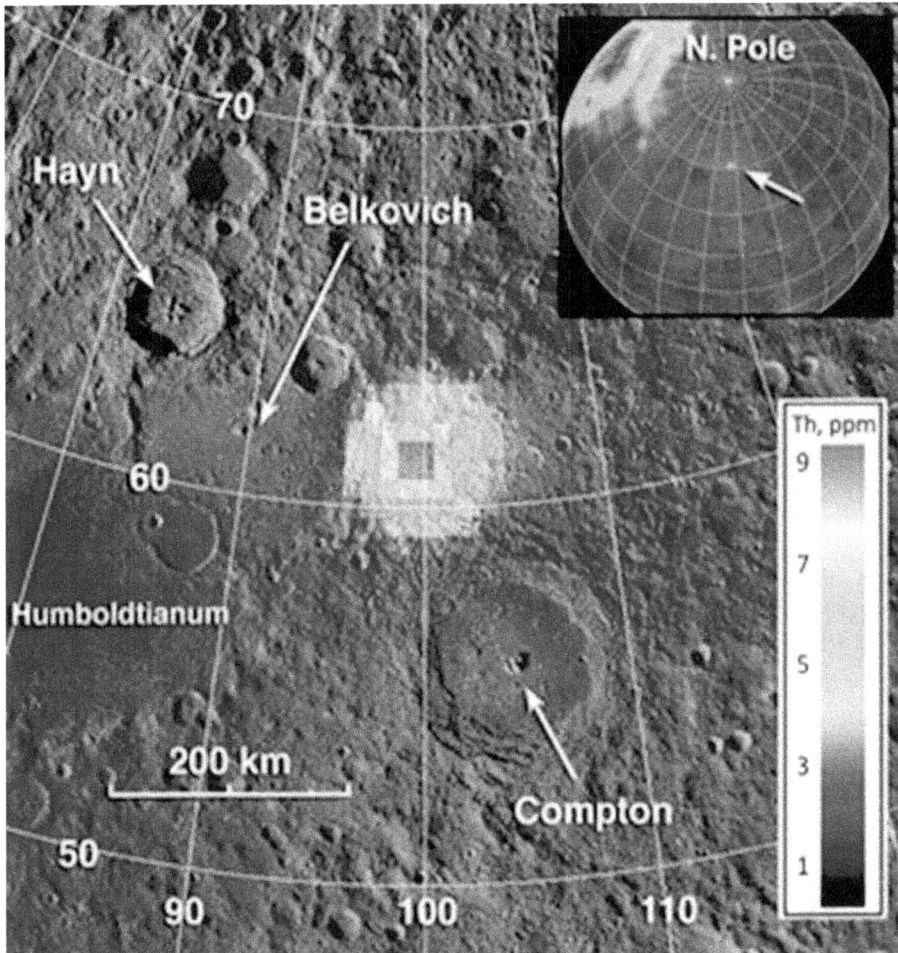

Figure 1.2. The **Compton–Belkovich Volcanic Complex** or **Compton–Belkovich Thorium Anomaly** is an unusual, volcanic region on the far side of the Moon that is perplexingly warm. It emits heat at about 180 mW m^{-2}, which is around 20 times higher than the background lunar highlands. That corresponds to a temperature nearly 90 °F warmer than elsewhere at a depth of around 2 m below the surface (but still a frigid −10 °F). The heat appears to be stemming from radioactive decay of thorium, uranium and potassium-40 in an underlying gigantic granitic slab (a batholith) that was possibly the product of a volcanic eruption 3.5 billion years ago. Such felsic (non-basaltic) volcanism is generally considered rare on the Moon (or anywhere in the Solar System aside from Earth). Multiple cycles of melting and re-solidifying probably distilled the radioisotopes in the frozen granitic magma present here. While it is 900 km (560 mi) from the Procellarum KREEP Terrane, the region may be inherently rich in KREEP (potassium (K), rare-earth elements (REE) and phosphorus (P)). KREEP is noteworthy because of its concentration of 'incompatible' elements (that is, elements that partition into the liquid phase during partial melting into magma) including radioactive potassium, thorium and uranium. The radioactivity in the Compton–Belkovich Volcanic Complex is primarily due to thorium. Elsewhere, data from the von Karman crater on the far side of the Moon (provided by the German-built Lunar **Lander Neutrons and Dosimetry** experiment aboard China's **Chang'E 4 lander**) indicate that the total absorbed dose rate on the lunar surface averages about 13.2 ± 1 μGy h^{-1} with a neutral particle dose rate of 3.1 ± 0.5 μGy h^{-1} and charged particle dose rate of 10.2 ± 1.1 μGy h^{-1}. Using a quality factor Q

(or more technically correct, a weighting factor W_R) of 4.3, this amounts to an equivalent dose rate of about 57 μSv h^{-1} or 1369 μSv d^{-1} on the surface of the Moon. For comparison, thanks to some shielding provided by our geomagnetic field, the reported dose equivalent onboard the International Space Station (as measured with the DOSIS 3D DOSTEL instruments) is about 2.6 times lower, averaging about 731 μSv d^{-1} (with contributions from galactic cosmic rays amounting to 523 and 208 μSv d^{-1} from Van Allen belt protons while crossing the South Atlantic Anomaly). Image from NASA/Goddard Space Flight Center/Arizona State University/Washington University in St. Louis. Reprinted from [6], Copyright (2011), with permission from Nature.

with an average surface pressure less than 1% of what we have at sea level. This will do little to attenuate the incoming cosmic rays from both the Sun and beyond the Solar System. In fact, the disproportionately thin atmosphere is basically the product of unrelenting attrition caused by the solar wind. As mentioned, this susceptibility to erosion by the solar wind is because Mars does not possess much of a deflective geomagnetic field. Thus, undeflected charged particles continue inward and, over millennia, continue to wear away Mars' naturally thin atmosphere.

While it was once believed that the absence of a protective planetary magnetic field was because Mars' iron core froze solid early in its life history, it now appears that the Martian core remains at least partly liquid. And seismic data from NASA's InSIGHT lander hints that the Martian core is larger but less dense than previously believed. This has led to a new model of the Martian core in which the composition is not mostly iron-nickel as in Earth's core but rather iron with a relative abundance of sulfur and hydrogen. This particular chemical composition could have led to immiscibility issues that prematurely terminated the needed convection to sustain a permanent magnetic field—even if the core remains partly liquid. Irrespective of whether its core is frozen solid or still molten, Mars is an essentially dead planet both geologically and biologically. Unlike Earth, with its numerous volcanoes, earthquakes and slow but steady migrating tectonic plates, Mars shows no such signs of activity today (despite hosting what is often touted as the biggest (but extinct) volcano in the Solar System, Olympus Mons). The present reality is that Mars' core is not generating much of a protective planetary magnetic field. Thus, future astronaut visitors and colonizers will have to cope with unadulterated incoming charged particle radiation from space.

Measurements made by the Radiation Assessment Detector (RAD) (on NASA's Mars Science Laboratory Curiosity rover) during a 300 d period coinciding with solar maximum, revealed a good estimate of what to expect on the surface. Note that while radiation from the Sun is higher during solar maximum, the increased solar activity fends off incoming galactic cosmic rays, so the radiation from galactic cosmic rays is lower during solar max. The radiobiological consequences of this tradeoff are uncertain, but it is believed that the reduction in HZEs (found in galactic cosmic rays) might make solar max generally less biologically hazardous. (On the other hand, the largest SPEs caused by solar flares and coronal mass ejections occur more often during solar max, so we are never truly safe!) The RAD instrument measured an absorbed dose of 76 mGy (or 7.6 cGy or rads) per year. (Radiation dose units and related basic physics is reviewed in chapters 4, 5, and 6.) This is far higher than the average annual natural background radiation dose in much of the continental United States (just under 1 mrem per day for an annual total of ~310 mrem or 3.1 mSv). The estimated annual dose of 7.6 cGy (which is the equivalent of 76 mSv if we assume a quality factor of 1) was alarming. However, one might recall that in parts of Guarapari, Brazil; Yangiang, China; Akraroola, Australia; Kerala, India and Ramsar, Iran, there are some areas that naturally receive high background doses that sometimes approach this figure (and some record-high regions that more than double this Martian figure), yet adverse health effects are not clearly manifest. On the other hand, the interplanetary dose incurred from traveling to and from the Red Planet must be factored in. The interplanetary radiation dose is higher than what is found on the Martian surface thanks to attenuation provided by its thin but not non-existent atmosphere. Based on the RAD readings, a round-trip mission consisting of 180 d each way plus 500 d on Mars itself during solar max would yield a total dose equivalent of roughly 1 Sv. This is a highly controversial finding because NASA presently limits its astronauts to a 3% lifetime risk of getting a fatal radiation-induced cancer. According to some models (such as the highly controversial linear–no threshold model), a 1 Sv dose from a round trip to Mars might confer a 5% risk of contracting cancer. Of course, all such models have their limitations, and the figures calculated by such flawed models must be tempered by the observed reality of a roughly 30% risk of dying from cancer for inhabitants of the United States—without ever venturing into space!

Further reading

[1] Yokoo S, Hirose K, Tagawa S, Morard G and Ohishi Y 2022 Stratification in planetary cores by liquid immiscibility in Fe–S–H *Nat. Commun.* **13** 644
[2] Jolliff B, Wiseman S and Lawrence S *et al* 2011 Non-mare silicic volcanism on the lunar farside at Compton–Belkovich *Nat. Geosci.* **4** 566–71
[3] Siegler M A, Feng J and Lehman-Franco K *et al* 2023 Remote detection of a lunar granitic batholith at Compton–Belkovich *Nature* **620** 116–21
[4] Shirley K A, Zanetti M, Jolliff B, van der Bogert C H and Hiesinger H 2016 Crater size-frequency distribution measurements and age of the Compton–Belkovich volcanic complex *Icarus* **273** 214–23
[5] Zhang S *et al* 2020 First measurements of the radiation dose on the lunar surface *Sci. Adv.* **6** eaaz1334
[6] Jolliff B L *et al* 2011 Non-mare silicic volcanism on the lunar farside at Compton–Belkovich *Nat. Geosci.* **4** 566–71

Chapter 2

Space radiation beyond Mars

2.1 Beyond Mars: Jupiter and its Galilean satellites

Beyond the Red Planet, attractive destinations for astronauts, space colonizers and unmanned scientific missions include the Galilean moons surrounding Jupiter—the large, innermost Jovian satellites Io, Europa, Ganymede and Calisto (in order of distance from Jupiter). Unlike Jupiter itself, which is a big ball of astrobiologically uninteresting gas, the Galilean moons are surprisingly earthlike. For its size, Io is the most volcanically active body in the entire Solar System. Io is relatively small (mean radius of 1131.7 miles or 1821.3 km, making it just slightly larger than our Moon) and therefore does not have much residual accretion heat from its initial formation. Additionally, Io cannot boast of an inordinate supply of radionuclides to power its volcanic activity. But it has something that Earth does not—a heat source from gravitational tidal friction. As mighty Jupiter tugs and twists its innermost moon during its slightly eccentric orbit, it generates a good deal of frictional internal heat. This tidal heating is what powers all of Io's numerous volcanoes. As mentioned, Io lies within the Jovian equivalent of our Van Allen belt and so is in a high-radiation environment. With no oxygen or liquid water, but fiery volcanoes and ionizing radiation galore, Io might not be at the top of anyone's list of vacation spots.

There is some tidal heating as well on Europa, the next moon out. The structure of Europa, with a thick crust of ice surrounding a sub-glacial liquid ocean of water, along with the heat generated by Jupiter's gravity makes Europa extremely interesting astrobiologically. If life evolves as readily as some scientists believe, there is a good chance that Europa's vast ocean (ten times as deep as our ocean at about 60 miles or 100 km and maybe 2–3 fold as voluminous) is teeming with microbial or more sophisticated life of some sort or another. Europa does have all the basic astrobiological prerequisites: ample liquid solvent (with water, the 'universal solvent'), an adequate energy source (not photons from the Sun but rather heat from friction caused by Jupiter's gravity) and an adequate amount of biochemically suitable elements (and in this author's opinion, nothing competes with

doi:10.1088/978-0-7503-5444-8ch2

carbon and the associated versatility of organic chemistry). Upcoming missions to Europa should be very revealing regarding the validity of the assumption that anywhere life can evolve, it will evolve. The discovery of even simple microbial life in the water of Europa would be a truly monumental moment in scientific history, and the absence of any such life could be equally revealing.

But one must keep in mind that Europa is located smack in the middle of Jupiter's intense radiation belts. As one might expect based on Jupiter's size and magnetic field strength, its radiation belts are both intense and extensive. The sources of the ions include the solar wind, just as in Earth's Van Allen belts, but in addition, charged particles arising from Jupiter's ionosphere as well as from its innermost moon Io contribute to the Jovian radiation belts. As mentioned, Io is extremely volcanically active. It belches nearly 1 ton per second of noxious gases such as sulfur dioxide into outer space, where the molecules are ionized and the covalent bonds are broken. The resulting ions become trapped and contribute heavily to Jupiter's unique radiation belts. This results in strong electrical currents between Io and the ionosphere of Jupiter, which incites intense Jovian auroras much like our northern lights (aurora borealis) and southern lights (aurora australis) but far more extensive. There is a radiation entity filled with energetic charged particles called the **Io plasma torus**, which circles Jupiter. The charged particles mostly originate from Io's volcanoes. When these charged particles are accelerated to even higher energies by Jupiter's forceful magnetic field, they collide with each other (and with the surfaces of Io and Europa) to create x-rays through bremsstrahlung (which is described further in chapter 6).

Clearly, manned missions to Io, Europa and Ganymede would face substantial radiation challenges thanks to the huge, high-dose radiation belts surrounding Jupiter. NASA's planned Europa Clipper spacecraft, which will orbit Jupiter and make close approaches to Europa about 50 times during its mission, should gather a great deal of valuable information. Of the four Galilean moons orbiting Jupiter, Callisto might be the best bet when it comes to radiation safety, as it lies just beyond the perimeter of Jupiter's radiation belts.

An astrobiologically fascinating point is that Europa's liquid water is believed to be devoid of oxygen. Thus, unlike our lakes, streams and oceans, Europa's subsurface ocean could be a 'reducing environment'. Biochemically speaking, this might initially seem like an attractive environment for the origin of life, wherein primitive life forms could harvest the ample chemical energy of reduced compounds in the Europan ocean water. But to do so would require an adequate supply of oxidants. Here on Earth, with ample molecular oxygen, this is not an issue. But what about Europa and other worlds bereft of oxygen? Here is where all that ionizing radiation might ironically come in handy. By definition, ionizing radiation removes electrons from atoms and molecules. In other words, it is an oxidizing agent— arguably the most powerful form of oxidizing agent as it can put practically any chemical oxidants to shame. The net result could be that, unlike Earth where aerobic respiration allows harvesting of the chemical energy in reduced compounds by exploiting the abundance of the powerful oxidant, diatomic molecular oxygen, on Europa there could be a completely different metabolism based on harvesting chemical energy through the use of abundant radiation-derived oxidized products. Again, the upcoming Europa Clipper mission could prove to be most important astrobiologically since Europa has all the necessary ingredients for life: water, organic compounds and an energy source. But in addition to the heat energy provided by tidal forces, the high-radiation environment might also be providing the needed missing ingredients to get life going. So, while the intense radiation on Europa's surface would likely sterilize any life forms by breaking covalent bonds in biological chemicals, there could be some extremely interesting chemistry (biochemistry?) beneath the surface, safely shielded by the ice layer (which might be as thick as 15 miles or 25 km in some places). On the other hand, there are some extremophilic organisms right here on Earth (such as *Deinococcus radiodurans*, tardigrades (water bears) and some rotifers (wheel animalcules)) that are incredibly radioresistant and could perhaps withstand the doses below Europa's surface. Of course, that average surface temperature of only 110 K (−160 °C or −260 °F) presents its own difficulties!

2.2 Saturn and its moons

Moving beyond Jupiter, Saturn's largest moon, Titan, is a most fascinating world from a variety of perspectives. Arguably it is the most earthlike body in our Solar System. For one, it is the only moon in the entire Solar System (of >150 and counting) with a substantial atmosphere. And its atmosphere is not wimpy—it is about 1.5 times as dense as Earth's with a surface atmospheric pressure 50% higher than at sea level here. And like Earth, Titan has an atmosphere mostly made of nitrogen. With such a pronounced atmosphere, one might expect Titan to shine in fluorescent x-rays just as Earth, Mars and Venus do when solar x-rays interact with their atmospheres. But curiously, Titan does not shine in fluorescent x-rays. The explanation lies in the fact that Titan's atmosphere is a very good absorber of x-rays. This fact was discovered when the Chandra X-ray Observatory glimpsed Titan while it was crossing in front of the Crab Nebula, a known x-ray source. Titan's atmosphere effectively blotted out these x-rays, thanks to its effective x-ray-absorbing properties. And like Earth's atmosphere, Titan's even thicker atmosphere is able to filter out much of the charged particulate cosmic rays despite the absence of an intrinsic deflecting magnetic field. (Although flybys performed by NASA's Cassini spacecraft demonstrated the presence of a transient, induced magnetic field thanks to its proximity to Saturn and its extensive magnetosphere.)

And while it is smaller than Earth (radius of 1600 miles or 2630 km vs 3950 miles or 6357 km), it is 50% larger in diameter and 80% more massive than the Moon. Titan is the second largest moon in the Solar System (behind Ganymede), and its diameter exceeds that of the innermost planet, Mercury. It is far less massive than Mercury and thus its density is far lower as well. This low density leads to a surface gravity similar to the Moon's (about 83.5% that of the Moon). The Cassini-Huygens mission provided a great deal of data on Titan, including the 2010 discovery of what appeared to be volcanic activity. But unlike the lava spewing volcanoes of Earth, those on Titan appear to be cryovolcanoes—'ice volcanoes' that spit out water, ammonia, methane or other volatile compounds into an environment that is colder than their freezing points. Titan is not the only moon in the Solar System with suspected cryovolcanism; Cassini also photographed another moon of Saturn, Enceladus, with erupting geysers. These cryogeysers are responsible for the 'tiger stripes' near Enceladus's south pole and feed into Saturn's E ring (which is an outer member of Saturn's several rings; the prominent A and B rings are separated by the so-called Cassini Division and are more proximal to the parent planet than the E ring). Other moons with suspected cryovolcanic activity include Jupiter's Europa and Ganymede, Uranus's Miranda, and perhaps Neptune's Triton. Cryovolcanoes that spit out water imply that there is a reservoir of liquid water under the surface of Titan, just as is suspected for Europa. And since (at least on Earth) where there is water there is life, Titan is another astrobiologically intriguing prospect. But unlike Earth where volcanic activity involves molten rock and the heat engine stems from residual accretion heat and radioactive decay, or on Io where immense tidal forces stretch and compress the celestial body to provide the internal heat, the cryovolcan- ism on Titan is likely powered primarily through radioactive decay of primordial radionuclides within its mantle. Most fascinatingly, Titan appears to be the only Solar System body besides Earth with liquid lakes and seas. Furthermore, Titan's thick atmosphere seems to offer real weather. But this is where the similarities to Earth abruptly stop. The lakes and seas on Titan are not filled with liquid water but rather light, liquid hydrocarbons (liquid methane and ethane). And the rain on Titan that fills these lakes is made of methane/ethane raindrops. And of course, given the distance from the Sun, Titan is unwelcomingly cold. Typical surface temperatures hover about 94 K or −179 °C. Nevertheless, with an internal heat source and a possible subsurface liquid water ocean, it is possible that some form of life as we know it could exist below the Titanic surface. And one can have fun by speculating on life as we *don't* know it thriving in the liquid lakes of methane!

The Cassini space probe provided much information about Enceladus, another astrobiologically intriguing moon of Saturn. Its surface is highly reflective (i.e. it has a high albedo) thanks to its young coating of clean ice. This high albedo makes a cold place even colder (typical temperatures average 75.1 K or −198 °C). This highly reflective ice covering is likely a consequence of the cryovolcanic activity (cryogeyers) that shoot out water, ice, hydrogen and other volatiles with a composition similar to that of comets. These cryogeysers also feed into Saturn's E ring. The cryovolcanic activity suggests liquid water under the surface, and where there is water, there might be life. Thus, Enceladus is astrobiologically interesting, but from our angle, Enceladus is also radiologically interesting. The fact that Enceladus is still geologically active is probably partly due to a rich supply of short-lived radioisotopes at birth. In particular, Enceladus had a substantial sum of aluminum-26 and iron-60 that contributed to the early heating of Enceladus's interior. Radioactive decay will be covered in depth in chapter 5 but for now, aluminum-26 has a half-life of 717 000 years and decays through beta decay ('beta plus' or positron emission, specifically) as well as electron capture. It therefore releases energy through x-rays, Auger electrons and positrons (which undergo mutual annihilation with electrons to yield gamma rays). Iron-60 has a half-life of 2.6 million years and undergoes beta minus decay into cobalt-60, which itself is radioactive with a 5.27 year half-life. The decay of these short-lived primordial radionuclides contributed to the heating and melting of Enceladus's interior over and above the heat generation by its long-lived radionuclides. This kick-start allowed magma chambers to form that stretch and flex under Saturn's gravitational influence causing a tidal effect as on other moons. Additionally, Enceladus is in an **orbital resonance** with another Saturnian moon, Dione—Enceladus completes two orbits for every one that Dione does. This resonance also provides another tidal source of heating for Enceladus. Given that Enceladus is only about 500 km or 311 miles in diameter—smaller than the state of Texas and roughly the east-west width of England, Enceladus needs all the help it can get as far as internal heating goes. A body this small has a very unfavorable surface area:volume ratio when it comes to retaining heat, so for it to presently still boast any sort of cryovolcanic activity strongly suggests an unusual past history and an ongoing source of energy.

Saturn itself has a strong magnetic field, so, like Jupiter, it was expected to display x-ray emission near its magnetic poles. NASA's Chandra X-ray Observatory, however, revealed the highest x-ray brightness emanating from Saturn's equatorial region. Analysis of the x-ray spectrum showed a remarkable similarity to the Sun's x-ray spectrum leading to the conclusion that Saturn's equatorial x-rays are simply reflected solar x-rays. Jupiter also exhibits equatorial x-rays, but these are less dramatic or intense compared to its polar x-rays. Jupiter's north and south magnetic poles occasionally flare up in x-rays thanks to electrically charged particles from the Sun getting trapped in Jupiter's strong magnetic field and being accelerated towards the poles, where they then collide with atmospheric molecules and atoms to generate x-rays. (X-rays are produced in this fashion via a process called bremsstrahlung. This and other mechanisms of radiation production are reviewed in chapter 6.)

Other fascinating bodies in our Solar System include Uranus's Miranda and perhaps Neptune's Triton among many others. Uranus has a magnetosphere and radiation belts, as does Neptune, although Voyager 2 simply confirmed their existence; these radiation belts are yet to be fully characterized. Although no definite proof of extant cryovolcanoes have been confirmed yet, Miranda is interesting because it does show signs on its surface of past extensional tectonics—crisscross canyon patterns that might have been laid down by cracks in a frozen water surface as on Europa. Based on some surface features, this past cryovolcanic activity may have involved a viscous solution of ammonia and water or maybe ethanol. But closer observations will be required to corroborate these hypotheses. Miranda's parent planet Uranus has the curious feature of rotating on its side. With an axial tilt of 97.8°, Uranus is more like a rolling ball rather than a spinning top as it revolves around the Sun. (Although this is extremely odd, since Venus rotates 'backwards' relative to Earth and the rest of the planets, it is Venus, not Uranus, that wins the weirdness award when it comes to axial tilt. At a measured 177°, Venus's axial tilt is practically 180°, which is another way of saying it 'spins the wrong way' relative to its orbital path of revolution about the Sun.) And although Uranus's axial tilt is essentially orthogonal to its plane of revolution around the Sun, its strong magnetic field (about 50 times stronger than Earth's) is tilted 59° away from its spin axis. This misalignment causes Uranus's aurorae to be offset from its axial poles.

Incidentally, Earth's spin axis and magnetic axis are also tilted relative to one another but only by about 11°. And of course, Earth's spin axis (axial tilt or obliquity) is angled about 23.5° away from its orbital plane around the Sun.

Neptune's spin axis is also inclined relative to its orbital plane around the Sun. But rather than lying on its side like Uranus, Neptune's obliquity is 28°, which is comparable to Earth's 23.5° and Mars' 25° tilts. And Neptune has a strong magnetic field—approximately 27 times as strong as Earth's. But like many other planets, Neptune's magnetic field axis is misaligned from its rotational axis; the two are off by about 47°. Because of this misalignment, Neptune's magnetosphere displays marked variations as Neptune rotates (and it rotates quickly—a Neptunian day is only 16 h). Neptune has 14 known moons but the largest one, Triton, is quite quirky. It is the only large moon in the Solar System that orbits its parent planet in a direction opposite to that planet's spin. In other words, Triton has a retrograde orbit around Neptune. This strongly suggests that, rather than forming simultaneously with Neptune, Triton was a captured satellite. Also quite surprisingly, a 1989 Voyager 2 flyby glimpsed what could be cryovolcanic activity in the form of geysers blasting ice over 8 km (5 miles) up into Triton's thin atmosphere. Some of these geysers eject nitrogen that is heated into the gaseous phase by a 'solid-state greenhouse effect' in which solar radiation penetrates the transparent surface ice and heats the internal darker substrate until solid nitrogen sublimates, expands and erupts in the gas phase. Triton's western hemisphere is the home of the odd cantaloupe terrain, which is probably composed of dirty water ice and is unique to Triton. Furthermore, it is unknown why Triton's surface is so geologically young, at an estimated age of only ten million years. Whether there is a radioactive decay-driven source of internal heating that supplements this process and contributes to more conventional cryovolcanic activity on Triton is unknown. Hopefully, NASA's tentatively planned Trident mission will reveal some of Triton's secrets.

And finally, the dwarf planet/trans-Neptunian object/Kuiper belt object formerly known as planet Pluto recently revealed quite a surprise when it was found to be aglow in x-rays. Previously believed to be just a dull and dead rock in the region beyond Neptune, Pluto came up with a surprise as the New Horizons spacecraft discovered a thin atmosphere of nitrogen, methane and carbon monoxide surrounding this not-so-dull world. Following the discovery of an atmosphere, the Chandra x-ray telescope was then pointed towards Pluto and detected a handful of x-rays (just seven photons). These x-rays are not solar-like and therefore do not appear to be reflected x-rays from the solar corona. And since New Horizons has detected no vestiges of a magnetic field around Pluto, it is unlikely that the x-rays could be produced through captured charged particles accelerated and pointed towards the poles where they would strike atmospheric atoms and make x-rays and aurorae. New Horizons saw no signs of auroral activity near Pluto's poles. The most likely explanation for these x-rays is a charge exchange process between solar wind carbon, nitrogen and oxygen ions and neutral atoms at the edge of Pluto's exosphere. This discovery makes Pluto the most distant known source of x-ray radiation in our Solar System.

2.3 The edge of the Solar System

In order to speak meaningfully about radiation within, at the edge of, and beyond the Solar System, one must properly define the boundaries of the Solar System in the first place. Early definitions were simple and included the Sun and its orbiting planets and asteroids. But as more sophisticated observations were made, the definition of the Solar System grew to include the comets and one of their major sources, the Kuiper Belt. The Kuiper Belt is a vast donut-shaped region beyond the orbit of Neptune, thus the name trans-Neptunian objects or Kuiper belt objects for its inhabitants. And its inhabitants number in the hundreds of thousands, so the number of members of the Solar System per se increased dramatically with its discovery. Pluto, a dwarf planet, is also one of the largest and closest Kuiper belt objects. So, although it was recently demoted from its previous 'planet' status, perhaps Pluto gains some redemption in a Miltonian manner as the 'King of the Kuiper Belt'. (It now seems most ironic that its namesake, Pluto of Greek mythology, was the ruler of the underworld). But while the Kuiper belt is one important source of comets, there are others. The (still hypothetical) Oort cloud occupies a space between 2000 and 5000 AU (astronomical units—the average distance between the Earth and Sun, or 93 million miles or 150 million km) from the Sun. The inner region, or Hill cloud, is torus-shaped, while the outer Oort cloud is believed to be spherical. The Oort cloud is believed to be the home of billions or even trillions of bodies that can become comets that intermittently visit our neck of the woods. It appears that short-period comets (with periods up to 200 years or so) originate in the Kuiper belt whereas very long period comets (with periods ranging into millions of years) might originate from the outer Oort cloud. Incidentally, the Kuiper belt extends from the orbit of Neptune at about 30 AU out to approximately 50 AU from the Sun. In any case, one must arbitrarily decide where the Solar System ends and what is to be included or excluded. So, why not use radiation to provide the definition?

At the moment, our Solar System is defined by the extent of heliosphere, the outermost atmosphere of our Sun. The heliosphere in turn is described by the solar wind. The boundary where the solar wind is finally stopped by the interstellar medium and can no longer push back against the stellar wind of other stars is called the heliopause—the outer edge of the heliosphere and the outer extent of the Sun's magnetic field influence. Beyond this solar-wind-defined boundary lies interstellar space, but everything before this boundary can be considered part of the Solar System. And since the solar wind is a steady stream of energetic charged particles originating from the Sun (in other words, a form of radiation), we can rightly say that the borders of the Solar System are actually defined by radiation.

The heliopause is about 123 AU from the Sun. The Voyager 1 space probe, launched in 1977, crossed the heliopause and thus left the Solar System in August 2012. Also launched in 1977, the Voyager 2 spacecraft was actually launched before Voyager 1 but was overtaken. Nevertheless, it too has left the Solar System, joining its twin in crossing the heliopause on 10 December 2018.

2.4 Interstellar space radiation

Beyond the heliopause, which is the maximum reach of the Sun's magnetosphere, lies true interstellar space. Here, spacecrafts will no longer be bothered by the solar wind or sporadic solar particle events but are still at the mercy of a variety of radiation risks. For one, galactic cosmic rays will be unabated by our Sun's previously protective heliospheric deflection. And since it is the galactic cosmic ray contingent that contains the most energetic heavy charged particles, with the highest linear energy transfer, this now-unattenuated radiation will be free to exhibit its full potential on electronics and biological entities.

As will be explained in chapter 17, it is believed that galactic cosmic rays originate from supernova explosions and the ejected charged particles are further accelerated by a galactic magnetic field. But what if, while we are traveling in deep space, a nearby star were to actually go supernova? Supernovae will be described in greater depth in chapter 15, but in the unlikely scenario wherein a spaceship is blasted head-on with the radiation from a nearby supernova, it could be lights out, both literally and figuratively. Fortunately, we have a fairly good understanding about which stars will become supernovae and which of those might be in the vicinity, so appropriate evasive actions or shielding might be attempted. Such possible actions will be discussed in subsequent chapters.

Supernova explosions are potentially catastrophic but also are rare and predictable. Only stars in the final phases of their lives are capable of supernova explosions. These stars reveal their intentions to end their lives in a bang by becoming red supergiants. One such red supergiant star is Betelgeuse in Orion's shoulder. This one could blow any day now. (In fact, since it is an estimated 643 light years away, maybe it already has!) But supernovae are not the only way that stars can end their lives and emit radiation on their way out. After running out of fuel for fusion in about five billion more years, the Sun will go through its red giant phases and then will ultimately end its life as a white dwarf inside a planetary nebula. Although not nearly as dramatic as a supernova, ordinary novae occur in the last stages of life of intermediate-sized stars such as our Sun if they are part of a binary star system. Such binary systems, where one partner has already become a white dwarf, may alternatively become super-soft x-ray sources. Regardless of whether they are solitary or part of a binary system, during their last gasps, stars with masses like our Sun's blow off their outer layers of gas as their inner cores condense further and further into what is called degenerate matter (discussed in chapters 7 and 15). Photographs of beautiful planetary nebula have been provided by the Hubble Space Telescope among many others, including amateur astronomers. Among the more famous and beautiful examples are the Cat's Eye Nebula, the Helix Nebula and the Ring Nebula. But from a radiological perspective, we must remember that the magnificent display is because these are emission nebulae; they fluoresce in visible light after absorbing more energetic electromagnetic radiation of various wavelengths. And included among those wavelengths is intense ultraviolet blazing from the extremely hot central planetary nebula nucleus—a white dwarf star. The result is ionization and fluorescence of the previously ejected cloud of surrounding gas, which we see as a stunning planetary nebula. X-rays also abound in the central regions. Although these planetary nebulae are very short-lived from an astronomical perspective (a few millennia to tens of millennia), that is still a very long time from a human lifespan perspective! Thus, from a space traveler's perspective, the old adage of 'look but don't touch' might be applicable. One ought not get too close to their ionized and radiating beauty.

2.5 Gamma rays of death, doom and destruction?

Of all the sporadic, explosive sources of ionizing radiation in outer space, perhaps gamma ray bursts are the most notorious. As will be covered in chapter 31, gamma rays bursts signify the birth of a black hole. Some black holes are created by the collapse of an ultra-massive star, while others result from the merger of two neutron stars. Either way, their formation is accompanied by an incredible outpouring of high-energy gamma photons. The intensity of these gamma rays is quite impressive —they typically originate not from our Milky Way Galaxy but rather from distant galaxies millions or billions of light years away. NASA's Neil Gehrels Swift Observatory (previously called the Swift Gamma-Ray Burst Explorer) detected one gamma ray burst that was an incredible 12.8 billion light years distant. The fact that we can detect anything that was emitted from this unbelievable distance is a testament to the phenomenal power of gamma ray bursts. Thus, for space travelers and colonizers, a nearby, unmitigated gamma rays burst would certainly deliver a dose that could cause serious trouble. If one were wondering whether or not the energies of these gamma rays is high enough to penetrate typical spacecraft shields and harm human bodies by depositing a deep dose, the answer is a resounding yes. For instance, the Neil Gehrels Swift Observatory and the Fermi Gamma-ray Space Telescope were able to study a gamma ray burst in 2013 (GRB 130427A) that emitted gamma rays reaching as high as 94 GeV (94 billion eV). Considering that for cobalt-60 gamma photons (1.17 and 1.33 MeV) the tenth-value layer (i.e. the thickness of a barrier needed to attenuate the radiation intensity to one-tenth its initial value) is 21 cm of concrete, and for 18 MV bremsstrahlung photons used in radiation therapy, it is 45 cm of concrete, 11 cm of steel or 5.7 cm of lead, one might recognize that for gamma photons with energies in the tens of giga electron volt range, the barrier thickness would become inordinate. One cannot easily hide from a gamma ray burst. Fortunately, their aim is bad, meaning that as explained in upcoming chapters, the beam of gamma rays is relatively narrow and is ejected only from the axial poles of the newborn rotating black hole.

Another source of gamma rays in deep space are soft gamma repeaters. Initially believed to be another type of gamma ray burst, it is now known that soft gamma repeaters do not mark the beginning of a new black hole but instead are a completely different astrophysical phenomenon. As explained in more depth in chapter 33, soft gamma repeaters differ from gamma ray bursts in that the former are one-time events, whereas soft gamma repeaters (as the name would suggest) recur periodically. They are believed to be caused by 'starquakes' on a special type of neutron stars. Such starquakes are massive restructuring of the surface structure (the crust), and more importantly, the magnetic field, of unique, highly magnetized neutron stars. Neutron stars with these super-strong magnetic fields are called magnetars. Just how strong are the magnetic fields of magnetars? Recalling that 1 tesla (T) equals 10 000 gauss (G), our own geomagnetic field, which can move a compass needle, is roughly half a gauss or in the 30–65 μT range. A small toy magnet might have a strength of 100 G, while a strong neodymium magnet might have 2000 G or 0.2 T. A medical MRI unit, which can dangerously fling ferromagnetic objects across a room at high velocities, easily erase hard drives and wipe out credit cards, might have a magnetic field strength of 3 T. In comparison, magnetars might reach up to 100 billion T.

Rearrangements of their ultra-intense magnetic fields are probably what causes the neutron star's crust to quake, not vice versa. But when a magnetar undergoes the equivalent of a seismic shift in its magnetic field, not only does the neutron star's crust crack and quake, tremendous amounts of energy are also released in the form of x-rays and gamma rays. In most instances, the photons are of only modest energies—x-rays somewhere around 10 keV. But every once in a while, they can exhibit giant flares in which high fluences of hard gamma rays (up to tens of mega electron volts) are emitted.

Other sources of radiation within the Galaxy include cataclysmic variables, x-ray binaries, x-ray pulsars and gamma-ray binaries. These exotic sources of radiation will be touched upon in later chapters. As we journey even beyond the confines of our Milky Way Galaxy, even stranger sources of radiation can be encountered. Just as our planet has its associated Van Allen belts, the Galaxy as a whole has giant lobes of radiation emanating from its central axis region, orthogonal to the plane of rotation (the disc). These so-called Fermi bubbles have been detected by the eROSITA x-ray instrument (built by the Max Planck Institute for Extraterrestrial Physics and on the Spektr-RG space observatory). The Fermi bubbles protrude from both sides of the galactic bulge and contain an area rich in gamma radiation 20 000 light years across and even more extensive x-ray bubbles reaching 45 000 light years across. Given the diameter of the Milky Way Galaxy is an estimated 100 000 light years across but only about 1000 light years thick in the spiral arm disc, the size of the Fermi bubbles is quite impressive.

2.6 The Solar System sine wave

There are some strange sources of extragalactic radiation out there too. But we don't have to actually leave the Milky Way Galaxy to encounter some of the effects of extragalactic radiation. The Milky Way, like other spiral galaxies, rotates. It takes an estimated 225 million years for our Solar System (including our Sun and Earth) to take one lap around the Galaxy. But as the Solar System revolves around the Galaxy, it does not remain perfectly confined to the galactic plane; it bobs up and down in a sine wave as it circles around the center. In this manner, every 62 million years or so, our Solar System reaches a far point from the central galactic plane. We are presently comfortably located right near the center of the Milky Way's plane, but in 31 million years, we will reach a maximal distance 'south' of the plane. From there, in another 62 million years, the Solar System will be poking out a maximal distance to the 'north' of the galactic plane. Excursions beyond the central plane expose us to higher levels of radiation as we venture beyond the protective shielding of the Galaxy's magnetic field. This is akin to astronauts braving the realm of space beyond our own protective geomagnetic field and becoming vulnerable to the charged particulate radiation of space. It is of course worth remembering that only charged particles are affected by any magnetic field, be it from the Earth, the Sun or the Milky Way; electromagnetic radiation is unabated by such magnetism. It has been speculated that because of this sinusoidal pattern of travel around the galaxy, our biosphere might take a big hit every 62 million years when the sine wave reaches maximum amplitude. Paleontologists are examining the fossil record to determine if such predicted mass extinctions have been recorded in rocks.

Lastly, once we are well beyond the borders of our galaxy, we enter the realm of intergalactic space. In this deep vacuum, are we finally free from any forms of radiation? Absolutely not! Intergalactic space itself is filled with ionized hydrogen (that is, a plasma of free protons and electrons). Students are reminded about the difference between temperature and heat—something can have a very high temperature but still not possess much total heat energy if it is extremely rarefied. Intergalactic space fits this description well, as the free electrons and protons are flying about with staggering speeds (and since temperature is a reflection of the average kinetic energy or speed of the constituent particles, the temperatures are similarly staggering). The temperature of these intergalactic particles reaches up to 10 million K. Temperatures of the particles in between galaxies (in galactic clusters —the so-called intra-cluster medium) get even hotter. The electrons, protons and helium nuclei here can reach temperatures in the 10–100 MK range. Strong x-ray emissions abound as well in this intergalactic space.

2.7 Quasars and blazars

In addition to this background radiation in intergalactic space, there are intermittent flashes from gamma ray bursts (which as mentioned, sometimes originate from extremely distant galaxies as the gamma photons fly across the Universe) and radiation from the occasionally encountered quasar or blazar. Quasars and blazars are examples of active galactic nuclei. As such, they are both powered by supermassive black holes in the center of young galaxies. (And the adjective 'supermassive' is apt since some of these black holes can be tens of billions times more massive than the Sun.) As their black hole engines greedily consume massive amounts of matter, the infalling superheated material emits high-intensity electromagnetic radiation of all wavelengths. Some of this radiation illuminates the surrounding interstellar gas near the active galactic nucleus, allowing us to see objects over 13 billion light years away through glowing gas over 13 billion years old. Interestingly, ancient quasars illuminated and heated the gas and dust in their vicinities, which made that gas and dust less likely to condense under gravitational attraction into new stars and planets. Therefore, during the younger days of the Universe, star formation might have been somewhat impeded by the hot, highly active galactic nuclei found in early galaxies. Some valuable data on quasars has been gleaned through NASA's Wide-field Infrared Survey Explorer and the Southern African Large Telescope among other observatories.

The fact that light has a finite speed has fascinating consequences. If light had no speed limit, we would be unable to see quasars, since for the most part, quasars are very young galaxies and their bright lights have actually burned out billions of years ago. Nevertheless, we can see quasars and other truly ancient relics from the dawn of time, billions of years back, thanks to the set and limited speed of light. When we see objects that are 13 billion light years away, that literally means we are seeing the light emitted from that object 13 billion years ago. Thus, we are provided with a series of snapshots of the Universe at various ages. This is tremendously helpful when reconstructing the past history of our Universe. It is almost like paleobiologists and paleoanthropologists getting to see Neanderthals, dinosaurs and *Anomalocaris* in action by just peeking into a telescope. In essence, looking at quasars is looking into a time machine. And as explained in later chapters, we can deduce the distance and age of the various objects we see based on the Doppler effect and quantifying the red shifts of their light. The fact that older galaxies look quite different from younger galaxies is an indication that the Universe has been evolving. In contrast to the expectations of the 'steady-state hypothesis', the Universe does not appear static; it has been changing over time and continues to evolve. This is taken as evidence of the Big Bang theory, which will be covered in more detail in chapter 35.

While quasars emit intense radiation at a relatively steady rate, blazars are even more luminous but are highly variable in their output. Along with the various types of radio galaxies and Seyfert galaxies, quasars and blazars are all types of active galaxies (with active galactic nuclei) in which a spinning supermassive black hole is actively gobbling up matter at a prodigious rate. These supermassive black holes are surrounded by rotating accretion discs of infalling matter that, through friction and conversion of gravitational potential energy into kinetic energy and heat, have become scorching hot. In fact, it is now believed that these active galactic nuclei are all essentially the same thing, with the only difference being the angle that they are inclined to us. A radio galaxy has a rotational axis that is tilted nearly perpendicular to us, whereas a blazar has its axis of rotation directly aimed straight at us, like staring down the barrel of a rifle. And shooting out of the core, along the rotational axis, of any active galactic nucleus is a relativistic jet of ions. It turns out that the axial relativistic jets that generate gamma rays and other electromagnetic radiation are blazing straight at us in a blazar. Quasars also have their jets of ionized matter streaming from the axis of rotation somewhat aimed at us, but not as directly as blazars are. Accordingly, since it is less likely for an active galactic nucleus to be pointing straight at us, quasars are more common than blazars and radio galaxies are more common than quasars. By 2020, the quasar count was over 750 000 whereas there were only 66 blazars tallied. Among the closest blazars is Markarian 421 at 2.5 billion light years away, while Markarian 231 is a 'nearby' quasar 581 million light years from Earth. As explained later in the text, some of the enormous output of x-rays and gamma rays from quasars and blazars is due to heat generated by internal friction within the accretion disc of matter spiraling into the supermassive black hole at the core. This superheated matter can emit highly energetic electromagnetic radiation. And of course, the immense power of the central black hole's gravity converts potential energy into kinetic energy, and as the infalling particles gain more and more energy, they increase their temperatures to the point that they can emit what is called 'thermal', or 'blackbody', radiation on the energetic end of the electromagnetic spectrum. Additionally, since the spinning supermassive black holes are magnetically charged, their magnetic fields can accelerate charged particles along the lines of force. As explained in chapter 6, electrically charged particles that are accelerated emit radiation. When the charged particles are accelerated by a magnetic field, the emitted radiation is 'non-thermal' radiation (and specifically called synchrotron radiation in this case). In most instances, synchrotron radiation tends to be low energy (i.e. in the radio wave end of the spectrum) but as discovered by the Italian-Dutch BeppoSAX x-ray satellite, in certain circumstances, synchrotron radiation can be on the opposite, high-energy side of the spectrum. Finally, the relativistic jets of charged particles can also produce x-rays through a process called inverse Compton scattering. Essentially, this is the familiar Compton scattering process in reverse. In Compton scattering, x-rays interact with and eject atomic electrons and are then scattered in specific directions. Thus, the Compton effect is one of the main mechanisms whereby x-rays and gamma rays interact with matter. In inverse Compton scattering, a high-energy charged particle such as an electron interacts with a low-energy photon, resulting in an energy boost such that x-rays or gamma rays are generated. As the high-energy charged particles fly away from the active galactic nuclei in quasars and blazars, they will encounter the ubiquitous cosmic background radiation microwave photons, which can generate x-ray and gamma-ray photons through this inverse Compton effect. Furthermore, the relativistic jets of plasma from blazars, quasars and other active galactic nuclei are candidate sources for some of the highest energy galactic cosmic rays. If true, it implies that some 'galactic' cosmic rays are not from our own galaxy, but rather, trace their origins to far further galactic cores.

2.8 Hawking radiation

As we continue our tour of the Universe, we might encounter what is perhaps the most exotic source of radiation of all—Hawking radiation. At this moment, Hawking radiation remains hypothetical; its existence has not yet been verified. This theoretical form of radiation will be covered in chapter 34, but for now, we can note that its origin is exactly the opposite of a gamma ray burst. Gamma ray bursts mark the birth of a black hole; Hawking radiation derives from the death of a black hole. As will be explained further later, black holes might not be truly one-way streets after all. There is a possibility that, through quantum mechanical effects (explained in chapter 7), black holes may very slowly emit blackbody radiation. A popular explanation (that will be more properly refined and expounded upon in chapter 34) is based on the concept that the cold, empty vacuum of space is actually alive and seething with virtual particles and antiparticle pairs that transiently pop up and then undergo mutual annihilation into photon pairs. When this happens near the event horizon of a black hole, interesting things can happen. (The event horizon is the point of no return. Anything that enters is unequivocally trapped—doomed by the relentless gravity of the black hole.) These virtual particle–antiparticle pairs would ordinarily undergo mutual annihilation and become photons of energy by Einstein's $E = mc^2$. But every once in a while, rather than both particles and both photons falling into the black hole, one member of the pair falls in while the other falls out. In this way, the virtual particle that falls onto the outside of the event horizon becomes a real particle ('Pinocchio particles'). And then this real particle annihilates with a real antiparticle to create photons of energy. These photons may be thought of as the equivalent of blackbody radiation (which will be covered in depth in chapter 7). But if a body emits radiation, and radiation is energy, and energy is the equivalent of mass... this implies that black holes can gradually lose mass. And if something is emitting blackbody radiation, it implies that that body has a temperature. This is the so-called Hawking temperature.

If a black hole is not constantly being 'fed' matter, it is conceivable that the pace at which it radiates (i.e. loses mass) will exceed the rate at which it takes mass in. At this point, the back hole begins to 'evaporate'. As will be explained later, the black hole's evaporation rate is proportional to its temperature and the more mass it loses, the higher its temperature rises. Additionally, the evaporation rate is inversely proportional to the black hole's mass. Regardless of the details, the evaporation rate is not quick—a black hole the mass of the Sun would take around 10^{67} years to evaporate. A supermassive black hole, like the ones found in quasars and blazars, might take a googol (10^{100}) years or so! As they grow smaller and smaller, however, their rate of radiation release grows faster and more energetic. Towards the very end, tiny black holes might go out in a glorious blast of gamma rays. According to present models, some miniature black holes may have been formed around the birth of the Universe, shortly after the Big Bang. These primordial black holes could be reaching their explosive expiration dates at this time. Given the difficulty in identifying mini black holes, such a rapidly evaporating and exploding black hole could take us by surprise along our journey.

At this point, we will end our brief tour of the radiation-filled Universe and begin to study it in more depth. We will start with a short review of relevant physics and chemistry along with a more detailed review of basic applicable astronomy.

Further reading

[1] Schilling G 2001 Astrophysics. Quasars or blazars? It's all in the angle *Science* **292** 1985
[2] Ma F and Wills B J 2001 Discovery of hidden blazars *Science* **292** 2050–3
[3] Padovani P, Costamante L, Ghisellini G, Giommi P and Perlman E 2003 Synchrotron x-ray emission from flat-spectrum radio quasars. High energy blazar astronomy *Astronomical Society of the Pacific* 299 High Energy Blazar Astronomy *(Berkeley, CA, 17-21 June 2002)* ed L O Takalo and E Valtaoja (Piikio, Finland: Astronomical Society of the Pacific) 63–8
[4] Lisse C M *et al* 2017 The puzzling detection of x-rays from Pluto by Chandra *Icarus* **287** 103–9

IOP Publishing

Space Radiation
Astrophysical origins, radiobiological effects and implications for space travellers
James S Welsh

Chapter 3

Astronomy basics

3.1 Galaxy, clusters and superclusters

Our galaxy, the **Milky Way**, is a spiral galaxy that hosts over 100 billion (10^{11}) stars, perhaps as many as 400 billion. And our galaxy is but one of many billions of similar galaxies. The Milky Way is a fairly large galaxy, second to only the **Andromeda Galaxy** in our **Local Group**. The Local Group is an example of a **galaxy group**. A galaxy group is smaller than a **galaxy cluster**, which in turn is smaller than a **galaxy supercluster**. The Local Group contains over 50 galaxies, including ours.

A galaxy cluster is a large group—hundreds to thousands—of gravitationally bound galaxies. One example is the **Virgo Cluster**. Galaxy clusters, therefore, are larger than galaxy groups such as our Local Group with its 50–80 galaxies. Galaxy clusters can, in turn, be aggregated into even larger collections of galaxies known as superclusters. Thus, while the Local Group is not part of the Virgo Cluster per se, both the Local Group and the Virgo Cluster are members of the Virgo Supercluster. The Virgo Supercluster itself could be a member of an even larger **Laniakea Supercluster**.

3.2 Galactic size and motion

At just over 100 000 light years in diameter (excluding dark matter), the Milky Way is relatively large, as only the **Andromeda Galaxy** is comparable in size within the Local Group. Andromeda is actually far larger, with a diameter spanning more than 200 000 light years. Andromeda also hosts far more stars with its estimated one trillion inhabitants. Estimates of the number of stars in the Milky Way range from 100 billion up to 400 billion. (Incidentally the old mnemonic that there are about 100 billion stars in the Milky Way and 100 billion neurons in the human brain no longer appears valid, as the number of estimated stars increases while the number of estimated human brain cells is now down to only 86 billion).

The Milky Way's gravity affects its smaller nearby galactic neighbors. For instance, the Large and Small Magellanic Clouds, which are about 150 000 and 200 000 light

doi:10.1088/978-0-7503-5444-8ch3

years away, respectively, are influenced by the immense gravity of the Milky Way. Even more heavily influenced are the Sagittarius Dwarf and Canis Major Dwarf galaxies. These two 'small' galaxies are destined to collide with and be torn asunder by the Milky Way. The term 'small' is relative since they still contain over a billion stars apiece. Given the immense distances between stars, when this galactic collision occurs, it is quite unlikely that any individual stars will actually bump into one another.

Presently, Andromeda is about 2.48 million light years away from us. But it is headed in our direction. The two giants of the Local Group, the Milky Way and the Andromeda Galaxy, are projected to collide in 3–5 billion years. This will also be very unlikely to result in any 'stellar accidents'; most stars and planets will simply zip right on by one another at huge distances. Nevertheless, the vast clouds of interstellar gas and dust will undoubtedly crash, spawning a new nursery for star birth.

The entire Milky Way is itself racing through intergalactic space. We are presently hurtling towards a galaxy cluster called the Virgo Cluster at a linear velocity of about 150 km s^{-1}.

3.3 Super-sized galaxies

While the Milky Way may be a big shot within the Local Group, there are many galaxies beyond the Local Group that are far larger than the Milky Way or even Andromeda. One of the largest known spiral galaxies is UGC 2885 (Rubin's Galaxy), which spans the Milky Way and perhaps as much as 463 000 light years across. **UGC** stands for **Uppsala General Catalogue of Galaxies**, which is one of several galactic catalogues. The UGC presently lists over 12 900 galaxies. Another popular catalogue is the **Messier Objects Catalogue**, but this is a far smaller list (110 total) that contains not just galaxies but other objects such as nebulae. For instance, the well-known Crab nebula is Messier 1 (M1). The Andromeda Galaxy is also known as M31 or UGC 454 (and also goes by the name NGC 224 in the '*N*ew *G*eneral *C*atalogue of Nebulae and Clusters of Stars', which categorizes 7840 bodies). NGC 6872 appears to be even larger than Rubin's Galaxy, and presently holds the record for largest spiral galaxy at a whopping 522 000 light years in greatest diameter.

Elliptical galaxies, particularly those in galaxy cluster cores, achieve even larger sizes than spiral galaxies. M87 is the largest elliptical galaxy in the Virgo Cluster at 980 000 light years in diameter. But even larger still is the central galaxy of the Phoenix Cluster (**Phoenix A**) at a diameter of 2.2 million light years. Galaxies host a central **supermassive black hole**, but Phoenix A's central black hole, with an estimated mass of 100 billion solar masses (10^{11} M$_\odot$, where the M$_\odot$ symbol represents the mass of the Sun), falls into its own unique category of massive black holes nicknamed '**stupendously massive black holes**'. For reference, the Milky Way has a supermassive black hole at its center (corresponding to the radio source called **Sagittarius A***) that is an estimated 4 million solar masses.

Beyond even Phoenix A in terms of galaxy size is **IC 1101**, with an estimated 100 trillion (10^{14}) stars. IC 1101 is a member of the Abell 2029 cluster of galaxies in the constellation Virgo. At this time, IC 1101 is the most populated known galaxy and has an unmatched 6 million light year diameter. This is far larger than any spiral galaxy, or elliptical galaxy for that matter. Technically IC 1101 falls into a morphological class somewhere between the ellipticals and lenticular galaxy shapes. It should be noted that astronomers define the diameter of galaxies in various ways, such as the length that contains one-half the total light and the so-called isophotal diameter metric, among others.

While we have been focusing on the big galaxies here, one must recognize that most galaxies are smaller than the Milky Way. Most galaxies host ≤1% of the Milky Way's stars.

3.4 Radio giants

But star content is not the only metric for galactic measurement. Another yardstick for galactic extent relates to the radiation they emit. The supermassive black holes at the centers of galaxies spew out radiation in the form of charged particles. These charged particles gush out like two enormous geysers from the galactic center, along the axis of rotation and perpendicular to the galactic discs. This phenomenon, in which collimated, magnetized, charged particles (and electromagnetic radiation) come firing out in two opposite directions from a central structure along the axis of rotation is called **astrophysical jetting**. The rotational axis provides the easiest route out since the material does not have to pummel through the thicker lateral discs to fully escape. Astrophysical jets are the basis for exotic galactic oddities such as **quasars** and **blazars**, but they are not limited just to galactic phenomena. Astrophysical jets are also involved in strange stellar phenomena including **pulsars** and **gamma ray bursts**. Many astrophysical jets contain ejected ions traveling at nearly the speed of light. As such, these particles exhibit effects of special relativity and are sometimes called **relativistic jets** for this reason.

The ejected jets of charged particles emanating from a galaxy can be visualized through radio telescopes. That is, they are radio bright or radio loud. **Giant radio galaxies** exhibit two enormous lobes above and below the plane of their stellar discs. One notable example is the giant radio galaxy **Alcyoneus** (named after the giant in Greek mythology). The two gargantuan lobes of radio bright charged particles stem from the active, rotating, central supermassive black hole. These lobes extend 16 million light years into space. These lobes are the largest galactic structures yet measured.

3.5 Number of galaxies in the Universe?

Data from the Hubble Space Telescope suggests there are about 100–200 billion galaxies out there. But some estimates range up to one or two trillion galaxies. The recently launched James Webb Telescope should shed new light on this ongoing question. Although the Milky Way is larger than most galaxies at 100–400 billion stars, many galaxies host far more. As mentioned, the Andromeda Galaxy might be home to around a trillion stars. Curiously, one can do some quick math to obtain a very rough estimate of *nearly 1 mole of stars* in our Universe.

3.6 Milky Way anatomy

Being a part of the Milky Way, we cannot view it from afar and say anything conclusive, but the Milky Way is believed to be a **spiral galaxy.** Hence, like other spiral galaxies (such as Andromeda), it probably has **spiral arms**. The Galaxy has a nucleus with a central '**bulge**' of stars around the nucleus. Emerging from the bulge are two main bars. Thus, some texts consider the Milky Way to be a barred spiral galaxy. Some models argue in favor of two major spiral arms—the **Perseus Arm** and the **Scutum-Centaurus Arm** with several smaller minor arms or **spurs**. Others assert that the two main bars break up into smaller arms, leading to four main arms:

1. Norma
2. Sagittarius
3. Orion
4. Perseus

In this reconstruction, our Solar System is located on the inner edge of the **Orion arm**.

These spiral arms and spurs are part of a flat **galactic disc** surrounding the central bulge of stars. The diameter of the Milky Way is about 100 000 **light years**[1] across. The disc, however, is only 1000, or just a few thousand, light years thick, implying that the Milky Way is extremely flat. The whole disc is surrounded by a round, dim **halo** of stars. Most of these halo stars, however, are found in the approximately 200 **globular clusters** scattered throughout the halo. Nevertheless, the overwhelming majority of the Milky Way's stars reside in the disc itself.

3.7 The Solar System's location within the Galaxy

Copernicus showed us that the Earth is not the center of our Solar System. Further extending this Copernican revolution, Harlow Shapley was able to demonstrate that our Solar System is not the center of our Galaxy (based on the distribution of the globular clusters). Our Solar System is located in a spiral arm in the galactic disc, about 28 000 light years from the galactic center (roughly one-half to two-thirds of the way from the center). And just as Earth orbits the Sun, our Solar System orbits the center of the Milky Way Galaxy. Travelling at approximately 200 km s^{-1}, it takes about 225 million years for the Earth and Sun to complete one lap around the Galaxy. Incidentally, the spiral arms of our galaxy and other spiral galaxies are not static structures consisting of the same stars indefinitely. They simply represent the equivalent of stellar 'traffic jams' and stars may come and go through them as they travel at different speeds than the spiral arms do. For example, our Sun and Solar System travels somewhat faster than the Milky Way spiral arms do. This means that the Solar System moves in and out of these spiral arms as they catch up, move into, and then exit them as they both orbit the Milky Way's center. Given the age of Earth at about 4.567 Ga, since Earth was born we have taken about 18 revolutions around

---

[1] A light year is the distance that light in a vacuum would travel in one year. This is approximately 6 trillion miles (5.879×10^{12} miles). Another unit of distance occasionally used is the **parsec** or 'parallax-second'. A parsec is the distance from Earth defined by a parallax angle of one arcsecond as the Earth moves a distance of one Sun–Earth radius; a parsec equals 3.26 light years. This raises the question of what one average Earth–Sun distance is. This distance (the average semi-major axis of Earth's elliptical orbit) is called an **astronomical unit** or **AU**. One AU corresponds to roughly 93 million miles or 150 million km. Hence, a parsec is the distance from Earth corresponding to 1 s of parallax when the Earth has moved one astronomical unit.

the center of the Milky Way. Furthermore, just as our Earth's rotational axis is tilted 23.5° relative to the plane in which it orbits the Sun (the **ecliptic plane**), the whole Solar System's rotational axis is tilted relative to the galactic plane. In fact, at about 60°, this angle is far steeper than the 23.5° of inclination of Earth. (Incidentally, the term for inclination of a spinning object's rotational axis relative to its plane of revolution is called its **obliquity**.) Thanks to the combined tilts of the Earth and the Solar System, the center of the Milky Way Galaxy is nearly overhead at high latitudes in the southern hemisphere during the winter.

3.8 Star clusters

Stars are not randomly distributed throughout the galaxy but instead tend to bunch together in clusters. The two broad categories of star clusters are **globular clusters** and **open clusters**. The open clusters are generally loosely bound groups of stars, typically with a few dozen up to a few thousand constituents. Not to be confused with clusters of galaxies ('galaxy clusters'), open star clusters are sometimes called **galactic clusters**. There are about 1000 known open clusters in the Milky Way, with the Pleiades or Seven Sisters ('Subaru' in Japanese) being a familiar example that is visible to the naked eye.

In contrast to these loosely bound open clusters, globular clusters tend to be tightly bound, spherical agglomerations of stars with hundreds of thousands to millions of members. Globular clusters are usually found in the outer halo region of galaxies and are probably among the oldest part of any given galaxy. The Milky Way has around 170 globular clusters in its halo but some huge elliptical galaxies (e.g. M87) have been found to contain over 15 000.

Astronomers can estimate the age of star clusters. A given cluster contains stars that are all roughly the same distance from us and were formed around the same time. Despite these commonalities, clusters are likely to contain stars of various sizes. The most massive stars (spectral type O and spectral type B) are hot, luminous and blue—and they use up their fuel fast and have short lifespans. Thus, by determining the spectral type of the largest and most luminous main sequence stars in a given cluster (its so-called **main sequence turnoff point**), we can approximate the cluster's age. For instance, if there are no large and luminous stars (those of type O, B, A or F) in a cluster but we see smaller, cooler and less luminous stars (type G, K and M), we know that the cluster must be old enough to have burned out all its high-mass stars already. But it must still be younger than the lifespan of its mid- and light-weight members. In contrast, if the cluster has a lower main sequence turnoff point and is devoid of type G stars like our Sun, then we know it must be pretty old (our Sun will last approximately ten billion years). On the other hand, clusters that have very high main sequence turnoff points and still have some hot, luminous, blue stars (type O stars) must be extremely young (since type O stars only last a few million years). Incidentally, the smallest and dimmest main sequence stars (spectral type M) might live hundreds of billions or even over a trillion years.

The age of some globular clusters are amazingly old. For instance, one of the two closest globular clusters is NGC 6397. (The other nearby globular cluster is M4.)

NGC 6397 contains roughly 400 000 stars and can be seen with the naked eye on a good night. Despite being only 7800 light years away, astronomers have been able to calculate its age at 13.4 billion years old.

Both types of star clusters are useful to astronomers because all members of a given cluster are all roughly the same distance from Earth. By carefully noting the apparent brightness (**apparent magnitude**) of stars within a cluster and comparing those apparent magnitudes with **absolute magnitudes** (i.e. the true luminosity at a standard distance of 10 parsecs) of similar stars that we know the distance to, astronomers are able to calculate the distances of these star clusters.

3.9 Standard candles

By comparing an object's absolute luminosity to its observed luminosity, the distance to that object can be calculated through the inverse square law. But there are only a handful of objects that we know the absolute luminosity of. These astronomical objects are called standard candles. Within the Milky Way galaxy, **Cepheid variable** stars serve as a useful standard candle.

In 1912, while observing the Small Magellanic Cloud, Henrietta Leavitt discovered a period–luminosity relationship in certain stars called Cepheid variables that fluctuated in their brightness. In essence, the period of a Cepheid variable was a function of the absolute magnitude—the more luminous, the longer the period. This relationship is sometimes called Leavitt's law. Therefore, by measuring the pulsation period of these variable stars, one knew the inherent luminosity of that star and therefore could calculate the distance to that star. Another set of variable stars with a period-luminosity relationship are the **RR Lyrae** stars. These too can serve as standard candles in the same way as Cepheid variable stars. In comparison to Cepheid variables, however, RR Lyrae stars are more common but dimmer and are generally only useful for measuring distances up to globular clusters in the Milky Way halo. Another curious difference between Cepheids and RR Lyrae stars is that the latter do not exhibit a strict period-luminosity relationship in the visible wavelengths but do so in the infrared.

The high luminosity of Cepheid variables allowed Edwin Hubble to measure the distances to several galaxies. Hubble compared the distances to these galaxies against their spectral Doppler shifts (the relative shift of spectral lines of known elements in the gathered light towards the red or the blue ends of the electromagnetic spectrum). A redshift indicates recession of the light source whereas a blueshift indicates that the light source is moving towards us. The majority of galaxies are strongly redshifted, and the further away a galaxy is, the higher the degree of redshift. In this manner, Hubble was able to demonstrate that the Universe as a whole was expanding (despite some galaxies moving towards us). The observation that the further a galaxy is, the faster it is receding is known as **Hubble's Law** (or the Hubble–Lemaitre Law) and can be mathematically expressed as:

$$v = H_0 \times d$$

where v is velocity of recession, d is the distance to the galaxy in question, and H_0 is a constant known as **Hubble's constant**.

Aside from the use of standard candles, there are a few more direct means of measuring distance to heavenly bodies. One that works well for nearby objects such as planets, moons and asteroids in our Solar System is **radar ranging**. This simply employs radar (radio detection and ranging) to bounce radio waves off celestial objects and analyze the reflected data. In this way, the size, shape and distance of many objects in the Solar System can be directly measured. Another means of determining distance is through **parallax**. This measures changes in an object's apparent location relative to the fixed background of very distant stars as the Earth moves around the Sun. This requires knowledge of the Earth–Sun distance, which itself can be calculated via radar ranging of nearby planets and application of some geometry and trigonometry. As we attempt to measure the distances to objects far beyond the Solar System, we must resort to the use of standard candles such as the Cepheid variable stars and RR Lyrae stars. Other standard candles of importance in astronomy include:

1. Tip of the red giant branch
2. Planetary nebulae luminosity function
3. Type Ia supernovae
4. The Tully–Fisher relation
5. X-ray bursts.

The utility and relevance of some of these standard candles will be discussed in subsequent chapters.

3.10 The interstellar medium

The galactic disc is permeated with the scant **interstellar medium (ISM)**—extremely rarified gas and dust. Despite its low density, the vast expanse of the interstellar medium obscures much of the galaxy from visible light. Fortunately, examinations with other wavelengths have provided us with a far more comprehensive perspective than can be appreciated with the human eye. Some areas within the ISM can be denser than others. This inhomogeneity can lead to gravitational instability and collapse, culminating in star formation. Also, some of the denser regions can reach concentrations near 1×10^6 particles per cubic centimeter, which is sufficient for molecule formation. The resultant **molecular clouds** may span a distance of a few hundred light years and can lead to the creation of clusters of stars. **Star clusters** can contain thousands of stars. (Star clusters within galaxies must not be conflated with galaxy clusters).

Since meteors can burn up in our atmosphere (in the rarified mesosphere no less!), one might wonder if a speeding spaceship would similarly burn up if traveling through 'dense' regions in the interstellar medium. Recalling that at sea level (about one atmosphere of pressure (1.0325 bar, which in turn is 100 kilopascals)) and 0 °C, a mole of ideal gas (6.02×10^{23} particles) occupies 22.4 liters, students might enjoy calculating the difference between 'dense' portions of the interstellar medium (1×10^6 particles per milliliter) and our meteor-melting mesosphere. Approximately 99.9% of our atmosphere is in the lower regions (troposphere and stratosphere) so the mesosphere is only about 0.1% of the density at sea level. And the mesosphere reaches temperatures around −90 °C. (Answer: approximately 13 orders of magnitude.)

3.11 Nebulae

In addition to the rarefied interstellar medium, there are other collections of gas and dust in our Galaxy called nebulae. Most nebulae in our Milky Way are in the spiral arms. While some nebulae are the remains of dead stars (e.g. planetary nebulae such as the 'Eye of God' (Helix Nebula) and supernova entrails like the Crab Nebula), other nebulae are the birthplaces of new stars. For instance, the brightest nebula in the night sky is the Orion nebula (also called the Great Nebula in Orion). It is about 1500 light years away and has spawned several new stars over the last few million years.

One class of extensive, boundaryless gas/dust cloud is the **diffuse nebula.** These come in three varieties (with a good deal of overlap):

1. Emission nebulae
2. Reflection nebulae
3. Dark nebulae

Those that glow thanks to the light and radiation from newborn stars within them are called **emission nebulae.** Young, hot, massive blue stars (type O and B stars) emit an enormous amount of ionizing radiation that can cause their host nebula to glow. But smaller (Sun-sized) pre-main sequence, variable stars known as **T-Tauri stars** (that are in the process of evolving into hydrogen-fusing stars) are also found in emission nebulae and contribute to their glow. T-Tauri stars are in the infancy of their stardom at less than 10 million years old and have masses under 2–3 $M_\odot$. Some T-Tauri stars are intense sources of x-rays (as well as many other wavelengths) and produce powerful stellar winds that further brighten the emission nebulae they reside in. In fact, these special stellar winds go by the name of **T-Tauri winds.** The T-Tauri winds are akin to the solar wind, but the output is far higher, amounting to in some cases as much as 10^{-7} $M_\odot$ of ionized matter blasted away per year.

Ultraviolet output during the T-Tauri phase is often 50-fold higher than during the hydrogen-fusing main sequence phase. Their radiation comes not from hydrogen fusion in their cores but rather solely from the conversion of gravitational potential energy into light and heat during continuous contraction. The surface temperature of T-Tauri stars is similar to main sequence stars such as our Sun but are far brighter overall thanks to their larger sizes.

More massive stars (>3 $M_\odot$ but less than 8 $M_\odot$) have a very similar early stage of evolution during their pre-main sequence days. But these larger versions are not T-Tauri stars (which are named after the first example of their class, T-Tauri) but rather **Herbig Ae/Be stars** (named after the astronomer who first identified them, George Herbig). Fully mature stars in this size range correspond to what are known as spectral types A and B, thus the designation Ae/Be. While both T-Tauri and Herbig Ae/Be stars are variable, T-Tauri stars exhibit far greater swings in brightness ('magnitude'). Still more massive stars (>8 $M_\odot$, type O stars) have not yet been seen during their pre-main sequence stage, probably because they pass through this stage so quickly.

T-Tauri stars are in a phase of development where they are heated by the gravitational energy provided by contraction, but their core temperatures have not yet reached the fusion flashpoint. This stage of stellar evolution is known as the **Hayashi track** on the **Hertzsprung-Russell diagram** (**HR diagram**; a plot of luminosity versus temperature, which will be discussed later). While on the nearly vertical Hayashi track, the early stars maintain roughly the same surface temperature but decrease their overall luminosity because of gravitationally induced shrinkage; meanwhile, their cores continue heating from the compression. Following the Hayashi track, stars with greater than 0.5 solar masses may take one final track on the HR diagram before hydrogen fusion begins. This is called the **Henyey track** and is nearly horizontal on the HR diagram, unlike the vertical Hayashi track. Sun-sized stars will follow this pattern over a period lasting up to 100 million years. Thus, an early Sun-like star will begin to condense out of the dust in a nebula, continue contracting and transitioning from a **protostar phase**, onto the Hayashi track as a T-Tauri star, and then onto the Henyey track, until finally its core can initiate hydrogen fusion. This marks the dawn of the **main sequence** phase of the star's evolution, where it will spend most of its lifetime.

Returning to emission nebulae in general, the ultraviolet radiation can ionize or excite atoms in the clouds of gas and make them glow, as the electrons recombine with nuclei or fall back to lower energy states. Energetic photons can ionize hydrogen to yield **HII regions**. (Neutral hydrogen is found in **HI regions**). Sufficiently energetic photons from very hot stars might even be able to ionize helium within these emission nebulae.

Reflection nebulae also glow, but their shine is from another source. Rather than directly emitting photons because of atomic recombination or deexcitation, reflection nebulae shine because photons from the visible part of the spectrum reflect off small space matter particles ('dust'). Since this reflection is most pronounced for blue light, most reflection nebulae have a bluish hue.

Absorption nebulae, also known as dark nebulae, are denser and contain so much dust that they block any view of their visible light from Earth. Nevertheless, even more so than emission or reflection nebulae, dark nebulae are sites of active star formation. It is important to recognize that all three nebula types can co-exist in the same region. One might reasonably ask how we know dark nebulae are sites of star formation if we cannot see anything inside them. Fortunately, their inner workings are revealed to us through infrared and radio telescopes.

It should be noted that the name 'nebula' is also used for several other objects, including planetary nebulae (which are the remnants of medium-mass stars like our Sun after they expand through red giant stages and end as white dwarfs) and spiral nebulae (which are actually distant galaxies like our Milky Way). Additionally, the scattered remains of huge stars blasted into space after supernova explosions are also called nebulae (such as the famous Crab Nebula). Further adding to the confusion is the fact that there can be some overlap, such that both planetary nebulae and some diffuse nebulae can both be classified as emission nebulae (since they glow because of the ionizing ultraviolet radiation within them). Finally, there are reflection nebulae called **protoplanetary nebulae** (or **preplanetary nebulae**), which represent a brief period in a star's late stages between the asymptotic giant branch phase on the HR diagram and the subsequent planetary nebula phase after the star has ejected much of its mass into space as a glowing, expanding cloud of ionized gas. The name 'protoplanetary nebula' is a most unfortunate choice since it can be confused with something called a **protoplanetary disc**, which surrounds young, newly formed stars such as T-Tauri and Herbig Ae/Be stars. These protoplanetary discs are rotating, circumstellar collections of gas and dust orbiting a young star, which, according to the **nebular hypothesis** of stellar system evolution, will eventually coalesce into the planets orbiting a star.

Proplyds (ionized, glowing protoplanetary discs) are relatively recently discovered phenomena surrounding young stars in their T-Tauri or Herbig Ae/Be phases. Most proplyds have been discovered in the Orion Nebula. Some are close to their luminous parent star and glow because of the star's luminosity. Others are further from the parent star and appear as dark outlines. It is believed that some proplyds may develop into planetesimals, which in turn can lead to planets, moons and asteroids. Hence, proplyds may be embryonic planetary systems. Curiously, proplyds appear to be undergoing **photoevaporation**, a process in which radiation ionizes their gases, causing it to gradually scatter from the central star. In some cases, the proplyds appear to be moving thanks to shock waves caused by their host star irradiance.

3.12 Our place in the Galaxy

Our Solar System is located in the inner edge of the **Orion Arm** (or Orion-Cygnus Arm) of the Milky Way. Also within the Orion Arm is the **Local Bubble**, a 300 light year wide region (perhaps even 1000 light years in diameter) of decreased density in the interstellar medium. And within the Local Bubble is the **Local Interstellar Cloud** (also called the **Local Fluff**), an area of slightly higher density. Our Solar System is presently moving through this Local Interstellar Cloud.

Our closest known stellar neighbor is actually a triple-star system called **Alpha Centauri**. At about 4.4 light years away, the red dwarf **Proxima Centauri** orbits a pair of larger stars, **Alpha Centauri A** and **Alpha Centauri B**. Incidentally, an **exoplanet** has been discovered orbiting Proxima Centauri in the so-called **circumstellar habitable zone** (where water might exist as a liquid on the surface, given sufficient atmospheric pressure). This exoplanet, **Proxima Centauri b** orbits its parent star, Proxima Centauri, much closer than Earth orbits its Sun—roughly only 0.05 AU (approximately 7.5 million km). (Note that exoplanets are named after their parent star and adding a lowercase letter). This is in fact far closer to its host star than Mercury is to the Sun (semimajor axis = 0.39 AU). But Proxima Centauri is far dimmer and cooler than the Sun. So, despite the proximity to its parent star, it is conceivable that liquid water could exist on the surface of this presumably Earth-like planet.

However, despite it being relatively dim and cool, Proxima Centauri is a **flare star**—a variable star that sporadically undergoes transient but dramatic increases in brightness. The mechanism remains uncertain, but it is possible that these flares are akin to solar flares that release enormous amounts of energy due to magnetic phenomena. In any case, the emitted x-rays and other radiation might erode any atmosphere on Proxima Centauri b, and might pose severe challenges for life on this planet.

3.13 The galactic carousel and cosmic rays

The stars in the galactic disc (including our Sun and the Solar System) circle the Milky Way's center about once every 225 million years; this is known as a galactic year. The stars in the galactic disc revolve about the center in roughly circular paths, also remaining roughly—but not exactly—in the same plane. As stars orbit the galaxy, they do not stay perfectly confined to the galactic plane. From a distant side view, the appearance of the Solar System's orbit might be aptly described like a carnival carousel ('merry-go-round'). Like the horses on a merry-go-round bobbing up and down as they circle around, the Solar System similarly weaves an up-and-down sinusoidal orbit around the galactic center. The amplitude of each up and down sine wave might be around 1000 light years, and the period is tens of millions of years, maybe 60 million years for a full cycle. This periodic oscillation causes the Solar System to protrude slightly out of the galactic plane every 30 million years or so. Hence every 30 million years or so, we are protruding out of the galactic plane and exposing ourselves more to the intergalactic medium. As discussed elsewhere, these excursions above and below the galactic plane can subject the entire Solar System (including our Earth) to higher levels of radiation during this period of exposure.

Although the intergalactic medium is extremely sparce, it contains energetic charged particles—protons, electrons, and helium nuclei (alpha particles)—at concentrations of tens to hundreds of particles per cubic meter. The Milky Way's path through this plasma creates a bow shock. This bow shock in turn accelerates these subatomic particles to even higher energies. Just as Earth has its protective geomagnetic field and the Sun provides a solar magnetosphere that protects our entire Solar System, the galaxy also has a magnetic field that confers some shielding against these intergalactic particles. Venturing beyond this protective shielding may pose some radiation risks.

3.14 The structure of the Solar System

Hypothetically, on the outskirts of our Solar System lies a big ball of comets known as the **Oort Cloud**. Still unobserved, the Oort Cloud may contain up to a trillion small, icy **trans-Neptunian** (i.e. beyond the orbit of Neptune) objects and is thought to be the source of **long-period comets**. (Historically, comets with periods greater than 200 years were called long-period comets, those with periods shorter than 20 years were called Jupiter-family comets, and those with intermediate periods were called Halley-family comets).

Unlike the flattened disc of planets and asteroids, the Oort Cloud is a spherical structure surrounding the Solar System at roughly 50 000 AU (a bit under 1 light year) but possibly extending to 100 000 AU (1.58 ly). The Oort Cloud objects move very slowly but can be perturbed by approaching stars. If a nearby star does perturb the Oort Cloud, comets could come plummeting inwards, potentially threatening Earth.

More proximal to the Sun than the Oort Cloud is the **Kuiper belt**, another reservoir of trans-Neptunian objects that can become comets. Unlike the spherical Oort Cloud, the Kuiper belt is flatter and torus shaped. It extends approximately 30–50 AU from the Sun. Formerly considered a planet, Pluto is now catalogued as the largest (and second most massive, after Eris) of the many thousands of Kuiper belt objects. It was previously believed that the Kuiper belt was the source of periodic comets with orbital periods of less than 200 years (the Jupiter-family comets and the Halley-family comets). However, the Kuiper belt might be relatively stable, and the true home of periodic comets may be the **scattered disc**, a third group of trans-Neptunian objects. **Scattered disc objects** may have extremely eccentric orbits with steep inclinations to the **ecliptic** (the plane of Earth's orbit around the Sun).

Among the planets, there is a division between the **giant planets** (also known as Jovian planets or gas giants) and the **terrestrial planets**. In order of distance from the Sun, the terrestrial planets include Mercury, Venus, Earth and Mars. These also happen to be the four closest planets to the Sun and therefore constitute the **inner planets**. The giant planets include Jupiter, Saturn, Uranus and Neptune in order of distance from the Sun. The four giant planets constitute the **outer planets** in our Solar System. Between the two groups (that is, between Mars and Jupiter) lies the **asteroid belt**. Curiously, although there are well over 100 000 individual asteroids in the asteroid belt, the total mass of all of them is just around 3%–4% of the mass of our Moon. A useful mnemonic for recalling the approximate distances of the planets and asteroid belt from the Sun is provided by **Bode's Law**: $d = (n + 4)/10$ AU, where n is given by the series 0, 3, 6, 12, 24, 48, 96, 192, etc. Thus, the third planet from the Sun, our Earth, would be $(6 + 4)/10 = 1$ AU from the Sun. Presently considered just a numerological curiosity, Bode's Law (also called the Titius-Bode Law) fails for Neptune but provides good rough estimates for the remainder.

3.15 Kepler's Laws of planetary motion

In the early 17th century, Johannes Kepler formulated his three laws of planetary motion. These can be stated succinctly as:
1. The planets orbit the Sun in ellipses, with the Sun at one of the foci.
2. As a planet orbits the Sun, it will outline the same area of space between it and the Sun in the same amount of time, regardless of where it is in its orbit
3. A planet's orbital period is proportional to its distance (its semi-major axis).

An ellipse is a conic section defined by two focal points. One property of an ellipse is that the sum of the distances from the two foci to any point on the ellipse is a constant. The eccentricity of an ellipse ranges from zero for a circle to one for a parabola. If the eccentricity is greater than one, the conic section is a hyperbola. If the eccentricity approaches infinity, the ellipse collapses into a straight line. Since ellipses are not necessarily circular, we cannot simply define its diameter. Instead, we assign the name major axis to the longest axis, while the shortest axis is called the minor axis. The **semi-major axis** is one-half the length of the major axis. It is this parameter that we most often allude to when describing planetary motion. Because of its elliptical orbit, the distance between a planet and the Sun is constantly changing as that planet revolves around the Sun. A planet's **perihelion** is the point when it is nearest to the Sun during its elliptical orbit. The point of greatest distance to the Sun is called the **aphelion**.

Kepler's Second Law can be understood by drawing an imaginary line joining a planet and the Sun. As the planet orbits the Sun, a triangle is created by the line from time a to time b. The area of this triangle is always the same for any similar time interval, regardless of where in the planet's orbit it happens to be. What this amounts to is that a planet moves faster when it is closer to the Sun and moves slower when it is further from the Sun. Kepler's Second Law is just a quantitative way of stating this fact. Therefore, by Kepler's Second Law, a planet is moving fastest when it is at perihelion and slowest at aphelion.

Kepler's Third Law relates a planet's orbital period to its distance from the Sun. More specifically, it says that the square of the orbital period of a planet is directly proportional to the cube of its semi-major axis. Hence, the period of a planet increases the further away it is. For instance, Mercury takes only 88 d to orbit the Sun, whereas Earth takes 365 d for a revolution. Similarly, Mars requires 687 d for a revolution, while Saturn takes a whopping 10 759 d to take a lap. Mathematically, Kepler's Third Law can be expressed as:

$$\frac{a^3}{T^2} = c \approx 7.5 \times 10^{-6} \quad \text{AU}^3/\text{days}^2,$$

where a is the semi-major axis and T is the orbital period.

Alternatively, this can be rearranged as:

$$T^2 = k \cdot a^3,$$

k is just a different form of the proportionality constant. Kepler's Laws were instrumental in the development of Newton's Laws of Motion as well as his Universal Law of Gravitation. In fact, Kepler's Third Law is often written in a form derived by Newton that is often called **Kepler's Third Law in Newton's form**:

$$T^2 = \left(\frac{4\pi^2}{G[m_1 + m_2]} \right) a^3,$$

where G is the gravitational constant, m_1 is the mass of the Sun, and m_2 is the mass of the planet in question.

Because the mass of the Sun is so much greater than that of any planet (e.g. about 333 000 times the mass of Earth or about 1048 times the mass of Jupiter), we can ignore m_2 in the equation, reducing it to:

$$T^2 = \left(\frac{4\pi^2}{Gm_1} \right) a^3.$$

From this, one can calculate the mass of the Sun by plugging in some familiar numbers such as Earth's period of revolution being 1 year and its distance from the Sun being 1 AU. Students may enjoy showing that the Sun is about 1.99×10^{33} g.

Also, in our Solar System, the planets revolve around the Sun in relatively circular orbits. Note that this is often not the case with exoplanets orbiting distant stars in other planetary systems. But this fact allows us to provide a rough calculation of planetary velocity. If we say that a planet's orbit is circular, then the orbit is given by $2\pi r$, where r is the radius of the planet's orbit. Through 'distance equals rate times time', we see that $2\pi r = vT$. (Assuming not only a perfectly circular orbit but also a constant orbital velocity, v.) Thus, the period, T is given by $T = 2\pi r/v$. Plugging this into the equation for Kepler's Law in Newton's form, we see:

$$T^2 = (2\pi r/v)^2 = \left(\frac{4\pi^2}{Gm_1}\right)a^3.$$

Thus, rearranging and cancelling, one gets:

$$m_1 = v^2 r/G.$$

And from this and solving for v:

$$v = (Gm_1/r)^{1/2}.$$

So, regardless of the simplification and rearrangements, one can still see that the velocity of the orbiting planet is inversely proportional to the square root of the distance from the Sun:

$$v \propto (1/r)^{1/2}.$$

3.16 Using Kepler's Law to measure the mass of galaxies

There are two general methods of estimating the mass of a galaxy. The first approach is simply weighing by counting. Astronomers can basically add up all the stuff they see or know about, such as the stars and the interstellar medium. If we apply this technique to the Milky Way, we estimate that 85% of the mass is stars, while only 15% is the planets, moons, asteroids and comets plus the interstellar medium, gas and dust. Our Milky Way is mostly stars by the counting methods. Applying this approach to elliptical galaxies, however, reveals that they are nearly all stars, whereas small, irregular galaxies are nearly all gas and dust. However, Kepler's Third Law provides us with another, and more reliable, means of estimating the mass of a galaxy.

Recalling the equation, $m_1 = v^2 r/G$ for a planet orbiting the Sun, we can generalize the formula to calculate the mass of a galaxy. Rather than reflecting the mass of the Sun based on the velocity of planets orbiting at specific distances, we can now replace m_1 with the mass of the galaxy as a whole and calculate it based on the velocity of stars orbiting that galaxy at specific distances from the center. Kepler's Law says that if we know the mass, we should be able to calculate the velocity of an object orbiting that mass if we know its distance. (Note that we do not need to know the exact mass of that orbiting body as long as its mass is negligible compared to the central mass—and a single star's mass is certainly negligible when compared to its parent galaxy). Conversely, if we know the velocity of an orbiting star and how far that star is from the center of the galaxy, we can calculate the mass of that galaxy. It should be pointed out that the mass m_1 in the original application of Kepler's Law was the mass of the Sun—the central star of the Solar System. In this application however, m_1 represents the mass of the entire galaxy interior to the orbiting star at distance r from the center; it is not just the mass at the center of the galaxy. As the distance r from the center increases, there are more and more stars inside that radius. Based on observations, one can create a rotation curve—a plot of star linear velocity versus distance from the galaxy's center.

In our Solar System, the Sun contains nearly 99.9% of the mass. Therefore, planets revolving at long distances from the Sun will be traveling far slower than the inner planets. But when we take a close look at galaxies like Andromeda (M31), the rotational velocity remains fairly constant, even far from the center. And this oddity is not restricted solely to the optically visible stars, it holds true for interstellar hydrogen gas as well. Based on the velocity of the stars and interstellar gas, it appears that there is a lot more mass than meets the eye. If one does the same thing with the smallest spiral galaxy in the Local Group, the Triangulum Galaxy (also known as M33 or NGC 598), the discrepancy in the rotation curve is even more marked—it differs from calculations by 300%. The observed velocities of stars and hydrogen gas on the perimeter are consistent with a huge amount of mass beyond the visible disc of stars. The measured amount of hydrogen indicates that this gas (or any other gas) cannot come close to accounting for this additional but unseen mass. The invisible mass that is causing the stars and interstellar gas and dust to rotate far faster than predicted is now known as **dark matter**. A great deal of effort is being devoted to characterizing and understanding this mysterious material.

An alternative hypothesis to account for the observed motions of stars and gas near the perimeter of galaxies is called **MOND**, an abbreviation for *Mo*dified *N*ewtonian *D*ynamics. In essence, this hypothesis holds that at great distances from the central source of gravity, Newtonian physics and Kepler's Laws, such as the inverse-square relationship, no longer hold in the form that works for closer distances. Opponents of this model argue that MOND might explain the motion of stars in a galaxy but fails at explaining the motion of galaxies in galaxy clusters. This is another area of intense investigation presently.

Further reading

[1] Anglada-Escudé G, Amado P and Barnes J *et al* 2016 A terrestrial planet candidate in a temperate orbit around Proxima Centauri *Nature* **536** 437–40

IOP Publishing

Space Radiation
Astrophysical origins, radiobiological effects and implications for space travellers
James S Welsh

Chapter 4

Elementary physics, chemistry and biology

4.1 Ionizing radiation units

The basic unit of ionizing radiation absorbed dose is the **gray**. A gray (**Gy**) is defined as 1 joule (**J**) of energy absorbed per kilogram of matter:

$$1 \text{ Gy} = 1 \text{ Jkg}^{-1}.$$

One gray is the equivalent of 100 cGy (centigray) or, to use the older units of dose, 100 **rads**. Obviously a centigray is now defined as one-hundredth of a gray, but the centigray or rad was originally defined as 100 ergs of energy absorbed by 1 g of matter. Matter may be loosely defined as something that has mass and takes up space (i.e. has a volume).

4.2 Derived versus fundamental quantities and the SI units

The gray is an example of a **derived unit**, meaning that it is a unit that can be derived from other, more basic units, the so-called **fundamental units**. The four fundamental physical quantities and their associated units are mass (measured in kilograms), length (measured in meters), time (measured in seconds) and electric current (measured in amperes). With rare exceptions, throughout the text we shall try to stick with units used in the International System of Units (abbreviated SI in all languages, but tracing back to the French, Système International). The base SI units and their abbreviations include:

s for seconds as the unit of time
m for meters as the unit of length
kg for kilograms as the unit of mass
A for amperes as the unit of electric current
cd for candela as the unit of luminous intensity
mol for moles as the unit of amount (of atoms, molecules, etc).

doi:10.1088/978-0-7503-5444-8ch4

4.3 Work and energy

Using the fundamental SI units, the joule is the derived unit of energy or work. The joule is defined as one kilogram-meter per second:

$$1 \text{ J} = 1 \text{ kg} \cdot \text{ms}^{-1}.$$

Energy can be in the form of kinetic energy (KE), thermal energy, potential energy (PE) and other ways that will be discussed below. Work is the energy transferred to or from an object as it is displaced by a force. Thus, in its simplest form, work may mechanically be considered the product of a force (F) and the distance (d) over which force is exerted:

$$W = \vec{F} \times d \text{ or more generally, } W = \int_{a}^{b} \vec{F} \cdot ds \text{ where s represents displacement.}$$

Note that, depending on the context, the distance or displacement may variably be designated by x, l, r or s. Also note that the line over the force symbol ($\vec{F}$) indicates that force is a vector. Sometimes vectors are represented this way, while other times they are simply boldfaced. (This in part, depends on the ease or difficulty of representing vectors with the line over the symbol in the word processing program!)

If the force is exerted at an angle θ with respect to the displacement, the equation can be expressed as:

$$W = \vec{F} \, d \cos \theta.$$

Kinetic energy is classically defined as one-half of the product of mass and the square of the velocity:

$$\text{KE} = \tfrac{1}{2}\text{mv}^2.$$

4.4 Velocity and acceleration

Velocity (**v**) is a familiar vector quantity that is formally defined as the ratio of the change in distance (x) typically measured in meters to the change in time (t) measured in seconds:

$$\mathbf{v} = \Delta x/\Delta t.$$

The instantaneous velocity can be expressed using calculus notation as:

$$\mathbf{v} = =\frac{dx}{dt}.$$

The boldface **v** here is intended to indicate that velocity is a vector quantity, which means that it cannot be expressed as a single number; a vector is characterized by two independent properties that each have a magnitude. In the case of velocity, the vector is a composite of the magnitude of distance and the direction of distance. (And in the case of this word processing program, it has proven challenging to use the conventional arrow over the symbol, thus the use of the alternative boldface notation for vectors.)

Acceleration (**a**) is another vector quantity that is the ratio of the change in velocity to the change in time:

$$\mathbf{a} = \Delta v / \Delta t.$$

Since velocity is simply $\frac{dx}{dt}$, one can see that acceleration can be related to distance as the change in distance per unit time squared:

$$\mathbf{a} = dx / dt^2.$$

A special example of acceleration is the average acceleration due to Earth's gravity at sea level. This is symbolized by **g** and is approximately 9.8 m s^{-2} (or 32.2 ft s^{-2}). With this in mind, gravitational potential energy, PE near the surface of Earth is often expressed as:

$$\mathrm{PE} = m\mathbf{g}h,$$

where h is the height above Earth's surface.

4.5 Force, Newton's Laws of Motion and the Law of Universal Gravitation

With this background, we can return to the concept of force in more detail. Force (**F**) is another vector quantity. Force can be related to acceleration through Newton's Second Law of Motion:

$$\mathbf{F} = m\mathbf{a}.$$

Thus, force may be thought of as the cause of a change in acceleration of a massive object. Newton's First Law of Motion states that an object in uniform motion stays in motion and an object at rest stays at rest, unless acted upon by an outside force. This tendency to resist change in the absence of an applied force is called inertia. In such cases, where an object is either at rest or moving at a constant speed, the sum of the forces acting upon the body must be zero. Mathematically, Newton's First Law can be expressed as:

$$\sum \mathbf{F} = 0$$

Newton's Third Law of Motion is often stated as 'for every action there is an equal and opposite reaction'. In other words, any action or force results in an equal (in magnitude) force that is opposite (in direction). For instance, when a bird's wings push air downwards, the air exerts a similar force on the bird directed upwards, enabling the bird to fly. Similarly, when a rocket engine burns fuel, the hot, expanding exhaust gases flow out of the back of the rocket, which induces an equal and opposite force (called thrust) that propels the rocket forward. Sometimes Newton's Third Law is expressed mathematically as:

$\vec{F}_{1,2} = -\vec{F}_{2,1}$ where $\vec{F}_{1,2}$, is the action and $-\vec{F}_{2,1}$ is the reaction.

Newton's Second Law, $\vec{F} = m\vec{a}$ relates force to acceleration. But force can be expressed mathematically through other equations including the famous *inverse-square relationship* known as **Newton's Law of Universal Gravitation**:

$$F = G\frac{m_1 m_2}{r^2},$$

where m_1 and m_2 are the masses of two objects and r is the distance between the two massive objects. The constant G equals 6.67×10^{-11} m$^3 \cdot$ kg$^{-1} \cdot$ s^{-2} and is known as the gravitational constant, the universal gravitational constant, or the Newtonian constant of gravitation. This should not be confused with **g**, the gravity of Earth (*vide supra*). Sometimes the former, G, is called 'big G' while the latter, g is called 'little g'. From Newton's Second Law of Motion and the Law of Universal Gravitation, one may derive the formula for the force of gravity at sea level:

$$\text{Since} \quad \mathbf{F} = G\frac{m_1 m_2}{r^2} \quad \text{and } \mathbf{F} = m\mathbf{a}:$$

$$G\frac{m_1 m_2}{r^2} = m\mathbf{a} \text{ or } \mathbf{a} = G\frac{m_1}{r^2}.$$

In this case, m_1 is the mass of the Earth (also written as M_E), and r is the distance between Earth's surface and its center of mass in the core of the planet (i.e. Earth's radius), which are both constants, as is G. Also, the acceleration **a** happens to be the acceleration due to Earth's gravity which is given by **g**. Thus, the force due to Earth's gravity, **g**, can be expressed as:

$$\mathbf{g} = G\frac{M_E}{r^2}.$$

Students are reminded that 'weight' is not synonymous with mass; weight is a force whereas mass is an inherent characteristic of an object that reflects the amount of matter present. An object's mass is constant anywhere in the Universe, but its weight is dependent on the gravitational field it happens to be in. Thus, on the surface of Earth, the force we call weight may be expressed as:

$$\mathbf{W} = m\mathbf{g}.$$

In analogy with Newton's Law of Universal Gravitation, there are other equations that express force in a similar inverse-square fashion. One is **Coulomb's Law** that equates electrostatic force to the magnitude of stationary electrical charges (q, measured in coulombs) and distance (r) between those electrically charged objects:

$$F = k\frac{q_1 q_2}{r^2},$$

where k is Coulomb's constant, which equals 8.988×10^9 N$\cdot$m$^2\cdot$C^{-2}. Unlike gravity, which is always attractive, the Coulomb force can be attractive when the two charges are opposite or repulsive when the two charges are the same. A similar inverse-square equation may be written to describe magnetic attraction and repulsion between two magnetic poles:

$$F = \mu_0 \frac{m_1 m_2}{4 \pi r^2},$$

where $\mu_0 = 1.256\ 637\ 06 \times 10^{-6}$ m · kg · s^{-2} · A^{-2} and is called the magnetic permeability of free space. Coulomb's Law may be written in a directly analogous manner, replacing the Coulomb constant, k, with another constant, the permittivity of free space:

$$F = \frac{1}{\epsilon 0}\ \frac{q_1 q_2}{4 \pi r^2}.$$

In this case, ε_0 is the permittivity of free space (see more about permittivity later in this chapter), which is the permittivity in a vacuum. One can see that the permittivity of free space is related to the Coulomb constant, k, by $\varepsilon_0 = 1/4\pi k$. In media other than a vacuum, the permittivity of the medium in question is always a bit higher than ε_0.

In any case, force is measured in SI units of newtons. One newton is defined as one kilogram-meter per second squared:

$$1\ N = 1\ kg \cdot ms^{-2}.$$

And since energy (or equivalently, the work done by a force) may be measured in joules by $W = F \times d$, newtons can be related to the basic unit of energy or work (joules) through the identity:

$$1\ J = 1\ \text{newton-meter} = 1\ kg \cdot m^2 s^{-2}.$$

One relevant force outside the realm of classical mechanics is the Lorenz force (electromagnetic force), which is the net sum of electrical and magnetic forces on a charged particle in an electromagnetic field. Briefly, an electrically charged particle with charge q, moving at a velocity $\mathbf{v}$, in an electrical field $\mathbf{E}$ and a magnetic field $\mathbf{B}$ will experience a force given by:

$$\mathbf{F} = q\mathbf{E} + q\mathbf{v} \times \mathbf{B}.$$

The direction of the force imposed by $q\mathbf{v} \times \mathbf{B}$ is given by the 'right-hand rule' in which the index finger gives the direction of the electric current $q\mathbf{v}$, the middle finger (flexed 90° from the index finger) gives the direction of $\mathbf{B}$, and the upward-pointing thumb gives the direction of the generated force, $\mathbf{F}$. Note that the × in the Lorentz force equation is a vector cross product.

Finally, with the general background information between force and work or potential energy, one may recognize that in three dimensions, the work done by a force over a distance or volume can be described as:

$$W = \iiint_V F(v) \mathrm{d}V.$$

Equivalently, the potential energy Φ in a gravitational energy field (for example) may be tapped to produce a force described by:

$$F = -\nabla\Phi,$$

where Φ is the potential energy function and ∇ is the 'del operator' $\left(\frac{\partial}{\partial x} + \frac{\partial}{\partial y} + \frac{\partial}{\partial z}\right)$. Using the mathematical terminology, the force is the gradient of the potential (in some instances).

4.6 Power

Unlike force, which is a vector quantity since it must be described by both a magnitude and a direction, energy is not a vector but rather is a scalar; it only has an associated magnitude. A related physical quantity in mechanics is power (P), which is defined as the work done per unit time. Like energy, power is a scalar that only has an associated magnitude. In SI units, power is measured in watts (W):

$$P = \frac{dE}{dt} \text{ or E t}^{-1},$$

$$1 \text{ Watt} = 1\text{Js}^{-1}.$$

Although the SI unit of power is the watt, in mechanics, another unit of power, the horsepower, remains in common usage, especially with respect to motor vehicles such as cars. (Note that 'motor vehicle' is often a misnomer since, technically, a motor converts electrical energy into kinetic energy, whereas a gasoline-powered car uses a combustion engine that converts chemical energy into kinetic energy). Anyway, one imperial horsepower equals 745.7 W (or in yet another set of power units, 550 foot-pounds per second). In electricity, power is the rate at which electrical energy is transferred in an electrical circuit. In this context, power is quantitated in the usual units of watts, but since energy is related to power by $E = P \times t$, electrical energy is usually sold to consumers in units of kilowatt-hours.

When focusing on electrical power, in a circuit, the relevant equation is:

$$P = IV,$$

where V is the potential difference (voltage drop) measured in volts and I is the current in amperes. If the circuit component in question is a resistor, from Ohm's Law ($V = IR$, where R is resistance in ohms), the equation can be rewritten as:

$$P = I^2R = V^2/R.$$

Yet another example of power is radiant power, which is expressed by the formula:

$$P = 4 \pi r^2 I,$$

where I is the intensity of radiation (e.g. light) and r is the distance from the source. In this example, the source of the light or radiation is assumed to be spherical.

4.7 Frequency and hertz

As mentioned, power is energy per unit time, and the watt is defined as one joule per second or 1 J s^{-1}. But when something is regularly repetitive and measured in events per unit time, we are talking about a frequency. Along with their amplitudes, sound waves and light waves are characterized by their frequencies, which measure the number of cycles per second. For frequencies, the unit is the reciprocal second or s^{-1}, which is known in the SI system as the hertz.

$$1 \text{ Hz} = 1s^{-1}.$$

An equivalent way of expressing the same concept is through the period of a wave, which is simply the time it takes for one cycle to be completed. Frequency is typically abbreviated as ν or f, while period is symbolized by T.

Mathematically, frequency and period are related by:

$$T = 1/\nu$$

A frequency–period relationship that will frequently recur is that between the speed of light in a vacuum (c) and the wavelength (λ) and frequency (ν) of the light in question. This is mathematically expressed as:

$$c = \nu \, \lambda.$$

Of course, this relationship between the speed of a wave, its frequency and its wavelength is not restricted to light. It applies equally well to sound.

4.8 Curies and becquerels

In the radiation sciences, radioactivity is typically measured in disintegrations per unit time. However, since the number of disintegrations per unit time is not perfectly constant (due to both random fluctuations and the steady decay of the parent radioisotope over time), we do not typically refer to radioactive decay in terms of frequencies. The SI unit of radioactive decay is the becquerel (Bq), which is defined as one decay or disintegration per second. It is worth keeping in mind that disintegrations per second do not necessarily translate one-to-one into counts per second in a radiation detector, since there will be some degree of inefficiency in any radiation detector. Therefore, counts per second are always less than actual decays pers per second. An older unit of radioactive decay that is still frequently used in nuclear medicine is the curie (Ci). Historically, the curie was based on the amount of radioactivity in pure radium-226; it was defined as the number of disintegrations that one gram of radium-226 would undergo in one second. In modern units, 1 Ci equals 37 billion Bq. Stated another way, 1 Ci = 37 GBq (where GBq stands for gigabecquerels or 10^9 Bq.) Thus, a curie is 3.7×10^{10} Bq. As a quick review of prefixes used in this text:

yocto (y) = 10^{-24}
zepto (z) = 10^{-21}
atto (e) = 10^{-18}
femto (f) = 10^{-15}
pico (p) = 10^{-12}
nano (n) = 10^{-9}
micro (μ) = 10^{-6}
milli (m) = 10^{-3}

centi (c) = 10^{-2}
deci (d) = 10^{-1}
(one = 10^{0})
deca (da) = 10^{1}
hecto (h) = 10^{2}
kilo (k) = 10^{3}
mega (M) = 10^{6}
giga (G) = 10^{9}
tera (T) = 10^{12}
peta (P) = 10^{15}
exa (E) = 10^{18}
zetta (Z) = 10^{21}
yotta = 10^{24}.

4.9 Linear momentum

Another important quantity in physics is momentum. There are two forms of momentum, linear momentum and angular momentum. Both are vector quantities, with magnitudes and directions. Additionally, both linear momentum and angular momentum are conserved quantities, meaning that the net momentum is the same before and after any interaction. In this manner, momentum is similar to energy, another conserved, non-vector quantity. In fact, one way of expressing the First Law of Thermodynamics is that 'energy is neither created nor destroyed'. Energy can be transferred, and it can change form (e.g. from nuclear energy into electrical energy into kinetic energy for example), but it cannot be completely lost nor can it be created de novo. (Technically, since $E = mc^2$ by the Einstein relation, it is more correct to say that 'mass-energy' can be neither created nor destroyed). The Laws of Thermodynamics will be discussed in further detail in chapter 34 in the context of black holes, but for now, we can mention that the First Law of Thermodynamics is accompanied by the Second Law, which states that the entropy (which quantifies disorder) of the Universe as a whole, always increases in any spontaneous process. Then there is the Third Law of Thermodynamics, which can be stated simply as 'the entropy (S) of a pure, crystalline material at a temperature of absolute zero is zero.' In other words, the disorder of a substance at absolute zero (0 K) is zero. Since absolutely perfect order or zero entropy is impossible to achieve, the Third Law implies that it is impossible to actually cool a system down to 0 K. Finally, there is a Zeroth Law of Thermodynamics, which says that if any two bodies are in thermal equilibrium with each other (meaning there is no heat flow from one to another), then they have the same temperature. Stated another way, if two bodies, A and B, are in thermal equilibrium with a third body, C, then A and B are in thermal equilibrium with each other and have the same temperature.

Returning to momentum, linear momentum (p) is defined as mass multiplied by velocity and is given by the equation:

$$\mathbf{p} = m\mathbf{V}.$$

Students should be careful not to conflate the symbol for power (P) with the symbol for momentum ($\mathbf{p}$). The boldface, small $\mathbf{p}$ (or equivalently, an arrow over the symbol) indicates that momentum is a vector, whereas power (P) is not.

$$\text{Since } \mathbf{p} = m\mathbf{v} \text{ and } \mathbf{v} = \frac{dx}{dt},$$

$$\text{One can see that } \mathbf{p} = m\left(\frac{dx}{dt}\right).$$

With this in mind, another way of expressing Newton's Second Law of Motion ($\mathbf{F} = m\mathbf{a}$) that incorporates momentum is:

$$\mathbf{F} = m\frac{d\mathbf{v}}{dt} = \frac{d}{dt}(m\mathbf{v}) = \frac{d\mathbf{p}}{dt}.$$

Recall that Newton's First Law of Motion (the law of inertia) said that a body at rest will stay at rest and a body in uniform motion will stay in motion unless acted upon by a force. This basically says that the momentum of any object is constant until a force is applied. Conversely, a change in momentum implies that a force had to be applied to change the inertia. From the above rearrangement of Newton's Second Law, we can revisit Newton's First Law and rearrange again:

$$F = \frac{d}{dt}(m\mathbf{v}),$$

$$F\,dt = d(m\mathbf{v}).$$

Or after integrating both sides:

$$\mathbf{F}\Delta t = \Delta m\mathbf{v} = \Delta\mathbf{p}.$$

This quantity, $\mathbf{F}\Delta t$ is also called the impulse and is symbolized by $\mathbf{J}$. Impulse is measured in units of newton-seconds. Working backwards, one can now see that another way of expressing impulse, J, is:

$$\mathbf{J} = \int \mathbf{F}dt = \Delta\mathbf{p} = m\Delta\mathbf{v}.$$

Note that the last equivalence, $m\Delta\mathbf{v}$, only applies when the mass is constant. While this becomes important when objects are moving at relativistic velocities, one need not go to such extremes to find the relevance of changing mass. A rocket-propelled spacecraft or jet-propelled aircraft will be losing mass as it burns fuel to generate thrust. Thus, corrections for changes in mass are important in calculations involving impulse in aerospace applications, regardless of special relativity.

One final point regarding linear momentum is that since KE is given by $KE = \frac{1}{2}mv^2$, one may multiply $\frac{1}{2}mv^2$ by (m/m) to get $(mv)^2/2m$. In this manner we obtain another useful expression, which relates momentum to kinetic energy:

$$KE = \frac{1}{2}mv^2 = (mv)^2/2\,m = \mathbf{p}^2/2\,m.$$

This relationship indicates that a body cannot have kinetic energy without also possessing linear momentum and vice versa. But it should be noted that this classical equation between energy and momentum must be modified when the velocities become 'relativistic' (which essentially means that the velocity has become a sufficiently large fraction of the speed of light in a vacuum (c) that special relativity must be taken into account). More about special relativity will be covered in subsequent sections.

The SI units of linear momentum are kilogram-meter per second or $kg \cdot m \ s^{-1}$, which is equivalent to newton-seconds.

4.10 Angular momentum

Akin to linear momentum is angular momentum ($\mathbf{L}$). Like linear momentum, p, L is a vector quantity that is conserved. Angular momentum can allude to orbital angular momentum (for example, the momentum associated with Earth's orbit as it revolves around the Sun) or spin angular momentum (for example, the momentum associated with Earth's rotation about its spin axis). The general equation describing spin angular momentum for a rigid body rotating about a fixed axis is:

$$\mathbf{L} = I \ \omega,$$

where $\mathbf{L}$ is the angular momentum, I is the moment of inertia, and ω is the angular velocity measured in radians per second. The moment of inertia, I, is given by the general formula: $I = \Sigma \ m_i r_i^2$ or $\int r^2 \ dm$.

Just as linear momentum is a conserved quantity that will not change unless a force is acting on the body in question, so too is angular momentum conserved and unchanging unless a torque is applied to the system in question. As mentioned earlier, a change in momentum caused by a force is the impulse, $\mathbf{J}$. Similarly, in a rotating system, the change in angular momentum caused by a torque is also called impulse and is defined analogously:

$$\mathbf{J} = \int \mathbf{F} dt = \Delta \mathbf{p} = m \Delta \mathbf{v},$$

for linear momentum, and

$$\mathbf{J} = \int \mathbf{\Gamma} \ dt = \Delta \mathbf{L} = I \Delta \omega,$$

for angular momentum (where $\mathbf{\Gamma}$ is torque).

Angular velocity, ω is given by $2\pi\nu$, where ν is the frequency of rotation. The average angular velocity can be given as $\Delta\theta/\Delta t$, and in the limit, as Δt approaches zero, the instantaneous angular velocity becomes $d\theta/dt$. The boldface $\mathbf{L}$ and ω indicate that these quantities are vectors. The angular velocity ω can be related to the linear velocity of an orbiting body (or to a point on a rotating body) by the equation:

$$v = r\omega.$$

For a planet revolving about a star, the general formula of $L = I\omega$ does apply but can be simplified. Since $\omega = v/r$, and I in this case can be reduced to mr^2, the overall equation for **L** reduces to:

$$\mathbf{L} = I\omega = (mr^2)\omega = (mr^2)(v/r) = mvr.$$

Of course, the same equation applies to the Moon revolving around the Earth, where the conservation of angular momentum has some interesting consequences over time. The Earth–Moon system is more complicated than one might initially anticipate because of the fact that Earth is largely covered with water in the form of its many oceans. This magnifies the tidal torque exerted by the Moon on the Earth and vice versa. The result is a deceleration of Earth's rotation rate (by about 66 ns per day) and a slow but steady recession of the Moon away from the Earth (about 3.82 cm per year). The latter figure was obtained through reflectors left on the Moon by NASA's Apollo Mission. The gradual recession of any orbiting moon or planet is generally accompanied by a decrease in the rotation rate of the central parent body, thanks to the conservation of angular momentum, but the presence of surface oceans on Earth adds substantially to the complexity of this problem.

It should be noted that for a situation such as an orbiting planet or moon, the angular momentum is mvr, and the associated centripetal force, holding that moon or planet in a stable orbit, is given by:

$$F = m\mathbf{v}^2/r \quad \textbf{Centripetal force.}$$

In the example of planets moving around a star, the centripetal force is the gravitational attraction provided by the central star. In this case, the centripetal force is given by Newton's Law of Universal Gravitation. But the case may be generalized to include analogous but not identical scenarios such as a charged particle orbiting another oppositely charged object. In that case, the centripetal force is given by Coulomb's Law. Additionally, one might notice that the equation for centripetal force, $F = m\mathbf{v}^2/r$, is similar to Newton's Second Law of Motion, $F = ma$. In fact, from these two equations, one can correctly deduce that the acceleration of a planet a planet around a central star is given by:

$$\mathbf{a} = \mathbf{v}^2/r \quad \text{Linear acceleration of an orbiting body.}$$

A subtle but important difference between linear momentum and angular momentum is that in the former, there is no acceleration, but in angular momentum, there is. This is because of the fact that an object revolving about another (like the Earth revolving around the Sun) is constantly changing its direction—and by definition, a change in direction is an acceleration. This is true even when the object in question is rotating or revolving at an unchanging rate. Students are warned about the confusion stemming from the terminology. 'Angular acceleration' (α) alludes to a change in angular velocity (ω) or $d\theta/dt$. One can have a constant angular momentum, which implies an acceleration because of rotation, without having an *angular* acceleration.

For clarification, a non-zero angular momentum, α, means that ω (that is, $d\theta/dt$) is not zero. This occurs for example, when a torque is applied to a rigid rotating body. But even when $d\theta/dt$ is constant (i.e. when ω is unchanging and α is zero), we still have an 'acceleration' going on, although it does not meet the definition of 'angular acceleration'; it is a linear acceleration. Of course, this simple analysis does not hold for planets or moons in highly eccentric orbits. According to Kepler's Second Law, a planet will be moving faster when it is closer to the Sun. Therefore, ω is in fact changing even though L, the angular momentum, is constant. Because L is conserved and constant, as r decreases in an eccentric orbit, ω must consequently increase to maintain $L = mvr = (mr^2)\omega$. Hence, in the case of a moon or planet in an eccentric orbit, angular momentum will remain constant but there is both a linear acceleration as well as an angular acceleration.

4.11 Electrical units

All units in electricity can be derived from the fundamental SI unit of current, the ampere (A). Current is electrical charge per unit time in an electrical circuit. Specifically, an ampere is $6.241\ 509\ 074 \times 10^{18}$ electrons flowing per second. Charge (Q) may be considered as a derived electrical quantity defined by current (I) times time (t):

$$Q = I \cdot t.$$

In this manner, electrical charge has the units of ampere seconds, which are also called coulombs (C). Thus:

$$1 \text{ coulomb} = 1 \text{ C} = 1 \text{ ampere second} = 1 \text{ A} \cdot \text{s}.$$

From the above, it is evident that 1 C of electrical charge is the equivalent of $6.241\ 509\ 074 \times 10^{18}$ electrons.

Electrical potential may be conceptualized as the 'pressure' in an electrical circuit that pushes electrons around and causes current to flow. In an electric circuit, the potential difference may be supplied by a battery to provide unidirectional direct current. While the most familiar example of an electric current is through a circuit of wires, semiconductors, resistors and transistors in a circuit board, one should keep in mind that a beam of electrons or ions in a vacuum is also an example of a direct current. Therefore, charged particle radiation flow as in the solar wind, solar particle events, galactic cosmic rays, and radiation therapy electron beams and proton beams are also examples of direct currents.

In a classic circuit connected to a battery, work is done by the battery and the work done is directly proportional to the electric charge Q moved through the circuit. This may be expressed as:

$$\text{Potential difference } (V) = \frac{\text{Work done in the electrical circuit}}{\text{Charge passing through the electrical circuit}}.$$

Because the unit of work is the joule and the unit of electrical charge is the coulomb, the units for potential difference are joules per coulomb. Joules per coulomb are called volts.

$$1 \text{ volt} = 1 \text{ V} = 1 \text{ joule/coulomb} = 1 \text{ J C}^{-1}.$$

By rearranging these relationships, one can show that the work done by an electrical circuit can be given by:

Work (in joules) $= Q \cdot V$ (in coulombs $\cdot$ volts) $= ItV$ (in amperes $\cdot$ seconds $\cdot$ volts).

Students are encouraged to work through the units to prove the equivalence.

Capacitance is the ability of an object or device to store electrical energy by accumulating electrical charges on two close, but non-touching, surfaces separated by an insulator. Capacitance is measured by the change in electric charge stored in response to the applied potential difference in a circuit. A device that performs this function is called a capacitor or condenser. When a charge Q is placed on a capacitor, the potential is increased to V and the capacitance C is given by:

$$\text{Capacitance } (C) = \frac{\text{Charge stored in the capacitor } (Q)}{\text{Potentilal difference to which the capacitor is raised } (V)},$$

or upon rearranging, $Q = CV$.

As charge is quantified in coulombs and potential difference is measured in volts, the units of capacitance are coulombs per volt. This provides the definition of the **farad (F)**, which is the formal SI unit of capacitance:

$$1 \text{ farad} = 1 \text{ F} = 1 \text{ coulomb/volt} = 1 \text{ C V}^{-1}.$$

In practice, the farad is a very large and cumbersome unit to use, so most quantitative discussions of capacitance allude to microfarads, nanofarads and picofarads.

The insulating material between the plates of a simple parallel-plate capacitor strongly influence the overall capacitance. The material separating the two plates is called a dielectric. Dielectrics are distinguished by their relative **permittivity**, ε. Permittivity measures the electrical polarizability of the dielectric material in question. Permittivity is often represented by the relative permittivity, which is simply the ratio of the permittivity of the actual dielectric material in question and the permittivity of a vacuum, ε_0. The relative permittivity of the dielectric material is sometimes confusingly just referred to as the permittivity. Another, older and now disfavored measure of permittivity is the dielectric constant. In principle, pure water has a relatively high relative permittivity or dielectric constant and therefore should be an excellent electrical insulator. However, water is virtually never pure; water with dissolved ions in it is highly conductive and a very poor insulating material.

Another physical quantity of relevance is **resistance (R)**. When an electrical potential difference V is applied to the wires in a circuit, a current I, will flow. The exact amount of current is a function of several variables inherent to the wire in question such as its area, its chemical composition and its length. The tendency of a current to flow is called the **electrical conductance**, which is measured in units of **siemens (S)**. The reciprocal of conductance is resistance. The resistance to flow in a circuit is defined by the ratio of the potential difference V to the resulting current I in volts per ampere. Resistance is important in electrical circuits and is measured in volts per ampere, which are also called ohms (Ω).

$$1 \text{ ohm} = 1 \ \Omega = \frac{1 \text{ volt}}{1 \text{ ampere}} = 1 \text{ V/A} = 1 \text{ V A}^{-1}.$$

The defining equation for resistance is **Ohm's Law**: $V = IR$.

From the above discussion, one will note that work (i.e. energy) is the equivalent of the product of charge and potential. A unit of great importance in space radiation, particle physics and radiological science in general is the **electron volt**. This is the amount of kinetic energy gained by one electron as it is accelerated through a vacuum with an electric potential difference of one volt. The electrical charge of an electron (symbolized by $-1 \ e$) has a negative charge of $1.602 \ 176 \ 634 \times 10^{-19}$ C. Therefore, an electron volt is numerically equal to $1.602 \ 176 \ 634 \times 10^{-19}$ J. When dealing with ionizing radiation, kilo electron volts (keV), mega electron volts (MeV) and giga electron volts (GeV) are often encountered.

One MeV=1 MeV = 1×10^6 electron volts
$$= 1 \times 10^6 \times 1.602176634 \times 10^{-19} \text{J} \approx 1.602 \times 10^{-13} \text{J}.$$

4.12 Photometry: luminous intensity, luminous flux, emittance and illuminance

Luminous intensity, luminous flux, emittance, illuminance and related terms are often conflated but represent distinctly different concepts. These terms belong to the field of **photometry**, the branch of physics focusing on measurements of light but perceived as brightness by the human eye. Thus, there is an anthropogenically biased subjectivity to photometry. Such subjectivity is absent from the related but purer field of radiometry, which focuses on the measurement of photons of all wavelengths.

The reason behind the subjectivity in photometry is partly based on the fact that the human retina is not sensitive to all wavelengths of electromagnetic radiation. The human eye is only sensitive to electromagnetic radiation in the visible portion of the spectrum (which for the purposes of this chapter will be considered roughly 400–700 nm). More subtly yet more importantly, even within this restricted range of wavelengths, the human eye is not equally sensitive to all wavelengths of visible light. For instance, the human retina is more sensitive to yellow and green than it is to blue or red wavelengths. The units in photometry are selected to reflect this differential sensitivity through the use of 'wavelength weighting'. In the field of radiometry, no such biologically based, subjective preference is given to certain colored photons over others (although differences in photon energy do make a difference, as will be discussed in chapter 7 when we review thermal or blackbody radiation in further detail).

4.13 Luminous intensity

Luminous intensity alludes to a light source's intensity to the human eye and in a specific direction. The SI unit of luminous intensity is the candela. This quantifies the luminous flux per unit solid angle of a light source. It is (imperfectly) analogous to **radiant intensity**, but unlike radiant intensity, which includes the contributions of photons from every wavelength of the electromagnetic radiation in question, luminous intensity exclusively deals with the wavelengths that the human eye can detect.

Another important distinction about luminous intensity is that, unlike radiant intensity, which is measured in power per unit area and decreases as the square of distance, luminous intensity is flux per solid angle in a certain direction; hence, luminous intensity does not decrease with increasing distance. Thus, the analogy is imperfect between luminous intensity in the branch of physics called **photometry** and radiant intensity in the branch of physics dealing with radiant energy (**radiometry**).

One candela is approximately the luminous intensity emitted by an ordinary wax candle. Thus, the older unit of luminous intensity was the candlepower (CP), which is equal to 0.981 modern candelas. Conversely, 1 cd equals 1.02 CP. The SI unit the candela has replaced the previous luminous intensity unit of candlepower from a 'standard candle', and for that reason, the candela is sometimes called 'the new candle'. For those interested in the history of science, the standard candle or international candle was a light source from one-sixth of a pound of sperm whale wax (spermaceti) burning at the rate of 120 grains per hour. Spermaceti is the waxy substance from the head of a sperm whale that gives these animals their characteristic large-headed appearance and probably aids them with buoyancy and echolocation. And a grain was 0.065 g (or 64.798 91 mg to be more precise) and was based on the weight of a single, ideal seed of cereal.

The technical definition of the candela is the luminous intensity in a given direction of a light source that emits monochromatic electromagnetic radiation of 540 THz and has a radiant intensity in that direction of 1/683 W per steradian. The choice of 540×10^{12} Hz was based on the fact that the human eye is most sensitive to light with this frequency (which is the equivalent of 555 nm based on $c = \lambda \nu$). This specification is necessary in order to standardize and quantify the visual sensation of 'brightness'. For instance, the human retina is not equally sensitive to all wavelengths of visible light—an incandescent light bulb emitting 1 W of blue light seems less bright than a bulb emitting 1 W of yellow light to the human eye. The biologically based weighting of wavelengths is given by what is called a **standard luminosity function**. This standard luminosity function in turn provides a quantity known as the **luminous efficacy factor**, which is described in the next paragraph.

4.14 Luminous flux

Luminous flux (often symbolized by Φ) is measured in lumens (lm). Luminous flux (or luminous power) is the quantity of *perceptible* light that is emitted from a light source *in all directions*. The latter point distinguishes luminous flux from luminous intensity, which is the intensity in a specified direction. Luminous flux is specifically the energy per unit time (dQ/dt) radiated over the wavelengths that the human eye can see (roughly 400–700 nm). It is therefore slightly different from **radiant flux** or **radiant power** in that luminous flux is adjusted (i.e. weighted) to the specific sensitivity of the human retina to certain wavelengths of visible light, whereas the radiant flux or radiant power includes all wavelengths. (Further adding to the confusion, in astronomy, radiant flux is sometimes called '**luminosity**'. More properly, luminosity should allude to the total power output of a star in energy per unit time, and not the area-dependent flux).

Under daylight conditions, where the human eye is 'light-adapted' (so-called **photopic vision** that involves the cone cells of the human retina), the average human eye has a peak efficacy at 555 nm; 555 nm is assigned a conversion value of 683 lm W^{-1}. At a wavelength of 555 nm, the unitless **luminous efficacy factor** is given a value of 1.0, and all other wavelengths are slightly lower than this maximum value. Hence, luminous flux can be related to radiant flux by the formula:

Luminous flux(in lumens) = Radiant power(in watts) × (683 lumens/watt)
× (the luminous efficacy factor)

The lumen is derived from the basic SI unit, the candela. One lumen is the amount of light streaming outwards through a solid angle of one steradian from a light source with a luminous intensity of 1 cd. Mathematically, the relationship between luminous flux (Φ, measured in lumens) and luminous intensity (I_v, measured in candela) is given by:

$$\Phi = I_v \cdot \Omega$$

(where Ω is the solid angle spanned in steradians)

If the light source in question is perfectly isotropic, meaning that its luminous flux is uniform in all directions, since a sphere measures 4π steradians, the above equation reduces to:

$$\Phi = 4\pi I_v$$

Older readers might remember when light was referenced with respect to another standard candle of sorts—the 100-watt light bulb. These incandescent bulbs of the 20th century have given way to light-emitting diodes and compact fluorescent bulbs that waste far less energy as heat. The luminous flux of the good old 100 W lightbulb was about 1500 lumens. But this same amount of luminous flux is now also provided by compact fluorescent light bulbs and LED light bulbs at around one-quarter or one-fifth the power consumption in watts.

4.15 Illuminance and luminous emittance

Illuminance is a quantity related to luminous flux but rather than referring to light emanating from a source, it refers to the amount of light that falls onto a surface. When luminous flux (measured in lumens) strikes an object's surface, that object is said to be **illuminated**. Illumination intensity is analogous to the intensity of total electromagnetic irradiation, which is measured in power (watts) per unit area. But again, unlike total irradiation (which includes all wavelengths) in the field of radiometry, illumination in the field of photometry is restricted to visible light, and is wavelength-weighted by the luminosity function in order to correspond better to the human perception of brightness. The illuminance (sometimes symbolized by E_v) is **the luminous flux per unit area**, which is measured in **lux (lx)** in SI units. One lux is defined as one lumen per square meter:

$$1 \text{ lux} = 1 \text{ lm} \cdot \text{m}^{-2}$$

An outdated unit of illuminance that persists, especially in photography, is the foot-candle (fc). This non-SI unit equals one lumen per square foot. As a square foot is smaller than a square meter, the foot-candle is correspondingly more than a lux:

$$1 \text{ fc} = 10.76 \text{ lux}$$

The complement of illuminance is **luminous emittance** (also called **luminous exitance**). Luminous emittance is a measure of light output from a surface (that is, exiting light), whereas illuminance is the amount of light falling onto a specified surface area. But while illuminance is measured in lux (which is $\text{lm} \cdot \text{m}^{-2}$), one should not use lux for emittance; the unit is still $\text{lm} \cdot \text{m}^{-2}$ but technically is not properly alluded to as lux. Note also that luminous flux is slightly different from luminous emittance in that the former is luminous energy per unit time while the latter is luminous flux emanating from a surface. The analogous quantity to luminous emittance or exitance in radiometry is radiant emittance or radiant exitance, which is the radiant flux emitted by a surface per unit area and is measured in watts per square meter.

The terminology in this field can be quite confusing but students might benefit from the mnemonic assistance of noting that *i*lluminance refers to *i*ncoming or *i*ncident light, which all begin with the letter 'i'. Illuminance measures how much incident visible light (weighted by the wavelength-based human eye luminosity function) illuminates a surface. The analogous quantity to illuminance in radiometry is called irradiance, which is the amount of incoming radiant energy received, measured in watts per square meter. Again adding to the confusion in the already terminology-challenging discipline, in astrophysics, the irradiance is often alluded to as radiant flux.

4.16 Luminance

Luminous intensity per unit area within a specified solid angle is a light source's luminance. As luminance is quantified per unit area, as the source recedes, the luminance falls off as the square of the distance. Luminance is quantified in SI units of candelas per square meter ($\text{cd} \cdot \text{m}^{-2}$). One candela per square meter is also called the nit, but this is not an officially accepted SI unit.

4.17 Recap of chemistry

4.17.1 Atomic structure

The smallest possible unit of any given element is the atom. For instance, if one were to break a piece of gold in half, the two products are just smaller pieces of gold, but if one repeats the breakage again and again, eventually there will come a point that any further breakage will lead to something no longer considered gold. That smallest unit is a gold atom. If these units (atoms) are broken down any further, the constituent parts no longer have the characteristic chemical features of the element in question.

Atoms are built such that the central nucleus (composed of positively charged protons and neutral neutrons) is a tiny fraction of the overall size of an atom. The size of atoms is dictated by the extent of the orbiting cloud of electrons. For reference, atomic radii are typically about four to five orders of magnitude larger than nuclear radii. But while electrons orbit about the nucleus, they do not have liberty to take on any orbit they please; they are restricted to very specific orbital distances. In other words, their orbits are quantized. This discovery is the basis for the simplest quantum model of the atom, the Bohr atom.

According to classical mechanics, orbiting charged particles will emit electro-magnetic radiation (synchrotron radiation) and thereby continuously lose energy and momentum. Thus, they will eventually be unable to maintain their orbits and will come crashing into the nucleus. This is not what is observed of course. To account for the observed stability of the hydrogen atom (one electron orbiting one proton), Niels Bohr proposed a model in which the electrons orbit only in fixed, stable orbits that are separated in energy by discrete values. The electron normally inhabits the lowest energy level (the ground state), but upon absorption of a photon of light with the right energy, it can 'jump' up to another energy level, the so-called 'quantum leap'. In this manner, only certain energy photons can be absorbed since only certain energy levels (and thus energy differences between levels) are permitted. Similarly, when an electron jumps from a higher energy level to a lower one, a photon of a discrete energy is emitted. The energy of the photon corresponds to the difference in electron energy levels and given by:

$$E = h\nu,$$

where ν is the frequency of the light in question and h is Planck's constant.

This equation is a very important one in quantum theory and many related branches of physics and chemistry. To honor the crucial roles of the discoverers, $E = h\nu$ is sometimes called the **Planck–Einstein relation**, often simply called the Planck relation.

The postulated model of the hydrogen atom explains the observed spectral lines of hydrogen that are covered in further depth in chapter 7. According to this model, only certain energies (and thus frequencies or wavelengths) of light are absorbed or emitted, giving rise to the dark lines in an absorption spectrum or the bright lines of an emission spectrum, respectively. Bohr observed that the observed spectral lines of hydrogen (characterized by the so-called Balmer series, Lyman series, and others) may be explained by angular momentum values of the electron that happen to be integral multiples of Planck's constant divided by 2π, that is $h/2\pi$. This unit, $h/2\pi$ is often given its own special symbol $\hbar$ and is sometimes called the reduced Planck constant or **Dirac constant**. In other words, the only values allowed for the electron's angular momentum are integral multiples of $\hbar$:

Angular momentum of an electron in orbit about a nucleus $= n\hbar n = 1, 2, 3, \ldots$

From this, one can set the angular momentum of an electron orbiting in a circular path (mvr) equal to $n\hbar$:

$$mv\,r = n\hbar.$$

Similarly, one can set the classical centripetal force of an orbiting electron (mv^2/r) equal to the Coulomb force (kZe^2/r^2, where Z is the atomic number and e is the electric charge of the electron) holding that electron in place:

$$F = mv^2/r = kZe^2/r^2,$$

where k is the Coulomb constant.

Upon rearranging and solving for r, one gets:

$$r = \frac{kZe^2}{mv^2}.$$

Students may use further algebra to eliminate v between the two equations above to obtain:

$$r = \frac{n^2\hbar^2}{kZe^2m} = 0.529 \times 10^{-10}\left(\frac{n^2}{Z}\right) m.$$

Thus, one can see that the lowest possible orbit in the hydrogen atom, where $n = 1$ and $Z = 1$, is given by 0.529×10^{-10} m or 0.529 Å. This is known as the **Bohr radius**.

Eliminating r in the equations $m\,vr = n\hbar$ and $r = \frac{kZe^2}{mv^2}$ yields the allowable speeds of the electron:

$$v = \frac{kZe^2}{n\hbar} = 2.\,19 \times 10^6\left(\frac{Z}{n}\right) m\ s^{-1}.$$

One can see from this relation, that the highest electron speeds are attained when Z is high and n is low. When one plugs in some numbers, it turns out that the velocity of an electron in the ground state ($n = 1$) is approximately 1/137 times the speed of light. A generalization of this relationship allows one to estimate the speed of electrons in the first Bohr orbit (K shell) of any atom relative to the speed of light through:

$$v_{\text{rel}} = \left(\frac{1}{137}\right) Z\ m\ s^{-1},$$

where v_{rel} is the speed of the electron relative to c, the speed of light.

We will see this curious figure, 1/137 (more precisely 1/137.0360), repeatedly and in various situations throughout the text. It is given the special designation α and is called the **fine structure constant** or the **Sommerfeld constant**.

One may calculate the allowable kinetic energies and potential energies of an orbiting electron in the Bohr atom from $KE = \frac{1}{2}\,mv^2$ and $PE = -\frac{kZe^2}{r}$:

$$\text{KE} = \tfrac{1}{2}mv^2 = \frac{k^2 Z^2 e^4 m}{2\hbar^2 n^2}.$$

$$\text{PE} = -\frac{kZe^2}{r} = -\frac{k^2 Z^2 e^4 m}{\hbar^2 n^2}.$$

If one sums the kinetic and potential energies to obtain a total energy through $E = \text{KE} + \text{PE}$:

$$E = \text{KE} + \text{PE} = -\frac{k^2 Z^2 e^4 m}{2\,\hbar^2 n^2} = -2.18 \times 10^{-18} \left(\frac{Z^2}{n^2}\right) \text{joules}.$$

Recalling that 1 eV is 1.602×10^{-19} J, in terms of electron volts, this reduces to:

$$E = -13.6\left(\frac{Z^2}{n^2}\right) \text{eV}.$$

When solving problems, students are reminded about the need to use the appropriate units for the problem in question. For instance, when calculating the energy difference between the $n = 2$ and the $n = 3$ energy levels (which corresponds to the $n = 3 \rightarrow n = 2$ transition in the Balmer series of the hydrogen atomic spectrum, where $Z = 1$), one can use the above equation to obtain: $E2 = -13.6/(2)^2 = -3.4$ eV and $E3 = -13.6/(3)^2 = -1.51$ eV, so that the difference between the energy levels is $(-1.51)—(-3.40) = 1.89$ eV. But when asked what frequency or wavelength this photon has, one often must convert back to joules since most tables provide Planck's constant in units of joule-seconds. Thus:

$$1.89 \text{ eV} \times (1.602 \times 10^{-19} \text{ J/eV}) = 3.02 \times 10^{-19} \text{ J}.$$

Then one can use the Planck–Einstein relation $E = h\nu$ to obtain the frequency:

$$\nu = E/h = (3.02 \times 10^{-19} \text{ J})/(6.63 \times 10^{-34} \text{ J} \cdot \text{s}) = 4.56 \times 10^{14} \text{ s}^{-1}.$$

To obtain the wavelength, one uses the equation, $c = \nu\lambda$:

$$\lambda = c/\nu = (2.99 \times 10^8 \text{ m s}^{-1}) / (4.56 \times 10^{14} \text{ s}^{-1}) = 6.58 \times 10^{-7} \text{ m} = 658 \text{ nm}.$$

This particular spectral line (sometimes called Hα or H-alpha) corresponds to visible light in the red portion of the spectrum. A more precise calculation for H-alpha will yield 656.46 nm. Three other transitions in the Balmer series similarly correspond to visible light.

For hydrogen in the ground state, the total energy is therefore −13.6 eV. This means that it takes 13.6 eV to ionize a hydrogen atom (when it is in its lowest energy state). This is the **ionization** energy of hydrogen in its ground state. That is, 13.6 eV is the minimum energy required to separate a tightly bound electron from a proton. If the electron is already in an excited state, it can be removed with less energy. It should be pointed out that this model, the Bohr atom, only works well for hydrogen and single-electron ions such as He^+ and Li^{2+}. Thus, we can use it to calculate the amount of energy needed to remove the remaining electron from He^+ and create

He^{2+} from the equation by plugging in $Z = 2$ and $n = 1$ to obtain 54.4 eV. But for more complicated situations such as atoms with multiple electrons in various energy levels, the Bohr atom must be replaced with a fuller version that is based on quantum mechanics, which will be covered in a later chapter. Nevertheless, the basic principles of quantization of energy and a limited number of discrete energy or angular momentum values are demonstrated by the simple Bohr model.

Additionally, the spectral analysis of molecules introduces additional variables above and beyond the simple electronic energy level transitions of atoms. These additional complexities include lines associated with molecular bond vibrational transitions (in the infrared typically) and molecular rotational transitions (often in the microwave range). Finally, it should be mentioned briefly that just as the electrons around atoms occupy specific arrangements of shells and subshells with only a limited number of allowable states, so too are the arrangements of protons and neutrons within nuclei. **Nuclear shells** similarly have specific, quantized energy levels. Transitions of protons and neutrons ('nucleons') from one energy level to another release photons, just as electrons do when transitioning from one energy level to another. However, nuclear transitions tend to be far higher in energy and are in the gamma ray range of the electromagnetic spectrum. And just as atoms have characteristic spectral fingerprints of only limited photon energies (as in the Balmer and Lyman series of hydrogen), the gamma rays released during nuclear transitions are very specific and can reveal the identities of the isotopes in question.

Beyond the angular momentum associated with orbiting the nucleus as discussed in the development of the Bohr atom, electrons have an inherent angular momentum (spin). Unlike the infinite number of angular momentum values that macroscopic objects can possess, according to the principles of quantum mechanics, the spin values that electrons (and other) fundamental subatomic particles can take on are limited. The electron happens to be a fermion, meaning that it has a spin that is a half-integer multiple of Plank's constant. (Bosons constitute the other main category of particles and have integer multiples of Plank's constant.) In particular, the electron possesses an inherent spin said to be 'one-half'. All fermions, that is particles with half-integral spins, follow the **Pauli exclusion principle**, which dictates that no two electrons in a given system (an atom for instance) can occupy the same exact quantum state at the same time. (More generally, it says that no two fermions can occupy the same quantum states in a given quantum system simultaneously; this will be discussed further in chapter 7). This principle explains how electrons arrange themselves in atoms and molecules, and it allows one to predict where subsequent electrons will go as the atomic number increases. When we say that no two electrons can occupy the same quantum state in any given atom, that is the equivalent of saying that no two electrons in an atom may simultaneously have the same four 'quantum numbers'. For electrons filling atomic orbitals, the four quantum numbers are:

1. The principal quantum number (n)
2. The azimuthal quantum number (ℓ)
3. The magnetic quantum number (m_ℓ)
4. The spin quantum number (m_s).

The **principal quantum number**, n, describes the overall energy of the level and has allowed values given by $n = 1, 2, 3, \ldots$. In the Bohr atom, these correspond to primary electron **shells**. They go by letters so that the **K shell** corresponds to $n = 1$, the **L shell** corresponds to $n = 2$, the **M shell** corresponds to $n = 3$, etc. The principal quantum number dictates the total energy of the electron as well as the most probable distance of the electron from the nucleus. The higher the value of n, the further from the nucleus the electron is and the larger the atomic radius is. The total number of electrons that may reside in a shell of n equals $2n^2$. Thus, the K shell ($n = 1$) can hold 2 electrons; the L shell ($n = 2$) can hold up to 8 electrons; the M shell ($n = 3$) can carry 18 electrons; and the N shell ($n = 4$) may contain up to 32 total electrons.

The **azimuthal or orbital quantum number**, ℓ, corresponds to subshells or **orbitals** within the primary shells. ℓ describes the orbital angular momentum and dictates the shape of the orbital. ℓ can take on the integer values $\ell = 0, 1, 2, 3, \ldots$ up to $(n - 1)$.

The various values of ℓ correspond to different atomic orbitals called s, p, d and f. (These letters come from historical spectroscopic names—sharp, periodic, diffuse and fundamental.) For a given principal quantum number, there are n allowed values of ℓ. Thus, for the $n = 1$ level, there is only one allowed value of ℓ, corresponding to an s orbital. This would be called the 1s orbital. For the principal quantum number $n = 2$, there are two permitted values of ℓ, namely an s orbital and a set of p orbitals; these are called the 2s orbital and the 2p orbitals. For $n = 3$, there are three values of ℓ, the 3s orbital, a set of 3p orbitals and a set of 3d orbitals.

The **magnetic quantum number** (m_ℓ) can assume a total of $(2\ell + 1)$ different integer values ranging from $-\ell$ to $+\ell$, including zero. For example, $\ell = 1$ (the p orbitals), the $(2\ell + 1) = 3$ possible m_ℓ values are -1, 0 and $+1$. This tells us that there are three different p orbitals. These three p orbitals are identical in size, shape and energy but differ in the way their 'lobes' are oriented. The three possible orientations are orthogonal to each other such that there are p_x, p_y and p_z orbitals. Likewise, for the d orbitals, which correspond to $\ell = 2$, m_ℓ can take on the $(2\ell + 1) = 5$ integer values of $-2, -1, 0, +1$ and $+2$, indicating that there are five different d orbitals. Overall, the s, p, d and f subshells contain 1, 3, 5 and 7 orbitals each based on the $(2l + 1)$ rule (table 4.1).

For s orbitals, $\ell = 0$ and $(2\ell + 1) = 1$, indicating that there is only one s orbital for any given value of n. Thus, there is a 1s orbital (where the 1 indicates $n = 1$), a 2s orbital, a 3s orbital, etc. But for any given value of n, there is still only one s orbital.

Table 4.1. To summarize, the different orbitals in an atom may host different maximum numbers of electrons as follows.

Azimuthal quantum number (ℓ)	Orbital designation	Number of orbitals = $(2\ell + 1)$ (Possible values of magnetic quantum number (m_ℓ))	Maximum number of electrons = $2(2\ell + 1)$
0	s	1	2
1	p	3	6
2	d	5	10
3	f	7	14

This contrasts with the situation for p orbitals where there are zero 1p orbitals, three 2p orbitals, three 3p orbitals, three 4p orbitals, etc. For the d orbitals, there are no 1d orbitals or 2d orbitals, but there are five 3d orbitals, five 4d orbitals, etc.

Finally, the **spin quantum number** (m_s) can assume the two and only two values of $\pm\frac{1}{2}$. This means that each individual orbital can accommodate up to two electrons, paired through opposite spins. The Pauli exclusion principle said that no two electrons in a given atom can have the same four quantum numbers simultaneously. Thus if there are two electrons with the same values for n, ℓ and m_ℓ, the only way they can both occupy the same shell, subshell and orbital is if their spins are antiparallel. That is, their m_s values must be different.

Electrons, being fermions, are 'antisocial'—they do not like being crowded together and will not occupy the same orbitals with the same quantum numbers. This behavior contrasts them with bosons such as photons, who do like to be crowded together. This is the basis for lasers, for example. Basically, bosons do not obey the Pauli exclusion principle.

The antisocial behavior of electrons dictated by the Pauli exclusion principle determines how atoms build up their surrounding shells of electrons. As electrons are added to nuclei, they tend to follow very clearly outlined rules—the **Aufbau principle** or **Aufbau rule**. Simply stated, this calls for electrons filling an atom's ground state through the lowest available energy levels first and then moving onto higher available energy levels. For instance, the lower energy 1s orbitals are filled before the higher energy 2s orbitals are.

Hund's rule also plays a role when electrons are added to atoms. Hund's rule states that if multiple orbitals are available at the same energy levels (i.e. if they are **degenerate**), the newly added electrons must occupy different orbitals singly and with the same spin before they start pairing up in the same orbital with different (balanced or opposite) spins. This is called parallel spin, and it maximizes the number of orbitals occupied by electrons as well as the net spin. As an example, in atomic oxygen, the four valence electrons in the three 2p orbitals fill up as: [↑↓] [↑] [↑]. Thus, there are two p orbitals with single electrons of the same spin, which *maximizes the net spin of the atom*. This contrasts with alternative ways of filling the orbitals such as [↑↓] [↑] [↓] in which electrons are filling different orbitals but are balancing out the spins (thereby reducing overall atomic spin). Similarly, oxygen does not fill its p orbitals as [↑↓] [↑↓] [ ] since this does not follow Hund's rule of moving onto vacant orbitals of equal energy before beginning to pair up in individual orbitals. A mathematical way of stating Hund's rule is that for any given electron configuration, the term with the maximum **multiplicity** has the lowest energy and is therefore favored. Multiplicity is given by **2S + 1**, where S is the atom's total electron spin angular momentum. Numerically, the multiplicity can be calculated as the number of unpaired electrons plus one. Thus, the ground state of oxygen, with two unpaired electrons has a multiplicity of 3. Alternatively, one could have added the one-half spins of the two unpaired electron's spins to get S = 1 and plugged that into the 2S + 1 formula to calculate a multiplicity of 3. States with multiplicity of 3, like oxygen, are called triplets. Overall, states with multiplicities of 1, 2, 3, 4, etc are called singlets, doublets, triplets, quadruplets, etc.

Incidentally, the Aufbau principle holds not only for electrons in an atom, but also for protons and neutrons in a nucleus. Just as electrons fill the available atomic electron energy states starting from the lowest available energy first, so too do protons and neutrons fill the available nuclear energy states. Thus, just as electrons are organized into energy shells, protons and neutrons are also arranged in similar energy shells, in accordance with the **nuclear shell model**. However, the useful graphical mnemonic for electron orbital filling, **Madelung's rule**, does not apply for nuclear shell filling. Madelung's rule basically states that when the atomic orbitals are written in the following pattern, the electrons fill vacancies from the top down, following the arrows. Thus, the 1s orbital is filled before the 2s orbital and the 2p orbital is filled before the 3s. However, the 3p orbital is filled before the 4s, according to the diagram.

1s

2s   2p

3s   3p   3d   **Madelung's rule**

4s   4p   4d   4f

With a fundamental understanding of atomic structure, the various trends across the rows and columns in the periodic table, such as ionization potential, electronegativity and atomic size, make more sense and will be reviewed below.

4.18 Ionization energy

By definition, ionizing radiation interacts with atoms and molecules by ionizing them (i.e. by removing electrons). But not all atoms and molecules are as easily ionized. An inherent property of every atom in the periodic table is its ionization energy or ionization potential, which is defined as the minimum energy required to remove the most loosely bound electron of an atom (or ion or molecule) in its gaseous state. For an atom (as opposed to an ion), this is called the **first ionization energy** and is given by:

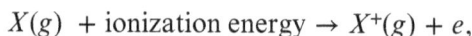

$$X(g) \; + \; \text{ionization energy} \rightarrow X^+(g) + e,$$

where $X(g)$ is an atom or molecule *in the gaseous state* and $X^+(g)$ is the resulting positively charged ion (a cation).

The second ionization energy alludes to the energy required to remove another electron from the resultant singly charged cation to generate a doubly charged cation:

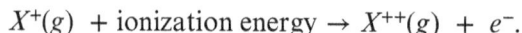

$$X^+(g) \; + \; \text{ionization energy} \rightarrow X^{++}(g) \; + \; e^-.$$

There can be a third ionization energy, and a fourth ionization energy, and so on depending on how many electrons the element in question possesses. For a given element, the first ionization energy is the lowest, the second ionization energy is the next lowest, etc. This is predictable since the positive charge on the atomic nucleus does not change with loss of electrons, and thus, with each ionization, the remaining negatively charged electrons are held more firmly by the positively charged nucleus.

For all neutral atoms, ionization is an endothermic process, meaning it requires the input of energy. Technically, an endothermic reaction is one that increases the enthalpy (H) of a system. This means that the reaction absorbs energy from its surroundings (or in other words, requires heat input). In chemistry, ionization energies are typically expressed as the energy required to ionize one mole of atoms or molecules, measured in kilojoules per mole ($kJ \cdot mol^{-1}$). However, in physics and radiation sciences, the ionization energy is often expressed in electron volts per atom or molecule. Recall that an electron volt is the kinetic energy gained by an electron as it accelerates through an electric potential difference of one volt when in a vacuum.

$$1 \text{ eV} = 1.602176634 \times 10^{-19} \text{ J}$$

Generally speaking, the farther away an electron is from its parent positively charged nucleus, the easier it is to remove. Conversely, the ionization energy is higher when the electrons are held closer to the nucleus. Two basic principles may be gleaned from an inspection of ionization potentials across the periodic table:
1. Ionization energies increase from left to right within a row ('period'). That is, as Z increases, ionization energy increases.
2. Ionization energies decrease from top to bottom in a given group or column.

Thus, the alkali metals have lower ionization potential than the halogens or noble gases in the same period. This is because all the elements in the same row are simply adding more electrons to the same shell as one moves from left to right. In this scenario, the atomic number continues increasing (meaning the number of protons and thus the positive electrical charge of the nucleus) from right to left, which implies that the electrostatic attractive force holding the electrons to the atom increases as well. But because the outer electrons are being added to the same shell (meaning the size of the atom is not substantially bigger), the overall effect is greater attraction of the positive nucleus on the negatively charged electrons.

This is different from what happens as one moves down a column. In this case, the decrease in ionization energy as one moves down a group is due to the newly added outermost electrons in the next member down belonging to an entirely new shell compared to the element above. As these outer electrons are in a completely different and higher-energy shell, they are further from the columbic attraction of the parent positively charged nucleus and are thus less electrostatically bound.

Overall, the ionization potential for a given element is a function of the collective influences of the electric charge of the nucleus, the size of the atom and the specific electronic configuration. Nuclear charge (that is, Z) is a major factor—as the positive charge increases, the electrons are held more tightly and the ionization energy increases accordingly. The number of electron shells (that is, atomic radius) is another major influencer—as atomic radius increases, the electrons are further away and thus less tightly held and the ionization energy correspondingly decreases. Another important factor is the **effective nuclear charge (Z_{eff})**. The outermost electrons are attracted by the positively charged nucleus but are also partially repelled by the negatively charged inner electrons. For a given actual Z, the Z_{eff} is dependent on the degree of electron shielding. The greater the degree of shielding (and thus the lower the Z_{eff}), the lower the ionization potential.

And the ionization potential is a reflection of just how eager or reluctant an atom is to engage in chemical reactions that require donation of electrons (e.g. reduction reactions) or ionization (e.g. salt formation). The heavier alkali elements such as Rb and Cs are easily ionized, whereas the smaller noble gases such as He and Ne are very difficult to ionize. When it comes to ionizing radiation, a standard element is hydrogen, and its ionization potential is 13.6 eV. Compared to other elements in Group 1, this is unexpectedly high. The reason is partly because of its single electron in a small electron shell (K shell) held closely and tightly by the nucleus. Additionally, there are no other electrons that shield the electrostatic attraction of the nucleus. Thus, the single electron feels the full effect of the positively charged nucleus. Another interesting but less significant factor is that in very high-Z elements, the outermost electrons are highly energetic, and their velocities are relativistic. The result is that relativistic considerations must be taken into account for the wave equations, and the end result is a smaller atomic radius and higher ionization energy.

4.19 Electron affinity

Electron affinity is a measure of the amount of energy released when an electron is added to a neutral atom or molecule. Because energy is typically released rather than required, the values of electron affinities technically should be negative (as is the custom in thermodynamics for exothermic reactions). Nevertheless, the present convention is to not use a negative sign for those with the greatest affinities. Students are advised to carefully note the sign convention being used when working problems in chemistry. In this text we shall adopt the use of positive signs for the situation where the atom, molecule or ion in question has a high affinity for an electron; thus, the higher the electron affinity, the more positive the value.

In a sense, electron affinity is the opposite of the ionization energy. And as with the case for ionization energy, the electron affinity is measured with the atom or molecule in question being in the gaseous state.

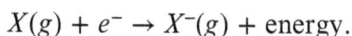

$$X(g) + e^- \rightarrow X^-(g) + \text{energy}.$$

Like ionization potential, there are predictable patterns of electron affinities across the periodic table. And since we mentioned above that electron affinity can be thought of as the opposite of ionization energy, it is unsurprising that the highest electron affinities are found in the halogens and they are typically low in Groups 1 and 2. As a general rule, electron affinity tends to diminish as one goes down a group. However, compared to the situation with ionization energies, this trend is far less obvious, and there are many exceptions to the 'rules'.

As one moves across a period, the trend is similarly imperfect—electron affinity may increase or decrease as one moves from right to left. The specifics depend on the electronic configuration of the atoms in question. For example, although nitrogen ($Z = 7$) has more electrons than carbon ($Z = 6$), the electrons in nitrogen have filled the three available p-orbitals. Therefore, addition of another electron means pairing up in one of the same three p-orbitals. In contrast, carbon has room for another electron in a vacant p-orbital, and for these reasons, the electron affinity of carbon is slightly higher than that of nitrogen. Nevertheless, the overall trend for electron affinity is for the value to increase from left to right across a given period up until the halogens are reached.

The halogens (Group VIIA in the traditional Mendeleev system or Group 17 in the modern International Union of Pure and Applied Chemistry (IUPAC) nomenclature), which need just one more electron to fill their valence shells and obtain the coveted 'octet' of outer electrons, have the greatest electron affinities. And as expected, the noble gases (Group 18 in the IUPAC system), which were born with that filled octet of outer electrons, have no desire to grab another electron and therefore have the lowest electron affinities.

One might predict that fluorine (F) would have the highest electron affinity of all elements, but in fact, chlorine (Cl) has an even higher electron affinity than F. The reason for this anomaly is that the electron–electron repulsion in the smaller 2p orbitals of row 2 elements (where F sits) is greater than the repulsion in the larger 3p orbitals of row 3 (where Cl is).

Although electron affinity is typically defined as the energy released upon addition of an electron to an isolated atom, another equivalent definition is the energy *required* to *remove* an electron from the singly charged anion (negatively charged ion; a positively charged ion is called a cation) in the gaseous state. This concept can be represented as:

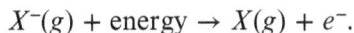

$$X^-(g) + \text{energy} \rightarrow X(g) + e^-.$$

When this convention is used, the electron affinity can be considered the same as the ionization energy of the −1 charged cation (or the zeroth ionization energy using the convention that the first ionization energy is removal of an electron from the neutral atom).

4.20 Electronegativity

Electronegativity (symbolized by X) is very similar to electron affinity in the sense that is represents an element's tendency to attract electrons. The difference is that electronegativity reflects this tendency when the atom is involved in a chemical bond. Therefore, unlike electron affinity and ionization energy, which are properties of isolated atoms alone, electronegativity relates to the behavior of atomic electrons in a molecule. Nevertheless, electronegativity does exhibit trends across the periodic table. For instance, electronegativity tends to increase from left to right across periods (rows) and tends to increase from bottom to top of groups (columns). Incidentally, this trend across the periodic table is the inverse of atomic radius —atomic radii decrease from left to right across periods and bottom to top in groups.

Incidentally, atomic radii are typically on the order of 30–300 pm or 0.03–0.3 nm. One nanometer is 1×10^{-9} m, and one picometer is 1×10^{-12} m. A non-SI unit that is still in common use when talking about these sizes is the **angstrom unit** or 1×10^{-10} m or one-tenth of a nanometer or 100 pm. The symbol for an angstrom is Å. Hence, atomic radii are about 0.3–3 Å. Atomic radii are around 10 000 (1×10^4) times the size of atomic nuclei. Nuclei typically have radii on the order of 1–10 fm. One femtometer is 1×10^{-15} m; this distance is sometimes called a **fermi**. Either way, both are abbreviated **fm**. Atomic radii have often been determined empirically through the technique of x-ray crystallography.

Electronegativity is quantified by the Pauling scale, using Pauling units. Historically, these ranged up to 4.0 for F (fluorine), the most electronegative of all elements, but today's scale assigns F a value of 3.98. Thus, F has the highest electronegativity and cesium has the lowest on the Pauling scale.

An important difference between electronegativity and the other three atomic metrics (ionization energy, electron affinity and atomic radius) is that the noble gases of Group 18 (generally) do not have any assigned electronegativity values at all. This is because of the definition of electronegativity as the tendency to attract electrons within a molecule (that is, while engaged in a chemical bond). Since noble gases do not normally form molecules and do not engage in chemical bonding, they cannot be assigned an electronegativity value. (Students of chemistry may recall that this generalization is not entirely true since the heavy noble gases (krypton, xenon and radon) can, under extreme conditions, form chemical bonds with very electronegative elements (e.g. F) and thus can be assigned electronegativity values.)

4.21 Chemical bonds

A chemical bond can simply be defined as an attraction between atoms or ions that allows at least transient formation of molecules or crystals. There are several forms of chemical bonds, including:
1. Ionic
2. Covalent
3. Metallic
4. London dispersion forces
5. Keesom bonds
6. Debye bonds
7. Hydrogen bonds.

The conceptually simplest bonds are caused by the purely electrostatic attraction between positively and negatively charged ions. Ions are simply atoms with an electric charge such that those with an excess of electrons and a net negative charge are called anions, whereas those with an electron deficiency and a net positive charge are called cations. As discussed above, some elements have very low ionization potentials and are easily ionized (meaning they lose an electron in this example). Sodium and potassium of Group 1 are good examples of such elements, and they tend to form singly charged cations. On the other side of the periodic table, F and Cl tend to gain electrons to form a complete outer octet of electrons. Thus, they are found as singly charged anions. When such cations and anions meet, their positive and negative electrical charges attract, and they form ionic bonds. Thus, sodium chloride (NaCl) crystals are held together by strong, attractive coulombic forces in a typical ionic bond. Ionic bonds are characterized by a large difference in electronegativity between the atoms involved. Quantitatively, when the electronegativity difference exceeds 2.0, an ionic bond is formed.

Like ionic bonds, metallic bonds arise from coulombic attraction. But instead of being between positively and negatively charged ions, metallic bonds materialize from the attraction between positively charged metallic nuclei and a cloud of free electrons surrounding them—'delocalized' electrons. Atomically, metals may be conceptualized as a scaffold of fixed, positively charged nuclei surrounded by a 'sea' of free electrons in which no given electron belongs to any particular nucleus. In this manner, metallic bonds may be considered as the sharing of valence electrons between several positively charged metal cations. This type of chemical bonding grants metals their characteristic properties such as malleability, electrical conductivity, thermal conductivity and reflectivity, among many others.

Parenthetically, the shiny yellowish ('golden') color of metallic gold is due to fascinating quantum and relativistic effects. Gold (Au) has atomic number 79, which is the highest of all naturally occurring metals (except for mercury (Hg, Z = 80), which as the only liquid metal at room temperature, has its own quirks). The high atomic number implies a high number of positively charged protons in the nucleus that exert a very strong electrostatic force on the orbiting electrons. This phenomenon results in what is called relativistic contraction of the orbitals; given orbitals are closer to the nucleus than they would be in a lighter atom. But to maintain their orbits in the presence of such force, these electrons must be travelling faster than similar electrons in smaller atoms. Using a simple approximation provided by the Bohr model, the speed of an electron may be estimated by:

$$v \approx cZ/137,$$

where c is the speed of light in a vacuum and Z is the atomic number.

From this, one can calculate that the speed of the electrons is about 58% the speed of light—certainly high enough to demand that relativistic effects be considered. What this means is that the mass of the electrons in question increases. Specifically, it increases by the factor γ, which is defined as:

$$\gamma = \frac{1}{\sqrt{1 - \frac{v^2}{c^2}}}$$

Because of the orbital contraction (especially the 6s orbital, which has greater proximity to the positively charged nucleus than the 5d orbitals in Au), the energy of photons absorbed by Au lies in the blue wavelengths of visible light (as opposed to the invisible ultraviolet wavelengths of most other metals like silver). The result is that Au absorbs blue wavelengths of visible light and reflects the other wavelengths, granting it its shiny yellowish color.

Covalent bonds are more conceptually complicated. They form because of an atom's drive to fill its outer electron octet. Because of this inherent thirst to complete their outer electron octet (that is, complete the shells of their valence electrons), atoms seem willing to 'share' electrons in order to attain this coveted state. The sharing of valence electrons leads to an attraction that is called covalent bonding. The sharing of one pair of electrons produces a single covalent bond. The sharing of two pairs of electrons creates a double bond, etc.

Although single bonds are the most common covalent bonds, many compounds possess double or triple bonds. For example, the carbon dioxide molecule (CO_2) has a central carbon with two oxygen atoms bound through double bonds. The carbon monoxide molecule (CO) has a triple bond between the carbon and the oxygen, meaning there are three pairs of electrons or six total that are being shared by the carbon and oxygen atoms. Other familiar examples include diatomic oxygen molecules, which have double bonds, and diatomic nitrogen, which has triple bonds.

Note that with covalent bonding, the term, 'molecule' is appropriate. Nitrogen gas molecules truly are discrete entities composed of two nitrogen atoms held together as a unit. This distinguishes covalent compounds from ionic compounds and metals. In ionic compounds, we should not think of the electrostatically attracted positively and negatively charged ions as genuine molecules. Rather, they are part of a large lattice or crystal composed of many oppositely charged units held together by the electrostatic force. Thus, in crystals of common table salt, NaCl, there are adjacent positively charged sodium ions and negatively charged chloride ions repetitively arranged in three dimensions. But these alternating sodium and chloride ions are not true molecules. However, NaCl can be called the 'formula unit' of the ionic compound. Similarly, we should not think of the individual cations swimming in a sea of detached free electrons as being molecules either. In contrast, covalent compounds may be considered to be composed of discrete, 'true' molecules. Nevertheless, there is some overlap that can lead to controversy. For instance, the bonds in silica (SiO_2) have characteristics of both covalent and ionic bonds and the question of whether quartz is an ionic or covalent compound is a subject of academic debate.

Covalent bonds result from electron sharing, but that does not mean the atoms share the electrons equally. Highly electronegative atoms will hog up the electrons for themselves, leading to a slight charge imbalance. In other words, an ionic bond character may be present. This unequal sharing of electrons means that the electrons are closer to one atom than the other and lead to a polarity such that the atom(s) with the electrons close by tend to be more negatively charged, while the atom(s) with the electrons pulled away from it tend to be more positively charged. This leads to so-called polar covalent bonds. Although not as pronounced as genuine ionic bonds, polar covalent bonds exhibit decidedly ionic chemical characteristics. They form when the electronegativity difference between the two atoms involved in the bond have an electronegativity difference of 0.5–2.0.

Given the electronegativity of oxygen is 3.44, while the electronegativity of hydrogen is 2.20, one would predict that water exhibits polar covalent bonds and the H_2O water molecule (which is shaped like a boomerang) is a polar molecule with the positivity on the two hydrogens and the negativity surrounding the oxygen atom. Such 'partial charges' on polar molecules are denoted by $\delta+$ or $\delta-$. Polar molecules like water behave like miniature electric dipoles, characterized by their partial charges $\delta+$ and $\delta-$. Such characteristics play a critical role in molecular biology, where a special subtype of chemical bond (hydrogen bonds) is frequently encountered (for example, in holding the two chains of the double helix of DNA molecules together).

In some situations, the sharing of electrons is more equitable, and this results in the formation of non-polar covalent bonds. When the electronegativity difference between two atoms is less than 0.5, non-polar bonds tend to form.

Because of its valence electron configuration, in order to obtain an octet, carbon atoms tend to form four total bonds with other atoms. These other atoms can be other carbons. When a carbon bonds to other carbon atoms, the bonds will be nonpolar (although the atom(s) attached to that other carbon may affect it somewhat). But carbon can also form strong covalent bonds to various other atoms, yielding polar or non-polar bonds. Carbon's capacity to create up to four strong covalent bonds confers a unique versatility among all other elements.

Carbon can form single bonds, double bonds or triple bonds (and even quadruple bonds in the unusual case of C_2). And with its ability to bond to four different or similar atoms, the chemistry of carbon is unparalleled. There are many millions of different carbon-based compounds, and the entire discipline of organic chemistry is dedicated to the study of carbon.

The field of organic chemistry is ill-defined. It was originally defined as the branch of chemistry that dealt with compounds obtained from living organisms—that is, chemical compounds that were somehow imbued with the 'elan vital' or force of life. But modern organic chemistry actually has surprisingly little overlap with the field of biochemistry nowadays. And although organic chemistry focuses on the chemistry of carbon, it excludes carbonates, oxides and carbides and also excludes the various allotropes of carbon such as diamond, graphite, graphene and the fullerenes. Thanks to this restriction, greenhouse gas carbon dioxide (which is the chemical foundation of all life through photosynthesis and chemosynthesis) and poisonous carbon monoxide (which is naturally produced in humans during the metabolic breakdown of heme and has recently been found to have a role as a neurotransmitter) are usually excluded from the study of organic chemistry. With the tens of millions of organic compounds along with the complexities of organic synthesis, detection and applications, students probably do not miss the carbonates, carbon oxides and carbides one bit.

This brief description of covalent bonding (based on individual atoms' quest to obtain an outer valance electron shell octet) does not do justice to the true complexity of covalent bonding. In order to properly describe the nuances of covalent bonding and explain phenomena such as why diatomic molecular oxygen is a triplet, one must resort to the quantum-mechanics-based molecular orbital theory. This will be touched upon briefly in chapter 9 when we go into greater detail on radiation chemistry.

In our list of the various types of chemical bonds, the last four (London dispersion forces, Keesom bonds, Debye bonds and hydrogen bonds) are quite different from ionic, metallic or covalent bonding. These four subtypes of bonds can collectively be called **Van der Waals bonds** since they all share a common mechanism of attraction between molecular dipoles called the Van der Waals force. Van der Waals forces are inter-molecular attractions (although they can also be inter-atomic, as in the attraction between noble gas atoms).

As mentioned above, some molecules have polar covalent bonds and result in polar molecules with permanent partial charges designated by δ+ or δ−. When such permanent molecular dipoles encounter one another, the δ+ on one dipole will be drawn to the δ− leading to a short-lived, weak electrostatic attractive force. These interactions are fittingly called **permanent dipole–permanent dipole interactions** or **Keesom interactions**. Although Keesom forces are conceptually similar to ionic bonds, the Keesom interaction between polar molecules is far weaker than ionic bonding. But it is stronger than the next interaction, the Debye force.

Another variation of the same underlying mechanism involves a molecule with a permanent dipole that encounters a non-polar molecule. Upon a close encounter, the δ+ of the permanent dipole will induce a transient partial charge δ− in the non-polar molecule (and of course, the δ− of the permanent dipole will induce a transient partial charge δ+ in the non-polar molecule). This phenomenon is called induced polarization and leads to a transient attractive electrostatic force. The resultant chemical bonding is the result of **permanent dipole–induced dipole interactions** or the **Debye force**. Not all non-polar molecules are equally susceptible to such transient induced dipole formation and thus do not participate in Debye forces equally. The Debye forces are weaker than Keesom forces but are stronger than the next interaction, the London dispersion force.

Non-polar molecules and atoms are, by definition, without a permanent dipole but nonetheless can have a non-zero instantaneous dipole moment. Random fluctuations in the electron clouds of such non-polar molecules (and atoms) lead to transient localizations of positive and negative partial charges that can result in weak electrostatic attractions. These weak attractive forces are known as **induced dipole–induced dipole forces** or **London dispersion forces**.

Students may recognize that for some of the above interactions (dipole–dipole and dipole-induced dipole) there can be a stronger version involving not a polar molecule, but an ion. Thus, there can be **ion-dipole forces** and **ion-induced dipole forces**. Because the net electrical charge on an ion is stronger than the partial charge on a polar molecule, these tend to be stronger bonds than the dipole–dipole Keesom forces or the dipole-induced dipole Debye forces. Historically, ion-dipole and ion-induced dipole forces are not called Van der Waals forces since technically, the definition of Van der Waals forces is limited to uncharged molecules and atoms, but not ions.

There is one final special subtype of intermolecular attraction (which thus qualifies as a Van der Waals force) between uncharged polar molecules that is very important in molecular biology. This is the **hydrogen bond**. Hydrogen bonds are an exceptionally strong form of dipole–dipole interactions and therefore might be considered an extreme case of Keesom bonding. Hydrogen bonding is limited to polar molecules that contain nitrogen, oxygen or fluorine atoms. When these very electronegative atoms are covalently bonded to hydrogen atoms, the partial charges ($\delta+$ and $\delta-$) in the polar molecules are very pronounced. This allows unusually strong intermolecular interactions to develop. In fact, hydrogen bonds have not only all the features of Keesom dipole–dipole interactions but also possess some features of covalent bonds. For example, hydrogen bonds possess a directionality; such directionality is not characteristic of conventional Keesom bonds. Furthermore, hydrogen bonds result in interatomic distances that are shorter than the calculated sum of their radii (technically, their Van der Waals radii), indicating that the attraction is strong. Hydrogen bonding gives water some of its very unusual properties, such as its oddly high boiling point, melting point, viscosity and surface tension. As with all intermolecular forces, hydrogen bonds are transient. In liquid water, hydrogen bonds within cells have a lifetime averaging around 10^{-11} s (10 ps).

Although we mentioned that Van der Waals forces are intermolecular interactions, for large biological macromolecules, such forces play important intramolecular roles as well. For example, hydrogen bonds play a prominent role in the characteristic secondary structures of proteins (such as the alpha helix and the beta pleated sheet). Hydrogen bonds and other Van der Waals forces are involved in the higher-order tertiary conformations and inter-subunit interactions that lead to quaternary structures of proteins as well. Student may recall that the two chains of the double helix of the genetic molecule, DNA, are linked together through hydrogen bonds. Specifically, the purine bases (adenine (A) and guanine (G)) in one chain are hydrogen bonded to the pyrimidine bases (thymine (T) and cytosine (C)) of the other chain. These GC and AT bonds are called Watson–Crick base pairs. Since there are three hydrogen bonds holding the G-C bases together compared to only two hydrogen bonds holding adenine and thymine together, sections of DNA that are GC-rich tend to hold together tighter than regions that are made primarily of AT base pairs.

4.22 Types of chemical reactions

In all examples the table 4.2, the chemical elements or compounds on the left are called **reactants**, and the ones on the right of the arrow are called **products**. (In biochemistry, when an enzyme catalyst is present to facilitate a reaction, the reactants are usually called substrates.)

The numbers in front of the elements or compounds are the coefficients for the balanced chemical reaction that indicate the number of molecules (or moles) of that particular reactant or product in the overall reaction. Chemical reactions that have the proper coefficients in front of the reactants and products are said to be stoichiometrically balanced chemical equations.

A basic principle in single displacement reactions is that an active metal will displace less active metals (or hydrogen) from their compounds when in an aqueous solution. (An aqueous solution is simply a solution in which water is the solvent.) For example:

$$Na(s) + H_2O \rightarrow NaOH(aq) + H_2(g).$$

Table 4.2. There are several ways to classify the countless chemical reaction types, and among the various subtypes, there can be significant overlap. Nevertheless, this is one simple system of categorizing chemical reactions.

Reaction type	General equation	Examples
Displacement	A + BC → B + AC	$Zn + CuSO_4 → ZnSO_4 + Cu$
Double displacement	AB + BC → AC + BD	$NaSO_4 + BaCl_2 → BaSO_4 + 2\,NaCl$
Synthesis	A + B → AB	$2\,Al + 3\,O_2 → Al_2O_3$
Decomposition	AB → A + B	$2\,H_2O_2 → 2\,H_2O + O_2$

In this example, the (s), (aq) and (g) indicate the phases solid, aqueous and gaseous, respectively.

The double displacement reaction is sometimes called a metathesis. When a product in such a reaction (or any chemical reaction for that matter) 'leaves the scene', meaning it is no longer available for the reverse chemical reaction, it pulls the reaction to the right. Such situations are encountered in double displacement reactions that yield gaseous products that evaporate away or solid products that precipitate out of solution such as:

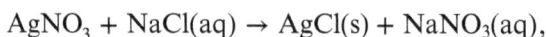

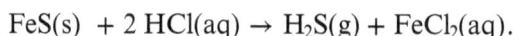

$$AgNO_3 + NaCl(aq) → AgCl(s) + NaNO_3(aq),$$

$$FeS(s) + 2\,HCl(aq) → H_2S(g) + FeCl_2(aq).$$

Synthesis reactions are sometimes called composition reactions or combination reactions.

Synthesis and decomposition reactions may be thermodynamically favorable (meaning that they should ultimately move in the direction of the chemical reaction as written) but often do not happen spontaneously under normal conditions. But the term normal must be properly defined. For any chemical reaction, we should specify the conditions, and one common set of conditions is called STP, for standard temperature and pressure. STP is defined as a temperature of 0 °C (293.15 K or 32 °F) and a pressure of 1 bar (100 kPa). Another commonly used standard is called NTP, for normal temperature and pressure, which is 20 °C and a pressure of 1 atmosphere (1 atmosphere or atm is 101.325 kPa). In any case, many chemical reactions do not occur spontaneously under either STP or NTP but instead require energy input (e.g. heat) to drive the reaction as written. When a reaction requires the input of heat, the reaction is called **endothermic** and the related metric, ΔH, is positive. Conversely when a reaction releases heat, the reaction is **exothermic** and ΔH is negative. The quantity ΔH is called the **heat of reaction** or **enthalpy** of reaction. The ΔH is calculated by the general concept of:

$$\Delta H = (energy\ of\ products) - (energy\ of\ reactants).$$

To ascertain if a chemical reaction will occur spontaneously under a given set of conditions (e.g. at STP), one needs to take into account a couple of additional variables—the **entropy** (S) and the Gibbs **free energy** (G). More specifically, one must consider the *changes* in these quantities (ΔS and ΔG) in order to properly perform the calculations. Under conditions of constant temperature and pressure (which is actually quite rare in nature) the general formula relating the three key variables is:

$$\Delta G = \Delta H - T\Delta S.$$

If the ΔG of a chemical reaction is negative, the reaction will be expected to occur spontaneously. One can see that in general, in reactions that release energy (that is, exothermic reactions where ΔH is negative), the likelihood of ΔG being negative is higher. Thus, exothermic reactions are generally more likely to occur spontaneously. But of course, the other factor $T\Delta S$ cannot be ignored. One can see that because of the negative sign, raising the temperature will further increase the chances that the overall ΔG will be less than zero. Additionally, one can see from the math that changes favoring an increase in entropy (conceptualizable as an increase in disorder) are also more likely to occur spontaneously.

Students of chemistry will recall that H, G and S are examples of so-called **state functions**, meaning that it is only the change (Δ) in value that matters, not the path one took to obtain that change. Other examples of state functions include pressure, temperature, mass and volume. This distinguishes state functions from quantities that do depend on the path taken, which are known as **path functions**. Work is proportional to the distance a force is applied, which means that it does depend on the path taken. Thus, work is a path function rather than a state function. State functions permit data tabulation of standard thermodynamic quantities at a given temperature and pressure. Students may look up such standard quantities such as the ΔG, ΔS and ΔH of formation (abbreviated ΔG_f°, ΔS_f° and ΔH_f°, respectively) in reference tables complied in the *CRC Handbook of Chemistry and Physics*, for instance, to perform calculations under standard conditions using the modified formula (to indicate standard conditions):

$$\Delta G_f^\circ = \Delta H_f^\circ - T\Delta S_f^\circ.$$

At chemical equilibrium, $\Delta G^\circ = -RT \ln K$, where K is the chemical equilibrium constant. From this one can obtain the final value of ΔG through: $\Delta G = \Delta G^\circ + RT \ln Q$, where Q is the molar concentrations of the products divided by the molar concentrations of the reactants (that is, $Q = \{[C]^c[D]^d/[A]^a[B]^b\}$ before the reaction proceeds to equilibrium). This will allow one to determine if a given reaction will occur spontaneously or not—if ΔG is negative, the reaction will proceed spontaneously as written.

One final point about spontaneous reactions is that even though a reaction may be thermodynamically favorable and expected to occur spontaneously (because ΔG is negative), that does not guarantee that the reaction will occur quickly. The rate at which a reaction proceeds is a function of temperature and another quantity called the activation energy (which is discussed in the next paragraph), but is generally independent of ΔG.

Chemical kinetics can be sped up substantially by a chemical catalyst. An example is the spontaneous decomposition reaction:

$$2\,H_2O_2 \rightarrow 2\,H_2O + O_2.$$

This reaction has a negative ΔG, so it will occur spontaneously over time. However, the rate at which the reaction occurs can be tremendously accelerated through the aid of a chemical catalyst such as manganese dioxide (MnO_2). Incidentally, this ability of manganese to serve as a powerful catalyst for the detoxification of hydrogen peroxide (which is an important product of radiolysis—the radiochemical breakdown water upon irradiation) will be discussed further when we review the mechanisms whereby certain extremely radiation resistant microorganisms cope with radiation. The way a catalyst works is not by affecting the overall thermodynamics of the chemical equation but rather by simply lowering the so-called activation energy of the reaction. The **activation energy (E_a)** can be thought of as a 'hump' in the road along a chemical reaction. Even if the road is downhill overall (meaning that the chemical reaction is thermodynamically favorable (ΔG is negative)), a large hump in the road may impede progress. Removing or reducing the height of the hump greatly accelerates progress. In this fashion, chemical catalysts (including protein enzymes in biochemical reactions) tremendously accelerate otherwise slow reactions. And the extent to which some enzymes can accelerate biochemical reactions is absolutely staggering. For instance, the enzymes orotidylate decarboxylase and adenosine deaminase allow biochemical reactions that would otherwise take millions of years, to occur in milliseconds. As another example, the Haber–Bosch process for synthesizing ammonia requires temperatures around 450 °C (840 °F) and pressures about 200 times normal atmospheric pressure even in the presence of a metal catalyst:

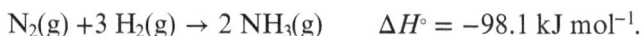

$$N_2(g) + 3\,H_2(g) \rightarrow 2\,NH_3(g) \qquad \Delta H^\circ = -98.1\ \text{kJ mol}^{-1}.$$

Although the reaction is exothermic (i.e. ΔH° is negative), ΔS° is also negative, making the reaction difficult under ordinary conditions. Nevertheless, the overall ΔG of the reaction is negative, indicating that it can occur spontaneously. Nitrogen-fixing bacteria, through the assistance of enzymes, can take inert atmospheric nitrogen and convert it into biologically useable ammonia, which enters the biosphere. And they do it quickly, under normal atmospheric pressures and at room temperature or below. Without this reaction and these nitrogen-fixing bacteria, life as we know it would likely not exist. These examples should make it clear that simply because a reaction is thermodynamically favorable and *can* occur spontaneously, does not imply that the reaction *will* proceed in a perceptible fashion. For instance, the conversion of carbon in the diamond allotrope (allotropes are the different physical forms in which an element can exist) can spontaneously transform into the graphite allotrope. Fortunately, this does not occur at a rate worth worrying about.

The rate at which a chemical reaction occurs is related to a parameter called the **rate constant, k**. If one starts with the general chemical reaction

$$aA + bB \rightarrow cC,$$

where the small letters (a, b, c) are the coefficients of the stoichiometrically balanced reaction representing the number of moles of each chemical reactant or product (A, B, C), then the rate of the reaction ($d[C]/dt$) can be related to the concentration of the reactants by:

$$\text{Rate} = d[C]/dt = k[A]^m[B]^n,$$

where the exponents m and n are called the **orders** of reaction.

The brackets around A and B above indicate the concentration of these reactants in **molarity**. Molarity is a common unit of concentration in chemistry and is defined as the number of moles per liter of solution. It is given the symbol **M**. Thus, a solution with one mole of solute in one liter of solution is said to have a concentration of 1 M or one molar.

Students should note that the stoichiometric coefficients a and b do not necessarily equate to the orders of the reaction, m and n. These orders must be determined experimentally, as must the rate constant, k. The reaction orders dictate the mathematical dependence of the rate on the concentration of the reactants. For instance, if a reaction is order m with respect to reactant A, and $m = 1$, the reaction is said to be 'first order with respect to A'. If the reaction is order $n = 2$ with respect to reactant B, the reaction may be said to be 'second order in B'. The overall reaction order is the numerical sum of all the reactant orders (that is, $m + n$ in the example above). For example, for the synthesis of water from hydrogen cations (H^+) and hydroxide anions (OH^-):

$$H^+ + OH^- \rightarrow H_2O.$$

The rate law is shown experimentally to be: rate $= k\,[H^+]^1\,[OH^-]^1$ indicating that the reaction is first order with respect to H^+, first order in OH^-, and second order overall.

Students should note that the reaction orders m and n are typically positive integers, but they do not have to be. Sometimes, these experimentally determined quantities are fractions, negative numbers or even zero. The subject of reaction order and the various rate laws will be touched upon again in the chapter on radioactivity.

A final point about chemical kinetics is that the rate constant k is a function of temperature and hence is sometimes written as $k(T)$. The relationship between temperature and chemical reaction rate is provided by the **Arrhenius equation**:

$$k = Ae^{-(Ea/RT)},$$

where R is the Universal Gas Constant and E_a is the activation energy of the chemical reaction. As a general rule of thumb, chemical reaction rates double for every 10 °C increase in temperature.

Returning to the various types of chemical reactions, there are several subtypes that are simply special versions of the main four groups. For instance, a **combustion reaction** is a special type of displacement reaction in which one of the reactants is diatomic molecular oxygen (O_2). Typically, one of the reactants contains carbon and one of the products is carbon dioxide, CO_2. (But if the reactants contain nitrogen atoms, diatomic nitrogen gas, $N_2(g)$[1], is generated instead of or in addition to carbon dioxide.) A familiar example is the combustion of the simplest hydrocarbon alkane, methane:

$$CH_4 + 2\,O_2 \rightarrow 2\,H_2O + CO_2 \qquad \Delta H^\circ = -890 \text{ kJ mol}^{-1}.$$

Similarly, combustion of octane, an eight-carbon alkane present in gasoline is given by:

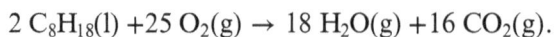

$$2\,C_8H_{18}(l) + 25\,O_2(g) \rightarrow 18\,H_2O(g) + 16\,CO_2(g).$$

The combustion reaction for ammonia is given by:

$$4\,NH_3 + 3\,O_2 \rightarrow 2\,N_2 + 6\,H_2O.$$

Notice that there is no carbon dioxide produced in this particular combustion reaction.

As is evident in fires and explosions, most combustion reactions are highly exothermic, meaning they release heat and ΔH is negative. For this reason, combustion is a key reaction for energy production in modern society. Automobile combustion engines may use gasoline (a mixture of hydrocarbons ranging from four carbons to 12 carbons) or diesel (a slightly heavier mixture with more long-chain hydrocarbons). Because of its higher average molecular weight, diesel is slightly less volatile than gasoline. Additionally, diesel is more efficient in its combustion, meaning that the reaction with oxygen is more likely to go to 'completion' and yield carbon dioxide; less biologically dangerous carbon monoxide is generated through 'incomplete' combustion. Interestingly, diesel does not require spark plugs for ignition, as the fuel spontaneously ignites when the piston pressure becomes high enough. Furthermore, diesel is more energy-dense than gasoline—it produces more energy per volume of fuel. However, there are some disadvantages to diesel including the heavier and more complicated engine. Incidentally, when a combustion reaction is explosive, and the gaseous products generates expand at beyond the speed of sound, the loud reaction is sometimes called a **detonation**. Subsonic speeds of combustion result from **deflagrations** and **diffusion flames**, in which the reactants are premixed or unmixed, respectively.

The involvement of molecular oxygen as one of the reactants in a combustion reaction introduces another category of chemical reactions—**oxidation**. By definition, when a substance is oxidized, it has lost electrons. The complement of oxidation is **reduction**, the gain of electrons. Chemicals that oxidize other chemicals are called **oxidants** or **oxidizing agents**; substances that reduce other chemicals are **reductants** or **reducing agents**. Oxidation and reduction are always coupled in the

---

[1] As an aside, along with diatomic molecular nitrogen and oxygen, there are five other diatomic molecules, bringing the total to seven. The other diatomic molecules are hydrogen, fluorine, chlorine, bromine and iodine. While most of the seven diatomics are gases at room temperature and normal atmospheric pressure, bromine is a liquid and iodine is a solid. Note that just because the most familiar chemical form of the pure element might be a diatomic molecule, this does not mean that other forms are not possible. For instance, molecular oxygen can exist in the common diatomic gas form or in the triatomic gas, ozone. These different forms are examples of oxygen allotropes, just as diamond and graphite are allotropes of carbon.

sense that if something has been oxidized in a chemical reaction, something else must have been simultaneously reduced. Because of this coupling, such reactions are collectively called **redox reactions**, which is short for 'reduction–oxidation'. The combustion reactions described above also fit into the category of redox reactions, demonstrating the overlap between the various subcategories of chemical reactions. Along with ionization, redox reactions are among the most important chemical reactions in radiation chemistry and radiobiology.

During redox reactions, the so-called 'oxidation state' (also called **oxidation number**) of the reactants change. For instance, when metallic iron (Fe) is oxidized by oxygen to form rust, the oxidation state of Fe (which is zero, as it is for all atoms existing as pure elements) increases to the +3 oxidation state. To indicate the fact that the new Fe product is in the +3 oxidation state, it is often written as Fe(III). The overall reaction may be written as:

$$4Fe(s) + 3O_2(g) + 6H_2O(l) \rightarrow 4Fe(OH)_3(s).$$

$4Fe(OH)_3$ is 'hydrated' iron oxide, which can then become Fe_2O_3 through the release of its chemically bound water:

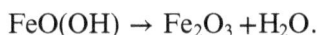

$$Fe(OH)_3 \rightarrow FeO(OH) + H_2O,$$

$$FeO(OH) \rightarrow Fe_2O_3 + H_2O.$$

In this final state, iron oxide (Fe_2O_3) contains iron in its +3 oxidation state (that is, Fe(III)), and to attain a net balance of zero, oxygen must be in its −2 oxidation state. By multiplying three oxygen atoms in their −2 state to get −6, one can see that the two Fe atoms in Fe_2O_3 must each have an oxidation state of +3 to yield +6, since the Fe_2O_3 product is uncharged and must have a net oxidation state of zero. This observation leads to some of the redox rules, including the fact that when in a compound, oxygen is normally in its −2 oxidation state.

Some of the basic rules for oxidation numbers include:

1. The oxidation number of any pure, free element is zero. Thus, the oxidation state of elemental, metallic copper is zero. Similarly, the oxidation numbers of nitrogen and oxygen in N_2 and O_2 are both zero.
2. The oxidation number of a monatomic ion is equal to the charge. For instance, the oxidation state of Cu^{++} would be +2.
3. The oxidation number of hydrogen in most compounds is +1. Thus, in water (H_2O), the two hydrogen atoms each have an oxidation number of +1 to yield a total of +2; for electrical charge neutrality and balance of oxidation number, this means that the sole oxygen must have an oxidation number of −2.
4. Hydrogen has an oxidation number of −1 when it is combined with an element that is less electronegative than itself. For example, lithium at 0.98 is less electronegative than hydrogen (2.20). Thus, in lithium hydride, LiH, hydrogen has an oxidation number of −1, while lithium has an oxidation state of +1.

5. The oxidation number of oxygen in most compounds is −2. Important exceptions arising when radiation interacts with matter include peroxides (oxidation state of −1) and superoxides (oxidation number of −½), which are discussed in subsequent chapters.

6. The oxidation number of elements in Group 1 (the alkali metals) is always +1 when in a compound. For example, the oxidation number of K^+ is +1.

7. The oxidation number of elements in Group 2 (the alkaline earth metals) is always +2 when in a compound. For example, the oxidation number of Ca^{++} is +2.

8. The oxidation number of F in a compound is always −1. The halogens as a whole (Group 17, formerly called Group 7A) usually, but not uniformly, have an oxidation state of −1 when in compounds.

9. The oxidation state of a compound anion (such as the sulfate anion, SO_4^{2-}) equals the net charge. Thus, the individual component atomic oxidation states must sum to the composite ion. In this case, the oxidation state of the sulfate anion is −2, and the four oxygens have −2 apiece for a total of −8, meaning that the single sulfur atom must have an oxidation number of +6.

10. In any electrically neutral chemical compound, the net sum of oxidation numbers must always be zero. For instance, in H_2SO_4 (sulfuric acid), the two hydrogens have oxidation numbers of +1 apiece, the four oxygen atoms have oxidation numbers of −2 each. Since the sum of the molecule as a whole must equal zero, the oxidation number of the single sulfur atom must be +6. In H_2SO_3 (sulfurous acid), the two hydrogens still have oxidation numbers of +1 apiece, and the three oxygen atoms still have oxidation numbers of −2 each. Thus, the oxidation number of the single sulfur atom in this case must be +4. In general, when there are two possible oxidation states (as in sulfur, iron and nitrogen), the higher oxidation state is given the suffix '-ic', whereas the lower oxidation state is given the suffix '-ous'. Hence, 'ferric' for Fe^{+3} and 'ferrous' for Fe^{+2}.

A special type of redox reaction, in which a substance is simultaneously both oxidized and reduced is called a **disproportionation** reaction. Disproportionations are sometimes also called **dismutation** reactions. In such reactions, a compound with atoms in an intermediate oxidation state is converted into products in which the atoms now have different oxidation states, with one higher and one lower than at the start. A classic example is the ultraviolet radiation-induced disproportionation of mercury (I) chloride Hg_2Cl_2:

$$Hg_2Cl_2 \rightarrow Hg + HgCl_2.$$

In this case, the initial oxidation state of mercury is +1, but the final products have mercury in the zero oxidation state (elemental mercury) and +2 oxidation state (in mercuric chloride or mercury (II) chloride). Another example, one that plays a crucial role in radiation biology, is the disproportionation of hydrogen peroxide:

$$2H_2O_2 \rightarrow 2H_2O + O_2.$$

In this case, oxygen starts out with an oxidation number of −1 in the parent hydrogen peroxide molecule, but in the products, the oxidation number of oxygen in water is its typical value of −2, and of course in its elemental form of O_2 its oxidation number is 0.

Redox reactions are especially important in radiation chemistry since ionizing radiation (by definition) involves the removal of electrons from atoms and molecules. Therefore, although one does not typically refer to radiation as an oxidizing agent, in a strict sense, interactions between radiation and matter do involve oxidations. Naturally, when an electron is removed (and becomes a 'solvated electron' roaming about in an aqueous solution, for example), it will inevitably attach itself to another molecule and reduce it. This disruption can have profound chemical and biological consequences, and forms the basis for radiobiology. Radiation-induced ionizations, with the concomitant release of powerful reductants such as solvated electrons start off chain reactions that can have biological sequelae. Along the way, free radicals play an important part of this chain reaction.

This will be discussed in greater depth in chapter 9 but for now, it is useful to recall that the definition of a **free radical** is an atom, ion or molecule with one or more unpaired electrons. Free radicals tend to be extremely reactive and can wreak havoc on biochemical systems if they are present in abundance in unwanted locations and remain unchecked. On the other hand, some of our immune cells such as neutrophils actually kill invading microbes by generating reactive oxygen species including superoxide anion, singlet oxygen and various free radicals and pouring them onto the intruders.

Although we tend to think of free radicals as rare and dangerous, one should recall that molecular oxygen itself is a radical. Specifically, it is a 'bi-radical' species in that it has two unpaired electrons. This explains oxygen's unusual reactivity and involvement in combustion reactions and other redox reactions. To understand why diatomic molecular oxygen is a biradical with two unpaired electrons, one must examine the **molecular orbital diagram** of diatomic oxygen. This demonstrates that in its ground state, diatomic molecular oxygen has two unpaired electrons, each following Hund's rule and residing in their own orbitals with parallel spins (rather than pairing up) with opposite spins. Recall that Hund's rule simply says that electrons will arrange themselves in separate orbitals with their spins parallel to one another before they start pairing up into the same orbitals with antiparallel spins. Parenthetically, diatomic molecular oxygen, O_2, is in a triplet state, meaning that it exhibits a threefold splitting of spectral lines. In contrast, highly-reactive singlet oxygen is in an excited energy state above the ground state and exhibits only a single spectral line. Molecular orbital theory explains why the ground state of oxygen is a triplet (with two unpaired electrons in two separate molecular orbitals), whereas singlet oxygen (with all its electrons paired up) is in an excited state since the electrons have violated Hund's rule. Curiously, without any unpaired electrons, singlet oxygen is not a free radical whereas triplet oxygen is, underscoring the complexity of this and related radiochemistry topics. We will cover this further in chapter 9.

Hydrolysis is another subtype of chemical reaction in which water, specifically, is involved in the breakdown of chemical bonds. Hydrolysis reactions are particularly important in biochemistry, in which large biological macromolecules such as polysaccharides, fats and proteins are broken down into their constituent monosaccharides, fatty acids and amino acids during digestion or other physiological processes. A classic example of hydrolysis is saponification in which a fat molecule (triglyceride) is broken down in an aqueous solution (that is, a water solution) in the presence of a powerful base such as lye into free fatty acids and glycerol—an elementary form of soap. Another example is the cleavage of amide bonds in proteins called peptide bonds, catalyzed by enzymes called proteases.

The reverse of hydrolysis is a form of synthesis reaction that can be called **condensation** or **dehydration synthesis.** As a couple of biologically important examples of dehydration synthesis, larger biological macromolecules such as nucleic acids (e.g. DNA and RNA) or proteins are built up from their fundamental building blocks of nucleotides and amino acids, respectively. In the biosynthesis of protein molecules, the amino group of an amino acid is esterified to the carboxyl group of another amino acid. The resulting amide bond is given the special name of **peptide bond**, and the growing chain of amin acids is called a polypeptide. One the full sequence of amino acids is assembled, the mature polypeptide chain may remain as a 'monomer' or may join others to form dimers, trimer, tetramers, etc. The resulting protein molecule might theoretically assume a variety of different three-dimensional conformations, but under physiological conditions of an aqueous solution with a very specific temperature, pH, and ionic milieu, the protein will normally only assume one very particular and fully functional conformation. Under unusual conditions, these exquisitely designed proteins may assume an abnormal three-dimensional conformation, leading to malfunction and occasionally serious disease (for instance, 'mad cow disease' or Creutzfeldt-Jacob disease, which are examples of the so-called spongiform encephalopathies that leads to irreversible and relentlessly progressing brain damage and death in bovines and humans, respectively).

As a very general rule in chemistry, the breakage of bonds requires energy, whereas the formation of new bonds releases energy. But one must keep in mind that this applies to covalent (and ionic chemical bonds), but in physiological chemistry, other weaker bond types (such as hydrogen bonds and other van der Waals interactions) frequently come into play, and the overall free energy change (ΔG) is what will determine whether a reaction proceeds or not. Thus, in molecular biology, the formation of bonds in the construction of large, complicated biological macromolecules does not proceed spontaneously with the release of energy but rather requires input of a good deal of chemical energy via the conversion of the universal biological 'commerce molecule' ATP (adenosine triphosphate) into ADT (adenosine diphosphate). Similarly, the adage that 'bond breakage requires energy' is not readily apparent in molecular biology and biochemistry, in which the breakage of covalent bonds in complicated molecules may occur relatively easily. Nevertheless, when one adds excess energy into a biological system via ionizing radiation, the disruptive net outcome is often abundant bond breakage. In fact, the ultimate molecular consequences of ionizing radiation are rooted in chemical bond breakage in key macromolecules such as DNA.

4.23 Acids and bases

A final subcategory of chemical reactions are **acid–base reactions**. Acid–base reactions maybe thought of as a special subset of double displacement reactions. For instance, a reaction between a strong acid and a strong base will lead to the products of water plus a salt. As a familiar example, one can consider the reaction between the strong acid, hydrochloric acid and the strong base, sodium hydroxide:

$$HCl + NaOH \rightarrow H_2O + NaCl.$$

The products include water and ordinary table salt, NaCl. The definition of an acid varies according to a few different conceptual models. In the simplest model, the **Arrhenius theory**, an acid is defined as a chemical compound that produces hydrogen ions (H^+) in aqueous solution. In other words, an acid increases the concentration of H^+ in a solution. The hydrogen ions come from the dissociation of water:

$$H_2O \rightleftharpoons H^+ + OH^-.$$

This leads to the most common means of measuring the acidity of a solution—the pH scale. The pH of a solution is given by:

$$\mathbf{pH = -log[H^+]},$$

where the brackets indicate that the concentration of hydrogen ions in question is in molarity or moles per liter.

One should note that the pH scale is logarithmic and negative. Thus, a solution that has a pH of 3 is ten-fold more acidic and has ten times the molarity of hydrogen ions than that of a solution with a pH of 4. Furthermore, it is possible (though not frequently encountered) for a solution to have a pH less than zero (a negative pH) if the solution has a concentration of hydrogen ions greater than 1 M. Since the log of 1 is zero, any solution with $[H^+] > 1$ mol l^{-1} would have a negative pH. Neutral pH is defined as the pH at which the concentration of H^+ equals the concentration of OH^-. Under familiar circumstances of normal atmospheric pressure and temperatures, neutral pH is 7.0.

The dissociation of water into its ions is an endothermic reaction, meaning ΔH is positive. Therefore, providing heat will favor the reaction in the left-to-right direction. This can lead to a bit of confusion since as heat is added (that is, the temperature is increased), the dissociation reaction will tend to proceed to the right, which means more H^+—but also more OH^-. Thus, as the temperature increases and the dissociation of water increases, there will be more $[H^+]$ *in the still neutral solution*. However, by the definition of pH, the pH has dropped. Therefore, the pH of a perfectly neutral solution without any dissolved solutes will decrease as the temperature increases. However, in the real world, where dissolved gases are impossible to exclude, the situation grows more complex. Carbon dioxide is a weak acid that upon dissolution becomes carbonic acid, H_2CO_3. Because gases are more soluble in water at lower temperatures, there will be more dissolved carbon dioxide in an aqueous solution at a cold temperature compared to a warm temperature. Therefore, in reality when one takes into account the inevitable presence of dissolved atmospheric gases, the actual pH becomes a complicated function of temperature. Rising temperature will increase water dissociation and hydrogen ion concentration, tending to drive the pH down, while the solubility of acidic carbon dioxide diminishes as the temperature rises, tending to drive pH up. The net result depends on the actual concentration of dissolved carbon dioxide and the dissociation of consequent carbonic acid.

While an Arrhenius acid creates hydrogen ions in aqueous solution, an Arrhenius base is a chemical compound that dissociates in water to create hydroxide ions (OH^-). Recall that another name for base is alkali. Thus, a solution that is high in hydroxide ion concentration can be said to be basic or alkaline. In this manner, a general equation between an acid and a base may be expressed as:

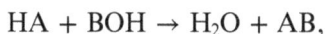

$$HA + BOH \rightarrow H_2O + AB,$$

where HA is the Arrhenius acid, BOH is an Arrhenius base, and AB is a salt like NaCl.

One should note that the Arrhenius definition of acids and bases is restricted to water—aqueous solutions. It does not apply to substances dissolved in non-polar organic solvents, for example. This restrictive definition led to other, more encompassing models of acids and bases, such as the Brønsted–Lowry theory and the Lewis theory.

Addition of hydrogen ions, H^+, to a solution is another way of saying that free protons are becoming available. Free protons interact with water molecules to cause 'protonation' of water or formation of **hydronium ions** (H_3O^+). It is in fact these hydronium ions that are detected by most pH indicators. In the **Brønsted–Lowry theory**, the concept of protonation is critical. The Brønsted–Lowry definition of an acid is a chemical that can donate protons (hydrogen ions) to other compounds. Likewise, a Brønsted–Lowry base is a chemical compound that can accept donated protons. Upon donation of a proton, the Brønsted–Lowry acid becomes a '**conjugate base**', and upon acceptance of a proton, a Brønsted–Lowry base becomes a **conjugate acid**. The Brønsted–Lowry theory is solvent independent, meaning that it is not restricted to aqueous solutions. A general equation for Brønsted–Lowry acid–base reactions would be:

$$HA + B \rightarrow A^- + BH^+,$$

where HA is the Brønsted–Lowry acid, B is a Brønsted–Lowry base, A^- is the conjugate base of HA, and BH^+ is the conjugate acid of B.

As a specific example:

$$HCl + H_2O \rightleftharpoons H_3O^+ + Cl^-.$$

The removal of H+ from the Brønsted–Lowry acid, hydrochloric acid (HCl), generates the chloride anion Cl^-, which is the conjugate base. Simultaneously the addition of H^+ to water (a Brønsted–Lowry base in this example) generates the hydronium ion, H_3O^+, which is the conjugate acid of the base, water.

Water is an example of an **amphoteric** substance, meaning that depending on the circumstances, it can act either as an acid or as a base. Such behavior is evident in the Brønsted–Lowry model:

$$H_2O + H_2O \rightleftharpoons H_3O^+ + OH,$$

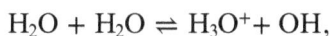

where H_2O is both the Brønsted–Lowry acid and the Brønsted–Lowry base in the reactants, while H_3O^+ and OH^- serve as conjugate bases and conjugate acids, respectively.

Given the Brønsted–Lowry definition of acids as hydrogen ion (proton) donors, it is evident that like the Arrhenius theory, the Brønsted–Lowry theory is limited. For instance, boron trichloride (BCl_3) is considered to be an acid but cannot fit the description according to the Brønsted–Lowry theory. Thus, the Lewis definition of acids and bases was developed to be more universal. A **Lewis acid** is a compound that can accept an electron pair, whereas a **Lewis base** is a chemical compound that can donate an electron pair.

The effect of ionizing radiation on pH of aqueous solutions does not appear to be very significant, even at doses far beyond what any human (or most pathogenic microbes) could tolerate. In food radiation sterilization experiments using gamma radiation on tomatoes and measuring the pH of the tomato juice with doses of 1, 2, 3 and 4 kGy, Gyimah et al showed a statistically significant but small increase in pH as a function of dose (e.g. 4.43–4.49 between 0 and 4 kGy for tomatoes in cold storage for 25 d), whereas Yussef et al found no statistically significant difference in tomato juice pH upon irradiation of tomatoes at doses of 1.5, 3 and 4.5 kGy. Thus, it does not appear that gamma or x-ray radiation affects pH very much.

On the other hand, charged particle radiation such as cosmic rays from the Sun or from beyond the Solar System might produce a different effect on the subsurface liquid water ocean on Jupiter's moon Europa. Europa orbits within Jupiter's radiation belts (the equivalent of our Van Allen belts but far larger and more intense). The constant bombardment of Europa by charged particle radiation might be generating a profusion of oxidants, including hydrogen peroxide and oxygen, on the icy surface that work their way down to the ocean below. If so, these oxidants could provide an energy source for extraterrestrial life. But these radiation-produced oxidants might be reacting with sulfides to generate sulfuric acid in abundance. Some calculations suggest the resulting pH could be as acidic as 2.6. Such acidity can dissolve the shells of calcareous organisms (sea life that uses calcium carbonate in their exoskeletons such as clams, snails, mussels, corals, etc) and would similarly wreak havoc on bone (which is made mostly of calcium phosphate in the form of hydroxyapatite). One might speculate that large underwater lifeforms in Europa's ocean could employ a different, acid-resistant, chemistry for their skeletons. One peculiar possibility is blue ironstone or vivianite. Crystals of this blue iron phosphate mineral are known to grow on human skeletal remains as well as many other fossils. Although it is pure speculation, organisms using vivianite for their skeletons (rather than calcium carbonate or calcium phosphate) might be better able to withstand the strong acidity of Europa's ocean. On the other hand, as will be discussed in subsequent chapters, there are organisms right here on Earth called extremophiles that are capable of withstanding extremes of pH, radiation exposure, temperature and various other conditions. For some of these acidophiles, a pH of 2.6 poses no major problem. In fact, the pH of gastric juice in the human stomach is normally in the range of 1 to 3, making it quite acidic (and capable of killing many pathogenic microroganisms). Nevertheless, some pathogens such as *Helicobacter pylori* tolerate, and even thrive in this acid. Archean microbes of genus *Picrophilus* have been found to tolerate pH as low as 0. So, perhaps the acidity of Europa's subsurface ocean might not be an insurmountable barrier to life as we already know it. Of course, we really do not know the pH of Europa's ocean at present—and we do not even know with certainty that Europa truly has a liquid water ocean beneath its icy surface. Hopefully NASA's Europa Clipper and the ESA's (European Space Agency) Juice (Jupiter Icy Moons Explorer) will answer some of these fascinating questions.

Further reading

[1] Shaik S, Danovich D and Wu W *et al* 2012 Quadruple bonding in C_2 and analogous eight-valence electron species *Nat. Chem.* **4** 195–200
[2] Youssef K A, Hammad A I, Abd El-Kalek H H and Abd El-Kader R M 2011 Ensure microbial safety and extending shelf-life of tomato juice by γ irradiation *Nat. Sci.* **9** 154–63
[3] Gyimah L A, Amoatey H M, Boatin R, Appiah V and Odai B T 2020 The impact of gamma irradiation and storage on the physicochemical properties of tomato fruits in Ghana *Food Qual. Saf.* **4** 151–7
[4] Pasek M A and Greenberg R 2012 Acidification of Europa's subsurface ocean as a consequence of oxidant delivery *Astrobiology* **12** 151–9
[5] Schleper C, Puehler G, Holz I, Gambacorta A, Janekovic D, Santarius U, Klenk H P and Zillig W 1995 *Picrophilus* gen. nov., fam. nov.: a novel aerobic, heterotrophic, thermoacido-philic genus and family comprising archaea capable of growth around pH 0 *J. Bacteriol.* **177** 7050–9

IOP Publishing

Space Radiation
Astrophysical origins, radiobiological effects and implications for space travellers
James S Welsh

Chapter 5

Fundamentals of radioactivity

5.1 Kinetics of radioactivity

Radioactive decay follows first-order kinetics and is described by the equation:

$$A = A_0 e^{-kT},$$

where A is the activity of the sample at time t, A_0 is the activity at time zero, k is the rate constant, and t is the time.

As a first-order process, the rate of radioactive decay is dependent on the concentration or absolute amount of only one reactant, namely the amount of parent radioisotope present.

To get a better understanding of this, a brief review of chemical kinetics might be helpful. Kinetics is crucially important in biochemistry (especially enzyme kinetics), as well as in general chemistry and radioactive decay. Generally speaking, the rate of a chemical reaction is proportional to the initial concentration of reactant(s) or absolute amount of a substance raised to some power. If that exponent is zero, the reaction is said to exhibit **zero-order kinetics**, if the exponent is one, the reaction is said to exhibit **first-order kinetics**, etc. The general expression of a rate equation may be written as a differential equation:

$$\frac{dA}{dt} = kA^x,$$

where A is the amount of reactant in question, t is time, and k is the rate constant.

Often the amount of reactant, A, is decreasing over time as the amount of product, B, increases. In such situations, k is negative. This the case in radioactivity. Thus, the differential equation is often written as:

$$\frac{dA}{dt} = -kA^x.$$

doi:10.1088/978-0-7503-5444-8ch5

This can be rearranged to:

$$\frac{\mathrm{d}A}{A^x} = -k \ \mathrm{d}t.$$

This can in turn be integrated from time zero to time t:

$$\int_{A_0}^{A_t} \frac{\mathrm{d}A}{A^x} = \int_0^t - k \ \mathrm{d}t.$$

The general solutions to this integral as a function of x are:

$$x = 0 \text{(zero order):} \quad A_t - A_0 = -kt$$
$$x = 1 \text{(first order):} \quad \ln(A_t) - \ln(A_0) = -kt$$
$$x = 2 \text{(second order):} \quad 1/A_t - 1/A_0 = -kt$$

These are collectively called the **integrated rate laws**. They can inform us about the amount of reactant(s) that will be consumed at any given time. As such, they are more powerful than the **basic rate law equation**, which is:

$$\text{Rate} = kA^m B^n.$$

A and B are the concentrations or absolute amounts of reactants A and B; exponents m and n are empirically derived constants that may or may not be integers.

This applies to the general chemical equation:

$$aA + bB \rightarrow cC + dD.$$

This general formula is based on several assumptions including that it does not run in reverse and that it does not have a complicated multi-step mechanism. It is an elementary reaction in which a mole of reactant A reacts with b moles of reactant B to yield c moles of product C and d moles of product D. In such a simplified situation, the reaction rate is 'first order with respect to A' if the exponent a is 1, 'second order with respect to A' if the exponent a is 2, etc. Similarly, the reaction rate is first order with respect to B if the exponent b is 1, second order with respect to B if the exponent b is 2, etc. The overall order of the rate equation is the sum of the exponents, $a + b$. For instance, if the reaction rate is first order with respect to A ($a = 1$) and second order with respect to B ($b = 2$), the overall reaction rate is 3, making it a third-order rate equation.

Students are advised that this simplification only applies to elementary reactions. When the reaction mechanism is more complex, the exponents m and n do not necessarily correspond to the stoichiometric coefficients a and b of the balanced chemical equation. In fact, for practical purposes, it should be assumed that they never work out that way. These exponents must be empirically determined. Once these exponents are properly ascertained, the integrated rate equations may be applied. These powerful integrated rate formulas can allow calculation of the amount of reactant left after 1 min, 1 h, 1 year, etc.

Relatively few examples of zero-order kinetics are found in biochemistry or pharmacology. Some examples include the elimination of alcohol (ethanol), aspirin (acetylsalicylic acid) and the chemotherapeutic agent cisplatin. The biochemical elimination process for these drugs is fully saturated, and their elimination rates are constant (for example, at a hypothetical elimination rate of a steady 1 mg h^{-1}). In contrast to the relatively rare zero-order kinetics, 95% of drugs are eliminated via first-order kinetics, wherein the rate is proportional to the drug concentration, for example, at a rate of 1% per minute where the higher the concentration, the higher the drug metabolism rate. First-order elimination kinetics is commonly observed when the enzymes involved in biochemical metabolism are not fully saturated. Because pharmacologists typically plot the logarithm of plasma drug concentration versus time, first-order drug elimination kinetics are often called 'linear kinetics'. On the same semi-log plots, drug elimination via zero-order kinetics will deviate from linearity. The rate appears to decrease at higher concentration. Thus, zero-order kinetics is sometimes called 'non-linear kinetics' in pharmacology. Of course, if not plotted on a semi-logarithmic graph, zero-order kinetics does appear linear, which can be a source of confusion for those unaccustomed to the nomenclature and traditional semi-log format.

5.2 Radioactivity revisited

Returning to radioactive decay, we already mentioned that this follows first-order kinetics, implying that the integrated solution will be of the form $\ln(A_t) - \ln(A_0) = -kt$. This implies that the decay rate is dependent on the amount of only one substance (namely, the radioactive parent isotope) raised to the first power. But we can start from experimental observations with an empirically derived basic rate equation:

$$\text{Rate} = kA.$$

The rate in this case is defined as the **activity** of a sample (e.g. measured in **disintegrations per second** or **becquerels** (Bq)). Hence, this rate equation can be rewritten in a general form as:

$$\frac{dA}{dt} = kA^x,$$

where $x = 1$ because it is first order. It is observed that the activity decreases over time. Since the activity monotonically decreases in a pure sample over time (ignoring for now radioactive daughter products), the rate constant, k is negative. This means that the equation can be expressed as:

$$\frac{dA}{dt} = -kA,$$

or upon rearranging:

$$\frac{dA}{A} = -kt.$$

Upon integration over a time interval, this yields:

$$\int_{A_0}^{A_t} \frac{dA}{A} = \int_0^t -k \, dt$$

or

$$\ln A_t - \ln A_0 = -kt + c,$$

which, upon taking the logarithms of both sides and normalizing to the amount of activity at time zero, can be rearranged to the familiar equation:

$$A = A_0 e^{-kT}.$$

A special case is given by the situation in which the activity A has fallen to $1/e$ of its original value. That is, $A = 1/e\,(A_0)$ or simply $A = A_0/e$. At this point in time, $A = A_0 e^{-kT} = A_0 e^{-1}$, which means that $kt = -1$. From this, one can state that $t = t_0$, where t_0 is the time taken for the activity to fall to $(1/e)$ times the original activity. This is given the special name of **mean lifetime** of the radioactive sample:

$$t_0 \equiv \text{time for a sample to fall to } 1/e \text{ of its initial activity.}$$

A more familiar version of the same concept is the time required for the sample to decay to ½ its initial activity. That is the **half-life** ($t_{1/2}$) of the sample:

$$t_{1/2} \equiv \text{time for a samplet of all to } 1/2 \text{ of its initial activity.}$$

From this, one can see that if $A = \frac{1}{2} A_0$, then $A/A_0 = \frac{1}{2}$ and $e^{-kT} = -\frac{1}{2}$. Taking the natural logarithms of both sides gives $-kt = \ln(\frac{1}{2})$, which is -0.693. Thus, the radioactivity rate constant k is related to the half-life by:

$$t_{1/2} = k/0.693$$

or

$$k = t_{1/2}/0.693.$$

In many texts, this rate constant is denoted by λ rather than k.

For special cases in which the time t happens to be an integral multiple of the half-life, we can apply the simple formula:

$$A = A_0 \times (1/2)^n,$$

where n is the number of half-lives.

Radioactivity may be quantified by the disintegrations per second (dps). One dps is one becquerel using SI units. A popular non-SI unit is the **curie (Ci)**, which is 37 billion (3.7×10^{10}) Bq. The curie was originally defined as the number of disintegrations per second in one gram of pure radium-226.

5.3 The valley of stability

For any given number of protons, that is, for a given atomic number (Z), there exists an ideal number (or numbers) of neutrons to ensure stability of the nuclide. Those with too many or too few neutrons are unstable and undergo radioactive decay. Such nuclides are called radionuclides or radioisotopes.

It is observed that as Z increases, the number of neutrons required to ensure stability increases. For example, the most abundant stable isotopes of low-Z elements such as carbon and oxygen (carbon-12 and oxygen-16) both have the same number of protons as neutrons. As one climbs higher in atomic number to calcium, the most common of the six stable isotopes is calcium-40, which has the same number of neutrons as protons. Beyond calcium-40 however, a trend develops with an ever increasing number of neutrons required for stability, up to $Z = 82$, lead. In simple terms, the increasing number of protons means an increasing electrostatic repulsion, and this requires an increasing number of neutrons to mitigate the repulsion. The most common isotope of lead is lead-208 with an abundance of 52.4%, but all the common stable isotopes of lead have roughly 1.5 times the number of neutrons as protons.

It was previously believed that bismuth-209 was the heaviest stable isotope, but it is now known that this decays with a half-life of 2.01×10^{19} years, making it the longest-lived alpha-emitting radionuclide. (Tellurium-128 presently holds the overall title of longest measured half-life at 2.2×10^{24} years; it decays via double beta emission).

Thus, all the elements with an atomic number over 82 are unstable. Apparently, no number of neutrons is capable of holding a nucleus with more than 82 protons together forever.

When stable nuclides are plotted out with Z on the abscissa, a 'zone of stability' appears with an ideal ratio of neutrons to protons. Such plots of protons versus neutrons are sometimes called Segrè plots or Segrè charts[1].

One can predict what type of radioactive decay mode a radionuclide will use depending on where it lies relative to the zone of stability. For instance, when a nucleus has too few neutrons, it can convert one of the excess protons into a neutron to come closer to the zone of stability. Thus, such a radionuclide might decay via positron emission or electron capture, the two common mechanisms of converting a proton into a neutron.

For heavy nuclei, alpha emission increases stability of nuclides because of the stipulation that the higher the value of Z, the more neutrons are needed. Thus, although the emission of two protons and two neutrons packaged together in the form of an alpha particle decreases the total number of neutrons, it also drops Z by 2 and thereby simultaneously diminishes the optimal number of neutrons required for stability. (There are other means of reducing the proton:neutron ratio, such as proton emission and double proton emission, but these are extremely rare.)

5.4 Alpha decay

Alpha particles are nothing more than helium nuclei—a set of two protons and two neutrons bound together via the strong nuclear force. Alpha particles thus have a mass number of 4 and may be represented as ^{4}He, where the 4 represents the mass number and the atomic number of 2 is implied by the symbol He for helium. Alternatively, one can spell out the mass numbers as superscripts and the atomic numbers as subscripts: $^4_2$He.

---

[1] https://upload.wikimedia.org/wikipedia/commons/8/80/Isotopes_and_half-life.svg.

When a radionuclide $_Z^M X$ with mass number M and atomic number Z undergoes alpha decay, it decreases its mass number by 2 and its atomic number by 2 as follows:

$$_Z^M X \rightarrow _{Z-2}^{M-4} Y + _2^4 He.$$

The **disintegration energy**, or **Q value**, of a radioactive decay is the amount of energy released through the decay reaction and is given by:

$$Q = (M_x - M_y - M_\alpha)c^2.$$

The Q value is often given in parentheses after the equation. Thus, substituting $_2^4 He$ with $_2^4 \alpha$, the same equation can be written as:

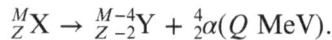

$$_Z^M X \rightarrow _{Z-2}^{M-4} Y + _2^4 \alpha (Q \text{ MeV}).$$

When working with such equations, note that the mass number on both sides must balance for conservation of mass, and the number of protons (Z) must balance as well for conservation of charge. A well-known example is:

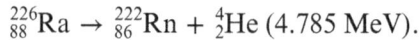

$$_{88}^{226} Ra \rightarrow _{86}^{222} Rn + _2^4 He \text{ (4.785 MeV)}.$$

The emitted alpha particles are monoenergetic with a kinetic energy of 4.785 MeV in this case. In alpha decay (and other forms of radioactive decay), there can be more than one pathway from parent to daughters. For example, in the decay of radium-226, 95% of the time, the decay proceeds as above, but in about 5% of decays, the emitted alpha particle has a lower energy of 4.60 MeV. In such cases, the interim radon nuclide is left in an excited state and promptly sheds the excess energy (4.785 MeV − 4.602 MeV = 0.186 MeV) in the form of a gamma photon. The whole process can be depicted in a decay diagram as follows (figure 5.1).

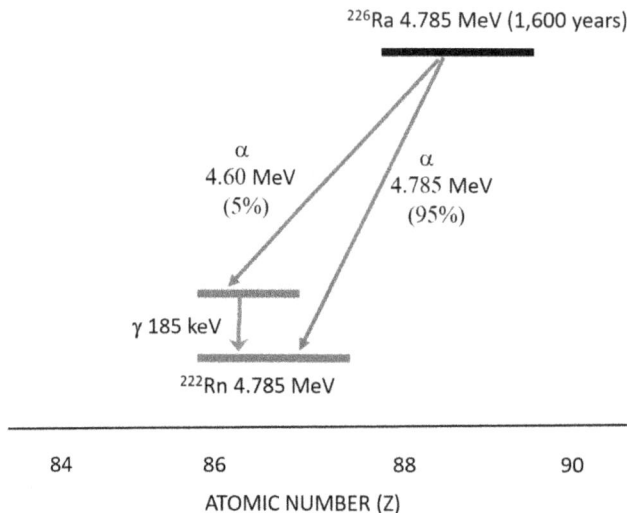

^{226}Ra 4.785 MeV (1,600 years)

α
4.60 MeV
(5%)

α
4.785 MeV
(95%)

γ 185 keV

^{222}Rn 4.785 MeV

84   86   88   90

ATOMIC NUMBER (Z)

Figure 5.1. Radioactive decay scheme of radium-226. Because the daughter products have less energy than the parent, they are lower on such schemes. Also, because the daughters have lower atomic numbers than the parent, they are to the left of the parent. See text for further details.

Note that the daughter is slightly to the left of the parent in the above diagram, signifying a decrease in Z. These decay scheme diagrams often provide the branching ratios, which are the ratio of the rate constant for a particular decay pathway to the rate constant for the total reaction.

All nuclei heavier than bismuth-209 can decay by alpha-particle emission (although other modes may dominate).

According to classical physics, alpha decay should not be possible. The energy barrier is such that no matter how many times the alpha particle attempts to surmount the escape energy barrier, it never will succeed. Nevertheless, we do observe alpha decay in nature. As discussed in chapter 7, the key is quantum tunneling.

5.5 Beta decay

When an unstable nuclide has a suboptimal number of neutrons, it can get closer to the zone of stability by several mechanisms. If the radionuclide has a relative overabundance of neutrons, it can convert one of its neutrons into a proton via standard beta decay, that is, emission of an electron. If, on the other hand, the radionuclide has a relative underabundance of neutrons, it can convert a proton into a neutron via two alternative mechanisms: beta-plus decay (also known as positron emission) or electron capture. All three mechanisms are generally lumped together as variations of beta decay, and the ejected charged particles are called beta particles. 'Beta-minus' particles are simply electrons, whereas 'beta-plus' particles are antimatter positrons. In all three scenarios, the mass number M remains unchanged. However, the atomic number Z, reflecting the number of protons, is altered:

$$\ _{Z}^{M}X \rightarrow \ _{Z+1}^{M}Y + \ _{-1}^{0}e^{-} + \ _{0}^{0}\bar{\nu}_{e}.$$

Neglecting the mass of the electron and the neutrino ν, the energy released by beta-minus decay can be approximated by its Q-value:

$$Q = (M_{x} - M_{Y})c^{2}.$$

As can be seen above, a neutrino, ν, is also produced in the reaction. This is necessary for conservation of momentum and energy since beta particles, unlike monoenergetic alpha particles, come out with a spectrum of energies. The neutrino carries away the 'missing' energy and momentum. Additionally, the neutrino produced in beta-minus decay as above is an *anti*neutrino for balance of matter/antimatter on both sides of the equation; this is indicated by the bar over the neutrino symbol. Furthermore, it should be noted that the antineutrino is specifically an *electron antineutrino* to maintain balance of electron lepton number on both sides; this is indicated by subscript $_e$. It should be pointed out that in contrast to alpha decay, neither the beta particle nor its associated electron antineutrino exists within the nucleus prior to beta decay; rather they are created in the decay process. Through beta decay, unstable atoms obtain a more stable ratio of protons to neutrons.

Neutrinos and antineutrinos have negligible mass but still carry enough energy and momentum to ensure conservation. In contrast to alpha particles, beta particles come out with an array of energies ranging from zero up to a defined maximum. The average beta particle energy is typically about one-third of the nominal maximum energy; the most probable energy is slightly less than the average energy. Standard beta decay is an intranuclear conversion of a neutron into a proton and an electron (plus an electron antineutrino).

The reaction can thus be portrayed as:

$$\mathrm{{}^1_0 n} \rightarrow \mathrm{{}^1_1 p} + \mathrm{{}^0_{-1}e^- e^-} + \mathrm{{}^0_{-0}\bar{\nu}_e}.$$

Sometimes, instead of writing $\mathrm{{}^0_{-1}e^-}$, the electron is written as β^-:

$$\mathrm{n} \rightarrow \mathrm{p} + \beta^- + \bar{\nu}_e,$$

where the mass and atomic numbers are understood but not spelled out.

An example of beta-minus decay is provided by cobalt-60:

$$\mathrm{{}^{60}_{27}Co} \rightarrow \mathrm{{}^{60}_{28}Ni} + \mathrm{{}^0_{-1}e^-} + \mathrm{{}^0_0\bar{\nu}_e}.$$

In the above example with cobalt-60, almost all decays emit a beta particle with a maximum energy of 331 keV followed by two prompt gamma photons of 1.1732 and 1.3325 MeV. However, in about 0.12% of decays, a higher-energy beta electron with a maximum energy of 1.48 MeV is emitted, and then only a single 1.3325 MeV gamma photon is released. Note that the gamma photons are monoenergetic with discrete nominal energy values, whereas the beta electrons have a spectrum of possible energies ranging from zero up to the nominal maximum value (figure 5.2).

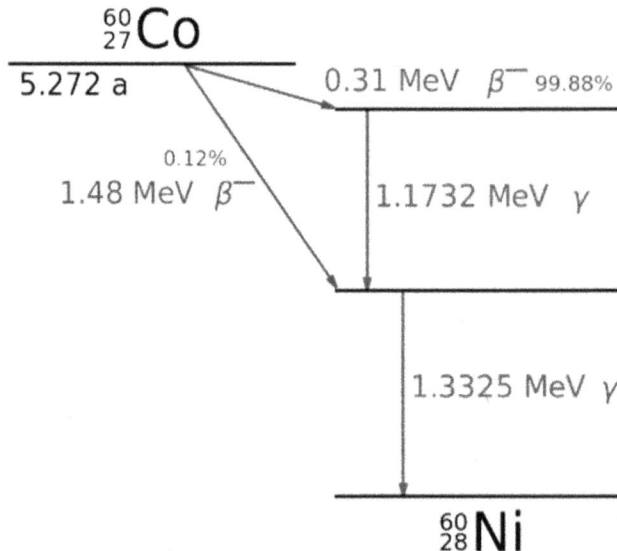

Figure 5.2. This cobalt-60 decay scheme image has been obtained by the author(s) from the Wikimedia website, where it is stated to have been released into the public domain. It is included within this book on that basis. It is attributed to Inductiveload.

Another familiar example of relevance in space radiation is the beta decay of radioactive carbon-14 in the atmosphere:

$$\ce{^{14}_{6}C} \rightarrow \ce{^{14}_{7}N} + \ce{^{0}_{-1}e^-} + \ce{^{0}_{-0}\bar{\nu}_e}.$$

Carbon-14 is useful in radiometric dating of organic material. As we will discuss later, carbon-14 is a **cosmogenic radionuclide** that is produced by cosmic ray-induced spallation of stratospheric and tropospheric molecules, which leads to free neutrons. The neutrons then react with nitrogen-14 atoms to create carbon-14:

$$\ce{^{14}_{7}N} + \ce{^{0}_{1}n} \rightarrow \ce{^{14}_{6}C} + \ce{^{1}_{1}p}.$$

Beta decay is a classic example of the weak nuclear force in action. (Students should notice that whenever neutrinos are present in a reaction, the weak force is at play.) As will be covered in greater detail in chapter 8, beta decay may be understood in terms of the constituent quarks of the protons and neutrons involved. Thus, beta-minus decay occurs when a proton (made of 2 up quarks and 1 down quark, symbolized as uud) converts one of its negatively charged ($-1/3$) down quarks into a positively charged ($+2/3$) up quark via a W-weak boson:

$$d \rightarrow u + W^-.$$

The W^- immediately decays into an electron an electron antineutrino:

$$W^- \rightarrow \ce{^{0}_{-1}e^-} + \bar{\nu}_e.$$

In terms of quarks and weak vector bosons the reaction is:

$$(udd) \rightarrow (udd) + W^- \quad \text{followed by} \quad W^- \rightarrow \ce{^{0}_{-1}e^-} + \bar{\nu}_e.$$

A Feynman diagram of the reaction with space on the abscissa and time on the ordinate is (figure 5.3).

As is customary in Feynman diagrams, the electron antineutrino (being anti-matter), has the arrow pointed in the opposite direction (that is, towards the initial state rather than towards the final state), which can be interpreted as matter traveling backwards in time. The transformations of quarks into different 'flavors' with emission of various vector bosons falls under the subject of quantum chromodynamics, and is covered in slightly further detail in chapter 8.

5.6 Positron emission (beta-plus decay)

When a radionuclide has a relative deficit rather than an excess of neutrons, it can gain stability by emission of a positive electron (a positron or beta-plus particle) thereby creating a neutron from a proton.

$$p \rightarrow n + \beta^+ + \nu_e.$$

Rather than an electron antineutrino being generated as in beta-minus decay, in this case, a regular electron neutrino is created for balance of matter/antimatter and lepton number.

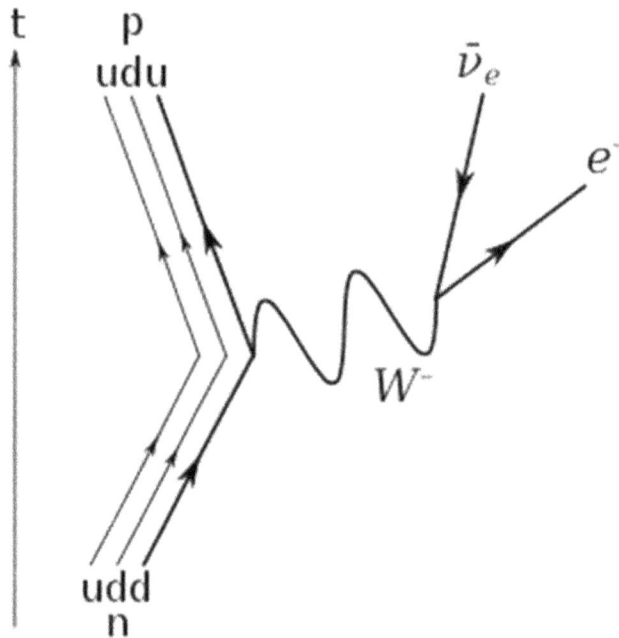

Figure 5.3. This Feynman diagram has been obtained by the author(s) from the WikiWand/Wikimedia website, where it is stated to have been released into the public domain. It is included within this book on that basis. It is attributed to Joel Holdsworth.

The overall reaction involves an intranuclear conversion of a proton into a neutron. Since the proton is lighter than the neutron (roughly 938 versus 939.5 MeV), this reaction is possible only in certain nuclei that meet specific requirements. One such requirement is that the available energy must exceed twice the rest mass of the electron (i.e. $\geqslant 2 \times 511$ keV/c^2 or $\geqslant 1.022$ MeV). What appears to be happening is:

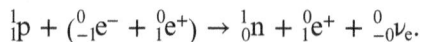

$$\,^1_1\text{p} + (\,^0_{-1}\text{e}^- + \,^0_1\text{e}^+) \rightarrow \,^1_0\text{n} + \,^0_1\text{e}^+ + \,^0_{-0}\nu_e.$$

The negative charge of the electron combines and cancels the positive charge of the proton, and a neutron is created. Thus, positron emission results in a daughter nucleus with the same mass number but one less proton (i.e. M is unchanged, but Z is reduced by 1). Neglecting the mass of the positron and the neutrino, the energy released by beta-plus decay can be approximated by its Q-value:

$$Q = (M_X - M_Y - 2m_e)c^2.$$

As in beta-minus decay, the emitted positrons have a spectrum of energies ranging from zero up to a specified maximum energy. The electron neutrino carries the balance to conserve energy and momentum. Beta-plus decay was first discovered by Irene and Frederic Joliot-Curie in 1934. One example is:

$$\,^{30}_{15}\text{P} \rightarrow \,^{30}_{14}\text{Ni} + \,^0_{+1}\text{e}^+ + \,^0_0\nu_e.$$

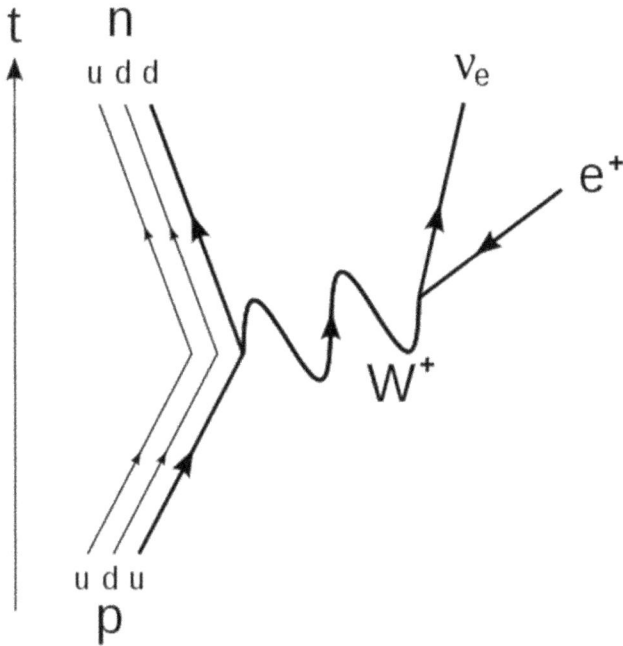

Figure 5.4. This Feynman diagram of positron emission has been obtained by the author(s) from the WikiWand/Wikimedia website, where it is stated to have been released into the public domain. It is included within this book on that basis. It is attributed to Pamputt.

Of historical interest, phosphorus-30 was the first artificially synthesized radio-isotope. The above reaction was discovered by Irène Joliot-Curie and her husband Frédéric who shared the 1935 Nobel Prize in Chemistry for artificially synthesizing this radioactive isotope. The radionuclide was created by bombarding aluminum with alpha particles.

Another familiar example that is often used in nuclear medicine for positron emission tomography scans is:

$$\ce{^{18}_{9}F} \rightarrow \ce{^{18}_{8}O} + \ce{^{0}_{+1}e^+} + \ce{^{0}_{0}\nu_e}.$$

Beta-plus decay is another classic example of a weak interaction and can be depicted by the Feynman diagram (figure 5.4).

Note that a $+2/3$ positively charged up quark is converted into a $-1/3$ negatively charged down quark mediated by a W^+ intermediate vector boson. Also note that the positron, (depicted by e^+) is antimatter, and therefore, the arrow is pointed in the reverse direction (i.e. towards the initial state rather than the final state). Again, this can be interpreted as matter traveling backwards in time.

5.7 Electron capture

When a radionuclide has a relative excess of protons, a proton can be converted into a neutron via beta-plus decay or an alternative method. Discovered by Luis Alvarez in 1938, this alternative involves the capture of an orbiting electron from the atomic electron cloud. Because such capture is more likely to involve an electron from the K-shell than the L-shell, and an L-shell electron is more readily captured than one from the M-shell, etc, the process is sometimes called K-capture:

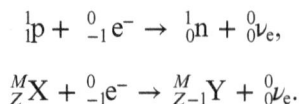

$$\,^1_1p + \,^0_{-1}e^- \rightarrow \,^1_0n + \,^0_0\nu_e,$$

$$\,^M_Z X + \,^0_{-1}e^- \rightarrow \,^M_{Z-1}Y + \,^0_0\nu_e.$$

The atomic number goes down by one unit, and mass number remains unchanged. Note that the neutrino in this situation is, as in beta-plus decay, an ordinary matter electron neutrino for conservation of lepton number. As both mechanisms achieve the same end, electron capture can theoretically compete with beta-plus decay. Electron capture may be the only option when the energy difference between parent and daughter does not satisfy the requirement of >1.022 MeV for positron emission. Additionally, thanks to electrostatic attraction between the positive nucleus and the negative electron cloud, as Z increases, electron capture may be more favorable. Unlike emission beta decay (with readily detected electrons and positrons), electron capture could be difficult to detect since by the above equations, apparently the only radiation emitted is the electron antineutrino. However, electron capture is often manifested by the x-rays given off as higher-shell electrons cascade downward to fill the vacancy created in the K shell. As discussed in the next chapter, Auger and Coster-Kronig electrons might be emitted as an alternative to these characteristic electrons. Furthermore, the nucleus is typically left in an excited state after electron capture, and a gamma photon may be emitted. But whenever a gamma photon can be emitted, alternatively, a so-called conversion electron may be ejected instead. Thus, the radiation emitted by electron capture process can be complex.

A curious phenomenon might occur with radionuclides that must decay by pure electron capture (for example, if their Q-value does not meet the 1.022 MeV requirement of beta-plus decay). In such circumstances, one can speculate about what might happen if the radionuclide is fully ionized and stripped of all its atomic electrons. Such isotopes might naturally exist upon formation via the r-process (which will be discussed in the sections on nucleosynthesis). Despite being unstable, these radionuclides might be unable to decay via their preferred method until they can acquire some orbital electrons for capture.

Electron capture is an important mode of decay in clinical radiation therapy where small, encapsulated sealed sources or 'seeds' containing radionuclides such as iodine-125, palladium-103 and cesium-131 are used for permanent implant brachytherapy. As will be discussed later, electron capture (and neutrino production) is critically important in the final phases of massive star collapse when protons and electrons are compressed together by overwhelming gravitational pressure in the formation of neutron stars. In such situations, it is the tremendous energy release through the formation of enormous numbers of neutrinos that is the primary driver of supernova explosions. Thus, neutrinos are the unsung heroes that provide the true force behind supernovas.

In terms of Feynman diagrams, electron capture can be depicted on the left side of the diagram (in figure 5.5) as the creation of a (+1) charged W+ boson through a (+2/3) charged up quark converting into a (−1/3) charged down quark, and that

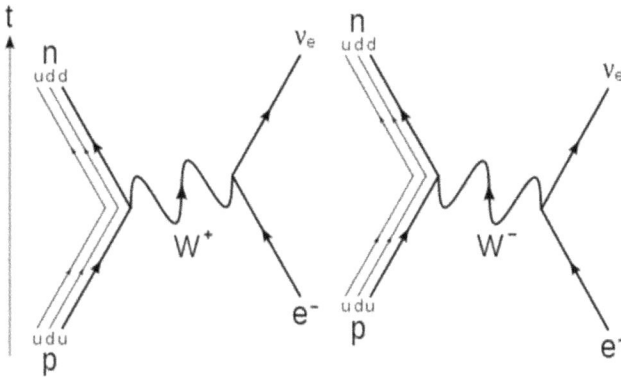

Figure 5.5. This set of electron capture Feynman diagrams depicts (on the left) a proton converting into a neutron and emitting a W+ boson, which then interacts with and neutralizes an electron, releasing an electron neutrino. The electron capture process illustrated on the right depicts the electron emitting a W− boson, which then interacts with the proton, converting it into a neutron. One can see the obvious similarities between the Feynman diagram on the left and the positron-emission Feynman diagram earlier in this chapter. This image has been obtained by the author(s) from the Wikimedia website where it was made available under a CC BY-SA 4.0 licence. It is included within this book on that basis. It is attributed to Servinjesus1.

W+ boson then interacting with an electron to neutralize its charge and generate an electron neutrino. On the right side of figure 5.5, an electron is creating an electron neutrino and a W− boson, which then interacts with an up quark in the proton, converting it into a down quark and thereby converting the proton into a neutron. Both diagrams are equivalent representations.

It could be pointed out that classic beta decay (beta-minus decay), positron emission (beta-plus decay) and electron capture all involve parent and daughter nuclides with the same total number of protons plus neutrons, i.e. mass numbers (A). Although the atomic numbers are altered in the various processes, when A is unaltered, we are dealing with **isobars**. Isobars, by definition, are nuclides that have the same total number of protons plus neutrons (i.e. A is constant). **Isotones** are defined as nuclides that have the same number of neutrons, and of course, **isotopes** have the same number of protons.

The **decay scheme of potassium-40** (figure 5.6) shows the various modes of decay available and the 'branching ratios' or relative probabilities of each pathway. Potassium-40 is unusual in that it has three known modes of decay: beta-minus, beta-plus, and electron capture. All three are considered types of beta decay. Most often (89.25%), potassium-40 decays via conventional beta decay ('beta-minus' decay), in which an ordinary (matter) electron and an antimatter electron antineutrino are emitted when a neutron is converted into a proton. The atomic number accordingly increases from 19 (potassium) to 20 (calcium). In decay scheme diagrams, this increase in Z is depicted by a move to the right; thus, the daughter, calcium-40, is lower and to the right of the parent, potassium-40, indicating an increase in atomic number and a decrease in energy. The decay energy (Q) in this beta-minus decay (1.31 MeV) is shared between the emitted electron and the electron antineutrino. In general, the beta particle gets roughly 1/3 of the energy and the

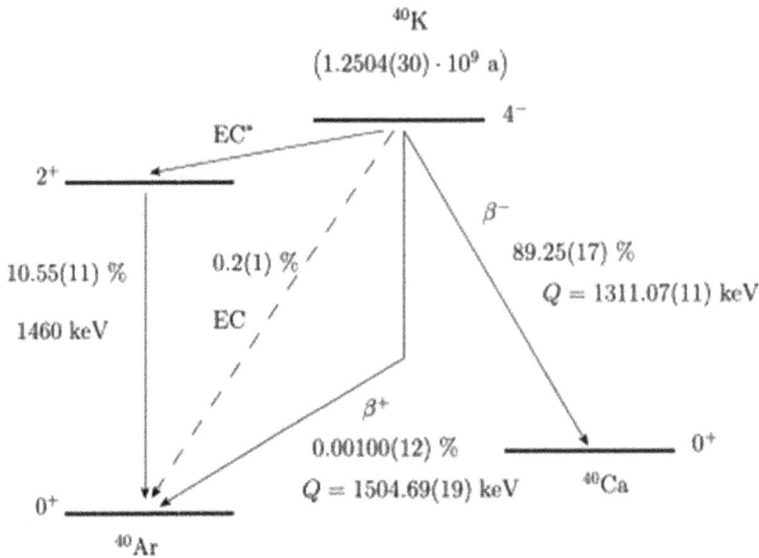

Figure 5.6. This decay diagram also depicts another rare mode of electron capture (dotted line) in which the parent potassium-40 decays straight to argon-40 without an excited intermediate. The energy released through this special decay mode is carried entirely by the emitted neutrino and no gamma ray is emitted. A small amount of energy (about 3 keV) might appear as Auger electron emission and x-ray fluorescence upon K-shell capture [6]. Reproduced from [6]. CC BY 3.0.

antineutrino gets 2/3; the recoil decay product gets so little energy that it can be ignored. The observed average beta electron energy is a bit higher than a third of the maximum energy (in this case, about 0.52 MeV).

Alternatively, about 10.55% of the time potassium-40 decays through electron capture. In this case, for lepton conservation and matter-antimatter balance, an electron neutrino is generated, which carries away some of the decay energy. The capture of an electron leads to an excited state of argon-40, which then promptly releases its energy via gamma decay to the ground state of argon-40. Because the argon-40 daughter has one less proton in its nucleus, Z has decreased by 1, and thus, as in all electron capture depictions, the daughter is lower and to the left of the original parent radionuclide. The total difference in energy, Q, between potassium-40 and the ground state of argon-40 is 1.505 MeV, but the initial electron capture yields an excited state of argon-40 with an energy level of 1.460 MeV. The 0.45 MeV difference is carried off by a neutrino and leaves only 1.460 MeV for the gamma photon, rather than the full 1.505 MeV.

Yet another, albeit very rare, third option for potassium-40 decay is positron emission (also called 'beta-plus' decay). This should be anticipated given that the competing mechanism of electron capture occurs. In general, positron emission wins out in low-Z radionuclides, whereas electron capture usually dominates in higher-Z radioisotopes. In potassium-40, positron emission occurs about 0.001% of the time. In positron emission, a proton is converted into a neutron plus a positron (and an electron neutrino for conservation of lepton number and matter-antimatter conservation). The atomic number decreases, thus the daughter product is down and to the left of the parent. The daughter produced this way is the same as if electron capture occurred (argon-40 in the case of potassium-40 decay). For positron emission to occur, the energy difference, Q, between the initial and final states must exceed 1.022 MeV. This is the threshold energy to create a positron plus an electron. Since the new daughter *atom* as a whole contains one fewer electrons, the net atomic equation yields both a positron and an electron: $^{40}_{19}K \rightarrow {}^{40}_{18}Ar + e^+ + \nu + e^-$.

In potassium-40, since $Q = 1.505$, this 1.055 Mev requirement leaves only 0.483 MeV available, which is shared between the positron and the electron neutrino.

Potassium-40 is one of the three primordial radionuclides (along with thorium-232 and uranium-238) that contribute to the radiogenic heating of Earth's deep interior, which, along with other sources such as leftover accretion heat, keeps the outer core molten and makes magma in the mantle. Potassium-40 has a half-life of 1.25×10^9 years. The decay of potassium-40 into argon-40 is the basis for potassium–argon dating of geological specimens. As argon is a noble gas, when a mineral is first formed it will be completely devoid of argon. However, if the mineral contains potassium (as orthoclase feldspars do), decay of the potassium-40 (which would have been significantly higher in the distant geological past) will lead to some argon-40 that will remain trapped in the mineral. By measuring the ratio of potassium-40 and argon-40 atoms remaining in the mineral, one can estimate the age of the mineral.

The natural abundance of potassium-40 at this time in Earth's history is about 120 atoms per million or 0.012% of all natural potassium. This might not seem like much until we consider that 1 g of potassium contains about 1.5×10^{22} atoms of potassium, with about 1.8×10^{18} of them being radioactive potassium-40. Given the numerous essential roles of potassium in physiology, potassium is quite plentiful in biological organisms, and despite its low abundance, potassium-40 is the main reason that life is radioactive (another reason is the cosmogenic radioisotope, carbon-14). Assuming an overall concentration of 2.5 g kg^{-1} in human tissues, there are about 175 grams of potassium in a 70 kg human body. And at a potassium-40 abundance of 0.012% or an observed activity of 31 Bq g^{-1}, this potassium load will constantly generate: (175 g) × (31 Bq g^{-1}) = 5425 Bq.

Potassium-40 is the basis for the 'banana equivalent dose', which allows comparisons between the radiation dose from eating one banana versus other sources of radiation. It may be calculated that the effective dose from consuming one banana is around 0.1 microsievert (based on the ICRP [International Commission on Radiological Protection] suggested estimate of 6.2 nSv/Bq for ingested potassium-40). Since an average banana contains about 500 mg of potassium, at 31 Bq g^{-1}, it will exhibit an average banana activity of about 15.5 Bq (along with a few beta particles per second from carbon-14 decay).

In radionuclides where the 1.022 MeV energy threshold of positron emission is not met, the only option might be electron capture. But in unusual circumstances, such as in extremely energetic galactic cosmic rays, the atoms are fully stripped of all electrons and fly through the Universe as bare atomic nuclei. Thus, the electron capture option is unavailable. In this case, the excited nucleus cannot decay and appears stable. Such is the situation with some beryllium-7 cosmic ray nuclei.

5.8 Inverse beta decay

Confusingly, electron capture is occasionally called inverse beta decay. Here, we will reserve the name inverse beta decay (often abbreviated IBD) for another phenomenon, namely the capture of a neutrino by a nucleus, resulting in the emission a beta particle. In this type of inverse beta decay, the neutrino happens to be an antineutrino and beta particle emitted turns out to be a beta-plus (a positron). A low energy neutrino interacts with a nuclear proton, converting that proton into a neutron. In this neutrino-capture process of converting of a proton into a neutron, the neutrino involved must be an electron antineutrino. The reaction may be written as:

$$p^+ + \bar{\nu}_e \rightarrow e^+ + n^0.$$

To be even more precise in the terminology, this form of inverse beta decay is an inelastic scattering event (meaning kinetic energy is not conserved) involving an electron antineutrino with a nuclear proton that yields a neutron and a beta-plus particle (positron).

There is an energy threshold for inverse beta decay—the incident electron antineutrino must possess a kinetic energy of >1.806 MeV. This can be predicted given the relative masses of the reactants (a proton and neutrino) compared to the products (a neutron and positron).

It was via this reaction that Reines and Cowan first discovered the neutrino in 1956. (Note that this first discovery actually involved antineutrinos rather than ordinary neutrinos.) Neutrinos have an extremely small cross section of interaction with matter—it is estimated that a typical solar neutrino might penetrate nearly a light year of lead before being stopped (in other words the half-value layer of lead for solar neutrinos is approximately one light year). For this reason, along with their conjectured existence based purely on scientific principles but no empirical evidence, it was considered these 'ghost' particles might remain elusive. Therefore, the Reines and Cowan quest to find the neutrino was nicknamed Project Poltergeist.

They conducted their work at a nuclear reactor that in principle should have been producing copious quantities of antineutrinos. They chose the *Savannah River Site* in South Carolina, because better shielding against cosmic rays was achievable there compared to other sites they considered. The experimental setup was 11 m from the reactor core and 12 m underground. They created tanks containing 400 l of water and about 40 kg of dissolved cadmium chloride. Antineutrinos then collided with protons in the water producing neutrons and positrons via the inverse beta decay reaction. The positrons then produced two gamma photons of 511 keV each via pair annihilation (a process which is reviewed in the next chapter) when encountering an electron; this process is measure in nanoseconds. These simultaneously detected photons constituted the first detected signal in liquid scintillation detectors. The other product, a neutron, is rapidly moderated in energy by water down to thermal energies. Natural cadmium contains about 12% cadmium-113, which has a very large cross section for capture of thermal neutrons and emits prompt gamma photons totaling ~9 MeV. This process is measured in microseconds. This constitutes the second signal detected by the liquid scintillators. Thus, when the first signal of two 511 keV annihilation gamma photons is followed microseconds later (a 'delayed coincidence') by a second signal (a third gamma photon emanating from cadmium-113(n,γ) with the appropriate energy) it was taken as evidence of a neutrino-induced inverse beta capture event. In this manner, neutrinos were first experimentally proven to exist.

As in electron capture (and in beta-plus decay for that matter) the net reaction is the same: a proton is converted into a neutron.

Another variation of inverse beta decay involves capture of an ordinary electron neutrino by a neutron to create a proton and an electron (beta-minus):

$$^1_0\text{n} + {}^0_0\nu_e \rightarrow {}^1_1\text{p} + {}^0_{-1}\text{e}^-.$$

A specific and important example of this reaction is the one in the Homestake experiment in which Davis and Bahcall first detected and quantified the solar neutrinos predicted to be generated by nuclear fusion within the core of our Sun:

$$^{37}_{17}\text{Cl} + {}^0_0\nu_e \rightarrow {}^{37}_{18}\text{Ar} + {}^0_{-1}\text{e}^-.$$

The detection of neutrinos in this fashion was strong supporting evidence for the proposed model of thermonuclear fusion of hydrogen in the core of the Sun.

5.9 Gamma decay

Gamma rays may be defined as photons given off when an excited nucleus decays to a lower-energy state. With this definition of nuclear origin, the energy is irrelevant. It is this definition that is frequently adopted by medical physicists and physicians in radiation oncology and nuclear medicine. In the radiotherapy parlance, x-rays are created by processes involving electrons, whereas gamma rays emanate from the nucleus. By this definition, gamma rays can have lower energy than x-rays. For example, the 140.5 keV 'gamma' photons from the decay of technetium-99m are far lower in energy than the 18 MeV 'x-rays' commonly used in 3D conformal radiation therapy, which are generated through bremsstrahlung in a medical linac (18 MV$_p$; 18 MeV maximum but approximately 6 MeV most probable photon energy). In astrophysics however, high-energy electromagnetic radiation is called x-rays rays, while even higher-energy photons are considered gamma rays, irrespective of mechanism of production. Thus, photons from a soft gamma repeater might have more energy than x-ray photons from a cataclysmic variable but both could be less energetic than the 23 MV$_p$ photons occasionally used in 3D conformal radiation therapy. Here, we shall stick with the convention of defining any photon emerging from a nucleus as a gamma ray (regardless of energy). We also include any photons with energies above an arbitrary threshold of 511 keV (the rest mass–energy of an electron) as gamma rays.

Just as characteristic x-rays emerge from the atomic electron cloud when electrons drop from higher-energy shells (principal quantum numbers) to lower ones, gamma photons emerge from the nucleus when nucleons drop from higher-energy nuclear shells down to lower ones.

Gamma emission does not change Z, N or A. The emitted photon has an energy of hν equal to ΔE between the initial and final nuclear energy levels, that is, the Q-value.

5.10 Isomeric transition

Isomeric transitions yield gamma photons when an excited metastable nuclear state relaxes to a lower-energy state. As the name indicates, nuclear isomers do not experience any change in atomic number or atomic mass upon decay. Thus, nuclear isomers are analogous to chemical isomers that have the same elemental composition and identical numbers of electrons. Nevertheless, nuclear isomers, just as with chemical isomers can be in excited states and therefore tend to decay to a more stable ground state. They do so via the emission of a gamma photon with discrete energy.

The nuclear shells relax from a higher-energy state to a lower-energy state and release photons of energy upon the shell rearrangement. This reshuffling of nucleons does not affect their total numbers. Thus Z, A and N remain unchanged as would be expected by the name isomeric transition. Nuclear isomeric transitions from an excited nuclear state to the ground state (or just a lower-energy state) with the emission of gamma photons are analogous to electronic transitions from an excited atomic state to the ground or lower-energy state with the emission of fluorescent photons. The differences of course lie in the origin and energy of the photons involved.

The term 'metastable' is presently loosely defined but can refer to nuclear isomers with measurable half-lives (measurable in minutes, hours or years). This distinguishes metastable gamma emission from 'prompt' gamma emission, which occurs on the order of picoseconds. However, students should be aware that sometimes the term metastable can sometimes refer to states with half-lives measured in nanoseconds.

When used in the traditional sense of the term, metastable nuclei have longer half-lives than anticipated due to **'forbidden transitions'** that slow their decay. Such forbidden transitions violate the so-called **selection rules** for allowed transitions. An example of such a forbidden transition could involve a large amount of nuclear spin transition from parent to daughter.

One example that is familiar to those in nuclear medicine is the decay of technetium-99m. (The m after 99 indicates a metastable state.) This metastable radionuclide decays via emission of a 140 keV gamma photon with a half-life of approximately 6 h. As an isomeric transition, the daughter product is technetium-99. The reason technetium-99m is metastable is because it has a spin of $\frac{1}{2}$, whereas the daughter, technetium-99, has a total spin state of 9/2. This large difference in total spin is tantamount to a large difference in spin multiplicity, S. The spin selection rules forbid transitions between states with $\Delta S > 0$; thus, this is a forbidden transition that takes far longer to occur than expected otherwise.

Nuclear isomeric transitions may lead to gamma emission as described above, but any time there is a potential gamma emission, there could be a competing alternative: internal conversion.

5.11 Internal conversion

Internal conversion is the process wherein decay of an excited nucleus leads to the emission of an electron rather than a gamma photon. During internal conversion, the relaxation energy from nuclear de-excitation is not emitted as a gamma photon but is instead used to accelerate one of the orbiting inner atomic electrons. The excited electron then exits the atom at high speed. As the electron originates in the atomic electron cloud rather than the nucleus itself, this is technically not a beta particle. The ejected electron is called a **conversion electron**. Hence, internal conversion is sometimes called electron conversion. It typically occurs following an electron capture, alpha emission or beta emission that has left the nucleus in a still-excited state that can relax further.

Internal conversion is possible whenever gamma decay is possible (except of course if the atom is fully ionized, and no orbital electrons are available for conversion; this rare situation might be encountered when the nucleus in question happens to be a high-energy galactic cosmic ray). As in gamma decay, internal conversion results in no change in the atomic number or mass number of the parent radionuclide.

Internal conversion competes with gamma emission. The relative ratio of internal conversions to gamma photons is quantified as the **conversion coefficient**, α.

α = (number of de−excitations via conversion electron emission)/(number of de−excitations via gamma photon emission)

Or more succinctly, $\alpha = e/\gamma$.

Using an example familiar thanks to permanent implant low dose-rate brachytherapy with iodine-125 'seeds' for prostate cancer, electron capture by iodine-125 results in the formation of tellurium-125. This tellurium-125 is still excited and can emit a gamma photon of 35.5 keV. However, gamma emission is observed only 7% of the time (i.e. 7% abundance). In 93% of decays, the decay energy is released in the form of conversion electrons. Thus, the conversion coefficient, α, of tellurium-125 is 93/7 = 13.3.

Although not entirely correct, conversion electron emission may be conceptualized as an 'internal photoelectric effect' wherein a gamma photon vanishes as it interacts with one of the atomic electrons and transfers all its energy. Although the internal photoelectric effect concept may be helpful in understanding the phenomenon, what truly happens is that inner atomic electrons (if residing in orbitals without a node at the nucleus, as is the case for s orbitals) may intermittently penetrate the nucleus thanks to their non-zero probability of being there. Upon entry into the nucleus, these electrons are subjected to intense electric fields generated by the protons of the nucleus as they rearrange themselves during relaxation. Transfer of nuclear energy to these electrons leads to their ejection from the atom as conversion electrons.

As in the real photoelectric effect (which is covered in chapter 6), the electron receives all the photon's energy but must expend some of that energy to free itself from the binding force of the atom. In other words, just as in the photoelectric effect, part of the expelled electron's energy goes into a work function. Thus, the kinetic energy of the conversion electron is numerically less than the energy of the competing gamma photons. The difference depends on the shell from which the electron originated. Because of the quantum nature of the atomic electron cloud (which is described in terms of shell structure), the electron's binding energy reflects the shell it belonged to. The kinetic energy of the conversion electron is the energy of the gamma photon minus the characteristic binding energy of the electron's original shell.

Since both the gamma photon energy and the electron shell binding energies have well-defined values, the conversion electrons exhibit well-defined values as well. These discrete energy values can therefore reveal the shell from which the characteristic electron originated. The highest probability of conversion electron origin is for the 1s-orbital of the innermost shell (that is, the K shell). The probability diminishes rapidly as one moves to outer layers. The simplest explanation for this is that the s-orbitals of the K-shell have the highest electron density at the nucleus, whereas higher-energy outer shells and orbitals have significantly lower electron density probabilities at the nucleus. It is worth remembering that although these higher-energy outer shells have a lower probability of an electron at the nucleus, when a conversion electron does originate from such levels, its binding energy is lower, and the resultant conversion electron will thus carry away more kinetic energy than if it originated from the usual tightly bound inner K shell.

The discrete, unique energies of conversion electrons clearly distinguish them from the electrons of beta decays whose energies exhibit a broad spectrum that varies between zero and a maximum value, with the balance of the decay energy being carried away by a neutrino.

The orbital electron expulsions in internal conversion leave a vacancy. This vacancy is filled by electrons cascading down from higher-energy levels, which is followed by a reorganization of the electron cloud. This reorganization is accompanied by the emission of characteristic x-rays, Auger electrons or both (as discussed further in chapter 6).

5.12 Double electron capture

Double electron capture results in the transformation of a radionuclide $^A_Z X$ into a daughter $^A_{Z-2} X$. It is represented by ($\varepsilon\varepsilon$).

It is similar to conventional electron capture except that instead of just one electron, two orbital electrons are captured by two protons. These two protons are converted into two neutrons and two electron neutrinos. Double electron capture has been experimentally confirmed in the decay of only three nuclides: $^{78}_{36} Kr$, $^{130}_{56} Ba$ and $^{124}_{54} Xe$. One reason is that the probability of double electron capture is stupendously small; the half-lives for this mode lie above 10^{20} years. A second reason is that the only detectable particles created in this process are x-rays and Auger electrons that are emitted by the excited atomic shell. In the range of their energies (~ 1–10 keV), the background is usually high. Thus, the experimental detection of double electron capture is challenging.

If the mass difference between the parent and daughter nuclides is >1.022 MeV (the rest mass of an electron-positron pair or two electrons), the Q-value is sufficient to allow another rare mode of radioactive decay—electron capture with positron emission. This process can occur in competition with double electron decay.

Finally, if the mass difference exceeds 2.044 MeV (the rest mass of two electron-positron pairs or 4 total electrons), yet another competing mode of decay (double positron decay) becomes possible. As one might recognize, double positron decay is a form of double beta decay, which will be discussed next.

5.13 Double beta decay

Double nuclear beta-plus and beta-minus ($\beta\pm$) decay was predicted by Nobel Laureate Maria Goeppert Mayer in 1935. Abbreviated $2\nu\beta\beta$, double beta decay is a rare type of nuclear transmutation in which two neutrons are simultaneously transformed into two protons. Two beta particles along with two electron anti-neutrinos are simultaneously released. Double beta decay allows an unstable radionuclide to efficiently move closer to its optimal ratio of protons and neutrons. Depending on the radionuclide, double beta decay may produce two electrons (due to conversion of two neutrons into two protons), or it may produce two positrons (due to conversion of two protons into two neutrons). The first is symbolized by $2\nu\beta^-\beta^-$, whereas the second is abbreviated $2\nu\beta^+\beta^+$.

In $2\nu\beta^-\beta^-$, an initial nucleus with proton number Z and total nucleon number A (Z,A), decays to $(Z+2,A)$. As A remains constant, this is a transition between isobars. Despite being isobars, there are minute mass differences between the isobars involved. A necessary requirement for double beta decay to occur is $m(Z,A) > m(Z+2,A)$ *and* single β-decay must be forbidden or suppressed because of mass conservation, for example, if $m(Z,A) < m(Z+1,A)$. This may be the case in certain 'even-even' nuclei with an even Z and even neutron number (N). The situation wherein both Z and N are even leads to increased stability due to spin coupling.

Thus, a typical $2\nu\beta\beta$ candidate is an even-even nucleus (Z,A both are even), which is more tightly bound than its $(Z+1,A)$ neighbor but less so than the nearby $(Z+2,A)$ nuclide. If single beta decay is permissible, double beta decay is obscured. And if alpha emission is allowed, double beta decay is also unlikely and difficult to detect (although it has been observed in uranium-238). Double beta decay has been observed in several radionuclides and has a half-life on the order of 10^{19} to 10^{21} years. Among naturally existing radionuclides, 35 are known to possess the specific requirements for double beta decay. Among them are calcium-48, germanium-76, zirconium-96, molybdenum-100, tellurium-128, xenon-138 and uranium-238. The first observation of double beta decay was made in a sample of selenium-82 by Elliott, Hahn and Moe in 1987 at the University of California, Irvine.

5.14 Neutrinoless double beta decay

In the still hypothetical neutrinoless double beta decay, abbreviated $0\nu\beta\beta$, only electrons would be emitted. It should be emphasized that this process (which would possibly violate the Standard Model since lepton number would not be conserved;

lepton number is covered further in chapter 8) has never been observed. Theoretically, neutrinoless double beta decay might be possible without a violation of lepton number if neutrinos were **Majorana fermions**—that is, particles that are their own antiparticles. Thus, the search for $0\nu\beta\beta$ is of great interest. Observing the process would prove that the neutrino can be its own antiparticle—a Majorana particle.

5.15 Induced fission

The nuclear binding energy curve of the elements reaches its maximum at a mass number, A, of 56. Thus, iron-56 is among the most stable of all nuclides. Among nuclides with greater values of A, fission could release energy and increase stability. Fission may be induced by neutrons (as in a nuclear reactor or bomb), or it may occur spontaneously in very heavy nuclei. Fission typically leads to gamma photons and more neutrons (e.g. 2 or 3 neutrons in the case of uranium-235 fission). These neutrons may induce nuclear fission in adjacent nuclei. Thus, a self-sustaining nuclear chain reaction becomes possible when the amount of parent radionuclide is sufficient. This amount is called the critical mass. When a critical mass of uranium-235 is amalgamated, spontaneous fission can initiate—and sustain—a chain reaction.

Nuclei that can fission upon bombardment with free neutrons are termed **fissionable**. Radionuclides that very readily undergo fission when bombarded by low energy (thermal) neutrons are termed **fissile**. Examples of fissile radioisotopes include uranium-233, uranium-235 and plutonium-239. Such isotopes are capable of self-sustaining nuclear chain reactions when assembled in amounts beyond their critical masses.

Uranium-238 is an example of an isotope that is fissionable but not fissile. Uranium-238 can undergo induced fission when bombarded with energetic neutrons (>1 MeV). However, upon fission of uranium-238, too few of the daughter neutrons produced are energetic enough to induce further fissions of uranium-238 nuclei. Hence, no chain reaction occurs with pure uranium-238.

Fission, whether spontaneous or induced, typically produces two daughter nuclei (fission products) of roughly equal atomic mass. This distinguishes it from alpha emission or **cluster decay** (in which one of the daughter products is far smaller than the other). The fission products themselves are usually unstable and tend to be relatively rich in neutrons. Thus, many fission products are radioactive and undergo beta decay.

Although it is said that the daughter fission products are roughly equal in mass, this is not precisely correct and there is a slightly asymmetrical mass distribution among the daughter products. Symmetrical fission, wherein the fission product masses really are at approximately one-half of the parent, is less favored. Thus, the daughter product abundance plot looks more like a two-humped Bactrian camel back than a bell curve and is called an M curve. For example, in the fission of uranium-235, the two peaks center around $A = 90\text{--}100$ and $A = 135\text{--}145$.

Of note, there are subtle effects of neutron energy on the shape of this curve. As the inciting neutron energy increases, the trough in the M curve becomes shallower, and the curve more closely approximates a Gaussian. If the incoming neutron is too energetic, the capacity to induce fission diminishes.

Finally, about 0.2%–0.4% of uranium fissions are ternary fissions; that is, they yield a third light nucleus. In ternary fissions, the third product is typically quite small and can be an alpha particle or a tritium nucleus (i.e. a triton).

5.16 Spontaneous fission

Spontaneous fission can occur in very heavy elements with $Z > 90$ and $A > 230$. The fission products of spontaneous fission are the same as with induced fission. The parent splits into two lighter nuclei with atomic masses described by an M-curve; neutrons are also released along with gamma radiation.

Although it does occur in very small branching ratios for naturally occurring radioisotopes including thorium-232, uranium-235 and uranium-238, it is more likely to occur in artificially produced elements such as plutonium-240, curium-250 and californium-252. Among these heavy radionuclides, those with lower mass numbers generally have longer half-lives for spontaneous fission. For example, uranium-238 has a spontaneous fission half-life of about 10^{16} years, while fermium-256 has a half-life of only around 3 h.

In elements capable of decaying via spontaneous fission, alpha emission is also possible. The relative frequency of the two competing decay modes is related to atomic mass. For instance, in uranium-238, alpha emission is two million times more likely than spontaneous fission, while in fermium-256, spontaneous fission occurs in 97% of all decays, compared to only 3% of decays occurring through alpha emission.

Californium-252 is somewhere in between these two extremes and can serve as a useful, portable neutron source. Californium-252 has a half-life of 2.645 y and undergoes decay by alpha emission (96.9%) and through spontaneous fission (3.1%). Californium-252 has been used as a 'sealed source' radioisotope for high-LET brachytherapy in cancer radiotherapy.

A small but important amount of spontaneous fission occurs in plutonium-240. This importance is evident in nuclear weapons design. Plutonium-240 is created through neutron capture by plutonium-239 and is thus a contaminant. Some plutonium-240 is inevitably present in samples of plutonium-239. This plutonium-240 contamination makes plutonium fission bombs far more complicated than the simple 'gun-type' explosives that are possible with uranium-235 enriched uranium. Even small amounts of plutonium-240 in a plutonium-239 bomb can lead to 'fizzle' and meltdown (rather than kaboom) during 'assembly' of two subcritical plutonium-239 masses using a gun-type design. This is because the spontaneous fission of plutonium-240 increases neutron flux, which in turn can initiate the chain reaction prematurely and release too much energy; this in turn stymies core implosion and detonation. Therefore gun-type designs will not work with plutonium-239. Also, for this reason, even in an implosion-triggering device, the maximum allowable amount of plutonium-240 for a nuclear weapon is 7%.

5.17 Natural reactors

As an aside, as predicted in 1956 by Paul Kuroda and discovered by Francis Perrin in 1972, natural nuclear reactors have existed on Earth. Approximately 1.7 billion years ago in Oklo, Gabon, a natural reactor went critical and remained intermittently active for approximately 300 000 years, leaving behind about 5 tons of telltale fission products. The slightly decreased percentage of uranium-235 in the natural uranium at the Oklo mine was initially a cause for alarm and sparked an investigation that led to the discovery. There, the uranium ore did not have a normal uranium-235 content of 0.7202%. Instead, the uranium ore was inexplicably depleted, containing only 0.7171%. While this might superficially seem like a tiny and inconsequential discrepancy, given the central role of uranium-235 in nuclear weapons, it was enough to trigger an intense investigation.

For exercise, students might want to calculate the relative abundance of uranium-235 when these natural reactors went critical 1.7 billion years ago, remembering that the half-life of uranium-235 is ~700 million years and the half-life of uranium-238 is 4.5 billion years.

(Answer: ≈3%; At the dawn of Earth's history, approximately 4.567 billion years ago, the relative abundance of uranium-235 was ≈17%.)

5.18 Proton radioactivity and neutron emission

When one examines a plot of atomic number against the number of neutrons for the stable nuclides, it becomes obvious that just any old arbitrary combination of protons and neutrons will not necessarily yield a stable nucleus. There is a zone of stability, which deviates from the $Z = N$ line as atomic number increases on a Segrè plot. A pattern emerges revealing that, as Z increases, to attain stability, the number of neutrons, N, must increase disproportionately relative to the number of protons. This requirement for stability is achieved for 251 different combinations of protons and neutrons. These natural nuclides can be found on Earth. Additionally, another 26 nuclei possess a 'quasi-stable' configuration, meaning that they decay with a half-life comparable to, or longer than, the age of the Earth. Thus, these nuclides may also still be present on Earth. But beyond this zone of stability, the resultant nuclides are unstable; they are radioactive and decay via various different mechanisms.

By definition, isotopes have a constant Z, but a variable number of neutrons, N. For high-Z isotopes, N must increase significantly beyond the value of Z. But for the most neutron-rich isotopes of each element, there is a limit beyond which additional neutrons are no longer accepted. Any additional neutrons will not remain bound. This limit, beyond which any additional neutron will not be bound, is called the **neutron drip line**.

Radioisotopes residing near but not beyond the neutron drip line may decay via beta decay (which converts a neutron into a proton) or alpha decay (which may create a daughter nucleus that is closer to the zone of stability). But once the number of added neutrons has exceeded the neutron drip line, another option emerges— **neutron emission**. In neutron emission, a neutron is directly ejected from the nucleus. Two radioisotopes known to decay via neutron emission are helium-5 and beryllium-13. Some very neutron-rich radioisotopes can decay through the emission of two or more neutrons simultaneously. For instance, hydrogen-5 and helium-10 decay through the emission of two neutrons, while hydrogen-6 can decay through the emission of three neutrons.

There are no naturally occurring radioisotopes that decay through neutron emission. But some can be created in the laboratory, and others are known (by circumstantial evidence) to have been created in stars. Note that when neutron emission occurs, Z does not change. Therefore, the daughter product is an isotope of the same parent element.

Just as with neutron emission and the neutron drip line, a similar situation is observed on the other side of the zone of stability. For radionuclides with N:Z ratios that have too few neutrons for the number of protons, electron capture and positron emission can convert protons into neutrons and help attain stability. But beyond an extreme limit called the **proton drip line, proton emission** comes into play.

In proton emission, a proton is ejected from the nucleus. This phenomenon is also called **proton radioactivity**. Proton emission may occur because of an overabundance of protons in the isotope in question or it may occur following beta-plus decay (that is, positron emission), in which case the phenomenon is called **beta-delayed proton emission**.

Cobalt-53m was the first radionuclide found to be capable of decay through proton emission (in 1.5% of decays). Early in the 21st century, observations of double proton decay were made for iron-45 and zinc-54.

5.19 Halo nuclei

Most atomic nuclei are extremely compact and tightly bound. However, when there is an excess of neutrons or an excess of protons (as is the case with nuclei near the neutron drip line or proton drip line, respectively), the nuclei take on an odd configuration. These nuclei develop an inner core surrounded by a peripheral 'halo' of nucleons.

Some nuclei appear to consist of an inner core surrounded by a 'halo' of protons and/or neutrons. Such **halo nuclei** are larger than one would expect. (The classical radius of a nucleus is roughly proportional to the cube root of the mass number, A.) But halo nuclei, are larger than this estimated radius:

$$r > 1.2A^{1/3} \text{ femtometers.}$$

Halo nuclei are unstable and reside near but not beyond the neutron or proton drip lines on a chart of the nuclides. They are unstable and decay with very short half-lives but not through proton emission or neutron emission. Some halo nuclei can be induced to undergo proton emission by intense x-ray lasers, however.

5.20 Halo nuclei in stellar interiors and in supernovae

Although addition of multiple neutrons to nuclei is not a natural process here on Earth, neutrons are added in the cores of certain stars and during some stellar explosions. One such neutron adding process is called the **s-process** (for slow neutron capture). During the s-process, the addition of neutrons makes the new, heavier nucleus unstable, and it typically will decay via beta emission. The new daughter is a different element since a neutron has been converted into a proton through the beta decay process. In the stellar environment, additional neutrons can be added to this new element to form yet another isotope that might decay through alpha or beta emission into yet another element. In this fashion, the s-process may manufacture a whole slew of new elements beyond iron. The s-process of neutron capture occurs in certain types of large, bright stars (especially in **asymptotic giant branch stars**) and probably also in intermittent violent stellar phenomena such as **novae** (during which a sudden thermonuclear detonation occurs when a companion star dumps material onto its white-dwarf partner).

However, a different form of neutron capture can occur in even more violent explosions such as **Type Ia supernovae** (where so much material is poured onto the white dwarf from its red giant companion that the whole white dwarf explodes in a tremendous thermonuclear blast). This is the rapid neutron capture process or **r-process**. In the r-process, neutrons are being added so quickly to nuclei that they do not have time to undergo the radioactive decay that happens in the s-process. Thanks to the r-process, heavy exotic nuclei can be produced, which then decay into the rare but familiar elements we see today such as gold, thorium and uranium. The r-process may also occur during neutron star mergers (NSMs). As will be covered in chapter 31, the merger of two neutrons stars may lead to the creation of a black hole and an accompanying gamma ray burst. Theoretically, protons and neutrons can be shuffled together to form over 6000 possible permutations. As mentioned, fewer than 300 of them are stable and are naturally found on Earth. Nonetheless, physicists have been able to create about 3000 different unstable nuclei in the laboratory setting.

But as discussed above, when excessive amounts of neutrons are added to elements, we approach the neutron drip line. And as we approach the neutron drip line, the nuclear shell model begins to falter. In such nuclei, the inner protons and neutrons form a central core, but the loosely bound neutrons on the perimeter no longer remain in defined shells but form a halo. It therefore appears that these halo nuclei can and probably do form during supernovae, and their decay contributes to the variety of bizarre radiation in the supernova environment.

5.21 Conservation in radioactive decay

In the decay of any radionuclide, the following conservation laws must be obeyed:
1. Conservation of total mass/energy
2. Momentum
3. Electric charge
4. Matter/antimatter
5. Total number of nucleons (protons plus neutrons)
6. Baryon number and quark number
7. Lepton number
8. Lepton subtype number.

Students are advised to double check that all conserved properties are balanced on both sides of the equation when working problems.

Whenever there is an observed conservation law, there is an underlying 'symmetry' and vice versa. This is a fundamental principle of physics that plays a very important role in not only radioactive decay but in any reactions involving the weak force. For that matter, symmetry plays a crucial role in the physics behind the strong force, electromagnetism and even gravity. In other words, it is an inherent and important principle in all of physics, but its role was most noticeable in the development of the Standard Model of particle physics. In honor of Emmy Noether, who first enunciated it, this principle is called **Noether's Theorem**.

5.22 The Oddo–Harkins rule

The Oddo–Harkins rule states that an element with an even atomic number will be more abundant than both elements with the adjacently larger and smaller odd atomic numbers. As a classic example, oxygen ($Z = 8$) is more abundant in the Universe than nitrogen ($Z = 7$) or fluorine ($Z = 9$); in fact, oxygen is the third-most abundant element in the Universe on a molar basis. Thanks to the Oddo–Harkins rule, when one plots abundance of the nuclides on the ordinate against atomic number, Z on the abscissa, an obvious saw tooth pattern emerges. This sawtooth pattern is far more pronounced when absolute number is plotted but is even evident when abundance is given logarithmically. Also evident on such a plot is a general decrease in abundance as atomic number increases from left to right. Iron tends to stand out—it is three orders of magnitude more abundant than its odd-numbered neighbor cobalt. This is because iron has the maximum binding energy per nucleon; it is the most compact nucleus that can be made through thermonuclear fusion in stars or through neutron capture in supernovae. Fusion of elements heavier than iron ($Z = 26$) or nickel ($Z = 28$) is not exergonic but endergonic.

One notable exception to the Oddo–Harkins rule is beryllium ($Z = 4$). Although it has an even number of protons, it is rarer than its odd-numbered neighbors, lithium ($Z = 3$) and boron ($Z = 5$). The explanation is that most of the Universe's beryllium was not made through stellar nucleosynthesis (or Big Bang nucleosynthesis) but instead via cosmic ray spallation. Furthermore, beryllium has only one stable isotope, whereas its odd neighbors have two stable isotopes.

5.23 Neutrinos

It was thanks to the requirements for conservation of energy and momentum that in 1930 Wolfgang Pauli proposed the existence of a previously undiscovered, uncharged particle that accompanies the electron upon beta decay. Enrico Fermi named this hypothetical ghost particle the 'little neutral one' or neutrino. Since the emitted electrons displayed a spectrum of energies rather than discrete energies as in alpha decay, it was assumed that this hypothetical particle must carry the energy and momentum that the electron was missing. Actual detection of the neutrino was a daunting task that took two decades to accomplish, since the neutrino 'cross section' of interaction is so low that they could pass through light years of matter without interacting. The discovery of neutrinos first took place in 1956 by a group led by Clyde L Cowan and Frederick Reines. They studied the theoretical phenomenon of 'inverse beta decay' in which a proton interacted with an electron antineutrino to become a neutron and a positron.

$$p + \bar{\nu} \rightarrow n + e^+.$$

Thus, the first observation of these hypothesized particles actually involved antineutrinos.

The details of this reaction are fascinating. Using a uranium fission reactor (Savannah River Site reactor in South Carolina), 'Project Poltergeist' exploited (the assumed) high flux electron antineutrinos derived from the beta decay of abundant fission products.

$$n \rightarrow p + e + \bar{\nu}.$$

In theory, the antineutrino flux was on the order of 10^{13} cm^{-2}s^{-1} in the vicinity of the core. They placed nearby tanks of aqueous cadmium chloride (CdC$_{12}$) with liquid scintillators to detect the proposed reactions. The immediate reaction involved capture of electron antineutrinos by protons:

$$\bar{\nu} + p \rightarrow n + e^+.$$

Within nanoseconds, the positron would meet and mutually annihilate with an electron to yield two gamma photons of 511 keV apiece directed approximately 180° apart from each other. The reaction would be documented by coincidence counters.

$$e^+ + e^- \rightarrow 2\gamma.$$

But in contrast with this fast reaction, the neutron could yield additional gamma photons at a slightly slower pace through another mechanism. The neutrons would slow down via interactions with hydrogen nuclei in the water until reaching thermal energies (~0.025 eV) over a matter of microseconds. Cadmium-113 has a very large cross section for the capture of such thermal neutrons. Cadmium neutron capture results in nuclear disintegration (discussed in a later chapter) with emission of gamma photons and a total energy of 9 MeV. The team would thus seek this second gamma signal a few microseconds after the first coincidence. Through this experimental setup, they were able to count approximately three events per hour, with perhaps one event per hour attributable to background and one that nicely matched the expected description.

In addition to finally detecting the hypothetical neutrino, the mass limit deduced from the measurements was <2 eV. (Compare this to the 511 keV of the lightest known charged lepton, the electron!) The above equations and analysis imply that the neutrino, like the electron, is a fermion with spin ½. Whether it had any mass at all was not answered by this set of experiments. But further experiments (which resolved the **solar neutrino deficit problem**) showed that neutrinos could change flavor en route from the Sun to the Earth—and a change in flavor implies that neutrinos could not be truly massless. The reason for this is simple: if neutrinos were massless, they would travel at the speed of light, c. If they traveled at c, by special relativity, time would stop for them. If time had stopped for these massless neutrinos, they could not change flavor over time. Therefore, the discovery of neutrino flavor oscillation implies that neutrinos are not moving at the speed of light and are not massless.

Further reading

[1] Blank B and Płoszajczak M 2008 Two-proton radioactivity *Rep. Prog. Phys.* **71** 046301
[2] Pfützner M, Badura E and Bingham C *et al* 2002 First evidence for the two-proton decay of Fe-45 *Eur. Phys. J.* A **14** 279–85
[3] Zuber K 2004 Double beta decay *Contemp. Phys.* **45** 491–502
[4] Wu B and Liu J 2022 Proton emission from halo nuclei induced by intense x-ray lasers *Phys Rev* C **106** 064610
[5] Elliott S R, Hahn A A and Moe M K 1987 Direct evidence for two-neutrino double-beta decay in 82Se *Phys. Rev. Lett.* **59** 2020–3
[6] Pradler J, Singh B and Yavin I 2013 On an unverified nuclear decay and its role in the DAMA experiment *Phys. Lett.* B **720** 399–404

Chapter 6

Basic radiation science

6.1 Directly ionizing and indirectly ionizing radiation

Ionizing radiation species may be categorized as **directly ionizing** or **indirectly ionizing**. Directly ionizing entities are electrically charged and virtually instantly rip electrons off atoms and molecules as they interact with matter. Examples of directly ionizing radiation species include electrons, muons, charged pions, protons, alpha particles and heavier nuclei.

Indirectly ionizing entities are uncharged and first interact with atoms or molecules in the medium. These interactions subsequently eject directly ionizing particles such as electrons or protons. Examples of indirectly ionizing radiation species include high energy photons, neutrons and uncharged pions.

6.2 Electromagnetic radiation

The electromagnetic spectrum spans energies ranging from radio waves to gamma rays. Depending on the part of the spectrum in question, scientists address the electromagnetic radiation in question in different ways. For instance, when talking about radio, people typically describe this radiation in frequencies of megahertz and gigahertz. Visible light in contrast is usually described by wavelength in nanometers. Radiation scientists typically describe x-rays and gamma rays in terms of energies measured in kiloelectron volts and megaelectron volts.

Regarding terminology and definitions, the **optical** part of the spectrum is generally defined to encompass electromagnetic radiation with wavelengths in the range from 10 nm to 10^3 μm or frequencies in the range from 300 GHz to 3000 THz. The **optical spectrum** is sometimes considered to be the same as the visible spectrum (which is roughly between 400 and 700 nm), but some authors define the term more broadly, to include the ultraviolet and infrared parts of the electromagnetic spectrum as well.

The wavelength of a photon is given by the equation:

$\lambda = hc/E_p$ where E_p is the energy of the photon, h is
Planck's constant, and c is the speed of light.

This can be rewritten for photon energy in eV and wavelength in nm as:

$$E(\text{eV}) = 1240/\lambda \ (\text{nm}).$$

For the more relevant energies in the range of ionizing radiation, the equation may be expressed as:

$$\lambda(\text{Å}) = 12.4/E \ (\text{keV})$$

Where the wavelength is given in non-SI units of angstroms and the photon energy is in keV. Furthermore, photon energy and temperature may be (albeit imperfectly, if one adheres to the thermodynamically correct definition of 'temperature') related through Boltzmann's constant by the equation:

$$E = kT.$$

The conversion factor is 11 605. In other words, 1 eV corresponds to 11 605 K and 1 keV corresponds to 11.6 million Kelvin. Similarly, 10^{-10} m (100 pm or 1 Å) corresponds to about 12 400 eV (12.4 keV) according to:

$$E_{\text{photon}} = h\nu = hc/\lambda.$$

6.2.1 X-rays vs gamma-rays

Gamma rays have the highest energies and shortest wavelengths among electro-magnetic waves. They have shorter wavelengths than x-rays, but the cut-off is fuzzy and inconsistent. One definition holds that, 'with frequencies above 30 EHz (3×10^{19} Hz), gamma rays impart the highest photon energy'. Alternatively, gamma (γ) radiation may be said to consist of electromagnetic waves with wavelengths usually smaller than 10^{-11} meters or 10 pm, or another alternative definition holds that it is the electromagnetic radiation with the highest energy ($\geqslant$200 keV). There are several other arbitrary dividing lines between x-rays and gamma rays.

In various disciplines, x-rays and gamma rays may be distinguished by their origin. For instance, in medical physics, nuclear medicine and radiation oncology, gamma rays are defined as photons created by nuclear decay, whereas x-rays originate from interactions involving electrons outside the nucleus such as bremsstrahlung. In astrophysics, gamma rays are conventionally defined as photons with energies over 100 keV, while radiation below 100 keV is classified as x-rays. Others, however, may use the rest mass–energy of the electron 511 keV as an arbitrary dividing line.

Gamma photons with energies above tens of giga electron volts have sufficient energy such that a small fraction of them might be capable of penetrating the entire thickness of Earth's atmosphere and reach the surface. The Crab nebula is the brightest persistent (i.e. non-sporadic or non-periodic) source of gamma rays in space and has a small flux of photons with energies reaching beyond an astounding 1 PeV. The IACT (Imaging Atmospheric Cherenkov Telescope), which can detect ultra-high gamma ray photons, has documented some extremely energetic gamma photons coming from the Crab Nebula and other sources (which are discussed elsewhere).

And scientists from China's Large High Altitude Air Shower Observatory (LHAASO) recently reported the detection of gamma photons reaching an incredible 1.4 PeV.

6.3 Dose, Kerma and terma

Radiation absorbed **dose** is measured in gray (Gy), which is simply energy absorbed per unit mass or joules per kilogram:

$$D = dE/dm.$$

Kerma is an acronym for '**kinetic energy released per unit mass**' or alternatively, '**kinetic energy released to matter**'. It is a measure of energy *transferred* from indirectly ionizing radiation to matter as opposed to the energy absorbed in that matter. Kerma is thus related to, but not exactly the same as, absorbed dose. Nevertheless, somewhat confusingly, kerma is measured by the same SI unit, the gray (joules per kilogram). Kerma measures the amount of energy that is transferred from photons (or other forms of indirectly ionizing radiation) to electrons (or other secondary particles) per unit mass at a certain position. These electrons may or may not stay put. Absorbed dose, on the other hand, measures the energy deposited in a unit mass at a certain position by liberated particles that have stayed in the vicinity and deposited their energy locally.

Kerma is formally defined as the sum of the *kinetic* energies of all *charged* particles liberated in an amount of material by *uncharged* ionizing radiation (i.e. indirectly ionizing radiation such as photons and neutrons), divided by the mass of that material. Mathematically, this is:

$$K = dE_{tr}/dm \quad \textbf{Kerma}$$

where E_{tr} is the (kinetic) energy transferred from the radiation to the material.

At modest radiological energies, transfer and deposition of energy is virtually equal; thus, kerma and dose are essentially the same. However, at higher energies, a photon may interact with tissue in one position and create an electron that possesses enough energy to escape that local region and deposit energy at a location distant from the interaction point. Hence, as energy increases, the discrepancy between kerma and dose becomes more pronounced.

There are two components to kerma—**collisional kerma** and **radiative kerma**. Collisional kerma is the portion of the *kinetic energy of the secondary charged particles*, which is caused by collisions that result in ionization and excitation of atoms in matter. Technically, it is the expectation value of the net energy transferred. Radiative kerma on the other hand is the fraction of the initial kinetic energy of the secondary charged particles, which is *converted into photon energy* (i.e. bremsstrahlung). This photon energy often does not stay locally, whereas collisional kerma usually does.

Kerma is maximal at the skin (or surface of inanimate material) and then falls off with depth. This is because of the progressive decline in photon energy fluence with depth. Note that this superficial location of maximum kerma is quite different from the depth of maximum absorbed dose. Especially for very high-energy photons, absorbed dose may be relatively low at the surface but then increases with depth (to a depth called d_{max}, which is the depth below the surface where absorbed dose is highest;

$d_{\max}$ is a function of energy). Absorbed dose falls off exponentially with depth beyond $d_{\max}$. The difference is because, unlike kerma (which is energy transferred), absorbed dose (which is transferred energy that stays locally) requires electron 'buildup'. The electrons contributing to this buildup may be very energetic and can travel a distance from the surface before maximum buildup is achieved. Generally speaking, kerma is higher than absorbed dose at shallow depths, but beyond a depth of $d_{\max}$, absorbed dose may be slightly higher than kerma.

This subtle difference between kerma and absorbed dose is also due to the fact that photons (or any other indirectly ionizing radiations) interact with matter in a two-step process. The first step is the transfer of energy from the radiation beam to the medium. This is the conversion of the energy in photons (or neutrons) into kinetic energy of electrons (or protons in the case of neutron radiation). The second step occurs when the secondary electrons or protons slow down and deposit their energy in the medium. In this way, the first step is kerma, and the second step is dose.

In low atomic number materials (such as biological tissues, where Z can effectively be taken as approximately 7), the majority of the kinetic energy of secondary electrons set in motion by photons is spent in ionization and excitation of atoms. Only a relatively small fraction is spent in radiative collisions with nuclei that generate bremsstrahlung. In other words, for human tissues, kerma is mostly collisional kerma, and relatively little is radiative kerma.

Terma is an acronym for the **total energy released per unit mass**. It measures the total loss of radiation energy from uncharged radiation (e.g. photons or neutrons) as these particles interact with a medium. That lost energy may stay local and contribute to absorbed dose, or it may travel distantly. Kerma is the product of photon energy fluence times the mass energy *transfer* coefficient. This differs slightly from terma, which is the product of photon energy fluence times the mass energy *attenuation* coefficient. One way to view the difference is that kerma addresses the transfer of radiation energy from a beam to kinetic energy of charged particles in a medium that the beam is traversing, whereas terma focuses on the total loss of energy from that beam while it travels through the medium. One subtle point is that kerma does not include energy losses due to coherent scattering (such as Rayleigh and Thompson scattering which are covered later in this chapter); terma does include coherent scattering.

6.3.1 Exposure

Exposure (X) is a measure of the ability of a photon beam to ionize dry air. Exposure thus quantifies the electric charge set free by x-ray or gamma ray radiation in a specified volume of air divided by the mass of that air. As such, exposure is a quantity akin to collisional kerma when high-energy photons interact with air.

Mathematically exposure may be represented as:

$$X = dQ/dm,$$

where m is the mass and Q is the total electrical charge of the ions produced in air *of one charge* (i.e. positive *or* negative) (since if one added all the charges of both signs, the sum would amount to zero).

Note that strictly speaking, the air must be dry, and all liberated electrons must be stopped within the mass of air in question.

Exposure is more readily expressed as the mean energy spent in air to form an **ion pair** (**W**). *W* can be quantified as the kinetic energy consumed in ionization and excitation divided by the sum of the ion pairs produced in air. It turns out that the mean energy needed to create an ion pair in (dry) air is virtually constant over all energies at **33.97 eV ion^{-1} pair**. Since the charge of an electron (e) is 1.602×10^{-19} C, and one electron volt is 1.602×10^{-19} J, *W*/e also equals **33.97 J/C**.

Exposure is measured in electric charge per unit mass. In SI units, exposure is thus measured in coulombs per kilogram (C·kg^{-1}), but this unit does not have an alternative name. A non-SI legacy unit of exposure, the **roentgen (R)**, denotes the amount of radiation that is required to produce one electrostatic unit of charge in 1 cm^3 of air under standard conditions of pressure, temperature and humidity. The roentgen in SI units is:

$$1R = 2.580 \times 10^{-4} \text{C} \cdot \text{kg}^{-1}.$$

Conversely, an exposure of 1 C kg^{-1} = 3876 roentgens. Although exposure usually closely matches absorbed dose, the two are not the same and can be quite different numerically, depending on the circumstances. For example, one roentgen deposits 0.008 77 Gy (0.877 cGy or rads) of dose in dry air but about 0.0096 Gy (0.96 rad) in soft tissue. And one roentgen of x-rays or gamma rays may deposit an absorbed dose anywhere from 0.01 to 0.04 Gy (1.0 to 4.0 rads) in bone depending on the beam energy.

Exposure is the ionization equivalent to collisional kerma when energetic photons interact with dry air. Thus, exposure may be calculated from the collisional kerma if one has the ionization charge generated per unit energy deposited by the incident photons. Since this is a known quantity (33.97 J/C), collisional kerma in air (K_{col} air) is related to exposure (*X*) by:

$$(K_{col} \text{ air}) = \text{WX} = 33.97 \cdot X \text{ when using SI units}$$

or

$$(K_{col} \text{ air}) = \text{WX} = 0.877 \times 10^{-2} X,$$

if *X* is in roentgens and kerma is in gray. A rule of thumb regarding exposure and exposure rates from the radioisotope, radium-226, is that the exposure rate at 1 meter from a 1 curie source of Ra-226 is about 1 roentgen per hour.

6.4 *W* values and specific ionization

Energetic charged particle radiation plays a disproportionately important role in space radiation. For directly ionizing charged particles (such as electrons, protons and alpha particles) traversing materials, including tissue, there are a couple of quantitative measures worth reviewing. First, the *W* **value** is defined as *the average energy lost by a charged particle per ion pair produced*. *W* depends on the type of charged particle and the material that is being ionized. Alpha particles and beta particles (electrons) both lose an average of approximately 22 eV per ion pair produced in water. In air, however, beta particles typically lose 34 eV per ion pair produced, while alpha particles lose an average of 36 eV per ion pair.

Specific ionization is a related concept that is defined as the *average number of ion pairs produced per unit distance* traveled in a medium by a charged particle. As such, specific ionization is closely related to **linear energy transfer (LET)**. High LET radiation will have a high specific ionization, whereas low LET radiation has a low specific ionization. Specific ionization and LET depend on the quality of the radiation (that is, the type of charged particle in question), the energy of the charged particle, and the specific medium being traversed. Depending on the exact energy, alpha particles (which are generally considered to be high-LET particles) might produce 20 000–60 000 ion pairs per centimeter in air. Beta particles (which are generally considered to be low-LET particles) have lower specific ionization and might yield around 100 ion pairs per centimeter in air.

Charged particles can ionize or excite the atoms and molecules in a medium. In addition, rapidly-moving charged particles racing through a medium can manufacture electromagnetic radiation (i.e. photons) through two mechanisms—bremsstrahlung and Cerenkov radiation. In general, for reasons discussed elsewhere, electrons are the only real sources of bremsstrahlung and Cerenkov radiation in most terrestrial forms of charged particle radiation. While protons and heavier charged particles can in theory generate such radiation, for the most part, these two types of electromagnetic radiation (bremsstrahlung and Cerenkov radiation) are generated in practice mainly by the lightest charged particles—electrons. Nevertheless, at the extraordinary energies of some cosmic rays (e.g. those in the TeV, PeV and EeV range) Cerenkov radiation is indeed created as these cosmic rays interact with atmospheric atoms. In fact, this is one means by which these exceptionally energetic cosmic rays are detected by ground level imaging atmospheric Cerenkov telescope systems such as the High Energy Stereoscopic System (H.E.S.S.) in Namibia.

Uncharged photons and neutrons interact with matter very differently. By interacting with atoms in a medium, these indirectly ionizing forms of radiation liberate electrons and protons, which then travel onwards and ionize matter just as primary electrons and protons do.

6.5 Half value layers and radiation lengths

The **half-value layer (HVL)** of a material is the thickness of that material required to reduce the radiation intensity of an x-ray or gamma ray beam to one-half its original value. Alternatively, it can be quantified as the thickness of a material required to reduce the **air kerma rate** of a radiation beam to one-half its initial value.

The HVL reflects the penetrating power of a polychromatic x-ray beam and is a surrogate for beam energy. The penetrating power of a beam is sometimes referred to as beam quality or penetrance. A lower HVL indicates lower photon energies while a large HVL indicates high photon energies. For historical reasons, HVL is often reported in millimeters of aluminum.

A related concept is the **radiation length, X_0**. Like the HVL, the radiation length is a characteristic of a given material and the energy and quality of the radiation beam used. Whereas the HVL is the thickness of a material required to decrease the radiation intensity to 1/2 of its initial value, the radiation length is defined as the thickness of attenuating material required to decrease the intensity to $1/e$ of its original value. Thus, HVLs are akin to half-lives in radioactivity, while radiation length is akin to mean lifetime of radioactivity. The radiation length or **radiation path length** is thus sometimes called the **mean length**. Other synonyms for the radiation length are the **attenuation length** or **mean free path**. Technically, the mean free path is the average distance a particle will travel in a given material before being absorbed. Half-value lengths are often used when describing photon beams, whereas radiation lengths are more often used when describing particle beams.

When dealing with single particles rather than beams of particles, the probability of finding a particle at a depth x in an attenuating medium may be estimated through **Beer's law** (also called the **Beer-Lambert law**):

$$P(x) = e^{-x/l},$$

where l is the radiation length.

Another related parameter is the **tenth-value layer (TVL)**, which is the thickness of an attenuator needed to diminish the radiation intensity to one-tenth of its initial value. Thus, the TVL is the amount of material needed to absorb 90% of the incident radiation. Mathematically, the TVL should be around 3.32 HVLs (ln 10/ln2); however, thanks to beam hardening which is discussed below, the measured TVLs tend to be greater than 3.32 HVLs.

Strictly speaking, 'HVL' refers to the *first* HVL. In a *polychromatic* x-ray or gamma ray beam (as from the bremsstrahlung used in radiology or from gamma ray bursts), the first half-value thickness is always the thinnest. Subsequent HVLs (e.g. second HVL, third HVL, etc) refer to the thickness of a specified material that will reduce the radiation intensity (or air kerma rate) by one-half after it has been attenuated by previous half value layers. The reason the second, third, etc HVLs are thicker than their predecessors is because of **beam hardening**.

6.6 Beam hardening

Beam hardening of a radiation beam occurs as lower-energy photons are filtered out through selective attenuation as the beam continues traveling through a material. After passage through any material, the lower-energy photons are preferentially absorbed first and will no longer contribute to the ongoing beam. Thus, at any given depth within a medium, the remaining beam will have a higher average energy than when it first entered that medium.

After filtration by one HVL and removal of lower-energy photons, the subsequent HVLs will be thicker since the filtered photon beam becomes primarily populated with photons of higher energy. In other words, thicker material is required to attenuate the now more penetrating beams.

Hence, as a polyenergetic beam traverses matter, it grows 'harder' with depth. And subsequent HVLs thus grow thicker. Thanks to beam hardening, a tenth-value layer (TVL) is always greater than or equal to 3.32 HVLs with bremsstrahlung beams, as they are polychromatic. But if the radiation beam were perfectly monoenergetic, one TVL would exactly equal 3.32 HVLs. Although it is obvious, it is worth stating that although a polyenergetic photon beam grows harder (i.e. more energetic) with depth, it also becomes less intense (that is, the flux or the number of photons is diminished).

6.7 Homogeneity factor

A metric called the **homogeneity factor (HF)** may be used to quantify the polychromatic nature of an x-ray or gamma ray. The homogeneity factor is given by:

$$HF = HVL1/HVL2,$$

where HVL1 is the first half value layer and HVL2 is the second.

Since in any polychromatic radiation beam the first HVL is always the thinnest, HF is always less than one. A very low HF indicates a very polyenergetic radiation beam. In contrast, a monoenergetic radiation beam (as from a radioisotope gamma emitter) will have an HF equal to one since the photon energy does not change after filtration. The higher the HF (that is, the closer to one), the closer to monoenergetic the radiation beam is.

It should be stated that HVLs only apply strictly to narrow beam ('pencil beam') geometry. With broad-beam geometry (a more realistic and general scenario), a higher amount of scattered radiation will reach a radiation detector. Measurements made under broad beam geometry tend to underestimate the amount of attenuation and overestimate the half-value thickness.

HVLs must specify the radiation and the absorbing material. For example, the approximate HVL for gamma rays from iridium-192 in lead is 4.8 mm, whereas it is 2.8 mm in uranium. And these HVL figures would be proportionately higher with the more energetic gamma rays from cobalt-60 (about 12.5 mm in lead and 6.9 mm in uranium). Just as a space radiation-related factoid, the HVL for energetic neutrinos in lead is approximately one light year! Incidentally, here on Earth, thanks to their 'unstoppability', about 60 billion solar neutrinos pass through each square cm per second (the size of a thumbnail).

6.7.1 Bremsstrahlung

When decelerating in the electric field of a nucleus, fast-moving, light, charged particles such as electrons and positrons lose energy by emission of electromagnetic radiation. This radiation is called 'braking radiation' or bremsstrahlung. The bremsstrahlung emission probability (cross section) is proportional to the inverse square of the particle mass:

$$\sigma \propto \ {\sim} m_e^{-2}.$$

As a consequence, because muons weigh about 200 times more than electrons, energy loss due to bremsstrahlung is 40 000 times smaller for muons than it is for electrons. Of course, one may look at this the other way and state that electrons are 40 000 times more efficient at generating bremsstrahlung than muons are. And since protons weight about 1836 times more than electrons, the bremsstrahlung associated with protons is around 3.4 million times less than with electrons.

The greater the charge in the nucleus (atomic number) of the target material, the greater the deflection of the electrons and the greater the intensity of the bremsstrahlung. Hence, in addition to the mass of the charged particle, the probability of making bremsstrahlung is also proportional to Z^2 (where Z is the atomic number of the absorber). For this reason, the target materials in most x-ray tubes and linear accelerators are made of high-Z metals like tungsten (which provides a good blend of bremsstrahlung production and resistance to melting).

Students should be aware that most medical x-rays are highly 'filtered', meaning that an added layer of absorbing material such as 0.5 or 1 mm of aluminum is placed in the beam path. The effect of such filtration is a hardening of the beam because the filter removes a large fraction of the very lowest energy x-rays without much effect on the high-energy portion of the spectrum. The result is a harder beam that is enriched in high-energy photons and largely devoid of low-energy ones.

But, in the absence of filtration, **Kramer's equation** applies:

$$I_E = KZ(E_M - E),$$

where I_E is the intensity of photons with energy E, Z is the atomic number of the target, E_M is the maximum photon energy, and K is a constant.

As dictated by the **law of Duane and Hunt** (which is also discussed elsewhere in this chapter), the maximum energy that a bremsstrahlung photon can have is equal to the energy of the incident electrons. Thus, the maximum energy of the bremsstrahlung photon in mega electron volts or kilo electron volts is equal to the potential difference, which is the peak voltage in megavolts or kilovolts (MVp or kVp, respectively). Notice that according to Kramer's equation, the actual intensity of photons with this theoretical maximum energy approaches zero. That is, when $E = Em$, $I_E = 0$.

The probability of bremsstrahlung production is proportional to the square of the atomic number of the target (Z^2). However, the **efficiency** of bremsstrahlung production is a function of the electron energy and Z to the first power. Since electron energy is proportional to the voltage across an x-ray tube, this can be expressed as:

$$\text{Bremsstrahlung efficiency} = 9 \times 10^{-10} \, ZV,$$

where V is the tube voltage (in volts).

Efficiency is the ratio of the emitted x-ray output energy to the input energy by electrons. One can see from this equation, just how miserably inefficient x-ray production is. Even using a high-Z material like tungsten ($Z = 74$) with electrons accelerated across a peak potential of 100 kVp yields an x-ray efficiency of under 1%. The balance of the input energy of electrons (>99%) is lost as heat.

From the Duane–Hunt law ($\lambda_{min} = hc/eV$) to calculate the minimum wavelength in angstroms when given V in volts one may use:

$$\lambda_{(min)} = 12.4 \times 10^3 \text{ volts} / V_{(peak)} \text{ (in angstroms)}.$$

The same equation, when the peak voltage is given in kVp is:

$$\lambda_{(min)} = 12.4 \text{ kV/kVp (in angstroms)}.$$

Working in reverse, to get the energy in kilo electron volts when given wavelength in angstroms:

$$E = 12.4 \text{ keV}/\lambda \text{ (in angstroms)}.$$

Of course, one may convert to nanometers through the fact that 1 nm = 10 Å.

When a beam of energetic electrons smashes into a target to generate x-rays, the fraction of electron energy that is actually converted into bremsstrahlung is tiny. Compared to collisional energy losses by electrons, a formula is:

$$\text{Radiative energy loss/collisional energy loss} = E_k \times Z/820\ 000,$$

where E_k is the kinetic energy of the incident monoenergetic electrons in kilo electron volts and Z is the atomic number of the target material.

This might be also expressed as:

$$f = 0.00012\ Z \times E,$$

where f is the fraction of energy of monoenergetic electrons that is converted to bremsstrahlung, Z is the atomic number of the material, and E is the kinetic energy of the monoenergetic electrons (this time in mega electron volts).

6.8 Cerenkov radiation and anomalous refraction

Although nothing can travel faster than c, the speed of light in a vacuum, charged particles traveling through a medium may travel faster than the speed of light in that particular medium. It is well known that the speed of light can be considerably slower in certain media than c; this is quantitatively given by the **index of refraction**, *n*. The index of refraction or refractive index is defined as:

$$n = c/v,$$

where v is the velocity of light in the medium with a refractive index of n.

In optics, the refractive index is well known through **Snell's Law**, which relates the angle of incidence (θ_i) to the angle of refraction (θ_r):

$$n = \sin \theta_i / \sin \theta_r.$$

This equation applies when a light wave is travelling in a vacuum and enters a medium of refractive index n at an angle θ_i. For the more general case, where light is travelling in a medium of refractive index n_1 and then enters another medium of refractive index n_2, Snell's Law can then be written as:

$$n_1 \sin \theta_1 = n_2 \sin \theta_2.$$

The light is bent is towards the normal when it enters a medium of higher refractive index. Thus, when a beam of light enters any transparent medium from a vacuum above that medium, the beam of light will be bent towards the normal. As will be discussed below, this is not always the case for x-rays and gamma rays.

The refractive index for visible light can be quite pronounced for certain transparent materials. One classic example is diamond, where $n = 2.417$. This implies that there is a substantial degree of bending of an entering light wave, which in turn confers much of diamond's attractive optical features. Another interesting related optical phenomenon is **dispersion**, which is due to slight differences in the refractive index of a medium for different wavelengths. In other words, the refractive index is not a constant for a given medium but instead is better represented as a function of frequency or wavelength such as $n = n(\lambda)$. In lenses for telescopes and other applications, dispersion can be problematic because it leads to a phenomenon called **chromatic aberration**.

Optical dispersion can be quantified in different ways such as the **Abbe number** (V_d) of a medium. The Abbe number is defined as:

$$V_d = (n_D - 1)/(n_F - n_C),$$

where n_D, n_F and n_C are the refractive indices at wavelengths of 656.3, 587.6 and 486.1 nm, respectively. (These are historical spectral lines called Fraunhofer lines C, D and F.)

There are other ways to quantify dispersion, but it is a gemstone's dispersion that gives it its '**fire**'. Diamond does not have the highest dispersion but ranks near the top in this category (as well as in refractive index).

As a general statement, the refractive index of x-rays is slightly *less than* 1.0. Hence, an x-ray beam entering a slab of ice or a pane of glass from air will be bent *away* from the normal, unlike a ray of light, which will be bent *toward* the normal. This is called **anomalous refraction**.

This leads to an enigma since the equation $n = c/v$ would indicate that the velocity of x-rays in glass and in other materials is greater than its velocity in empty space. But since x-rays in empty space travel at c, and according to the theory of relativity nothing can travel faster than c, there is a serious paradox. The paradox is resolved by recalling that according to relativity, no *information* can travel faster than the speed of light in vacuum. The **phase velocity** of light (e.g. x-rays) does not carry information and can travel faster than c. And since the refractive index alludes to the phase velocity of light, there is no violation of special relativity.

The phase velocity is the speed at which the crests of a wave move; this may be faster than c. On the other hand, the **signal velocity** of x-rays in a medium always remains less than c. Hence, the observed refractive indices less than unity for x-rays are allowed. In reality, refractive indices of x-rays are often lower than, but very close to, one. For instance, 30 keV x-rays have a refractive index of 0.999 999 74 in water. X-rays thus tend to travel very straight through biological tissues and other materials, which is very useful when creating high-resolution, accurate medial images. Incidentally, this contrasts x-rays from charged particles such as protons, which as discussed below, may deviate from straight lines due to electrostatic repulsion from nuclei in the medium being traversed.

Another example of anomalous refraction or a refractive index less than unity is provided by radio waves in the ionosphere. Since the refractive index of the ionosphere is less than one, electromagnetic waves propagating through the ionosphere are bent away from, rather than towards, the normal. This unusual phenomenon may be exploited for long-distance radio communications since it allows radio waves to be refracted back toward, rather than away from, Earth.

When a charged particle moves through a medium faster than the speed of light in that medium, electromagnetic radiation is emitted. This electromagnetic radiation is called **Cerenkov radiation** (also spelled Cherenkov, after 1958 Nobel Laureate Pavel Cherenkov). Conceptually, it is akin to the sonic boom associated with supersonic movement.

Cerenkov radiation may be seen as the eerie blue light emanating from a radioactive source in water such as the core of a nuclear reactor. Patients undergoing radiation therapy for head and neck tumors often say they see flashes of light during treatment. This may be due to Cerenkov radiation occurring in the vitreous humor of the eye.

Technically, for Cerenkov radiation to be emitted, the particle must be electrically charged (e.g. an electron) and must be moving faster than the wave velocity of light in that medium. Furthermore, the medium must be a dielectric (meaning it can be polarized like water). That is:

$$v_p > c/n,$$

where v_p is the velocity of the particle and n is the refractive index of the medium.

The Cerenkov radiation does not emanate from the moving particle in arbitrary directions but instead is radiated at a specific angle from the direction of the moving particle. If we define β as the ratio of the speed of the particle in the medium (v_p) to the speed of light in a vacuum (c), then the Cerenkov radiation is emitted at an angle given by:

$$\cos \theta = 1/n\beta \text{ (\textbf{Cerenkov radiation emission angle}).}$$

The number of photons emitted per unit path length and per unit wavelength through the Cerenkov effect is given by the following formula:

$$\frac{d^2N}{d\lambda \, dx} = \frac{2\pi z^2 a}{\lambda^2}\left[1 - \frac{1}{\beta^2 n^2 \lambda}\right].$$

The energy loss due to emission of Cherenkov photons is small compared to the energy loss induced by inelastic collisions with atomic electrons. Nevertheless, it is possible to quantify beta emitters by measuring the intensity of their Cerenkov radiation

6.9 Interactions between energetic photons and matter

X-rays and gamma rays interact with matter through three broad basic mechanisms:
1. Elastic scattering
2. Inelastic scattering
3. Absorption.

In the first two categories, the incident photon does not vanish but rather is bounced about by the atoms and molecules in the medium. Examples include the Compton effect, Rayleigh scattering and Thompson scattering. In absorption, the photon does vanish as it transfers all its energy into the atoms and nuclei of the medium. Examples of photon absorption include include the photoelectric effect, pair production and photonuclear disintegration.

Compton scattering (also called inelastic or incoherent scattering) involves loosely bound (essentially free) electrons and involves the transfer of a fraction of the incident x-ray energy to the electron. Absorption on the other hand typically occurs when the x-ray photon transfers *all* of its energy to an atom, thereby ionizing the atom (although nuclear interactions can and do occur as well).

6.9.1 Coherent scattering

Coherent or elastic scattering of photons is caused by two basic processes:
1. Rayleigh scattering
2. Thomson scattering.

Very simply, Thompson scattering is due to single atomic electrons, whereas Rayleigh scattering occurs from multiple bound electrons in an atom or molecule that act together in a cooperative manner.

6.9.2 Rayleigh scattering

This type of scattering does not lead to any absorbed dose in the traversed material. However, the individual photons are deflected. Rayleigh scattering thus leads to a broadening of the radiation beam and decreased intensity.

Rayleigh scattering is sometimes called **molecular scattering**. Rayleigh scattering is most effectively achieved by molecules, atoms or particles that have much smaller diameters than the wavelength of the incidence radiation. Mathematically, α must be $\ll 1$, where the parameter alpha is given by $\alpha = 2\pi r/\lambda$ and r is the radius of the scatterer and λ is the wavelength of the radiation.

In daily living, Rayleigh scattering is most commonly illustrated by the atmospheric scattering of sunlight. Rayleigh scattering leads to our familiar blue sky. The atmospheric scattering is due mostly to the two most abundant molecules in air—diatomic nitrogen and diatomic oxygen. The scattering intensity is proportional to λ^4.

Since blue light has a wavelength of $\lambda = 0.46$, while red light has a wavelength of around $\lambda = 0.66$, one can see from $(0.66/0.46)^4 = 4.24$ that blue light scatters sunlight over four-fold more efficiently than red light. This accounts for the blue daytime sky.

Regarding x-rays and gamma rays, since there is no ionization and hence no loss of an electron, there is no dose deposition due to Rayleigh scattering. It just reduces intensity and broadens the penumbra of a beam. Most x-rays and gamma rays are scattered in the forward direction by Rayleigh scattering. Without ejection of an electron, the causative atom as a whole does not experience any significant recoil and thus, most incident photons continue moving in the forward direction with a small angle, theta. There is essentially no energy loss in the incident photon after Rayleigh scattering. Thus, it is a form of elastic scattering.

For x-rays, the probability of Rayleigh scattering (σ_{coh} or σ_R) increases with Z of the medium and decreases with photon energy. In soft tissue, the probability of Rayleigh scattering is low (maybe ~5% of all scattering events) because of the low effective atomic number of soft tissues (Z_{eff} = ~7.5) and the relatively high energy of x-ray photons.

Rayleigh scattering is a form of coherent or elastic scattering that occurs because of the electric polarizability of the scattering particles. The oscillating electric field of an incident light wave acts on the electrical charges within the atoms of the scatterer, causing them to then oscillate at the same frequency. The scattering particle becomes a tiny radiating dipole, which then radiates a photon of the same energy as the one that excited it. In other words, Rayleigh scattering occurs by temporarily raising the energy of the electron without removing it from the atom or molecule. The transiently excited electron returns to its initial energy level by emitting a photon of equal energy but with a slightly different direction.

6.9.3 Thompson scattering

Thompson scattering is another form of coherent or elastic scattering of photons. Thus, as in Rayleigh scattering, the energy of the incident photon does not change but the direction can. Hence, Thompson scattering does not deposit dose. While Rayleigh scattering involves bound electrons, Thompson scattering involves free electrons. In Rayleigh scattering, the incoming photon is scattered by bound atomic electrons, without causing overall atomic excitation or ionization. In essence, Rayleigh scattering is a type of scattering by the atom as a whole while Thompson scattering is scattering is caused by an isolated, single electron.

Thompson scattering can occur when the incident photon energy is $\ll$ electron rest mass (511 keV) and when the involved electron's speed is non-relativistic ($v \ll c$). In Thompson scattering (as in Rayleigh scattering), the wavelength and frequency of the photon is unchanged, that is $\lambda' - \lambda \approx 0$.

More descriptively, Thompson scattering is the non-relativistic limit of Compton scattering of photons by free electrons. This occurs when both the electron and photon momenta have magnitudes that are small compared with $m_e c$ (i.e. $\ll$511 keV).

The cross section for scattering of electromagnetic radiation by an electron is given by the **Thomson cross section**,

$$\sigma_T = 8\pi/3 r_e^2,$$

where $r_e = e^2/4\pi\varepsilon_0 \, m_e c^2 = 2.82 \times 10^{-15}$ m is the classical electron radius.

Hence, the probability of Thompson scattering is proportional to the square of the classical electron radius. Another way of formulating this is in terms of alpha, the fine structure constant:

$$\sigma_T = 8\pi/3(\alpha\hbar c/\mathrm{mc}^2)^2 = 0.665 \text{ barns}.$$

6.10 Compton scattering

Photon scattering by free electrons can be described by Thomson scattering in the low-energy limit and by Compton scattering in the high-energy, relativistic limit. But unlike Rayleigh and Thompson scattering, which are elastic, *Compton scattering is inelastic*. This means that energy is transferred in the process and absorbed dose is deposited. Hence, Compton scattering is also called **inelastic photon scattering** or **incoherent scattering**.

The Compton effect can be considered as a case of conservation of energy and momentum involving three participants:

1. An incoming photon
2. An outgoing (scattered) photon
3. An ejected electron.

The Compton effect considers both relativistic and quantum mechanical aspects of these three participants. The energies, momentum and directions of the scattered products are calculated through conservation of the momentum and energy (which all comes from the incident photon).

Importantly, in the Compton effect only part of the energy of the incident photon is transferred to an electron. Thus, the photon is not entirely absorbed and eliminated. This contrasts Compton scattering from the photoelectric effect, which will be discussed below.

Within the photon energy range of 10–150 keV, the probability of Compton scattering is nearly independent of energy. At higher energies, the Compton cross section decreases approximately as $1/E$.

At lower photon energies, a relatively large fraction of the incident energy is carried away by the Compton scattered photons. As the photon energy increases, however, a greater proportion of the total energy is transferred to the recoil electron. At a photon energy of ~1.5 MeV, the energy is shared equally between the outgoing scattered photon and local energy absorption (that is, radiation dose transferred to the ejected electron).

In sharp distinction from coherent scattering, the frequency (i.e. energy) of the inciting photon is decreased after interaction with an electron during Compton scattering. Mathematically, for conservation of energy and momentum, the incoming photon energy and momentum must be taken into account and must balance the outgoing energy and momentum of the products.

Taking basic principles of special relativity into consideration, the electron has an initial energy of $E_e = m_e c^2$, its rest mass. Its final energy is:

$$E_e = \sqrt{m_e^2 c^4 + p^2 c^2}.$$

Regarding momentum, the electron starts with an initial momentum of zero ($p_{initial} = 0$) and ends with a final momentum of $p_{final} = m_e v$. As discussed below, the electron's mass should take into account relativistic changes. Sometimes these values of electron momentum are written as p_e and p'_e, respectively.

The energy of the deflected photon $h\nu'$ (E' or E_{final}) is reduced from that of the incident photon energy $h\nu$ (E or $E_{initial}$) according to:

$$h\nu' = h\nu/1 + \{(h\nu/m_e c^2)(1 - \cos \phi)\}$$

in which $m_e c^2 = 511$ keV is the rest mass energy of an electron (or positron) and ϕ is the angle of the scattered photon. Note: some texts use ϕ as the angle of scatter for the photon, whereas others use θ. This can be the source of great confusion!

This equation can also be written in terms of wavelengths of the photons:

$$\lambda' - \lambda = \Delta\lambda = \lambda_c(1 - \cos \phi) \text{ (\textbf{Compton effect}),}$$

where λ' and λ are the wavelengths of the deflected and incident photons (sometimes written as λ_{final} and $\lambda_{initial}$, respectively).

In this formula, λ_c is known as the **Compton wavelength**. It is a constant given by

$$\lambda_c = h/m_e c.$$

The Compton wavelength is equal to 0.024 26 Å or 2.43×10^{-12} m.

In the low energy extreme, where the wavelength of the incident photon is much larger than the Compton wavelength (that is, where $\lambda'-\lambda \approx 0$), elastic scattering (i.e.

Scattered photon

$\frac{h\nu}{c}$

$\frac{h\nu}{c} \sin \varphi$

Target electron

$E=h\nu$
$p=\frac{h\nu}{c}$

φ

$\frac{h\nu}{c} \cos \varphi$

Incident photon

$E=h\nu$
$p=\frac{h\nu}{c}$

$E=m_e c^2$
$p=0$

θ

$\frac{h\nu}{c}$

θ

$p \cos \theta$

$p \sin \theta$

Scattered electron

$E=\sqrt{m_e^2 c^4 + p^2 c^2}$
$p=p$

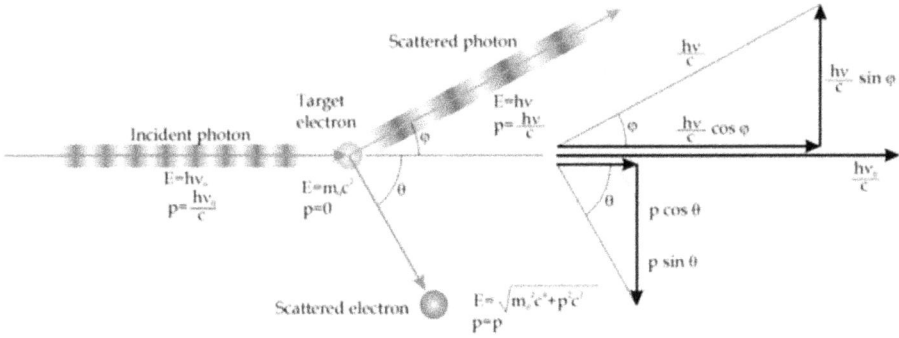

Figure 6.1. In this idealized illustration of the Compton effect, the incident photon approaches from the left and strikes an isolated stationary electron. The electron is then deflected downward at angle θ, while the photon, now reduced in energy and longer in wavelength, is scattered upward at angle ϕ. Through basic principles of quantum mechanics and special relativity, along with conservation of linear momentum and energy, one may obtain the essential equations of the Compton effect. This Compton scattering image has been obtained by the author(s) from the Wikimedia website, where it is stated to have been released into the public domain. It is included within this book on that basis. It is attributed to Astarte.Mauro.

Thomson scattering) is a reasonable approximation. In such cases, the photon wavelength does not change much (nor does the energy).

On the high-energy extreme, it is obvious that the maximum possible change in photon wavelength is two Compton units, i.e. $\Delta\lambda = 2\lambda_c$ when $\cos \theta = -1$. *This corresponds to a photon being backscattered at an angle of $\theta = 180°$.* At this angle, no matter how high the incident photon energy was, *the backscattered photon energy E' can never exceed $m_e c^2/2 = 255.5$ keV* (figure 6.1).

6.11 Derivation of the Compton effect equations

The energies of the incident photon ($E_{initial}$ or just E) and scattered photon (E_{final} or E') are given by the Planck relation:

$$E_{initial} = h\nu,$$

$$E' = h\nu'.$$

Regarding the scattered electron (whose kinetic energy all came from a transfer of energy from the incident photon), the final energy minus the initial energy before the collision, the energy can be calculated by:

$$E_e = E - E' \quad \text{or} \quad E_{initial} - E_{final}.$$

The initial energy and momentum of the electron are taken as zero. Additionally, the electron is considered 'free' or unbound, and therefore, there is no energy expended in overcoming its binding energy (as is needed for analysis of the photoelectric effect). Thus, the final energy value of the electron is equal to the amount of energy lost by the photon and is given by:

$$E_e = \Delta h\nu = h\nu - h\nu' \text{ or } (h\nu_{\text{initial}} - h\nu_{\text{final}}).$$

Moving to momentum, the momentum of the ejected electron is simply:

$$p = m_0 v,$$

where m_e is the mass of the electron and v is its velocity.

Of course, for a more accurate calculation, this should be $\gamma m_e v$, where γ of special relativity is given by:

$$\gamma = \frac{1}{\sqrt{1 - \beta^2}}.$$

(And $\beta = v/c$). From the **de Broglie relation** of elementary quantum mechanics, we know that the momentum of the incoming photon is given by:

$$p = h/\lambda = h\nu/c = E/c \quad \textbf{(photon momentum).}$$

For conservation of linear momentum, the total linear momentum before and after the collision must be balanced. In the direction of the incoming photon (which is taken as the x-direction) this calls for:

$$h\nu/c + 0 = h\nu'/c \cos(\phi) + p \cos(\theta) \quad \textbf{(momentum along } x\textbf{)}, \tag{6.1}$$

where $h\nu/c$ is the momentum of the incident photon, 0 is the initial momentum of the electron, $h\nu'/c \cos(\phi)$ is the momentum in the x-direction of the scattered photon, and $p \cos(\theta)$ is the momentum in the x-direction of the scattered electron.

But in addition to this simple before-and-after consideration in the x-direction, the linear momentum must balance in the y-direction. Using rectilinear analysis, the initial momentum of the incident photon along the y-direction is taken as zero (since the angle of incidence is taken as zero by convention). Thus, the resultant products must also have a net momentum in the y-dimension of zero. Photon momentum is again given by the general equation of:

$$p = h/\lambda = h\nu/c = E/c.$$

The scattered photon is defined here as being deflected at an angle of phi (ϕ); again, it is essential that students take into consideration whether the photon scatter angle is being called ϕ or θ! Hence, the scattered photon carries a linear momentum in the y-direction of:

$$p_y(\text{photon}) = h/\lambda' \sin(\phi) = h\nu'/c \sin(\phi). \tag{6.2}$$

This must exactly balance the momentum of the ejected electron in the y-direction. The electron is scattered at an angle of theta (θ) in this treatment. Thus, the electron momentum in the y-dimension is given by:

$$p_y(\text{electron}) = m_e v \sin(\theta). \tag{6.3}$$

Setting these quantities (equations (6.2) and (6.3)) equal and then multiplying both sides by c yields:

$$pc \sin(\theta) = h\nu' \sin(\phi) \ \textbf{(momentum along } y\textbf{).}$$

Then, returning to the momentum perpendicular to this (that is, along the x-direction) and multiplying equation (6.1) through by c as well, we get:

$$h\nu + 0 = h\nu' \cos(\phi) + pc \cos(\theta),$$

or

$$pc \cos(\theta) = h\nu - h\nu' \cos(\phi).$$

Hence, the two expressions for conservation of momentum are:

$$pc \cos(\theta) = h\nu - h\nu' \cos(\phi) \ \textbf{(momentum along } x\textbf{),}$$

$$pc \sin(\theta) = h\nu' \sin(\phi) \ \textbf{(momentum along } y\textbf{).}$$

Students may remember that $\tan(\theta) = \sin(\theta)/\cos(\theta)$ and the **law of cosines**: $c^2 = a^2 + b^2 - 2ab \cos(C)$. Taking this into account and squaring both sides and recombining allows elimination of θ to yield:

$$p^2 c^2 = (h\nu)^2 - (h\nu)(h\nu')\cos(\phi) + (h\nu')^2.$$

Recalling some basic special relativity physics where $E_{\text{total}} = \text{KE} + m_e c^2$ and $E_{\text{total}} = \sqrt{m_e^2 c^4 + p^2 c^2}$.
We obtain:

$$(\text{KE} + m_e c^2)^2 = m_e^2 c^4 + p^2 c^2,$$

or upon rearranging:

$$p^2 c^2 = KE^2 + 2 m_e c^2 \text{KE}.$$

But recalling that the electron kinetic energy is equal to the difference between the energies of the incoming and outgoing photons (that is, $\text{KE} = h\nu - h\nu'$), we get:

$$p^2 c^2 = (h\nu)^2 - 2(h\nu)(h\nu') + 2 m_e c^2 (h\nu - h\nu')$$

or

$$2 m_e c^2 (h\nu - h\nu') = (h\nu)(h\nu')(1 - \cos \phi).$$

These equations may be less cumbersome in terms of wavelength, λ:

$$\lambda' - \lambda = (h/m_e c)(1 - \cos \phi) \ \textbf{(wavelength based equation for the Compton effect).}$$

The factor $h/m_e c$ is known as **the Compton wavelength or λ_C**. For an electron, $\lambda_C = 2.426 \times 10^{-12}$ m ≈ 2.43 pm. For photon scattering off particles other than electrons, the Compton wavelength λ_C will be less. In terms of λ_C, the equation becomes:

$$\lambda' - \lambda = \lambda_C (1 - \cos \phi) \ \textbf{(Compton effect equation in terms of } \lambda_C\textbf{).}$$

This equation describes the change in wavelength for a photon that is scattered through the angle ϕ by a free electron of rest mass m_e (or more generally, by any particle of rest mass, m_o). This difference in wavelength is known as **the Compton shift**.

Note that this change in photon wavelength is independent of the initial wavelength λ of the incident photon. Also, the greatest possible change in photon wavelength corresponds to $\phi = 180°$ or equivalently when $\cos \phi = -1$. That is, when the photon comes straight backwards. In this situation, the photon wavelength change will be twice the Compton wavelength, $2\lambda_C$. Since $\lambda_C = 2.426$ pm for an electron, the maximum allowable wavelength change in Compton scattering is 4.852 pm. This also corresponds to the maximum allowable transfer of energy to the electron.

Another parameter often used in the Compton effect equations is α, which is defined as:

$$\alpha = h\nu/m_e c^2.$$

This metric is the ratios of the initial energies: $h\nu$ is the initial energy of the incoming photon, and $m_e c^2$ is the initial energy of the electron, that is, its rest mass. Using α, a few additional valuable equations may be written:

$$E = \frac{h\nu\,\alpha(1 - \cos \phi)}{1 + \alpha(1 - \cos \phi)},$$

where E is the energy of the ejected electron;

$$h\nu' = \frac{h\nu}{1 + \alpha(1 - \cos \phi)},$$

where $h\nu'$ is the energy of the deflected photon;

$$\cot(\theta) = (1 + \alpha)\tan(\phi/2),$$

where θ is the angle at which the electron leaves.

From these equations, one can understand and calculate the results of some special circumstances such as when the photon comes directly backwards, when the photon continues moving straight ahead, and when the photon leaves at a 90° angle.

When a photon is directly backscattered ($\phi = 180°$ and $\cos \phi = -1$), regardless of how energetic the initial photon was, the reflected photon energy E' can never exceed $m_e c^2/2$ or 255.5 keV. Again, this scenario represents the maximum transfer of photon energy to the electron. In other words, when the photon comes straight back, E is E_{max}:

$$E_{max} = \frac{h\nu\,\alpha(1 - \cos \phi)}{1 + \alpha(1 - \cos \phi)} = E_{max}\frac{h\nu 2\alpha}{1 + 2\alpha} \quad \left(\textbf{photon scattered 180°}\right).$$

And $h\nu'$ is $h\nu'_{min}$ meaning that the scattered photon has the lowest mathematically allowable value:

$$h\nu' = h\nu'_{min} = \frac{h\nu}{1 + 2\alpha} \quad (\textbf{minimum allowable scattered photon energy}).$$

From $\lambda'-\lambda = \lambda_C$ (1−cos ϕ), one can see that when $\phi = 180°$ and cos $\phi = -1$, the photon wavelength change will be twice the Compton wavelength (i.e. $2\lambda_C$). Since $\lambda_C = 2.426$ pm for an electron, the maximum allowable wavelength change in Compton scattering is 4.852 pm. Again, this is approximately $h\nu/2\alpha$, which is 255.5 keV or 0.255 MeV. In essence, when the photon makes such a 'direct hit' and bounces straight backwards, its energy is virtually always 0.255 MeV, and the electron has gained the maximum allowable kinetic energy. The electron will move straight ahead at $\theta = 0$. The odds of this maximal energy transfer to the electron increase as incident photon energy increases.

On the other extreme, the minimum energy transfer from photon to electron occurs for a 0° photon deflection. In this case, there really is no interaction. Thus, the 'scattered' photon has the same energy as the incident photon as it moves straight ahead, ($\phi = 0$ and cos $\phi = 1$) and $h\nu' = h\nu$ and $E = 0$. The electron moves off at a right angle ($\theta = 90°$) with zero kinetic energy.

$h\nu' = h\nu$ **(photon scattered 0°)**

$E = 0$ **(minimum allowable electron energy)**

$\theta = 90°$ **(maximum allowable electron scattering angle)**

All combinations in between are possible ranging for photon angles of $\phi = 0$ to $\phi = 180$ and electron angles ranging from $\theta = 0$ to $\theta = 90$.

In general, the angle of the outgoing electron (θ in this analysis) is given by:

$$\cot \theta = [1 + h\nu/\{m_e c^2\}] \tan(\phi/2).$$

Another interesting case is presented by a 90° Compton scattered photon, which has an energy given by:

$$h\nu' = h\nu/(1 + \alpha) \text{ (photon scattered 90°).}$$

In this case, the 90° scattered Compton photon energy will always be <511 keV (which equals $m_e c^2$) regardless of how high the initial photon energy was. The ejected electron will have an energy of $E = h\nu - h\nu' = h\nu$ ($\alpha/(1 + \alpha)$), and it travels in a direction that depends on the incident photon energy.

6.12 The Klein–Nishina formula

The probability of a photon interacting with an electron via the Compton effect is given by the Klein–Nishina equation, which provides the differential cross section for the scattering of incident photons off an electron. The cross section for Compton interactions per electron (σ_C) varies tremendously over the relevant energy range, from 66.3×10^{-30} m^2/electron at 1 keV to 0.82 at 100 MeV.

$$\frac{d\sigma}{d\Omega} = \frac{1}{2}r_e^2 \left(\frac{\lambda}{\lambda'}\right)^2 \left[\frac{\lambda}{\lambda'} + \frac{\lambda'}{\lambda} - \sin^2(\phi)\right] \text{ (Klein–Nishina formula).}$$

Where $d\sigma$ is the cross section for scattering of the photon into solid angle $d\Omega$, when it leaves at an angle ϕ. The Klein–Nishina formula can be used to calculate the probability of energy transfer to an ejected electron energy as a function of photon energy. For instance, with 1 MeV photons, electrons may be liberated with kinetic energies ranging from 0 up to 800 keV, but the mean energy of a released electron is 440 keV.

When the incident photon energy is low, relatively little energy is transferred to the medium in the form of energetic electrons; most of the energy remains in the scattered photons. In other words, the exiting photons have about the same energy as the incoming photons. In the limit, as the energy of the incident photons approaches zero, Compton scattering reduces to the coherent (classical) forms of scattering discussed earlier.

Hence, to absorb a radiation beam with low energy, many Compton interactions are needed. But if the incident radiation beam has high energy (e.g. 10–100 MeV), the opposite is seen—most of the photon energy is transferred to recoil electrons, while very little remains in the scattered photons. Of course, this discussion ignores the real situation wherein the photoelectric effect is very much the dominant (ionizing) interaction at low photon energies.

The above analysis of Compton scattering assumes the electron is free (not bound in an atom) and ignores any atomic binding energy. Of course, this is unrealistic since essentially all electrons are bound in tissues and most materials. A more complete and accurate analysis should factor in the binding energies of the electrons involved. But unlike in the photoelectric effect where tightly bound *inner* electrons may be dislodged, in Compton scattering, the electrons involved are typically loosely bound *outer* (valence) electrons. Additionally, although one might conclude that calculations assuming completely free electrons with low x-ray photon energies (e.g. below 10 keV) would be slightly inaccurate, one must keep in mind that at these low radiation energy levels, most of the interactions between photons and matter are through the photoelectric effect and the Compton effect plays a relatively small role.

6.13 Compton effect summary

To summarize, the Compton cross section is nearly independent of Z; it is dependent on **electron density** (which is related to the number of electrons in the outer shell). That is, the probability of the Compton effect reflects the number of electrons per gram in the absorbing material. Electron density may be calculated by: $N_e = N_A Z/A$, where N_e is the electron density and N_A is Avogadro's number. In most cases, this means the probability of Compton scattering is directly related to the physical density of the material. For most elements, this value is approximately constant at roughly 3×10^{23}. The 'exception element' is hydrogen (which has no neutrons in its nucleus) and has an electron density around two times as high as all other elements (approximately 6.02×10^{23}).

The Compton effect is weakly dependent on photon energy; it is relatively constant from 10–600 keV and then slowly decreases with increasing energy. That is, it is proportional to 1/E. The fraction of energy transferred from the photon to the emitted electrons increases with photon energy. It is the dominant radiation interaction in soft tissues for photons between 100 keV and 10 MeV. In human soft tissues, for photons over the energy range between 100 keV and 10 MeV, the Compton effect is the dominant means of photon interactions and dose deposition.

As some broad generalizations, for water (which can serve as a surrogate for soft tissues), up to 50 keV the photoelectric effect is the dominant interaction. Between 60 and 90 keV, the photoelectric effect and Compton scattering are both important. From 200 keV to 2 MeV, the Compton effect is the dominant interaction. From 5 to 10 MeV, pair production and the Compton effect are both important interactions. Above 50 MeV, pair production becomes the dominant photon interaction. These generalizations allude to monoenergetic photons. It is helpful to remember that bremsstrahlung radiation beams are polyenergetic, typically with a mean energy of one-third the accelerating potential. Thus, a 100 keV diagnostic x-ray beam would be the equivalent of a 33 keV monoenergetic photon beam and interactions would largely be photoelectric. This allows maximum contrast between soft tissues and bones. In contrast, for a 10 MVp bremsstrahlung photon beam used for radiotherapy, the monoenergetic equivalent would be around 3.3 MeV and most interactions with soft tissues would be through the Compton effect, with some contribution through pair production.

6.13.1 The photoelectric effect

Another important interaction between radiation and matter is the photoelectric effect. In this mechanism, the incident photon is completely absorbed by the struck atom; it no longer exists after the interaction. This contrasts with photon scattering phenomena such as the Compton effect and coherent scattering (figure 6.2).

The photoelectric effect ionizes an atom after the energy of the incoming photon is fully absorbed. Electrons liberated via the photoelectric effect are called **photoelectrons**. The photon energy goes into liberating the electron as well as providing the kinetic energy of the ejected electron. Thus, the photoelectric effect equation may be represented as:

$$h\nu = W + \text{KE} \text{ (photoelectric effect)},$$

where $h\nu$ is the energy of the initial photon, KE is the energy of the liberated electron, and W is the **work function.**

W is the so-called **work function**, which is the energy required to overcome the binding energy of the involved electron. That is, it is the ionization energy for the electron in question. Unless the incident photon has sufficient energy, it cannot liberate an electron through the photoelectric effect. Thus, there is a minimum threshold energy to set a tightly bound electron free. The work function varies from one element to another and depends on the particular energy level or electron shell that the photoelectron is knocked out of. For example, it takes far more energy to liberate a tightly bound inner K shell electron from uranium than an outer N shell electron from potassium.

From this it follows that the formula for the maximum kinetic energy of the ejected electrons is:

$$\text{KE} = h\nu - W.$$

Since kinetic energy is always a positive value, the requirement is that $h\nu$ must exceed W for the photoelectric effect to occur.

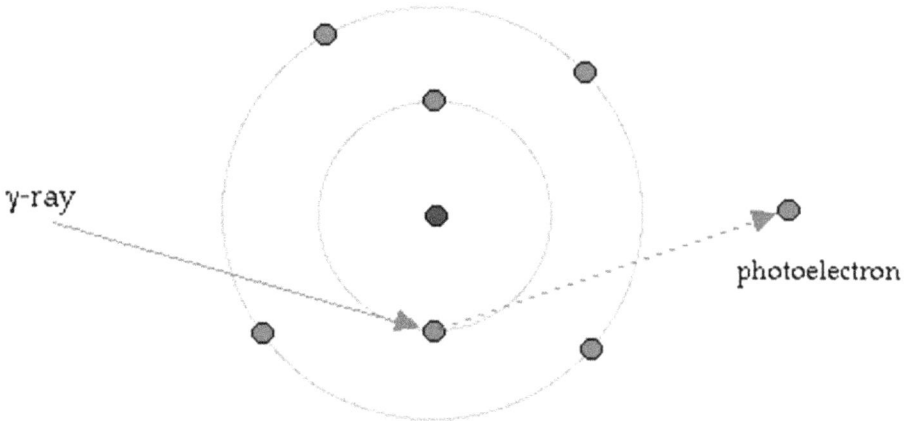

Figure 6.2. A diagram of the photoelectric effect in which an incident x-ray or gamma ray interacts with an inner electron (in the K shell in this case) to eject an electron. The ejected electron is called a photoelectron, and its kinetic energy equals the energy of the incident photon minus the binding energy of the electron to the atom. As with pair production, a threshold energy (given by the Planck relation $E = h\nu$) is needed by the incident photon; this is required to overcome the binding energy and dislodge the electron. Also, as with pair production (but in contrast to the Compton effect), the incident photon vanishes in the photoelectric effect. It was Albert Einstein's analysis of the photoelectric effect that truly ushered in the era of quantum theory and earned him his Nobel Prize. This Illustration of the Photoelectric Effect image has been obtained by the author from the Wikimedia website, where it is stated to have been released into the public domain. It is included within this book on that basis. It is attributed to KieranMaher.

Importantly, it is the photon energy (through the Planck relation $E = h\nu$), that is the key parameter. If the photon energy (that is, the frequency of the electromagnetic radiation) is not sufficient, electrons will not be ejected no matter how intense the radiation or light is. Above the threshold frequency however, the number of electrons ejected (that is, the current created) becomes a function of light intensity.

Furthermore, the kinetic energies of the ejected electrons are a function of the energy (frequency) of the incident light—but not the light intensity. The intensity of the radiation beam dictates the yield of ejected electrons (that is, the current generated or the number of electrons liberated). As mentioned however, the kinetic energy of the released electron is independent of the light intensity; regardless of how intense or how dim the light is, the kinetic energy of the released photons remains unchanged as long as the frequency of the incident light is the same. Another important observation is that, in contrast to expectations from wave theory, there is no delay between irradiation and electron emission, even when the light or x-ray beam is very dim. If the minimum energy threshold is exceeded, electrons are ejected promptly, (albeit at a low current if the x-ray beam is dim). There is no delay.

The photoelectric effect offers strong proof that light behaves like a particle. This was a monumental discovery, because ever since Thomas Young's double slit experiment, light was considered to be a wave, and only a wave. Heinrich Hertz is credited with discovering the photoelectric effect, but it was Albert Einstein who, in 1905, properly analyzed the photoelectric effect and unequivocally explained it in terms of the particle-like behavior of light. To some science historians, Einstein's discovery marks the true dawn of quantum physics. The Planck relation, $E = h\nu$ does come naturally from Max Planck's analysis of blackbody radiation, but it was Einstein's treatment of the photoelectric effect that definitively showed the existence of particle-like light photons or **quanta**. It is these quanta that were capable of ejecting electrons. For this reason, Einstein's work is sometimes alluded to as the **corpuscular theory of light**. Incidentally, it was for his thorough explanation of the photoelectric effect that Albert Einstein was awarded the 1921 Nobel Prize in Physics. He did not receive the Nobel Prize for either his special or general theories of relativity.

The various elements have different thresholds for the photoelectric effect. For some metals with very loosely bound outer electrons, the threshold energy might lie in the ultraviolet or even visible range. For other elements, very-high-energy x-rays might be needed to eject electrons from the tightly bound inner electron shells and orbitals.

6.13.2 Probability of the photoelectric effect as a function of atomic number

In general, the probability of the photoelectric effect *per atom* is proportional to Z raised to a power (n) that typically ranges from 3–5. For high-Z materials, $n \approx 4$, whereas for low-Z materials, $n = 4.8$. But since each atom has Z electrons, the coefficient *per electron* varies as Z^3 for high-Z materials and $Z^{3.8}$ for low-Z materials. However, it should be recalled that the number of electrons per gram of material is largely independent of Z except for hydrogen. Hence, on a *per gram* basis, the mass coefficients for the photoelectric effect goes (roughly) as Z^3 for high-Z materials and $Z^{3.8}$ for low-Z materials.

6.13.3 Probability of the photoelectric effect as a function of photon energy

Additionally, the probability of the photoelectric effect is inversely proportional to the incident photon energy raised to another exponent (provided that the photon has more than the minimum threshold energy required for the work function). For water, the mass coefficient (τ/ρ) is proportional to $1/E^n$, where n is empirically found to be just a bit above 3. The same roughly $1/E^3$ relationship is generally true also for high-Z materials like lead. But at higher photon energies (>1 MeV), the relationship changes such that the mass coefficient is inversely proportional to E to the first power ($\propto 1/E^1$). Thus, at higher photon energies, doubling the energy is likely to reduce the photoelectric attenuation coefficient two-fold.

Hence, for the probability of photoelectric effect interactions, one may write:

$$\sigma_{PE} = \tau = Z^n/E^m,$$

where n is an empirically derived exponent that varies with Z and m is another exponent that varies with energy in a nonlinear fashion.

The photoelectric effect is most likely to occur when the photon energy is *just above the binding energy of the electron*. As mentioned, photon energies below the binding energy threshold are incapable of dislodging the electron. Thus, a plot of photoelectric effect cross section as a function of energy gives a complicated curve with discontinuities and sharp peaks just above the binding energies. These sharp peaks correspond to the binding energies of the various electron shells. Thus, because the binding energy of electrons in the K-shell of lead is 88 keV, photons with less than 88 keV cannot eject a K-shell electron while photons just above 88 keV exhibit a dramatic five-fold increase in absorption. This is the so-called **K-edge** or **K-break**. Maximal photoelectric absorption (as exhibited by these peaks) occurs when the incident photon has just above the K-edge energy. In addition to the K-edge peak, there are three equivalent, but less pronounced absorption peaks corresponding to the L_i, L_{ii} and L_{iii} shells and five peaks associated with the M_i, M_{ii}, M_{iii}, M_{iv} and M_v shells.

6.13.4 Characteristic x-rays, the Auger effect and conversion electrons

High-frequency electromagnetic radiation such as x-rays and gamma rays can, through the photoelectric effect, eject tightly bound inner electrons near the atomic nucleus. When such inner electrons are removed, higher-energy electrons residing in outer electron shells and orbitals quickly drop down to fill the newly formed vacancies. When electrons from higher-energy shells fall into the lower-energy vacancies, energy is released. This energy can be in the form of monoenergetic electromagnetic radiation, in which case the photons are called **characteristic x-rays**. Alternatively, this energy can be released in the form of so-called **'radiation-less' transitions**, in which another ejected electron is released. In such cases, the additional electron is called an **Auger electron**, a **Coster–Kronig electron**, or a **super Coster–Kronig electron**. When electrons are released instead of characteristic x-rays, the phenomenon is called the **Auger effect** in general. The name Auger effect applies regardless of whether the emitted electron in question qualifies as an Auger, Coster–Kronig, or super Coster–Kronig electron (defined below). Just as the monoenergetic characteristic x-ray photons have characteristic unique values, the ejected electrons are also only allowed to have discrete values. This is a basic tenet of quantum mechanics.

Characteristic x-rays and Auger electrons are collectively referred to as **atomic radiations** since they both arise from the atomic electron cloud (rather than the nucleus). Characteristic x-ray emission and Auger effect emission are competing processes. The relative rates of each depend on the atomic number, the electron shells involved, and the electron configuration of the atom that has lost an electron through the photoelectric effect.

Of course, characteristic x-rays and Auger electrons may be emitted not only after an electron is released through the photoelectric effect, but also after radioactive decay involving electron capture. Both processes leave a vacancy in the atom's electron cloud and that newly created vacancy must be filled somehow. Similarly, these two competing atomic radiation processes (characteristic x-rays vs Auger electrons) may also arise when an electron vacancy is created through other radioactive decay modes such as an isomeric transition that has led not to gamma emission but instead to the ejection of a **conversion electron** (that is, **internal conversion** has occurred). Internal conversion leaves a vacancy in a lower-energy electron shell, just as in the case of the photoelectric effect. This vacancy must then be addressed through characteristic x-ray production or Auger electron emission.

6.14 Fluorescence yield

Characteristic x-rays are also called **fluorescence x-rays**. Given the general phenomenon of fluorescence, wherein incident light of higher energy leads to emission of light with lower energy, the name 'fluorescent x-rays' does make sense since the incident x-ray (which was absorbed through the photoelectric effect) has a higher energy than the emitted characteristic x-ray. Hence, the ratio of the probability of characteristic x-ray production to the sum of the probability of characteristic x-ray production plus the probability of Auger electron emission is called the **fluorescence yield**. The fluorescence yield (also called the **yield of characteristic radiation**) is symbolized as **f** or **ω**. The fluorescence yield may simply be considered to be the probability of characteristic photon emission as opposed to Auger electron emission, after the creation of an electron shell vacancy through the photoelectric effect. That is:

$$\omega = p_r/(p_r + p_A) \text{ (\textbf{fluorescence yield}),}$$

where p_r is the probability of emitting x-ray radiation and p_A is the probability of releasing an Auger electron (or Coster–Kronig electron).

The fluorescence yield of a given electron shell (K, L or M) may be defined as the number of fluorescence (characteristic) photons emitted per vacancy in the given shell. It may be abbreviated as ω_K, ω_L and ω_M for K, L and M electronic shells, respectively. These shell specific fluorescence yields (ω_K, ω_L and ω_M) increase as a function of both Z and mass number (A). An empirically derived formula for fluorescence yield is:

$$\omega_i = Z^4/(A_i + Z^4),$$

where A_i is around 10^6 for the K shell and 10^8 for the L shell.

Conversely, the probabilities for electron emission are the complementary values —that is, $1-\omega_K$, $1-\omega_L$, $1-\omega_M$, etc. The quantity $(1-\omega)$ may be called the **electron yield**.

For vacancies created by ejection of an innermost electron (as happens with the photoelectric effect using x-rays), a plot of the electron yield, $1-\omega_K$, against absorber Z reveals a sigmoidal curve. This curve has values of $1-\omega_K$ ranging from 0 for elements with low atomic number ($Z < 10$), a midpoint of $1-\omega_K = 0.5$ for $Z \approx 30$, and up to $1-\omega_K = 0.96$ for elements with very high Z. Thus, for the K-shell at least, Auger electron emission dominates characteristic x-ray emission for absorbers made of low atomic number elements such as human tissues. But for the heaviest elements, characteristic x-ray production dominates.

Consistent with the formula for fluorescence yield $[\omega_i = Z^4/(A_i + Z^4)]$, the relationship for L-shell vacancies is slightly different. For elements with L-shell vacancies, the fluorescence yield is $\omega_L = 0$ at $Z < 30$, then it rises with Z to reach a highest value of $\omega_L = 0.5$ at $Z = 100$. For M-shell vacancies, ω_M is zero for all elements with $Z < 60$, but for $Z > 60$ elements it rises slowly with increasing Z to attain a value of only $\omega_M \approx 0.05$ for very-high-Z elements. This shows that fluorescence emission from the M shell (and higher-level electron shells) is virtually negligible for all elements, even those with very high Z. This is in stark contrast to the case of vacancies in the K shell with low-Z absorbers, where Auger emission dominates.

6.15 Auger electron energy

Although the highest energy Auger electrons originating from K-shell transitions can reach up to 100 keV, the yield of such high energy Auger electrons per decay is low—typically under 10%. The majority of Auger electrons have lower energies (<25 keV and occasionally just a few electron volts). Given this observation, most of the dose deposited by radioactive Auger emitters is over very short (nanometers to micrometers) distances in tissues. This short range is accompanied by high linear energy transfer or LET. For instance, for Auger electron energies <1 keV, the LET is rather high at around 26 keV μm^{-1}. In comparison to beta particles, and even alpha particles (which are generally considered to have short ranges), Auger electrons have a much shorter range in tissue and other materials. This makes radioactive Auger emitters potentially attractive for cancer radiation therapy (that is, if they can somehow be localized within the cell nucleus, and preferably immediately adjacent to the target molecule, DNA).

6.16 Classification of characteristic x-rays

Characteristic x-rays may be classified through a traditional type of x-ray terminology known as **Siegbahn notation**. When an electron falls from the L shell (principal quantum number, $n = 2$) into the K shell ($n = 1$), the resultant characteristic x-ray is called a K-α x-ray. When an electron falls from the M shell ($n = 3$) down to the K shell, the characteristic x-ray X-emitted is called a K-β x-ray. Similarly, if an electron falls from higher energy levels down to a vacancy in the L shell, the characteristic x-rays are called L-α (for a transition from the M to L shell), L-β (for a transition from the N to L shell), etc.

The transition energies between atomic electron energy levels and the resulting characteristic x-rays may be approximated through **Moseley's law**:

$$E_{K-\alpha} = (3/4)R_y(Z - 1)^2 = (10.2 \text{ eV})(Z - 1)^2,$$

where Z is the atomic number and R_y is the Rydberg energy.

This is the formula for the K-α x-rays. It should be mentioned that this approximation works reasonably well for atomic numbers up to $Z = 26$ (iron) but fails for higher Z. Another way of expressing Mosely's Law is:

$$\nu = a(Z - b),$$

where $a = 4.971\,07$, $b = 1$, ν is the frequency for K-α lines, and Z is atomic number.

6.17 Auger, Coster–Kronig, and super Coster–Kronig electrons

As an aside on the Auger effect, there are three variations:
1. The standard Auger effect
2. The Coster–Kronig effect
3. The super Coster–Kronig effect

The subtle distinctions between these three slightly different versions of the Auger effect are as follows:

1. In the standard Auger effect, the primary transition occurs between two different electron shells (that is, between two different principal quantum numbers). For instance, transitions between the K shell and the L shell (that is, between the $n = 1$ and $n = 2$ energy levels in terms of principal quantum numbers). In the standard Auger effect, the transition energy between electron shells is transferred to an orbital electron from the initial shell that had the vacancy or to a higher-level shell. The emitted electron is called an **Auger electron**.

2. In the Coster–Kronig effect, the transition energy originates from two subshells within the same shell. That energy is then transferred to an electron in another (higher-level) shell. For instance, a vacancy in the L shell (i.e. $n = 2$) may be resolved by rearrangements within the L shell, but the ejected electron comes from the M shell ($n = 3$). In such cases, the released electron is called a **Coster–Kronig electron**.

3. In the super Coster–Kronig effect, the transition energy also originates from two subshells within a given electron shell. However, the transition energy is then transferred to a subshell electron within the same shell that the primary transition occurred. For example, a vacancy in the M shell ($n = 3$) may be resolved by rearrangements within the electronic hierarchy within the M shell, and the electron ejected is also from the M shell. When this occurs, the emitted electron is called a **super Coster–Kronig electron**.

The terminology is the same regardless of how the original electron vacancy was created. Thus, an electron emitted after filling a hole in the K shell caused by the photoelectric effect is called an Auger electron just as an electron emitted after filling a hole caused by electron capture is called an Auger electron. Again, creation of electron shell holes or vacancies can come about through a variety of mechanisms including the photoelectric effect, electron capture, internal conversion, the Compton effect and **pair production** and **triplet production** among a few other causes.

6.18 Pair production

In both the photoelectric effect and Compton effect, the energy of the incident electromagnetic radiation is transferred to atomic electrons. In the photoelectric effect, aside from the work function to break the binding energy of the electron, all the incident photon energy is converted into electron kinetic energy. In Compton scattering, a variable fraction of the incident photon energy is converted into electron kinetic energy, with that fraction increasing as incoming photon energy increases. But in pair production, the incident photon energy is converted into the creation of matter and antimatter, as well as the kinetic energy of those produced particles.

Just as with the photoelectric effect, in the pair production mechanism, the incoming photon is eliminated. It vanishes completely as its energy is converted into matter and antimatter through the Einstein relation, $E = mc^2$. Also akin to the photoelectric effect, there is an energy threshold. But unlike the photoelectric effect (where the threshold varies with the electron binding energy of the particular element in question) in pair production the threshold is a constant. It is two times the rest mass of the electron m_e (or equivalently, the sum of the rest mass of an electron and a positron). This amounts to 1.022 MeV/c^2. That is:

$E^3 2m_e c^2 \geq 1.022 \text{MeV}/c^2$ (**threshold for electron–positron pair production**).

It should be stated that while the product of pair production is ordinarily an electron and its antimatter counterpart, a positron, there can in principle be other matter-antimatter pairs created as well if the photon energy is sufficiently high. This can occur if the incident photon has more than the threshold energy required to create pairs of matter-antimatter out of the photon energy, hν. If the inbound photon energy is extremely high, it might be capable of materializing different pairs of matter and antimatter, such as a muon and an antimuon or a proton and an antiproton. In very general terms then, *pair production is the creation of a matter-antimatter pair of particles from the energy of a neutral gauge boson.*

6.19 Pair production probability and photon energy

The probability of pair production (annotated by the pair production cross section symbol σ_κ) increases rapidly with photon energy above the 1.022 MeV threshold limit. Thereafter, it slowly rises with increasing energy, (approximately) as the natural logarithm of the photon energy. That is, $\sigma_\kappa \propto \ln E$.

This leads to the curious observation that a very-high-energy photon beam might be more readily stopped by a barrier than a lower-energy radiation beam. This reduction in penetration as a function of energy is markedly different from the radiation interactions via the photoelectric and Compton processes (since their interaction probabilities decrease with increasing photon energy, meaning the higher the energy, the deeper the penetration). Such is not the case when pair production dominates (figure 6.3).

6.20 Pair production probability and atomic number

The pair production probability coefficient σ_κ approximately varies *per atom* as Z^2 but varies as Z^1 *per electron* in the medium traversed. (And because the number of electrons per gram of most materials is similar, it also varies with Z^1 *per unit mass*). Hence, since the atomic number of lead is about ten times the atomic number of oxygen ($Z = 82$ versus $Z = 8$), the dose deposited in a gram of lead would be ten times the dose deposited in the same mass of oxygen (if it were all due to pair production).

An equation for the overall probability of pair production is:

$$\sigma_\kappa = \alpha r^2 Z^2 f(Z, E),$$

where r is the nuclear radius, α is the fine structure constant, and $f(Z,E)$ is a function of atomic number and photon energy.

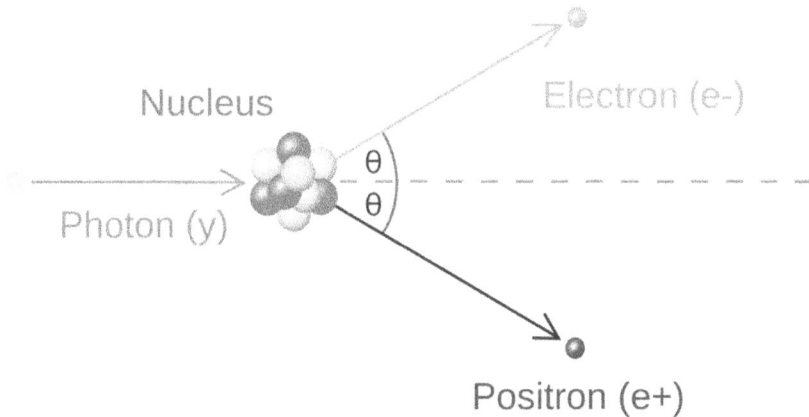

Figure 6.3. An illustration of pair production with the incoming gamma ray approaching from the left and interacting with an atomic nucleus. Via the Einstein relation, $E = mc^2$, photon energy is converted into mass in the form of a matter-antimatter pair (an electron–positron pair in this case). The incident photon must have an energy that exceeds the threshold for electron–positron pair production, namely 1.022 MeV, which is twice the rest mass of an electron (or positron). If the incident photon is sufficiently energetic, matter–antimatter pairs heavier than electrons and positrons can be created such as muon–antimuon or pion–antipion pairs. Although pair production tends to occur in the electric field of the nucleus, it can in principle also occur in the field of an atomic electron. In such cases, the electron is also ejected along with the created electron–positron pair, amounting to three total particles. This is called triplet production, and, as explained in the text, the threshold energy is $4mc^2$ or 2.044 MeV in the case of electrons. This pair production image has been obtained by the author from the Wikimedia website where it was made available under a CC BY-SA 3.0 licence. It is included within this book on that basis. It is attributed to Davidhorman.

6.21 The need for the nucleus in pair production

The inbound photon must be near a nucleus for pair production to occur since linear momentum must be conserved. An electron–positron pair produced in free space cannot satisfy conservation of both energy and momentum. Hence, when pair production occurs, there is a tiny bit of recoil imparted to the involved atomic nucleus. When the photon encounters the strong field of an atomic nucleus, it can disappear as a photon but the energy rematerializes as a pair of particles. And of course, since electric charge must be conserved in pair production, a negatively charged electron must be accompanied by the appearance of a positively charged electron (a positron).

Just as in the Compton effect, momentum must be conserved in both the incident photon's direction (taken as the x-direction) and perpendicular to that direction. But unlike the substantial angles that the scattered photon and ejected electron can assume in Compton scattering, the created electron and positron in pair production are emitted in very nearly exactly opposite directions. That is, they are nearly colinear.

6.22 Energy and momentum in pair production

Ignoring the nucleus for now, the kinetic energies of the newly created pair of electrons and positrons are given by:

$$KE = h\nu - 2m_ec^2.$$

Another way of formulating this is:

$$h\nu = 2m_ec^2 + KE(e^+) + KE(e^-),$$

where $KE(e^+)$ is the kinetic energy of the positron and $KE(e^-)$ is the kinetic energy of the electron.

Generally speaking, this energy is roughly shared equally between the electron and positron although it does not have to be so. The kinetic energy might go all to the electron with zero going to the positron, or all may go to the positron with zero going to the electron, or any distribution in between these extremes. In most cases, the positron has slightly more kinetic energy than the electron because of the electrostatic repulsive 'push' from the positive nucleus near which it forms.

Of course, for a more thorough and accurate calculation, the energy gained by the nucleus must be taken into account. Similarly, when calculating the momentum of the outgoing electron-positron pairs from the incoming photon momentum, the linear momentum transferred to the nucleus can be taken into account. But for most applications, the energy and momentum transferred to the nucleus can be ignored, especially when the produced pair is an electron and a positron. As always, the incoming photon energy is given by:

$$p = E/c = h\nu/c = h/\lambda \quad \textbf{(photon momentum).}$$

As photon energy increases, pair production becomes progressively more important, and it eventually overshadows the Compton effect. By approximately 100 MeV for most materials, pair production becomes the dominant mechanism whereby radiation interacts with matter.

From an overall perspective, as the photon energy increases, the dominant interaction mechanism between radiation and matter shifts from the photoelectric effect to Compton scattering to pair production. Given that they operate on opposite ends of the spectrum, rarely do the photoelectric effect and pair production compete at any given energy in any material. Compton scattering, however, at relatively low energies, does compete with the photoelectric effect. And at relatively high energies it competes with pair production. In lead for example, the dominant radiation interaction at <100 keV is essentially all photoelectric; between 100 keV and 2.5 MeV both photoelectric and Compton processes occur, along with some pair production. Between 2.5 and 100 MeV, Compton scattering and pair production compete as the dominant interaction; and beyond 100 MeV, pair production becomes the main mode of interaction.

6.23 The fate of the positron

Being antimatter, the positron will eventually encounter its ordinary matter counterpart, the electron, and the two will undergo **mutual annihilation**. 'Positronium' is the name of the short-lived electron-positron entity that exists briefly when both are orbiting each other with virtually zero kinetic energy.

Just how far the positron will have to travel before it comes to rest and encounters an electron depends on its kinetic energy. As mentioned, the kinetic energy of the positron comes from the incident photon (minus the energy required to create the matter–antimatter pair of particles) and is shared with the energy of the electron created during the production of the pair. The kinetic energy is not necessarily equally shared and the positron energy can range from zero to all of the kinetic energy. Very-high-energy incident photons can generate some very-high-energy positrons, which could in principle, travel a fair distance before they slow down, come to rest and annihilate an electron. Regardless of how far the positron has travelled, when this encounter does happen, the combined rest energy of the positron-electron pair (1.022 MeV) will be reconverted into electromagnetic energy in the form of gamma rays. Typically, this will be two gamma photons travelling at 180° apart.

However, (as reviewed in chapter 8), thanks to the spin of the photon being 1 as a vector boson, other possible outcomes that preserve angular and linear momentum may occur, such as three photons emerging in the direction of the angles of an equilateral triangle. In ~0.1% of cases, three photons come off rather than two, if the positronium had a value of l (angular momentum) not equal to zero. This can be understood by the fact that electrons (whether positively or negatively charged) are fermions with spin quantum number $J = \frac{1}{2}$ (which means they can assume spin states of $+1/2$ or $-1/2$ in units of $\hbar$). In contrast, photons are gauge bosons with spin quantum number $J = 1$, with potential nominal values of $+1$, 0 or -1. The only requirement, as always, is that the outgoing angular momentum of the photons must match the angular momentum of the initial participants.

6.24 PET scanning

The fact that electron–positron mutual annihilation usually leads to two photons of 511 keV apiece traveling in diametrically opposed directions is the basis for **positron emission tomography** or **PET scanning** in nuclear medicine and diagnostic radiology. PET scans are particularly useful in diagnostic imaging for cancer patients. Thanks to the **Warburg effect** (the fact that cancer cells usually prefer **fermentation** over **aerobic respiration** even when oxygen is available), cancer cells take up glucose far more greedily than most normal cells. One can label the glucose analogue, deoxyglucose, with the positron emitter, fluorine-18, to obtain fluorine-18 deoxyglucose (FDG). Because cancer cells readily take up and trap this agent, FDG selectively accumulates in cancer cells and provides a signal (two gamma ray photons of 511 keV each reaching the detector nearly simultaneously 180° apart) that can be perceived by the proper equipment.

As mentioned, electron-positron mutual annihilation is much more likely when the participants have low kinetic energies. Although in most cases the positron is unlikely to be captured and annihilated by an electron until it has depleted its kinetic energy and virtually comes to rest, it can in principle annihilate while still in motion. In such rare situations, the extra energy is shared between the produced photons.

6.25 Cascade production

At sufficiently high energies (e.g. around or above 100 MeV), an interesting phenomenon called **cascade production** (or **shower production**) can occur. In this situation, a very-high-energy photon may lead to pair production, and the created electrons and positrons can then make new photons through **bremsstrahlung**. If sufficiently energetic, these bremsstrahlung photons may then make new pairs of electrons and positrons, and so forth, until the energy is degraded to the point where new pair production and additional bremsstrahlung is no longer possible. Another type of cascade might be initiated by very high energy gamma photons creating, not electron–positron pairs, but rather, muon–antimuon pairs. This can happen if the photon has an energy of over twice the rest mass of a muon (that is, twice 105.7 MeV or 211.4 MeV). The antimuon might then annihilate with a matter muon and generate photons with sufficiently high energy to then make electron–antielectron pairs. In more general terms, because electrons and positrons (and muons and antimuons) are leptons, these sequences may be called **leptonic electromagnetic cascade showers**. As discussed below in the context of photonuclear reactions, there can be an equivalent cascade involving pairs of pions and antipions (or even protons/antiprotons) rather than electrons/positrons, and that would be the **extranuclear hadronic cascade**.

Because of the possibility of producing bremsstrahlung radiation by the fast moving, newly created electrons and positrons, some of the absorbed radiation energy may be lost (rather than remaining locally). That is, the difference between dose and kerma becomes more pronounced. Thus, when considering *transferred* energy versus *locally absorbed* energy, there may be substantial discrepancies at the very high energies involved in pair production. *Absorbed radiation dose will be less than the transferred energy* since some of that transferred energy eventually escapes as new photons made via bremsstrahlung. The percentage of transferred energy that is then lost as bremsstrahlung radiation is calculated by:

$$100 \times (E_{tr} - E_{ab})/E_{tr},$$

where E_{ab} is the average energy absorbed and E_{tr} is the average energy transferred out.

This fraction is generally only significant with photons above 10 MeV. For a 10 MeV photon beam traversing carbon, the E_{ab} (7.04) is approximately 3% less than the E_{tr} (7.30). But at 100 MeV, the difference is 95.62 versus 71.9, which amounts to almost a 25% difference. In other words, at this high energy, nearly a quarter of the transferred energy is lost as outgoing bremsstrahlung.

6.26 Pair production and electron cloud vacancies

In the section above on the photoelectric effect, we digressed a bit about how the ejected electron leaves a hole or vacancy in the electron cloud into which higher-energy electrons might fall. Upon falling into that electron vacancy hole, they emit energy that could be in the form of photons (characteristic x-rays) or electrons (Auger, Coster–Kronig, or super Coster–Kronig electrons). Holes in the atomic electron cloud can be formed by other processes including electron capture or conversion electron emission (which competes with gamma emission from the nucleus). Electron vacancies or holes may also arise through pair production if the positron annihilates an inner atomic electron. Additionally, an electron cloud hole will appear in **triplet production** (discussed later in this section) as the third electron is emitted along with the electron–positron pair.

6.27 Internal pair production

Another fascinating phenomenon can occur in nuclear physics called **internal pair production**. Just like internal conversion, internal pair production is another variation on gamma decay. In normal gamma decay, a photon is emitted from the excited nucleus. In internal conversion, rather than a photon emanating from the nucleus, an electron is ejected from the atomic cloud. But in the case of internal pair production, instead of just an orbital electron being ejected, an electron–positron pair is formed. This can occur if the gamma photon energy exceeds the minimum threshold energy of 1.022 MeV. And just as in ordinary pair production, the electric field of a nucleus is required. In internal pair production, the required nuclear field happens to be associated with the unstable nucleus that has just decayed.

6.28 Pair-instability supernovae

Pair production is invoked as the mechanism behind so-called **pair-instability supernovae**. In this hypothetical type of stellar explosion, when temperatures are sufficiently high and photon energies exceed the threshold, pair production suddenly starts. And the odds of such pair production rise dramatically as the temperature (energy) further increases. Since electron–positron pairs do not contribute to the externally directed **radiation pressure** that is fighting off gravity, the commencement of pair production drastically and calamitously lowers the pressure inside a supergiant star. The result is a sudden core collapse. This in turn is accompanied by a great increase in compression of the core. This causes a further increase in temperature that culminates in a catastrophic thermonuclear detonation. This causes the entire stellar remnant to explode, leaving nothing behind. While still unproven, pair-instability supernovae should only be able to occur in extremely massive stars with an initial mass ranging between 130 to 250 solar masses, and low to moderate metallicity. Supernova SN 2006gy may have been a pair production type supernova.

6.29 Pair production and Hawking radiation

As will also be discussed in chapter 34, pair production is also often invoked as an explanation of the still-hypothetical **Hawking radiation** associated with evaporating black holes. Basically, this model asserts that virtual quantum pairs (that is, electrons and positrons) materialize in the vacuum of space incessantly, but when this happens near the event horizon of a black hole, one member of the pair (with negative energy) might fall in, while the other member (with positive energy) escapes and becomes 'real'. In this manner, real particles are escaping from isolated black holes. As nothing is supposed to be able to escape from a black hole, this is a real paradox. But if it does happen, it means that black holes are emitting particles—and losing mass (or slowly 'evaporating'). And a black hole that is losing mass is a shrinking black hole. As the black hole continues to shrink, its rate of mass loss progressively increases. Eventually, the miniature black hole will explode in a blast of gamma rays. The mechanism is a bit more convoluted than this simple description and is covered a bit further in a separate chapter. (Note—these unobserved and still hypothetical exploding black holes are different from the regularly observed **gamma ray bursts**, which are also covered in a separate chapter.)

6.30 Alternative pairs

Although we have focused on electron-positron pair production, countless other particle pairs can be produced through the $E = mc^2$ Einstein mass–energy relation, if the threshold energy is exceeded. For instance, if the photon energy is greater than the sum of the rest masses of a muon-antimuon pair, these leptons may materialize. Muons have a rest mass of 105.66 MeV, so the threshold energy for muon-antimuon pair production would be around twice that, or approximately 212 MeV. Note that muons, as the second lightest leptons are approximately 206.77 times the mass of the electron; thus, the threshold energy for their pair production is commensurately increased (that is, about 207 times higher than the threshold for electron-positron pair production). And if the photon energy is greater still, pion-antipion pairs might form. Pions are the lightest mesons and are also the lightest hadrons (these terms are reviewed in more detail in chapter 8 for those unfamiliar with these subatomic particles). The rest mass of a charged pion is 139.57 MeV, so the threshold for charged pion pair formation is twice that value, or about 280 MeV. Incidentally, besides the positively and negatively charged pions, there is a third, uncharged, form of pion. This neutral pion has a slightly lower rest mass of 134.98 MeV. When the incident photon energy exceeds the threshold of twice that rest mass, or about 270 MeV, two neutral pions can appear (since the neutral pion is its own antiparticle). This is an example of pair production in which the two products are uncharged (although they are still antiparticles to each other).

The same principles apply for proton-antiproton pair production; in this case, the threshold is twice the rest mass of a single proton (938 MeV × 2, or about 1.879 GeV). Neutron–antineutron pair production occurs at a slightly higher energy threshold.

In particle physics, many massive particles decay into particle–antiparticle pairs. For example, the most common decay mechanism for the Higgs boson is into a bottom quark and an anti-bottom quark. The concept is that when sufficient energy is available, that energy will yield mass via the Einstein relation. And the resulting mass must be in the form of a matter–antimatter pair of particles.

6.31 Triplet production

As mentioned, pair production must occur in an electric field, and such fields are found in proximity to an atomic nucleus. And for conservation of linear and angular momentum, besides the incoming photon and the outgoing electron–positron pair, a third body is required in the pair-production process. If that body is a heavy nucleus, there is negligible recoil energy. Therefore, the threshold is just twice the rest energy of the electron ($E \geqslant 2m_ec^2$).

Since even a single proton is about 1836 times the rest mass of an electron (938 versus 0.511 MeV), one can neglect the recoil momentum and energy imparted to the relatively heavy atomic nucleus in pair production. However, in rare instances pair production may occur in the field of an atomic electron. In this case, thanks to its small mass, the involved electron will experience considerable recoil and energy. And this must be accounted for in all calculations. In fact, the imparted energy and momentum will dislodge the electron from the atom. Thus, observers will see *three* particles produced: the electron and positron pair along with the displaced electron. Hence, the phenomenon is called **triplet production**.

The energy threshold for the triplet production process is twice as high as for pair production or four times the rest energy of an electron. That is:

$$E \geqslant 4m_ec^2. \ \textbf{(Threshold energy for triplet production)}$$

This can be understood when one examines the basic aspects of the overall reaction. The overall energy equation for the pair production process is:

$$h\nu_{(\text{init})} = mc^2_{(\text{electron})} + mc^2_{(\text{positron})} + KE_{(\text{electron})} + KE_{(\text{positron})} + KE_{(\text{nucleus})}.$$

As discussed previously, the kinetic energy given to the atomic nucleus is negligible compared to the rest energies of the electron and positron and the kinetic energies they are given. The same holds for the momentum transfer to the nucleus versus the electron–positron pair. Thus, nuclear recoil is ignored when doing calculations for pair production.

In contrast, when pair production occurs in the electric field of an atomic orbital electron, considerable recoil energy is imparted to that electron. This recoil energy causes the ejection of the orbital electron (in addition to the appearance of the electron–positron pair). In this case, three particles are detected. The overall threshold for pair/triplet production may be written:

$$h\nu \geqslant 2m_ec^2[1 + (2m_ec^2)/(2Mc^2)],$$

where M is the mass of the third particle involved (usually the nucleus but occasionally an electron).

One can see that if M is the mass of a nucleus (M_{nucleus}) then $M \gg m_e$ and $(2m_ec^2)/(2Mc^2)$ is negligible. In this case, pair production ensues, and the equation reduces to:

$$h\nu \geqslant 2m_ec^2.$$

But if M is the mass of an electron, then $M = m_e$ and $(2m_ec^2)/(2Mc^2) = 1$. The equation becomes:

$$h\nu \geqslant 2m_ec^2[1 + 1] = 4m_ec^2.$$

Thus, when triplet production ensues, one can see from the equations that the threshold for triplet production is $4m_ec^2$.

Although the average kinetic energy of the three particles would be predicted to be around one third of the total available energy, measurements indicate that the positron tends to be slightly more energetic than predicted. This is attributed to the extra little repulsive 'kick' it gets from the positively charged protons in the nucleus.

The total pair-production cross section is the sum of the two components: nuclear (regular pair production) and electronic (triplet production). As mentioned above, these cross sections depend on the energy of the incident photon as well as the atomic number of the absorber.

6.32 Photonuclear disintegration

As discussed above, x-ray and gamma photons usually interact with atomic electrons, or they are affected by the nuclear field without penetrating it. However, in some situations, highly energetic photons can indeed penetrate the nucleus and lead to its disintegration. The most familiar example involves neutron emission, which can be represented as $^{x}A(\gamma,n)^{x-1}A$. This is called the 'gamma-n reaction'. But more broadly, through **photonuclear reactions** high energy photons may provoke the emission of neutrons (γ,n), protons (γ,p), deuterons (γ,d), tritons (γ,t), α particles (γ,α) or heavier nuclear fragments. These products of photonuclear interactions can be categorized as **photon-induced nucleon emissions** on one extreme and **photofission** (γ,f) on the other, with a host of spallation options in between. In many cases, especially after neutron emission, the product nucleus is radioactive. This formation of radioactive nuclides via high-energy photon bombardment is called **photoactivation**.

It should be mentioned that the cobalt-60 and carbon-137 radioisotopes used in food irradiation to sterilize the food do not generate gamma rays with sufficient energy to induce photonuclear reactions. Thus, food irradiation does not make food radioactive through photoactivation. On the other hand, patients undergoing radiation therapy with high-energy photons from a linac may become transiently radioactive thanks to rare photonuclear reactions. This is of no clinical importance.

The electromagnetic radiation energy threshold for photonuclear reactions (such as production of neutrons) varies with the target element. For example, the threshold for iron-53 is 14.2 MeV, while the threshold for carbon-12 is 18.7 MeV. In general, the threshold is lower for heavy nuclei than for light nuclei. Although the threshold energies for photonuclear interactions are generally high (with most elements requiring photon energies in the range of 6–19 MeV) two neutron-releasing, (γ,n) reactions occur at relatively low photon energy. These are:

1. $^{2}H(\gamma,n)^{1}H$ (threshold = 2.226 MeV)
2. $^{9}Be(\gamma,n)^{8}Be$ (1.666 MeV)

This is because *deuterium and beryllium-9 have the lowest neutron separation energies* among all known nuclides. This allows the manufacture of portable neutron sources that capitalize on these γ,n reactions. For example, in a **SbBe neutron source**, a combination of antimony-124 and beryllium-9 serves as a relatively high-yield photoneutron source. The energic gamma rays from antimony-124 strike the adjacent beryllium target, resulting in a (γ,n) reaction. This is possible because the

1.69 MeV gamma photons from Sb-124 (at a respectable 50% abundance) exceed the 1.66 MeV threshold for photoneutron production in beryllium-9.

A few other examples of photonuclear reactions include:

$$^{10}B(\gamma, d)^{8}Be$$
$$^{11}B(\gamma, t)^{8}Be$$
$$^{12}C(\gamma, \alpha)^{8}Be$$

Note in these examples how the same product, beryllium-8, can be generated through a variety of different nuclear reactions. Also, note that beryllium-8 does not really exist. It is extremely unstable and decays in under a femtosecond. Because of this extreme instability, beryllium-8 is considered an 'unbound resonance' rather than a radionuclide.

Another set of photonuclear disintegration reactions is illustrated by **photofission**. These are abbreviated as (γ, f) reactions. **Fissile radionuclides** such as uranium-235 are inherently unstable, so it is not surprising that high-energy photon irradiation can increase that instability to the point of fission. But other heavy radionuclides can also be induced to undergo photofission if the incident radiation exceeds the threshold. For example, the (γ, n) photofission thresholds (in mega electron volts) of a few representative actinides are as follows:

$$^{230}Th\ (5.40\ MeV)$$
$$^{233}U\quad(5.18)$$
$$^{235}U\quad(5.31)$$
$$^{238}U\quad(5.08)$$
$$^{239}Pu\ (5.31)$$

Note that among those heavy actinide radioisotopes that are susceptible to this type of nuclear interaction, the photofission threshold energies do not vary much from nuclide to nuclide.

6.33 The giant dipole resonance

The probability cross sections for photonuclear reactions display broad peaks called **giant dipole resonances** or, more simply, **giant resonances**. The giant resonance corresponds to the most probable energy for a photonuclear interaction. For lead, it occurs at around 15 MeV; for oxygen, it lies at 25 MeV. The giant resonance energy is proportional to $r^{-1/2}$ and $A^{-1/6}$, where r is the nuclear radius and A is the mass number. In general, the giant resonance maximum occurs at energies between 12–25 MeV, and the span of the giant resonance varies from 4–10 MeV wide.

In some ways, the process of photonuclear disintegration is the opposite of the gamma decay process. In gamma decay, an excited nucleus relaxes through the emission of a gamma photon; in photonuclear disintegration, a gamma photon causes excitation of a nucleus. Photonuclear disintegration may also be considered the inverse of certain **radiative capture reactions** as discussed next.

6.34 Radiative capture

Some nuclei can capture neutrons, alpha particles or protons; become excited; and then emit **prompt gamma** photons. This type of capture reaction is called **radiative capture**, and the photons produced are called **capture gamma rays**. This concept is reviewed further in the section on neutron interactions, but as a quick example, a lithium-7 nucleus can capture an alpha particle and emit a capture gamma ray in an (α,γ) reaction:

$$^{7}\text{Li} + {}^{4}\text{He} \rightarrow {}^{11}\text{B} + \gamma.$$

Or using the nuclear shorthand notation:

$$^{7}\text{Li}(\alpha,\gamma)^{11}\text{B}.$$

Another example of radiative capture, this time a **thermal neutron capture reaction** involving the most common isotope of gold, is:

$$^{197}\text{Au} + \text{n} \rightarrow {}^{198}\text{Au} + \gamma$$

or

$$^{197}\text{Au}(\text{n},\gamma)^{198}\text{Au}.$$

The created gold-198 is a beta emitter with a half-life of 2.7 d and a maximum beta particle energy of 1.37 MeV. This isotope of gold has been used in cancer brachytherapy.

6.35 Gadolinium neutron capture therapy

Another radiative neutron capture reaction is the basis for a novel cancer therapy called **gadolinium neutron capture therapy**. The nuclear reaction is as follows:

$$^{157}\text{Gd} + \text{n} \rightarrow {}^{158}\text{Gd} + \gamma$$

or

$$^{157}\text{Gd}(\text{n},\gamma)^{158}\text{Gd}.$$

In this case the capture gamma photons lead to Auger and Coster–Kronig electrons that have high LET (linear energy transfer) and high RBE (relative biological effectiveness) against cancer cells. The very short ranges of these electrons make this an attractive oncological option, and research is going on. But returning to the basic mechanisms of photonuclear disintegration, one can see that these radiative capture reactions (that release gamma rays promptly upon capture of neutrons or other particles) are the inverse of photonuclear reactions (in which gamma rays release neutrons or other particles).

Above energies of 25 MeV, the cross section for photonuclear reactions may be modeled as a photon being absorbed by a **proton–neutron pair** within the nucleus. As a proton–neutron pair is a deuteron, this is called the Levinger **quasi-deuteron** or **qd model**. This theoretical model may be useful in estimating neutron (and proton) ejection from nuclei upon absorption of high-energy photons via photonuclear interactions. And at photon energies more than 280 MeV, the overall cross section for photoneutron production again increases because the threshold for **photo-pion production** is exceeded. (Pion production is a form of pair production just like electron–positron pair production is). This introduces another potential source of neutrons through the so-called **extranuclear hadronic cascade shower** (the hadronic equivalent of the **leptonic electromagnetic cascade** described above with regards to pair production). In this cascade however, high-energy photons may create pion pairs rather than electron–positron pairs. These matter–antimatter pion pairs then self-annihilate to create new photons that can then undergo additional photonuclear reactions. Hence, through this cascade, an individual photon gets a 'second chance' at generating neutrons (which are of great radiobiological importance).

6.36 Attenuation (absorption) coefficients

As mentioned earlier, HVLs are used to quantitatively describe polyenergetic beams. But for genuinely monoenergetic radiation beams, the linear and mass **attenuation coefficients** (μ) may be useful. Attenuation coefficients quantify the degree of attenuation of a radiation beam through a given material as a function of penetrating distance through that material. Attenuation coefficients reflect the amount of absorbed dose deposited by a radiation beam and therefore are also called absorption coefficients. But the absorbed dose is not entirely synonymous with the stoppage of photons, nor is it always equal to the amount of energy transferred to the medium by the radiation beam. The concepts of kerma and terma more accurately quantify these differences.

But for monoenergetic beams, the HVL can be related to the **linear attenuation coefficient** (μ) through the following formula:

$$\mathrm{HVL} = \ln 2/\mu = 0.693/\mu.$$

The linear attenuation coefficient is a constant that describes the fractional attenuation of incident radiation (in a monoenergetic beam) per unit thickness of a given material. It is expressed numerically in units of reciprocal centimeters (cm^{-1}). It includes all the potential interactions between radiation and matter including coherent scattering, Compton scattering, the photoelectric effect, pair production and photonuclear reactions. The complement of radiation attenuation is the **transmission**, which is just the transmitted fraction of the beam.

The linear attenuation coefficient increases with atomic number and physical density of the absorbing material. Attenuation coefficients generally decrease with increasing photon energy (except right at photoelectric K-edges and at the extremely high energies where pair production and photonuclear effects dominate). It may be worth mentioning here that although most radiation science texts focus on the moderate energies at which the photoelectric and Compton effects dominate, in space radiation, the extremely high energies at which pair production and photo-nuclear interactions dominate are often encountered.

The attenuation coefficient allows mathematical treatment of the probability that a photon will interact and be absorbed or attenuated in a material of thickness dx:

$$dN/N = \mu \, dx.$$

This may also be rearranged to expressed as the loss of photons per unit thickness:

$$dN/dx = -\mu N.$$

This, of course, is the equivalent of the number of photons lost per unit thickness:

$$dN = -\mu N \, dx.$$

One will immediately recognize the similarities to the equations of radioactive decay. Indeed, one may solve these differential equations and write formulas that are very familiar such as:

$$N = N_o e^{-\mu x},$$

where N_o is the number of photons entering an absorber and N is the number of photons remaining after traversing a thickness x.

While these equations that quantify numbers of individual particles work well for particulate radiation such as neutron beams, for photons, **intensity** is typically used instead. Intensity may be considered the *energy* per unit time crossing a unit area that is perpendicular to the beam. Intensity can be calculated by multiplying the energy per photon ($h\nu$) by the number of photons (n) crossing the unit area per unit time. The planar **photon flux** (ϕ) is defined as the number of photons crossing the unit area that is perpendicular to the beam per unit time. (Note: The term flux refers to a *rate* whereas **fluence** (ψ) is the time integral of a flux. Hence, photon fluence is the total number of photons crossing an area as opposed to flux, which is photons per second per unit area.) Thus, fluence (ψ) is dn/da, whereas flux (ϕ) is $d\psi/dt$. Students are cautioned about the confusing interchange of symbols (ϕ versus ψ) for these two entities; one must be careful to know exactly which concept is being discussed rather than relying on the symbols ϕ or ψ.

Based on these principles, one may write:

$$\phi = \phi_o e^{-\phi x} \quad (\textbf{photon flux } (\phi)),$$

where ϕ_o is the initial photon flux and ϕ is the photon flux at depth x in the medium.

$$I = h\nu\phi \quad (\textbf{photon beam intensity}),$$

where is the energy of the individual photons in the beam.

Thus, the intensity of a beam at a distance x within an attenuating material may be calculated through the following equation:

$$I_x = I_o e^{-\mu x},$$

where I_x is the intensity at depth of x cm, I_o is the original intensity, and μ is the linear attenuation coefficient. One may rearrange and take the logarithms of both sides to obtain an equation for μ:

$$\mu = \ln (I_o/I_x)/x \quad (\textbf{linear attenuation coefficient}).$$

The distance x is variably given in units of length (e.g. centimeters), length times density (g · cm^{-2}), length times the number of atoms per cubic centimeter (atoms · cm^{-2}) or length times the number of electrons per cubic centimeter (electrons · cm^{-2}). These four attenuation coefficients are respectively called the **linear, mass, atomic and electronic attenuation coefficients**. The linear and mass absorption coefficients are the ones most commonly used.

The **mass attenuation coefficient** is a normalization of the linear attenuation coefficient per unit density of a given material. This yields a value that is constant for a given element or compound. This means that, unlike the linear attenuation coefficient, the mass attenuation coefficient is independent of the density of the material. The linear attenuation coefficient is expressed in units of reciprocal centimeters (cm^{-1}), while the mass attenuation coefficient is expressed in square centimeter per gram (cm^2/g).

Attenuation coefficients may be related to the **mean free path** (ℓ) of a pencil beam of monoenergetic photons. As with attenuation coefficients, the mean free path depends on the material and the energy of the photons. It is given by:

$$\ell = \mu^{-1}.$$

Attenuation coefficients are very important for calculations in radiation shielding. The overall absorption of radiation depends on the energy of the radiation as well as the chemical composition of the absorbing material. And the overall attenuation coefficient is a composite of the individual attenuation coefficients. Thus, μ (the overall attenuation coefficient) is the sum of the individual coefficients: σ_{coh} (for coherent scattering), τ (the coefficient for the photoelectric effect), σ_C (Compton effect), π (the coefficient for pair production) and η (the coefficient for photonuclear events):

$$\mu = \sigma_{coh} + \tau + \sigma_C + \pi + \eta \quad \textbf{(overall attenuation coefficient)}.$$

In the way of example, for monochromatic high-energy photons traversing biological tissues with an effective atomic number (Z_{eff}) of 7, the absorption of photons is dominated by the photoelectric effect below 30 keV; the Compton effect dominates above 30 keV and up to about 24 MeV; and above 24 MeV, pair production becomes the dominant interaction. For other materials with different Z, the energy levels at which the various mechanisms dominates will vary (figure 6.4).

Again, attenuation coefficients are truly valid only when the radiation is monoenergetic. In real circumstances with polyenergetic beams, the calculations are more complicated, and the attenuation coefficient is not a constant but is a function of energy. As such, it changes as a function of depth in a material because of beam hardening. For simplicity, the highest energy of a polychromatic beam is often used as a surrogate for calculations extrapolating to a monoenergetic beam. But in reality, there may be very little radiation with this maximum nominal value in a bremsstrahlung beam.

6.36.1 Electron interactions with matter

For most purposes, 'radiation' means energetic photons—x-rays and gamma rays. However, in space radiation, particulate radiation plays an important role. And aside from neutrinos (which for all practical purposes do not play much of a role in radiation science unless you happen to be too close to a supernova!), electrons are

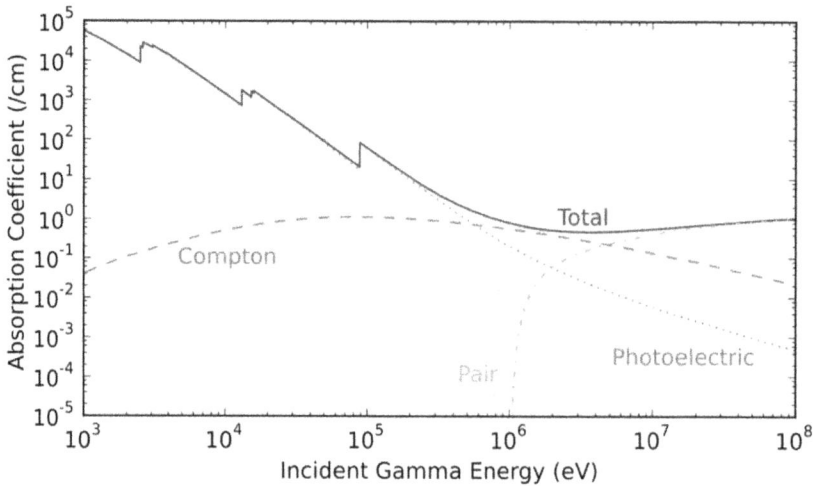

Figure 6.4. A plot of the absorption coefficient of energetic photons in lead as a function of energy in electron volts. At lower energies (up to around 100 keV), the photoelectric effect (green dotted line) dominates. At intermediate energies (around 100 keV–100 MeV), the Compton effect (red dashed line) dominates. At very high photon energies (e.g. >100 MeV), pair production (blue dashes) becomes the dominant interaction. Note that pair production has a threshold energy of 1.022 MeV, below which it does not occur. The solid blue line represents the composite or total absorption coefficient. Note the sharp discontinuities in the lower-energy region that represent photoelectric effect K-edges and L-edges. This lead-gamma-xs image has been obtained by the author from the Wikimedia website where it was made available under a CC BY-SA 3.0 licence. It is included within this book on that basis. It is attributed to Joshua Hykes.

the lightest form of particulate radiation. Additionally, they are electrically charged, which dominates their interactions with matter. The general way in which all electrically charged particles interact with matter is similar but because electrons are so small and light, they are far more susceptible to scattering. Because of the multiple scattering events that any individual electron is likely to experience, an electron radiation beam does not exhibit the **Bragg curve** that is characteristic of heavier charged particles. Thus, electrons are typically discussed separately from other heavier leptons such as muons and are also discussed separately from charged hadrons such as pions, protons, alpha particles heavier nuclei, and galactic cosmic rays. Electrons predominantly interact with atoms in matter through electrostatic or Coulomb collisions. Such collisions may be elastic or inelastic. Elastic collisions cause the electron to change direction without losing energy. Inelastic collisions can also cause the electron to change direction but the electrons do lose energy. Some inelastic interactions also lead to the production of electromagnetic secondary radiation such as bremsstrahlung and synchrotron radiation.

The rate of energy loss of an electron beam increases with the energy of the incident electron and the atomic number of the medium. And on a *per energy basis*, electrons penetrate deeper in tissue and other materials than protons or alpha particles. This may seem strange since in clinical radiation oncology, electron beam radiotherapy is used to treat superficial conditions. In contrast, proton beam therapy is used to treat either superficial or deep tumors. But this is because a clinical proton beam may be produced by an isochronous cyclotron, a synchrocyclotron or a synchrotron and have kinetic energies ranging up to 250 MeV; electron beam radiotherapy on the other hand is typically delivered by a linac with energies usually ranging from 4–20 MeV.

Charged particles of the same energy and charge but of different masses will penetrate different distances into matter. As a generalization, the lighter the particle, the deeper the penetration. Hence, alpha particles with about 4 MeV will penetrate less deeply into tissue than 4 MeV protons, and these will penetrate less deeply than 4 MeV electrons. Students must avoid confusion about heavy charged particles in the form of galactic cosmic rays. These are exceptionally energetic and have tremendous penetration power despite their high mass. But just as clinical proton beams penetrate deeper than clinical electron beams, the deep penetration of galactic cosmic rays is only because of the vast differences in their kinetic energies. The remarkable penetration of cosmic rays is simply due to their extraordinary energies. Some primary cosmic rays have energies in the tera electron volt, and even peta or exa electron volt, realm.

6.37 Coulombic scattering of electrons

Thanks to their extremely light weight, electrons are likely to experience **multiple coulombic (electrostatic) elastic scattering events** upon encounters with the electromagnetic fields of orbital atomic electrons and nuclei. This type of scattering event does not cause excitation or ionization of the medium being traversed, nor do the electrons lose energy. But it does cause the electrons to change direction. And given the light weight of electrons, this directional change can be quite significant. Hence, electron beams tend to have rather broad **penumbras** (meaning the beam has 'fuzzy' edges and broadens with depth of penetration rather than remaining tightly collimated).

6.38 Elastic and inelastic scattering with nuclei and atomic electrons

As with other charged particles, electron scattering encounters may be classified as either **elastic** (resulting in no loss of electron kinetic energy) or **inelastic** where there is a transfer of energy from the incident electron to the medium. Another way of thinking about electron interactions is based on what they interact with. Incident electrons may interact with the nucleus or with orbital electrons of an atom. In this context, there are four possible scenarios:

1. Elastic interactions with the orbital electrons of an atom
2. Elastic collisions with the nucleus
3. Inelastic interactions with orbital electrons
4. Inelastic collisions with the nucleus.

Multiple coulombic elastic scattering (whether from other electrons or from nuclei) does not cause excitation or ionization of the medium, and the scattered electrons do not lose much energy. Elastic electron interactions are therefore akin to Thompson and Rayleigh scattering of photons in that they do not result in transfer of energy or deposition of dose, but they do lead to attenuation of an electron beam because the primary electrons are deflected and no longer continue moving in the forward direction. Additionally, this scattering changes the directions of individual electrons and thereby widens the beam with depth. Hence, electron beams tend to have rather broad penumbras, meaning the beam edges grow 'fuzzier' with depth of penetration. This contrasts with a tightly collimated carbon ion beam made of much heavier charged particles.

Electrons also experience inelastic encounters, which do cause atomic excitation and ionization as well as result in loss of electron energy in the beam. Inelastic interactions with orbital electrons result in transfer of energy from the incident electron to an orbital electron. Even if the inelastic collision does not result in ionization, the orbital electron may be raised to a higher energy level. When this energy transfer causes the electron to jump to a higher shell it is called **excitation**. In this case, the vacated orbital in the lower-energy shell is rapidly filled by another orbital electron from a higher energy level. This may thereby generate characteristic x-rays (if the energy difference between the shells is sufficiently large, as is often seen with high-Z elements such as lead or tungsten) or visible light when the energy difference is lower.

If the transferred energy is sufficiently high, it can lead to ejection of an orbital electron, i.e., **ionization**. The ionized atom may be highly reactive and can oxidize adjacent atoms and molecules to regain stability. During ionizing encounters, if the ejected electron is sufficiently energetic, it may go on to produce another ionization track of its own. In such cases, the ejected electron is called a **delta ray** or simply a **secondary electron**.

6.39 Interactions between electrons and nuclei

When elastic electron scattering occurs with a nucleus, media with higher atomic numbers result in more scattering. The amount of scattering that occurs is dependent on Z^2 of the nucleus. Also, electrons can interact inelastically with the essentially immovable electromagnetic field of the nucleus and suffer dramatic directional changes or decelerations, including coming to a sudden stop. Because any accelerating electric charge emits electromagnetic radiation (and since changing directions or coming to a sudden stop is a form of acceleration), inelastic collisions between electrons and nuclei result in the generation of radiation called **bremsstrahlung** (German for 'braking radiation'). The energy that is lost by the negatively accelerated electron is released in the form of bremsstrahlung photons. When this occurs, a good deal of the electron's energy may be converted into photons. The rate of energy loss of an electron beam increases with the energy of the incident electron and the atomic number of the medium. Hence, tungsten targets ($Z = 74$) are often used for generating bremsstrahlung radiation in diagnostic medical x-ray units and radiotherapy linear accelerators.

The conversion of electron kinetic energy into electromagnetic radiation may be quantified through the **Law of Duane and Hunt**, which relates the maximum energy (or maximum frequency) of the manufactured bremsstrahlung radiation to the accelerating potential of an x-ray tube. The maximum potential of an x-ray tube in turn represents the maximum energy of electrons in the contained beam. Mathematically, the Duane–Hunt law may be written:

$$\nu_{max} = eV/h$$

or

$$\lambda_{min} = hc/eV,$$

where e is the charge of the electron and V is the electric potential (accelerating voltage).

Using familiar units of kilovolts and picometers, the Duane–Hunt law may be expressed as:

$$\lambda_{min} = 1240\, pm/V (\text{in kV}).$$

The law of Duane and Hunt may be rearranged into:

$$h\nu_{max} = eV = E_{max}.$$

This simply says that the maximum energy of the bremsstrahlung x-rays (E_{max}) equals the maximum kinetic energy of the incident electrons.

While bremsstrahlung can be made by any charged particle that is accelerated, electrons produce far more bremsstrahlung than protons do. The intensity of bremsstrahlung is proportional to Z^2 of the target material (hence, tungsten is a favorite target). But additionally, it is also proportional to z, the charge on the incident particle. Therefore, one might naively imagine that heavier nuclei, which have higher electrical charges, would be more effective at generating bremsstrahlung. But another factor is that bremsstrahlung intensity is inversely promotional to particle mass. Hence, the formula includes a charge-to-mass ratio: a (z/m) factor. Since protons are 1836 times more massive than electrons, the bremsstrahlung intensity is correspondingly higher for electrons. And for all (stable) nuclei other than hydrogen (a bare proton with $Z = 1$), the addition of neutrons makes the charge-to-mass ratio (z/m) even worse than it is for hydrogen. Hence, decelerating protons make more bremsstrahlung than decelerating alpha particles or carbon nuclei do, but electrons make far more bremsstrahlung than any of these heavier particles do.

It is worth mentioning again that, unlike characteristic x-rays (which are monoenergetic and of unique energy levels that correspond to energy level differences in the electron cloud), bremsstrahlung is polyenergetic with a broad spectrum. In medical applications, the peak of the bremsstrahlung spectrum is located at roughly one third of the nominal maximum incident electron energy, and the average photon energy is roughly one half of the maximum energy. But one must recall that in clinical practice, the low-energy photons are filtered out. This filtering of low energy x-rays means the clinically used x-ray beam is **harder** (i.e. enriched in high energy photons) than it is inherently.

Although bremsstrahlung is a very important interaction, such inelastic nuclear interactions (also known as **radiative interactions** because of photon emission) only account for a small fraction of electron interactions with matter overall—perhaps on the order of 2%–3% of all inelastic interactions (and the probability of inelastic interactions is only around 10% of the probability of elastic interactions). The low probability of bremsstrahlung makes sense when one considers that it is an interaction between inbound electrons and an atomic nucleus. Since the nucleus is so small in relationship to the atom, the odds of any type of interaction are far greater for the relatively voluminous atomic cloud of electrons than for the tiny, shielded nucleus. An equation to estimate the relative contributions of radiative interactions (which produce bremsstrahlung) to collisional energy losses is:

$$\text{Radiative energy loss/collisional energy loss} = E_k \times Z/820\,000,$$

where E_k is the kinetic energy of the incident electrons in kilo electron volts and Z is the atomic number of the target material.

For the electron energy levels usually used in diagnostic radiology, only a small minority of the incident energy leads to bremsstrahlung. For example, if 100 keV electrons collide with a tungsten target where $Z = 74$, the approximate ratio of radiative to collisional losses is $(100 \times 74)/820\,000 = 0.009$ or 0.9%. Only 0.9% of the incident energy goes into generating x-rays, while about 99.1% of the incident energy is released as heat.

Thus, besides the four interactions discussed above, electrons can interact with matter to produce heat. Thermal energy can be transferred to the media when electrons interact with it. This intense concentration of heat energy in a very small volume can be very useful in a variety of applications of electron beam technology. On the other hand, this production of heat can lead to premature degradation of the metal targets used to generate photons in x-ray tubes and linear accelerators.

6.40 Proton interactions with matter

Like electrons and all other forms of charged particle radiation, protons can interact with matter via elastic and inelastic collisions with atoms in the medium they traverse. Protons transfer energy to matter primarily through inelastic electro-magnetic interactions with atomic electrons; this is the main means through which protons deposit dose. But compared to electrons, protons have a relatively higher probability of inelastic nuclear reactions. Like electrons, protons can also interact with matter without transferring energy, i.e. via elastic interactions. The nuclear interactions which protons undergo include elastic scattering (which is also called **Rutherford scattering** or **multiple Coulomb scattering**), inelastic nuclear scattering ('hard scatters') and other nuclear reactions.

As with all charged particles traveling through a medium, the way that protons penetrate matter may be quantitatively described by the **stopping power**. An equivalent way of thinking about this is not from the proton beam's perspective, but rather from the perspective of the medium being traversed. In this case, the medium has a characteristic stopping power. Stopping power, or *dE/dx*, may be defined as the *average amount of energy lost per unit length* as a particle travels through matter. Stopping power will be covered in greater detail later in this section, but for now, at a given energy, say 4 MeV, the stopping power of electrons is lower than the stopping power of 4 MeV protons, and the stopping power of these protons is lower than that of 4 MeV alpha particles. Since stopping power relates inversely to depth of penetration, in a given medium, electrons will travel further than protons, and protons will penetrate deeper than alpha particles *of the same energy*. Examination of the stopping power, *dE/dx*, as a function of depth of penetration leads to a characteristic curve known as the **Bragg curve** (figure 6.5).

The initial part of a Bragg curve is called the **entry region** or **plateau** portion. Here, as protons progressively lose energy through Coulombic interactions with electrons in the medium, they slow down. And as they slow down, the number of ionization events rapidly increases and reaches an apex called the **Bragg peak**. Beyond the Bragg peak, the number of ionizations quickly diminishes to zero because the protons lose all their kinetic energy and come to a stop. The width of a Bragg peak broadens as the energy of the incident protons (and thus the penetration in tissue) increases.

Furthermore, despite calculations, the real range of a proton beam in tissue or any medium is not an absolute constant. This is because energy transfer to the medium is a stochastic process with some protons losing more, and others losing less energy, than an average proton. This phenomenon is known as **range straggling**. The result is that any two protons with the same starting energies may have different path lengths. In practice, the effect of range straggling is about 1.1% of the range of the beam. Furthermore, a real proton beam rarely has all protons starting with exactly the same initial energy; this slight variation also contributes to range straggling.

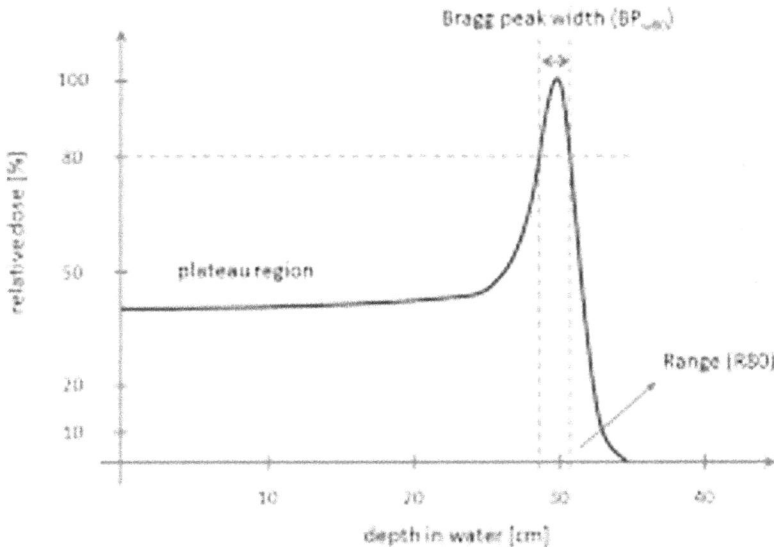

Figure 6.5. Reprinted from [5], Copyright (2019), with permission from Elsevier. The Bragg curve depicts the deposited dose of a proton beam as a function of depth of penetration in a medium. Notice the relatively low dose upon entry and the so-called plateau region, followed by a region of high dose deposition called the Bragg peak. Following the Bragg peak, the protons abruptly stop. Heavier charged particles may have a pronounced 'tail' on the distal end of the Bragg peak caused by spallation (break up of atomic nuclei caused by the charged particle). This tail radiation may be of high-LET and high biological effectiveness.

6.41 Nuclear interactions between protons and the medium

As with electrons, there are two main inelastic mechanisms by which protons transfer energy to a medium. First, as discussed above, energy may be lost through contact with atomic electrons. This is the principal explanation for the limited range of protons in matter. While inelastic interactions with electrons are the main mechanism whereby protons interact with matter and deposit dose, another important source of proton energy loss is through interactions with nuclei in the medium.

It should be first stated that not all nuclear scattering events are inelastic and dose-depositing. Some nuclear interactions are purely elastic and just cause deviations from a straight line. Such bends in the path of a proton beam caused this way is called **multiple Coulomb scattering (MCS)** or **Rutherford scattering**. In low-Z media, multiple Coulomb scattering may also occur due to atomic electrons. Students will recall that the deflection due to repulsive electrostatic forces from positively charged nuclei is governed by **Coulomb's Law**:

$$F = \frac{kqQ}{r^2},$$

where q and Q are the electrical charges in question, r is the separation between them, and k is Coulomb's constant ($k \approx 8.988 \times 10^9$ N·m^2·C^{-2}).

In clinical proton beam therapy for cancer, MCS is generally insignificant at shallow depths but becomes more pronounced with increasing depth as the protons slow down and grow more susceptible to electrostatic deflections. This scattering of protons contributes to the lateral dose distribution. In fact, broadening of a proton beam (i.e. the lateral penumbra) is mostly due to MCS. This is analogous to the scattering of electrons in an electron beam, but electrons are far more readily scattered than protons are. Furthermore, atomic electrons themselves are the primary cause of elastic scatter with electron beams. Atomic electrons generally do not cause as much elastic scattering of the much heavier protons.

In contrast, MCS via nuclei does make the proton beam penumbra broader with depth. As a rule of thumb, the contribution of MCS to the lateral width of a proton beam at the depth of the Bragg peak (σ) is approximately 2% of the range in water.

$$\sigma \approx 0.02 \times R.$$

Thus, as energy (and thus range) of a proton beam increases, so does the lateral penumbra at depth. The width of a proton beam at the Bragg peak (σ) is sometimes called the spot size.

6.42 Characteristics of a Bragg curve

The characteristic shape of a proton Bragg curve is due to several factors. One is the increasing energy loss per path length as the beam slows down. This is based on the so-called 'continuous slowing down approximation' (CSDA) theory. This is due mainly to electromagnetic interactions with atomic electrons and causes a slightly increasing slope along the plateau region. This is also the cause of the rapidly decreasing fluence near the end of range, which results in the Bragg peak. Finally, as protons run out of energy and eventually stop, the distal portion of the Bragg peak is created with a relatively abrupt end. Along the plateau portion, in addition to interactions with atomic electrons, the primary proton beam may interact with atomic nuclei. These nuclear interactions may generate secondary protons and, in so doing, decrease the fluence of the primary proton beam.

6.43 Stopping power and restricted stopping power

In plain English, the **stopping power** or the **LET** (linear energy transfer) is the average energy lost by a charged particle per unit distance traveled. From the medium's perspective, the stopping power is a measure of a material's ability to slow and stop charged particles traveling through it. Mathematically this is *dE/dx*. The Bragg curve describes dose versus depth, and *dE/dx* represents the loss of proton beam energy with increasing depth of penetration into a medium. In other words, *dE/dx* is the stopping power of a medium. Notably, *dE/dx* means different things to different people depending on the perspectives of their specific disciplines. To a medical physicist, *dE/dx* might represent the deposition of radiation dose as a function of depth. To a particle beam physicist, *dE/dx* might represent the loss of energy of a particle beam as a function of depth. To a medical dosimetrist, *dE/dx* might represent proton stopping power of various tissues, whereas to a radiation safety officer or aerospace engineer, *dE/dx* might represent the stopping power of specific shielding material. Finally, to a radiobiologist, *dE/dx* is synonymous with **LET**. LET will be discussed in greater depth in subsequent chapters.

Some might argue that there are subtle differences between stopping power and LET. Stopping power describes the total energy lost by the charged particle, whereas LET more properly describes the energy lost by the charged particle *that is locally absorbed* in the material the particle is traveling through. LET can thus be referred to as the **restricted stopping power**. When discussing LET, the assumption is that any secondary electrons (or other particles) deposit their energy locally and do not travel outside the volume of interest. This would therefore exclude interactions that produce long-range delta rays or bremsstrahlung. In this sense, stopping power is akin to **kerma** (kinetic energy released in matter or kinetic energy released per unit mass) or **terma** (total energy released in matter or total energy released per unit mass), whereas LET is akin to absorbed radiation dose. If no restriction is placed on the energy of the interactions, the **unrestricted LET** is indicated as **LET$_\infty$**. In this case, LET$_\infty$ is thus fully synonymous to stopping power.

6.44 The Bethe Bloch equation

To first order, $dE/dx \approx 1/v^2$, where v is the velocity of the proton at a given time and location within the medium it is passing through. As a rough first approximation, the range of a proton beam may be estimated using the **Bragg–Kleemann rule**:

$$R(E) = \alpha E^p,$$

where α is a medium-specific constant, E is the initial energy of the proton beam, and the exponent p considers the dependence of the proton's energy or velocity.

In human tissue or water, this can be approximated as:

$$R(E) = E^{1.75}.$$

In medical practice, this equation is rarely used. Rather, rough estimates can be obtained via tabulated data, or one can use the more precise **Bethe-Bloch equation**. To understand the Bethe-Bloch equation, it is helpful to review the history behind its development. Among the earliest to tackle the problem was Niels Bohr who considered dE/dx as the product of a **kinematic factor** (A) and a **stopping number** (B). Thus, he wrote:

$$dE/dx = (A) \times (B).$$

Including his kinematic factor, the Bohr equation is thus:

$$dE/dx = (4\pi Z^2 e^4 N/mv^2) \times (B),$$

where Z is the atomic number of the incoming particulate radiation (e.g. a proton) and N is the atomic density of the medium (in atoms/volume). Hans Bethe then further developed the equation by introducing a modified stopping number (B):

$$B = Z_2 \ln(2mv^2/I),$$

where Z_2 is the atomic number of the absorbing medium and I is the mean excitation potential.

With this development, the equation becomes:

$$dE/dx = (4\pi Z^2 e^4 N/mv^2)[Z_2 \ln(2mv^2/I)].$$

Further modifications that took special relativity and quantum mechanical considerations into account led to the present form of the Bethe-Block equation:

$$\left(\frac{dE}{dx}\right)_0 = \frac{2\pi n_e r_e^2 m_e c^2 \, z^2}{\beta^2} \left[\ln\left(\frac{2 m_e c^2 \beta^2 T_{max}}{I^2 (1 - \beta^2)}\right) - 2\beta^2 + 2z L_1(\beta) + 2z^2 L_2(\beta) - 2\frac{C}{Z} - \delta + G \right].$$

- I: mean excitation energy, material-dependent
- δ: density correction
- C: the shell correction which accounts for the fact that atomic electrons in the medium are not stationary, important at low energies
- T_{max}: maximum energy transfer to an electron
- L_1: Barkas–Anderson correction (Z^3)
- L_2: Bloch (Z^4) correction
- G: Mott correction.

It could be mentioned that the Bethe-Block formula is strictly appropriate only for a limited range of proton kinetic energies (specifically from around 500 keV to around 10 GeV where the proton (or other charged particle) does not carry any atomic electrons with it and at energies where the inequality $\gamma m/M \ll 1$ holds. (γ is the

Lorenz factor discussed in the next section.) Below this energy range, the so-called **Anderson-Ziegler approximation** might be more accurate, and at lower energies still, the **Lindhard–Scharff approach** is more precise. And at very-high-energy levels, **Fermi's density correction** must be added.

6.45 Special relativity and proton stopping power

Students might wonder if such elaborate adjustments are really necessary. Recalling from relativity that 'beta' is the ratio of a particle's velocity to the speed of light in a vacuum:

$$\beta \equiv \frac{v}{c}.$$

And the **Lorenz factor** γ is given by

$$\gamma = \frac{1}{\sqrt{1 - \beta^2}}.$$

One can see that as v increases and approaches c, γ approaches infinity, and β approaches 1.

A few rules of thumb may be useful for students:
1. When $\beta < 0.3$, $\gamma \approx 1$ (to within 5%).
2. When $\gamma > 7$, $\beta \approx 1$ (to within 1%).
3. Departures of >1% occur when $\gamma > 1.01$.
4. Similarly, departures of >1% occur when $\beta > 0.14$ (that is, when $v > 14\%$ of the speed of light).

Some additional equalities of value from special relativity stem from the Einstein relation:

$$E = mc^2.$$

Or more completely, for protons, other particles (or any object for that matter) with rest mass, m_0 moving at high speeds:

$$E = \gamma m_0 c^2.$$

In this case, E represents the total energy (rest energy plus kinetic energy). And equivalently, for momentum:

$$p = \gamma m_0 V.$$

Rearranging and equating yields:

$$\gamma = \frac{E}{m_0 c^2} = \frac{p}{m_0 V}.$$

Substituting $v = c\beta$ and rearranging yields:

$$\gamma \beta = \frac{p}{m_0 c}.$$

Another basic relationship in special relativity is:

$$E^2 = (m_0 c^2)^2 + (pc)^2 = P^2 c^2 + m_0 c^4.$$

Coupling this with $p = mv = \sqrt{\frac{2m}{E_k}} = \frac{2E_k}{v}$, where E_k represents kinetic energy, one can derive:

$$pc = \frac{\sqrt{E^2 - m_0^2 c^4}}{E_k + m_0 c^2}.$$

Thus:

$$\beta = \frac{v}{c} = \frac{pc}{E} = \frac{\sqrt{E^2 - m_0^2 c^4}}{E_k + m_0 c^2} = \frac{\sqrt{E_k^2 + 2E_k m_0 C^2}}{E_k + m_0 c^2}.$$

Simply put, the relativistic momentum ($p = \gamma\, m_0 v$) deviates by 1% from the Newtonian calculations when $\gamma = 1.01$ (that is, a 1% threshold). As mentioned, this occurs when $\beta = 0.14$ (when $v = 14\%$ of c). For protons, this 1% threshold occurs at a modest energy of 9.38 MeV. For comparison, the 1% rule for electrons is even more modest at only 5.11 keV, but for far heavier alpha particles, it is 37.27 MeV. Alpha particles from radioactive decay do not attain such energies. Thus, beta electrons from radioactive decay are relativistic but far heavier alpha particles from alpha decay are not.

Since most galactic cosmic rays have energies between 100 MeV (corresponding to a velocity for protons of 43% of the speed of light) and 10 GeV (corresponding to 99.6% of the speed of light), it is obvious that relativistic corrections are needed. In fact, clinical proton beam radiation therapy may employ protons up to 230–330 MeV, so even in the radiotherapy clinic, special relativity must be kept in mind. Again, considering that some primary galactic cosmic rays can possess energies on the order of many thousands, millions or even billions of giga electron volts, it is apparent that corrections for special relativity are indeed appropriate.

6.46 Bragg's rule for molecules

The above analysis of proton stopping power has all been explicitly worked out only *for atoms*. **Bragg's rule** states that the stopping number *of a molecule* is the sum of the stopping numbers of all the atoms composing the molecule. For most molecules Bragg's rule applies with reasonably high precision (within a few percent) although molecular hydrogen (H_2) and nitrous oxide (NO) are notable exceptions.

6.47 Proton-induced nuclear reactions

In addition to the interactions between primary protons with electrons, protons may also undergo inelastic nuclear reactions with target nuclei in the medium. As a rule of thumb, *about 1% of protons in a proton beam are lost per centimeter* of tissue or water through inelastic nuclear interactions. Thus, for a proton beam that has a range of 10 cm in tissue, by the time the Bragg peak is reached, $\sim$10% of the primary protons have been lost to nuclear interactions. In addition to attenuating the beam through the loss of primary protons, nuclear interactions can create **secondary protons**. For example, protons may strike a nucleus and dislodge a proton. The proton may then continue onward as a secondary proton. The reaction may be expressed as:

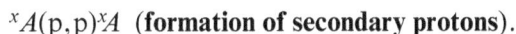

$$^xA(p,p)^xA \text{ (\textbf{formation of secondary protons}).}$$

In other situations, a neutron may be ejected or perhaps both a proton and a neutron can be knocked out. In such cases, the nucleons may be ejected independently, or they may come out as a packet, that is, as a deuteron:

$$^xA(p,pn)^{x-1}A \text{ (\textbf{proton--induced deuteron ejection}).}$$

As a specific example, oxygen-12, which is quite plentiful in normal biological tissues as well as water, may undergo a (p,pn) reaction:

$$^{16}O(p,pn)^{15}O.$$

Regardless of whether the proton and neutron are ejected separately or as a deuterium unit, the created oxygen-15 is radioactive. The process of creating radioisotopes through proton bombardment is called **proton activation**. *For light nuclei, proton activation often generates positron emitters*; oxygen-15 is a positron emitter with a half-life of about 2 min. Similarly, the most common isotope of carbon, carbon-12, undergoes a (p,pn) reaction to create the positron emitter, carbon-11:

$$^{12}C(p,pn)^{11}C.$$

Carbon-11 has a half-life of about 20.3 min. Another important (p,pn) reaction in human tissues is the reaction between incoming protons and the most common isotope of nitrogen, nitrogen-14:

$$^{14}N(p,pn)^{13}N.$$

Nitrogen-13 has a half-life of approximately 10 min and is another positron emitter. But the (p,pn) reaction is not the only proton nuclear reaction that can lead to nitrogen-13. For instance, protons can strike and react with oxygen-16 nuclei and lead to nitrogen-13 along with the ejection of an alpha particle:

$$^{16}O(p,\alpha)^{13}N.$$

Analogously, nitrogen-14 may undergo a nuclear reaction with an incident proton to yield carbon-11:

$$^{14}N(p,\alpha)^{11}C.$$

Hence, there may be several different pathways that lead to the same product radionuclide, and there may be several reactions between protons and nuclei that might lead to different product nuclei.

As mentioned, proton activation of light nuclei often yields positron emitters. These positron emitters typically have short half-lives that are measured in seconds to minutes. The energy of the emitted positrons (and thus their ranges in tissues) is inversely proportional to the half-life of the parent. For example:

Carbon-11 has a half-life of 20.3 min, and its positrons have a maximum energy of 0.96 MeV.

Nitrogen-13 has a half-life of 9.96 min, and its positrons have a maximum energy of 1.19 MeV.

Oxygen-15 has a half-life of 2 min, and its positrons have a maximum energy of 1.72 MeV.

Importantly, the range of positrons in tissues is directly related to their maximum energy, and therefore, the shorter the half-life, the longer the range. Long-range positrons do not make ideal PET scanning radioisotopes since the long distances they travel before encountering an electron and generating the two 511 keV gamma photon signals reduce the anatomical resolution of the PET image.

Carbon-11 has a half-life of 20.3 min, and its positrons have a maximum range of 4.11 mm.

Nitrogen-13 has a half-life of 9.96 min, and its positrons have a maximum range of 5.39 mm.

Oxygen-15 has a half-life of 2 min, and its positrons have a maximum range of 8.2 mm.

The presently popular radioisotope in PET imaging is fluorine-18, which has a half-life of 109.7 min, a maximal positron energy of 635 keV and a range in water or tissue of about 2.39 mm. Thus, unlike the undesirably long 8.2 mm range of oxygen-16, the annihilation gamma photons of fluorine-18 originate from a site not far from their parent radionuclide. Fluorine-18 can be made via a (p,n) reaction from ordinary oxygen:

$$^{16}O(p,n)^{18}F.$$

It is apparent that there may be several different pathways that lead to the same product radionuclide, and there may be several reactions between protons and the same nuclei that lead to different products. The energy thresholds for the incoming proton differ for the various reactions. For example, the production of carbon-11 through the $^{14}N(p,\alpha)^{11}C$ reaction has a proton energy threshold of 3.1 MeV, whereas the $^{12}C(p,pn)^{11}C$ reaction has a threshold of 20.3 MeV.

Aside from protons, neutrons, deuterons and alpha particles, other nuclear fragments may be ejected after proton interactions such as tritons, helium-3 and heavier nuclei. Additionally, the inelastic nuclear reactions between protons and nuclei can lead to transient excitation that is relieved by the nearly instantaneous emission of a gamma photon. The reaction can be written as (p,γ). For lack of a better term, these reactions are called proton capture reactions, and the photons are called **prompt gammas** or **proton capture gammas**. An example is:

$$^{7}Li(p,\gamma)^{8}Be.$$

The created beryllium-8 is initially in an excited state that then decays to the ground state through emission of a 17.6 MeV photon, or it can emit a 24.4 MeV gamma photon and two alpha particles. It should be noted that beryllium-8 is highly unstable and disintegrates into two alpha particles in well under a femtosecond, and the latest estimates are 8.19×10^{-17} s. In fact, with such a high degree of instability, beryllium-8 is sometimes alluded to as an **unbound resonance** rather than a radio-isotope of beryllium. This extreme instability is of great importance in both Big Bang and stellar nucleosynthesis, because new heavier nuclei cannot be readily made from the instantly collapsing beryllium-8 building blocks. The bottleneck for further nucleosynthesis posed by the instability of beryllium-8 is called the **beryllium barrier**.

A proton capture reaction that leads to prompt gammas and happens to be of relevance to irradiation of human tissue is the radiative interaction between protons and oxygen-16:

$$^{16}O(p, p\gamma)^{16}O.$$

In this case, the incoming proton excites the oxygen-16 nucleus, and a secondary proton is ejected, often at an angle relative to the original primary proton. Also, in a timeframe measured in picoseconds, a gamma photon is emitted. Such reactions are sometimes called **radiative nuclear reactions** since gamma radiation is emitted.

Although Compton interactions are normally thought of as photon-electron interactions, photons may similarly interact with protons as well under certain circumstances. In such Compton scattering events, the photon is deflected and leaves with reduced energy while an ejected proton travels off as well. The maximum energy transferable to a proton by a photon of energy $h\nu$ is:

$$E_{\max} = h\nu[2\xi/(1 + 2\xi)],$$

where $\zeta = h\nu/Mc^2$ and M is the mass of the proton.

6.48 Alpha particles

As charged particles, alpha particles interact with matter in a manner similar to how electrons and protons do but thanks to their increased mass, they tend to be less easily scattered. Also, their range will be shorter than the range of protons or electrons of equal initial energy. In air, alpha particles with energies of 4–8 MeV (which applies to almost all alpha emitters) have a range approximated by a form of the Bragg–Kleeman rule:

$$R(\text{in cm}) = 0.324 \, E^{3/2}$$

This amounts to approximately 4 to 5 cm in air for most alpha particles. The range in air of alpha particles with energies below 4 MeV follows a simpler rule of thumb:

$$R(\text{cm}) = 0.56E.$$

In tissue, most alpha particles will have a range of 20–70 μm. This amounts to one, two or three cell diameters depending on type of tissue in question—recall that a typical human red blood cell diameter is about 7.5 μm; lymphocytes are more variable with slightly larger diameters, usually ranging from about 6–18 μm and averaging around 12 μm; neutrophils are typically 12–14 μm in diameter; and epithelial cells are also highly variable and might average 25 μm depending on subtype.

Alpha particles are densely ionizing and thus have high LET. As such, they have a high **RBE** (relative biological effectiveness) and low **OER** (oxygen enhancement ratio) for most biological endpoints. But it is important to remember that alpha particles tend to have moderate energies—usually in the 3–10 MeV range. As a particle's energy increases, its range increases, but it becomes less densely ionizing. Because of the increased speed, the distance between ionization events becomes greater. Therefore, although alpha particles emanating from atomic nuclei during radioactive decay (with their moderate kinetic energies) are high LET, very energetic helium nuclei (in the hundreds of mega electron volts or above from particle accelerators or in cosmic rays) are not densely ionizing and do not have LET values as high as alpha particles of just a few mega electron volts. As mentioned, electrons emitted from radioactive isotopes by beta decay are relativistic, but relativistic effects may be ignored for alpha particles.

Of course, for more accurate estimates of the ranges of alpha particles and heavier charged particles in various media, the Bethe–Block equation can be used. Of interest and importance, for alpha particles ($Z = 2$) and all heavier charged particles with more total protons, as the particle slows down, another phenomenon occurs—it changes total charge as the nucleus picks up electrons. For example, as alpha particles slow down in matter, they capture electrons and become elemental helium atoms.

This change in charge as the charged particle slows down can be ignored for proton interactions but must be accounted for in carbon ion radiotherapy. And of course, when tackling detailed calculations involving HZEs (high-Z, high-energy galactic cosmic rays), such progressive changes in the particle's electrical charge must be considered as the particle slows down.

6.49 Neutron interactions with matter

6.49.1 Neutron energies

Neutrons are often categorized based on their kinetic energies or velocities. Although the boundaries are poorly defined, one such classification is:

High energy	>10 MeV
Fast	10 keV–10 MeV
Intermediate	100 eV–10 keV
Slow	0.03 eV–100 eV
Epithermal	~1eV
Thermal	0.025 eV
Cadmium	>1 eV

There are obvious overlaps in the classification, and different texts use different definitions for these categories. For instance, the neutron beam at Fermilab had a spectrum of neutron energies with an average neutron energy of 25 MeV but was often alluded to as a fast neutron beam.

Cadmium neutrons are those capable of passing through an absorber made of cadmium. This is because cadmium has a very high absorption cross section for low-energy neutrons (<0.4 eV) with an absorption peak around 0.176 eV. Above about 1 eV, however, cadmium becomes largely transparent to neutrons. Incidentally, this is largely thanks to one particular isotope, cadmium-113, which has an abundance of about 12.26%. Curiously, this isotope has been found to be radioactive with a half-life of 7.7×10^{15} years. Given the Universe is only about 13.8 billion years old, we have not yet approached one half-life of this primordial radionuclide, so we can continue exploiting its high capture cross section for billions of years to come.

As neutrons travel through matter, they lose energy and eventually reach a thermal equilibrium with their surroundings. A neutron that has reached such equilibrium with its surroundings at room temperature is called a thermal neutron. Although the mean kinetic energy of such neutrons is 0.025 eV, the actual energy distribution is a Maxwellian distribution around this mean. Students may enjoy calculating the energy and velocity of a thermal neutron (by setting $E = kT = mv^2/2$) and getting 2200 m s^{-1}. This seems surprisingly fast for something that does not even qualify for the definition of 'slow' by some classifications!

A general formula relating neutron energy and temperature is:

$$T = 1.159 \times 10^4 \, eV.$$

An equivalent formula for velocity is:

$$v = 13.83\sqrt{E},$$

where v is in kilometers per second, E is in electron volts, and T is in Kelvin.

6.49.2 Interactions of neutrons with matter

Photons and protons interact predominantly with atomic electrons. In contrast, neutrons interact predominantly with nuclei. In so doing, they often change the identity of the struck nucleus. Hence, in a very literal sense, neutrons are alchemists.

As with other particle interactions, neutron collisions may be classified as elastic or inelastic. Elastic scattering of neutrons is in many ways similar to elastic scattering of other particles in that after a series of such direction-changing collisions, an originally unidirectional beam of neutrons will cease to exist. As neutrons may be deflected at angles from 0° to 180°, the initially forward pointing beam grows broader with depth and eventually is attenuated to zero. In a way, the initial neutron beam may be considered as a neutron current, with a specified number of neutrons and a certain amount of total energy flowing across a unit area. This amounts to a neutron energy current density and number current density (with the two related by the energy per particle).

The net rate of neutron flow across a given unit area is called the **neutron flux**. It can be quantified as either an *energy* flux or a *number* flux. The formula for number flux (F) may be expressed as:

$$F = F_0 e^{-\sigma n x},$$

where σ is the cross section for absorption, n is the number of atoms per cubic centimeter in the absorber, and x is the thickness of the absorbing medium.

As with other forms of radiation, the total cross section (σ) for neutron interactions is made of several components:

$$\sigma(\text{tot}) = \sigma(\text{elastic}) + \sigma(\text{inelastic}) + \sigma(\text{absorption}).$$

The cross sections are strong functions of neutron energy. Hence, when specifying the cross section, it is imperative that the energy is clearly stated. A typical atomic nucleus will present a neutron cross section of around one barn (10^{-24} cm^2). If the neutron energy is less than a few mega electron volts, the effects of special relativity may be ignored, allowing an equation to be written relating neutron energy to its de Broglie wavelength:

$$\lambda = 2.87 \times 10^{-9} / \sqrt{E} \text{ cm.}$$

This shows that for fast neutrons, λ is about 10^{-12} cm, and the neutron cross section is a few barns. For thermal neutrons, in contrast, the de Broglie wavelength is far longer at around 1.8×10^{-8} cm. This is quite large compared to the radius of the nucleus. Hence, neutron cross sections correlate inversely with energy in general. Of course, resonance effects associated with specific isotopes play a very important role and do not follow this general rule.

6.49.3 Neutron diffraction

Low-energy neutrons have relatively large de Broglie wavelengths that are close to the interatomic distances in crystal lattice. Therefore, neutron **diffraction** can be useful in analyzing structures of organic compounds and other materials containing a large amount of hydrogen (since neutrons have a high scattering cross section for hydrogen, whereas x-rays have a low scattering probability for hydrogen). Hence neutron and x-ray diffraction can be complementary. Both use the same familiar formula known as **Bragg's Law**:

$$n\lambda = 2d \sin \theta.$$

In addition to its utility in determining structure, neutron diffraction also allows formation of low intensity but monoenergetic neutron beams. This is because the neutron diffraction process is highly selective and diffracts only specific wavelengths at specific angles.

6.50 Neutron absorption resonance peaks

Returning to cadmium as an example, natural cadmium has a neutron absorption cross section of 7000 barns (at an energy of 0.176 eV), but this is largely due to one individual isotope, cadmium-113. Cadmium-113 is present at an abundance of about 12.26% but has a huge cross section of 20 000 barns at this energy. Thus, one can see that for neutron absorption, there are very pronounced **isotope effects**.

Neutron absorption curves may exhibit areas with unusually high capture cross sections, called **resonance peaks**. Resonance capture effects are isotope dependent as well as energy dependent. For instance, cadmium has a resonance peak at 0.176 eV, whereas uranium has a resonance peak at 6.7 eV, cobalt has a resonance peak at 120 eV, and manganese shows a peak at 300 eV. Gadolinium exhibits a resonance peak at just about thermal neutron energies—0.028 eV. Each one of these elements have their resonance peaks due to a particular isotope of that element.

Upon absorption of a neutron at the resonance energy, the new nucleus will be in an excited, radioactive state. The excitation may be relieved through various de-excitation modes or channels. For example, the new, compound excited nucleus may de-excitate through the release of a photon (n,γ), a proton (n,p), an alpha particle (n,α) or another neutron (n,n) or two neutrons (n,2n) among various other channels.

6.51 Elastic scattering of neutrons

As with other particle interactions, neutron collisions may be classified as elastic or inelastic. But elastic scattering for neutrons is a bit different from other scattering scenarios. Elastic collisions involving photons, electrons and protons cause directional changes without transferring much energy. The particles bounce like a racquetball off an immovable wall. However, since neutrons are about the same mass (slightly more, actually) than protons, elastic collisions between neutrons and protons are akin to billiard ball collisions; the protons are set in motion too. These protons then contribute dose to tissue. Hence, for neutrons, substantial dose is deposited through scattering. It should be recalled that protons are synonymous with hydrogen atomic nuclei, which are the most abundant atoms in tissue. In fact, roughly 70%–80% of fast neutron dose is contributed via the recoiling protons created in this fashion. Curiously, upon a collision with a proton, the neutron's energy is reduced to about 1/e, or 37%, of its initial value.

In elastic neutron scattering, the scattered nuclei (usually hydrogen nuclei) dissipate their energy in the immediate vicinity of the primary neutron interaction. That is, the scattered secondary protons do not travel far, and they deposit their radiation dose nearby. The locally absorbed dose is called the **first collision dose**.

Of course, for neutrons, being uncharged and around 1839 times the mass of an electron, these elastic collisions are with atomic nuclei rather than with orbital electrons. Elastic scattering of neutrons is one type of **recoil reaction** involving neutrons. With repeated collisions, fast neutrons eventually slow down as they travel through a medium via multiple elastic collisions in which kinetic energy is transferred without ionization or excitation. Although neutrons also experience elastic collisions off heavier nuclei, elastic neutron collisions occur primarily with hydrogen (i.e. protons) because of similarity in mass. For this reason, wax and polyethylene (which contain a relatively large proportion of hydrogen) are good absorbers. In contrast, lead is not as efficient at absorption of neutrons as one might expect based on its effectiveness in absorbing other forms of radiation. After multiple scattering events, the fast neutrons eventually 'thermalize' (meaning they lose kinetic energy until they come to equilibrium with the surrounding temperature) and get 'captured' (as will be discussed later).

The protons set in motion then contribute dose to tissue. Hence, for neutrons, substantial dose is deposited through elastic scattering. It should be noted that the definition of 'elastic' in this context is that the overall kinetic energy before and after a collision is conserved. This is the distinction between elastic and inelastic interactions in general. During inelastic collisions the kinetic energy before and after do not equal since some of the incoming particle's energy goes into nuclear excitation, overcoming the binding energy of an electron, etc.

6.52 Dose deposition by neutrons

Since hydrogen is basically a proton, neutrons heavily interact with materials containing a lot of hydrogen (such as biological issues). As far as transfer of energy goes, the majority of absorbed dose from neutron radiation is contributed by recoil protons. In this case, the incoming neutron collides with a similarly sized proton (in the form of a hydrogen atom in a molecule). Because of the relatively higher concentration of hydrogen in fat and other lipids, the dose from neutrons in fat is approximately 20% greater than dose in muscle. And since the brain is also largely lipids, brain neutron dose is also higher than many other tissues.

6.52.1 Moderators

The average neutron energy before ($E1$) and after ($E2$) collisions with nuclei of atomic mass A is given by the following equation:

$$E2/E1 = e^{\wedge} - \zeta \text{ where } \zeta = 1 - \left\{ (A-1)^{\wedge}2/2A \right\} \times \ln\{(A+1)/(A-1)\}.$$

Therefore, fractional energy loss per collision *increases with decreasing A*. In other words, neutrons transfer their kinetic energy most efficiently to hydrogen. (ζ is given a value of 1 in these equations for hydrogen). From these equations, one can see why wax and polyethylene (which contain a lot of hydrogen atoms) are good absorbers, but lead and tungsten are not. The equations also explain why the average energy of a neutron after a collision with a proton is about 37% of its initial value.

Materials that are good at slowing neutrons without capturing them and undergoing nuclear reactions are called **moderators**. In nuclear reactors, graphite, ordinary water ('light water'), and heavy water (deuterated water or HDO) are useful moderators.

6.53 Elastic and inelastic scattering of neutrons

6.53.1 Nuclear disintegrations

Aside from recoiling protons, neutron-induced nuclear disintegrations are another important mechanism for depositing dose. *Approximately 30% of the total dose through neutrons in biological tissues is due to nuclear disintegrations.* Upon collision with a heavy nucleus, neutrons may induce a variety of reactions. The products of such reactions may include nuclear fragments (e.g. protons, deuterons, alpha particles, larger nuclear fragments), more neutrons and gamma rays. This neutron-induced nuclear fragmentation is called **spallation**.

6.53.2 Neutron capture

Unlike electrons, protons and other charged particles, uncharged neutrons do not undergo any Coulombic scattering. Neutrons do not experience nuclear repulsion the way protons do. Therefore, they enter atomic nuclei far more readily. Upon such entry, they may be 'captured'. And capture reactions can have a variety of consequences. In most cases, neutron capture leads to a new radioactive nucleus. This creation of radioactive nuclei upon neutron irradiation is called **neutron activation**. Nuclear reactors, which provide enormous neutron fluxes, allow vast amounts of artificial radionuclides to be created through neutron activation.

The production of radioactive isotopes through a neutron flux F may be calculated by:

$$R = (F\sigma_{act}N)/k \times (1 - e^{-kt})e^{-kt},$$

where t is the bombardment time and t' is the time from end of bombardment to measurement, F is the neutron flux, k is the radioactive decay constant of the newly manufactured radionuclide, N is the number of target atoms , and σ_{act} is the cross section for the nuclear reaction. N is the number of target atoms. One can see that the rate of neutron bombardment (that is the flux) must be fast enough to compensate for radioactive decay of the product radionuclide if one is to obtain any meaningful yield.

The activity generated will be given by:

$$A = Rk = (F\sigma_{act}N) \times (1 - e^{-kt})e^{-kt}.$$

The resulting radioactive nucleus might undergo immediate ('prompt') decay, or it might be metastable and decay over a measurable amount of time.

As neutrons slow down and their kinetic energy approaches zero, they experience an *attractive* **strong nuclear force** leading to capture as they enter a nucleus. And although their kinetic energy may be zero, upon capture the reaction adds about 8 MeV of energy to the newly formed nuclide. This is about the binding energy of almost all but the very lightest nuclei. This new, excited nucleus may then release its excitation energy through various means such as:

$$(n,\gamma)$$
$$(n, p)$$
$$(n,\alpha)$$
$$(n,f)$$

6.54 Radiative capture

The (n,γ) reaction is when a neutron is captured and a gamma ray is promptly emitted. This **prompt gamma photon** is also called a **capture gamma**, and the whole phenomenon is called **radiative capture**. Radiative capture tends to predominate at *lower* neutron energies. The simplest example involves the simplest nucleus, hydrogen:

$$^1H + {}^1n \rightarrow {}^2H + h\nu$$

or

$$^1H(n,\gamma)^2H.$$

In rough agreement with the neutron binding energy, radiative capture of neutrons by hydrogen leads to gamma rays with a maximal energy in the range of 6–10 MeV. One or several photons may be emitted. For hydrogen, radiative capture dominates at low neutron energies. Another very important radiative capture reaction is the ^{113}Cd(n,γ) reaction.

Cadmuim-113 absorbs neutrons very effectively at low energies (thermal or 'slow' neutrons). With remarkably high probability, neutrons with energy below the **cadmium cut-off** will be absorbed. However, those above the cut-off will be transmitted. The 'cadmium cut-off' is about 0.5 eV, and low-energy neutrons with kinetic energies just above 1 eV are called **cadmium neutrons**. Neutrons below this energy level are variously labeled as **slow** or **thermal neutrons**, to distinguish them from intermediate and fast neutrons.

6.55 (n,p) Proton Emission Reactions

In addition to radiative capture reactions, neutron capture may alternatively lead to charged particle emission (such as proton or alpha emission). Ordinarily, (n,p) and (n,α) reactions require fast neutrons. And the threshold energy required for (n,α) reactions is typically higher than for (n,p) reactions. As with the (n,2n) process, over a hundred (n,p) and (n,α) examples have been experimentally verified, and the created nuclei are usually radioactive. Hence, neutron irradiation may be monitored through the creation of radioactive daughter products which can be quantified. Very broadly speaking, the thresholds for (n,p) reactions are in the range over up to 3 MeV, whereas the thresholds for (n,α) reactions lie in higher energies of 9–20 MeV. Aside from charged particle emission upon neutron capture, sometimes additional neutrons can be ejected. These (n,2n) reactions typically require an incident neutron energy exceeding 9 MeV.

A prominent neutron capture reaction is the (n,p) proton emission reaction. In contrast to most radiative capture reactions (n,γ), which are stimulated by slow neutrons, this **proton emission reaction** usually requires fast neutrons. Nevertheless, there are three important (n,p) reactions that do occur at low neutron energies. These involve:

$$^3He$$
$$^{14}N$$
$$^{35}Cl$$

The ^{14}N(n,p)^{14}C is a well-studied and scientifically important (n,p) reaction that occurs in the upper atmosphere where cosmic ray interactions with atmospheric atoms produce abundant free neutrons. The formation of radioactive carbon-14 allows radiocarbon dating as explained elsewhere. Because carbon-14 gets incorporated into the food chain and is present at a steady-state in virtually all living materials, it is of great relevance to our main topic as a cosmogenic radionuclide. Additionally, this particular (n,p) reaction can happen *in vivo* upon neutron irradiation since biological organisms have a great deal of nitrogen in their cells in the form of proteins and nucleic acids.

6.56 (n,α) Alpha Emission Reactions

The (n,α) reaction, like the (n,p) reaction typically requires fast neutrons, and the threshold for such reactions usually exceeds that for prompt proton emission. But again, there are notable exceptions. Two such exceptions that can undergo (n,α) reactions with slow neutrons involve boron-10 and lithium-6.

Boron-10 has a natural abundance of around 19%. It has a very high cross section of 3840 barns for thermal neutrons (of 0.025 eV). In comparison, the capture cross section for boron-11 is only about 0.005 barns for thermal neutrons. Additionally, unlike cadmium-113, the boron-10 isotope has a high cross section for the (n,α) reaction across the entire neutron energy spectrum, not just for slow neutrons. For fast neutrons, its cross section is on the order of barns. The neutron capture cross sections of most other elements become very small at high energies. (Including cadmium, which is an important neutron absorber in nuclear reactors). The boron-10 reaction is

$$^{10}B(n,\alpha)^7Li.$$

More precisely, this boron neutron capture reaction has two branches: one that releases 2.79 MeV of kinetic energy in its two hadronic daughter products and another that provides 2.31 MeV of kinetic energy along with a 480 keV gamma photon. The latter reaction is more abundant at 94%, while the former reaction occurs only 6% of the time. Thanks to the gamma emission, this boron neutron capture reaction is used in neutron detection instruments and in nuclear reactors. It also forms the basis for the cancer treatment known as **boron neutron capture therapy**.

Another (n,α) reaction that occurs with slow neutrons is:

$$^6Li(n,\alpha)^3H.$$

This reaction serves as an effective means of making tritium in a nuclear reactor.

Aside from charged particle emission upon neutron capture, sometimes additional neutrons can be ejected in an (n,2n) reaction. In general, the incident neutron energy for this must exceed 9 MeV.

6.57 Inelastic scattering of neutrons

For low-Z elements with low neutron energies (below 1 MeV), neutrons interactions are mostly through elastic scattering (and capture events). The same holds true for high-Z elements with neutrons up to 100 keV. But above these energies, several types of inelastic neutron interactions can occur. With elastic scattering of neutrons, there is a significant transfer of kinetic energy, especially to hydrogen (i.e. protons). Nevertheless, as with all elastic scattering events, the total kinetic energy of the incoming neutron is equal to the sum of the kinetic energies of the outgoing products. And in addition to elastic scattering, neutrons may disappear (e.g. through neutron capture). Alternatively, neutrons may inelastically bounce off with less kinetic energy but leave target nucleus in excited state: $a + A \rightarrow a + A^*$

As in all inelastic collisions, the kinetic energy of the outgoing neutron is less than the incoming neutron. The energy difference goes into exciting the struck nucleus, leaving it in an excited state that may relax promptly by emitting a gamma ray, particle or nuclear fragment. Alternatively, the lost kinetic energy may lead to a to a radioactive state that may decay after an observable time has elapsed. The excited nucleus releases its excess energy via a host of possibilities:

(n,γ) prompt gamma reactions
(n,2n) reactions
(n,n) reactions
(n,nγ) reactions
(n,p) reactions
(n,α) reactions
Other charged particle emissions
Spallation
Fission.

In biological tissues with neutrons below 6 MeV, neutron interactions are mostly elastic, but above 6 MeV, inelastic interactions begin. Inelastic scattering begins at relatively high energy for the low-Z elements common in biological materials (e.g. the threshold is 4 MeV for carbon). However, the threshold energy is much lower (around 100 keV) for high-Z elements. Thus in heavy elements, inelastic scattering predominates with fast neutrons.

Neutron scattering with fast neutrons is largely inelastic. One example is the (n,2n) process. In most such cases, the activated product is a positron emitter. The $^{12}C(n,2n)^{11}C$ reaction can occur in human tissue but requires energetic neutrons with over 20 MeV of kinetic energy. The residual carbon-11 nucleus decays via 'beta-plus' emission with a half-life of 20.5 min. This positron emission theoretically enables monitoring for high-energy neutron exposures through this reaction. In general, (n,2n) reactions require energetic neutrons with thresholds typically in excess of 9 MeV.

6.58 Neutron decay

Finally, unlike photons and protons, free neutrons are unstable. They undergo beta decay into protons, electrons and electron antineutrinos with a characteristic half-life. Fascinatingly, this half-life is still uncertain. There are two basic experimental methods for determining the half-life, called the 'bottle' and 'beam' techniques. Without going into details, the bottle approach confines a specified number of neutrons and measures how many remain after a while, whereas the beam techniques count the number of protons in a neutron beam after a period of time. Recent bottle experiments have shown the mean lifetime of a free neutron to be 877.75 +/− 0.28 s. (Recall that the mean lifetime is the time required to fall to 1/e of the initial value, which is 1.44 times the half-life.)

Despite this high precision measurement, it is still about 10 s shorter than meticulous measurements made with the beam technique. The explanations for the bottle versus beam discrepancies range from allegations of experimental imprecision to the discovery of **dark matter**. For instance, the beam methods would lead to incorrect conclusions if there were a decay mode that did not produce a proton. It has been suggested that in perhaps 1% of cases, a neutron can decay into a dark particle rather than a proton, and these dark matter particles are not counted among the protons measured in beam counting methods. Thus, the beam techniques lead to a longer lifetime than the bottle techniques. Although this is a fascinating idea, it remains unproven and the discrepancy between the bottle and beam results remains an area of contention.

The beta decay of the neutron is of great interest in particle physics as well as cosmology, since the lifetime of the neutron plays an important role in the timing of events during the first few minutes after the Big Bang (which is covered in a separate chapter). From a more practical perspective, only in a vacuum will a neutron exhibit radioactive decay. In any medium, a free neutron will be thermalized and captured long before decay.

Further reading

[1] Tendler I I et al 2020 Experimentally observed Cherenkov light generation in the eye during radiation therapy Int. J. Radiat. Oncol. Biol. Phys. **106** 422–9

[2] Anirban A 2022 Precise measurement of neutron lifetime Nat. Rev. Phys. **4** 9

[3] Gonzalez F M, Fries E M and Cude-Woods C et al UCNτ Collaboration 2021 Improved neutron lifetime measurement with UCNτ Phys. Rev. Lett. **127** 162501

[4] Fornal B and Grinstein B 2018 Dark matter interpretation of the neutron decay anomaly Phys. Rev. Lett. **120** 191801

[5] Obodovskiy I 2019 Passing of charged particles through matter Radiation Fundamentals, Applications, Risks, and Safety (Elsevier) 103–36

[6] Cao Z, Aharonian F A and An Q et al 2021 Ultrahigh-energy photons up to 1.4 petaelectron-volts from 12 γ-ray Galactic sources Nature **594** 33–6

[7] Newhauser W D and Zhang R 2015 The physics of proton therapy Phys. Med. Biol. **60** R155–209

[8] Gottschalk B 2018 Physics of proton interactions in matter Proton Therapy Physics 2nd edn ed H Paganetti (CRC Press)

IOP Publishing

Space Radiation
Astrophysical origins, radiobiological effects and implications for space travellers
James S Welsh

Chapter 7

Elements of quantum mechanics for space radiation

7.1 Basic quantum mechanics

Quantum mechanics deals with the behavior of matter and energy at very small scales. It is the realm of molecules, atoms and the subatomic world. As ionizing radiation comes in the form of electromagnetic waves and subatomic particles, the wave–particle duality concept of quantum theory underscores the central importance of quantum mechanics in modern radiation science. This is but one of the many topics in quantum theory that are highly relevant to the study of space radiation. While we will not go into much detail or cover very much of this broad and fascinating subject here, we will delve into a few basic concepts that are relevant to our main topic.

7.2 The tunnel effect

Quantum tunneling is a key concept in space radiation. For instance, alpha decay occurs thanks to the tunnel effect. Also, thermonuclear fusion in the Sun's core is dependent on quantum tunneling. In the way of example, we shall briefly go through the role of quantum tunneling in the proton–proton reaction (p–p chain), which is the first step in how our Sun generates energy via thermonuclear fusion of hydrogen into helium. As discussed in greater depth elsewhere, the first step in the p–p chain is the fusion of two protons into deuterium:

$$^1\text{H} + {}^2\text{H} \rightarrow {}^1\text{H} + \text{e}^+ + \nu_e.$$

This is an extremely slow reaction for several reasons, including its mediation via the weak force. Furthermore, the electrostatic repulsion of two positively charged protons would seem to make it prohibitively unlikely at all but the most extreme temperatures and pressures. Does the Sun, at a mere 15 million K in its core have sufficient temperature and pressure to make this reaction feasible? Well, since the reaction does occur, the answer must be yes. But let's take a look at the specifics.

As two positively charged protons are brought closer and closer together, their electrical potential energy (E_e) due to coulombic repulsion increases as:

$$E_e = k_e e^2 / r \ \ \textbf{(Coulomb's law)},$$

where k_e is Coulomb's constant, e is the electric charge on each proton, and r is the distance between them.

Recall that the strong force is approximately 137-fold stronger than the electro-magnetic force, but its very short range limits its ability to overcome electrostatic repulsion. In the Sun, only when the two protons get within about 1.5×10^{15} m or 1.5 fm (or Fermis) does the strong force take over and dominate the electromagnetic repulsion. But what is the height of this potential energy barrier? To calculate this, we can replace r with 1.5 fm and rearrange the equation to get:

$$E_e = k_e e^2 / 1.5 \text{ fm} = 1 \text{ MeV}.$$

In other words, to get two protons close enough for the strong force to take effect and allow fusion, the protons must possess at least 1 MeV of kinetic energy. If a proton approaches with less than this threshold energy, theoretically the repulsion will slow the proton down and halt it before it arrives at the 1.5 fm distance. So, do the protons in the Sun's core have the theoretical requisite kinetic energy for fusion? At any given temperature, the average kinetic energy is given by:

$$<E> = 3/2 \, k_B T,$$

where k_B is Boltzmann's constant (which is often symbolized as just k).

The Sun's core is approximately 15 million K. So, plugging in 15×10^6 K into the above equation, one arrives at 0.002 MeV or 2 keV. This is far below (by nearly three orders of magnitude!) the expected requirement of 1 MeV. Therefore, at the temperature in the Sun's core, no two protons should have a chance of overcoming the electrostatic repulsion pushing them apart if we use a classical approach; their positive charges should force them apart before reacting every single time. But quantum mechanics tells us that all moving objects have both particle-like and wave-like properties. The **wave–particle duality** is a central concept in quantum theory. According to the de Broglie hypothesis, a photon has a momentum which can be calculated via $p = E/c = h\nu/\lambda$, and moving particles have an associated wavelength given by $\lambda = h/p$. So, when treated as waves rather than particles, might the electrostatic repulsion be overcome?

The wave properties are governed by a wave function, $\Psi(x,y,z)$, given by the **Schrödinger equation**, which describes a probability distribution over a region of space. Essentially the Schrödinger equation is a **wave equation** that relates potential energy (V) plus kinetic energy (KE) to the total energy (E) in the general form of:

$$KE + V(x,y,z) = E.$$

Or since kinetic energy can be related to momentum (p) through $KE = p^2/2m$, this can be expressed:

$$p^2/2m + V(x,y,z) = E.$$

In quantum mechanics, the total energy can be described in terms of the Hamiltonian operator **H**, while the potential energy operator is $V(x,y,z)$ and the kinetic energy can be replaced by a differential operator (specifically including the Laplace operator ∇^2 which is just the second partial derivative in the three spatial dimensions of x, y, and z) that incorporates the de Broglie wave–particle duality: $-h^2/2m \nabla^2$.

Using these symbols, the general equation can be expressed:

$$-h^2/2m\nabla^2 + V(x,y,z) = \mathbb{H}.$$

From this and spelling out the Laplace operator ∇^2, Schrödinger wrote his time-independent wave mechanics equation as:

$$-\frac{\hbar^2}{2m_e}\left(\frac{\partial^2\Psi}{\partial x^2} + \frac{\partial^2\Psi}{\partial y^2} + \frac{\partial^2\Psi}{\partial z^2}\right) + V(x,y,z)\Psi = E\Psi.$$

Since the wave function Ψ, can vary in any given position over time as well as in position at any given point in time, a more general form can be written with $\Psi(x,y,z,t)$. The time-dependent Schrödinger equation is:

$$-\frac{\hbar^2}{2m_e}\left(\frac{\partial^2\Psi}{\partial x^2} + \frac{\partial^2\Psi}{\partial y^2} + \frac{\partial^2\Psi}{\partial z^2}\right) + V(x,y,z)\Psi = i\hbar\frac{\partial\Psi}{\partial t}.$$

For a system with constant energy, this has the form:

$$\Psi(x,y,z,t) = \Psi(x,y,z)\exp\left(-\frac{iEt}{\hbar}\right).$$

Some of these concepts are familiar to students of chemistry in the description of electron orbitals in atoms and molecules. The relevant point here is that waves, unlike classical particles, do not have to absolutely halt when encountering a barrier. Thus, if a potential energy barrier is encountered by a proton, if treated as a pure particle, the proton must stop, but if it is treated as a wave, it can leak through.

The probability of finding a particle at any given location in space depends on the amplitude of the wave function (ψ) at that particular location. More specifically, it depends on the square of ψ (that is, $|\psi|^2$) in accordance with **Born's rule**, which states that the probability of finding an electron (for instance) at a given location in an atom or molecule is approximated by $|\psi|^2$ at that given location. Similarly, the probability of finding the proton on the other side of a seemingly impenetrable potential energy barrier is given by the value of ψ^2 at the particular location beyond the barrier. Solutions to the Schrödinger equation show that ψ diminishes exponentially beyond the barrier, indicating that the probability of finding the proton on the other side of the potential energy barrier is tiny—but not zero. The finding of the proton beyond the apparently insurmountable potential energy barrier is called the **tunnel effect** or **quantum tunneling**. It is thanks to this small, but non-zero, probability of finding the proton on the other side of the classically-insurmountable electrostatic potential barrier that the proton–proton reaction proceeds. Quantum tunneling can only be understood when one accounts for the dual wave–particle behavior of entities in the quantum realm.

7.3 The wave–particle duality

In the macroscopic world, there are clear distinctions between waves (such as ocean waves) and particles (such as baseballs). But for centuries, the nature of light was debated. In the early 19th Century, physician and physicist Thomas Young performed the original double slit experiment, which unequivocally established the wave nature of light. This experimental set up involved a light that was shined upon a surface with two small slits in it, and the resulting pattern of light emerging from the slits was analyzed. That emerging pattern clearly displayed evidence of diffraction and interference, phenomena that can only be explained if light behaves as a wave. In 1905, Albert Einstein analyzed the photoelectric effect and showed that light must also behave as a particle. Thus, was born the wave particle duality for photons. Einstein wrote in his book (co-authored by Leopold Infeld), *The Evolution of Physics*, 'We have two contradictory pictures of reality; separately neither of them fully explains the phenomena of light, but together they do.'

Expanding further on this bizarre split-personality behavior of the photon, Louis de Broglie postulated that at sufficiently small enough levels, matter (that is, particles) behave like waves. According to the **de Broglie hypothesis**, as with any wave, a particle must have an associated wavelength. Such particle behavior has been clearly demonstrated for electrons, which have an associated wavelength. This is the basis for the electron microscope as discussed later in this chapter.

While particles have wave-like properties that are characterized by the Schrödinger equation, another important metric is the wavelength of the particle. The wavelength of a moving particle is given by the **de Broglie equation**:

$$\lambda = h/\mathbf{p},$$

where λ is the de Broglie wavelength, h is Planck's constant, and $\mathbf{p}$ is the momentum.

Conceptually, the **de Broglie wavelength** λ corresponds to the spatial extent of a moving particle's cloud of probability. For example, in classical mechanics a particle should be exactly where one would expect it to be, while in quantum mechanics, the particle would be within about 1λ of the expected location. Returning to the example of proton fusion in the Sun, what this means is that the two protons do not have to approach precisely within the originally calculated 1.5 fm of each other. Instead, they can interact if they get within one de Broglie wavelength of one another. So, what is that value?

$$\text{Since } \mathbf{p} = m\mathbf{v} \text{ and therefore KE} = \tfrac{1}{2}\, m\mathbf{v}^2 = \mathbf{p}^2/2m$$

and

$$\mathbf{p} = \sqrt{2mKE},$$

one can substitute the momentum in the de Broglie equation with this identity to obtain:

$$\lambda = h/\mathbf{p} = h/\sqrt{2mKE}.$$

In light of this, the correct energy calculation for how close the two protons need to get to one another is:

$$E_e = k_e e^2 / \lambda_p = k_e e^2 \sqrt{2mKE} / h .$$

$$\sqrt{KE} = k_e e^2 / h \sqrt{2m} .$$

$$KE = 2m(k_e e^2 / h)^2 .$$

Upon plugging in the values, one can see that if the proton's kinetic energy equals or exceeds 0.0025 MeV (2.5 keV), there is a probability of tunneling through the potential energy barrier. This value is far closer to the average kinetic energy of solar core protons as calculated above and is far lower (about three orders of magnitude) than the classically predicted threshold of 1 MeV.

While we will not examine radioactive decay via alpha emission in the same depth as p–p fusion, we can say that the basic principles are similar. Just as there was a formidable electrostatic potential barrier to overcome for proton fusion, there is a daunting obstacle to alpha emission, as the strong force and a Coulomb barrier must be overcome to set the alpha particle loose. The energy of this barrier is on the order of 20–25 MeV. The quantum tunnel effect must come into play to overcome this barrier.

For a given alpha-emitting radionuclide, there is a characteristic lifetime (τ) or equivalently, a half-life ($t_{1/2}$). Equally characteristic of any given alpha decay is the liberated energy.

The probability of quantum tunneling is related to this energy and the half-life. Rutherford noted that the most energetic alpha particles emanated from the radionuclides with the shortest half-lives. For instance, polonium-212 has a very short half-life of 3×10^{-7} s and emits an 8.95 MeV alpha particle, whereas Th-232 has a 1.4×10^{10} year half-life and emits an alpha particle of 4.0 MeV.

It turns out that the transition energy between most parents and daughters (and thus the energies of the emitted alpha particle) tends to fall in the relatively small range of 2–8 MeV. In other words, E varies only by a factor of about 4 among alpha-emitters. In contrast, measured values for $t_{1/2}$ vary enormously from microseconds to hundreds of billions of years. In fact, bismuth-209, which was previously believed to be stable, has recently been found to undergo alpha decay with a half-life of 2.01×10^{19} years. Thus, the range of half-lives is on the order of 10^{26}. This means that the relation between energy (or as measured in the laboratory, range) and half-life must be an extremely strong function of energy.

Geiger and Nuttall expressed this relationship between range half-life in 1912 through what is now called the **Geiger–Nuttall rule**, which relates the half-lives of alpha emitters to the ranges of the released α particles.

log R(α) = A + B log λ The Geiger-Nuttall law,

where A and B are empirical constants that vary with Z of the radionuclide, $R(\alpha)$ is the range of the ejected alpha particle in air (which is a surrogate for energy of the ejected particle), and λ is the decay constant of the parent radionuclide. (Recall that $\lambda = 0.693 / t_{1/2}$.)

The relationship also shows that half-lives are exponentially dependent on decay energy, so that very large changes in half-life make comparatively small differences in decay energy or alpha particle energy. In practice, this means that alpha particles (from essentially all alpha-emitting isotopes across many orders of magnitude of difference in half-life) all have roughly the same decay energy.

Stated simply, short-lived alpha-emitting radioisotopes emit more energetic alpha particles than long-lived ones. Gamow, Condon and Gurney developed a more complete theory of alpha decay that incorporates quantum tunneling and showed good correlation with the Geiger–Nuttall rule. Given the enormous energy barrier for an alpha particle to overcome the attractive strong nuclear force, it would be challenging to explain this correlation through any theory that did not incorporate quantum tunneling.

7.4 Degeneracy pressure

Another highly instructive example of quantum mechanics is provided by the concept of degeneracy pressure, which is something we shall discuss repeatedly when dealing with the internal structure of stars. Stars are simply colossal agglomerations of matter that are held together through gravity. Gravity can be described by **Newton's Law of Universal Gravitation**:

$$F = G\frac{m_1 m_2}{r^2}, \quad G = 6.67 \times 10^{-11}\,\text{N} \cdot \text{m}^2/\text{kg}^2,$$

where F is the gravitational force of attraction, G is the gravitational constant, m_1 and m_2 are the masses involved, and r is the distance between them.

As the mass increases, the gravitational force of attraction increases. And as a huge star contracts (i.e. as r decreases), the gravitational force increases as the square of this decreasing distance. In other words, gravitational attraction continually squeezes the matter tighter and tighter together. There must be some form of resistance or pressure to prevent gravity from simply crushing the matter out of existence in the form of a **black hole**. As will be discussed later, sometime gravity is victorious in its battle with this outward pressure, and the matter is indeed squeezed tighter and tighter until a singularity of zero volume is achieved. In other words, a hole in spacetime is created. But under less extreme circumstances, to prevent gravity from crushing the matter within a star into oblivion, various externally-directed forms of pressure are called into play to fight gravity. Thermal pressure from heat produced through thermonuclear fusion is one source of such outwardly directed pressure. This can withstand gravity up to a point, but beyond this point, other forces must come into play. One other form of pressure that is sometimes called upon to stave off gravity is **quantum degeneracy pressure**.

We described the characteristics of **bosons** and **fermions** in another chapter. Bosons are particles with integral spin, whereas fermions possess half-integer spin values. The bosons follow statistical mechanics known as **Bose–Einstein statistics**, whereas fermions follow **Fermi–Dirac statistics**. Another feature is that the wave function ψ for bosons is symmetric, whereas the wave function for fermions is antisymmetric with respect to an exchange of any pair of particles. To understand this, we should note that the complete wave function of a system of n non-interacting and indistinguishable particles is given by:

$$\Psi(1, 2, 3, \ldots n) = \Psi(1)\Psi(2)\Psi(3)\ldots \Psi(n).$$

Mathematically this is represented:

$\Psi(2, 1) = \Psi(1, 2)$ (**symmetric wave function** for bosons),

$\Psi(2, 1) = -\Psi(1, 2)$ (**antisymmetric wave function** for bosons).

Wave functions that are symmetric are unaffected by the exchange of indistinguishable particles, and this is a feature of bosons. Any number of bosons can co-exist in the same quantum state of a system. In stark contrast, wavefunctions that are antisymmetric with respect to exchange of indistinguishable particles result in a reversal of sign, and this is characteristic of fermions. Only one fermion can exist in a particular quantum state in a given system. This is another way of stating that fermions obey the **Pauli exclusion principle**.

The Pauli exclusion principle imposes an inherent quantum resistance to crowding by fermions. Macroscopically, this manifests as a resistance to compression of matter into more compact volumes. Electron degeneracy occurs at densities of about 10^6 kg m^{-3}. A classic example of this resistance to compression is given by **white dwarfs**. These are stellar remnants that reach a static equilibrium with the inward directed pressure of gravity, not through heat generated by thermonuclear fusion, but rather through **electron degeneracy pressure**. This is simply the resistance to further compression by gravity into more compact volumes that is caused by the natural reluctance of electrons to crowd one another. It is electron degeneracy pressure that prevents further collapse of a white dwarf into another form of matter with even higher density.

But electron degeneracy pressure can only withstand gravity up to a certain limit. This maximum limit is known as **Chandrasekhar's limit**. Chandrasekhar's limit varies with the elemental composition of the white dwarf remnant, but for our purposes, we will use the value of 1.44 solar masses here. If a white dwarf has more mass than Chandrasekhar's limit, even electron degeneracy pressure cannot fight off gravity, and the stellar remnant will collapse further still. When electron degeneracy pressure can no longer withstand the immense squeeze of gravity, against their will, the electrons are forced into the space of the nucleus. In the presence of such incredible pressure, the electrons are physically forced into proximity of the nuclear protons and are captured, forming neutrons (plus electron neutrinos). This next step in stellar evolution is a **neutron star**.

As mentioned in the discussion about electron capture in radioactivity, when neutrons are created through electron capture, neutrinos accompany the reactions for balance of lepton number. Thus, a flood of neutrinos is generated the moment when gravity wins its war with electron degeneracy pressure. (But as explained elsewhere, this is not the only, nor the primary source of the profusion of neutrinos in a supernova)

Students may wonder what prevents gravity from just crushing the neutrons into oblivion just as it overpowered the electrons. The answer lies in the fact that neutrons are also fermions. And, as such, they too obey the Pauli exclusion principle and exert a form of quantum degeneracy pressure of their own—**neutron degeneracy pressure**. Neutron degeneracy pressure is far stronger than electron degeneracy pressure and can withstand the gravitational contraction of stellar cores even beyond Chandrasekhar's limit of 1.44 solar masses. Nevertheless, if the relic star is sufficiently massive, neutron degeneracy pressure too can be beaten by gravity. The weight limit for a neutron star, called the **Oppenheimer–Volkoff limit**, is less certain than the Chandrasekhar limit, with present estimates at about 2.17 solar masses.

Degenerate matter can be modeled as a **Fermi gas**, which is a collection of non-interacting fermions such as electrons. As with electrons in an atom, such particles can only occupy discrete quantum energy states and therefore take up a finite volume. When thermal energy is minimal, the electrons will occupy the lowest energy states. When all the lowest energy quantum states are occupied, the condition is called full degeneracy. An important characteristic of fully degenerate matter is that it will exhibit degeneracy pressure regardless of the temperature—only the density matters. Theoretically, even at absolute zero, fermionic matter will exert a degeneracy pressure manifested as a resistance to further compression. Thus, in stellar evolution, during the constant battle against inward directed gravity, the outward directed degeneracy pressure and the outward pressure generated by heat from thermonuclear fusion are completely independent allies.

Another way to look at degeneracy pressure is from the perspective of the **Heisenberg uncertainty principle**. Heisenberg's uncertainty principle basically states that there is an inherent uncertainty surrounding certain pairs of physical attributes of particles such position and momentum or between energy and time. Such pairs of variables are known as **complementary variables**. Measurement uncertainties aside (as well as the consequences of taking a measurement, such as measuring an electron's position by using a photon that affects that electron's position), the uncertainty principle asserts that there is a *fundamental* limit to the accuracy with which one can simultaneously know the position and momentum of a particle. Mathematically it is usually represented as:

$$\Delta p \Delta x \geqslant h/4\pi \quad \textbf{(Heisenberg uncertainty principle)},$$

where h is Planck's constant, Δp is the uncertainty in momentum, and Δx is the uncertainty in position

Using the **reduced Plank's constant** ($\hbar$, which is $h/2\pi$ and is also called Dirac's constant) this can be written as:

$$\Delta p \Delta x \geqslant \hbar/2.$$

Another form of the uncertainty principle incorporates the complementary variables of energy and time:

$$\Delta E \Delta t \geqslant \hbar/2.$$

As gravity continues to squeeze a star's interior, unless thermal pressure can be generated through thermonuclear fusion to combat gravity, the core will be compressed. If it has run out of fuel for fusion or simply does not have the required temperature and pressure to fuse what fuel it does have, gravity will gain the upper hand, and the end product might be a white dwarf. But as compression continues, the volume in which electrons reside diminishes. In other words, Δx, the position, becomes more and more restricted and therefore more precisely known. From the uncertainty principle, Δp, the uncertainty in momentum must correspondingly increase. Since momentum is related to velocity, this means the velocity must be increasing as the position becomes increasingly bound. But from special relativity, we know that velocity has an upper limit, c. As the pressure mounts and the volume decreases, the position becomes progressively more precisely defined but the velocity continues to grow faster and faster. But this presents an inherent obstacle thanks to special relativity's speed limit. Hence, an intrinsic quantum mechanical resistance to further compaction emerges from the exclusion principle.

This leads to a paradoxical behavior of degenerate matter. Normally, temperature is a reflection of particle kinetic energy, but in highly compressed degenerate matter, even when the temperature is extremely cold, some of the particles are moving at extremely high speeds because of the decrease in positional uncertainty. The consequence of the high speeds and momenta of these particles is an immense force (i.e. pressure). Mathematically, this can be understood by first starting with a review of ideal gas behavior and then moving on to ideal Fermi gas behavior. Classical gases can be modeled through the Ideal Gas Law:

$$PV = nRT,$$

where P is pressure, V is volume, n is the number of moles of the gas, R is the universal gas constant, and T is temperature.

This is the familiar form of the equation that dictates behavior of gases on a macroscopic level, that is, large collections of particles. When we are focusing on the individual particles rather than moles of particles, the Ideal Gas Law can be rewritten as:

$$P = kNT/V = kT(N/V),$$

where k is Boltzmann's constant and N is the number of particles rather than the number of moles, n. Thus, n is simply N multiplied by Avogadro's number (6.02×10^{23}). One can now see that the pressure is proportional to the 'number density' (N/V) in an ideal gas. For an ideal Fermi gas (a gas of non-interacting fermions) at low densities, the pressure can be described by:

$$P = (3\pi^2)^{2/3} \hbar^2 / 5m (N/V)^{5/3}.$$

This is analogous to the ideal gas pressure with the only differences being the constants and the exponent on (N/V). As the density of this Fermi gas increases (for instance, in a stellar interior), where some of the particles are forced into quantum states with very high energies (and therefore relativistic velocities), the equation must be modified slightly into:

$$P = C(N/V)^{4/3}$$

C is a slightly different constant that is a function of the particular particles constituting the Fermi gas. But the concept is the same: pressure is proportional to (N/V) raised to some power, in this case 4/3.

In any case, this quantum degeneracy pressure imposes a limit to just how tightly squeezed the electrons can get in a white dwarf. This quantum phenomenon becomes macroscopically observable as the electron degeneracy pressure—the pressure that resists further gravitational contraction in a white dwarf.

Similar phenomena occur in neutron stars (and in proposed strange quark stars). Namely, a gravity-defying quantum degeneracy pressure is created by the constituent fermions obeying the Pauli exclusion principle and refusing to be crowded into the same quantum states or because the Heisenberg uncertainty principle demands that the positions of these particles cannot become ever more well-defined. This weird quantum mechanical pressure permits some stellar corpses to rest in peace as neutron stars in a sort of truce in their timeless battle between inward gravity and outward pressure. In subsequent chapters, we shall see that the battle does not always end in a stalemate.

7.5 Statistical mechanics

As mentioned, bosons are subatomic particles with integer spin values, whereas fermions have half-integer spins. Additionally, bosons do not obey the Pauli exclusion principle, whereas fermions do. Bosons, having symmetrical wave functions, follow a statistical behavior called Bose–Einstein statistics, while fermions, with their antisymmetric wave functions, follow Fermi–Dirac statistics. Mathematically, the distribution functions $f(E)$ are correspondingly different:

$$f(E) = 1/[e^{\alpha}e^{E/kT} - 1] \text{ (\textbf{Bose–Einstein distribution function}),}$$

$$f(E) = 1/[e^{\alpha}e^{E/kT} + 1] \text{ (\textbf{Fermi–Dirac distribution function}),}$$

where α is a parameter that depends on the properties of the particular system and may be a function of temperature.

Just as half-life is a useful concept when dealing with radioactivity, the value of $f(E) = \frac{1}{2}$ can be a helpful parameter in Fermi–Dirac statistics. This special case yields the **Fermi energy**, E_F:

$$E = -\alpha kT = E_F.$$

In terms of the Fermi energy, the Fermi–Dirac distribution function becomes:

$$f(E) = 1/[e^{(E-E_F)/kT} + 1].$$

All of the above discussion refers to particles that are identical and inherently indistinguishable such as photons and electrons. Their wave functions overlap. But for classical particles such as gas molecules or atoms, spin is irrelevant. And such particles are in practice identical but hypothetically distinguishable. These classical particles can be described by yet another distribution function, the Maxwell–Boltzmann distribution function. **Maxwell–Boltzmann statistics** apply to particles that tend to be far enough apart that their wave functions do not overlap.

$$f(E) = Ae^{-E/kT},$$

or when written in a form analogous to the others above:

$$f(E) = A \, 1/[e^{E/kT} + 0] \text{ (\textbf{Maxwell–Boltzmann distribution function})}.$$

7.5.1 Kirchhoff's laws and blackbody radiation

Two key concepts in astronomy are:

1. **Hubble's Law** (also known as the Hubble–Lemaître law), which states that the Universe is expanding and that the farther away a distant galaxy is, the faster it is receding from us.
2. One can determine the chemical composition of remote stars.

Both essential concepts trace their scientific histories to spectral analysis of light. The fact that very remote galaxies are rapidly receding was determined through detection of a red shift of their emitted light. By analyzing the light spectrum coming from such galaxies, scientists could see a Doppler shift caused by their relative motion. And by scrutinizing the spectrum of light produced by stars (including our Sun), one could determine their chemical compositions. But in order to do all this, one must examine certain dark lines in their spectra, the so-called **Fraunhofer lines**. Although the presence of these lines was known for decades, it was not until quantum theory was properly developed that a clear understanding of their meaning was possible.

In the 19th Century, it was known that an opaque body, when heated sufficiently, will glow in the visible portion of the spectrum. In fact, we now know that any body with a temperature above absolute zero (which limits the discussion to everything) emits electromagnetic radiation. This emitted radiation is called **thermal radiation**. A blackbody is an idealized, opaque body that reflects no incident light but absorbs all of it, and emits its own radiation. In the real world, such entities do not exist, but the theoretical treatment of these blackbodies and their emitted radiation (**blackbody radiation**) is closely approximated by the thermal radiation given off by certain heated objects, including the Sun, the stars, and incandescent light bulbs. In the following discussion, thermal and blackbody radiation will be treated synonymously.

Simply stated, **Kirchhoff's first law** says that any object with a temperature gives off a spectrum of thermal electromagnetic radiation. Just where this particular spectrum lies and peaks along the overall span of electromagnetic wavelengths depends on the exact temperature. For example, an afebrile human body at 98.6 °F (37 °C) emits radiation mainly in the infrared part of the spectrum, mostly at about a wavelength of 12 μm. But while humans do not perceptibly glow in the dark, they do emit tiny amounts of visible light and even shorter wavelengths. If considered a blackbody, humans would theoretically even give off some gamma rays and radio waves. The rate at which such photons are emitted is so low as to be undetectable, of course. (It is important to not confuse this theoretical blackbody emission of gamma rays with the *real* radiation emitted by humans that stems from the decay of potassium-40 and carbon-14 that naturally exists in all living entities). Nevertheless, the Sun and stars emit a full spectrum of thermal radiation, with much of it in the visible part of the spectrum. In fact, the Sun's most probable wavelength of emission, λ_p, lies around 500 nm.

The way to calculate the most probable wavelength, λ_p, is through **Wein's law**:

$$\lambda_p = b/T,$$

where b is Wein's constant $(2.898 \times 10^{-3} \text{ m} \cdot \text{K})$.

Wein's law is also called **Wein's displacement law** because it calculates the displacement of the peak of a **Planck curve**, which is the plot of emitted thermal radiation on a graph of intensity versus wavelength. Wein's displacement law mathematically shows that the wavelength of peak emission (λ_p) is inversely proportional to the temperature of the blackbody (figure 7.1).

Historically, Planck curves are created by plotting **spectral radiance** versus wavelength (but the abscissa can be frequency, wavenumber or their angular equivalents). Spectral radiance is the power emitted from an emitting surface per unit area of emitting surface, per unit solid angle, per spectral unit (typically wavelength, but again this can be wavenumber, frequency or their angular equivalents). Instead of spectral radiance, the ordinate can be luminosity, power, emittance, etc. In the typical format where the abscissa is wavelength, Wein's law says that the Planck curves are moved to

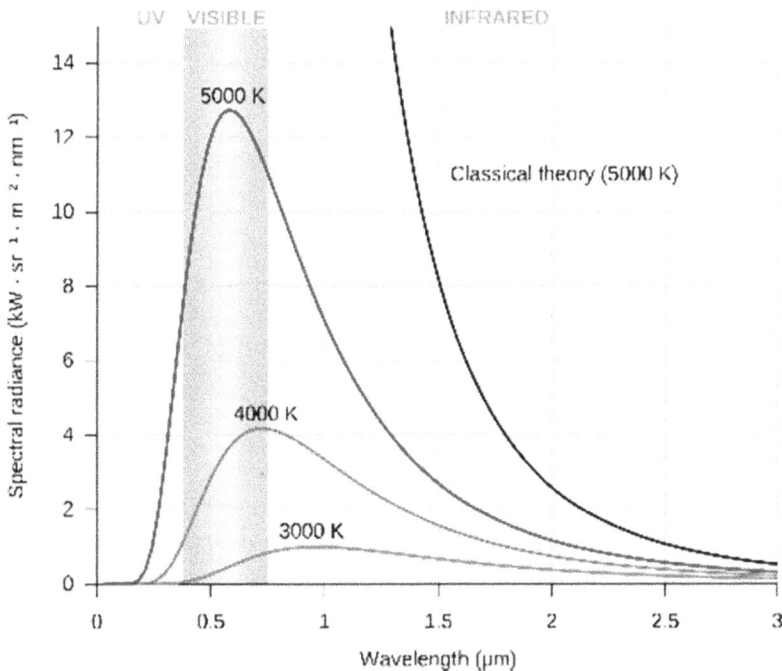

Figure 7.1. A series of Planck curves at various temperatures (3000 K, 4000 K and 5000 K) are shown in this plot of spectral radiance versus wavelength. As the temperatures increase, the peaks of the curves are shifted to the left in accordance with Wein's law. Also, the areas under the curves increase with temperature in accordance with the Stefan–Boltzmann law. Also shown is the incorrect curve predicted by classical theory that continues rising indefinitely as temperature increases and wavelength decreases; this is known as the ultraviolet catastrophe. This blackbody image has been obtained by the author(s) from the Wikimedia website, where it is stated to have been released into the public domain. It is included within this book on that basis. It is attributed to Darth Kule.

the left (i.e. to shorter, more energetic wavelengths) as the temperature of the emitting object increases. Obviously, since $c = \nu\lambda$, there are a number of different ways the same concept can be quantified. Students should just be aware of what is being plotted against what when examining Planck curves.

Along with the shape and peak of the curve itself, another important quantitative feature of Planck curves is the area under the curve. If the curve is plotted as E_λ (the energy radiated per unit volume for a given wavelength) versus wavelength interval (λ to $\lambda + \Delta\lambda$), the Planck curve can be expressed by the equation:

$$E_\lambda = \frac{8\pi hc}{\lambda^5} \times \frac{1}{\exp(hc/kT\lambda) - 1}.$$

This mathematical expression is known as **Planck's Law** (and should not be confused with the simpler and more familiar **Planck–Einstein relation**, $E = h\nu$ which is often simply called the **Planck relation**). One can see that as wavelength decreases, according to Planck's Law, the total amount of energy emitted falls off dramatically (to the fifth power).

Another useful way of characterizing these Planck curves emerges when one examines the shape of the curve as a function of temperature. If E is the energy radiated from a unit area per unit time (that is, the power per unit area, which is the time-averaged energy flux or intensity), and T is the absolute temperature, then:

$$E = \sigma T^4 \quad \textbf{(Stefan–Boltzmann law)}.$$

This is known as the Stefan–Boltzmann Law. More precisely, the equation should be written as:

$$E = e\sigma T^4,$$

where e is the **emissivity** of the radiating body. Emissivity ranges from zero for a perfect reflector that does not radiate at all to one for a perfect blackbody. σ is the **Stefan–Boltzmann constant**, which is a constellation of different numbers and constants given by:

$$\sigma = \frac{2\pi^5 k^4}{15h^3 c^2}.$$

The Stefan–Boltzmann law means that the 'spectral radiance' or total energy radiated from a blackbody is directly proportional to temperature of the blackbody raised to the fourth power. This is illustrated by the areas under the Planck curves at various temperatures in figure 7.1. One should notice that E, the power per unit area, is the same as energy per unit area per unit time. This quantity is known as the energy flux. In astronomy, the flux or power per unit area can be related to the overall luminosity of a star since stars are spherical. The surface area of a sphere is given by:

$$A = 4\pi r^2.$$

Thus, the total **luminosity** (L) is simply the flux (luminosity per unit area) multiplied by the total area:

$$L = AE = (4\pi r^2)(\sigma T^4).$$

Students should be aware that the symbols used in the above derivation are not universally applied. So, one should be alert about potential confusion stemming from simple differences in the symbols chosen for quantities such as luminosity and flux.

Although the philosopher August Comte famously asserted in 1835 that, 'the true temperature of the stars will necessarily always be concealed from us', it was not very far in the future that Comte was proven wrong through a combination of Kirchhoff's first law plus Wein's law. All one must do is examine a star's blackbody continuous emission spectrum, plot out a Planck curve and record the wavelength that is brightest. Wein's displacement law then enables a calculation of the temperature.

Another assertion made by 19th Century philosophers was that we could never know the chemical composition of the stars. However, if these philosophers acquainted themselves with the work of Joseph von Fraunhofer (1787–1826), they would have realized that the clues were already available to them way back then.

Kirchhoff's first law focused on *opaque* bodies, which upon heating emitted a full spectrum or rainbow of radiation. Kirchhoff's second and third laws focus on *transparent gases*. Specifically, **Kirchhoff's second law** states that a heated transparent gas will not emit a full spectrum of light (like a heated opaque body does) but, instead, only emits specific discrete wavelengths. These discrete wavelengths constitute what is now called an **emission spectrum**. Complementing this is **Kirchhoff's third law**, which states that a transparent gas placed in front of a full spectrum light source will introduce dark lines in the spectrum, disrupting the rainbow. The dark lines superimposed on a broad spectrum of light constitute what is now called an **absorption spectrum**. The bright lines on an emission spectrum correspond to the same wavelengths as the dark lines on an absorption spectrum for any particular transparent gas so tested. In a very real sense, these spectral lines constitute a 'fingerprint' that reveal the chemical identity of the gas in question. Fraunhofer studied the Sun's spectrum and noticed that there were dark absorption lines superimposed on the otherwise full spectrum rainbow of light produced by the Sun. These absorption lines are sometimes called **Fraunhofer lines** in honor of their discoverer (figure 7.2). (Actually, it was William Hyde Wollaston (discoverer of the elements palladium and rhodium) who first noticed the dark lines in the solar spectrum in 1802, but Fraunhofer systematically studied the exact locations of these lines and determined their corresponding wavelengths in 1814. In any case, they reveal to us the various gases present in the solar atmosphere.)

Kirchhoff's first law states that an opaque body will emit a continuous spectrum of light (thermal or blackbody radiation) when sufficiently heated. Kirchhoff's second law states that a transparent gas that is sufficiently heated will emit light in the form of discrete lines of specific wavelengths, and Kirchhoff's third law states that a transparent gas placed in front of a continuous spectrum will result in dark absorption lines in that spectrum. The wavelengths of the emission lines and the absorption lines are the same and are characteristic of the elements in the gas. Students should not confuse Kirchhoff's laws of spectroscopy with Kirchhoff's rules in electrical circuits. (Kirchhoff's first rule— the junction rule: The sum of all currents entering a junction must equal the sum of all currents leaving the junction. Kirchhoff's second rule—the loop rule: The algebraic sum of changes in potential around any closed-circuit loop must equal zero.)

Continuous Spectrum

Emission Lines

Absorption Lines

Figure 7.2. Kirchoff's first law states that a heated opaque body will emit radiation. If the temperature is sufficently high, that radiation might be in the visible range and would be a continuous spectrum as at the top. If a transparent gas is heated, it does not emit a full spectrum but instead only emits specific emission lines. This is Kirchoff's second law. Kirchoff's third law states that if a transparent gas is placed in front of a light source with a continuous spectrum, dark absorption lines become evident. These specific emission lines and absorption lines are characteristic of the element in question. As an interesting historical aside, helium was discovered in the Sun through spectral analysis before it was discovered on Earth. This spectral lines image has been obtained by the author(s) from the Wikimedia website, where it is stated to have been released into the public domain. It is included within this book on that basis. It is attributed to Jhausauer.

Thanks to our modern understanding of quantum mechanics, we now know that these spectral lines are caused by electrons making jumps from one quantum energy state to another in atoms. The fact that these emission and absorption lines are always in the same place for a given element reveals that the electrons in that particular element are confined to very specific energy levels and jumping from one discrete energy level to another must emit or absorb the exact energy difference between those two levels. For that reason, the wavelength of emission or absorption is always in the same location and corresponds to a very precise amount of energy. This energy corresponds to a so-called quantum leap from one distinct energy level to another. There is no 'in-between' possibility in electron energy levels; only very specific energy transitions are allowed by theory and observed by experiments. These electron energy levels are separated by a precise amount of energy in an absorbed or emitted photon, and that energy is given by the Planck–Einstein relation $E = h\nu$.

In fact, when analyzing thermal radiation and striving to derive the right formula for the spectral radiance versus wavelength plots, Planck had to postulate that light was emitted in discrete energy units or quanta, characterized by $E = h\nu$. This was needed in order to circumvent the 'ultraviolet catastrophe' that befell his predecessors Wein, Rayleigh and Jeans, who all used classical approaches to the problem. Some science historians argue that this remarkable insight made by Max Planck was the dawn of quantum theory. This postulate was a revolutionary change because ever since Young's famous double slit experiment in 1801, which clearly showed constructive and destructive interference patterns, light was considered to be a wave. Planck showed that if light was emitted from a blackbody as a particle (a quantum as he called it, or a photon as we now call it), the correct formula for blackbody (thermal) radiation could be obtained. Other historians might argue that Planck never truly asserted that light was a particle. He only argued that if that the energy of the light emitted from a blackbody was quantized in units of $h\nu$, the correct formula for blackbody radiation could be obtained. It was Albert Einstein in 1905 who conclusively proved that light behaved as a particle through his analysis of the photoelectric effect. Thus, Einstein may arguably be considered the father of quantum theory.

Today, we know that light exhibits both wave and particle characteristics, one of the hallmarks of quantum theory. Furthermore, consistent with the de Broglie hypothesis of wave behavior for particles, the double slit experiment can be duplicated with electrons. Thus, just as with photons, electrons behave like particles in some situations and as waves in others. The wave-like behavior of electrons is the basis for the **electron microscope**. The resolution of any microscope is inversely proportional to the wavelength of the light or electrons used to examine the object in question. Recall that the de Broglie wavelength for an electron is indirectly related to its momentum p (which equals $m_e v$) by:

$$\lambda = h/p = h/m_e v.$$

Plugging in $h = 6.6 \times 10^{-34}$ kg $\cdot$ m^2 s^{-1}, the mass of an electron ($m_e = 9.1 \times 10^{-31}$ kg) and the velocity of the electrons (given modestly as $v = 10^6$ m s^{-1}), one can calculate that $\lambda \approx 10^{-10}$ m or an Angstrom. This is just about the diameter of an atom, which is why electron microscopy is so powerful. (Recall that the wavelengths of visible light are in the 400–750 nm range (1 nm = 10^{-9} m), so visualizing anything at the molecular level is not feasible with visible light). Another point to recall is that electrons are so light that it does not take much energy to get them up to relativistic velocities (where one must make corrections for **special relativity**). Instead of using the standard $KE = 1/2\ mv^2$ equation, one might have to employ:

$$KE = \gamma mc^2 - mc^2,$$

where $\gamma \equiv \dfrac{1}{\sqrt{1 - \frac{v^2}{c^2}}}.$

As a rule of thumb, kinetic energy and momentum will deviate from classical calculations by around 1% when gamma = 1.01. This corresponds to a velocity $v = 0.14c$ or 14% of the speed of light. Using this 1% threshold for calling special relativity corrections into play, this amounts to only 5.11 keV for electrons or 9.38 MeV for protons (1% of the rest mass for each particle, respectively). Therefore, it is clear that when working out problems with electrons and positrons created at the rest mass energies of 511 keV apiece during pair production, relativistic corrections should be applied if one wants maximal accuracy.

One final aspect of the electron version of the double slit experiment is worthy of mention. In the **Copenhagen interpretation** of quantum mechanics, a mixed-state quantum particle such as the electron does not exist in one state or another but rather in all possible states of the wave function simultaneously. However, upon observation, the wave function 'collapses' into a single observed state—a so-called pure state. This phenomenon is best illustrated by the double slit experiment using electrons in which a device is created to keep track of which slit the electrons have gone through. In the absence of such measurement, the electrons pass through the two slits and demonstrate a classic wave-like (interference-pattern) distribution on the detector. However, when a device is activated to keep track of which slit the electron went through, the wave-like interference pattern is eliminated. Only two bright spots are seen on the detector. This strange phenomenon is not due solely to the interaction of the electrons with the measuring method. In one version of the experiment, the electrons interact with photons to yield a signal indicating which slit they traversed. One might be tempted to attribute the electron–photon interaction as the cause of the loss of wave characteristics and the appearance of only particle characteristics. However, this interpretation is shown to be wrong by another version of the experiment in which only one of the two slits has a photon-based detector. After all, if there are only two slits and we have a detector on one that measures whenever an electron went through that slit, we do not really need such a detector on the other slit—if it didn't go through the slit with the detector, it must have gone through the slit without the detector. Nevertheless, the interference pattern vanishes if any sort of measurement methodology is imposed. And the explanation for this wave function collapse is not because the photons used for measurement have somehow meddled with the electrons; it is inherently associated with knowing which slit the electron has used. Another illustration of this fact is based on diminishing the intensity of the light photons used to determine which slit the electron went through. As the light intensity is decreased, there is initially no impact on the electron distribution pattern—until one reaches the point where the intensity of the light is so low that it is no longer capable of telling us which slit the electron truly traversed. At that point, the electron interference pattern miraculously reappears. The same thing is found when one decreases not the intensity of the light but, instead, the energy of the photons. Lower-energy photons will disturb the electrons less, and thus, one might expect that red photons would have less effect on the electrons than blue light. But as the wavelengths of the photons are increased further and further, at one point they will no longer be able to tell us which slit the electron 'chose'. And it is at this photon wavelength that the electron interference pattern again remerges. It appears that anything that allows one to know which slit was selected by the electrons collapses the wavefunction and the associated wave-like behavior. This weird quantum phenomenon may be related to the uncertainty principle. By knowing exactly which slit the electron has gone through (i.e. its position), we must have greater uncertainty in the momentum of the electron (represented by the wave-like interference pattern).

7.6 The Balmer series and the Bohr atom

It was a careful study of spectral lines that ultimately led Niels Bohr to come up with the model of the hydrogen atom known as the Bohr atom in 1914. Elemental hydrogen displays characteristic absorption and emission spectral lines across the electromagnetic spectrum. In the visible portion of the spectrum is the **Balmer series**, and in the ultraviolet is the **Lyman series**. The Lyman series represents electron energy transitions from the $n \geqslant 2$ principal energy levels down to the $n = 1$ level (the ground state). The lines of the Balmer series originate from transitions from excited levels where $n \geqslant 3$ back down to the $n = 2$ level. Although all the Lyman series lines lie in the ultraviolet, most of the Balmer series lines fall into the visible (with some in the ultraviolet with a lower limit of $\lambda = 364.5$ nm). Other series of spectral lines are evident in hydrogen besides the Lyman and Balmer series. In order of increasing wavelengths (decreasing energy) these are the **Paschen, Brackett, Pfund and Humphrey series**, which correspond to electron energy transitions back down to the $n = 3$, 4, 5 and 6 levels, respectively. The location (wavelengths) of these spectral lines can be predicted through the **Rydberg formula**:

$$\frac{1}{\lambda} = Z^2 R \left(\frac{1}{n_1^2} - \frac{1}{n_2^2} \right),$$

where λ is the wavelength, Z is the atomic number, R is Rydberg's constant, and n represents the principal quantum numbers involved in the transition.

Bohr recognized that the success of the Rydberg formula for hydrogen spectral lines indicated that electrons must be bound to the nucleus (which, in the case of hydrogen, is a single proton) only in very specific energy levels. These quantized energy levels are fixed in energy, and electrons cannot exist anywhere in between these allowed levels. As introduced in chapter 3, the energy of these electron levels can be given by:

$$E_n = -\frac{m_e e^4}{2[4\pi \in \hbar]^2} \frac{1}{n^2} = -13.6 \left(\frac{1}{n^2} \right) \text{eV}.$$

For example, when an electron in the $n = 2$ state drops down to the $n = 1$ energy level, a photon of 121.7 nm is emitted. Likewise, when a photon of 121.7 nm is absorbed by a hydrogen atom, an electron jumps from the $n = 1$ state up to the $n = 2$ energy level. This explains the emission and absorption spectral lines, respectively, and implies that only certain energies of photons can be emitted or absorbed. This is indeed what is observed. The Bohr model works well only for hydrogen. For multi-electron atoms and molecules, a more comprehensive description based on electron wavefunctions provided by the Schrödinger equation is needed. Nevertheless, the qualitative concepts are the same, and this is the essence of quantum mechanics behind atomic and **molecular orbital theory**. Discrete emissions at various but specific wavelengths caused by electronic transitions in various atoms and molecules account for various phenomena covered in this text including fluorescence, the beautiful glow of **planetary nebulae** and the **auroras**, as well as **characteristic x-rays**.

Notably, the constant in the Bohr atom energy equation, 13.6 eV, is the ionization energy of hydrogen. Thus, this figure is taken by some as the minimum energy defining 'ionizing radiation'. The values of n in the above equation can take on values of $n = 1$, 2, 3, etc. Today, we call n the principal quantum number, but in the

Bohr model, this parameter corresponds to electron shells, familiar to students of chemistry as the K, L, M, N, etc shells. When an electron transitions from one shell to another, characteristic spectral emission lines are evident that can be calculated by a different expression of the Rydberg formula. These wavelengths fit the equation:

$$\lambda = hc/(E_i - E_f).$$

E_i is the energy of the initial energy state, and E_f is the energy of the final electron energy state. One can see that this denominator can also be written as ΔE. This happens to be the energy of the emitted photon. From here is a short intellectual step to the energy in the Planck–Einstein relation, given by $E = h\nu$ (which of course is the same as $E = hc/\lambda$, mathematically proving the equality).

The energy transitions clarified by the Rydberg formula were foundational in the history of quantum mechanics but were developed from a study of the simplest atom, hydrogen. Nevertheless, analogous energy level transitions are important in modern astronomy and radiation science. A classic transition is the **21 cm line of hydrogen**, the so-called **H1 line**. A wavelength of 21 cm (or equivalently, 1420 MHz) falls into the radio region of the spectrum, just below the microwave range. It is emitted during **hyperfine transitions** between electrons in the 1s ground state of atomic hydrogen. In this case, the electron is not jumping from one energy level to another but rather is undergoing a far more subtle energy transition—it is flipping its spin from 'up' to 'down' relative to the spin of the proton. For this reason, the transition is therefore sometimes called the **spin-flip line**. It plays an important role in radio astronomy with applications in cosmology and even in the search for extraterrestrial intelligence, because it was depicted in messages on the Pioneer and Voyager spacecrafts (hopefully not too cryptically for any aliens who eventually find them!)

7.6.1 Hubble's Law

As a final illustration of the importance of spectral analysis in astronomy, Hubble's Law (also called the **Hubble–Lemaître Law**), which states that the Universe is expanding and that galaxies that are farther away are moving away faster than those that are closer to us, was determined through a close look at the spectral lines in distant stars. Quantitatively, this says that galaxies are moving away from us at a rate (v) proportional to their distance:

$$v = H_0 d \quad \textbf{Hubble's Law,}$$

where H_0 is **Hubble's constant** (which is not a true 'constant' in that it has changed over time).

As mentioned above, the emission and absorption lines for any given element are in characteristic locations, corresponding to very specific wavelengths. It was noticed that the light from most galaxies have spectral lines that are markedly Doppler shifted towards the longer wavelengths (i.e. **redshifted**) compared to their normal locations. This indicates motion away from us; motion toward us would be a **blueshift**. Doppler shifts are quantified in units denoted by the metric **z**, with

z-values	Time the detected light left its source
$z = 0.5$	≈5 billion years ago
$z = 1.0$	≈7.7 billion years ago
$z = 1.5$	≈9.3 billion years ago
$z = 2$	≈10.3 billion years ago
$z = 3$	≈11.5 billion years ago
$z = 4$	≈12.1 billion years

redshifts being positive and blueshifts being negative. The formula for calculating z is:

$$z = (\lambda_{observed} - \lambda_{emitted})/\lambda_{emitted},$$

or equivalently:

$$1 + z = \lambda_{observed}/\lambda_{emitted}.$$

Given that the speed of light is finite, when we see heavily redshifted, very distant galaxies, we are not seeing them as they are today but instead as how they appeared in the distant past. Because of the finite value of c, when we look at galaxies that are extremely far away, we are actually looking back in time. This allows us to propose plausible models for **galactic evolution**.

Among the objects displaying the largest redshifts are very distant galaxies and gamma ray bursts. One can tabulate so-called **lookback tables** in which z-values are correlated to the number of years ago the light left the source. (table 7.1)

Presently, the largest known redshift is $z = 10.6$ (for galaxy GN-z11). The light that we are seeing now left that galaxy well over 13 billion years ago! It is now ≈31 billion light-years away from us and came into existence around 400 million years after the Big Bang. (The distance an object actually is from us today is called its 'proper distance'.) Students are encouraged to ponder how a galaxy can be this far away if the Universe is only about 13.8 billion years old and nothing can travel faster than the speed of light, c. Of note, the recently launched James Webb Space telescope is expected to detect galaxies that will break this record for highest redshift.

Further reading

[1] Cho A 2018 A weight limit emerges for neutron stars *Science* **359** 724–5

Chapter 8

Quarks, leptons and radiation: the Standard Model

The purpose of this chapter is to introduce some very basic concepts behind the **Standard Model** of particle physics and **quantum chromodynamics** and to review the nomenclature of the various particles and force carriers, which are repeatedly alluded to throughout the text. We will not be delving into the very complicated mathematical underpinnings of the Standard Model, which are beyond the scope of this text.

8.1 Fundamental forces and particles

The four fundamental forces or interactions include:

1. Gravity
2. Electromagnetism
3. The strong nuclear force
4. The weak nuclear force or interaction.

Aside from gravity, these forces have been shown to be mediated by force-carrying particles. These force carriers fall into the category of particles known as bosons, which are defined below. Force-mediating bosons are called **gauge bosons** or **field bosons**. Both the gravitational force and the electromagnetic force are infinite in range and their strength falls off inversely to the square of distance. The force carriers of these infinite-range fields are the hypothetical **graviton** and the **photon**, respectively. Much of the material covered in this book on space radiation deals with high-energy photons, i.e. electromagnetic radiation. In contrast to the infinite range of gravity and electromagnetism, the weak force is a short-range interaction and the gauge bosons carrying the weak force are the W^+, W^- and Z^0 particles, sometimes referred to as **intermediate vector bosons**. The bosons mediating the strong force (which is another short-range interaction) are called **gluons**.

8.2 Bosons

Particles may be categorized in various ways, and one such way is by inherent angular momentum or **spin**. Particle spin follows the rules of quantum mechanics, meaning that unlike the classical situation where any value is allowed, particles can only exhibit a limited number of specified values. That is, particle angular momentum is 'quantized'. Particle spin can be experimentally determined by a **Stern-Gerlach apparatus**, which measures the deflection of a particle in a nonuniform magnetic field. From such experiments, it was observed that the electron can only display two spin values: $\pm\frac{1}{2}$ in units of $h/2\pi$, often symbolized as $\hbar$. This is Planck's constant, h, divided by 2, also sometimes called the **Dirac constant** and pronounced 'h-bar'. In particle physics, this unit is often understood and not written out. The electron is thus said to have a spin of $\frac{1}{2}$.

Bosons are defined as particles that possess integer spin values. Thus, any particle with a spin of 0, 1, 2, etc would be classified as a boson. In this manner, there are indivisible **elementary bosons**, and there are **composite bosons** made of other particles. Examples of the former include the Higgs boson and the force-carrying gauge bosons (e.g. photons). Examples of composite bosons include **mesons**, which are defined and discussed further in this chapter. But even atomic nuclei can be bosons—for example, the alpha particle, which is simply a helium nucleus, is a composite boson with a net spin of 0. Naturally, if there are particles with integral spin values (bosons) there must be another class of particles with non-integral spins; these particles are called **fermions**. Fermions are thus defined as particles possessing half-integer spin values.

Electromagnetism, the weak force and the strong force are conveyed by known mediators. These force-mediating particles are bosons called **gauge bosons** or **field bosons**. Specifically, electromagnetism is conveyed via photons, the strong force is conveyed by gluons, and the weak interaction is mediated by the W^-, W^+ and Z^0 bosons. Gravity is proposed to similarly be mediated by a gauge boson called the **graviton** with a spin of 2. It is worth emphasizing that, while all force-mediating particles are bosons, not all bosons are force mediators.

All the (known) force-mediating gauge bosons belong to the category of bosons known as **vector bosons**. This designation is dictated by their spins of 1. Photons, gluons, and the W and Z bosons are thus vector bosons since they all have a spin of 1. Obviously, the terminology can become confusing since we just mentioned above that these are all force-mediating bosons, which means they are gauge bosons. In contrast, the **Higgs boson** (with spin = 0) is a **scalar boson**. The proposed, but yet undiscovered, graviton allegedly has spin = 2 and would be a **tensor boson**. The **Higgs boson** is not a force-carrying gauge boson but is the mediator of something called the **Higgs field**. The Higgs field confers mass to certain particles that interact with it. In order for the Higgs boson to interact with a given gauge boson, the Higgs must have a 'charge' that is by the gauge boson in question. For example, for the Higgs to interact with and confer mass to the photon, it would have to have an electrical charge, since the photon is the mediator of electromagnetism. The Higgs does not possess electric charge, however, and thus does not interact with the photon. Hence, the photon remains massless because it does not interact with and gain mass through the Higgs field. Similarly, the Higgs boson does not possess **color charge** (color is explained in more detail below) and thus does not interact with the colored gluons that mediate the strong force. For this reason, gluons do not acquire mass through the Higgs mechanism, and gluons are massless (despite the very short range of the strong force they mediate). On the other hand, the Higgs boson does possess **weak charge** and does interact with the mediators of the weak force, namely the W and Z bosons. It is through this interaction that the W and Z bosons have acquired mass. Note that the presence or absence of mass in the force-mediating gauge bosons is not determined by whether the force they convey is long range or short range. Instead, it has to do with whether or not the gauge boson in question interacts with the Higgs field. Thus, the mediator of the infinite-range electromagnetic force, the photon, has zero mass and the very short-range strong force carriers, the gluons, also have zero mass. In contrast, the W and Z bosons, which carry another short-range force, the weak force, do have mass.

The statistical behavior of large numbers of bosons is described by **Bose–Einstein statistics** as described in another chapter. The behavior of large numbers of fermions is dictated by a different mathematical model called **Fermi-Dirac statistics**.

8.3 Fermions

Fermions are defined as particles with half-integer spins. Thus, electrons, with their spins of ½ are fermions. All six **quarks** (the fundamental particles that make up protons and neutrons as discussed below) have spins of ½ and therefore, like electrons, are also fermions. Fermions, such as the electron, obey the **Pauli exclusion principle** and follow Fermi-Dirac statistical mechanics.

As covered below, the fundamental fermions (the quarks) can be put together to make larger composite particles. Such compound particles composed of quarks include protons and neutrons and are collectively called **hadrons**. As expounded upon further later in this chapter, hadrons are composite particles that can interact through the strong force. Since they are made of quarks, which are fermions, one might naively guess that these composite hadrons would also be fermions. However, one must harken back to the basic definition of a fermion, which is possession of half-integral spin. Adding the spins of particles with integral and half-integral spins is just like addition of even and odd numbers. Therefore, **mesons** (which are composite hadrons made of a quark and antiquark pair) are not fermions but instead are bosons due to their net spins of zero. As mentioned above, alpha particles and other atomic nuclei with an even number of protons and neutrons can have integral spins (and therefore be bosons rather than fermions). Thus, some atomic nuclei are bosons while others are fermions, despite the fact that they are all made up of fermion building blocks. It all depends on whether the nucleus in question has an even or odd total number of protons and neutrons.

An alpha particle is not a force mediator and thus is not a gauge boson. But is a composite particle that happens to have a net spin of zero. And because zero is an integer, the alpha particle is a boson. Furthermore, since the net spin is zero, the alpha particle also fits the definition of a scalar boson. But atomic nuclei can be bosons or fermions based on their net spins. For instance, the most abundant isotope of nitrogen, nitrogen-14, is a boson with a quadrupole magnetic moment, whereas the less-abundant (at 0.36%) nitrogen-15 is a fermion. This subtle difference in spin can make a significant difference in techniques such as nuclear magnetic resonance spectroscopy.

8.3.1 Fundamental particles

In the early 20th century, atoms were believed to be composed of *all* the fundamental particles—electrons, protons and neutrons. If the world truly were that simple, detailed classifications schemes would not be so necessary. However, with the vast proliferation of particles identified by the profusion of colliders in the middle of the 20th century, classification schemes of the ever-expanding 'particle zoo' became desperately needed. Among the earliest indications that such classi-fication schemes could be devised was when Heisenberg suggested that the proton and neutron could be considered essentially the same particles except for electric charge. Thus, they could be collectively lumped together as the **nucleons**. Upon the introduction of **isotopic spin** or **isospin** by Wigner, other natural groupings of hadrons (particles that feel the strong force) began to emerge. For example, some semblance of sanity surfaced when the panoply of particles was arranged on simple plots where the abscissa and ordinate quantified certain characteristics such as electric charge, hypercharge, isospin or strangeness. Such arrangements, whimsically called the **Eight-Fold Way**, yielded certain natural groupings. The Eight-Fold Way thus hinted at some underlying order, reminiscent of the order conferred by organizing chemical elements in the Periodic Table. But just as the underlying order of the Periodic Table required quantum mechanics for a more complete under-standing, the Eight-Fold Way required the quark model and **quantum chromody-namics** for a more complete understanding. Building upon the quark model, the fundamental particles of nature can be classified in various ways.

8.3.2 Quarks

Based on initially perplexing results of high-energy electron–proton deep inelastic scattering experiments, it was inferred that, unlike the electron, which appears to be a truly indivisible fundamental particle, the proton is a composite particle with an internal substructure. Further experiments confirmed the existence of these point-like internal structures, which now go by the name quarks. Thus, we now know that the fundamental building blocks of nucleons (protons and neutrons) are quarks. While nucleons are composed of only **up quarks** and **down quarks**, further research identified additional '**flavors**' of quarks so that there are a total of *six quarks in three generations* that constitute nuclear matter. (Incidentally, as will be discussed below, the counter-parts to the quarks, the **leptons**, are also found in six flavors organized in three generations.) Arranged in mass order (lightest to heaviest), these quark flavors are:

1. Up (u)
2. Down (d)

3. Strange (s)
4. Charmed (c)
5. Bottom or beauty (b)
6. Top or truth (t).

A peculiarity of quarks is that they are the only particles known to possess fractional electric charges. For instance, the up quark carries an electric charge of +1/3, while the down quark carries a −2/3 charge. And all quarks are spin one-half particles, indicating that they are fermions and thus obey the Pauli exclusion principle and follow Fermi-Dirac statistical mechanics. The quarks are tabulated below based on charge and mass. Note that charge and mass do not correlate the way one might anticipate. For instance, in the first generation, the lighter up quark has a charge of +2/3 while the heavier down quark has a charge of −1/3 but in the second generation, the lighter strange quark has a charge of −1/3, while the heavier charmed quark has a charge of +2/3. The quarks, with their masses and charges are tabulated below (table 8.1).

Each quark has an antimatter counterpart with the same mass but the opposite electrical charge. Thus, we have matter up quarks with a charge of +1/3, and antimatter antiup quarks, which would have a charge of −1/3. Similarly, we have down quarks and antidown quarks with −2/3 and +2/3 charges, respectively.

The quarks are the only elementary particles that can experience all four fundamental forces. Unlike **leptons** (which are covered below) such as the electron, quarks possess **color** (or **color charge**) and thus can participate in interactions involving the strong nuclear force. The strong force is carried by vector gauge bosons called gluons. Given that the electromagnetic force is infinite in range and the corresponding force-carrying boson, the photon, is massless, it is somewhat surprising that gluons, the mediators of the ultra-short-range strong force are also massless. In fact, the photon and the eight gluons (as well as the W and Z particles before 'symmetry breaking') are all *inherently* massless. But it is through interaction with the Higgs field (or lack of such interaction) that certain particles gain mass or do not. It has nothing to do with whether the force the boson mediates is long-range or short-range.

Table 8.1. A table of the six flavors of quarks in their three generations, along with their masses and electrical charges.

Flavor	Symbol	Generation	Mass (MeV/c^2)	Electrical charge
Up	u	I	2.4	+2/3
Down	d	I	4.8	−1/3
Strange	s	II	104	−1/3
Charmed	c	II	1270 (1.27 GeV)	+2/3
Bottom or beauty	b	III	4200 (4.2 GeV)	−1/3
Top or truth	t	III	171 200 (171.2 GeV)	+2/3

8.3.3 Color

As fermions, quarks must obey the Pauli Exclusion Principle, which in simple terms states that no two fermions in a bound state can have the same exact set of quantum numbers. An example familiar to students of chemistry is how electrons in an atom will fill orbitals in a specific fashion (following the **Aufbau principle**, meaning lower energy orbitals are filled first), but they avoid ever having any two electrons with the exact same four quantum numbers in a given atom. Quarks initially appeared to be violating the Pauli exclusion principle when bound together in protons, neutrons and other particles. To resolve the apparent discrepancy, it was postulated that quarks must possess some other, previously unknown 'charge' of sorts. This new charge was dubbed **color charge** or simply **color**. In that fashion, the quarks in a proton, for instance, would not have the same exact set of quantum numbers—the two up quarks must have different colors. These colors are called **red**, **blue** or **green**.

8.4 Gluons and quantum chromodynamics

There are eight gluons that mediate the strong interaction, and each gluon carries a certain color combination. Specifically, gluons carry a color and an **anticolor** (antired, antiblue and antigreen). Quarks stick together via the attractive strong force. The way they interact with each other through the strong force is via the incessant *exchange of* **virtual gluons**. Upon interaction of a quark with a colored gluon, there is a color change in both the quark and the gluon. For example, if a green quark emits a green-antiblue gluon that quark will turn from green to blue. If a red quark absorbs a blue-antired quark it will turn blue too. Because of these 'colorful' interactions, the mathematically complicated branch of physics that deals with strong interactions is called **quantum chromodynamics**.

Since there are nine possible permutations combining the color-anticolors (between red, blue and green and their respective 'anticolors') we might expect nine different gluons. Unfortunately, there is no intuitive explanation for the fact that there are only eight. It is a mathematical consequence of the **SU(3) symmetry** characteristic of quantum chromodynamics, which is briefly discussed later in this chapter.

As would be anticipated, heavier quarks decay into lighter quarks, ultimately culminating in only the lightweight first generation up and down quarks. Although the heavier quarks naturally decay under ordinary (that is, low energy) conditions, the lightweight first-generation up and down quarks appear to be completely stable. (Although up and down quark flavors can be interconverted via weak interactions during radioactive decay within nuclei.) Another fine point worth stating is that although the strong force, via gluon exchange, can change the *color* of quarks, only the weak force can change the *flavor* of quarks. Hence, if students encounter a reaction in which an up quark becomes a down quark, the flavor change is an unmistakable telltale sign that the weak force was at play. Additionally, a neutrino or antineutrino was probably involved, which is another giveaway that the weak force was involved.

Among the various conservation laws to be respected during radioactive decay and other particle physics interactions is *conservation of quark number*. As will be shown in a later section, this is essentially the same as the more specifically observed *conservation of baryon number*.

These six quarks are the fundamental building blocks of larger particles known as hadrons. Thus, all hadrons, including the nucleons, are made of smaller quark constituents. As discussed below, hadrons include two broad classes of subatomic particles, the baryons and mesons.

8.4.1 Leptons

Aside from the quarks, there is another category of elementary subatomic particles, the leptons. Like the quarks, the leptons have spin ½ and thus are fermions. Also, like the quarks, they come in six flavors categorized into three main families or generations. These are:

1. Electron (e)
2. Electron neutrino (ν_e)
3. Muon (μ)
4. Muon neutrino (ν_μ)
5. Tauon or tau lepton (τ)
6. Tau neutrino (ν_τ).

The first two (electron and electron neutrino) constitute the **first generation of leptons**, which along with the first generation of quarks (up and down), make up essentially all matter today. The three generations of leptons are each made up of a charged lepton along with a neutrino partner. Hence, the three generations are electron and electron neutrino, the muon and muon neutrino, and the tauon and tau neutrino. The electron, muon and tauon each carry a charge of −1, while their corresponding neutrinos are neutral. The neutrinos have no significant mass and for quite a while were believed to be entirely massless. However, as will be discussed below, the phenomenon of neutrino **flavor oscillation** indicates that they must possess at least some minute amount of mass. Electrons are the lightest of the charged leptons at approximately 1/1836 mass of the proton. The ratio of proton rest mass (938.272 MeV/c^2) to electron rest mass (0.511 MeV/c^2) is confusingly some-times given the symbol μ. Muons, with the suitable symbol μ, are the next most massive of the leptons. They are approximately 207 times the rest mass of an electron (105.66 MeV/c^2 versus 0.511 MeV/c^2). The most massive lepton is the tau lepton (τ), which, at approximately 1777 MeV/c^2, is around 3477 times the mass of the electron. In table form, the masses of the charged leptons are:

$$m_e = 0.5110 \text{ MeV}$$
$$m_\mu = 105.7 \text{ MeV}$$
$$m_\tau = 1776.86 \text{ MeV}$$

Like quarks, each lepton has an antimatter counterpart that is essentially identical to the matter variety lepton except that the electrical charge is reversed. Of course, since neutrinos do not possess electrical charge, their antineutrino counterparts are similarly uncharged. Nevertheless, antineutrinos constitute antimatter, and this must be factored in when performing calculations and respecting conservation laws. Unlike the quarks however, the charged leptons do not have fractional charge—they are either electrically neutral (neutrinos) or have a charge of -1. Also, unlike quarks, which aggregate into hadrons (next section), leptons do not form composite structures. Electrons, for example, do not combine with other electrons or other leptons to make larger compounds. (Unless of course one considers molecules, which are atoms held together via electron-mediated chemical bonds, but this is a qualitatively different concept from composite hadrons composed of quarks.) Finally, but importantly, unlike quarks who can experience all four of the fundamental forces, leptons can experience only gravity, electromagnetism and the weak force—they do not feel the strong force.

There is a conservation law for leptons (the conservation of **lepton number**, L) such that the total number of leptons in any weak reaction is constant. For instance, in beta decay, the classic example of a weak force mediated reaction, if an electron of $L = +1$ is created, to ensure that L remains balanced on both sides of the equation, an antilepton of $L = -1$ must also be created. Therefore, in beta decay, an electron antineutrino must be formed along with the beta electron. Similarly, in electron capture, an electron ($L = +1$) is consumed by a proton, so in order to maintain balance, a neutrino of $L = +1$ must be created on the other side of the equation. This would be an electron neutrino (not an antineutrino), since L is always $+1$ for matter leptons and -1 for antimatter.

In addition to the requirement that overall lepton number must be conserved, there is a law of conservation of lepton subtype. This refers to the family or generation of leptons in question, such that electrons and electron neutrinos constitute family or generation I, muons and muon neutrinos are family or generation II and so on. Thence, the number of electrons and electron neutrinos is accounted for separately from the number of muons and muon neutrinos, for example. Therefore, in a beta plus decay where a positron ($L = -1$) is created, an electron neutrino ($L = +1$) must also be created for overall conservation of lepton number, but because of conservation of lepton subtype, the formation of a muon neutrino or tau neutrino will not fit the bill.

8.4.2 Hadrons

The name 'hadron' is derived from the ancient Greek for strong, sturdy or stout. ('Lepton' on the other hand is derived from the Greek for slender, small, light or delicate). *Hadrons can be defined simply as those particles that feel the strong force.* The corollary of this concept is that the strong force is limited in its effect only to certain particles, and those particles are the hadrons. Protons and neutrons are affected by the strong force and therefore are hadrons. In contrast, leptons like electrons are unaffected by the strong force; they are not hadrons.

All hadrons are composite particles composed of smaller, more fundamental units called quarks. Hadrons come in two basic categories:

1. **Baryons**
2. **Mesons**.

Baryons are combinations of *three different quarks of different colors*. The nucleons—the proton and the neutron—are the classic example of baryons. A **meson** in contrast is a combination of two quarks, specifically a *quark-antiquark pair*. These various arrangements into baryons and mesons must respect the need for **color neutrality** in hadrons. This means that baryons must be made of three quarks of three different colors (i.e. red, green and blue) to yield 'white', while mesons are pairs of colored and 'anti-colored' quarks (e.g. a red and 'anti-red' pair of quarks and antiquarks). Both situations lead to the required color neutrality in the composite hadrons. Both baryons and mesons are composite hadrons made of fermion components (namely quarks and antiquarks). But because of differing possible combinations of their constituent spin ½ quarks, **baryons are fermions**, whereas **mesons are bosons**.

While there is a requirement for color charge neutrality in composite hadrons, there is no such requirement for electrical neutrality. Baryons and mesons may be electrically charged (e.g. positively charged protons and positively (or negatively) charged pions) or neutral (e.g. neutrons and neutral pions).

Besides the need for color neutrality in composite hadrons, another phenomenon observed in hadrons is **quark confinement** or **color confinement**. Confinement is simply the observation that isolated quarks are never detected in nature. Instead, they are always 'confined' to compound particles and found in the form of baryons and mesons; this is also called **hadronization**. Only under extreme conditions (e.g. ultra-high temperatures) can quarks escape their hadronic jailhouses and exist as isolated individuals. The temperature at which hadrons 'melt' is called the **Hagedorn temperature** (T_H). Above the Hagedorn temperature, hadrons can dissociate into a **quark soup** or **quark-gluon plasma**. That is, above a certain temperature, quarks can be **deconfined**. Conversely, as the available energy or ambient temperature falls below T_H, hadronization commences, and quarks become confined in hadrons.

Such extreme conditions (i.e. $T > T_H$) can be found in some high-energy physics labs and were also in effect during the first moments after the Big Bang. At these ultra-high energies, the quark's vibrational energy exceeds the strength of the gluon-mediated strong force that confines the quarks together in a hadron. Physically, confinement can be explained by the fact that the strong force, unlike the other fundamental forces, *grows stronger with distance within a given hadron*. This strange phenomenon (which is further explained below in the section on gluons) is because the carriers of the strong force, gluons, themselves possess color charge. In contrast, the photons that convey the electromagnetic force do not possess electrical charge.

As with electric charge, mass, lepton number and various other quantities, there is a **law of conservation of baryons**. This stems from a more fundamental **conservation of quark number**. What this amounts to is that 'protons are forever'. Despite predictions by the theory of **supersymmetry**, to date, there is no indication that the proton ever really decays. The requirement for conservation of quarks and baryons leads to a **conservation of baryon number**. Specifically, **baryon number** (B) can be defined as:

$$B = \frac{(n_q - n_{\bar{q}})}{3},$$

where n_q and $n_{\bar{q}}$ represent the number of quarks and antiquarks.

Just as an aside about the smallest and apparently completely stable baryon, recent data indicates that the proton is actually about 5% smaller than was assumed in the 1990s and 2000s. The latest estimate is 0.84–0.87 fm. For practical purposes, one may still assume the proton's diameter is just under 1 fm. Just as 10^{-10} m is called an Angstrom unit (Å), a femtometer or 10^{-15} m is also called a fermi, which is conveniently abbreviated fm.

8.4.3 Gluons

Although photons are the carriers of the electromagnetic force, they themselves are not electrically charged. Thus, they do not interact with themselves. On the other hand, gluons, being colored themselves, can interact with one another. Gluons can potentially absorb or emit other gluons. This ability leads to a phenomenon called **asymptotic freedom**.

When two quarks are in close proximity within a bound state, they exchange gluons and create a **color force field**. This very powerful color field is what binds the quarks together. A unique feature about this force field is that it gets stronger as the quarks get further apart. As quarks are pulled further and further from one another, the attractive force grows (in stark contrast to the familiar inverse-square falloff of strength in gravity and electromagnetism). Mathematically, this can be expressed as a variant of **Hooke's Law** ($F = kr^x$), where r is distance between the quarks (but in this case the exponent, x, is larger than 1).

What is happening is that as the color field is stressed by the stretching (that is, the input of energy) as one attempts to pull quarks apart from one another in a hadron, more and more gluons are spontaneously generated from that energy via $E = mc^2$ to abet the attractive force. Conversely, as quarks are drawn asymptotically closer together, the attractive force grows progressively weaker since fewer and fewer gluons are exchanged. Thus, in the limit, there is **asymptotic freedom** of the quarks; the attractive force asymptotically approaches zero as the distance between them decreases. This accounts for the paradoxical phenomenon in which the attractive force between quarks grows vanishingly small as quarks come closer and closer together.

If the force trying to separate two quarks exceeds a certain energy threshold, instead of producing free isolated quarks, the energy is converted into **pair production** of new quark-antiquark pairs, which then combine with the quarks that were forced apart to yield new hadrons. This process is known as **hadronization**. Note from the previous section that the name 'hadronization' is used in two different contexts (the combination of quarks and gluons in hadrons vs the creation of new hadrons from liberated quarks).

8.5 Symmetry and the Standard Model

The mathematical underpinnings of the Standard Model are quite complex and beyond the scope of this text. Suffice it to say the demonstration of **conservation laws** implies an underlying **mathematical symmetry**. In particle physics, numerous conservation laws are patently operational such as the conservation of electrical charge, mass, momentum, lepton number, baryon number, etc. The existence of these conservation principles indicates an associated inherent symmetry. **Noether's Theorem** provides the mathematical proof of the relationship between symmetry and an associated conservation law, and forms the basis for the Standard Model.

In the Standard Model, symmetries may be expressed in terms of mathematical group properties. Here we use the terminology where **U** and **SU** stand for symmetry groups of '**unitary**' and '**special unitary**' groups, respectively. These designations are followed by a number in parentheses, which specifies the **dimensionality** of the groups. Thus, we can say that electromagnetism is depicted by **U(1)** symmetry, the weak force is described by **SU(2)** symmetry, and the strong force is expressed by **SU(3)** group symmetry. SU(3) may be described as 'special unitary group of dimension 3'.

The *number of dimensions* is in parentheses and given by the quantity n; the corresponding number of **generators** (**N**) for a group of dimension n is given by:

$$N = n^2 - 1 \text{ when } n > 1.$$

When $n = 1$, there is but a single generator. And the number of generators dictates the number of force-mediators there are for the force in question.

Following this convention, consistent with its U(1) symmetry, electromagnetism is mediated by a single gauge boson, the photon. The weak force, exhibiting non-Abelian SU(2) symmetry, is mediated by $N = n^2 - 1 = 3$ different gauge bosons. These are the W−, W+ and Z^0 bosons. These are known as **intermediate vector bosons**. (But the other known force-carrying gauge bosons technically also happen to be vector bosons because they possess a spin of 1.) The strong force involves quarks of three different colors. Transitions between colors emit spin 1 gauge bosons, which we know of as gluons. Since the transitions involve triplets, the gauge group is SU(3). Because quantum chromodynamics exhibits SU(3) symmetry, there are $N = n^2 - 1 = 8$ force-carrying bosons. Thus, rather than the naively anticipated nine gluons, there are only eight.

8.6 Sea quarks and valence quarks

Unlike the electron, the proton (and other hadrons) are composite particles consisting of a number of '**valence quarks**' (q_v). In the case of the proton, these valence quarks are two up quarks and one down quark; the neutron is made of one up valence quark and two down valence quarks. Similarly, mesons are made up of valence quark and antiquark pairs. In addition to these valance quarks, however, hadrons contain an indeterminate number of **virtual quarks** or '**sea quarks**' (q_s) bound by gluons.

It is the valence quarks that contribute to the quantum numbers of hadrons; sea quarks do not influence a hadron's quantum numbers. In a manner analogous to pair production of electrons and positrons when sufficient energy is input, sea quarks form when a gluon splits. But the process can also work in reverse as the mutual annihilation of two sea quarks creates a gluon. The result of these two ongoing processes is a constant flux of gluon splits and creations, colloquially called '**the sea**'.

Sea quarks, as virtual particles, are naturally far less stable than their valence counterparts. Within the sea inside a hadron, they typically annihilate each other right after formation. Despite this transient nature, under certain circumstances, the virtual sea quarks can occasionally hadronize into real baryons or mesons. The extreme conditions for such hadronization abounded in the first fractions of a second after the Big Bang.

8.7 Neutrinos

Neutrinos can be thought of as the nearly massless 'ghostly' counterparts of the negatively charged massive leptons, the electron, muon and tauon. They come in three flavors that pair up with their partners: the electron neutrino, the muon neutrino and the tau neutrino. The term ghostly is apt given their tremendous difficulty in being seen. Verification of their existence only came decades after they were initially hypothesized by Wolfgang Pauli. Being electrically neutral, traveling at close to the speed of light, c, and originally being assumed massless, it was evident from the start that their detection would be daunting. Further complicating their experimental verification is their incredibly low probability of interaction—in other words, they have an extremely small **cross section**. The **half-value layer** (i.e. the thickness of an attenuator needed to decrease the number of incident neutrinos to one-half the initial number) in solid lead is estimated to be approximately 1 light year (roughly 6 trillion miles)! Since they all have a spin of ½, neutrinos are fermions.

Incidentally, their high velocity has taken them out of the running as candidates for the elusive **dark matter** since dark matter is considered 'cold' or slow moving, whereas neutrinos zipping about at nearly the speed of light would be 'hot' dark matter.

Unlike their partner leptons, (electrons, muons and tauons, which are essentially immutable and do not readily interconvert), neutrinos do seem to change flavors relatively easily. This **flavor oscillation** contributed to the **solar neutrino deficit problem** since the early experiments were designed to find only electron neutrinos (ν_e) rather than muon neutrinos ν_μ or tau neutrinos ν_τ.

Neutrino flavor change is conversion from ν_α to ν_β [where $\nu_\alpha \in (\nu_e \ \nu_\mu \ \nu_\tau)$] and is generally dependent on time and energy. Specifically, it is proportional to t/E, where t is time and E is neutrino kinetic energy. As distance is proportional to time, another way of expressing this is d/E. This propensity for flavor variation over distance (from the Sun for instance) leads to progressive flavor transitions from ν_e to ν_μ to ν_τ and then back and forth again; this phenomenon is known as flavor 'oscillation'. Because of flavor oscillation, neutrinos may be viewed as **wave packets** with a certain mixture of flavors rather than immutable individual particles. Importantly, flavor oscillation implies that neutrinos have mass. This contrasts with previous assumptions that neutrinos, like photons, were completely massless. Only particles with a definite rest mass can oscillate from one flavor into another. The Nobel Prize in Physics 2015 was awarded jointly to Takaaki Kajita and Arthur B McDonald for their discovery of neutrino flavor oscillations.

Further reading

[1] Lin Y-H, Hammer H-W and Meißner U-G 2022 New insights into the nucleon's electro-magnetic structure *Phys. Rev. Lett.* **128** 052002
[2] Bezginov N *et al* 2019 A measurement of the atomic hydrogen Lamb shift and the proton charge radius *Science* **365** 1007–12

IOP Publishing

Space Radiation
Astrophysical origins, radiobiological effects and implications for space travellers
James S Welsh

Chapter 9

Radiation chemistry

9.1 Radiochemistry versus radiation chemistry

The scientific discipline that examines the chemical effects of ionizing radiation on mixtures, solutions and other materials is **radiation chemistry**. This must be distinguished from the discipline of **radiochemistry**, which focuses on the chemical attachment of radionuclides to substrates for scientific studies (e.g. studies of metabolism in biochemistry) or medical applications (such as the creation of radiolabeled diagnostic tracers and unsealed radionuclide-based cancer therapies). Most aspects of radiation chemistry of interest in this book deal with aqueous solutions, but solvents aside from water (e.g. organic solvents) are of great importance in general radiation chemistry. Also, although human biology is largely liquid based (thus, the focus on radiation chemistry of water and aqueous solutions), the radiation chemistry of materials in the gas and solid phases are also of relevance to space radiation. For instance, the chemical deterioration of solid materials exposed to space radiation is of relevance to space missions. Constant exposure to radiation in space can seriously test the integrity of solid-state electronics. For instance, **sputtering** is a process in which atoms, ions and molecular species in the surface of a target material are ejected because of radiation. This is most often due to bombardment with very energetic particle radiation such as that found in galactic cosmic rays. More information on the effects of space radiation on electronic devices is covered in chapter 23. Also, the effects of large doses of radiation on planetary atmospheres via rare events such as gamma ray bursts fall into the realm of radiation chemistry in the gaseous phase.

9.1.1 Radiation chemistry

Radiation chemistry is characterized by the interactions of ionizing radiation with solvent and solute molecules. These interactions can very rapidly result in highly reactive species. These highly reactive chemical species may then recombine with one another through quenching reactions, or they may interact with and alter normal, unexcited molecules. Biological molecules are among the normal molecules that may

doi:10.1088/978-0-7503-5444-8ch9

be targeted by reactive, radiation-generated species. Such reactions between biological molecules and radiation-generated active species in turn lead to radio-biological phenomena.

Although water is the solvent of greatest concern in radiobiology, there is much more to radiation chemistry than just aqueous reactions. Radiolysis, or breakdown of molecules by radiation, can occur to solute molecules in either aqueous or non-aqueous solvents. And radiolytic reactions involving non-aqueous solvents are of great importance in organic chemistry. Furthermore, radiation chemistry is not limited to liquid solutions. As discussed in several areas throughout the text, our atmosphere (and other planetary or lunar atmospheres) is subjected to radiation effects that fall under the general rubric of radiation chemistry. For example, the feared dissolution of our stratospheric ozone layer after intense atmospheric irradiation from a gamma-ray burst would involve multiple radiation-initiated chain reactions that ultimately target and destroy ozone molecules.

9.2 Aqueous radiation chemistry

In liquid water and aqueous solutions, ionizing radiation generates secondary electrons that progressively slow down or '**thermalize**' to energies below 7.4 eV (which is the threshold for electron transitions in liquid water).

For a primary charged particle and the secondary electrons it may create, this thermalization process is simply a transfer of energy from the radiation to the medium. Although stopping power ($-dE/dx$), which is a very important parameter in the physics of radiation (and in radiobiology under the synonym of linear energy transfer (LET)), treats the slowing down process of a particle in a medium as a continuous function, the energy transfer from radiation to an aqueous medium actually proceeds in a sequence of discrete events. Thus, stopping power, and other treatments of thermalization as a continuous process, are sometimes alluded to as '**continuous slowing down approximations**'.

Depending on the amount of energy transferred to an electron in the solution, the water molecules and solutes (including biological macromolecules) can cause simple thermal transfer (increases vibration, rotation and translational motions), atomic/molecular excitation or full-blown ionization. Of course the thresholds for these different processes vary. The ionization threshold in water is ≈ 13 eV, whereas the excitation threshold in water is lower, at ≈ 7.4 eV.

9.3 The timescale of radiation chemistry

The various discrete steps involved in the transfer of energy from radiation to a medium occur with a characteristic timescale. For example, the time that it takes for a directly ionizing particle to traverse a molecule is measured in attoseconds (10^{-18} s), while the time interval between successive ionizations in an aqueous medium is measured in femtoseconds (10^{-15} s). Radiation-excited water molecules (and other molecules) may dissociate into ion pairs in tens of femtoseconds, and these ions may begin chemically reacting with molecules on the same timescale. By around 100 fs, the secondary electrons traversing a medium will have thermalized and will cease to be capable of inducing ionizations in the medium. The free radicals generated through ionizations and excitations diffuse through the solution at around a picosecond (10^{-12} s). By about 10 ps, aqueous (solvated) electrons are formed and reactive. By about a nanosecond or so, all radiochemical reactions that depend on rapid diffusion will have been completed. By about 10 ns (10×10^{-9} s), biologically relevant molecular products will have been formed. Also around this time, excited singlet states will undergo radiative decay to their ground states. At about 10^{-5} s (10 μs), reactive species will be captured and neutralized provided there are such free radical quenchers available. At about a millisecond, damage to biological macromolecules (especially in the presence of oxygen) will be '**fixed**', meaning the damage is now permanent (unless somehow later reversed through intricate molecular biological mechanisms). Students are cautioned about the unfortunate terminology in radiation chemistry where the names 'fixed' or '**fixated**' mean that the molecular damage is now chemically permanent—these words do not correlate with the colloquial meaning of 'fix' as in repair. In essence, the same words are antonyms! By 1 s, most radiochemical reactions will have been completed. Nevertheless, certain reactions, especially in non-aqueous organic media can continue for hours to days. Biochemical enzymatic reactions that address macromolecular damage such as DNA strand breaks may proceed on the order of milliseconds to several seconds or minutes.

As mentioned, the time scale of radiation chemistry spans many orders of magnitude, ranging from the extremely short time required for a photon or fast electron to traverse a molecule (measured in attoseconds, 10^{-18} s) to the relatively long time (measured in hours) required for neutralization processes in viscous media (typically about 3 h). Thus, in radiation chemistry, the **pt scale** is occasionally utilized. As with pH in the pH scale, the pt is defined as the negative (base 10) of the time in seconds. That is, **pt $= -$log10 t**.

9.4 Initial physico-chemical actions of irradiation

Among the earliest physical chemistry events in radiochemistry is the transfer of energy to the medium. The amount transferred in aqueous radiochemistry is between 7 and 100 eV to the medium (a solution in water) per event. Depending on the exact amount of energy transferred, this energy may excite or ionize one or more water molecules. Transfer of radiation energy to the medium in cells typically involves ionization of water molecules; this leads to the so-called **indirect effect** of ionizing radiation. The excited products of such ionizations may then go on to damage biological molecules. It should be noted that the indirect effect relies on the ionization of the solvent molecules as intermediaries and is diffusion limited. As such, the dose response relationship can be quite complicated.

Alternatively, radiation might directly excite or ionize a biological macromolecule, leading to breakage of covalent bonds and chemical alterations. This is called the **direct effect** of ionizing radiation. Much of basic radiation biology focuses on nuclear DNA as the principal target of radiation, but damage to other macromolecules in sufficient amounts can certainly have significant biological effects. Incidentally, the reason that DNA is so important radiobiologically is because unlike all other macromolecules, it is found in only two copies per cell (in a normal diploid human cell). Also, unlike lipids in membranes and protein enzymes, DNA is capable of radiation damage repair. This damage repair can be quite extensive and quite impressive. It is in stark contrast to the general absence of repair in other biological macromolecules such as proteins, which are typically (but not always) replaced, rather than repaired, after radiation damage.

Ionizations and excitations caused by the passage of a charged particle through a biological medium produce three important radiochemical species in the local vicinity of the particle track, namely:

1. A free electron (e^-)
2. A positively charged water radical ion ($H_2O\cdot^+$)
3. Or an excited water molecule (H_2O^*)

Radiation may lead to ionization of an H_2O molecule (or in general, any molecule), which then yields a radical ion and a free electron (called a **free subexcitation electron**). Such free subexcitation electrons in water usually have a kinetic energy less than 7.4 eV. In water, along with the liberated electron, a water radical ion is also made. These products may recombine into innocuous water molecules or they may proceed to initiate biochemically significant reactions.

$$H_2O \overset{\otimes}{\longrightarrow} H_2O\cdot^+ + e^-$$

(where the $\overset{\otimes}{\longrightarrow}$ symbol just indicates irradiation of the sample).

And of course, energy transfer from ionizing radiation can simply leave a water molecule in an excited state:

$$H2O \overset{\otimes}{\longrightarrow} H_2O^*$$

(where the * indicates an excited state).

As mentioned, these initial reactions occur very quickly—the time scale for the formation of these species is on the order of 10^{-16} s.

9.5 'Pre-chemical' reactions of radiation chemistry

Following these first physicochemical events, in the next phase (pre-chemical reactions), these three initial species (free electron, the positively charged water radical ion and the excited water molecule) may then diffuse and react with each other or with other molecules in the medium. Some of these reactions can produce **free radicals** (sometimes just called radicals). Simply stated, radicals are atoms, ions or molecules that contain at least one unpaired electron. In general, radicals are chemically extremely reactive.

Among the reactions that involve the free electron is capture by water molecules. During this capture, through dipolar interactions, the free electron can become solvated. Thereafter, it is referred to as an **aqueous electron** (e^-_{aq}, also called a **solvated electron**):

$$e^- + H_2O \rightarrow e^-_{aq}$$

The solvated electron is thus surrounded by a 'cage' of water molecules, which means it may be considered a clathrate. Alternatively, the free electron may react with a hydrogen ion (H^+ or 'hydron') to yield a **hydrogen radical (H·)**:

$$e^- + H^+ \rightarrow H\cdot$$

Among the reactions involving the second product (the water radical ion ($H_2O\cdot^+$)), is its spontaneous dissociation to yield a **hydroxyl free radical** and a hydrogen ion:

$$H_2O^+ \rightarrow H^+ + HO\cdot$$

The hydroxyl radical is among the most reactive of the various active species generated by radiation in water and is considered to be the most radiobiologically important.

Notably, it requires approximately 5 eV to break an O-H bond. And as mentioned earlier, the formation of the last product (an excited water molecule (H_2O^*)) requires at least 7.4 eV. So, an excited water molecule (H_2O^*) may dissipate its excess energy through bond cleavage, generating hydroxyl radicals and **hydrogen radicals**:

$$H_2O^* \rightarrow HO\cdot + H\cdot$$

Thus, through various 'pre-chemical' reactions, the three initial species made via the interaction of radiation with water (H_2O^*, $H_2O\cdot^+$ and e^-) can and do react further to produce other chemically active and radiobiologically important species: $HO\cdot$, $H\cdot$ and e^-_{aq}.

9.6 Concentration and potency of radiation products

It should be emphasized that the absolute concentrations of radiation-induced radical species are quite small, especially when compared to the products of other common chemical reactions such as the dissociation of water. Water naturally dissociates into positively charged hydrogen ions and negatively charged hydroxyl ions as a function of temperature. The concentrations of ions present from this dissociation are tiny—but they still dwarf the concentrations of free radicals made via radiation in most situations. The pH scale is defined as the negative base 10 logarithm of the molar concentration of hydrogen ions [H^+]. Neutral pH under standard conditions is a pH of 7. Under such conditions of neutrality, the [H^+] = 10^{-7}. This figure grossly eclipses the concentrations of radiation-produced reactive radicals. Thus, even though some radiation chemical reactions may lead to H^+, irradiation does not affect pH of the medium much under most circumstances.

This observation highlights the fact that radiation can have profound chemical and biological effects despite the relatively tiny amounts of energy actually absorbed. For instance, *it would require a radiation dose of about 10 000 Gy to raise the temperature of a solution by just a few degrees Celsius.* Yet, an acute dose of only 1 Gy may have significant biochemical consequences that could possibly kill a cell.

Also, a total body dose of about 4 Gy of x-rays acutely absorbed by a human body is lethal in about half of the individuals exposed (i.e. this is the LD50/60 or lethal dose to 50% of individuals in 60 days' time). This radiation dose represents the absorption of energy of only about 67 calories, assuming the body is a 'standard man' weighing 70 kg. This tiny amount of energy, if converted into heat, would represent a temperature rise of 0.002 °C. Obviously, this temperature increase would be biologically harmless. It is about the same amount of heat energy absorbed by taking a sip of warm coffee. Therefore, the substantial chemical and biological effects of radiation are disproportionate to the absolute amount of energy absorbed.

Thus, the effects of radiation are clearly not mediated by the deposition of thermal energy. Similarly, a radiobiologically significant dose of energetic protons (despite literally being H^+ ions) does not have much of an effect on the pH of a solution until many thousands of grays are deposited. Students are invited to calculate the pH change induced by an absorbed dose of 1 Gy of proton radiation to see how little the pH is affected. However, it should be mentioned that the pH of a solution can affect the concentration of specific radiolysis products made. For instance, under highly acidic condition (low pH), the H+ ions can scavenge the ·OH radicals. And at very high pH, the redox potential of hydroxyl radicals is decreased. Also, key catalysts involved in radiation chemistry, such as Fe^{2+}/Fe^{3+}, can be strongly influenced by pH, thereby altering yields of certain radiation chemical reactions. Nevertheless, although the absolute temperature rise of a solution that has absorbed a biologically relevant dose of radiation does not change perceptibly, at a microscopic level (that is, at the radiation track level), there can indeed be profound temperature increases, as will be discussed below.

9.7 Linear energy transfer

LET may be defined as the amount of energy released by a particle or wave of radiation over the length of its track in a given medium. **Specific Ionization** is a related concept and is defined as the number of ion pairs produced per unit track length.

The dividing line defining 'high' LET radiation is not clear, although an arbitrary cutoff that is often used is around 10 keV/μ (although some authors use an even lower figure of 3 keV/micron). Radiation that is more densely ionizing than this figure is often classified as 'high' LET. High-LET radiation is densely ionizing but often is not very penetrating. For instance, alpha particles may have LET values between 50 and 200 keV but a range in tissue of only a few cell diameters or about 14–77 μm (depending on the alpha particle energy and tissue type). Similarly, although energetic electrons (that is, those with energies measured in MeV) are low LET and have relatively long ranges in tissue measured in centimeters, low-energy electrons, such as Auger and Coster-Kronig electrons, have high LET but extremely short ranges. The range of these electrons is usually less than a single cell diameter and often less than the size of a cell nucleus. Thus, for them to have a biologically significant effect, they must be within range of the biological target molecule, DNA (ignoring bystander effects, which are discussed elsewhere). An obvious exception to this 'high LET = short range' rule is the densely ionizing radiation of heavy galactic cosmic rays or high atomic number (Z), high-energy (E) particles (abbreviated as

'HZEs'). Not only are these very high LET, but, thanks to their extreme energies, they are also capable of penetrating deeply.

9.8 LET and radiation chemistry

Low-LET radiation such as gamma rays causes radiolysis (that is, it ionizes and dissociates water molecules) over a relatively long track length. In the case of low-LET radiation, the ionizations along the track are relatively far apart, and the radiation is described as **sparsely ionizing**. When ion pairs are made in water (or any medium), the created ions and radicals are adjacent and of the opposite type. For instance, a single pair of H· and OH· radicals generated through radiolysis of water may reunite and reform harmless, inert H_2O.

In contrast, when many ion pairs are created along the radiation track, that radiation is called densely ionizing. Densely ionizing, high-LET radiation (like alpha particles and low-energy electrons such as Auger electrons) leads to multiple ionization events that lie close to one another over a very short track. When this happens, the multitude of immediate radiolysis products, such as H· and OH· radicals, may mix, match and recombine into products other than water, such as hydrogen peroxide, H_2O_2. These products in turn may precipitate further reactions that ultimately lead to oxidative damage. Hence, irradiation of water with photons or beta electrons does not yield much hydrogen peroxide, whereas irradiation of the same sample with alpha particles does.

Of course, this general phenomenon can occur in solvents other than water. Regardless of what the solvent is, microscopic regions in an irradiated solution that contain a high concentration of reactive species after absorption of radiation are referred to as **spurs**. A spur is defined as a small, spherical region in irradiated material, where the absorbed radiation energy is deposited. Energy deposition of about 100 eV along the tracks leads to these spurs in an aqueous medium. It is estimated that 95% of the energy deposition events from x-rays and γ-rays occur in spurs. As a rule, spurs are about 4–5 nm in diameter and contain up to three ion pairs. This contrasts spurs with **blobs**, which are larger versions of the same phenomenon. Blobs in water typically require 100–500 eV for formation and are typically around 7 nm in diameter and contain an average of 12 ion pairs. Recalling that the diameter of the double helix of DNA is 2 nm, one can see that spurs and blobs are more than adequate in size to induce DNA damage. Finally, joining spurs and blobs is another microdosimetric category of ions and reactive species set in motion by the radiation-produced charged particles traveling through a medium; these are called 'short tracks'. Of course, the distinction between short tracks, spurs, and blobs is purely arbitrary and there is a continuum of such entities.

In a medium irradiated with low-LET radiation, the spurs are sparsely distributed across the track and are usually unable to interact with one another. But for high-LET radiation, the spurs may overlap, enabling inter-spur reactions. These inter-spur reactions lead to different yields of reactive products when compared to the same medium irradiated with the same dose of low-LET radiation. Furthermore, high-LET radiation is more capable of producing blobs than low-LET radiation is.

9.9 Radiation-induced heating

Bridging the space between radiation physics and chemistry is simple heating of an irradiated sample. As one might surmise from the definition of absorbed dose (amount of energy absorbed from the radiation in a given mass of irradiated material), heating is one of the elementary consequences of irradiation. However, for low-LET radiation, the actual heating is, for most purposes, negligible.

Recall that a spur is a small, spherical region in irradiated material in which the absorbed radiation energy is deposited. The temperature rise, ΔT, of a spur at a point located a distance r, away from the spur at time t, is given by the equation:

$$\Delta T(r, t) = \Delta T_{max}(1 + 4\gamma t/a^2)^{-3/2} e^{\char`\^} - (r^2/(a^2 + 4\gamma t),$$

in which a is the initial spur-size parameter, γ is the thermal diffusivity of the medium (that is, heat conductivity divided by the density times specific heat at a constant volume), and ΔT_{max} is the initial maximum temperature rise at the center of the spur.

Using some realistic values for energy deposition (30 eV) and spur size (20 Å or 2 nm) in water (which has a density equal to 1 g cm^{-3} and a specific heat of 4184 J·kg^{-1}·K^{-1}), ΔT_{max} may be estimated to be 30 °C (54 °F), and the temperature will fall off exponentially from that place and over time. But despite this surprisingly high value (albeit to a small volume), the time to cool off is astonishingly fast. The cooling half-life ($t_{1/2}$), which is the time required for the central temperature to drop to half its initial value, may be calculated by solving for $t_{1/2}$ in the equation:

$$(1 + 4\gamma t_{1/2}/a^2)^{3/2} = 2.$$

Using a typical thermal diffusivity (γ) of 10^{-3} cm^{-3} s^{-1}, $t_{1/2} = 6 \times 10^{-12}$ s.

The net result is that the local temperature increase caused by low-LET radiation is too small and too brief to have any appreciable physico-chemical impacts. As mentioned above, the LD50/60 for total body radiation of an average human is about 4 Gy, which causes a temperature rise of 0.002 °C overall. Even though there is a local temperature increase caused by the ionization events, the overall effect is negligible. It is akin to a person at 98.6 °F (37 °C) jumping into the frigid Arctic Ocean—although you are far warmer than the ocean, the overall effect on the ocean's temperature is negligible. Furthermore, the actual temperature rise will be slightly less than what is calculated because some of the absorbed radiation energy is consumed by ionization and chemical bond breaking.

The case is different for high-LET radiation. When high-LET particles travel through a medium they produce, not small spherical spurs, but rather cylindrical ionization tracks. The equation for temperature rise due to high-LET radiation is slightly different:

$$\Delta T(r, t) = \Delta T_{max}(1 + \gamma t/a^2)^{-1} e^{\char`\^} - (r^2/(a^2 + 4\gamma t),$$

where in this case a is the initial size parameter for the track cylinder and ΔT_{max} is the maximum initial temperature rise on the cylinder axis.

For high-LET fission fragments with LET (approximately 500 eV per angstrom) and with a value of $a = 20$ Å, ΔT_{max} for water turns out to be 1.6×10^4 K. This calculation is a slight overestimation for the same reasons that apply for low-LET radiations above, but it is still likely that the microscopic temperature rise truly is high —perhaps on the order of 10 000 K. The time for this temperature to drop to half the initial value is given by $t_{1/2} = a^2/4\gamma$, which in this case is estimated to be about 10^{-11} s.

This time is not much larger than the corresponding time for survival of an isolated spur secondary to low-LET radiation but thanks to the much higher local temperature, the reaction time for radiation-generated chemical intermediates also becomes much shorter. Recall that the rate constant (k) for a chemical reaction is given by the Arrhenius equation:

$$k = Ae^{-(E_a/RT)} \text{ (\textbf{Arrhenius equation}),}$$

where R is the Universal Gas Constant and E_a is the activation energy of the chemical reaction.

Given the high value of T in this situation, the chemical kinetics are different for high-LET radiation. Thus, even though the heat pulse survives only a short time, the high temperature means that the short time is still long enough to allow radiation-induced chemical reactions involving short-lived intermediates. These radiolytic products can then go on to cause indirect biological damage. Hence, even though high-LET radiation predominantly causes direct radiation effects, there are also important indirect radiation effects that contribute about a third of the overall radiobiological impact.

9.10 Reactive species made by radiolysis of water

As mentioned above, water radiolysis leads to highly reactive HO· and H· radicals. These radicals are far more chemically reactive and biochemically damaging than the hydroxide ions (HO^-) or H^+ from ionic dissociation in acid-base chemistry.

The hydroxyl radical (HO·) especially, is a very potent oxidizing agent and is extremely reactive chemically. In radiation biology, it has been experimentally verified (via studies involving specific scavenging agents) that the hydroxyl radical is the principal agent involved in DNA damage as well as other biological macro-molecular damage.

Recall that oxidation is the loss of electrons from an atom or chemical compound. The electrons are transferred to the oxidizing agent, which then becomes reduced. Reduction is the addition of electrons. Reduction may involve the addition of an electron only, or (especially in biochemistry) it may involve the addition of hydrogen together with an electron. While hydroxyl radicals are powerful oxidants, the hydrated electron is an exceptionally powerful reducing agent. Hence radiation chemistry and the radiobiological effects are largely mediated through redox reactions.

9.11 Chemical reactions involving the species made through irradiation

In many cases, excitation fails to culminate in any further significant chemical consequences as the excited molecules dissipate energy without inciting chemical reactions. In contrast, ionization typically does lead to further chemical effects. Ionization may cause a wide range of processes involving the outgoing electron and the resulting positive ion. Among the reactions involving the positive ion are **parent ion fragmentation**, **ion–molecule reactions** and **charge neutralization reactions**. The products of such processes can, in turn, precipitate further chemical reactions.

9.12 Parent ion fragmentation

Regarding positively charged **parent ion fragmentation**, various pathways of decomposition are often possible from the same parent ion. These fragmentation channels compete with one another for dominance. The excited parent ion usually has sufficient energy for bond breakage, and thus, subsequent bond breakage allows different relaxation (decomposition) options including:

1. **Atom elimination**
2. **Molecule elimination**
3. **Radical elimination**.

Thus, parent ion fragmentation involves de-excitation of the parent ion radical through creation and ejection of a single atom such as atomic hydrogen (which is highly reactive), a molecule such as hydrogen gas (which is not as reactive) or the decomposition into other radical ions (such as a methyl cation, which can also be highly reactive). Parent ion fragmentation is more common in the gas phase; the next set of reactions (**ion–molecule reactions**) are more prominent in the solid state.

9.13 Ion–molecule reactions

Another outcome of the ions made by radiation may be **ion–molecule reactions**. These types of reactions occur between the positively charged ion formed from irradiation of the medium (such as a positively charged water ion, H_2O^+) and a neutral medium molecule (such as a native, ground-state H_2O molecule). In water, the classic example is thus:

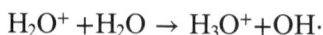

$$H_2O^+ + H_2O \rightarrow H_3O^+ + OH\cdot$$

Of course, the positively charged parent ion does not have to be from the solvent, and the neutral molecule does not have to be the solvent itself, but that is naturally the most common situation.

The products of ion–molecule reactions are the results of charge exchange, such as, in the case of aqueous irradiation, H_2O^+ to H_3O^+ with the production of $OH\cdot$. Although ion–molecule reactions are generally more important in the solid phase, in liquid water, this particular ion–molecule interaction is a very important reaction channel for the parent ion (H_2O^+).

It should be noted that the same positively charged parent ion may undergo either fragmentation or ion–molecule reaction under different circumstances. A classic example is the irradiation of methane (CH_4) by gamma rays. The products include a free electron, and a methane ion CH_4^+. This positively charged parent ion may then undergo various means of fragmentation or may undergo ion–molecule reactions. A final fate category of the positively charged parent ion is charge neutralization. It is worth recognizing that although the excited parent ion may become electrically neutral through this process, the uncharged products themselves may still be in an excited state and be chemically reactive.

9.14 Interactions of the ejected electron

An ejected electron from the initial ionization event may excite or ionize other atoms and molecules along its path, thereby precipitating further radiochemical reactions. The electrons set free by radiation may have sufficient energy to cause further ionizations (meaning additional ejected electrons, or secondary electrons), which in turn may ionize other molecules and liberate additional electrons (tertiary electrons). This domino effect type process can continue until the energy falls below the ionization potential of the atoms and molecules encountered.

Such an ejected electron may cause chemical changes independently through ionization or **dissociative attachment** (wherein the electron attaches to an atom, creating a negative ion that then dissociates from the parent molecule). The negative ions made through dissociative attachment (such as H^-, O^-, etc) are often highly reactive.

There are myriad other reactions provoked by ionizing radiation beyond this brief introduction. Each generated product may then serve as the progenitor of additional excitation and ionization reactions. As a general principle, the distance such chemically active products may travel is a function of the radiation energy, the type of radiation and the particular medium.

9.15 The chemical reaction phase of radiation chemistry

In aqueous radiochemistry, the phase following the 'pre-chemical' phase of radiation reactions is the 'chemical phase'. After $\sim 10^{-12}$ s, the new chemically reactive species will still be located in the vicinity of the original three radiolytic products (H_2O^*, $H_2O^{\cdot +}$ and e^-) of the initial physico-chemical events. Three of the new species created in the pre-chemical phase are radicals: $HO\cdot$, $H\cdot$ and e^-_{aq}.

These free radicals now begin to randomly diffuse through the solution away from their initial positions. As this diffusion proceeds, individual radical pairs may come close enough together to react with one other. A host of radical recombination reactions are possible near the tracks of the charged particle such as:

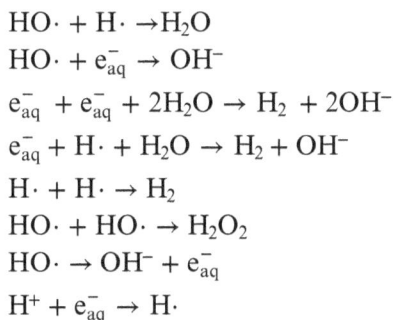

$$HO\cdot + H\cdot \rightarrow H_2O$$
$$HO\cdot + e^-_{aq} \rightarrow OH^-$$
$$e^-_{aq} + e^-_{aq} + 2H_2O \rightarrow H_2 + 2OH^-$$
$$e^-_{aq} + H\cdot + H_2O \rightarrow H_2 + OH^-$$
$$H\cdot + H\cdot \rightarrow H_2$$
$$HO\cdot + HO\cdot \rightarrow H_2O_2$$
$$HO\cdot \rightarrow OH^- + e^-_{aq}$$
$$H^+ + e^-_{aq} \rightarrow H\cdot$$

The first five of the above reactions may be considered **quenching reactions**. They are representative of most radical reactions, and they remove chemically reactive species from the system, leaving biologically inoffensive molecules such as water, hydroxide ions (OH⁻) and hydrogen gas. The last three examples above convert reactive species into different reactive species.

Within a very short time (that is, by about 1 μs), the radiation-generated reactive species will have diffused far enough away from each other that further such reactions are improbable. Thus, the chemical phase of radiochemical track development terminates by 10^{-6} s.

9.16 Reaction radius

The distance λ, that a reactive chemical species will diffuse over a given time τ, with a diffusion constant D is quantitatively given by:

$$\lambda = \sqrt{\frac{D}{6\tau}}.$$

The **reaction radius**, r, is the distance within which a chemical reaction is likely to occur if two entities approach one another. Hence, the reaction radius is a measure of the reactivity of the individual species in question. If a reactive species diffuses within the chemical reaction radius of a target molecule, the two will react. Part of the reason that the hydroxyl radical is so reactive is because it has a relatively long reach—its reaction radius is significantly larger than for most other reactive species. The diffusion distance of the hydroxyl radical is about 9 nm over a timespan of 2.5 ns.

9.16.1 *G* values

The G value is the empirical chemical yield or number of a particular species produced (or eliminated) per 100 eV of energy input by a charged particle and its secondaries when it stops in a medium. In other words, the G value is the 100 electron volt yield of a radiochemical reaction. For example, irradiation of cyclohexane with cobalt-60 gamma rays yields approximately 5.6 hydrogen molecules per 100 eV input, which can be written $G(H_2) \approx 5.6$.

For the same amount of radiation absorbed dose in different materials, the lower the G-value, the more radioresistant or radiation insensitive the material is.

But when giving G values, it is important to state the type of radiation and the energy. Different types of radiation give different G values. For instance, protons give higher G values for the creation of hydrated electrons (e_{aq}^-) or hydroxyl radicals (HO·) than alpha particles of the same energy or velocity. Unsurprisingly, G values also vary with the energy of the particles in question.

As mentioned above, by about 1 μs (10^{-6} s), all chemical development of the track is over. Therefore, G values do not change much after that time point.

9.16.2 'Fixation' of damage by oxygen

The damaging effects of free radicals are significantly magnified in the presence of oxygen. In fact, molecular oxygen in the medium may cause radiochemical damage to become fixed. As mentioned, this is an unfortunate term that in chemistry, means 'made permanent', whereas in common colloquial conversation, fixed means 'repaired'—the exact opposite. **Molecular oxygen**, O_2 is somewhat unusual in that it has two unpaired electrons making it a stable biradical. Regardless, O_2 is a powerful oxidizing agent.

Molecular oxygen readily reacts with radiation-generated free radicals and converts them into other reactive species (collectively called **reactive oxygen species (ROS)**). Irradiation of water in the presence of air, for example, will produce the superoxide radical ($\cdot O_2^-$) as well as its protonated form, the perhydroxyl radical ($HO_2\cdot$), instead of the hydrated electron and hydrogen atoms, which dominate in the absence of oxygen. The creation of these new species decreases the odds that the initial free radicals from the radiolysis of H_2O will recombine into nontoxic compounds such as water or hydrogen gas.

For example, oxygen can combine with the hydrogen radical ($H\cdot$) to form ROS such as the hydroperoxyl radical ($HO_2\cdot$):

$$H\cdot + O_2 \rightarrow HO_2\cdot$$

In the presence of molecular oxygen, chemical repair of radical-induced damage is inhibited by the transformation of organic radicals into peroxyradicals, represented by the general equation:

$$R\cdot + O_2 \rightarrow RO_2\cdot$$

In this manner, if free oxygen reacts with a DNA radical before it is repaired, the damage becomes harder, if not impossible, to repair. DNA hydroperoxides, for instance, cannot be repaired by chemical restitution. In other words, the DNA damage is chemically fixated.

O_2 reacts readily with excited biochemical species such as organic radicals. In this example, DNA serves as the excited biochemical species in question:

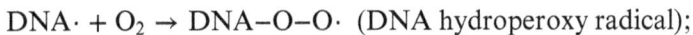

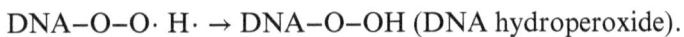

$$DNA\cdot + O_2 \rightarrow DNA-O-O\cdot \quad (DNA \text{ hydroperoxy radical});$$

$$DNA-O-O\cdot \ H\cdot \rightarrow DNA-O-OH \ (DNA \text{ hydroperoxide}).$$

In both cases, these final products, created through the consolidative effects of oxygen, are far harder for biochemical repair than the original damage was. Hence, such damage may be considered fixed in place. It should be pointed out, however, that even extensive damage to DNA may be sometimes repaired through the host of complicated molecular repair mechanisms found in humans and other organisms. These DNA repair mechanisms will be covered elsewhere.

Another effect of oxygen is that, through the generation of new ROS, it increases the overall duration and range of the biochemically reactive entities. As discussed above, the time frame of creating and quenching radiolysis products is extremely short. But by creating new species with their own half-lives, oxygen in effect increases the lifespan and, thus, physical reach of reactive products. The effective reaction range is thereby increased by oxygen. Enhanced creation of the extremely reactive and relatively long-range hydroxyl radical is an example of how oxygen can augment the effects of radiation.

Oxygen augments the biological impact of ionizing radiation, especially for low-LET radiation. Using cell killing as the biological endpoint, at a given dose, the percentage of cells killed is significantly higher when cells are growing under full oxygenation compared to the same scenario under hypoxic conditions. This effect of oxygen is called the **oxygen enhancement ratio (OER)**. The OER is defined as the ratio of doses administered under hypoxic to aerated conditions required to accomplish the same biological outcome:

OER = [radiation dose without oxygen]/[radiation dose in the presence of oxygen].

The OER for survival after irradiation of human cells is approximately 2.5–3.5 for sparsely ionizing (i.e. low LET) radiation such as x-rays, gamma rays, and high-energy electrons and protons. However, for densely ionizing (i.e. high LET) radiation such as neutrons, alpha particles, heavier charged particles and low-energy protons, the OER is significantly lower. With alpha particles, the survival curve is virtually identical irrespective of the concentration of oxygen; the OER is unity, meaning there is no enhancement due to oxygen. For fast neutrons (which are high LET but have lower ionization density than alpha particles), the experimentally observed OER for cell survival is also quite low and usually amounts to around 1.6.

There is some variation of the OER with the phase of the cell cycle. For instance, cells in G1 tend to have a lower OER than cells in S phase. Since G1 cells are more radiosensitive, they dominate the low-dose region of the survival curve in rapidly growing cells *in vitro*. Thus, the OER of an asynchronous population of rapidly dividing cells in culture is slightly smaller at low doses (around 2.5) than at high doses.

It is noticed that the amount of oxygen required to bring about the oxygen enhancement effect is quite modest. There is a marked increase in radiosensitivity of cells from oxygen concentration of 0%–5% (~30–40 mm Hg partial pressure). At an oxygen concentration of 5%, the cell survival curve becomes virtually indistinguishable from that obtained under full oxygenation. A relative radiosensitivity halfway between full oxygenation and anoxia corresponds to an oxygen tension of about 3 mm Hg or 0.4% oxygen, indicating a steep, sigmoid-shaped response curve. Note that in terms of radiosensitivity (i.e. exhibition of an oxygen enhancement effect), there is essentially no difference between pure oxygen (100% oxygen), normal air (21% oxygen) and 5% oxygen (which corresponds to the oxygen concentration in venous blood). For reference, our atmosphere is 21% oxygen and has a pressure of 760 mm Hg (millimeters of mercury as a unit of barometric pressure). This amounts to 160 mm Hg oxygen. Fully oxygen-saturated arterial blood has a partial pressure of between 75–100 mm Hg. In contrast, after the arteries deliver oxygen to the tissues, the returning venous blood only has an oxygen tension of 30–40 mm Hg. But as mentioned, even this oxygen-depleted venous blood has enough oxygen to exhibit a full oxygen enhancement effect for low-LET radiation.

Finally, the timing of oxygenation needed to demonstrate an OER sheds light on the radiochemical and radiobiological mechanisms of oxygen. The OER is evident when oxygen is present at the moment of irradiation. However, if it is added afterwards, the oxygen effect may be abrogated. However, there is a very brief interval

(<5 ms) after irradiation that the addition of oxygen might still allow expression of the oxygen effect. This observation is consistent with the proposed mechanism involving the production of ion pairs (with lifetimes on the order of 10^{-10} s or 100 ps) and free radicals (which have lifetimes on the order of 10^{-5} s or 10 μs). Thus, it is the longer-lasting free radicals that do more of the biochemical damage.

If molecular oxygen is present, the formation of organic peroxides (represented as $RO_2 \cdot$) from these radicals is possible. These organic peroxides represent a form of irreversible chemical change in targeted biological molecules. For example, the product of DNA attack in the absence of oxygen may be represented as DNA·; this type of damage is restorable. In contrast, when oxygen is present, the DNA product may be a peroxyl radical, DNA-OO·, which is considered irreparable. In this manner, oxygen 'fixes' radiation-induced chemical damage, meaning that it makes such damage permanent. This mechanism is called the **oxygen fixation hypothesis**. As explained later, even extensive 'permanent' DNA molecular damage may be repaired under certain circumstances.

9.16.3 Radicals

As mentioned above, much of radiation chemistry is mediated through the production and subsequent reactions of free radicals made through the ionization and excitation of solvent and solute molecules.

Free radicals may serve as either oxidizing agents or reducing agents when they attack other chemical moieties, including those in biological macromolecules. Radicals react in a variety of different ways. Some classic examples are hydrogen abstraction (RH + ·OH → ·R· + H_2O), ·OH addition (R + ·OH → ROH·) and radical elimination.

Chemical repair of radical-caused damage to biological organic molecules (i.e. **restitution**) can occur via radical recombination (e.g. R· + H· → RH) or, more commonly within cells, through hydrogen donation from **thiol** compounds, symbolized as R-SH. Thiols are chemical compounds that contain a sulfhydryl group (-SH). Thus, thiols are good free radical scavenging agents that are critically important in biochemistry.

$$R - SH + R \cdot \rightarrow RH + RS \cdot .$$

This restitution reaction results in much less reactive and less damaging **thiyl radicals**.

Thiol SH bonds are significantly weaker (about 20% weaker) than CH bonds. Thus, when thiols are abundant, free radicals preferentially attack them, rather than damaging the CH bonds in organic biological molecules. Hence, thiols are good radiation protectors because they scavenge free radicals and minimize the indirect effects of radiation on DNA and other important biological molecules.

9.17 Initiation, propagation and termination of radical reactions

Chain reactions involving free radicals can usually be divided into three distinct processes. These are initiation, propagation and termination.

Initiation reactions are those that result in a net increase in the number of free radicals. They may involve the formation of new free radicals from stable species, or they may involve reactions of free radicals with stable species to form more free radicals.

Propagation reactions are those reactions involving free radicals in which the total number of free radicals remains the same.

Termination reactions are those reactions resulting in a net decrease in the number of free radicals. Typically, termination involves two free radicals combining to form a more stable species, for example: $2Cl\cdot \rightarrow Cl_2$.

9.18 Reactivity of radical species and ease of formation

Generally, radicals requiring more energy to form are less stable than those requiring less energy. In other words, high-energy radicals are inherently more chemically reactive. Additionally, the more substituted the radical center is, the more stable it is. Thus, formation of a tertiary organic radical ($R_3C\cdot$) are favored over secondary organic radicals ($R_2HC\cdot$), which are favored over the very reactive primary radicals ($RH_2C\cdot$). Another observation is that radicals next to functional groups such as carbonyl, nitrile and ether groups are more stable than tertiary alkyl radicals.

Furthermore, sometimes radical formation is spin-forbidden, presenting an additional energy barrier. Hence, significant energy is often needed to create such reactive species; radiation is often a highly effective means of generating these reactive radicals. Although radical formation (that is, the initiation process) may be chemically difficult, the propagation steps are very exothermic and thermodynamically favorable.

9.19 Ion radicals

Although radical ions do exist, most species are electrically neutral. One notable example is the superoxide anion ($\cdot O^{-2}$). In contrast, the **hydroxyl anion** (HO^-), the **oxide anion** (O^{-2}) and the **carbenium cation** (CH^{+3}) are not real radicals, since the seemingly 'dangling' or unattached covalent bonds are in fact resolved by the addition or removal of electrons and there are no unpaired valence electrons.

9.20 Combustion as a radical reaction

A familiar free radical reaction is combustion. The diatomic oxygen molecule is unusual in that it is a stable diradical, meaning it has not just one, but two unpaired electrons. It may be represented as $\cdot O\text{–}O\cdot$. In accordance with Hund's rule, the two unpaired electrons go into two separate available molecular orbitals with their spins parallel, thereby maximizing the overall spin of the molecule. This unusual situation is the ground state of oxygen, but nevertheless, molecular oxygen is a relatively reactive entity. However, molecular oxygen can exist in a far more reactive state in which the two electrons are not in separate orbitals but instead spin-paired in the same orbital. This is the singlet state, singlet oxygen. In combustion reactions, the triplet state is converted into the more reactive singlet state, with heat supplying the activation energy. Combustion then ensues via a host of free radical chain reactions initiated by the excited singlet oxygen molecule. Interestingly, singlet oxygen is more reactive than triplet oxygen, but singlet oxygen is not a free radical, whereas triplet oxygen is (a diradical, in fact). The combustibility (flammability) of a given substance depends on the concentration of free radicals that must be obtained before the subsequent propagation reactions dominate.

As a historical aside, the combustion engine, which operates through combustion reactions in a gasoline and air (i.e. oxygen) mixture, is susceptible to preignition and knocking. The latter is due to spontaneous combustion of unburnt residues in the reaction chamber after the car has been turned off. This uncontrolled knocking can persist for several seconds after the ignition is turned off and can seriously damage the engine. Because lead can deactivate free radicals in the gasoline-air mixture, tetraethyl lead was once routinely added to gasoline. This lead would thereby prevent unwanted preignition or post-ignition combustion. Unfortunately, the lead added to gasoline to fight preignition and knocking would get mixed in with the combustion products and contaminate the environment.

9.21 Non-reactive radicals

Although most free radicals are very chemically reactive, there are a few free radical species such as **melanin** that are not chemically reactive. In fact, despite being a radical chemically, melanin serves as a potent ROS scavenger and quencher. As discussed in chapter 14, this can be important biologically. For instance, the highly radiation resistant fungi found growing and thriving in very high radiation regions such as the Chernobyl reactor are melanotic species (meaning they are dark due to their enrichment in melanin).

9.22 DNA damage reversal

This topic will be addressed also in the radiobiology chapters but for now, the basic chemical reactions will be introduced:

1. **Recombination**: This is the reaction of a nearby radical with the DNA radical. The result is the regeneration of the original unblemished DNA. The timescale is $<10^{-11}$ s.
2. **Restitution**: Restitution is the chemical restoration of damaged or excited DNA. No enzymes are involved in restitution. There are several intracellular reducing agents that can react with DNA radicals. Among the most important is **glutathione (GSH)**. The sulfhydryl group on GSH can donate an H· to the DNA radical, thereby yielding a restored DNA molecule and a less reactive sulfur radical (thiyl radical). The time scale for DNA restitution is under a millisecond ($<10^{-3}$ s).

$$DNA· + GSH \rightarrow DNA + GS·$$

3. **Repair**: There are multiple intracellular DNA repair enzymes that recognize and repair various forms of DNA damage through a number of different mechanisms. The time scale for this type of DNA damage repair is measured in minutes to hours.

9.23 Clustered DNA damage

In passing, it may be mentioned that clustered DNA damage (also known as **multiply damaged sites**) is more difficult, if not impossible, for cells to repair. Multiply damaged sites consist of closely spaced single lesions in proximity to one another (meaning within one or two helical turns of the DNA molecule). Recall that there are about 10.4 base pairs per helical turn of B-DNA, the form described by Watson and Crick and the form most often found in cells. Thus, clustered DNA damage is defined as multiple molecular lesions within around 10–20 base-pairs of the DNA molecule. The types of damage often encountered in multiply damaged sites include strand breaks, oxidative base damage and abasic sites (which means that the pyrimidine or purine moiety of the base pair has been removed). Clustered DNA damage is one of the main reasons (if not the main reason) that high-LET radiation is more damaging to cells than low-LET radiation. It appears that clustered DNA damage is more mutagenic and cytotoxic than isolated DNA damage and therefore is very biologically important. Clustered DNA lesions are divided into two major groups: double-stranded DNA breaks and non-double strand break clusters (also known as **oxidatively induced clustered DNA lesions (OCDLs)**). OCDLs may involve either two opposing DNA strands or all be on the same DNA strand. Low-LET radiation generates three to four times more non-double strand clustered DNA lesions than double strand breaks, but the proportion and complexity of clustered DNA lesions increases with increasing LET. Regardless of these specifics, clustered DNA lesions are difficult to repair and are a hallmark of radiation-induced damage.

9.23.1 Scavengers and quenchers

Scavengers are chemicals that can react with reactive species like free radicals and reactive oxygen/nitrogen species, thereby making them less reactive. They are related to free radical quenchers, which not only make free radicals less reactive, they completely neutralize them. In essence, when a quencher reacts with a free radical, the products are not free radicals. Thus, scavengers and quenchers may block the indirect effects of radiation by neutralizing such species before they can damage DNA. Some neutralizing scavengers are more specific for certain reactive species than others. For example, melatonin scavenges superoxide and peroxynitrate anions. And it also scavenges hydroxyl and peroxyl radicals and quenches singlet oxygen, so it is not very limited in scope. Ascorbate (vitamin C) and nitrate anions on the other hand are relatively more selective and restricted in scope than melatonin and preferentially scavenge hydroxyl radicals and hydrated electrons, respectively.

To simplify the highly complex radiation chemistry in solutions or cells, selective scavengers, which preferentially react with a limited set of radicals, can be added to an experimental system. In this fashion, specific scavengers that focus on a particular reactive species have helped to define the detailed radiation chemistry in each situation.

Quantitative experiments involving scavengers provide a means of assessing the relative contributions of direct and indirect effects of radiation. Radiobiological experiments with scavengers suggest that 60%–70% of the radiation damage to cells exposed to low-LET radiation is due to indirect effects of radiation, with the majority caused by HO· radicals. This contrasts with the situation involving high-LET radiation, where 60%–70% of the damage is due to direct effects on DNA and other biological molecules.

9.23.2 Glutathione

Glutathione or **γ-glutamylcysteinylglycine** or GSH is a natural free radical scavenging agent and antioxidant found in humans and other animals, as well as plants, fungi and some prokaryotes. As the name γ-glutamylcysteinylglycine suggests, it is a tripeptide made of glutamate, cysteine and glycine. It contains an unusual **gamma peptide linkage**. In the way of review, the amino acids found in biochemistry that serve as the building blocks of peptides (and larger chains of peptides called proteins) are alpha-amino acids. That is, the amino group is attached to the alpha carbon; the carboxyl group is also attached to the alpha carbon. And in regular **peptide bonds**, the alpha amino group of one amino acid is linked to the alpha carboxyl group of another amino acid through an amide link. Glutamate (glutamic acid) contains two carboxyl groups—one attached to the alpha carbon, and another attached to the gamma carbon. In glutathione, the glutamate carboxyl group is linked to the cysteine amino group via its gamma carboxyl rather than its alpha carboxyl. This gamma amide linkage renders glutathione increased stability through resistance to hydrolysis by peptidases. Thanks to its thiol group (–SH) on the cysteine moiety, it is a powerful antioxidant. Because it is a thiol, it is sometimes abbreviated GSH. Upon reaction with a radical, GSH becomes oxidized to glutathione disulfide (GSSG). This oxidized state can be reduced back into the active reduced form, GSH, by the enzyme **glutathione reductase**, using nicotinamide adenine dinucleotide phosphate (NADPH):

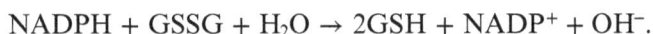

$$NADPH + GSSG + H_2O \rightarrow 2GSH + NADP^+ + OH^-.$$

This capacity for regeneration to the active form is important for several reasons, including the fact that GSH serves a role in reducing other oxidized antioxidants back into their active (reduced) forms. Thus, regeneration of active GSH can result in regeneration of many other key antioxidants in cells. Although it is not a co-factor in glutathione reductase, **selenium** is an essential component of two important antioxidant enzymes that work through GSH: **glutathione peroxidase** and **phospholipid hydroperoxide glutathione peroxidase**. Several other principal antioxidants and quenchers of reactive oxygen and nitrogen species are also 'selenoprotein' enzymes. These selenoproteins contain the amino acid **selenocysteine**, in which the sulfur atom of cysteine is replaced by a selenium atom instead. In such cases, instead of a sulfur-containing thiol (-SH) functional group doing the radical scavenging, an equivalent, selenium-containing functional group (-SeH) is the reducing agent.

In its reduced state, GSH is a good general free radical scavenger that can react with hydroxyl radicals, hydrogen atom radicals and hydrated electrons. GSH mitigates radiation damage to intracellular components including DNA, enzymes and membranes by neutralizing ROS, including singlet oxygen, along with various free radicals, peroxides and lipid peroxides *in vivo*.

9.24 Haber–Weiss and Fenton reactions

Thanks to the central role of the hydroxyl free radical in radiation chemistry and radiobiology, no review of radiation chemistry would be complete without a discussion about the Haber–Weiss and Fenton reactions. These reactions occur in aqueous solutions and yield hydroxyl radicals. The **Haber–Weiss reaction** is the overall equation for the iron-catalyzed conversion of superoxide and hydrogen peroxide into oxygen, hydroxide ions and the all-important hydroxyl radical:

$$\cdot O_2^- + H_2O_2 \rightarrow O_2 + OH^- + \cdot OH$$

This net reaction can be broken down into two individual components:

$$\cdot O_2^- + Fe^{3+} \rightarrow O_2 + Fe^{2+}$$

and

$$Fe^{2+} + H_2O_2 \rightarrow OH^- + \cdot OH^- + Fe^{3+}$$

The second reaction, the conversion of peroxide into hydroxide anions and hydroxyl radicals (with the oxidation of ferrous iron into ferric iron) is part of the Fenton reaction. Thus, the catalyst, Fe^{3+} or iron(III) is regenerated through the Fenton reaction. And the Fenton reaction can be considered part of the overall Haber–Weiss reaction. Conversely, the Haber–Weiss reaction may be called the **superoxide-driven Fenton reaction**. Although the Fenton reaction does appear to occur *in vivo*, the real role of the Haber–Weiss reaction in generating free radicals and leading to oxidative stress within cells is debated.

The full Fenton reaction is a bit more complicated than just the second half of the Haber–Weiss reaction. There are actually two steps to the complete Fenton reaction, and besides a hydroxyl radical, a hydroperoxyl radical and a proton (H^+) are also made.

In the first step, ferrous iron (Fe^{2+} or iron(II)) becomes oxidized into ferric iron (Fe^{3+} or iron(III)) by hydrogen peroxide into a hydroxide anion and the hydroxyl radical:

$$Fe^{2+} + H_2O_2 \rightarrow OH^- + \cdot OH + Fe^{3+}.$$

In the second step, the Fe^{3+} is reduced back into Fe^{2+} by additional hydrogen peroxide with the production of a hydrogen ion (proton or H^+) plus a hydroperoxyl radical:

$$Fe^{3+} + H_2O_2 \rightarrow H^+ + HOO\cdot + Fe^{2+}.$$

The hydroxide anion (OH^-) made in the first step pairs up with the hydrogen cation (H^+) in the second, to form water (H_2O). Hence, the net result of the complete Fenton reaction is:

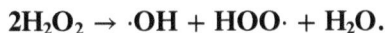

$$2H_2O_2 \rightarrow \cdot OH + HOO\cdot + H_2O.$$

Chemically, this amounts to a disproportionation of hydrogen peroxide because it is both reduced by iron(II) in the first reaction and oxidized by iron(III) in the second.

Although the Fenton reaction is of great relevance to radiobiology because it does occur *in vivo*, the reaction was originally the basis for the so-called **Fenton reagent**, which can be used as a decontaminant and means of oxidizing hazardous organic compounds, including waste waters for sewage treatment; it was developed by Henry Fenton in the 1890s. The Fenton reagent is basically a solution of peroxide with ferrous iron sulfate.

Although ferrous/ferric iron is the classic catalyst in the Fenton reactions, 'Fenton-like reactions' may be catalyzed by other metals with multiple oxidation states such as copper, cerium, cobalt, chromium, manganese, aluminum and ruthenium. Therefore, rather than alluding just to the Fenton reaction, chemists occasionally allude to the entire repertoire of reactions under the broader umbrella of **Fenton chemistry**.

Further reading

[1] Tomusiak-Plebanek A, Heczko P, Skowron B, Baranowska A, Okoń K, Thor P J and Strus M 2018 Lactobacilli with superoxide dismutase-like or catalase activity are more effective in alleviating inflammation in an inflammatory bowel disease mouse model *Drug Des. Dev. Ther.* **12** 3221–33

[2] Koppenol W H 2022 A resurrection of the Haber–Weiss reaction *Nat. Commun.* **13** 396

[3] Bukowska B and Karwowski BT 2018 The clustered DNA lesions - types, pathways of repair and relevance to human health *Curr. Med. Chem.* **25** 2722–35

IOP Publishing

Space Radiation
Astrophysical origins, radiobiological effects and implications for space travellers
James S Welsh

Chapter 10

Radiation biochemistry

In this chapter, we shall conduct a brief review of the major enzymes that are involved in inactivating the chemical products of radiation before they can cause macromolecular damage. Such enzymes neutralize reactive oxygen species (ROS), reactive nitrogen species (RNS), singlet oxygen and the host of biologically damaging free radical species. Additionally, we shall include in this chapter some of the non-enzyme antioxidants that are of relevance to radiation response.

DNA is the most important target of ionizing radiation, largely because it contains the blueprint for making all other components in the cell but also because, unlike other biological molecules, there are usually only one or two copies of this critical molecule. The repair of radiation-induced DNA damage is extremely complex and is the subject of molecular radiobiology, which is covered in chapter 11.

Biological macromolecules are large molecules found in cells that carry out critical functions. For instance, the **enzymes** that speed reactions up many million-fold are **proteins**, which are one major class of macromolecules. Biological macromolecules include:

1. Proteins (which are made of amino acid subunits)
2. Polysaccharides (which are made of monosaccharide (sugar) subunits)
3. Nucleic acids such as DNA and RNA (which are made of nucleotide subunits)
4. Lipids (which, in the case of triglycerides, include three fatty acid chains).

Many biological macromolecules are **polymers** that are composed of **monomers**. For instance, the energy storage molecules of glycogen and starch are polysaccharides that are made of repeating units of monosaccharides (sugars). Fat (triglyceride molecules) is a type of lipid that is made of monoglyceride components. And RNA (ribonucleic acid) is a nucleic acid that is composed of multiple ribonucleotide subunits. The resulting macromolecular polymer may be relatively simple (as in the

doi:10.1088/978-0-7503-5444-8ch10

case of polysaccharides made of identical repeating units of glucose, like cellulose) or extremely complex (as in the 3D conformation of large multimeric protein enzymes, which are composed of countless permutations of 20 different amino acids). In the context of proteins, 'multimer' alludes to the combination of multiple units into a cohesive final 'quaternary' 3D structure. Hence, multimeric proteins (dimers, trimers, tetramers, etc) or multi-subunit proteins are macromolecules that have more than one subunit. And, in the case of proteins, each subunit may be composed of long chains of amino acids that fold upon themselves in an intricate manner that is essential to their proper functioning. A classic example is the oxygen-carrying hemoglobin molecule, which is a tetramer composed of four individual globular protein subunits called globins, along with four iron-containing heme groups.

After radiation damage, most biological molecules are simply removed and replaced rather than actively repaired. DNA is the obvious exception to this rule, and the various DNA repair mechanisms that have evolved are quite complicated, yet often amazingly efficient. Some types of protein damage, such as disulfide bridges caused by oxidation, are also repaired by specific mechanisms, as are some types of membrane lipid damage.

10.1 Interactions of radiation with biological macromolecules

The interactions of radiation-generated free radicals with biological macromolecules depend largely on the chemical composition of these biological compounds. Recall that the aliphatic amino acids (alanine, glycine, isoleucine, leucine, proline and valine) behave hydrophobically and dictate protein folding. The aromatic amino acids are tyrosine, phenylalanine and tryptophan, and the sulfur-containing amino acids incorporated into proteins include cysteine and methionine. Radiation-induced hydroxyl radicals typically attack three major reaction sites in a protein molecule:

1. The carbon-hydrogen and carbon–carbon bonds in the main peptide backbone
2. The aliphatic amino acids
3. The aromatic and sulfur-containing amino acid residues.

Hydroxyl radicals initiate several types of reactions with proteins such as addition, electron transfer, and hydrogen abstraction. The main reaction with the polypeptide 'backbone' begins with hydrogen abstraction at the α-carbon. This is followed by a reaction with molecular oxygen to give a peroxyl radical. This ultimately causes cleavage of the protein backbone, leaving amide and carbonyl fragments. The hydroxyl radical reacts more readily with aromatic amino and sulfur-containing amino acid moieties than with aliphatic amino acid residues in proteins. The higher reaction rates with aromatic and sulfur-containing residues therefore mean that a relatively higher percentage of these amino acids a given protein will dominate the radiation chemistry in that protein molecule. Radiation-induced disulfide crosslinking is an important reaction in proteins that contain any substantial amounts of cysteine. Radiation-induced crosslinking can also occur between aromatic amino acid residues in proteins. Besides inducing protein crosslinking, radiation and radicals may attack peptide bonds since peptide bonds are highly reactive with hydrated electrons (e_{aq}^-). But in contrast to the rapid reactions between hydroxyl radicals and aromatic amino acids, the reaction rates of hydrated electrons with aromatic amino acids are relatively slow; hydrated electrons react with aromatic amino acids around an order of magnitude *slower* than they do with histidine and cysteine. Thus, the radiation chemistry of hydrated electrons with proteins is dominated mainly by reactions with histidine and cysteine moieties and with peptide bonds (in contrast to the reactions with aromatic and sulfur-containing amino acids which characterize the protein radiation chemistry of hydroxyl radicals).

The indirect effects of radiation on DNA and RNA are due to the chemical reactions between nucleic acids and the radiolysis products of water (hydroxyl radicals, superoxide anion, solvated electrons, and free hydrogen atoms). As mentioned earlier, some of these reactions are extremely fast. In fact, hydroxyl radicals react with free nucleosides and nucleotides at or near the 'diffusion limit', meaning the rate approaches the theoretical maximum. Recall that bimolecular chemical rate constants have a maximum value that is determined by how frequently the two molecules can encounter one another and collide. The fastest this can happen is limited by diffusion. Thus, in aqueous solutions at room temperature, a bimolecular rate constant has an upper limit of $k_2 \leqslant \sim 10^{10}$ $M^{-1}s^{-1}$.

Hydrogen atom radicals react with nucleic acid bases slower than hydroxyl radicals do (a bit over ten-fold slower rate). Nevertheless, these bimolecular rate constants ($\sim 10^8$ M^{-1} s^{-1}) are still faster than many other radical reactions.

When hydroxyl radicals and free hydrogen atom radicals react with pyrimidines in DNA, the predominant pathways involve organic addition reactions to the double bonds in the pyrimidine ring. Hydroxyl radicals prefer to add to the C-5 carbon, whereas hydrogen atoms, being more nucleophilic than hydroxyl radicals, demonstrate different **regioselectivity** and preferentially add to the C-6 carbon of the pyrimidine ring (figures 10.1 and 10.2).

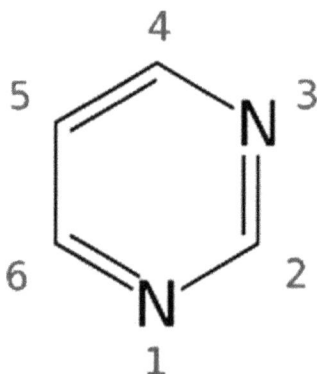

Figure 10.1. Pyrimidine molecular numbering. This Pyrimidine 2D numbers image has been obtained by the author(s) from the Wikimedia website (https://commons.wikimedia.org/w/index.php?curid=25262171), where it is stated to have been released into the public domain. It is included within this book on that basis. It is attributed to Jynto.

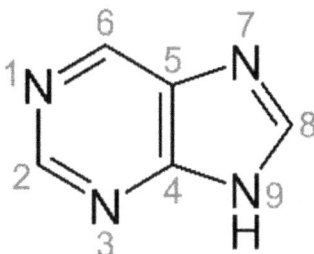

Figure 10.2. Purine molecular numbering. This Purin num2 image has been obtained by the author(s) from the Wikimedia website (https://commons.wikimedia.org/w/index.php?curid=4542866), where it is stated to have been released into the public domain. It is included within this book on that basis. It is attributed to NEUROtiker.

In addition to these indirect effects mediated through radicals, radiation can interact directly with pyrimidine bases in DNA and RNA. **Nucleobase radicals** are the major intermediates produced by both the direct (e.g., dA·+) and indirect (e.g., dA·) effects of radiation in aqueous solutions. Direct interactions lead to ionized pyrimidine and purine bases, which chemically are **alkene cation radicals**. Such pyrimidine alkene cation radicals may then suffer nucleophilic attacks by water molecules. The major product of this nucleophilic attack on pyrimidines is hydroxylation at C-6. Note that this radiochemical product of the direct effect differs structurally from the two main products of the indirect effect mentioned above. Another possible outcome of the direct radiation effect ultimately yields a C-6 hydrogen atom adduct; this is identical to the major product of hydrogen atom radical attack on a pyrimidine base.

10.2 Radiation countermeasures

Radiation countermeasures, defined here as chemical agents that are given in an effort to minimize the negative effects of high-dose, high dose-rate radiation, fall under three general categories:

1. Radiation protectors
2. Radiation mitigators
3. Radiation therapeutics.

Radiation protectors or **radioprotectors** are typically administered *before* ionizing radiation exposure; hence, they are sometimes called **radiation prophylactic agents**. They are given in an effort to protect people from imminent radiation-induced injuries. Radiation protectors work through several mechanisms such as scavenging the free radicals that are created through the initial radiochemical events.

Radiation mitigators are agents that are administered *after* radiation exposure but before the onset of radiation-specific symptoms. Mitigators work by promoting recovery and repair from radiation-induced injuries or by accelerating the removal of radioactive elements from the body after internalization. Thus, **decorporation** and **chelating agents** that aim to reduce damage caused by internalized radionuclides fall into this category; they promote the removal of certain radionuclides from the body. '**Blocking agents**' that minimize uptake of radionuclides by organs is another class of mitigator. An example of this latter group is **potassium iodide** to reduce uptake and subsequent damage to the thyroid by ingested radioisotopes of iodine such as iodine-131.

The final category of radiation countermeasures are the **radiation therapeutic agents**. These agents are administered after the onset of radiation-related symptoms such as **acute radiation syndrome**. Radiation therapeutic agents act by promoting healing and regeneration of tissues that have suffered radiation injury. These agents aim to treat and accelerate recovery from acute radiation exposure.

Not all experts use this classification system of radiation countermeasures. For instance, others may consider 'radioprotectors' in a far broader fashion than just agents given before radiation exposure to prevent injury. In this sense of the word, radioprotectors may be given before or after exposure and thus include the above category of radiation mitigators.

10.3 Radioprotectors and radiosensitizers

In the broad sense of the term, radioprotectors are chemical agents that reduce adverse reactions to radiation. Most radioprotectors work by modifying the indirect effects of radiation by scavenging the formed free radicals and other reactive species. The **dose modifying factor (DMF)** of radioprotectors is defined as:

DMF = (Dose to produce an effect in the presence of a radioprotector)/
(Dose to produce same effect without the radioprotector)

In contrast, chemical agents that augment the biological response to radiation are called **radiosensitizers**. Radiosensitizers are used in cancer radiation therapy and may work by promoting either the direct or indirect effects of radiation (or both). A classic example are the **halogenated pyrimidines** (such as iododeoxyuridine). These molecules intercalate between the strands of DNA, thereby inhibiting repair. Thus, the normal repair mechanisms that restore DNA after irradiation are impaired. These halogenated pyrimidine analogues are classified as **non-hypoxic radiosensitizers** since they augment the activity of the indirect effects of radiation in the presence of oxygen. Other radiosensitizers are **hypoxic cell radiosensitizers**, which mimic the effects of oxygen in its absence (i.e. under hypoxic conditions).

A related concept involves drugs that mimic the effects of ionizing radiation. Such drugs are called **radiomimetics**. A classic example of radiomimetic drugs are the nitrogen mustards (a class of chemotherapy agents that were originally produced as chemical warfare agents).

As mentioned, the term 'radioprotector' may be defined in a very limited sense to include only agents given before radiation exposure to ward off potential injury, or the term may be more inclusively defined to also include radiation mitigators that are given after exposure to reduce damage. When used in this more inclusive way, there are four main mechanisms whereby radioprotectors function. These are:

1. Competition for the strong oxidizing agents and free radicals formed upon the radiolysis of water, solvents and other media. In such competition,

radioprotectors may win and thereby quench the activity of potentially damaging free radicals.

2. Protection of radiosensitive enzymes, hormones and protein molecules from the damaging effects of radiation.
3. The chelation and sequestration of heavy metal ions and other cations that possess several oxidation states. (Such ions can promote the damaging effects of radiation by engendering new radicals if not complexed and inactivated.)
4. The termination of oxidation chain reactions. By donating electrons and becoming relatively innocuous free radicals, some radioprotectors can inhibit or terminate otherwise hazardous chain reactions.

10.4 Three lines of intracellular radiation defense

There are several different reactive species generated by radiation and other causes (including natural endogenous processes related to use of oxygen in aerobic respiration). Such reactive species have the potential for intracellular oxidative stress and oxidative damage. Some examples include genuine **radicals** such as the **hydroxyl radical, superoxide anion, singlet oxygen radical, peroxyl radical, nitric oxide radical** and **lipid peroxy radicals,** as well as **non-radical reactive species** such as **hydrogen peroxide, peroxynitrate, trichloromethane** and **hypochlorous acid,** which have high potential to generate free radicals. The term **ROS** (reactive oxygen species) alludes to chemically active entities regardless of whether they are true radicals or not. The same applies to **RNS** (reactive nitrogen species).

While ionizing radiation can lead to intracellular reactive entities such as free radicals and non-radical reactive species, many such entities are engendered endogenously during respiration, wound healing and repair processes. Among the sources are **mitochondria,** the organelles where oxidative phosphorylation and its associated electron transport chain occurs. Mitochondria manufacture various ROS when molecular oxygen is reduced via the **electron transport chain** on the inner mitochondrial membrane. **Phagocytes** (white blood cells such as **neutrophils,** which engulf and destroy microbial invaders) also produce abundant ROS through enzymatic activity of nicotinamide adenine dinucleotide phosphate or **NADPH oxidase** (e.g. superoxide generation via $O_2 + e^- \rightarrow O_2 \cdot^-$) in their efforts to kill consumed pathogens. Oxygen-derived radicals may also be produced as intermediate metabolites in cascades of various enzyme catalyzed reactions. One such source of ROS and cellular oxidative stress is the **cytochrome P450 (CYP) system.** CYP enzymes function to transform toxic chemicals into less toxic metabolites. They do not always succeed in this effort, however, and on occasion, the system backfires, and a relatively non-toxic entity can be converted into a more dangerous one. Even in their normal activity, they may manufacture ROS such as superoxide anions, hydrogen peroxide and hydroxyl radicals. Overall, ionizing radiation is, by far, the lesser of the two sources of ROS when one compares natural intracellular endogenous production and normal background radiation.

Hence, it is not surprising that there are many mechanisms of defense against free radicals and ROS, whether they are generated by radiation or produced endogenously through various metabolic activities.

10.5 Biochemical defenses against radiation

Cellular **antioxidant** molecules such as ascorbic acid (vitamin C), alpha-lipoic acid, thioredoxin, glutathione, melatonin, coenzyme Q, beta carotene, retinoids (vitamin A and derivatives), alpha-tocopherols (vitamin E), polyphenols and bioflavonoids (such as apigenin), as well as antioxidant enzymes including superoxide dismutase (SOD), catalase, glutathione-peroxidases, glutathione reductases, glutathione-s-transferases and thioredoxin reductase have been studied for their mechanisms of action as well as for their potential in prevention and treatment of diseases resulting from oxidative damage (including radiation injury). Some agents such as **WR-1065** and **WR-2721** (**amifostine**) were developed specifically to prevent and address radiation-related injury. (WR stands for Walter Reed National Military Medical Center for the United States Army.)

Thus, cells have multiple layers of defense against such reactive entities. Three general levels of defense include: 1. formation prevention, 2. free radical scavenging and 3. radical-induced biomolecular damage repair. Regarding free radical prevention, antioxidants are a critical participant. Three categories of antioxidants are described: first-line defense antioxidants, second-line defense antioxidants and third-line defense antioxidants.

10.5.1 First-line defense antioxidants

The category of first-line defense antioxidants constitutes a collection of antioxidants that act to prevent or diminish the formation of free radicals and other reactive species in cells. Such antioxidants are incredibly quick in neutralizing chemical entities that have the potential of forming free radicals. They also quickly quench newly formed free radicals before they promote the production of other radicals. Among first-line defenses are **neutralizing enzymes**. Three key enzymes (**superoxide dismutase** or **SOD, catalase** and **glutathione peroxidase**) are the most well-studied on this list of enzymes that respond to radiation and other potential sources of reactive species. In addition to these neutralizing enzymes, first-line defense antioxidants also include certain metal ion-binding proteins. Classic examples include **transferrin** and **ceruloplasmin**, which chelate and sequester free iron and copper, respectively. In so doing, these ion-binding proteins consequently prevent metal ions from forming free radicals. An exogenous metal ion chelator, **gallic acid** (3,4,5-trihydroxybenzoic acid) is a **phenolic acid phytochemical** (discussed further below) found in sumac, tea, garlic and witch hazel among other plants. It and related entities represent a category of natural products that are capable of attenuating cellular damage caused by radiation.

10.5.2 Second-line defense antioxidants

This class of defense agents includes the free radical scavenging antioxidants. These antioxidants scavenge existing free radicals, thereby preventing chain initiation and halting chain propagation. These antioxidants neutralize or scavenge free radicals by functioning as electron donors. In the act of electron donation, they become free radicals themselves. However, these antioxidant molecules become free radicals with far less potential for biomolecular damage.

This is the largest class of antioxidants and includes molecules that are both hydrophilic and lipophilic. Among the hydrophilic agents are **ascorbic acid** (**vitamin C**), uric acid and **glutathione**. Classic lipophilic antioxidants include **alpha tocopherol** (**vitamin E**) and ubiquinol. Some of these antioxidants such as **alpha lipoic acid** are amphiphilic (meaning they have both hydrophilic and lipophilic properties).

10.5.3 Third-line defense antioxidants

The entities in this category of defense are not literally active antioxidants, because they repair the damage done by radiation-induced reactive species. They comprise a group of enzymes that repair damaged DNA, proteins and lipids. Thus, they come into play after the damage is done, for the purposes of repair rather than prevention. This category of defense includes enzymes that repair biomolecular damage caused by free radicals. For example, such repair enzymes may be involved in DNA damage repair, protein damage repair (such as reduction of disulfide bridges) and restoration of damaged membranes. Examples of this line of defense include the various DNA repair enzyme systems such as DNA polymerases, glycosylases, nucleases and more complicated mechanisms of DNA damage repair including the non-homologous end joining and homologous recombination processes, which are covered in the next chapter. It also includes certain proteolytic enzymes (proteinases, proteases and peptidases) that may be involved in protein restitution.

10.6 Radical-inactivating enzymes

First-line defense antioxidants include specific enzymes that are geared towards inactivating potentially harmful reactive species that are made via radiation or other means. These enzymes play a fundamental role in the overall antioxidant armamentarium by averting macromolecular damage through disarming free radicals, singlet oxygen, ROS, RNS and other biochemically reactive radiation-created entities. The three most important enzymes in this category are catalase, SOD and glutathione peroxidase.

10.6.1 Catalase

Hydrogen peroxide (H_2O_2) is a nonradical ROS that, in excess, can be damaging. Nevertheless, hydrogen peroxide at low levels, is also required for optimal cellular functioning. Numerous enzymes neutralize or regulate hydrogen peroxide levels including catalase, glutathione peroxidase, cytochrome c peroxidase and NADH peroxidase.

One of these enzymes, catalase, is ubiquitous in all aerobic organisms. In fact, it is not only found in true aerobes (that is, organisms that use aerobic respiration as a means of metabolism for generating useable energy through catabolism of organic chemicals like carbohydrates and fats); catalase is found in essentially all organisms that are ever even exposed to oxygen in their environments. In fact, the presence or absence of catalase serves as a basis for microbial classification (through the **catalase** test), with most microorganisms being **catalase-positive** but some pathogenic bacteria being facultative or obligate **anaerobes** that are **catalase-negative**. Among other anti-microbial defenses, neutrophils engulf pathogens and douse them with hydrogen peroxide and other more potent oxidizing agents. While this can cause some catalase-negative anaerobic bacteria to succumb, catalase-positive microbes can ward off this attack mode. The antibacterial oxidants made in neutrophils and other phagocytic cells of the immune system stem from a multi-step process starting with the enzyme **NADPH oxidase**. This enzyme generates superoxide anions in the neutrophil's **phagosome** (the little intracellular vacuole that contains the engulfed microbe). The superoxide is then converted into the potent antimicrobial oxidizing agent, **hypochlorite** via hydrogen peroxide, chloride ions, and the myeloperoxidase enzyme. People with **chronic granulomatous disease** have an inborn deficiency of NADPH oxidase, and their phagocytes cannot produce peroxide at an adequate rate to ward off certain infectious pathogens, especially those that are catalase-positive. In contrast, individuals with a deficiency in catalase itself (because of homozygous loss of function mutations in the gene for catalase production, *CAT*) have a condition known as **acatalasemia** or **acatalasia**. Such individuals are phenotypically normal and mostly asymptomatic, although they may be more prone to developing oral ulcers and gangrene. People with autosomal recessive myeloperoxidase deficiency also are phenotypically normal and tend to be asymptomatic. However, especially if they have concurrent diabetes, they may exhibit immunodeficiency, especially for fungal infections with *Candida albicans*.

Catalase is a key enzyme in protecting against oxidative damage by ROS made by radiation or other means. Therefore, it is highly conserved and evolutionarily dates back billions of years to the time when oxygen first became abundant in the environment (i.e. around the **great oxidation event** around 2 billion years ago). Catalase contains iron in the form of **heme** in its active site; the iron undergoes transitions in its redox state as it catalyzes reactions involving hydrogen peroxide. Heavy metals may act as non-competitive inhibitors of catalase.

Catalase is not only important for neutralizing hazardously high levels of hydrogen peroxide. It is a key enzyme for regulating and maintaining *ideal levels* of hydrogen peroxide for intracellular signaling processes. In most eukaryotic cells, including human cells, catalase activity is highest in cytoplasmic membrane-bound organelles known as **peroxisomes**. In humans, the organ with highest catalase activity is the liver, although practically every organ expresses catalase to some degree.

Catalase is responsible for neutralization of H_2O_2 through decomposition of two hydrogen peroxide molecules into one oxygen molecule and one water molecule.

$$2H_2O_2 \rightarrow O_2 + H_2O.$$

It proceeds in a two-step process that oxidizes and reduces iron in its enzymatic active site. Because the peroxide is both oxidized and reduced, the overall reaction is a **disproportionation** or **dismutation**.

The molecular efficiency of catalase is phenomenal. One catalase molecule can catalyze the conversion of millions of hydrogen peroxide molecules into water and oxygen in one second. This incredible enzymatic turnover rate is one of the highest known and appears to defy classical explanations through the so-called 'lock and key' mechanisms of enzyme action. Thus, catalase and other phenomenally fast enzymes are of great interest in **quantum biology**. Some enzymes appear to achieve their uncanny turnover rates through **quantum tunneling**. Quantum tunneling also appears to play a role in the biochemistry of photosynthesis and electron transport.

A fascinating use of catalase is seen in the entomological realm in the form of **bombardier beetle** weaponry. Bombardier beetles can blast their enemies with hot **hydroquinones** and **quinones**. The beetle has two sets of glands (pygidial glands), sequestering hydrogen peroxide and hydroquinones. When ready to fire upon their foes, these beetles mix the contents of the two compartments in the explosion chamber (reaction chamber) where catalase and peroxidase enzymes promote powerful chemical reactions between these compounds. Catalase creates oxygen from hydrogen peroxide. The newly formed oxygen gas provides propulsive force and also oxidizes the hydroquinones into quinones, catalyzed by the peroxidases. This oxidation reaction is highly exothermic and heats the aqueous solution nearly to the boiling point. The explosive ejection of the hot liquid contents is accompanied by a loud, intimidating pop. As the 1,4-benzoquinone products are strong irritants (even at low temperatures), this is a powerful deterrent to would-be predators—including beetle collectors!

10.6.2 Superoxide dismutase

SOD is an enzyme that promotes the **dismutation** of superoxide free radical anion into hydrogen peroxide and gaseous molecular oxygen.

$$O_2^{\cdot-} + 2H^+ \rightarrow H_2O_2 + O_2$$

The overall reaction catalyzed by SOD converts the potentially hazardous superoxide anion into two relatively innocuous species.

The superoxide anion, O_2^- is a reduced form of molecular oxygen (O_2) with an extra electron. It is an ROS that happens to be an **ion radical**. Superoxide radicals are capable of denaturing proteins (including enzymes), oxidizing lipids in membranes and fragmenting DNA. Additionally, superoxide is known to react with the important signaling molecule and biological second messenger, **nitric oxide**. Nitric oxide synthesis is catalyzed by nitric oxide synthase enzymes from oxygen, arginine and NADPH. Nitric oxide is important in causing vasodilation through blood vessel smooth muscle relaxation. The drug sildenafil (Viagra) works by increasing nitric oxide activity by inhibiting the enzymatic activity of PDE_5 (cyclic guanosine monophosphate (cGMP) phosphodiesterase type 5). PDE_5 breaks down cGMP, which is an important mediator of the vasodilating effects of nitric oxide. Hence, inhibition of the degradation of this key molecule is important in maintaining the actions of nitric oxide. Nitric oxide is a radical. As such, it is preferentially acted upon by superoxide radicals since superoxide reactions with non-radicals is spin-forbidden and thus slower than the reactions with radical species. When superoxide encounters nitric oxide, it generates the highly reactive and genotoxic **peroxynitrate** anion.

Superoxide and other ROS are produced in human phagocytes (white blood cells, such as neutrophils that engulf and eliminate pathogenic microbes). These cells of the immune system attempt to destroy the invading bacteria by using the enzyme **NADPH oxidase** to generate superoxide anions. Some bacterial species are able to withstand this attack by using their own SOD and other detoxifying enzymes to defuse the ROS.

Chemically, a **dismutation** (also called **disproportionation**) is a type of redox reaction in which the parent compound is converted into two daughter products of different oxidation states, with one higher and one lower than the original reactant's oxidation state. In essence, the parent compound undergoes both auto-oxidation and reduction. The reverse of dismutation or disproportionation is known as **comproportionation** (or **synproportionation**). In such synproportionation reactions, a compound of intermediate oxidation state is produced from reactants of lower and higher oxidation states.

Dismutation of superoxide involves both oxidation and reduction of the two oxygen atoms in the superoxide anion. The initial oxidation state of both oxygens in superoxide is $-1/2$, but it becomes -1 in hydrogen peroxide and 0 in diatomic molecular oxygen. (Recall that all diatomic elements have an oxidation state of 0.) The subsequent conversion of hydrogen peroxide into water and oxygen, which can be catalyzed by the enzymes catalase or glutathione peroxidase, is another classic example of a radiobiologically important disproportionation reaction, since the oxygen in hydrogen peroxide initially has an oxidation number of -1 but becomes -2 and 0 in the final products of water and molecular diatomic oxygen, respectively:

$$H_2O_2 \rightarrow H_2O + O_2.$$

SOD is not a single enzyme but rather is a class of different enzymes with different structures and subcellular locations. In other words, there are several **isozymes** of SOD. Practically, all aerobic cells have some form of SOD. In humans, three forms of SOD exist.

SOD1 is located primarily in the cytoplasm, **SOD2** is in the mitochondria, and **SOD3** is extracellular. Curiously, these different forms of SOD use different metals in their active sites. Human SOD1 uses copper and zinc ions in its reaction center, SOD2 uses manganese, and SOD3 employs copper and zinc. Other organisms use different redox metal co-factors in their active sites; **trypanosomes** for instance (which are unicellular parasites that can transmit parasitic diseases such as **Chaga's disease**) use iron in their SOD. In fact, there are three distinct families of SOD, which are categorized based on the metal they employ—the copper/zinc family, the iron and manganese family (found in mitochondria and bacteria), and the nickel family (exclusively in bacteria and archaea). **Amyotrophic lateral sclerosis (ALS**, also known as **Lou Gehrig's disease**) is usually sporadic, meaning there it is not inherited. However, there are some familial forms of this disease and germline mutations in

SOD1 have been linked to this, suggesting that the antioxidant actions of SOD are mechanistically important in this disease.

There is some data suggesting that among the various antioxidant enzymes involved in neutralizing radiation-related radicals and reactive species, manganese-based SODs may play the most crucial role and may be slightly more important than the copper/zinc utilizing SODs and glutathione peroxidases.

Just as an interesting aside, *Lactiplantibacillus plantarum* is an example of a bacterial species devoid of SOD. It is normally a fermenter but is tolerant of oxygen. However, under the right circumstances (e.g. the presence of heme and menaquinone in its environment or medium), it can engage in aerobic respiration and put oxygen to good use. But in the absence of heme and menaquinone, the free oxygen is converted first into hydrogen peroxide and then water via NADH peroxidase. It appears that the bacteria then use the manufactured hydrogen peroxide against other oxygen-sensitive competing bacterial species. But instead of real SOD, *L. plantarum* accumulates large quantities of inorganic manganese (in the chemical form of manganese polyphosphate), which serves the same chemical function. The manganese is also used to neutralize other reactive oxygen species via a 'pseudo-catalase' mechanism. Curiously, because these antioxidant mechanisms are disrupted by iron ions, *L. plantarum* contains virtually no iron. It is one of the very few biological entities to entirely eschew iron. Another fascinating tidbit about this species is that in its fermentation, it produces both **D- and L-stereoisomers** of lactic acid. Humans only make L-lactate under normal circumstances. Under pathological circumstances, however, such as short bowel syndrome, harmful D-lactate may be produced by certain bacteria in the patient's gut.

Like catalase, SOD is an incredibly fast enzyme. It has a turnover rate of 2×10^9 $M^{-1}s^{-1}$. At present, SOD has the highest known $\mathbf{k_{cat}/K_M}$ **ratio** (which is a measure of enzyme catalytic rate) of any known enzyme. In effect, the activity of SOD approaches the theoretical maximum, meaning it is diffusion-limited and is another example of an enzyme of great interest in quantum biology. It should be stated that because the reactions of the superoxide radical with non-radical biological molecules are spin-forbidden, in general, the SOD reaction outcompetes these damaging reactions. Also, the reactions between superoxide and itself (i.e. 'self-dismutation') and with other radicals found in cells outcompete the interactions with biomolecules. However, because the 'naked' (i.e. uncatalyzed) self-dismutation reaction of superoxide requires two superoxide anions to react with one another, it is second-order with respect to concentration of superoxide. Therefore, in the absence of SOD, the chemical lifetime of the superoxide radical is short at high concentrations but relatively long at low concentrations. In contrast, the interaction of the superoxide anion with the SOD enzyme is first-order with respect to the concentration of superoxide anion and is therefore faster.

It should also be mentioned that superoxide anions may be intracellularly neutralized by means other than just SOD. For example, the nitroxide, 4-hydroxy-TEMPO or **TEMPOL** (4-hydroxy-2,2,6,6-tetramethylpiperidin-1-oxyl) can catalyze the disproportionation of superoxide, facilitate hydrogen peroxide metabolism and inhibit **Fenton chemistry** (which was covered in chapter 9).

Biochemically, TEMPOL is thus an SOD mimic. For these reasons, TEMPOL and related nitroxides are undergoing investigation as antioxidants and non-thiol radioprotectors.

Furthermore, some antioxidants stimulate and/or work in conjunction with SOD to enhance their radioprotective effects. **WR-1065** is an **aminothiol** that is the active metabolite of the **phosphorylated aminothiol amifostine (WR-2721)** that has been widely used in clinical oncology to reduce the side effects of radiation therapy and chemotherapy. WR-1065 has a number of different effects (sometimes called pharmacological pleiotropy) including effects on DNA topoisomerase II enzymes, activation of the redox-sensitive nuclear transcription factor NFκB, and subsequent expression of manganese SOD. Therefore, in addition to its thiol-mediated radiation mitigating mechanism, WR-1065 may have additional, delayed radioprotective effects through these other mechanisms.

10.7 Glutathione peroxidase

Glutathione peroxidase (GPx) functions like catalase to deactivate hydrogen peroxide, but also reduces **lipid hydroperoxides** in cellular membranes to their corresponding alcohols. Unlike catalase however, glutathione peroxidase uses glutathione as the reducing agent and the reaction does not liberate free oxygen. It may be represented as:

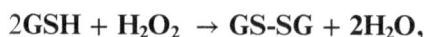

$$2GSH + H_2O_2 \rightarrow GS\text{-}SG + 2H_2O,$$

where GSH is glutathione and GSSG is the oxidized dimeric form, **glutathione disulfide**.

Unlike the disproportionation reaction catalyzed by catalase, this reaction is solely a reduction of hydrogen peroxide into water, with glutathione serving as the reducing agent. The oxidized dimeric form of glutathione, GSSG, can then be converted back into the active, monomeric, reduced form of glutathione via the enzyme **glutathione reductase**, which is discussed below.

Glutathione peroxidase, like SOD, is not just a single enzyme but comes in various forms (isozymes) that are located in different cellular locations, expressed differentially in various organs and coded for by different genes. At this time, eight different glutathione peroxidase isozymes have been discovered in humans. Glutathione peroxidase 1 is found intracellularly and is expressed in virtually all tissues; glutathione peroxidase 2, on the other hand, is an extracellular enzyme that is expressed mainly in the gastrointestinal tract. Glutathione peroxidase 4 (GPx4) is known for its unusual function of detoxifying lipid peroxides in the cell membrane. Cytosolic GPx4 is also an important inhibitor of ferroptosis (a non-apoptotic form of lipid peroxidation-induced cell death).

Glutathione peroxidase utilizes selenium in the form of a **selenocysteine** amino acid moiety in its active site. The selenium in selenocysteine functions somewhat similarly to sulfur in that it is attached to a hydrogen atom and can donate that hydrogen in biochemical reduction reactions. In other words, instead of a thiol (-SH), a **selenol** (-SeH) group is exploited. Selenols are also called **selenothiols** or

selenomercaptans. Given the central role of glutathione peroxidase in maintaining the integrity of cellular (and intracellular) membranes, selenium is a critical element for minimizing radiation-related (and other oxidative) threats to membranes.

10.7.1 Glutathione reductase

GSSG reductase or **glutathione reductase** is an enzyme that catalyzes the reduction of inactive, oxidized glutathione (GSSG) into two molecules of the active, reduced form of glutathione, GSH. Glutathione reductase uses **flavin adenine dinucleotide (FAD)** as a redox-active enzyme cofactor. Hence, adequate dietary **riboflavin** (vitamin B2), which is a precursor of FAD, is essential for proper functioning of glutathione reductase. The reaction itself uses NADPH as the reducing agent to reduce GSSG:

$$\text{GS-SG} + \text{NADPH} + \text{H}^+ \rightarrow \text{2GSH} + \text{NADP}^+.$$

As mentioned, glutathione plays a key role in warding off membrane damage by radiation-produced (and other) oxidants. Aside from its central role in neutralizing hydrogen peroxide, glutathione also scavenges hydroxyl radicals, singlet oxygen, and various hazardous electrophiles. Hence, it should come as no surprise that glutathione reductase is found in all Kingdoms of life and is structurally highly conserved.

As a medical digression, G6PD deficiency is an X-linked inborn error of metabolism caused by mutations in the glucose-6-phosphatate dehydrogenase (G6PD) gene. The G6PD enzyme catalyzes the first reaction in the pentose phosphate pathway and produces NADPH, which is critical for maintaining adequate levels of glutathione through the glutathione reductase reaction. Red blood cells lack any other source of NADPH; they are entirely dependent on the pentose phosphate pathway. Hence, G6PD deficiency leaves erythrocytes with no defense against oxidative damage. Patients with G6PD deficiency may develop favism—an acute episode of hemolytic anemia after they eat fava beans or are exposed to certain drugs or oxidants. Although such patients might be expected to exhibit radiation sensitivity, radiation therapy for cancer in G6PD deficiency patients has not revealed any untoward effects.

10.8 Thioredoxin reductase

Thioredoxin reductase is an NADPH-dependent enzyme that is related to glutathione reductase in that it uses FAD as a redox cofactor in its prosthetic group. In other words, thioredoxin reductase, like glutathione reductase, is a **flavoprotein**. And like glutathione peroxidase, thioredoxin reductase is a selenoprotein that contains **selenocysteine** and similarly uses NADPH as a substrate and source of reducing equivalents. But whereas glutathione reductase reduces GSSG into active glutathione, thioredoxin reductase reduces **thioredoxin** back into its active state. Thioredoxin is important in reducing disulfide bonds in biological molecules, which is one of the consequences of radiation and other forms of cellular oxidative stress.

Thioredoxin is a small protein, which in humans contains 105 amino acids and has a molecular weight of 12 kilodaltons. And like glutathione, thioredoxin switches back and forth between an active, reduced form and an oxidized dimeric form containing a disulfide bridge. In its reduced form, thioredoxin is a potent protein **disulfide oxidoreductase**, which means it repairs disulfide bridges in proteins caused by oxidation. It serves several functions besides its role in addressing oxidized proteins, including acting as a hydrogen donor for **ribonucleotide reductase**, an essential enzyme in DNA synthesis. Hence, it is not surprising that the thioredoxin reductase system has a long evolutionary history and is ubiquitous in living cells. Curiously, experiments have confirmed thioredoxin's potential as a radiation mitigator. Administration of thioredoxin can save the lives of mice even if given up to 24 hours after an otherwise lethal, total-body radiation dose. Obviously, this type of research is relevant to astronauts who might be unexpectedly exposed to large radiation doses through solar particle events or other avenues.

10.9 Non-enzyme antioxidants

Aside from intracellular enzyme proteins and peptides such as glutathione, non-protein/peptide bioactive chemical compounds are another important class of antioxidants with radioprotective properties. Such compounds may be administered as drugs, or they may be regularly eaten in the form of foods or taken as supplements. For instance, vitamins C, E, A and D have been investigated for their radioprotective capabilities, with vitamins C and E demonstrating significant activity.

Plants (both marine and land) are a rich source of such antioxidants that may be invaluable radioprotectors. Such compounds are called radioprotective **phytochemicals**. Among these phytochemicals are polyphenolic compounds including **anthocyanins, flavonoids, tannins** and **lignins**. Of these, the flavonoids presently appear to have the highest potential as practical radioprotective antioxidants. These radioprotectors may interact with various intracellular radiation signaling molecules such as Cox2, Nrf-2, NF-κB, TNF-α, STAT-3, NO, FOXO3, AP-1, Bax-1 and IL-6, among others. These various signaling molecules serve as mediators of biological damage repair following radiation exposure and other sources of cellular stress.

Radiation produces reactive species and depletes intracellular alpha-tocopherol (vitamin E), ascorbic acid (vitamin C) and selenium. This has been observed in normal cells and tissues of cancer patients undergoing radiation therapy. Additionally, such antioxidants have been found to decline in patients receiving ROS-producing chemotherapy agents such as doxorubicin. This raises the question about whether supplementation with antioxidants should be used by cancer patients during radiation therapy or chemotherapy. Because radiation therapy and certain chemotherapies specifically aim to damage and destroy cancer cells through oxidation, this question has not been fully answered—it is unclear whether antioxidant supplements might protect the malignant cells from the effects of treatment. At present, cancer patients are advised to avoid use of high-dose antioxidants during their radiotherapy or chemotherapy, but this is a soft recommendation that is now under reevaluation. In contrast, for people who have been exposed to high doses of radiation through nuclear accidents or other avenues, antioxidants may be a valuable medical treatment.

As mentioned earlier in this chapter, 'second-line defense antioxidants' work by scavenging free radicals and other reactive species. By serving as electron donors, they stymie free radical chain initiation and halt chain propagation. They constitute the largest class of antioxidants and may be hydrophilic, lipophilic, or amphiphilic. Both natural and synthetic antioxidants (such as amifostine, captopril and N-acetylcysteine) may display dose-modifying effects on DNA damage and cell survival if they are present at the time of irradiation. This prompt protective effect is conferred by the scavenging of radicals and reactive species such as peroxides. Several natural antioxidants have been experimentally found to reduce DNA damage, including caffeine, melatonin, bioflavonoids, polyphenols and other phytochemicals.

When considering the overall efficacy of such non-enzyme antioxidants, their bioavailability must be factored in. The absorption, cellular uptake, and biodistribution of antioxidants (used for treatment of radiation exposure, for instance) affects their radiation-mitigating effects. This is also critically important when considering such antioxidants as side effect mitigators in the radiotherapy clinic. **Amifostine (WR-2721)**, for instance, displays differential uptake between tumors and normal tissues, with normal tissues being more likely to accumulate this agent. This is because of an active transport mechanism found in normal cells that is generally absent in malignant cells (which can still take up the agent, albeit more slowly via passive diffusion). Additionally, and more importantly, there is a higher concentration of **alkaline phosphatase** enzymes in normal cells than in malignant cells. Amifostine is a precursor molecule that must be **dephosphorylated** by alkaline phosphatase into its active form, **WR-1065**, to carry out its function. Thus, amifostine is activated more in normal cells than it is in cancer cells. Hence, the protective effects of amifostine are largely limited to normal cells; cancer cells are not afforded such protection. (The fact that cancer cells contain less alkaline phosphatase than normal cells should not be too surprising since one of the 'hallmarks of cancer' is **aberrant metabolism**. There are many examples where cancer cells are markedly different from their normal counterparts, such as aerobic glycolysis (the Warburg effect), glutamine addiction, and reverse Krebs cycle to name a few.)

Amifostine is a clinical radioprotector that has found a role in the radiation therapy clinic but was originally developed as part of a program to develop agents that could offer protection to military personnel when in a radiation environment.

10.9.1 Selenium

Through selenocysteine, the micronutrient selenium is an integral part of glutathione peroxidase, which, like catalase, destroys peroxides. Interestingly, dietary selenium intake tends to increase levels of this antioxidant enzyme in normal cells but not in many malignant cells (such as leukemic cells). This raises the possibility of selenium being a clinically useful radioprotector during radiation therapy. In normal cells exposed to radiation, selenium inhibits malignant transformation and stimulates DNA repair in cells with functional p53 (that is, in normal cells with wild type p53; many cancer cells do not have normal p53 activity). Selenium is also important in the form of selenocysteine in the enzyme thioredoxin reductase.

Along with seafood and organ meats, Brazil nuts are a good source of selenium. But there is controversy about how many Brazil nuts per day is safe. This is because too much selenium can be toxic. An average Brazil nut may contain 91 micrograms of selenium, which is over 165% of the recommended daily allowance of selenium (which is 55 micrograms per day for adults). Selenium intake should be limited to 400 mcg per day, and with each Brazil nut containing between 70 to 100 mcg, intake obviously should be limited. Furthermore, Brazil nuts are radioactive, especially if grown in their native environments. This is thanks to the deep roots of the Brazil nut tree, *Bertholletia excelsa*, which reaches and extracts radium in the soil. Of course if the soil they grow in does not contain much radium, the nuts will not contain it either. Penna-Franca *et al* [11] showed that Brazil nuts may contain as much as 3.6 pCi/g Ra-226 and 3.6 pCi/g Ra-228 in the edible part of the nut.

10.9.2 Vitamin C

Vitamin C or **ascorbic acid** is a water-soluble ketolactone vitamin with two ionizable hydroxyl groups bound to a furan ring. The double bond between the two hydroxyl or alcohol groups (enols) is easily oxidized, leaving ketone groups rather than enol groups. This makes ascorbic acid an excellent donor of electrons and hydrogen in biochemistry. Vitamin C reduces reactive species such as superoxide and hydroxyl radicals, while becoming oxidized itself. In its oxidized form (**dehydroascorbic acid**), it is a radical but this radical is far less potentially damaging than the reactive species that it has reduced and quenched.

Although plants and the majority of animals can synthesize ascorbate from glucose, humans, other primates, guineapigs and a few species of fruit-eating bats cannot synthesize ascorbate. Hence, it is a true 'vitamin' for these organisms.

As discussed above, radiation promotes oxidative damage to specific amino acids in protein molecules. Cysteine, methionine, tryptophan and tyrosine residues are the most radiosensitive amino acid moieties in proteins. This damage may compromise the protein's functionality if the resulting conformational change is too severe. Denaturation of a protein's secondary structure (alpha-helix and beta-sheet) may ensue after irradiation, and this alteration in secondary structure will naturally affect tertiary structure (3D folding) and overall function of the protein molecule. Vitamin C and vitamin E can protect proteins from such structural changes.

10.9.3 Vitamin E

Vitamin E is the generic name for all biologically active **tocopherols** and **tocotrienols**. In contrast to water-soluble ascorbic acid, these compounds are lipid-soluble and carry out much of their functions in cellular membranes. Vitamin E compounds are potent antioxidants that scavenge radicals and reactive species thanks to the hydroxyl group on their **chroman rings**, which is an effective hydrogen donor and reducing agent in cells.

Alpha tocopherol can synergize with selenium in protecting cells against radiation-induced malignant transformation. Pretreatment of cells with vitamin E and selenium increases intracellular levels of glutathione, glutathione peroxidase and catalase. The observed result is a doubling of the breakdown of peroxide and a reduction in radiation-induced cellular transformation into cancer cells. It should be mentioned that alpha-tocopherol, while the most-studied form of vitamin E, is not the only chemical form of it. The vitamin E family consists of eight different 'vitamers', including four tocopherols (α, β, γ and δ), which are saturated analogs, and four tocotrienols (also α, β, γ and δ), which are unsaturated analogs. These various forms are being investigated for their radiation countermeasure effectiveness.

10.9.4 Melatonin

N-acetyl-5-methoxytryptamine or **melatonin** is an antioxidant produced by the **pineal gland**. Like serotonin, melatonin is an amino acid derivative that is structurally related to tryptophan. Its antioxidant actions result in a reduction of **nitric oxide** formation, which in turn, decreases the inflammation following radiation exposure. Before or after total body irradiation, melatonin administration can increase the overall survival rate of exposed animals and reduce signs and symptoms of acute irradiation syndrome. Melatonin may protect mitochondria by increasing the efficiency of oxidative phosphorylation. In so doing, it reduces the 'leakage' of free electrons from the electron transport chain (that might otherwise go on to form ROS from these electrons and damage the mitochondria). Additionally, melatonin induces the production of higher levels of antioxidant enzymes, such as glutathione peroxidase. It also stimulates the activity of glutathione peroxidase as well as the activity of glutathione reductase. Finally, melatonin increases intracellular levels of glutathione. All of these actions are important in reducing levels of oxygen radicals and peroxides in cells following radiation exposure.

10.10 Phytochemicals

A broad array of plant-derived organic compounds (phytochemicals), including **polyphenols**, **carotenoids** and **organosulfur compounds**, are potent antioxidants that have been shown to be radioprotective in lab experiments. For this reason, there has been a good deal of interest in these phytochemicals in conjunction with radiation therapy as a means of reducing side effects. Not only are many phytochemicals potential radioprotectors, but some also seem to have anti-cancer activities of their own. For example, **S-allylmercaptocysteine**, a water-soluble organosulfur compound that is found in aged garlic, induces **programed cell death** (**apoptosis**) in human prostate cancer cells, breast cancer cells, and colon cancer cells. Hence, research is underway to investigate the possible role of phytochemicals in conjunction with radiation therapy for certain cancers. But presently, the polyphenols appear to have the most important role in terms of general radiation mitigation.

10.10.1 Polyphenols: flavonoids and non-flavonoids

Tea (especially green tea), dark chocolate and red wine are often touted for being rich sources of **polyphenols**. Polyphenols are known for their antioxidant properties but most of these compounds have only been studied *in vitro*. It is presumed that they are made by plants for protection against the damaging effects of ultraviolet radiation and to fight off pathogens and perhaps some herbivores.

Chemically, polyphenols are a group of natural organic compounds containing multiple **phenol groups**. Members of this very heterogenous group of compounds differ significantly in their chemical and biochemical stability, bioavailability, and physiological functions. Over 8000 polyphenolic structures have been discovered or synthesized, and among them, over 4000 fall under the group of polyphenols called **flavonoids**.

The flavonoids in turn, can be categorized into several broad chemical classes:

1. Flavonols
2. Flavones
3. Anthocyanidins
4. Proanthocyanidins
5. Isoflavenoids
6. Neoflavenoids
7. Chalcones.

Using this chemical classification, **genistein** (an antioxidant and **phytoestrogen** found in soybeans and other plant sources) is an isoflavone isoflavonoid flavenoid polyphenol. **Epigallocatechin gallate (epigallocatechin-3-gallate (EGCG)**, which is found in green tea) is a type of **catechin**, which means it belongs to the subgroup of polyphenols called flavonoids and more specifically to the group called **flavonols**; technically, ECGC is thus a catechin flavonol flavonoid polyphenol. EGCG is structurally related to the parent chemical compounds, catechin and epicatechin. These compounds in turn are the building blocks of the **proanthocyanidins**, a type of condensed **tannin**. The first two listed classes of flavonoid compounds (flavanols and flavones) are sometimes lumped together as **anthoxanthins**.

One of the most well-known anthoxanthins is **quercetin**, which is found in certain fruits and vegetables, including capers, red onions, fennel, radicchio and kale. The name is derived from the oak tree genus, *Quercus*. Although the various flavonoids have been most studied for their antioxidant properties, some display additional ways to detoxify harmful substances. For instance, quercetin is one of the most effective inducers of **phase II detoxification enzymes**. Such substances help the liver convert harmful substances into less harmful agents that are more soluble and therefore more readily excreted and eliminated.

10.10.2 Non-flavonoid polyphenols

Although most polyphenols are flavonoids, of the more than 9000 different polyphenol structures identified, not all are flavonoids. The other major group, simply called **non-flavonoids**, include **phenolic acids, xanthones, tannins** and **lignans**.

Resveratrol, which is a powerful antioxidant found in blueberries, blackberries, grape skins, and red wine, is a well-studied stilbene.

Flavonoids and non-flavonoids mainly act as antioxidants, free radical scavengers, and anti-inflammatory compounds thereby providing cellular protection (cytoprotection) against radiation in *in vitro* experiments. Furthermore, some of these compounds downregulate several pro-inflammatory cytokines. Among those with potential are the **phenylpropanoids**, especially **caffeic acid phenylethylester**, **curcumin**, **thymol** and **zingerone**. Their radioprotective effects are mediated by a host of different mechanisms that reduce intracellular stress.

There is some overlap in the chemical classification system, with proanthocyanidins traditionally being considered condensed tannins but other bioorganic chemists putting them in the flavonoid category.

10.10.3 Carotenoids

Carotenoids are richly colored molecules (i.e. pigments) that confer the yellow, orange and red colors of many fruits and vegetables. Thus, bright fruits and vegetables are the source of most of the 40–50 known carotenoids found in the human diet. α-Carotene, **β-carotene**, β-cryptoxanthin, beta-cryptoxanthin, **lutein**, zeaxanthin and **lycopene** are the most common dietary carotenoids. Foods containing carotenoids include brightly colored fruits and vegetables including corn, pumpkins, carrots, orange/red bell peppers, tomatoes, watermelon, pink grapefruit, cantaloupe and broccoli as well as some leafy greens such as spinach and kale. In addition to plants, carotenoids are also found in some bacteria, algae and fungi. When consumed by animals, they can grant them bright colors—as in the pink colors of phytoplankton-eating flamingos. The pink color of wild salmon is also due to carotenoid consumption by these fish.

If people eat too many carrots or other sources of beta-carotene, they may develop **carotenemia**, which casts their skin in an orange-yellow hue. Although carotenemia is benign, it can be confused with **jaundice** (which is often a serious condition associated with liver disease). A quick distinction may be picked up on physical examination since carotenemia spares the sclera of the eyes and the mucosal membranes, whereas jaundice does not (in the sclera it is called **icterus**).

Chemically, carotenoids are **terpene** hydrocarbons, which means they are made up of **isoprene** subunits. Carotenoids are divided into two large chemical classes, the **carotenes** and the **xanthophylls**. One difference between these two classes of carotenoids is that xanthophyll contains oxygen, whereas carotenes do not. Lutein and zeaxanthin are two well-studied xanthophylls. The various carotenes (α-carotene, β-carotene, γ-carotene) and lycopene (which is abundant in tomatoes) are in the carotene class since they contain no oxygen atoms. Carotenoids function as quenchers of ROS, including the singlet oxygen made in plants during photosynthesis.

Some pathogenic microbes use carotenoids against us. For example, *Staphylococcus aureus* (which gets its name *aureus* from its golden color) possesses a carotenoid called **staphyloxanthin**. This antioxidant allows the bacteria to fend off the oxidative burst of ROS made in our defensive phagocytic neutrophils.

10.10.4 Glucosinolates and isothiocyanates

Cruciferous vegetables of botanical family Brassicacea contain glucosinolates. These glucosinolates, naturally found in common foods such as cabbage, broccoli, cauliflower and Brussel sprouts may be metabolized by the enzyme myrosinase into indoles and isothiocyanates. Some of these compounds have substantial antioxidant activities. One isothiocyanate, sulforaphane, also induces antioxidant enzymes after consumption. The induction of such antioxidant enzymes appears to be through the Nrf2 pathway (which is further discussed below).

10.11 Miscellaneous antioxidant agents

The **peroxiredoxins** are a family of antioxidant enzymes that serve to inactivate hydrogen peroxide and other intracellular organic and inorganic peroxides, such as alkyl peroxides and peroxynitrate. They serve an important role in maintaining the stability and integrity of red blood cells (erythrocytes); **peroxiredoxin 2** constitutes the second-most abundant intracellular protein in erythrocytes after hemoglobin. Of interest, radiation-induced suppression of **Nrf2** activity was restored by **peroxiredoxin 6**, even when it was administered after radiation exposure. Nrf2 or nuclear factor erythroid 2-related factor 2 is a key regulator of the radiation-related antioxidant cellular response.

Ferulic acid is a polyphenol that was originally derived from the giant fennel plant *Ferula communis*, thus the name ferulic acid. Supplementation with ferulic acid has recently been shown to attenuate total body irradiation-mediated bone marrow damage, as well as stem cell senescence, and general hematopoietic injury. It seems to work by enhanced antioxidant cellular defense and may have significant potential as a radiation countermeasure.

Lactoferrin is a multifunctional glycoprotein in the **transferrin family**. It is widely present in mammalian secretory fluids such as tears, saliva and milk and is also present in neutrophil granules where it plays an important role in the **innate immune system**. In a study of mice exposed to total body irradiation, human lactoferrin increased survival from 28% to 78%, reduced mouse weight loss and increased leukocyte counts among other physiological parameters. The potential of lactoferrin as a radiation countermeasure is under further study.

10.11.1 Apigenin

Apigenin (4′,5,7 trihydroxyflavone) is a dietary substance found in various aromatic herbs, cereal grains, onions, apples, celery, chamomile, tulsi, and other vegetables. It is a natural **flavone** which means it is a flavonoid polyphenol chemically. Apigenin displays potent free radical scavenging activity as well as antioxidant, anti-inflammatory, anti-microbial and anti-neoplastic activities. Apigenin protects human lymphocytes from gamma radiation *in vitro* and intraperitoneal apigenin diminishes inflammation in the bone marrow of gamma-irradiated mice. Curiously and perhaps very importantly, this agent may work through mechanisms beyond its antioxidant and anti-inflammatory properties. Significant differences in gut

microbial diversity were seen after space-like irradiation using silicon-28 ions to simulate the HZEs in cosmic rays. A three-fold increase in the **Firmicutes: Bacteroidetes ratio** (F:B ratio) was seen in samples from irradiated mice not on dietary apigenin compared to irradiated mice on an apigenin diet, sham controls or mice that received apigenin without being irradiated. Generally speaking, gut microbiome studies suggest that an increased F:B ratio is an unfavorable finding that is associated with various pathological conditions in humans such as diabetes and obesity. On the other hand, a very low F:B ratio may be linked to inflammatory bowel disease. When there is an imbalance between these two major phyla of gut bacteria, the condition is called gut microbial dysbiosis. Irradiation with silicon-28 ions induced an imbalanced duodenal microbiome in irradiated rats, but apigenin restored balance to the microbiota. Also, silicon-28 ions decreased the relative abundances of **probiotic bacteria** (**Lachnospiraceae** and **Bifidobacteriaceae**), but apigenin effectively restored the relative bacterial abundances. Daily dietary apigenin consumption restored balance to the 'gut microbiome' (more correctly, the **microbiota balance** in the gut tissues) of irradiated mice, hinting at a significant radiation countermeasure potential—especially for radiation in an outer space environment. Silicon-28 ion-induced radiation injuries to hematopoietic and gut tissues were minimized by apigenin-related anti-inflammatory activity, reduction of gut microbiota dysbiosis and an increase in probiotic bacteria. Hence, it appears that apigenin is a viable, novel candidate for space radiation countermeasures.

10.11.2 Antioxidant supplements versus foods?

The above discussion about the effects of apigenin raises the possibility that radiation countermeasure agents may work, not only through their direct antioxidant and anti-inflammatory actions but also through a previously poorly understood and under-appreciated impact on the host microbiome. Thus, the ongoing debate about the efficacy of supplements versus foods that contain such substances may be blurred by the previous failure to account for the impact of such supplements and foods on the host microbiota. For example, it remains possible that foods that contain a certain repertoire of antioxidants will work differently in human space travelers from how the same set of antioxidants would work in the form of supplements thanks to fiber and other items in foodstuffs that are not present when given as supplements. And if this proves to be true, the overall impact on the gut microbiota could be a key difference. Future work, including radiation mitigation for acute radiation syndrome and space travel, may focus more on the previously unrecognized role of the gut (and other sites) microbiome.

Further reading

[1] Tomusiak-Plebanek A, Heczko P, Skowron B, Baranowska A, Okoń K, Thor P J and Strus M 2018 Lactobacilli with superoxide dismutase-like or catalase activity are more effective in alleviating inflammation in an inflammatory bowel disease mouse model *Drug Des. Dev. Ther.* **12** 3221–33

[2] Yu Y, Lin X and Feng F *et al* 2023 Gut microbiota and ionizing radiation-induced damage: is there a link? *Environ. Res.* **229** 115947

[3] Novoselova E G, Sharapov M G and Lunin S M *et al* 2021 Peroxiredoxin 6 applied after exposure attenuates damaging effects of x-ray radiation in 3T3 mouse fibroblasts *Antioxidants* **10** 1951

[4] Wagle S, Sim H-J and Bhattarai G *et al* 2021 Supplemental ferulic acid inhibits total body irradiation-mediated bone marrow damage, bone mass loss, stem cell senescence, and hematopoietic defect in mice by enhancing antioxidant defense systems *Antioxidants* **10** 1209

[5] Rithidech K, Peanilkhit T, Honikel L, Liu J, Li J, Zimmerman T and Welsh J 2022 *Radiation Research Society's 68th Annual Meeting Apigenin: a promising countermeasure for space-radiation-induced damage (Hawaii, October 15–9)*

[6] Zarei H, Bahreinipour M, Sefidbakht Y, Rezaei S, Gheisari R, Ardestani S K, Uskoković V and Watabe H 2021 Radioprotective role of vitamins C and E against the gamma ray-induced damage to the chemical structure of bovine serum albumin *Antioxidants* **10** 1875

[7] Parsons B J 2012 Sterilisation of healthcare products by ionising radiation: sterilisation of drug-device products and tissue allografts *Sterilisation of Biomaterials and Medical Devices* ed S Lerouge and A Simmons (Sawston, England: Woodhead Publishing) 8 212–39

[8] Maiorino M, Conrad M and Ursini F 2018 GPx4, Lipid peroxidation, and cell death: discoveries, rediscoveries, and open issues *Antioxid. Redox Signal.* **29** 61–74

[9] Chung W P, Hsu Y T, Chen Y P and Hsu H P 2019 Treatment of a patient with breast cancer and glucose 6-phosphate dehydrogenase deficiency: A case report *Medicine (Baltimore)* **98** e14987

[10] Sundaramoorthy P, Wang Q and Zheng Z *et al* 2017 Thioredoxin mitigates radiation-induced hematopoietic stem cell injury in mice *Stem Cell Res. Ther.* **8** 263

[11] Penna-Franca E, Fiszman M and Lobao N *et al* 1968 Radioactivity of Brazil Nuts *Health Phys.* **14** 95–9

[12] Rithidech K N, Peanlikhit T and Honikel L *et al* 2024 Consumption of Apigenin prevents radiation-induced gut dysbiosis in male C57BL/6J mice exposed to silicon ions *Radiat. Res.* Epub ahead of print

Chapter 11

Molecular radiobiology

Bridging the gap between radiation chemistry and classical radiobiology is the emerging field of molecular radiation biology. This field focuses on the biomolecular responses to radiation exposure along with the molecular biology behind the various repair mechanisms used to address radiation-induced molecular lesions. Thus, radiation-induced DNA damage repair falls under this topic.

First enunciated in 1906, the **law of Bergonie and Tribondeau** is an essential concept in radiobiology and is an underlying principle behind radiation therapy for cancer. This law states that the radiosensitivity of a given biological tissue is directly proportional to the mitotic activity of its cells and inversely proportional to their degree of differentiation. Rapidly dividing cells are most radiosensitive, and highly differentiated cells are most radioresistant. Hence, poorly differentiated, rapidly-growing cancer cells tend to be more sensitive to ionizing radiation than their normal counterparts are. This principle also has relevance to general radiation safety concerns since acute, high doses to the entire body (or much of it) affect the body's dividing cells most. Thus, acute radiation syndromes affect the hematopoietic, gastrointestinal, and endothelial (blood vessel linings) systems since the cells in these tissues are mitotically active.

Cancer cells divide more frequently and are less differentiated than their benign neighbors. It is as though they have taken a step backwards, towards a more primitive cellular state. In essence cancer cells are atavistic, but this makes them more sensitive to radiation and vulnerable to judicious application of well-targeted ionizing radiation. In addition to being geographically targeted, radiation therapy is often given in a fractionated fashion (meaning small doses on a daily basis spread over a period of time, as opposed to all at once). This fractionation capitalizes on some inherent radiobiological differences between normal cells and cancer cells. It thereby augments cancer cell killing and diminishes normal cell damage. Fractionation of radiation therapy also often improves the oxygenation of cancer cells, which in turn makes them even more sensitive to radiation therapy. This re-oxygenation of originally hypoxic and radioresistant cancer cells in the central core of a tumor mass converts them into oxygenated and radiosensitive cancer cells. Reoxygenation in clinical radiotherapy occurs when the cells at the perimeter of a tumor (which is populated by oxygenated cells near capillaries), are selectively killed by each daily fraction of radiation, thereby allowing the capillaries to bring oxygen to the previously hypoxic cells in the core of a tumor and thereby grant them radiosensitivity.

Cancer cells usually harbor a host of mutations that promote cell division and tumor growth. The number of somatic mutations in a malignant cell can be astonishing. While most cancer cells harbor thousands to tens of thousands of mutations, some cancer cells may possess hundreds of thousands of mutations. Some of the mutations in cancer cells may offer a selective growth advantage to the cancer cells, allowing them to out compete their normal neighbors (who tend to cooperate with each other rather than compete with one another). Such mutations are called 'driver mutations'. In addition to these driver mutations, however, cancer cells may also harbor many 'passenger mutations' that simply evolve because of the cancer cell's tendency to develop more mutations. Passenger mutations do not confer any selective advantages when it comes to cell growth; they are simply there for the ride, so to speak. Such multiple mutations often come with a price of decreased ability to withstand radiation because of diminished ability to repair DNA damage and a disregard of 'checkpoints' throughout the cell cycle. In normal cells, **cell cycle checkpoints** halt the cell cycle and allow inspection, identification and repair of damaged DNA, rather than just rushing headlong into cell division. Cancer cells focus more on getting to, and through, mitosis rather than addressing DNA damage. And this makes them vulnerable. In this chapter we will examine some of the molecular responses of cells to radiation.

In this chapter, with a few exceptions, we will follow the convention of italicizing and capitalizing human genes (e.g. *RB*) whereas the protein products of these genes will not be capitalized nor italicized (e.g. p53).

11.1 Deterministic versus stochastic effects of radiation

The adverse health effects of high dose radiation exposure are classically categorized into two broad subtypes:

1. **Deterministic effects** (also called **tissue reactions**). Tissue reactions are typically caused by death or malfunction of cells, tissues, and organs after high dose radiation exposure. Such reactions normally display a threshold below which no adverse effects are seen. Additionally, the severity of deterministic effects usually increases as a function of dose.
2. **Stochastic effects**. Examples in this category include radiation-induced cancer caused by mutation of somatic cells caused by mutations in reproductive (germ) cells.

Stochastic radiation effects are characterized by a probability of occurrence which increases with dose, while the severity is independent of dose. Additionally, stochastic events are traditionally considered to have no threshold. Radiation-induced cancer, teratogenesis, cognitive decline, and heart disease are among the stochastic effects induced by ionizing radiation. The mathematical probability of developing many of these alleged stochastic effects (such as radiation carcinogenesis) are presently described via a linear–no threshold (LNT) model. In the LNT model, there is no threshold—even a single, tiny exposure can theoretically lead to an adverse outcome. Furthermore, the probability of such an adverse effect is a linear function of dose.

The concept of stochastic effects of radiation in radiobiology is controversial. It has its historical roots in radiation physics and chemistry. But while electronic hardware, rubber hoses, and human cells can all sustain radiation damage, only cells can repair radiation damage. Thus, while an LNT model applies perfectly well when repair of damage is not involved, its applicability is doubtful in situations where repair is possible.

With this important caveat stated, the most well-studied example in the stochastic effects category is the induction of cancer. Radiation-induced cancer has a **latency period**, which may be years or decades. That is, the malignancy associated with radiation may only appear many years after exposure. In this model of radiation carcinogenesis, radiation exposure is the **initiator**. But for a cell to transform into a fully malignant cancer cell, other events must happen. These other events are caused by **promoters**. Thus, in principle, the radiation will have caused a mutation, but additional mutations are needed to allow the cell to progress from a normal but damaged cell into a full-blown cancer cell. These additional required mutations may or may not be introduced over time by the action of promoters. Thus, not all cells harboring unrepaired radiation damage develop into cancer cells.

This traditional initiator-promoter model is attractive for its simplicity, but other important effects such as the role of the immune system are ignored in this model. For instance, the concept of **immunosurveillance** holds that abnormal cells (that could someday evolve into cancer cells) arise regularly but are detected and destroyed by the immune system. This step is called **elimination** in the **immunoediting hypothesis**. There are '3 E's' of immunoediting:

1. Elimination
2. Equilibrium
3. Escape.

Equilibrium is a sort of prolonged truce between the immune system and the cancer cells where neither has the upper hand. The cancer cells are not entirely eradicated by the immune system, but they are not causing problems either. In such scenarios, the cancer does not progress into a clinically evident cancer. In principle, equilibrium can go on for many years. On the other extreme is immune **escape** in which the cancer cells have evolved mechanisms that allow them to evade immunity and finally escape from equilibrium. Once this has happened, the cancer cells can progress into clinically evident tumors.

Returning to the idea of radiation carcinogenesis as a stochastic event, the present quantitative models predicting the level of risk remain controversial. The most widely accepted model (the LNT model) posits that 'there is no safe dose' (meaning there is no threshold for the induction of cancer) and the incidence of cancers due to radiation increases linearly with effective radiation dose at a rate of 5.5% per sievert. As mentioned, the LNT model remains highly controversial and there is a need for better, more accurate models that take immunosurveillance into account along with other radiobiological phenomena that we will discuss, such as the bystander effect and the adaptive response.

11.2 Cellular consequences of irradiation

Irradiation of a cell may culminate in one of the following nine possible outcomes:

1. No effect. The radiation-induced damage is fully repaired and the cell continues to function normally.
2. Cell division delay: The cell cycle is temporarily (or permanently) halted and the cell is delayed from undergoing mitosis. Cells that are permanently arrested in the cell cycle may have experienced **cellular senescence**.
3. Apoptosis: The cell undergoes **programmed cell death** and dies by fragmentation into smaller bodies (apoptotic bodies), which are taken up by immune cells.
4. Mitotic cell death (reproductive failure): Also known as mitotic catastrophe, the cell dies when attempting mitosis (either the first time or during subsequent efforts) because of too much unrepaired DNA (i.e., chromosomal) damage.
5. Genomic instability: Cells may sustain DNA damage that leads to inadequate repair of DNA damage in future cell generations. In other words, the daughter cells tend to develop mutations at a higher rate than usual; this is the so-called 'mutator phenotype' that is the result of unrepaired damage in certain DNA repair genes. Reproductive failure in cellular descendants may result from genomic instability.
6. Mutation: The irradiated cell survives but contains a DNA mutation. Mutation would be the initiating event in the initiator–promoter model of carcinogenesis.
7. Transformation: The cell survives but the mutation (change in genotype) leads to a 'transformed' phenotype. Such transformed cells may lead to cancer. This would be the outcome after both an initiating mutation and further mutations by promoters in the initiator–promoter model.
8. Bystander effects: An irradiated cell can send signals (e.g. cytokines) to neighboring unirradiated cells and induce death, genetic damage or other changes in them.
9. Adaptive responses: The irradiated cell reacts to the damage and upregulates its repair mechanisms. It thereby reverses the sustained damage and may become more resistant to subsequent radiation challenges (or chemical challenges). Beyond just the cellular and tissue levels, an adaptive response may occur at the organismal level such that irradiated cells might provoke a cancer-attacking response by the host's immune system for example. This action-at-a-distance phenomenon is called the abscopal effect in oncology.

There are other variations on these general themes. One is the so-called 'rescue effect' wherein unirradiated adjacent normal cells can rescue nearby irradiated cancer cells. (Rescue in this sense means less cell death through apoptosis.) This is the opposite of the bystander effect in some ways, and it might have an impact on the efficiency of radiation therapy for cancer.

11.3 Cellular senescence and senescence-associated secretory phenotype

As mentioned earlier, cellular senescence is another possible consequence of irradiation. This is an active area of research since there is increasing evidence that radiation-induced senescence in both tumor cells and in adjacent normal tissues contributes to tumor recurrence, metastasis, and resistance to therapy. Furthermore, persisting senescent cells in irradiated normal tissues and organs may be a source of **late effects** (also known as chronic radiation toxicity or long-term radiation damage).

Cellular senescence is characterized by irreversible cell cycle arrest in response to radiation and various other stress stimuli. Senescence may cause resistance to apoptosis and result in the **senescence-associated secretory phenotype (SASP)**. Cells exhibiting SASP secrete high levels of inflammatory cytokines, growth factors, and immunological modulators. Initially, the SASP condition is immunosuppressive (characterized by increased expression of TGF-β1 and TGF-β3) and profibrotic (meaning that it may promote fibrosis, which basically is the formation of scar tissue). Over time, cells exhibiting SASP become proinflammatory (characterized by increased secretion of IL-1β, IL-6, and IL-8).

It is commonly stated that cellular senescence is 'a normal consequence of aging'. While senescence plays important roles in development and wound healing, it appears possible that outside those functions, senescence is not normal. And, as an abnormal state, cellular senescence may have dire consequences for neighboring normal cells and tissues. Whether low dose radiation exposure might help an organism overall by reducing the number of burdensome, abnormal senescent cells is an interesting area of radiation research.

11.4 Classification of radiation damage

Radiation damage to mammalian cells is divided into three main categories:

1. Lethal damage, which is irreversible, irreparable, and inevitably leads to cell death.
2. Sublethal damage, which can be repaired in hours unless additional sublethal damage is added (e.g. from another dose of radiation) that eventually leads to lethal damage.
3. Potentially lethal damage, which can be altered and reversed by repair mechanisms when the irradiated cells are allowed to remain in a non-dividing state.

Some texts include another category called non-lethal damage. Non-lethal damage does not prevent proliferation of irradiated cells but may affect the rate of proliferation.

11.5 Potentially lethal radiation damage

Potentially lethal damage (PLD) is radiation damage that can be modified by post-irradiation environmental conditions. PLD is 'potentially' lethal because under normal conditions it causes cell death, but this fate may be averted under different conditions. For example, if cells are incubated in a balanced salt solution rather than a full growth medium for several hours following irradiation, they may avoid cell death. But this is a highly unusual and artificial situation. The concept is that if mitosis is delayed by suboptimal growth conditions, DNA damage can be repaired. The fraction of cells surviving a given dose of x-rays is increased if post irradiation conditions are suboptimal for growth (such that cells do not try to attempt the complicated process of mitosis while their chromosomes are damaged). Having said this, the clinical relevance of PLD repair in radiation therapy remains uncertain. It has been hypothesized that the radioresistance of certain cancer cells is thanks to their enhanced capacity to repair PLD. While this sounds plausible, it has not yet been proven. Interestingly, there does not appear to be much (if any) PLD repair following exposure to high linear energy transfer (LET) radiation such as alpha particles or the HZE radiation in galactic cosmic rays.

11.6 Sublethal damage repair

Sublethal damage (SLD) repair describes the increase in cell survival that is observed when an otherwise lethal radiation dose is split into two or more fractions separated by a time interval. Basically, SLD, which is chiefly evident at low dose levels, is somehow reversed or overcome after the passage of a relatively short period of time in healthy normal cells.

The proposed mechanism behind the repair of SLD is the repair and rejoining of double-strand breaks before they can interact to form irreversible lethal lesions. In other words, SLD repair reflects double-strand DNA damage repair. When a radiation dose is split into two fractions separated by a time interval, some of the double-strand breaks will be repaired and rejoined before the second dose of radiation is given. Consequently, there are fewer opportunities for broken ends of DNA to interact and incorrectly rejoin. Hence, fewer irreversible, lethal lesions will be formed (such as **dicentric chromosomes** which are discussed later in this chapter).

From this brief introduction it is evident that, up to a limit, mammalian cells can repair radiation damage. Lethal damage is beyond such limits, but SLD and PLD are both theoretically reparable. The molecular mechanisms behind both SLD repair and PLD repair are highly complex, multi-step processes that involve a host of DNA repair enzymes and pathways. And as will be discussed in this chapter, the width of the 'shoulder' on the cell survival curve reflects the degree of sublethal damage, which incidentally is more readily repaired in the S phase of the cell cycle.

11.7 Radiation dose rate effects

For low LET radiation the dose rate at which a given dose of radiation is given makes a vast difference in biological outcome. Very generally speaking, as the dose rate is lowered, the biological effect is diminished. This is called the 'classic dose rate effect' and sublethal damage repair is the underlying mechanism. The degree to which the dose rate effect is seen varies substantially between various cell types. Cells that have a smaller 'shoulder' on the acute radiation survival curve (discussed in the next section) exhibit small dose rate effects; dose rate doesn't matter much. In contrast, cells that have large shoulders on their acute radiation cell survival curves exhibit correspondingly larger dose rate effects. The shoulder size on the cell survival curve and the magnitude of the dose rate effect may reflect differences in the predominant mode of cell death following irradiation. For example, in cells with broad shoulders and large dose rate effects (meaning that there is a very pronounced difference in outcome between high and low dose rates), apoptosis appears to not be a very important mechanism of radiation induced cell death. In contrast, for cells with small shoulders and a less pronounced dose rate effect, apoptosis appears to be an important mechanism of radiation-induced cell death. For cells exhibiting broad cell survival curve shoulders and pronounced dose rate effects, at very low dose rates the impact on such cells becomes barely perceptible or practically non-existent at low and moderate cumulative doses; low dose rate irradiation may not even be very evident at large cumulative doses in such cell types.

Some cells may exhibit a paradoxical phenomenon in which cell death increases at lower dose rates. This is not seen in all cells and for those that do exhibit such an **inverse dose rate effect**, the dose rates which can cause this may be different. It may be explained by a tendency for certain cells irradiated at a certain low dose rate to progress through the cell cycle up to late G2, where they become arrested. Late G2 is a radiosensitive phase of the cell cycle and additional radiation (even at a low dose rate) leads to the demise of these cells. At higher dose rates, cells are typically 'frozen' in the phase of the cell cycle they were in when the irradiation occurred. But at lower dose rates, the cells may continue onward in the cell cycle during irradiation. In this fashion, a population of asynchronous cells (which may be in sensitive or resistant phases of the cell cycle) all become synchronized at G2. Continuous radiation exposure, even at a low dose rate, might cause such cells to succumb. Thus, such cells exhibit a perplexing *low dose rate hypersensitivity* phenomenon.

External beam radiation therapy via linear accelerators has historically been delivered at dose rates measured in gray per minute, with 1 Gy min^{-1} being a fairly typical dose rate. This dose rate (around 1 Gy/min) is also used in certain forms of temporary brachytherapy and is called high dose rate (HDR) brachytherapy. A very different dose rate is encountered in permanent implant brachytherapy. This form of radiotherapy is delivered with implanted sealed sources of radioisotopes and is called low dose rate brachytherapy. Typical dose rates in this type of treatment are measured in cGy per hour, which is several orders of magnitude different from the Gy/min rates of HDR brachytherapy and LINAC-based external beam radiation therapy. Thus, the total doses prescribed are very different. For instance, in fractionated external beam radiation therapy for prostate cancer, 78 Gy in 39 fractions of 2 Gy apiece is a popular prescription. To achieve the same clinical outcome, 144 Gy is often prescribed when using I-125 permanent implant brachytherapy for a similarly staged prostate cancer (I-125 decays via electron capture with a half-life of about 60 d).

Another phenomenon which is presently under active investigation is seen with ultra-rapid radiation dose rates. Instead of dose rates measured in cGy/h or Gy/min, in **FLASH radiotherapy**, dose rates > 40 Gy per second are employed. The **FLASH effect** is still not fully understood but it holds great clinical promise since normal tissues may experience fewer side effects when tumoricidal doses of radiation are delivered at tremendous dose rates. In contrast, cancer cells do not seem to enjoy the same protection when radiation is delivered at these HDRs.

11.8 Cell survival curves and the linear-quadratic model

When the results of an irradiation experiment (using low LET radiation) are plotted with the logarithm of the fraction of surviving cells on the y-axis against radiation dose on the x-axis (that is, a log-linear plot), most cells exhibit a curve with a relatively flat 'shoulder' at low doses followed by a downward bend in the curve and a linear decline afterwards at higher doses. The shoulder on the curve may be thought of a means to quantify a sort of 'threshold' dose, after which the curve dramatically changes shape. The width of shoulder of the survival curve indicates the degree of SLD. It appears that this SLD is more readily repaired in the S phase. The width of this shoulder is very variable, depending on the cell lines studied. But a generalization that has emerged is that, in cells lacking a significant shoulder, apoptosis is an important mechanism of radiation-induced cell death. In contrast, for cells exhibiting a pronounced shoulder, apoptosis is absent or at least not a prominent mechanism. In this latter situation, **mitotic cell death** is the main mechanism. (Mitotic cell death or mitotic catastrophe is discussed elsewhere in this chapter.)

There are a few models through which the same cell survival curve may be mathematically described. The one which is presently most favored is the linear-quadratic formalism, which may be written as:

$$S[D] = e^{-(\alpha D + \beta D^2)}.$$

Using the linear quadratic model, the first component—the **linear component**—represents the dose range at which cell killing is proportional to the dose. Alpha (α) is the coefficient during this linear component. In the second component of the curve, the **quadratic component**, beta (β), is the coefficient, and here survival is related to dose squared. The linear and quadratic contributions to cell killing are equal at a dose equal to the ratio of alpha to beta (that is, when $D = \alpha/\beta$). At this dose:

$$\alpha D = \beta D^2$$

and

$$\alpha/\beta = D.$$

This α/β **ratio** is a clinically useful parameter that is used for **biologically effective dose (BED)** calculations that can compare the biological effectiveness of various dose-fractionation schemes used in clinical radiation therapy. The formula for this is:

$$BED = nd[1 + d/(\alpha/\beta)]$$

where n is the number of fractions of radiation therapy and d is the dose for each fraction.

Note that $nd = D$, where D is the total dose in the course of treatment. Therefore, one may alternatively write:

$$\mathbf{BED} = \mathbf{D}[1 + d/(\alpha/\beta)].$$

Variations derived from this formula are used to estimate the clinically equivalent dose to prescribe when using alternatives to the classic 1.8–2 Gy fraction sizes of conventional external beam radiation therapy. For instance, when calculating the total dose to prescribe when using fraction sizes different from the familiar 2 Gy dose-fractionation scheme, one may use the **equivalent dose in 2 Gy** (EQD2) equation:

$$EQD2 = D[(d + \alpha/\beta)/(2 + \alpha/\beta)].$$

This tells the radiation oncologist what the biological equivalent is when using a dose size of d rather than the common daily dose of 2 Gy. A more general way of comparing the biological effectiveness of two different dose-fractionation schemes (with total doses of D1 and D2, using fraction sizes of $d1$ and $d2$, respectively) may be provided by:

$$D1/D2 = (\alpha/\beta + d1)/(\alpha/\beta + d2).$$

There are many assumptions made in using such formulas including the idea that both courses (with total doses of D1 or D2 using daily fractions of $d1$ and $d2$) are comparable in overall duration. This may not be true, especially when daily fraction sizes are large; the overall course in this situation may be significantly shorter than the conventional 1.8 or 2 Gy per fraction course. This strategy of using fewer but larger than usual daily fractions is called **hypofractionation**. When smaller fractions are used more than once per day, the approach is known as **hyperfractionation**. And when a treatment course is shorter in overall duration, it is called **accelerated fractionation**.

In addition to being cheaper and more convenient to the patient, hypofractionated courses of external beam radiotherapy may be radiobiologically advantageous when the α/β ratio is low. Most cancers exhibit α/β ratios that are relatively high (e.g. around 10 Gy; technically, since α/β is a ratio, the units of Gy should be omitted), whereas most normal tissues have α/β ratios that are lower, averaging about 3. But for prostate cancer (adenocarcinoma), the α/β ratio appears to be anomalously low, perhaps as low as between 1–2. For this reason, hypofractionation makes sense for external beam radiation therapy for prostate cancer, according to the equations of the linear-quadratic model. At this time, the former approach of 78 Gy in 39 fractions of 2 Gy each has been largely replaced by hypofractionated dose-fractionation regimens with strong, level I supportive data provided by several phase III clinical trials.

Beyond the radiobiological advantages conferred by the low alpha/beta ratio, since the overall duration of the treatment course is shorter, hypofractionated courses may be radiobiologically advantageous since they avoid **accelerated repopulation**. Accelerated repopulation is the phenomenon in which cancer cells begin growing faster than they were initially. Obviously, if cancer cells have accelerated their rate of growth, there may come a point in which tumor cell repopulation can theoretically overcome the rate at which cells are being killed by radiation therapy. Hence, it is clinically prudent to finish a course of radiation therapy before cancer cells begin accelerated repopulation. The time at which cancer cells 'kick into high gear' and begin accelerated repopulation is sometimes called the **kickoff time** or T_k. Clinical strategies that specifically attempt to avoid the problems of accelerated repopulation are alluded to as accelerated fractionation approaches. While avoiding accelerated repopulation is an important goal, clinicians must be careful to avoid late toxicity of treatment. Unfortunately, one of the largest contributors to late toxicity (i.e. long-term side effects of radiation therapy) is the size of the daily fraction. Therefore, limiting the volume of normal, healthy tissue that gets full dose is critically important for limiting collateral damage in hypofractionated external beam radiotherapy approaches.

In the linear quadratic model, alpha (the linear component) is hypothesized to represent damage caused by a single event. An example of this would be a double-strand DNA break caused by a single radiation track. The quadratic component (beta) is thought to represent damage caused by two or more events. This might be represented by multiple, close, single-strand DNA breaks that are made by two separate radiation tracks but yield the radiobiological equivalent of a double-strand DNA damage.

11.9 The multi-target model

The very same survival curve that led to the presently popular linear-quadratic formalism may mathematically be represented through another model variably called the **multi-hit model, multi-target model,** or **two-component model.** In this model there are four key parameters: n (the **extrapolation number**), Dq, D_1 and D_0. D_0 characterizes the slope of the latter portion of the curve (the straight portion on the log-linear plot) while n, D_1 and Dq characterize the early part of the curve—the so-called shoulder region. Dq, the 'quasi-threshold dose' is a measure of the width of the shoulder. Some cells exhibit large values of Dq while others have very small values. As mentioned, a large shoulder and thus a large value for Dq, correlates with a tendency towards mitotic cell death and greater capacity for SLD repair. Cells exhibiting a low value of Dq have a greater tendency to undergo apoptosis upon irradiation.

The parameter n is the intercept on the y-axis of a straight line overlapping the final, high-dose portion of the survival curve projected back to the y-axis and was originally believed to be the number of targets in the cell. Early experiments suggested that n was 2, and this correlated well with the double-stranded nature of the DNA molecule. Later experiments showed that n was very variable and is no longer believed to correspond to DNA in any physically tangible way. D_0 represents the dose needed to decrease cell survival to $1/e$ or 37% of the initial value once the dose is on the straight part of the curve (i.e. beyond the shoulder). On the other hand, D_1 represents that dose required to drop the cell survival to $1/e$ of the initial value on the initial, 'curvy' part of the cell survival curve (that is, the shoulder region). Beyond the shoulder region, the survival fraction in this is given by the multi-hit multi-target model equation:

$$S = 1 - [1 - e^{-D/D_0}]^n \quad \text{Multi-hit multi-target model}$$

The three parameters that describe the 'straight' high-dose part of the cell survival curve can be related to one another through:

$$\ln n = D_1/D_0$$

Again, these models apply to low LET radiation since irradiation with high LET radiation often does not demonstrate a shoulder or Dq. With high LET radiation the cell survival curve is usually 'linear' right from the start. And again, survival fraction is on the ordinate (y-axis) and is logarithmic while the dose is plotted linearly on the abscissa (x-axis) on these plots.

11.10 The D_0 concept

The dose of radiation that induces, on average, one lethal event per cell leaves about 37% ($1/e$) of the irradiated cells still viable. When we are on the straight, high-dose part of the survival curve, this dose is known as the D_0 **dose**; it is known as D_1 if we are on the low dose, curved, shoulder region. Other metrics such as the dose that leaves only 10% of cells still viable (D_{10}) are also sometimes used. For mammalian cells, the D_0 with low LET radiation is typically on the order of 1 to 2 Gy.

Generally speaking, DNA is the primary target molecule of radiation when it comes to cell death, mutations, and carcinogenesis. Radiation causes multiple lesions in the cell's single DNA molecule (or double in diploid cells). Most of these lesions are quickly repaired, but if there are too many lesions in too short a time frame, the DNA repair mechanisms can be overwhelmed, leading to cell death or propagation of unrepaired damage. The number of DNA lesions per cell detected immediately after a D_0 dose leads to (very approximately):

over 1000 nucleobase chemical alterations,
about 1000 single-strand DNA breaks,
approximately 30 DNA–DNA inter-strand cross-links,
and about 40 double-strand breaks.

As discussed in the radiation chemistry chapter, oxygen augments the biological impact of ionizing radiation, especially for low LET radiation. Using cell killing as the biological endpoint, at a given dose, the percentage of killed cells is significantly higher when cells are growing under full oxygenation compared to the same cells under hypoxic conditions. This effect of oxygen is called the **oxygen enhancement ratio,** or **OER.** The OER for most human cells is approximately 2.5–3.5 for low LET x-rays, gamma rays, electrons, and protons. However, for high LET radiation such as neutrons, alpha particles, heavier charged particles, and low energy protons, the OER is significantly lower.

11.11 DNA structure overview

DNA (deoxyribonucleic acid), along with the various RNA (ribonucleic acid) subtypes, are nucleic acids, meaning they consist of multiple monomeric nucleotide subunits. Each nucleotide is made up of phosphoric acid, a five-carbon sugar unit, and nitrogenous base. In nucleic acids, the five-carbon sugars are ribose (in RNA) and deoxyribose (in DNA). The difference is that deoxyribose only has hydrogen at the 2' carbon, whereas ribose has a hydroxyl group which includes hydrogen and oxygen. Nucleotides are composed of a nitrogenous base, a five-carbon sugar, and a number of phosphate groups. The number of phosphates varies from one to three. For example, ATP is adenosine triphosphate, and AMP is adenosine monophosphate; both are nucleotides. If there is no phosphate attached, the compound is called a nucleoside (figure 11.1).

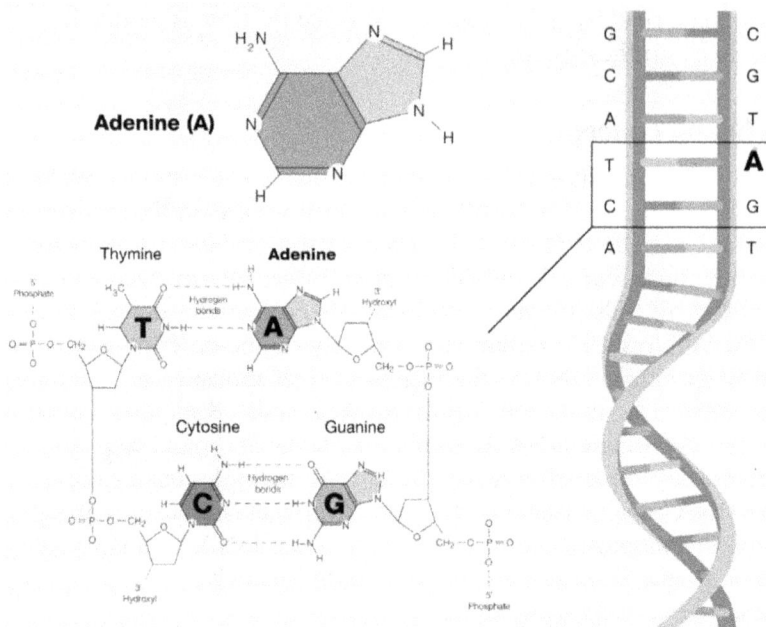

Figure 11.1. The nucleotides that make up DNA are base-paired in a specific manner such that purine bases (adenine and guanine) are always paired with pyrimidine bases (thymine and cytosine). Specifically, adenosine (A) base-pairs with thymine (T) while guanine (G) base-pairs with cytosine (C). Note that G≡C base pairs have three hydrogen bonds whereas A=T only have two. Credit: National Human Genome Research Institute.

The nucleotide units are assembled into a long chain that is connected from the phosphate group of one nucleotide to the 3′ hydroxyl group on the ribose or deoxyribose of the adjacent nucleotide. During the formation of these bonds, two phosphate groups are cleaved off the nucleotide unit during assembly, and this cleavage provides the energy needed for synthesis. The bond between the phosphate and the sugar is a version of bonding between an acid and an alcohol, which is called an ester bond. This particular type of ester is known as a phosphodiester, since the phosphate group (phosphoric acid) is linked to the 3′ hydroxyl group (alcohol functional group) to one sugar above it and to the 5′ hydroxyl group one sugar below it. This long chain of alternating phosphate and sugar units in DNA or RNA constitutes the phosphodiester backbone of the nucleic acid.

DNA and RNA differ in their nucleotide components. DNA has deoxyribose in its backbone whereas RNA has ribose. The nitrogenous bases in DNA and RNA are different as well. The bases found in DNA are adenine (A), guanine (G), cytosine (C), and thymine (T). Through hydrogen bonding, A complementary base pairs with T, whereas G base pairs with C. Note that there are two hydrogen bonds between A = T, whereas there are three between G ≡ C. The phosphodiester backbone is on the outside of the DNA molecule while the bases are on the inside facing one another, allowing hydrogen bonding. The two strands in the double helix are arranged antiparallel to one another with one oriented from 5′ to 3′ while its base-paired partner will then be oriented 3′ to 5′. Instead of thymine, RNA contains uracil. The nitrogenous bases adenine and guanine are in the purine family, whereas cytosine, thymine, and uracil are considered pyrimidines. Also, DNA differs from RNA in having a double helical configuration, while RNA is usually single-stranded (although it may double back on itself to create regions that are base-paired). DNA is normally found in the classic B form discovered by Watson and Crick, which is a right-handed helix, but beside the B form, there are other possible conformations, including a slightly more compact A form (which is also a right-handed helix, but the bases are not as perpendicular to the phosphodiester backbone as in the B form, where they are somewhat angled from the vertical). Additionally, there is a left-handed helix called Z-DNA. The role of Z-DNA in molecular biology is still uncertain, although this conformation may be associated with regulation of gene transcription. Also, the nucleotides in double-stranded DNA are typically in the form of classic Watson–Crick base pairs, but there are other possibilities, such as **Hoogsteen base pairing**. Although adenine still base pairs with thymine, and G base pairs with C, the geometry of the hydrogen bonding in Hoogsteen base pairs is different from Watson–Crick base pairs. This permits unusual conformations to arise, including segments of triple helix or triplex structures. Additionally, strange secondary structures of G-rich single-stranded DNA called **G-quadruplexes** (G4-DNA) may occur, especially in the telomere regions. Prior to mitosis (or meiosis in germ cells (gametes)), DNA is replicated. Specifically, DNA replication occurs in the S phase of the cell cycle. It is during M phase (mitosis or meiosis) that DNA is most susceptible to radiation-induced damage.

11.12 Molecular radiobiology

As the genetic material, DNA is considered the most important target of radiation since it is found in low copy number, is indispensable, and if damaged, can lead to cell death or problems that are propagated to daughter cells. Hence, it comes as no surprise that there are many means of preventing and addressing DNA damage. And in most cells, if the DNA is too severely damaged, the cell is commanded to take itself out of the gene pool through programmed cell death rather than allowing that badly damaged DNA to perpetuate as mutations. Alterations of the DNA molecule constitute mutations, which can come in various forms, both at the invisible molecular level and at the visible chromosomal level.

Radiation exposure can lead to a host of different DNA structure alterations. Among these are:

1. single-strand DNA breaks (ss DNA breaks or SSBs)
2. double-strand breaks (ds DNA breaks or DSBs)
3. chemically modified bases
4. abasic sites
5. bulky adducts
6. inter-strand crosslinks
7. intra-strand crosslinks
8. base-pairing mismatches.

These molecular alterations may have many direct and indirect consequences. For instance, incorrect repair of an abasic site (which is a nucleotide that has lost its purine or pyrimidine base) may lead to the conversion of one base into another (e.g. a guanine becoming an adenine). A change from one nucleotide base into another within the DNA chain is called a point mutation. We shall return to point mutations in greater detail later in the chapter, but for now we will review mutations in general.

11.13 Mutations and epigenetic changes

Mutations may be broadly defined as alterations in the DNA molecule. In the form of base-pair **triplets**, DNA carries the **genetic code** that converts the stored message in DNA (the 'information' molecule) into proteins (the 'action' molecules) (figure 11.2). Hence, alterations in the DNA may lead to genetic changes that result in different expression of proteins or different protein molecules altogether. Such mutated protein molecules may be less functional or non-functional (or in rare instances, better functioning than the original). These genetic changes, caused by alterations in the DNA itself must be distinguished from **epigenetic** changes. Epigenetic changes do not change the DNA, but rather alter the molecules that bind to and regulate DNA. For example, a change from a cytosine into a thymine base in DNA would be a mutation —a genetic change; methylation of that cytosine base would be an example of an epigenetic change. **DNA methylation** is a common form of epigenetic change. DNA methylation often results in **gene silencing** (meaning that the DNA is no longer transcribed into messenger RNA (mRNA) and the mRNA is no longer available to be translated into protein). Another common type of epigenetic change is **histone modification**. DNA in the cell nucleus of eukaryotes is not a single circular molecule as in prokaryotes, but rather is broken up into many DNA segments complexed to

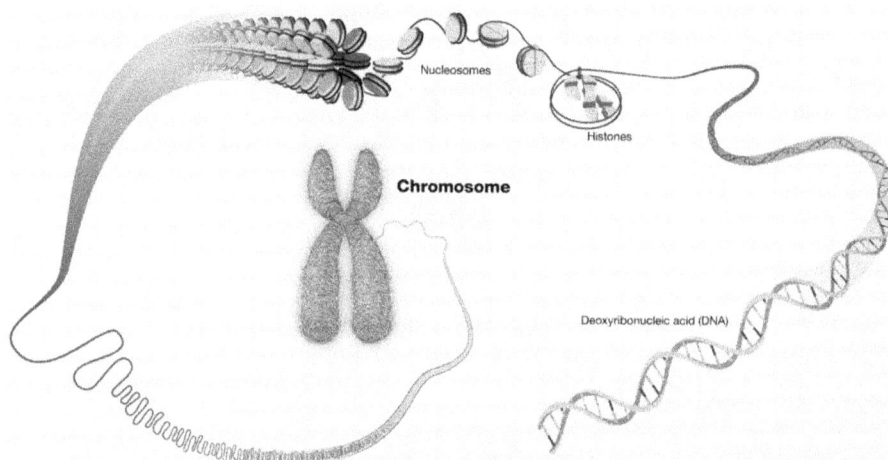

Figure 11.2. Chromosomes are readily visible under the microscope during mitosis and are the most condensed form of chromatin. Chromatin in turn is composed of DNA (deoxyribonucleic acid) complexed with proteins. The double helix of DNA is wrapped around proteins called histones, which are then organized into nucleosomes and compacted further. Credit: National Human Genome Research Institute.

proteins called **histones**. Thus, in human cells, the DNA is found in the form of **chromatin**—a complex of DNA and associated protein molecules. (Note that during cell division (mitosis and meiosis), the chromatin condenses or undergoes 'super-coiling' into much thicker discrete structures called **chromosomes**, which can be readily seen under the light microscope.)

The DNA double helix is wrapped around histones in the form of octamers called **nucleosomes**. These nucleosome octets contain two sets of four histone 'core' molecules (histones H2A, H2B, H3, and H4) and serve to anchor the DNA, which is wrapped around it. This yields the ultrastructural configuration of 'beads on a string'. The histones may be chemically modified through various processes such as methylation, acetylation, phosphorylation, adenylation, and ADP ribosylation. Such modification of the DNA-binding histone proteins alters their shapes and influences how they attach to the DNA, which in turn affects the three-dimensional conformation of the chromatin. For example, acetylation of H3 histone proteins may lead to a more 'open' chromatin structure, which in turn facilitates gene transcription. Drugs that inhibit the enzymes responsible for histone modification are becoming important in medical oncology. For example, histone deacetylase inhibitors prevent the removal of acetyl groups from histones, and thereby alter chromatin structure and the expression of some key cancer-related genes (figure 11.3).

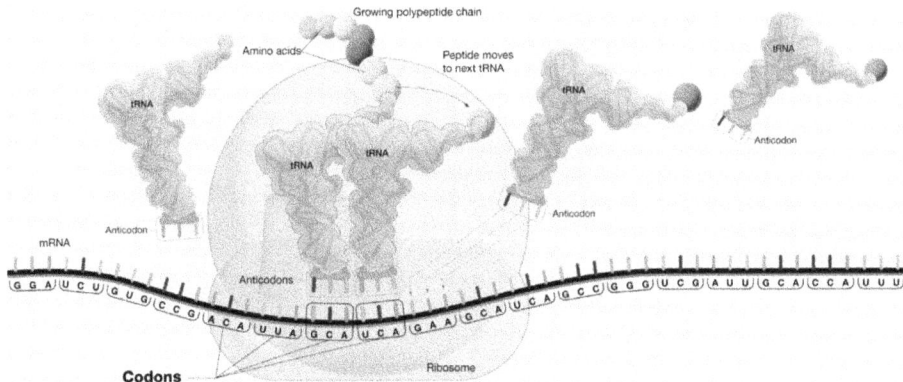

Figure 11.3. DNA is transcribed into mRNA, which is then translated into polypeptide chains that are then folded into protein molecules. During translation, the genetic code, which is written in the form of triplets called codons, is read such that specific triplets of bases represent individual amino acids. For example, the mRNA codon, GCA, signals the addition of the amino acid alanine to the polypeptide chain while the codon UCA calls for insertion of the amino acid tyrosine into the growing chain. The particular amino acids are brought to the mRNA molecule via transfer RNA (tRNA) carrier molecules, which have specific 'anticodons' that match the codons on the mRNA molecules. The entire assembly is carried out on structures called ribosomes. Ribosomes in turn are often attached to the endoplasmic reticulum or may be free-floating factories within the cytoplasm. Ribosomes contain their own form of RNA (rRNA), and these molecules catalyze the peptidyl transferase reaction, which creates the peptide bonds between the amino acids, thereby linking them together to form the polypeptide chain. Hence, not all biochemical catalysts are protein enzymes; in rare instances, RNA may serve as a catalyst. Credit: National Human Genome Research Institute.

11.14 Chromosomes and chromosomal aberrations

Chromosomes are condensed chromatin (DNA and associated proteins such as histones) that become visible under the light microscope during the **M phase** of cell division (mitosis or meiosis). Chromosomes may have a short arm (**p arm**) and a longer **q arm**. The arms are separated by a constriction called the **centromere**. Sometimes the arms are about the same length and the centromere is in the middle; such chromosomes are called **metacentric**. Chromosomes in which the centromere is not quite in the center but is slightly more towards one pole than the other are called **submetacentric**. Most human chromosomes are of this type. Chromosomes in which the centromere is near one end are called **acrocentric chromosomes**. Finally, if the centromere is right at the end of the chromosome, the chromosome is classified as **telocentric**; there are no telocentric chromosomes in humans (figure 11.4).

Chromosomes may be stained for identification. For example, **Giemsa stain** is used in **Giemsa banding** (**G-banding**) and the various chromosomes will have characteristic identification bands. Autosomal chromosomes are listed in size order with chromosome

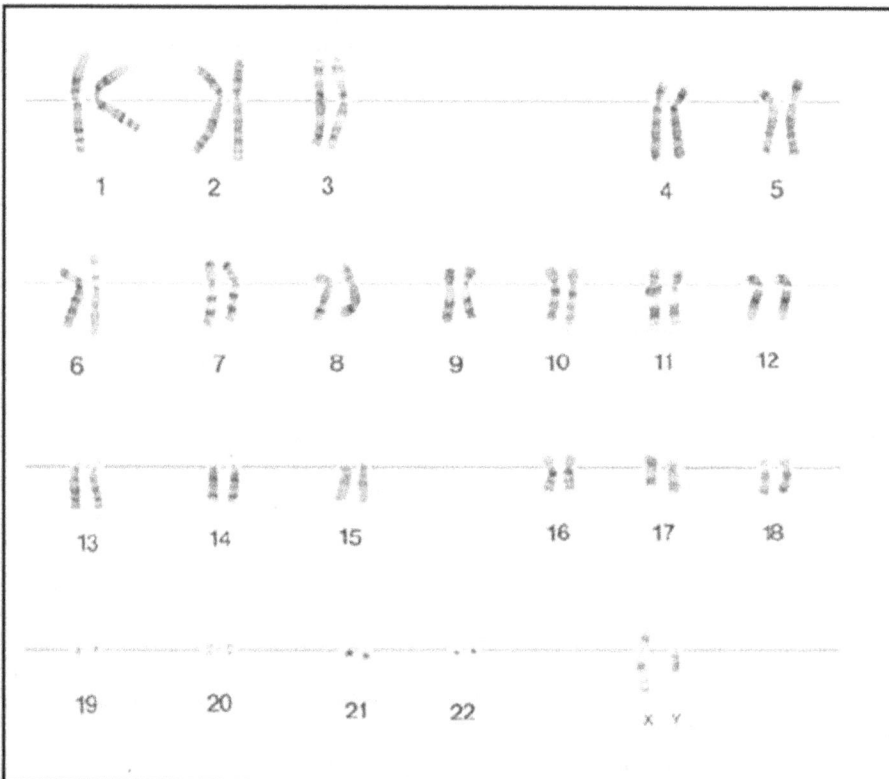

Figure 11.4. A human karyotype. In creating such lineups, cells such as lymphocytes are induced to undergo mitosis and then the 23 homologous pairs of chromosomes in such cells are photographed and literally cut and pasted in size order onto a background for photographing. Credit: National Human Genome Research Institute. Copyright © 2008, Genetic Alliance.

1 being the largest. In humans, there are 22 autosomal chromosomes or autosomes (which are found in both sexes) and one pair of sex chromosome (the X and Y chromosomes). Human females have two X chromosomes while males have an X and a Y. Radiation, especially high LET radiation, can cause double-strand DNA breaks, which may in turn be evident in the form of chromosomal breaks. These breaks may rejoin in various ways, sometimes normal and reconstituting the original chromosome, or at other times in an abnormal manner, leading to **chromosomal aberrations**. The scientific field of studying chromosomes is called **cytogenetics**.

Returning to mutations, there are three general types of DNA mutations: base substitutions (point mutations), deletions, and insertions. This terminology applies to both the invisible, molecular structure of DNA and also applies to the gross, visible form of DNA in chromosomes. Thus, there can be **chromosomal deletions** and **insertions**, as well as **inversions** in which a section of the chromosome has been removed and reinserted backwards. Entire chromosomes may be **duplicated**, resulting in **aneuploidy** (which is any deviation from the normal total number of chromosomes). For example, in humans the total number of chromosomes in a normal diploid cell is 46 (23 pairs of chromosomes). In Down syndrome, there is a duplication of one member of the pair of chromosome 21; there are three rather than two copies of this chromosome. Thus, Down syndrome is an example of aneuploidy and is specifically **trisomy 21**.

Entire sections of chromosomes may be transferred from one chromosome to another; this process is called **translocation**. If this transfer is an even swap of material, it is called a balanced or reciprocal translocation. However, many translocations are accompanied by a loss of chromosomal material. Many hematological cancers (leukemias and lymphomas) carry characteristic translocations. For instance, **follicular lymphoma** is characterized by a translocation between chromosomes 14 and 18. This is abbreviated as **t(14;18)** and causes the *BCL2* gene on chromosome 18 to be juxtaposed to the **heavy chain promoter** on chromosome 14. Since the heavy chain promoter on chromosome 14 is hyperactive in B lymphocytes (because lymphocytes make antibodies and the heavy chain is a part of immunoglobulin antibody molecules), this causes inappropriate hyperactivity of the *BCL2* gene. *BCL2* activation leads to over production of the Bcl-2 protein, which inhibits apoptosis. The result is that the affected lymphocytes carrying this abnormal t(14;18) chromosomal translocation do not undergo apoptosis when they should; they become 'immortal' (especially if the evasion of apoptosis is accompanied by **telomerase** activation which overcomes the naturally imposed restriction on the number of times a cell can replicate (the Hayflick limit). Cellular immortality and evasion of apoptosis are among the **hallmarks of cancer**.

Various other chromosomal aberrations can occur after radiation exposure. Among the most commonly observed chromosomal aberrations after radiation exposure are deletions, chromatid breaks, acentric fragments, and isochromatid fragments. Also often seen following irradiation are dicentric chromosomes, ring chromosomes and anaphase bridges. **Ring chromosomes**, where the two ends (telomeres) have fused together to form a circle, represent a classic radiation-induced chromosomal aberration. Anaphase bridges are due to a similar phenomenon in which the telomeres of sister chromatids fuse together and cause incomplete segregation of chromosomes during mitosis with a linkage that persists even after completion of mitosis as an 'interphase bridge'.

Another chromosomal aberration that is often seen after radiation exposure is the **dicentric chromosome**, in which a single chromosome has two centromeres. These so-called **dicentrics** are the product of two chromosomes that have been broken and inappropriately fused together. The result is a new chromosome that has two centromeres and another chromosome that had no centromere at all; this chromosome without a centromere is called an acentric fragment (figure 11.5).

Chromosomal dicentrics are a common finding in human lymphocytes following radiation exposure. In fact, dicentrics can appear after very low doses (e.g. 5 cGy) and are a useful bioassay and biodosimeter for radiation exposure (the **dicentric chromosome assay**). Dicentrics in lymphocytes are considered the most sensitive and most specific chromosomal alterations for assessing radiation dose. Increasing doses of radiation result in increasing numbers of detected dicentrics. Hence, the dicentric chromosome assay is potentially useful after acute radiation exposures even at very modest doses. However, it should be noted that these dicentrics are not necessarily permanent. In total body irradiation experiments involving male C57B1/6 mice, the number of lymphocytes with dicentrics falls to baseline after about 112 d.

The **Robertsonian chromosome** is a subtype of translocation. A **Robertsonian translocation** or Robertsonian chromosome (abbreviated ROB) is a chromosomal aberration in which the long arms (q arms) of two acrocentric chromosomes become fused to each other. Recall that acrocentric chromosomes are those with arms that are highly discordant in length. In other words, they have long q arms and very short p arms. The acrocentric chromosomes in humans include five autosomal chromosomes (13, 14, 15, 21, 22) and the Y sex chromosome. In a Robertsonian translocation, the two short arms (p arms) also fuse to form a smaller reciprocal chromosome, meaning there is no loss of genetic material—the chromosomal gains and losses are balanced. However, these products often contain only nonessential or redundant genes that are also present elsewhere in the genome. Such chromosomes, made only of the two fused tiny p arms of acrocentric chromosomes are usually lost within a few cell divisions without necessarily causing a loss of cell viability.

The Robertsonian type of translocation is readily visible cytologically and may reduce the total haploid chromosome number from 23 to 22 (or equivalently, the diploid number from 46 to 44).

In human evolution, because Robertsonian translocations result in a reduction in the number of chromosomes, it appears that a Robertsonian chromosomal fusion occurred in the ancestor of humans, reducing the 48 chromosomes (24 pairs) still seen in today's great apes down to the 46 (23 pairs) in modern humans. Specifically, the human chromosome 2 is apparently a Robertsonian chromosome derived from two separate chromosomes in apes.

Another chromosomal aberration that may occur after radiation exposure (but more often does not have an obvious etiology) is the **isochromosome**. An isochromosome is one in which the arms of a particular chromosome are mirror images of each other. This derived chromosome consists of either two copies of the long arm (q arm) or the short arm (p arm) of the original chromosome. Obviously, this leads to duplication of one part of a

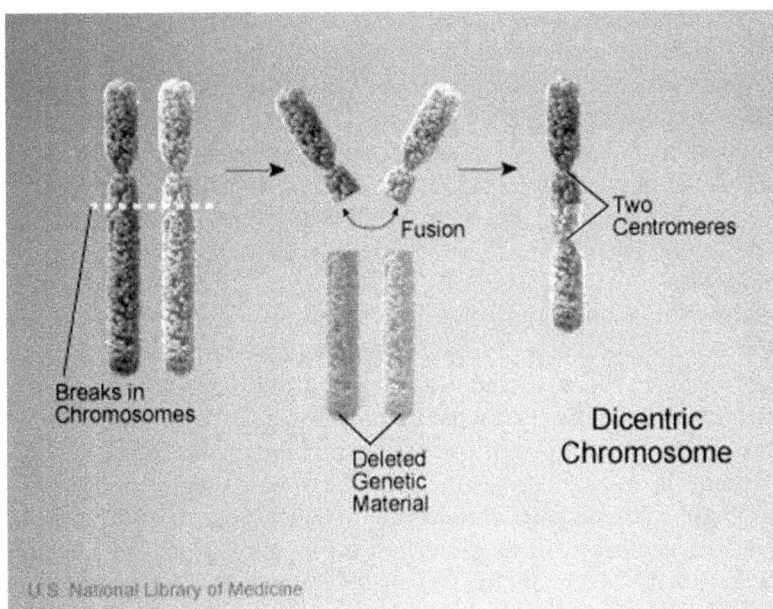

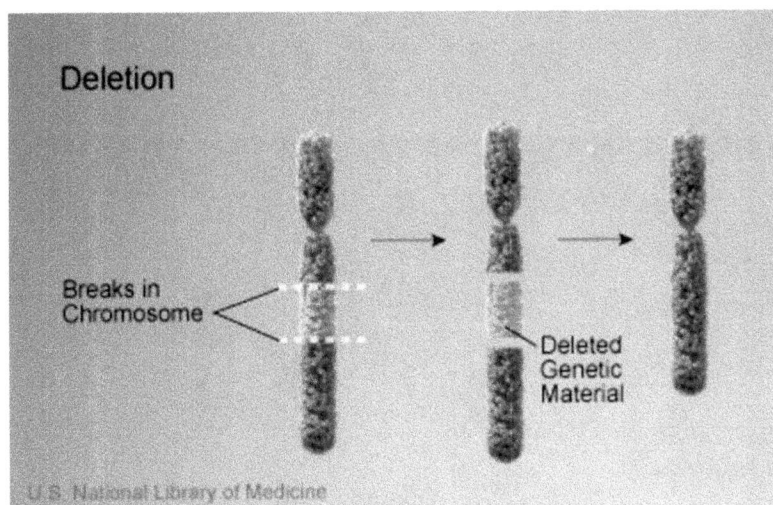

Figure 11.5. One type of chromosomal aberration that is often seen after radiation exposure is the **dicentric chromosome** (top image), in which a single chromosome has two centromeres. Dicentrics are the product of two chromosomes that have been broken and inappropriately fused together. The result is a new chromosome that has two centromeres and another chromosome that had no centromere at all; this chromosome without a centromere is called an acentric fragment and is often lost. Deletions (lower image) are another commonly seen chromosomal abnormality detected following irradiation. Point mutations and more subtle chromosomal mutations may be caused by a variety of genotoxins, but large-scale chromosomal rearrangements are caused by major events that shatter chromosomes (**chromothripsis**). Such breakage allows them to re-join in abnormal ways and are more often seen following high-LET radiation exposure. Chromothripsis may also be seen without irradiation in cancer cells. Cancer cells appear especially prone to cytogenetic disruptions and hence, hypermutability. This predisposition towards new mutations is one of the hallmarks of cancer. Credit: U.S. National Library of Medicine.

chromosome and deletion of another part. Hence, there is simultaneous **trisomy** of the genes present in the isochromosome and **monosomy** of the genes in the arm that is lost.

Isochromosomes are occasionally found in cancer cells. One example is an isochromosome derived from chromosome 17 that is a duplication of the two q arms. This is abbreviated as i(17q) and is not uncommonly found in medulloblastoma, a pediatric brain tumor. i(17q) is present in around 30%–50% of medulloblastoma cases. In these cases, the q arm of chromosome 17 is duplicated while the p arm is lost. It may be relevant that the critical tumor suppressor gene *TP53* is on chromosome 17p.

Chromosomal inversion is another example of a chromosomal aberration that may occur following radiation exposure. An inversion is a chromosomal rearrangement in which a segment of a chromosome is removed and reinserted backwards. Inversion formation requires two separate breaks in a chromosome. This is followed by the released segment flipping its orientation and then reinserting itself between the two break points in that same chromosome. The reversed segments in inversions can be as short as 100 kilobase pairs or as long as 100 megabase pairs. The chromosomal break points of inversions most often happen in regions rich in repetitive DNA nucleotide base pairs. Inversions are classified as **paracentric inversions** (which do not involve the centromere) and **pericentric inversions** (which do include the centromere) (figure 11.6). It is worth noting that although radiation can cause chromosomal aberrations, such aberrations can and do occur spontaneously.

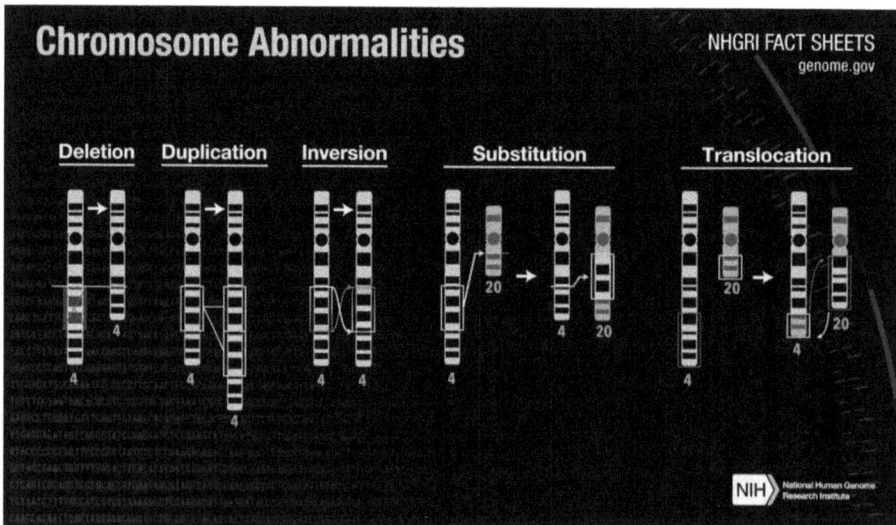

Figure 11.6. Cytogenetics is the subbranch of medical genetics that focuses on chromosomes. Chromosomes are long strands of DNA and associated proteins that contain most of the genetic information in a cell (organelles such as mitochondria and chloroplasts also carry their own DNA). Chromosomal aberrations come in various forms. Above are some classic chromosomal changes (deletions, duplications, inversions, substitutions, and translocations) that can occur from a variety of genotoxins, including ionizing radiation exposure. Credit: National Human Genome Research Institute.

11.15 DNA point mutations

Among the most obvious and simplest significant DNA alterations are point mutations. Point mutations are single base substitutions of one base for another. Point mutations can be **transitions** (wherein a purine is replaced by a purine, or a pyrimidine is replaced by a pyrimidine) or **transversions** (wherein a purine is replaced by a pyrimidine, or vice versa).

Point mutations that occur in DNA sequences encoding proteins can be classified as either **missense, nonsense, synonymous,** or **silent**.

Missense mutations occur when the base substitution results in a triplet codon that specifies a different amino acid, and hence leads to a very slightly different polypeptide sequence with a single changed amino acid. Depending on the particular amino acid substitution, a missense mutation is labeled as **conservative** or **non-conservative**. If the structure and functions of the protein containing the substituted amino acid are very similar to the original, the missense mutation is called conservative and has little biochemical impact. If, on the other hand, the substitution leads to a protein with a very different structure and properties, the missense mutation is classified as non-conservative. Most such non-conservative missense mutations are **deleterious**, which makes sense, given the millions of years that evolution has had to hone and perfect these molecules.

Nonsense point mutations occur when a base substitution produces a **stop codon** (TAA, TAG, TGA). This means the point mutation is a truncating mutation that halts translation. Such nonsense mutations most likely lead to a nonfunctional protein.

11.16 Silent and synonymous point mutations

The consequences of mutations on cells and organisms may be non-existent. Such non-consequential mutations are called **silent mutations**. Silent mutations do not affect the cell's function and therefore do not inhibit its reproduction. If silent mutations occur in the germ cells (gametes, such as sperm and egg cells), these silent mutations will be passed on to future generations. In fact, there is a baseline rate of silent mutations in many genes, and this rate is used in creating molecular genetic clocks. Molecular clocks allow the estimation of closeness of phylogenetic relationship between different organisms as well as an estimate of how long ago they diverged. For instance, a given human gene may be found to have 99% DNA sequence identity in the same gene in one animal, but only 89% sequence similarity in another animal. In such a scenario, it could be concluded that humans are more closely related phylogenetically to the first animal than to the second one. And, based on the genetic sequence divergence, one might be able to calculate the times in paleontological history the human lineage separated from these two animals provided that there are good 'calibration points' in the fossil record which allow estimations on how fast the molecular clock is ticking for the gene in question.

Curiously, if a base substitution occurs in the third position of a codon there is a good chance that a synonymous codon will be generated. This is because the third base in any given set of triplet codons that all code for the same amino acid tends to be most variable. For example, CCG, CCA, CCU, and CCC all code for proline. This phenomenon is part of the '**wobble hypothesis**' which asserts that the first two bases of a codon are more important for the specificity of that codon than the third base. In essence, the genetic code is degenerate because only the first two base pairings between the mRNA codon and the tRNA anticodon must be precise—the the pairing between the third bases of codon and anticodon may 'wobble' and vary.

This wobble permits a single tRNA molecule to recognize more than one codon. Hence, although there are 61 codons for amino acids, the number of tRNA types is less (around 40) thanks to wobbling. Thus, the amino acid sequence encoded by the gene is not changed and the mutation is said to be synonymous.

Note that *silent mutations are not the same as so-called **synonymous mutations***. Synonymous mutations lead to the same amino acid, thanks to the redundancy of the genetic code. In a **silent mutation**, the DNA sequence is different but there is *no biological consequence*. This may be because the point mutation was a missense mutation that has led to a different amino acid that does not affect the overall function of the protein molecule. Alternatively, the DNA mutation might happen to simply lead to the exact same amino acid. This would be a synonymous *and* silent point mutation.

Synonymous point mutations are defined as changes in the DNA sequence that do not lead to a different amino acid in the final protein product. This can happen because the triplets of DNA (codons) that constitute the genetic code are redundant. For example, the triplet sequences TTA and TTG both code for the same amino acid, leucine. In fact, leucine and arginine are coded for by six different triplet codons each. In contrast, methionine is coded for only by ATG, and tryptophan is coded only by TGG. All other amino acids have codon redundancy in between these extremes.

Thus, a synonymous point mutation may occur in the DNA sequence when TTA is converted into TTG; both codons lead to the same amino acid (leucine) in the final protein product. In contrast, a silent point mutation means that a different (or the same) amino acid is placed into the protein primary sequence but there are no obvious biological consequences.

11.17 Synonymous mutations that are not silent

It would seem that synonymous point mutations should always be silent since the exact same amino acid ends up in the same place within the protein, but most curiously, this is not true. Sometimes, one DNA base may be mutated, but the resulting triplet codon still codes for the *same* amino acid but the final protein product functions differently. In most cases, such a synonymous mutation would lead to no consequences and the mutation would be silent since the final primary protein sequence is unchanged. But, in very rare situations, such a point mutation can have biological consequences despite the protein having the exact same primary structure ('primary structure' refers to the amino acid sequence). This is because the tertiary structure of the folded final protein winds up different in the two cases. What appears to happen is that the ribosome 'stalls' or slows down when it must use a non-preferred codon rather than the usual, fast-reacting one. The different **tRNA** molecules that carry the same amino acid into the ribosome for protein synthesis may operate at different paces; one might be swift while the other is sluggish. In the case of a swift codon being mutated into a sluggish codon, the nascent polypeptide chain might fold up differently while waiting for the next amino acid to be added, leading to a different overall 3D conformation in the final product, despite having the exact same amino acid sequence! Hence, the molecular protein synthesis machinery does have its preferences. This unusual situation is called **codon bias**.

11.18 DNA deletions and insertions

Deletions appear to be the most commonly encountered mutations following radiation exposure. Radiation may result in combinations of insertions and deletions, leading to a variety of consequences.

The deletion of a DNA base may result in a **frameshift**, which means that the reading frame of the genetic code triplets has been altered. This can have a very serious effect on the coded-for protein in which all amino acids from that deletion onward are different. Frameshift mutations may lead to a completely garbled message in the transcribed mRNA and a non-functioning translated protein.

Frameshift mutations can occur if one or two bases are deleted but a deletion of three bases (or a multiple of three) may leave the reading frame intact. The end product of such a deletion might be a protein similar to the original but with one or more amino acids missing. Such deletion mutations may or may not be deleterious.

As with deletions, the insertion of additional base pairs into the DNA molecule may lead to frameshifts. This depends on whether or not the number of inserted bases is a multiple of three. As with deletions, if there is no frameshift, the final product protein may or may not be functional. Thus, insertions may or may not be deleterious.

11.19 DNA damage response

The DNA damage response (DDR) is in intricate, interconnected signaling network that serves to preserve genomic integrity against a variety of endogenous and exogenous genotoxic assaults including reactive oxygen species and ionizing radiation. In its most inclusive definition, the DNA damage response spans the events from the initial detection of DNA damage all the way through the various multi-factorial signaling pathways that address different types of DNA repair which are covered in more depth later in this chapter. Thus, the DDR ranges from simple **direct base repair** to the extremely complicated **homologous recombination** pathway. By this inclusive definition, the DDR also includes the **mismatch repair system (MMR)** for mistakenly paired bases, the **base excision repair system (BER)** for small base modifications, the **nucleotide excision repair (NER)** for intra-strand crosslinks and pyrimidine dimers, and the **single-strand break repair** and **double-strand break (DSB) repair pathways** (non-homologous end joining, microhomology-mediated end joining, and homologous recombination). Nevertheless, all descriptions of the DDR involve two basic branches: DNA damage repair and halting of the cell cycle (at designated cell cycle checkpoints).

The cellular response to DNA damage involves several key DNA damage sensor proteins, including:

1. ATM
2. ATR
3. DNA-PKcs
4. CHK1
5. CHK2
6. WEE1
7. PARP-1.

These molecules constitute the central regulators of the DNA damage cellular response. Several of these molecules are **kinases**, meaning they phosphorylate (that is, add phosphate groups to) other molecules. **ATM** (Ataxia telangiectasia mutated), **ATR** (ATM and RAD3-related), and **DNA-PKcs** (DNA-dependent protein kinase, catalytic subunit) are all protein kinases belonging to the family collectively called **PIKKs (phosphoinositide-3-kinase-related family of protein kinases)**. All three kinases have a predilection for phosphorylating their target proteins on serine or threonine residues that happen to be followed by a glutamine residue (i.e. **S/T-Q sequences**). Thus, these kinases share substrates and have overlapping functions. They also undergo autophosphorylation, which may be involved in their regulation. They are recruited to sites of DNA damage by other molecules: ATM is recruited by **NBS1**, while ATR is recruited by **ATR-interacting protein (ATRIP)**, and DNA-PKcs by **Ku80**.

Following their arrival at sites of DNA damage, the protein kinases ATM and ATR cooperate with various other proteins to initiate the DDR. It appears that double-strand DNA breaks preferentially activate ATM. ATM is mainly involved in the repair of DSBs sensed by the **MRN protein complex**, (which is discussed later in this chapter). ATR on the other hand, primarily responds to various types of DNA damage that involve the abnormal persistence of single-stranded DNA, with the assistance of the regulatory single-strand DNA binding proteins, ATR-interacting protein or **ATRIP**, and replication protein A (**RPA**). ATM and ATR then phosphorylate and activate the 'downstream' substrates **Chk1** and **Chk2**. Downstream simply refers to molecules and reactions that fall later in the sequence of a biochemical pathway than early or 'upstream' molecules and events. These in turn, phosphorylate other downstream effector proteins such as E2F, p53, and Cdc25 family members, which halt cell cycle progression and activate DNA repair systems. If the DNA damage is excessive, the p53 family members trigger programmed cell death or apoptosis.

11.20 The *TP53* gene and the p53 protein

The p53 protein plays a central role in DNA damage detection and repair. It is a general sensor of cellular stress and cytogenetic damage caused by radiation or other physical and chemical agents. Upon sensing DNA damage, a series of intricate pathways are initiated and coordinated by p53. Depending on factors such as the degree of DNA damage, p53 might initiate efforts to repair the DNA or dictate that the irreparably damaged cell be disposed of.

The locus for the human *TP53* gene is on the short arm of chromosome 17, specifically at 17p13.1. The *TP53* gene provides instructions for making the critically important protein called tumor protein p53 (or simply p53). *TP53* is a tumor suppressor gene, i.e. its activity stymies the formation of tumors. As with several other tumor suppressor genes, if a person inherits only one functional copy of the *TP53* gene, they are predisposed to cancer. Such germline mutations lead to the **Li-Fraumeni syndrome**, a classic cancer predisposition syndrome which can lead to multiple different cancers in various organs at relatively young ages. Although germline mutations in *TP53* are rare, sporadic somatic mutations in *TP53* are found in many, if not most, malignant cells in advanced human cancers. Somatic mutations in the *TP53* gene are thus among the most common genetic changes found in human cancers.

The p53 protein is so-named because of its apparent molecular weight of 53 kDa on sodium dodecyl sulphate–polyacrylamide gel electrophoresis (SDS-PAGE). In reality, the molecular mass of p53 is probably closer to 43.7 kDa. The discrepancy is due to the relatively high number of proline residues, which slow its migration on SDS-PAGE and confers a higher apparent molecular weight.

p53 is normally present in very low intracellular levels but, upon radiation exposure and damage to nuclear DNA, the *TP53* gene is upregulated. This increased translation of the gene (in conjunction with slowed p53 protein degradation), leads to p53 protein accumulation. p53 localizes and functions in the cell nucleus. There it directly binds to DNA and influences gene activity; in other words, it is a **transcription factor**. In fact, p53 regulates dozens of target genes that have a multitude of diverse biological functions. Thus, p53 is considered a master transcription factor and because of its central role in regulating DNA repair and cell division, it has been nicknamed the '**guardian of the genome**'. It deserves this moniker thanks to its key function in determining whether damaged chromosomal DNA will be repaired and the cell will be salvaged, or if the cell will self-destruct (i.e. undergo **apoptosis** or **programmed cell death**). Activated p53-mediated signaling pathways can cause cell cycle arrest and facilitate DNA repair (which promotes cell survival) or alternatively they can initiate the intrinsic pathway of apoptosis and cell senescence, promoting cell death. p53 clearly plays a crucial role in determining cellular fate following radiation exposure.

The pathway dictated by p53 depends on several variables, including the degree of DNA damage and the specific tissue type involved. For instance, activation of p53 typically results in increased interphase apoptosis in tissues that have a rapid turnover rate (such as the hematopoietic system and the epithelium of the gastro-intestinal tract). This partially accounts for the hematopoietic and gastrointestinal acute radiation syndromes following total body radiation exposure at sufficiently high doses. In contrast, in tissues with slower turnover rates such as bone, muscle, and heart, accumulation of p53 usually does not cause significantly increased apoptosis, but rather triggers molecular pathways that control cell-cycle checkpoints (such as **p21**) and DNA repair. Even within a specific tissue there can be varying effects of p53 depending on subtle differences in cell type. In the bone marrow, for example, p53 activation provokes a different response in hematopoietic stem cells compared to progenitor cells (descendants of stem cells that differentiate into specialized blood cells).

11.21 Mitotic cell death

DNA damage caused by radiation or other genotoxic insults normally leads to an effort to correct the damage. This is carried out by p53-mediated halting of the cell cycle and initiation of DNA repair pathways, or the cell may trigger apoptosis if the damage is deemed too severe. Another possible fate of an irradiated cell with extensive DNA damage is **mitotic cell death**. This means that the cell might survive the damage until it attempts to undergo mitosis, but upon attempting mitosis, the chromosomal damage will prohibit successful completion of chromosome replication and nuclear division (karyokinesis) and cell division (cytokinesis). This phenomenon is also called mitotic catastrophe. Mitotic cell death is a significant mechanism in radiation therapy of cancer. Additionally, mitotic cell death and other forms of cell death following extensive DNA/chromosome damage) are important since uncorrected DNA damage could lead to replication and propagation of the damaged DNA (i.e. perpetuation of mutations).

In addition to mitotic cell death, there are several other known mechanisms whereby cells die after radiation exposure (or other causes). In essence, various stressors including radiation can cause DNA damage, telomere shortening, intracellular metabolic changes, mitotic failure, and an immunological response to the damaged cells. These cellular stressors may lead to various responses that eventually culminate in cell death. Some of these include:

1. Mitotic cell death (mitotic catastrophe)
2. Necrosis
3. Apoptosis
4. Autophagic cell death (autophagy)
5. Senescence (which is often considered to be irreversible cell cycle arrest)
6. Immunogenic cell death (where the damaged cell is destroyed by the immune system)
7. Necroptosis
8. Pyroptosis
9. Ferroptosis.

Note that these mechanisms of cell 'death' do not always lead to outright destruction and elimination of the cell in question. In radiobiology, cell death may be achieved through the inability to reproduce. In this regard, the cell may be said to be inactivated. This is a very valid definition of death for a cancer cell. Many of these inactivation mechanisms require the p53 molecule. Hence, the loss of p53 function (which is commonly seen in cancer cells) leads to a loss of some means of cell death.

A final point that underscores the complexity of the interplay between cells and the various pathways they may take is that, although **senescent cells** are no longer capable of replicating, they may still be hazardous. This is because senescent cells may secrete pro-inflammatory cytokines, which may ultimately trigger tissue dysfunction and potentially be tumor promoting. Thus, these damaged but non-dividing cells may not be as innocuous as originally thought. Conversely, however, in other situations senescent cells seem capable of secreting tumor suppressing factors, which can counter carcinogenic insults such as radiation exposure.

As mentioned, p53 activation, in response to radiation-induced DNA damage, leads to cell growth arrest, allowing and stimulating DNA repair, or it may direct cellular senescence or apoptosis, thereby maintaining genome integrity by prohibiting propagation of cells with irreparably damaged DNA.

11.22 How does p53 pause the cell cycle?

Along with affecting DNA replication and promoting repair, p53 is an important regulatory transcription factor that controls the expression of several genes involved in the cell cycle. Among the myriad functions of p53 is the stimulation of the expression of genes encoding 'downstream' effectors such as **p21** and **Gadd45** (figure 11.7).

11.23 p21, cyclins, and cyclin-dependent kinases

Upon activation and the binding of p53 protein to DNA, one critical gene that becomes activated leads to the transcription of a protein called **p21**. The p53-dependent cell cycle arrest following DNA damage is principally mediated via this p21 protein. In other words, p21 is the primary downstream effector of p53-induced cell cycle arrest. Cells deficient in p53 are unable to cause a cell cycle arrest via p21; cells deficient in p21 cannot halt the cell cycle upon commands from p53.

To better understand the role of p21, it may be instructive to review the role of **cyclins** and **cyclin-dependent kinases** in the cell cycle (figure 11.7). Cyclin-dependent kinases (also called **cell division kinases**; both go by the abbreviation **CDK**) are *the primary regulators of the cell cycle*. They are **serine/threonine protein kinases** that require a separate subunit (a cyclin) for full enzymatic activity. Without the bound cyclin, CDKs have little kinase activity; only the complete cyclin-CDK complex is fully active.

Cyclins control progression through the cell cycle and initiate mitosis by activating the CDK enzymes. Many cyclins have been discovered, but one means of categorizing the main ones is based on when they function in the cell cycle. In this manner, the D, E, A and B cyclins operate during different phases of the cell cycle. After mitosis, cells may exit the cell cycle to enter a non-dividing phase called **G0**. Alternatively, the cell may continue dividing and enter '**interphase**'. The first stage in interphase is **G1 phase** of the cell cycle (G stands for gap). Cells grow during **G1** and assess the integrity of the DNA in preparation for replication. Following G1, the next phase is **S phase**. It is in S phase that DNA is duplicated. S therefore stands for synthesis. After S phase comes the **G2 phase**. During G2, organelles are replicated. After G2 phase, the cell moves out of interphase and begins mitosis.

Mitosis has several steps, which reflect what is happening with the chromosomes and **mitotic spindle** (a set of fibers made of **microtubules** that attach to chromosomes

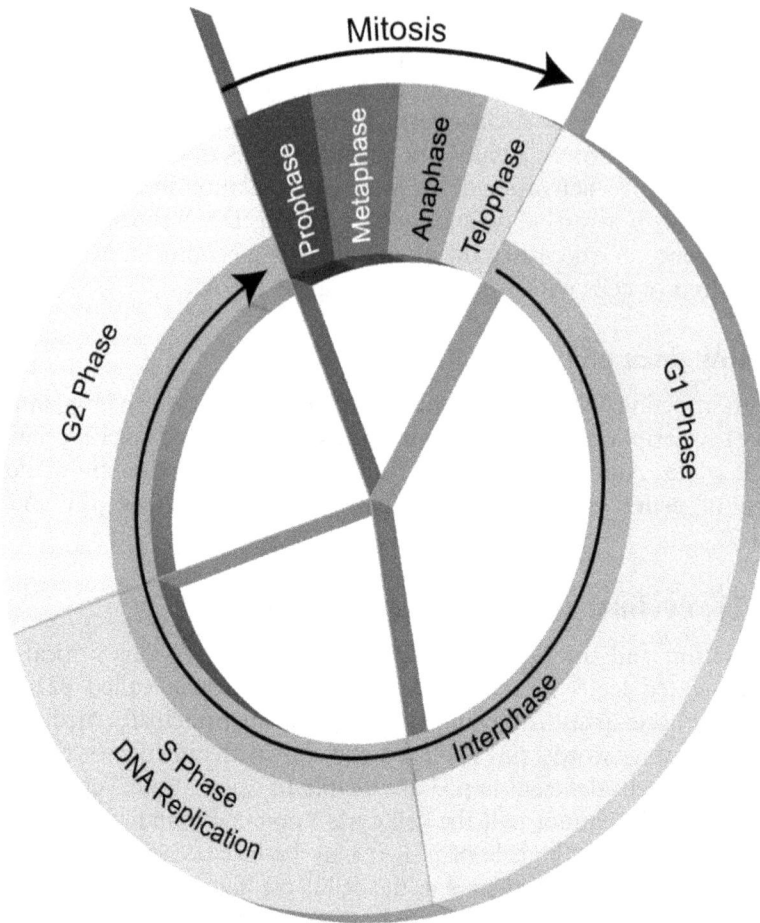

Figure 11.7. The cell cycle may be divided into mitosis and interphase. Interphase is then broken up into G1, S phase, and G2. G1 is the phase where the cell is preparing to divide. S stands for DNA synthesis and the S phase is when all the DNA is duplicated. After the DNA is replicated and a complete extra set of the genetic material now exists, the cell progresses through the cell cycle to G2. During the G2 stage the cell organizes the genetic material and prepares to divide. The next step, mitosis, is sometimes called M phase. Mitosis is divided into prophase, metaphase, anaphase, and telophase. These substages are based on the cytogenetic appearance of the chromosomes and the spindle apparatus that pulls the chromosomes apart and into the two daughter cells. During anaphase and telophase of mitosis, the nucleus divides (karyokinesis) and then cell division or cytokinesis occurs, leaving two daughter cells. Credit: National Human Genome Research Institute.

and pull them away from the middle and towards the poles of the newly forming daughter cells). The mitotic spindle connects the chromosomes to structures called **centrosomes**, which move towards the poles. Prophase is the first part of mitosis. This is when the mitotic spindle starts to form, and the chromosomes start to condense. Also, the nucleolus (a part of the nucleus where ribosomes are made) disappears. The nuclear membrane (nuclear envelope) breaks down in prophase as well. Towards the end of prophase, some of the microtubules of the spindle begin to attach to the forming centromeres of the condensing chromosomes (technically, the microtubules attached to the **kinetochore** of the chromosomes, which are in the centromere regions). In metaphase, the now-condensed chromosomes align at the so-called metaphase plate and the mitotic spindle firmly attaches to the kinetochores. The cell undergoes a **spindle checkpoint** to ensure that the sister chromatids will be evenly segregated between the two daughter cells upon cell division. If there is any detection of improper alignment, the cell will pause to correct the problem before proceeding with cell division. After metaphase comes anaphase. This is the part of mitosis when the microtubules of the spindles pull the chromosomes away from the metaphase plate and towards the centrosomes at the opposite poles of the cell. Finally, the cell enters telophase, the last stage of mitosis. In telophase, the mitotic spindle breaks down, the nucleoli reform (now one for each daughter cell) and the nuclear membrane(s) reform. The chromosomes revert from their condensed form into their invisible chromatin form. The cell also undergoes cytokinesis, the division of the cytoplasm into two separate daughter cells during the final phases of mitosis, anaphase, and telophase. The original cell becomes pinched in the middle; the pinch site is called the **cleavage furrow**. The two daughter cells then separate, each with a full set of chromosomes identical to their parent. Incidentally, cytokinesis is the division of a cell into two daughter cells; karyokinesis is the division of the nucleus into two new nuclei. Karyokinesis can occur without cytokinesis (leading to in some cases, multi-nucleated giant cells) but cytokinesis does not proceed without karyokinesis.

Returning to the role of cyclins in cell division, below are some very broad generalizations:

1. Cyclin D and cyclin E regulate progression from G1 to S phase.
2. Cyclin A functions during S phase.
3. Cyclin B regulates the transition from G2 to mitosis.

Although there is overlap, these cyclins tend to work with specific cyclin dependent kinases at particular phases of the cell cycle. Here are some examples:

1. Cyclin D pairs up with Cdk4 and to Cdk6 to regulate progression from G1 to S phase.
2. Cyclin E also binds to Cdk4 and to Cdk2 to help regulate progression from G1 to S phase.
3. Cyclin A works with Cdk2 during S phase.
4. Cyclin B interacts with Cdk1 to regulate progression from G_2 to M phase.

The cyclin dependent kinases are divided into cell-cycle-related subfamilies (Cdk1, Cdk4, and Cdk5) and the transcriptional subfamilies (Cdk7, Cdk8, Cdk9, Cdk11, and Cdk20). Regarding protein family affinities, CDKs belong to the *CMGC group of kinases* along with mitogen-activated protein kinases (MAPKs), glycogen synthase kinase-3 beta (Gsk3β), members of the dual-specificity tyrosine-regulated kinase family, and CDK-like kinases.

The p21 protein (the principal effector of p53-mediated cell cycle arrest) also goes by the name *CDK inhibitor 1*. It binds and inactivates certain cyclin dependent kinases, which results in cell cycle arrest. Just where in the cell cycle the arrest occurs depends on the specific cyclin dependent kinase involved. p21 can inhibit several cyclin/CDK complexes, although it is primarily an inhibitor of Cdk2. As mentioned above, Cdk2 is a principal effector of the G1/S checkpoint, also called the late **G1 restriction point**. Specifically, Cdk2 binds to, and is regulated by, two different cyclins, cyclin E and cyclin A. Cyclin E binds to Cdk2 during G1 phase. This cyclin E/Cdk2 complex is instrumental in the G1-S transition. In contrast, the binding of Cdk2 to cyclin A promotes progression through S phase. However, the p21 protein has numerous additional inhibitory functions and also binds to and inhibits the activity of other cyclin/Cdk complexes at the G1 and S phases.

Another role played by p21 is **PCNA inhibition**. PCNA is **proliferating cell nuclear antigen**. It is a **DNA polymerase accessory factor** that has a regulatory role in DNA replication (which occurs in S phase). PCNA is also instrumental in DNA repair, particularly through **nucleotide excision repair**, which is discussed further elsewhere. Curiously, the p21 protein appears to inhibit PCNA-dependent S phase DNA synthesis while not inhibiting PCNA-dependent nucleotide excision repair. Thus, when p53 activates p21, one of many effects is the halting of DNA synthesis, while simultaneously promoting DNA damage repair.

11.24 Proliferating cell nuclear antigen

Proliferating cell nuclear antigen, or **PCNA,** is a protein involved in DNA synthesis. PCNA encircles DNA to act as a base or scaffold onto which other enzymes involved in DNA repair, replication, and remodeling attach. This attachment to a DNA molecule is called **DNA clamping**; thus, PCNA is a **DNA clamp protein**.

Under normal physiological conditions, PCNA is **sumoylated** (which is the attachment of SUMO groups to the parent molecule; SUMO stands for small ubiquitin-like modifier proteins). When DNA is damaged by radiation or chemicals, the SUMO molecule is replaced by a different molecule called ubiquitin. Monoubiquitinated PCNA recruits DNA polymerases that can carry out DNA synthesis even on damaged DNA; but this process is very error-prone, possibly resulting in the synthesis of mutated DNA. Polyubiquitination of PCNA on the other hand allows it to engage in a less error-prone mutation bypass known as the **template switching pathway**. The template switching pathway allows DNA synthesis to continue even in the presence of DNA damage by utilizing a newly synthesized undamaged sister strand as a template.

11.25 p53, p21, and the G1 checkpoint

When p21 is complexed with cdk2, the cell cannot proceed to the next stage of cell division. From these observations, it is not surprising that irradiation of fibroblasts *in vitro* induces a p53- and p21-dependent cell cycle arrest, and this is probably mediated via cyclin E/Cdk2 complex inactivation, thereby causing a pause during G1. Specifically, the cell cycle is stalled at late G1. This is known as the G1/S **checkpoint** or the G1 **cell cycle restriction point**. Mutant forms of p53 might not effectively stimulate production of p21, and therefore the p21 protein would not be available to act as a 'stop signal' for cell division. Thus, cells with certain p53 mutations divide uncontrollably and form tumors.

11.26 Gadd45

Gadd45, like p21, is turned on by p53. Gadd stands for **growth arrest and DNA damage**. Gadd45 is one of the protein gene products of the *GADD* family of genes, which now includes *GADD45A, GADD45B,* and *GADD45G.* Curiously, overexpression of *GADD45* in the nervous system of fruit flies substantially prolongs lifespan. The likely mechanism is improved recognition and repair of DNA damage caused by radiation and various external causes, along with the spontaneous DNA damage that is constantly caused by oxidative metabolism. As one would predict, *GADD45* is strongly induced by a host of DNA damaging agents, including ionizing radiation.

The growth inhibiting function of the Gadd45 protein is due to its binding to and suppression of the cyclin B1/Cdk2 kinase activity in G2. Normally, the cyclin B1/Cdk2 complex plays a crucial role in regulating the progression of cells from the G2 phase of the cell cycle into mitosis (i.e. at the **G2/M checkpoint**). Cells with up-regulated cyclin B1/Cdk2 activity can enter mitosis, whereas cells with suppressed cyclin B1/Cdk2 activity due to Gadd45 binding remain arrested in G2. The various isoforms of the Gadd45 protein interact with numerous other proteins involved in the cellular stress and DNA damage repair pathway, such as PCNA, p21, MEKK4, and p38 MAP kinase.

11.27 Functions of p53 and the impact of *TP53* mutations

As mentioned, DNA repair is regulated by various sensing mechanisms, which detect DNA damage and activate a host of repair pathways. Among these repair pathways is activation of the p53 protein. It is important to recognize that radiation exposure may lead to a variety of different outcomes across different tissues and sometimes even within the same tissue type. This diversity of radiation response is partially thanks to the various functions of p53. For instance, in gastrointestinal epithelial cells and hematopoietic progenitors in the bone marrow, activation of p53 by radiation triggers the intrinsic pathway of apoptosis. However, in many other cells, activation of p53 by radiation does not lead to apoptosis.

In very general terms (since the specifics vary substantially from tissue to tissue), *if* the DNA damage can be repaired, p53 activates genes to halt the cell cycle and repair the damage. If the DNA damage is deemed too extensive and cannot be repaired, p53 protein prevents the cell from dividing and directs it to undergo apoptosis as a last resort. By preventing cells with badly mutated or damaged DNA from dividing and propagating, p53 helps avert malignant transformation (i.e. the development of tumors). Thanks to these and many more functions, p53 was honored with the prestigious Science Magazine Molecule of the Year award in 1993.

TP53 is one of the most frequently mutated genes in human cancer. Among its many roles, its main biological function is arguably the protection of DNA integrity. The *TP53* gene is considered a **tumor suppressor gene**. The tumor suppressor function of p53 is based on its abilities to induce transient cell cycle arrest to evaluate DNA integrity and repair it when possible, promote apoptosis if the damage is too severe, and also induce a permanent form of growth arrest known as cell **senescence** under certain circumstances. Researchers have recently discovered various other functions of p53 protein including **autophagy, ferroptosis,** immune system regulation, and global regulation of gene expression and micro-RNA species (miRNAs). The p53 protein is also involved in pathways leading to the generation of reactive oxygen species.

The p53 protein has several main domains including the trans-activation domain, the DNA binding core domain, the regulatory domain, the tetramerization domain, and a proline-rich region. The p53 protein binds to the **promoter** region of numerous genes to activate transcription. p53 binds to these DNA sequences via its **transcription-activation domain**, which is also known as **activation domain 1**. Upon binding to the gene promoters, various **transcription factors** are expressed. As the name indicates, transcription factors then stimulate the transcription of other genes (and by definition, p53 itself is a transcription factor). The p53 transcription-activation domain binds to a special sequence of DNA base pairs on the various target genes called the **p53 response element (p53RE)**. Although there are several variations on the theme, p53REs tend to have two repeats of a decamer motif: 'RRRCWWGYYY', separated by a spacer of 0 to 13 base pairs. Here 'R' represents purines (adenine or guanine), 'Y' represents pyrimidines (thymine or cytosine), and 'W' represents adenine or thymine.

It has long been assumed that p53 works *exclusively* as a **transcription factor** that turns on other genes that then execute activities, but in some of these newly identified pathways, it appears that p53 does not function as a transcription factor, but rather it directly interacts with other proteins to carry out these functions. This direct interaction of p53 to effect its functions may hold not just in these newly identified roles but also in some of its well-known traditional functions. For example, in activating the apoptosis pathway, initiation may occur via the direct interaction of p53 with anti-apoptotic proteins normally sequestered in the mitochondria.

11.28 Dominant negative p53 mutants

Unlike **oncogenes,** which derail cell cycle regulation by molecularly dominant gain of function (activating) mutations, tumor suppressor genes must be inactivated in order to cause trouble. In other words, they operate via loss of function and thus are molecularly recessive. Therefore, one would expect that, since the human genome is diploid, with two copies of any given gene (one maternally derived and the other paternally derived), that both alleles of the *TP53* gene must be inactivated to observe a deleterious biological effect. This pattern is observed for most p53 mutations involving the DNA binding domain. Nevertheless, there are some paradoxical **dominant negative** mutations in *TP53* that have deleterious effects even when present in only one chromosome. Such dominant negative mutations work by altering the way the p53 molecule interacts with itself.

Functional p53 protein is a tetramer. During assembly of this tetramer ('oligomerization'), if the tetramer incorporates just one component that is mutated in the tetramerization or **oligomerization domain** (OD domain), the whole molecule will malfunction, and activation of transcription is inhibited. Thus, OD mutations in the p53 oligomerization domain demonstrate a dominant negative effect. Only one copy of the gene needs to be mutated to affect the overall function in such unusual dominant negative circumstances.

11.29 Inherited mutations in *TP53*

Somewhat confusingly, the **Li-Fraumeni syndrome**, which is due to an inherited germline mutation in *TP53,* exhibits an **autosomal dominant** pattern of inheritance. In other words, inheriting a single defective *TP53* gene from the mother or father is sufficient to yield the affected phenotype (which in this situation is the development of cancer). One would normally expect autosomal dominant inheritance to be caused by a gain of function mutation rather than a molecular loss of function. Logically, a loss of function mutation would be expected to follow a recessive pattern of inheritance. This is because in the sporadic (non-hereditary) situation, normal cells have *two* functional copies of a given wild-type (i.e. non-mutated) tumor suppressor gene—one from each parent. In the process of tumorigenesis, one copy of a given tumor suppressor gene like *TP53* might acquire an inactivating mutation or be deleted. This is termed the first genetic 'hit'. But this first hit is insufficient for creating cancer and may have no obvious effects on cell function since the second copy of the tumor suppressor gene is still present and functioning. Tumor suppressor genes thus generally behave as recessive genes; they require both copies to be mutated for manifestation of loss of function. Only when the second hit occurs is the cell completely without that given tumor suppressor gene's function. Cells that have taken the second hit, and now harbor two mutated (or deleted) copies of the tumor suppressor gene, have total loss of function of the tumor suppressor gene's activity. This total loss of function contributes to carcinogenesis and is known as the **Knudson two-hit hypothesis**.

A classic example of the Knudson two-hit hypothesis is provided by the **Vogelstein model** of familial colon cancer progression. In this model, germline mutations in the **familial adenomatous polyposis gene** (*APC*, which is on chromosome 5q21) are inherited as the first hit, and subsequent somatic mutations (most often deletions) in the second copy of *APC* constitute the second hit. When both copies of this gene are inactivated, colorectal cancer development may ensue (after several additional steps are taken). This colonic adenoma-to-carcinoma progression is perhaps the best example of tumorigenesis involving a two-hit tumor suppressor gene as the initiator of a complicated pathway. Incidentally, this general pathway is not only followed by patients with familial polyposis; sporadic colorectal cancers often follow similar steps along the way from polyp to cancer.

The first hit in a tumor suppressor gene can be inherited in **cancer predisposition syndromes** that make patients susceptible to tumors such as the Li-Fraumeni syndrome or **hereditary retinoblastoma**. Some others are listed in the table (table 11.1). The exhibited autosomal dominant inheritance is not due to the above-mentioned occasional 'dominant negative' mutations in the gene. Rather, it

Table 11.1. Table of some autosomal dominant cancer predisposition syndromes.

Tumor Suppressor Gene	Syndrome	Cancers manifested	Protein gene product	Putative protein function(s)	Caretaker or Gatekeeper
BRCA1 or *BRCA2*	Hereditary breast and ovarian cancer syndrome (HBOC syndrome)	Female breast cancer, male breast cancer, ovarian, prostate, pancreatic	BRCA 1 protein BRCA 2 protein	Involved in repair of dsDNA breaks	Caretaker
TP53	Li-Fraumeni syndrome	Breast, soft tissue sarcoma, osteo sarcoma, leukemia, brain tumors, adrenocortical carcinoma and several others	p53 protein	DNA repair Cell cycle arrest Induction of apoptosis (And more)	Both
PTEN (**phosphatase** and **tensin** homolog)	Cowden syndrome (**PTEN hamartoma tumor syndrome**)	Breast, thyroid, endometrium (uterine lining),	PTEN protein	A lipid phosphatase that down regulates phosphatidylinositol-3,4,5- trisphosphate (PIP3) and thus regulates the Akt/PKB pathway	Gatekeeper
MLH1, MSH2, MSH6, PMS2, EPCAM	**Lynch syndrome** (also called **hereditary nonpolyposis colorectal cancer or HNPCC**)	Colorectal, endometrial, ovarian, renal pelvis, pancreatic, small intestine, liver, biliary tract, stomach, brain, breast	DNA mismatch repair enzymes	Repair of Watson-Crick DNA base-pair mismatches	Caretaker

Gene	Syndrome	Tumors	Protein	Function	Type
RB	Hereditary Retinoblastoma	Retinoblastoma, pinealoma, osteosarcoma, melanoma, soft tissue sarcoma	Rb (retinoblastoma) protein	Inhibits cell cycle progression from G1 to S phase in the cell cycle by binding to the E2F enzyme	Gatekeeper
VHL	Von-Hippel Lindau syndrome	Renal cell carcinoma, pheochromocytoma, hemangioblastomas	pVLH	E3 ubiquitin ligase that marks damaged proteins for degradation	Caretaker
MEN1	MEN 1(Multiple endocrine neoplasia syndrome type 1 or Wermer syndrome)	Pancreatic endocrine tumors, parathyroid, pituitary gland tumors	Menin	Uncertain, but appears to be involved in regulating cell division, DNA repair, and apoptosis	Both?
RET	MEN 2(Multiple endocrine neoplasia syndrome type 2)	Medullary thyroid cancer and pheochromocytoma	RET protein	Ret protein is a receptor tyrosine kinase for extracellular signaling molecules of the GDNF family	Gatekeeper
APC	Familial adenomatous polyposis	Colorectal cancer, small intestine, brain, stomach, bone, skin	APC protein	A negative regulator of beta-catenin which is involved in cellular adhesion	Gatekeeper

is because, if all cells in the patient's body are handicapped with one defective *TP53* allele since birth, the odds are quite high that someday, somewhere, one of the trillions of actively dividing cells in that person's body will acquire another mutation that will make inactive the other *TP53* alleles over the person's lifetime. Thus, the observed inheritance pattern appears to be autosomal dominant even though on a molecular level, both maternal and paternal copies of the *TP53* must be mutated to demonstrate the phenotype. Finally, another confusing terminology involves the phenomenon of **loss of heterozygosity (LOH)**. LOH alludes to the situation in which a cell is left with no normal functioning copies of an important gene such as *TP53*. In such situations, a heterozygous somatic cell might become homozygous for a mutant or deleted gene because the one remaining normal copy of the two alleles becomes mutated or is lost. Nevertheless, it is called loss of heterozygosity rather than gain of homozygosity.

11.30 The role of p53 in the G1 cell cycle checkpoint

Upon the binding of p53 to DNA, another gene is activated and transcribed into a protein called **p21** (also known as **CDK inhibitor 1** or **CDKI 1**). The p21 protein then interacts with a cell division-stimulating protein, **cdk2** (among others).

Recall that **cyclins** are a group of proteins associated with the cell cycle and initiation of mitosis. Cyclins control progression of a cell through the cell cycle by activating other proteins called **CDK enzymes** or cyclin-dependent kinases. When p21 is complexed with cdk2, the cell cannot pass through to the next stage of cell division. Specifically, the cell cycle is stalled at late G1. This is known as the **G1 checkpoint** or the **cell cycle restriction point**. Mutant p53 might not effectively stimulate production of p21, and therefore the p21 protein is not available to act as a 'stop signal' for cell division. Thus, affected cells may divide uncontrollably and form tumors.

11.31 p53 activation and regulation

The *TP53* gene is indirectly activated by radiation, and the p53 protein product is stabilized after radiation exposure by phosphorylation. The phosphorylation prevents mdm2-mediated degradation of p53 (which is discussed below). Given its myriad critical functions, the cellular concentration of p53 must be closely controlled. For example, while it can suppress tumors, an inappropriately high level of p53 may accelerate the aging process by excessive apoptosis. It appears that p53 regulation occurs at the transcriptional, post-transcriptional, and post-translational levels. Transcriptional regulation affects gene expression and post-translational regulation affects the function of the protein. But post-transcriptional regulation seems odd and inefficient since the DNA has been translated into mRNA. Nevertheless, it gives the cell 'one last chance' to turn off certain genes. This gene suppression is effected by microRNA molecules or miRNAs. miRNA molecules are small (averaging 22 nucleotides), non-coding, single-stranded RNA molecules that bind to specific mRNA molecules and actively destroy it by cleavage or destabilize it by shortening the poly(A) tails and allowing degradation. In any case, the process of p53 mRNA translation into p53 protein may be halted through miRNA mechanisms. At the transcriptional level, wild-type (that is, normal) p53 binds directly at **p53 response elements** on the DNA to regulate its own gene expression.

The main regulator of translated p53 protein is **Mdm2** (for **murine double minute 2**); mdm2 can trigger degradation of p53 by the **ubiquitin** system, which is reviewed below. Mdm2 itself is classified as an E3 ubiquitin ligase.

11.32 Mdm2

It turns out that the functional activity of p53 is diminished in nearly all tumors, either by mutations in the *TP53* gene itself, epigenetic modifications that decrease gene expression, or by molecular interactions that decrease the activity of the wild-type protein, such as an overabundance of the p53 repressor Mdm2. The gene product of the *MDM2* gene is a nuclear-localized **E3 ubiquitin ligase** (which is explained in greater depth below) that targets p53 protein for **proteasomal degradation** (i.e. dismantling of proteins in organelles called **proteasomes**). Unsurprisingly, given its role in incapacitating p53, overexpression or amplification of the *MDM2* gene locus has been observed in a variety of different human cancers. The *MDM2* gene is itself transcriptionally regulated by p53 in a negative feedback loop—high levels of p53 protein turn on the *MDM2* gene, thereby promoting degradation of the excess p53.

11.33 Post-translational modifications of p53 protein

The regulation of p53 protein is quite intricate with various post-translational chemical modifications known. Such modifications include:

1. Acetylation
2. Methylation
3. Phosphorylation
4. Neddylation
5. Sumoylation
6. Ubiquitination (also known as ubiquitylation or ubiquitinylation).

All of these covalent modifications of p53 have important downstream effects on p53 stability and activity as a transcription factor. Naturally, given the central role of p53, the spatial and temporal overlap of these p53 chemical modifications can profoundly impact cellular fate. Here we shall address just two post-translational modifications of p53, **ubiquitination** and **neddylation**.

11.34 Ubiquitination

Ubiquitination or **ubiquitylation** is the most well-studied pathway of molecular disassembly and degradation. In this process, the **ubiquitin peptide** (or small 'protein' depending on semantics; its molecular weight is 6.7 kDa and consists of 76 amino acids) becomes attached to p53 or another substrate protein. This process most commonly binds the last amino acid of ubiquitin (glycine 76) to a lysine residue on the substrate. **Deubiquitinating enzymes** oppose ubiquitination by removing ubiquitin from substrate proteins.

Ubiquitylation affects proteins in many ways, including changing their cellular location, altering their activity, and promoting or preventing various protein interactions. Here we shall focus only on the role of ubiquitination in marking them for degradation within proteasomes (the cellular equivalent of a trash disposal unit). Ubiquitination is a highly conserved pathway for the orchestrated and regulated degradation of cytosolic, nuclear, and membrane proteins in all eukaryotes. The first step, activation, involves creation of a bond between a cysteine of the **ubiquitin activating enzyme (E1)** and the carboxyl terminus of ubiquitin. The activated ubiquitin is next transferred to a **ubiquitin conjugase (E2)**. The final step is conjugation or linkage to a lysine residue within the target substrate (or to the N-terminal amino group) by an enzyme called a **ubiquitin ligase (E3)**. Additional ubiquitin molecules may be added onto the first, forming a **polyubiquitinated** adduct. This polyubiquitinated product is recognized by the **26S proteasome**, which then degrades the substrate in the proteasome. As Mdm2 is an E3 ubiquitin ligase, p53 is degraded in this manner via 26S proteasomes.

11.35 Proteasomes

Simply described, the 26S proteasome is a large (~2.5 MDa), protease complex that serves as the final degradation compartment of the ubiquitin system. It consists of two subunits, the **20S core particle** and a **19S regulatory particle**. 'S' here stands for **Svedberg units**, which are the sedimentation coefficients in ultracentrifugation and correlate to a particle's size, shape, and density. Note that Svedberg units do not add up arithmetically. Students might remember the similar situation in which 80S eukaryotic ribosomes are composed of 40S and 60S subunits. In man and all eukaryotes, proteasomes are found both in the cytoplasm and in the nucleus.

Collectively, the ubiquitin and proteasome degradation pathways are called the **ubiquitin-proteasome system** or **UPS**. Dysfunction of the UPS has reportedly been linked to several neurological diseases including Alzheimer's disease, amyotrophic lateral sclerosis, Parkinson's disease, Huntington's disease, and transmissible spongiform encephalopathies (prion-related neurodegenerative diseases such as mad cow disease (bovine spongiform encephalopathy), Creutzfeldt-Jakob disease and fatal familial insomnia).

11.36 NEDDylation of p53 and other proteins

NEDDylation or neddylation is another posttranslational modification of p53. NEDDylation diminishes p53 action by inhibiting the transcription factor functions of p53. NEDDylation is achieved by reversibly adding the small ubiquitin-like molecule **NEDD8** ('neuronal precursor cell-expressed developmentally down-regulated protein 8') to a lysine residue on p53 or other protein targets. Mdm2 and FBX011 promote p53 NEDDylation and inactivation of p53

Recent studies have highlighted the role of NEDDylation in p53 regulation and tumorigenesis. Inhibition of various neddylation pathways, in addition to directly acting on tumor cells through p53 manipulation, also affect components of the **tumor microenvironment**. Real tumors *in vivo* are not simply masses of tumor cells; they also include various immune cells (e.g. macrophages, neutrophils, and lymphocytes), cancer-associated fibroblasts, and cancer-associated endothelial cells along with their chemical products (cytokines). All of these players in the tumor microenvironment are crucial for tumorigenesis. Thus, there is increased attention being devoted to the role of NEDDylation in cancer biology and treatment. NEDDylation might also be involved in the pathogenesis of Alzheimer's disease where its activation appears to drive neurons into programed cell death (apoptosis).

11.37 Gatekeepers and caretakers

Tumor suppressor genes can be divided into two major categories: **caretakers** and **gatekeepers**.

1. Caretaker genes are responsible for maintaining the molecular integrity of chromosomal DNA. Thus, they promote stability of the genome. Caretakers include genes that make enzymes involved in DNA repair.
2. Gatekeeper genes inhibit cell growth or induce apoptosis.

Caretaker genes basically prevent genomic instability. Therefore, mutations in caretaker genes can, in principle, lead to accelerated acquisition of new mutations and conversion of a normal cell to a cancer cell. Several caretaker genes are required for proper maintenance of genome integrity.

A classic set of caretakers are the genes involved in **DNA mismatch repair** (**MMR**). These genes code for enzymes that are responsible for identifying and correcting mismatches in Watson–Crick base pairs (i.e. when bases are paired other than A-T or G-C). Among these MMR genes are *MLH1* (MutL homologue 2), *MSH2* (MutS homologue 1), *MSH6*, and *PMS2* (postmeiotic segregation increased homolog 2). As listed in the table of familial cancer predisposition syndromes (table 11.1), germline mutations in these genes is associated with Lynch syndrome. Additionally, deletions in a non-mismatch repair gene *EPCAM* (epithelial cellular adhesion molecule) is also known to be associated with Lynch syndrome. *EPCAM* functions to silence *MSH2* expression.

When the DNA mismatch repair pathway is inactivated either by mutation or epigenetic silencing, **microsatellite instability** or **MSI** manifests. **Microsatellites** are repeated sequences of DNA consisting of repeating units up to six base pairs in length. The most common microsatellite in human DNA is a simple dinucleotide repeat of cytosine and adenine nucleotides (CA). This dinucleotide may be seen tens of thousands of times across the human genome. Astonishingly, these non-functional microsatellites make up approximately 3% of the human genome. Although the length of these microsatellites varies from person to person (and contributes to an individual's 'DNA fingerprint'), each individual typically has microsatellites of a specified length. This length is usually constant within all the cells of that individual.

MSI is seen as abnormal variability in the lengths of a given individual's microsatellites from cell to cell and is due to DNA replication errors.

MSI is taken as cytogenetic evidence that the DNA MMR system is malfunctioning. The result is **genetic hypermutability** or a '**mutator phenotype**'. Hypermutability due to failure of DNA MMR results in cells that are prone to multiple mutations, which in turn predisposes a person to cancer. Germline mutation in any of the MMR genes is the cause of Lynch syndrome or **hereditary non-polyposis colorectal cancer syndrome**. (Note: 'non-polyposis' simply means the patient does not have a 'polyposis syndrome' in which up to many thousands of adenomatous polyps develop within the patient's colon. It does not mean that the cancers that such individuals get do not arise in polyps. As in sporadic cancers, they typically do.)

11.38 Hoogsteen base pairing and G-quadruplexes

Just for thoroughness, it may be mentioned that Watson–Crick base pairing is not the only form of base pairing observed in eukaryotes. **Hoogsteen base pairing** also links adenines to thymines and guanines to cytosines, but the geometry is quite different. It is characterized by a 180 degree rotation of the involved purine (adenine or guanine) in the base pair compared to the the typical orientation of purines in Watson–Crick base pairs. Hoogsteen base pairing allows integration of a third strand into the classic B-DNA double helix to create triple strand helical structures. Hoogsteen base pairs also permit formation of an unusual secondary structure of guanine-rich DNA called the **G-quadruplex** (also called G4-DNA or the G quartet because four guanine residues are associated together in these structures in a guanine tetrad). Such structures might be instrumental in regulating gene transcription as they are seen in transcriptional regulatory regions. G-quadruplexes have also been observed at **telomeres** (the ends of chromosomes). The presence of G-quadruplexes in telomeres seems to decrease the activity of the enzyme **telomerase**, which is responsible for maintaining the length of telomeres (and is active in around 85% of human cancers, thereby granting them immortality). Their presence at telomeres and in regions regulating oncogene expression in humans has led to the hope that G-quadruplexes might someday serve as targets for cancer therapy.

11.39 Gatekeeper tumor suppressor genes

In contrast to the caretakers who maintain the integrity of DNA, the gatekeepers regulate the cell cycle. Since uncontrolled proliferation is one of the key characteristics of cancer cells, the inactivation of gatekeepers is obviously involved in tumorigenesis. The classic example in this category is *RB*, the gene responsible for the pediatric eye tumor called retinoblastoma. The gene product of *RB* is Rb protein, which regulates the cell cycle by binding to another protein called E2F. E2F promotes the transition from late G1 into S phase. In other words, cells remain at the G1 checkpoint or restriction point until freed by E2F. However, like a brake Rb binds to and inhibits the function of E2F until the proper signals are received. Those cell cycle progression signals include activation of cyclin D/cdk4. This cyclin/cdk complex phosphorylates Rb. Upon phosphorylation, Rb releases E2F, which can then go about its business of stimulating cell cycle progression. (This is the classic function of Rb but it can bind to over 100 other proteins and probably has numerous presently unknown additional functions.)

Obviously, if this tightly regulated system is disrupted, cells may not remain at the G1 checkpoint for the amount of time required for DNA evaluation and repair; they may rush into S phase and replicate damaged DNA, thereby leading to mutation accumulation. Thus, germline mutations in *RB* lead to cancer (hereditary retino-blastoma) in most carriers, but sporadic mutations in *RB* are associated with various other cancer types including osteosarcoma, breast cancer, and small cell lung cancer.

Interestingly, certain cancer-causing viruses such as the HPV16 and HPV18 subtypes of the human papilloma virus (HPV) target Rb protein and degrade it, thereby disrupting cell cycle control. Specifically, the **E7 protein** of the cervical cancer–associated HVP16 and HPV18 subtypes binds to and promotes degradation of Rb. Another HPV protein called **E6** specifically binds to and degrades the p53 protein.

Another gene mutation that disrupts regulation of the Rb/E2F pathway involves the $p16^{INK4a}$ cyclin kinase inhibitor (usually just called **p16**). The molecular weight of the p16 protein is about 16 kilodaltons, just as the molecular weight of p53 is around 53 kilodaltons. The gene for this protein is ***CDKN2A***. As mentioned, Rb is activated by cyclin D/cdk4; the p16 protein controls cyclin D/cdk4 activity. Hence, loss of p16 activity leads to hyperactivity of cyclin D/cdk4, which leads to Rb phosphorylation, E2F release, progression through the G1 restriction point, and cellular proliferation. Therefore, the loss or diminution of of p16 activity is functionally equivalent to the loss of Rb. Inherited mutations of the $p16^{INK4a}$ gene have been associated with familial cases of melanoma, but sporadic mutations are often found in esophageal cancer, oropharyngeal cancer, cervical cancer, as well as non-familial cases of melanoma.

Numerous other examples of gatekeeper tumor suppressor genes are known, and some are included in table 11.1 on autosomal dominant cancer syndromes.

11.40 DNA damage repair deficiency syndromes

It was mentioned that caretaker tumor suppressor genes are involved in maintaining the integrity of the genome. With some of these caretakers, when just a single allele of these genes is mutated in germ cells, the individual who has inherited the defective gene has a high probability of manifesting the phenotype (that is, developing cancer). Hence, many of these cancer predisposition syndromes are passed along in what appears to be an autosomal dominant fashion, with less than 100% penetrance. (**Penetrance** is just the probability that a given gene will be phenotypically expressed.) However, several other genetic syndromes may be passed along in an autosomal recessive manner. These syndromes are similarly associated with care-taker tumor suppressor genes, but the individual must have two defective alleles to manifest the disorder. Some of these syndromes are associated with a marked cellular sensitivity to radiation and other DNA-damaging agents. Such conditions include **xeroderma pigmentosum, ataxia telangiectasia, Seckel syndrome, Nijmegen** breakage syndrome, **Bloom syndrome**, **Werner syndrome**, **Rothmund–Thompson syndrome**, **Cockayne syndrome**, and **Fanconi anemia** among others. Our understanding of the various enzymes and pathways involved in DNA damage repair is largely thanks to patients with these rare genetic disorders.

Patients with these autosomal recessive conditions have inherited deficiencies in key DNA repair genes or genes participating in the signaling pathways that are activated by DNA damage. This manifests as chromosomal instability and a pronounced predisposition to malignancy. For instance, patients with **xeroderma pigmentosum** are exceedingly sensitive to solar ultraviolet and exhibit an extreme predisposition to skin cancer. Because they are deficient in the **nucleotide excision repair pathway**, patients with xeroderma pigmentosum are unable to effectively address the pyrimidine dimers in their DNA routinely caused by ultraviolet radiation. Ultraviolet induces the formation of specific photoproducts in DNA called **cyclobutane pyrimidine dimers**, most often adjacent **thymine dimers**. The nucleotide excision repair pathway is covered later in this chapter.

Bloom syndrome is caused by mutations in the *BLM* gene, which belongs to the so-called RecQ DNA helicase family, which is highly conserved from *Escherichia coli* to humans. Patients with Bloom syndrome typically have short stature, a malar rash (red coloration across the cheeks) due to their sensitivity to sunlight, mild immunodeficiency, and a high risk of developing cancer (especially gastrointestinal tract cancers, leukemias, and lymphomas). Bloom syndrome appears to be associated with a shorter lifespan, with an average life expectancy of less than 30 years (mostly due to fatal cancer). The cells from patients with Bloom syndrome display a higher frequency of sister-chromatid exchanges, which can lead to genomic instability, chromosomal aberrations, loss of heterozygosity, and increased cancer risks.

Besides BLM, two other human RecQ helicase gene mutations are involved in the genomic instability diseases, Werner syndrome and Rothmund–Thomson syndrome. **Werner's syndrome,** or **adult-onset progeria**, is a rare autosomal recessively inherited genetic disorder that is characterized by premature aging, increased somatic mutation rates, and chromosomal aberrations. Unlike other progeria syndromes, Werner's syndrome is not evident during childhood and infancy, thus the name adult-onset progeria. It is linked to mutations in the gene called *WRN*, which encodes a **DNA helicase-endonuclease enzyme**. The associated deficiency in DNA repair capability confers an increased sensitivity to the effects of ionizing radiation. Like Hutchinson–Gilford syndrome, xeroderma pigmentosum, and Cockayne Syndrome, Werner's syndrome is a 'segmental progeria', meaning that patients exhibit some, but not all, aspects of aging. In people with Werner's syndrome, there is an unusually high incidence of soft tissue sarcomas, although other malignancies including skin, thyroid, colorectal, and pancreatic cancers, along with meningiomas and leukemia, have been reported.

Hutchinson–Gilford syndrome (or HGPS for Hutchinson–Gilford progeria syndrome) is a rare, genetic disorder that is caused by a point mutation in the gene on chromosome 1 known as *LMNA*, which makes a protein called **lamin A**. Lamin A is needed for holding the cell's nucleus together. Lamin A is a key component of the double-layered nuclear membrane (nuclear envelope). The abnormal lamin A protein produced in HGPS is called **progerin**. In addition to the diminished integrity of the nuclear envelope, there is abnormal epigenetic modification seen in patients with Hutchinson–Gilford syndrome. Among several epigenetic alterations is the abnormal methylation of histone H4. **H4K20me** is the methylation of the 20th lysine residue of the histone H4 protein. It is usually trimethylated and abbreviated as H4K20me3. It is critical for genome integrity and functions in DNA damage repair, DNA replication, and chromatin compaction. H4K20me3 is upregulated in HGPS cells. Interestingly, H4K20me3 marks telomeric heterochromatin, and an increase in H4K20me3 inhibits telomere elongation. (Recall that telomeres are chromosomal clocks that grow shorter after each cell division, thereby marking the age of a cell lineage.) The increased expression of H4K20me3 is consistent with the reported telomere dysfunction and accelerated senescence seen in the cells of patients with Hutchinson–Gilford syndrome.

Fanconi anemia is another rare hereditary chromosomal instability and radiation sensitivity syndrome. Patients with Fanconi anemia are prone to bone marrow failure (manifesting as pancytopenia or aplastic anemia) along with certain cancers (especially acute myeloid leukemia, hepatic tumors, and head and neck squamous cell cancers). Patients with Fanconi anemia have a 28% cumulative incidence of solid cancers by the age of 40. They typically exhibit short stature, hypoplastic thumbs, cardiac/renal abnormalities, and cutaneous café-au-lait spots. They also have high sensitivity to DNA crosslinking agents and ionizing radiation. Fanconi anemia patients have been found to experience a high rate of complications with radiation therapy. Unlike Bloom syndrome and Hutchinson–Gilford syndrome, which are caused by mutations in a single gene, mutations in over 20 genes (*FANC-A, B, C, D1, D2, E, F, G, I, J, L, M, N, O, P, Q, R, S, T, U, V, and W*) have been identified in Fanconi anemia. These *FANC* **gene products** act at various steps in the so-called **Fanconi anemia DNA damage response pathway** (or just the **FA pathway**), which serves to repair **DNA inter-strand crosslinks**, which stall DNA replication at replication forks. Such crosslinks can only be repaired during the S phase of the cell cycle in coordination with DNA replication. Genes in this pathway are involved in inter-strand crosslink and double-strand DNA break repair (DSB repair). Of note and clinical importance, the two breast cancer susceptibility genes, *BRCA1* and *BRCA2*, are involved in the FA pathway and are also known as *FANCS* and *FANCD1*, respectively. Hence, the FA pathway is sometimes called the **FA-BRCA pathway**. This key DNA repair pathway, along with other DNA repair processes such as homologous recombination, nucleotide excision repair, trans-lesion synthesis, and alternative non-homologous end joining (also known as microhomology-mediated end joining), forms an intricate DNA damage repair network (far beyond just the core inter-strand crosslink repair components) to correct diverse DNA lesions.

Ataxia telangiectasia is an autosomal recessive genetic disorder characterized by progressive cerebellar degeneration (which causes the ataxia), telangiectasia, immunodeficiency (often leading to recurrent sinopulmonary infections), radiation sensitivity, premature aging, and a strong tendency to develop cancer. Patients with ataxia telangiectasia (also called the **Louis–Bar syndrome**) have mutations in the ATM gene, which confers on them a very high incidence of lymphomas, often before the age of 20. In addition to being a cancer predisposition syndrome, ataxia telangiectasia may be considered a genome instability syndrome, a chromosomal instability syndrome, a DNA repair disorder, a DNA damage repair syndrome, and a radiation sensitivity syndrome. Ataxia telangiectasia is a complex disease and not all patients have the same collection of symptoms or laboratory findings (e.g. telangiectasia are not present in all individuals despite the name of the disease). This is an example of what is known as **variable expressivity** in medical genetics. The *ATM* gene codes for a serine/threonine protein kinase (called ATM protein), which is activated by double-strand DNA breaks. Thus, its activation may be an early response to radiation damage. Upon activation, ATM protein phosphorylates and activates critical DNA repair proteins involved in the DNA damage checkpoint, including p53. This, in turn, leads to either cell cycle arrest and DNA repair or apoptosis. ATM phosphorylates and activates target proteins other than just p53, however, including Chk2, BRCA1, NBS1, and H2AX. While deficiency of ATM leads to ataxia-telangiectasia, deficiency of ATR leads to another autosomal recessive DNA damage repair disorder known as Seckel syndrome.

11.41 ATM

The cell cycle has several DNA damage checkpoints during which cell cycle progression halts while the cell surveys for DNA damage and addresses it. The two main checkpoints are the G1/S and the G2/M checkpoints. The ATM protein plays an important role in cell cycle delay after DNA damage, especially after double-strand DNA breaks. ATM is recruited to sites of DSBs by **double-strand DNA break sensor proteins**, such as the **MRN complex**. The MRN complex is involved in the initial processing of double-strand DNA breaks prior to repair by the two main means of correcting them—**homologous recombination** or **non-homologous end joining**. The MRN complex binds to double-strand DNA breaks and seems to tether the broken ends together before repair by non-homologous end joining or it may initiate '**DNA end resection**' (which is 'shaving back' one of the two DNA strands at the break site so that just a 3' end is left hanging) prior to repair by homologous recombination. Thus, the MRN complex is involved in activation and recruitment of the ATM protein to the sites of DNA damage. After being recruited, the ATM kinase phosphorylates NBS1, along with other DSB repair proteins. These now modified mediator proteins then amplify the DNA damage signal. They transduce the signals to important downstream effectors such as CHK2 and p53. Activation of ATM is initiated by conformational changes induced by the MRN complex. This, in turn, causes autophosphorylation of ATM. ATM normally resides in cells as dimers; autophosphorylation dissociates the dimers into active monomers, which carry out the various functions.

11.42 γH2AX

Among the early but evident (and measurable) changes following cellular exposure to radiation is the modification of certain histone proteins. Recall that eukaryotic DNA is a double helix molecule that is then wrapped around a core of histone molecules in structures called **nucleosomes** (with 146 base pairs per nucleosome). Chromatin architecture, nucleosome positioning, and access to the DNA for gene transcription is partly regulated by histones. Each nucleosome is made of an octamer of two identical subunits, with four histones apiece: **H2A, H2B, H3, and H4**. The **H1** histone does not form part of the nucleosome but rather acts as the 'linker' histone to stabilize inter-nucleosomal DNA in chromatin. Histone proteins can be subject to post-translational modifications, which affect their interactions with DNA and regulate chromosomal functions. Some histone modifications disrupt histone-DNA interactions and encourage nucleosomes to unwind. This 'open' chromatin conformation is called euchromatin. DNA in **euchromatin** is more accessible to binding of transcriptional molecular machinery and gene activation. In contrast, other histone modifications might reinforce histone-DNA interactions, leading to a more tightly packed chromatin structure called **heterochromatin**. In compact heterochromatin the transcriptional machinery cannot as readily access DNA, resulting in gene silencing. In this manner, modification of histones alters the architecture of chromatin and affects gene activation. Post-translational modifications to histones include acetylation, methylation, phosphorylation, ubiquity-lation, crotonylation, sumoylation, among a few others. Histone modifications make up what is known as the **histone code**, which determines the transcriptional state of DNA in a particular region. The detailed histone modifications in a given segment of chromatin dictates gene activation states. Examination of histone modifications can reveal the locations of gene promoters, enhancers, and other regulatory elements scattered among the 6 billion base pairs in each diploid human cell.

Following radiation exposure, the histone subtype **H2AX** becomes phosphory-lated in response to double-strand DNA damage; this phosphorylated form of histone H2AX is called **γH2AX**. γH2AX is so-named because it was first discovered in cells after they were exposed to gamma rays. Phosphorylated histone H2AX foci can thus be used as a simple assay for DNA DSBs. The phosphorylation is performed by the kinases, ATM, ATR, and DNA-PKcs. γH2AX is a sensitive assay when looking for double-strand DNA breaks in cells. The amount and intranuclear location of γH2AX can be assessed and recorded in order to quantify the biological effects of ionizing radiation. It is important to remember that biological systems, in stark contrast to inanimate matter such as manmade equip-ment and structural support items, are capable of some degree of repair. Therefore, when estimating the potential degree of biological impact of γH2AX levels, the timing is essential. For instance, similarly high levels of γH2AX might be found for two different cell types 15 min after exposure but at 24 h post exposure, there might be vast differences in the amount of γH2AX between these different cells. In addition to γH2AX, other proteins known to form potentially detectable foci in response to DNA damage include ATM, RPA, RAD51, and BRCA1.

11.43 Molecular mechanisms of γH2AX

γH2AX is involved in early steps of DSB repair through homologous recombination as well as non-homologous end joining. In response to double-strand DNA breaks,

H2AX is phosphorylated by ATM, ATR, and DNA-PK to form γH2AX. The histone phosphorylation causes the chromatin to become less condensed. By sterically opening up space, this is believed to assist in the recruitment of the proteins necessary for repair of DSBs. γH2AX is phosphorylated on serine 139. It appears relatively quickly after radiation exposure that caused double-strand DNA breaks; after such irradiation it can often be detected within 20 s of exposure. The half-maximum amount of γH2AX typically appears within one minute with maximal phosphorylation around 30 minutes post-exposure. The number of base pairs associated with this phosphorylated histone product may be extensive. Usually, about two million base pairs surrounding the site of a DNA DSB become associated with γH2AX. The γH2AX does not directly cause chromatin de-condensation. The de-condensation is mediated by a protein known as **RNF8**, which is recruited by γH2AX. RNF8 does this by interacting with **CHD4**, a component of the nucleosome complex called **NuRD**.

γH2AX attracts the mediator of DNA damage checkpoint protein 1 (MDC1) to the damage sites. This γH2AX/MDC1 complex then orchestrates further interactions in DSB repair, including the binding of the ubiquitin ligases RNF8 and RNF168. These then ubiquitylate other chromatin associated proteins, which ultimately encourages the recruitment of either BRCA1 or **53BP1** to the γH2AX/MDC1 chromatin. If BRCA1 joins the MDC1 complex, homologous recombination will be initiated; if 53BP1 joins, non-homologous end joining proceeds. Depending on which pathway is initiated, various other proteins join the action, including the MRN complex (a protein complex of Mre11, Rad50, and Nbs1), RAD51, and ATM. Additional DNA repair proteins (such as RAD52 and RAD54) may rapidly but reversibly interact with these core components.

11.44 Formation of γH2AX

The initial formation of γH2AX is promoted by the phosphorylation of another chromatin associated protein, **heterochromatin protein 1 beta (HP1-beta)**, very shortly after double-strand DNA damage occurs (half-maximum within about one second). This phosphorylation causes HP1-beta to dissociate from chromatin. (It is normally bound to modified histone H3, which has been methylated on lysine 9 (abbreviated as **H3K9me**.) Disengagement of HP1-beta leads to conformational changes in chromatin, and this alteration in chromatin structure promotes H2AX phosphorylation by ATM, ATR, and DNA-PK. Thus, γH2AX is made. Each gray of low-LET radiation causes about 1% of the cellular H2AX to undergo phosphorylation. Analyses have contributed to the conclusion that there are around 35–40 double-strand DNA breaks per cell per gray of radiation.

11.45 DNA damage repair

It has repeatedly been stated that DNA is the most critical target molecule of radiation. This is certainly true when it comes to radiotherapy, where the aim is to kill malignant cells. But even in space radiation and the radiation safety setting, DNA is the most important target molecule. Especially at low and moderate radiation doses (where the damage may be repaired), DNA remains the key molecule. Thus, it is appropriate to review the ways in which DNA can be repaired following damage induced by ionizing radiation or other causes.

11.46 Mismatch repair

Sometimes during DNA synthesis, DNA polymerase will incorporate an incorrect nucleotide during strand synthesis, Normally, the inherent 3' to 5' exonuclease editing system will correct this, but in rare instances it fails. These Watson–Crick base pair mismatches may then be repaired by the DNA mismatch repair (MMR) mechanism. The MMR system also addresses single base insertions and deletions.

When a mismatch such as G-T is encountered, the question arises—how does the cell know if the G or the T is correct? Proper MMR relies on some sort of signal within the DNA to distinguish the original (parental) strand from the daughter strand, which contains the error. This signal (in prokaryotes) is differential DNA methylation of the sequence GATC. DNA replication is semi-conservative, meaning that every newly made double-strand DNA molecule consists of one original strand and one new complementary strand. The new daughter strand remains unmethylated for a brief interval following replication. This difference (methylated old strand versus unmethylated new strand) allows the MMR system to determine which strand contains the error. In *Escherichia coli*, MutS recognizes and binds the mismatched base pair; in man, the equivalent system does this (the MSH proteins (for MutS homologues)). Another protein, MutL (or in humans, the MLH proteins) then bind to MutS forming a MutL/MutS complex. Next, the unmethylated GATC sequence is recognized and bound by an endonuclease, MutH. The MutL/MutS complex then links with MutH, which nicks the unmethylated DNA strand at the GATC site. A DNA helicase called MutU then unwinds the DNA strand in the direction of the mismatch and then an exonuclease degrades the imperfect strand. Finally, DNA polymerase fills in the gap and DNA ligase seals the nick. Eventually, methylation of the GATC sequences occurs some time after DNA replication, after the system has had the chance to identify and correct any mismatched base pairs. The details of how this works in eukaryotic cells are being worked out.

Human cells possess a MMR system similar to that of *E. coli* and, as mentioned, defects in the human DNA MMR genes are associated with the cancer-predisposition syndrome known as Lynch syndrome or hereditary non-polyposis colorectal cancer. The seven identified human DNA MMR genes (or proteins) are MLH1, MLH3, MSH2, MSH3, MSH6, PMS1, and PMS2.

11.47 Base excision repair

The base excision repair (BER) system pathway is responsible for removing small, non-helix-distorting base lesions from DNA. This contrasts it slightly from the nucleotide excision repair (NER) pathway, which addresses bulkier, helix-distorting DNA lesions.

BER is initiated by **DNA glycosylases**, which recognize and remove specific damaged or inappropriate bases. Such bases are usually formed by deamination, oxidation, and alkylation. These chemically modified bases may hydrogen bond differently from the unmodified base, resulting in incorrect base-pairing. Left unaddressed, this may lead to point mutations. For example, guanine should normally base-pair with cytosine (G:C), but an oxidized form of guanine (8-oxoguanine) may base-pair with adenine instead. During DNA replication, this mistake can cause a G:C base pair to become a T:A base pair (a transversion point mutation). The glycosylase action removes the unnatural base, forming an **apurinic/apyrimidinic (AP) site** (also called an **abasic site**). These AP sites are then cleaved and removed by an enzyme called **AP endonuclease** called apurinic endonuclease I or APE1. Next, the created vacancy is filled by DNA polymerase beta (DNA pol β) and the site is joined together by DNA ligase III complexed with XRCC1. The resulting single-strand break is then processed by either 'short-patch BER' (where a single nucleotide is replaced) or 'long-patch BER' (where 2–10 new nucleotides are removed and replaced). Human cells have both options but which one is selected, and why, is still under investigation. Some lesions (such as radiation-induced oxidized or reduced AP sites) are better processed by the long-patch BER approach. Different enzymes are involved in the short-patch and long-patch BER mechanisms. While APE1 is involved in both subpathways, short-patch BER employs DNA pol β) and DNA ligase III-XRCC1, whereas long-patch BER employs a complex containing RFC (replication factor C), PCNA (proliferating cell nuclear antigen) and DNA polymerase delta/epsilon (DNA pol δ/ε). The overhanging 'flap' is then removed by FEN1 (flap endonuclease 1) and then DNA ligase I seals the gap.

Aside from 8-oxoguainine, other examples of base lesions repaired by the BER system include other oxidized bases, alkylated bases (e.g. 7-methylguanine), deaminated purines (such as hypoxanthine and xanthine), and inappropriately placed uracils (uracil is supposed to only be in RNA, not in DNA); uracil can also form *in situ* through deamination of cytosine. In addition to addressing base lesions, the BER pathway is also involved in the repair of single-strand breaks.

The DNA glycosylase gene *MUTYH* is important for making a glycosylase involved in BER. Mutations in this gene are linked to a rare, autosomal recessive disorder called ***MUTYH*-associated polyposis** that significantly increases one's susceptibility to colorectal cancer.

11.48 Nucleotide excision repair

As mentioned earlier, patients with xeroderma pigmentosum have germline mutations in their nucleotide excision repair (NER) genes, which makes them exquisitely sensitive to ultraviolet radiation. Nucleotide excision repair (NER) is an important mechanism for dealing with ultraviolet-induced cyclobutane pyrimidine dimers and bulky helix-distorting chemical adducts on DNA. The NER pathway in human cells is quite complicated with nearly two dozen different genes and proteins involved. Nevertheless, the NER pathway involves several basic steps:

- identification of the damaged DNA,
- making nicks in the damaged DNA strand on each side (5′ and 3′) of the lesion,

- excision of an oligonucleotide (24–32 nucleotides in length) containing the damaged DNA,
- filling in the created gap by DNA polymerase using the undamaged DNA strand as a template, and
- ligation of the DNA.

Defects in some specific proteins lead to certain diseases, and the gene/protein names are associated with the disease. For instance, the various *XP* genes (such as *XPA*, *XPB*, *XPC*, etc) are involved in the radiation sensitivity syndrome **xeroderma pigmentosum**. Similarly, the *CSA* and *CSB* genes are named for the disease they are linked to, **Cockayne syndrome**. Cockayne syndrome is similar to xeroderma pigmentosum in that patients with either disorder are extremely sensitive to ultraviolet radiation and cannot repair cyclobutane pyrimidine dimers made by radiation. *CSA* and *CSB* are also known as *ERCC*8 and *ERCC*6, respectively, which stand for excision repair (as in human DNA excision repair protein 8 or 6). The NER process begins with the formation of a complex of proteins (XPA, XPF, ERCC1, HSSB) at the lesion on the DNA. The multi-subunit transcription factor, TFIIH, then binds to that complex in an ATP-dependent reaction and makes an incision in the DNA. This opens the DNA double helix after damage is initially recognized and allows access for the rest of the repair enzymes. The resulting 24–32 nucleotide segment containing the damaged DNA is then unwound and removed, the gap is filled (by DNA polymerase), and the nick is sealed by DNA ligase. There are two subpathways of NER:

- Global genome repair (GG-NER)
- Transcription-coupled repair (TC-NER).

GG-NER works across the entire genome, regardless of whether the DNA is coding or non-coding. The TC-NER subpathway in contrast exclusively repairs lesions in the DNA of actively transcribed genes. The mechanisms of the two subpathways differ only in the initial detection of the helix-distorting DNA lesion. Thereafter, two subpathways are the same. This was elucidated by the observation that actively transcribed genes undergo NER far faster than regions of the genome not undergoing transcription. It appears that xeroderma pigmentosum is associated with global genomic NER, whereas Cockayne syndrome is associated with transcription-coupled NER (as is yet another photosensitivity syndrome called photosensitive trichothiodystropy). Although mutations in the NER genes confer ultra-sensitivity to ultraviolet, they tend not to also lead to extraordinary ionizing radiation sensitivity. Nevertheless, patients with defective NER genes exhibit significant sensitivity to chemotherapy agents such as alkylating agents that form bulky DNA adducts.

11.49 Direct repair of damaged DNA

Sometimes damage to a base in a DNA molecule can be directly repaired *in situ* by specialized enzymes without even having to excise the nucleotide. Thus, the phosphodiester DNA chain is never broken, there is no need for a template, and there is no new DNA synthesis required. It is quite simple and uncomplicated. Certain cells possess mechanisms to repair the most commonly encountered types of damage. For example, there are enzymes that directly remove alkyl groups from DNA bases. Most organisms, especially those exposed to large amounts of ultra-violet radiation such as plants, have enzymes that can directly repair cyclobutane pyrimidine dimensions such as thymine dimers without the complicated NER process. This is called **photoreactivation** or light repair. It uses photolyase, an enzyme that depends on visible light to directly break the bonds joining the thymine bases. While this seems like a wonderful mechanism, sadly, man and other placental mammals do not have photolyases or the ability to carry out this type of repair. Hence, humans must rely on BER and NER for reversal of radiation-induced thymine dimers.

On the other hand, humans do have **MGMT** (methylguanine methyltransferase), the enzyme that can directly remove methyl and other alkyl groups from guanine residues at the O6- position. This is important since methylation of guanine produces a structural change that makes it complimentary to thymine rather than cytosine, creating a high risk of G=C to A=T point mutations. MGMT can restore the original guanine by transferring the methyl group to its active site. It is a rare example of a 'one and done' enzyme that is inactivated after carrying out its mission. Such enzymes that are incapacitated after carrying out their functions only once are sometimes called **suicide enzymes**. MGMT is important in oncology because this enzyme can reverse the intentional damage done to DNA in tumor cells caused by chemotherapy with alkylating agents. If a tumor has active MGMT, it may not be susceptible to alkylating agents. However, some tumors naturally exhibit *MGMT* gene promoter methylation. Recalling that DNA methylation is an example of epigenetic modification that typically turns genes off, *MGMT* promoter methylation means the *MGMT* gene is silenced. Hence there is no *MGMT* enzyme that can directly reverse alkylated guanines and therefore alkylating agent chemotherapy is more likely to be effective in such cases. It is helpful to know if the *MGMT* promoter is methylated in the malignant brain tumor known as glioblastoma because it has both prognostic and predictive significance (meaning it may predict the relative efficacy of alkylating agent chemotherapy).

In addition to MGMT, another system of direct repair works in human cells—the ALKBH system. ALKBH stands for AlkB homologues and AlkB is the Alkylation B enzyme in *E. coli* that was found to directly reverse alkylation of DNA. In humans, removal of alkyl adducts from DNA is known to occur through three ALKBH proteins: ALKBH1–3. These can directly remove 1-meA (1-methyladenine) and 3-meC (3-methylcytosine) adducts.

11.50 Single-strand DNA break repair

Although ionizing radiation and other exogenous genotoxins can and does cause DNA damage including single-strand DNA breaks, it is important to recognize that such damage regularly occurs endogenously simply because of the 'hazardous' use of oxygen in aerobic respiration, which naturally and routinely generates reactive oxygen species. The net result is that single-strand breaks (SSBs) arise frequently—on the order of tens to hundreds of thousands per cell per day. Double-strand DNA breaks (DDBs) can also occur endogenously, but SSBs occur three orders of magnitude more often than DSBs.

This equates to an astonishing rate of one single-strand DNA break in every cell every 1–10 seconds or over 10 000 times per day. Single-strand DNA breaks are the most common form of DNA damage. Obviously, if this damage rate were not matched by a commensurate repair rate, DNA integrity would rapidly deteriorate and cellular viability would be compromised. Therefore, it should come as no surprise that SSB repair is a very rapid and efficient process.

When only one of the two strands of a DNA double helix has suffered damage, the other fully-intact complementary strand can be used as a template to guide the correction of the damaged strand. Hence, SSBs are far more readily repaired than DSBs, and DSBs have higher potential for causing catastrophic effects.

Single-strand DNA breaks are usually accompanied by the loss of a single nucleotide and by damaged 5'- and/or 3'-termini at the site of the break.

11.51 Direct versus indirect single-strand DNA breaks

Single-strand DNA breaks come in two varieties: direct and indirect. Some SSBs are made directly (e.g. from electrophilic attack (oxidation) of deoxyribose by free radicals and reactive oxygen species generated by radiation or other sources) while other breaks are made indirectly (via enzymatic cleavage of the phosphodiester backbone during formation of normal intermediates during the process of DNA BER or the relief of torsional strain by topoisomerase I for example). Note that the use of the terms 'direct' and 'indirect' here for single-strand DNA breaks is different from the use of these terms when alluding to the direct and indirect action of radiation (where direct action means the radiation directly damages the DNA molecule and indirect action means that free radicals generated by radiation damage the DNA).

But as far as single-strand DNA break repair goes, both are largely addressed by the same set of proteins. Also the sequence of steps is the same for both types of damage:

1. Damage detection and binding
2. End processing
3. DNA gap filling
4. Strand ligation.

SSBs are detected by the PARP enzymes poly(ADP-ribose) polymerase 1 & 2 (PARP1 and PARP2), which regulate chromatin remodeling, transcription, and the recruitment of SSB repair factors such as XRCC1 protein complexes to the site of damage. Once activated, PARP enzymes transfer ADP-ribose subunits to themselves and/or other proteins, consuming the key cofactor nicotinamide adenine dinucleotide (NAD+) in the process. Thus, excessive SSBs can theoretically deplete NAD.

A key role of PARP is the recruitment of **XRCC1** (x-ray repair cross-complementing protein 1) protein complexes that contain essentially all of the enzymes necessary for direct single-strand DNA break repair. (Recall that XRCC1 is also involved in the short-patch BER pathway.) It is evident that PARP is also needed for the repair of indirect SSBs that arise as intermediates of BER, since the loss of PARP activity dramatically decreases the efficiency of BER and increases cellular sensitivity to DNA base damage.

11.52 DNA damage binding

Many indirect SSBs are created during BER by AP endonuclease (APE1). APE1 itself then binds and takes the first step in indirect single-strand DNA damage repair, **DNA damage binding**, to rectify the indirect SSBs. The series of coordinated efforts of SSB intermediates from one enzyme to the next following the formation of an indirect SSB during BER is often likened to 'passing the baton' in a relay race. In contrast, during direct SSBs, PARP enzymes (rather than APE1) conduct the surveillance for such damage and bind to it.

11.53 DNA end processing

The next step is **DNA end processing**. Most SSBs have an abnormal 5'-terminus, 3'-terminus, or both. These abnormal ends must be reconverted to normal 3'-hydroxyl and 5'-phosphate moieties in order for repair to proceed. Abnormal 5' *indirect* SSBs are usually removed by the actions of Polβ. *Direct* SSBs usually possess an abnormal 3'-phosphate or a 3'-sugar fragment, both of which can be removed by APE1 in a reaction that may be stimulated by interaction with XRCC1. In fact, the scaffold protein XRCC1 is the key player in this stage of SSB repair. XRCC1 is a **molecular scaffold protein**, which means that it plays a central role in the binding, recruitment, and coordination of various polypeptides during SSB repair.

XRCC1 thus serves as the critical coordinator molecule, which interacts directly with most of the other components required for repair of SSBs, including the DNA end processing proteins PNKP, APTX, POLβ, and LIG3. Incidentally, the primary source of direct DNA SSBs is the oxidation of deoxyribose. This may be caused by free radicals or reactive oxygen species. Oxidized deoxyribose can lead to SSBs terminating in abnormal, dangling 3'-phosphoglycolate or 3'-phosphate endings. Oxidized deoxyribose can also indirectly lead to SSBs, following loss of the associated base and cleavage of the remaining oxidized abasic site by apurinic/apyrimidinic endonuclease (AP endonuclease) during DNA BER. Another common source of SSBs is incomplete action of DNA topoisomerases. Topoisomerase enzymes induce DNA breaks during their reduction of torsional stress and/or topological constraints during DNA replication and gene transcription. Topoisomerase-induced single-strand DNA breaks are thus more common during S phase and at transcriptionally active sites. Under normal circumstances, these breaks are ligated by the topoisomerase as part of its overall job, but on occasion, the intermediate cleavage complexes can become 'abortive' (i.e. they fail to finish the job and thereby leave SSBs behind) upon collision with RNA polymerases or DNA polymerases and/or by proximity to other DNA lesions. In human cells, there are two topoisomerases, and topoisomerase I is the primary source of such indirect SSBs. This is because topoisomerase I inherently creates cleavage complexes that harbor SSBs. Nevertheless, the topoisomerase II enzyme may also induce SSBs.

11.54 DNA gap filling

The DNA polymerase primarily responsible for the next step, **DNA gap filling** is called Polβ. This enzyme usually inserts just a single nucleotide (but on occasion can extend the gap by as much as 15 nucleotides).

11.55 DNA ligation

The final step of SSB repair is DNA ligation. During this process, the phophodiester backbone of the DNA is reconnected. DNA ligase IIIα (Lig3α) does this job during single nucleotide repair while DNA ligase I (Lig1) carries out the ligation during 'long patch' repair.

11.56 Means of correcting double-strand DNA breaks: homologous recombination and non-homologous end joining

Radiation, especially high LET radiation such as the HZEs in galactic cosmic rays, can cause double-strand DNA breaks. Left unrepaired, these DSBs can cause large-scale rearrangement of chromosomes (e.g. deletions and other chromosomal aberrations).

Cells essentially employ two major pathways for repairing DSBs: homologous recombination (HR) and non-homologous end joining (NHEJ). (There are two additional pathways of addressing double-strand breaks: microhomology-mediated end joining (MMEJ) and single-strand annealing (SSA); these are covered later in the chapter.) Non-homologous end joining involves the simple rejoining of the broken DNA ends, regardless of the DNA sequence. NHEJ has the advantage of being able to ligate broken DNA ends together without any requirement of a homologous sequence in a sister chromatid, but this process often occurs after some DNA processing (i.e. removal or addition of short stretches of nucleotides) near the ends. Therefore, this repair process can result in **illegitimate recombination**, meaning it tends to be error-prone, as small deletions or insertions may be introduced at the break site. For example, because a portion of the chromosome might have been lost following breakage, the mere joining of two chromosomal segments at the breaking site could lead to a mutation due to lost nucleotides. This mutation potential is averted when the repair is carried out by the more-complex but far more faithful **homologous recombination** mechanism. But whereas non-homologus end joining can take place at any time in the cell cycle, homologous recombination can only occur at limited phases in the cell cycle.

After replication has taken place to generate an identical copy of each chromatid (i.e. creation of sister chromatids), double-strand DNA breaks may be repaired by homologous recombination (HR). It should be noted that after S-phase, the normally diploid (2n) cell is now tetraploid (4n) and has twice the usual complement of DNA. In the HR pathway, a DSB present in one chromatid is repaired using its intact sister (homologous) chromatid as a template. The repair of breaks by HR is a very high fidelity process, which ensures that all the genetic information at the break site is retained.

11.57 HR or NHEJ?

Whether HR or NHEJ is used to repair a double-strand DNA break depends in part on when in the cell cycle the damage occurs. Although NHEJ operates throughout the cell cycle, the HR pathway only functions during S/G2 phase.

HR will repair a double-strand DNA break before the cell enters M phase. HR occurs during and shortly after DNA replication, that is, in the S and G2 phases of the cell cycle. These phases are when intact sister chromatids (that is, homologous DNA strands) are more easily available. It should be noted that unlike homologous chromosomes, which are similar but may have slightly different alleles, sister chromatids are an identical copy of a given chromosome. Hence, sister chromatids are an ideal template for HR.

In contrast to HR, NHEJ is predominant in G1 while the cell is growing but not yet ready to divide. It occurs less frequently after the G1 phase but maintains at least some activity throughout the cell cycle. It should not be interpreted that HR and NHEJ are mutually exclusive and competitive processes. They often function in a cooperative and overlapping manner. Nevertheless, in rare situations they paradoxically do compete with each other.

Deficiencies in HR are clearly associated with an increased predisposition to cancer. For instance, diminished capacity for HR causes inefficient double-strand DNA break repair which increases the risk of cancer. This increased risk is evident when mutations occur in *BRCA1* and *BRCA2*, the two separate genes associated with **HBOC syndrome** (hereditary breast and ovarian cancer syndrome). These two genes are **caretaker tumor suppressor genes**. Cells deficient in BRCA1 or BRCA2 exhibit decreased rates of HR and increased sensitivity to ionizing radiation. The fact that the only (known) function of *BRCA2* is to initiate HR strongly implicates HR in cancer prevention.

Additionally, the autosomal recessive cancer predisposition and radiation sensitivity syndrome **Fanconi anemia** is caused by HR defects. Cells from patients with Fanconi anemia are extremely sensitive to radiation and agents like the chemotherapy drug mitomycin C, which create **inter-strand crosslinks (ICLs)** since these cells are defective in ICL repair. ICLs are repaired through multiple steps involving several coordinated DNA repair proteins, including the HR repair mechanism. As mentioned, around 20 Fanconi anemia genes (**FANC genes**) have been identified, and these play roles in HR and repair of ICLs. In fact, it turns out that *FANCD1* is synonymous with *BRCA2*—they are the same gene.

Furthermore, the autosomal recessive cancer radiation sensitivity syndromes **Bloom syndrome, Werner syndrome,** and **Rothmund–Thomson syndrome** are caused by malfunctioning genes and proteins involved in the regulation of HR (specifically *BLM*, *WRN*, and *RECQL4*, respectively). These disorders are clearly linked to high rates of cancer (figure 11.8).

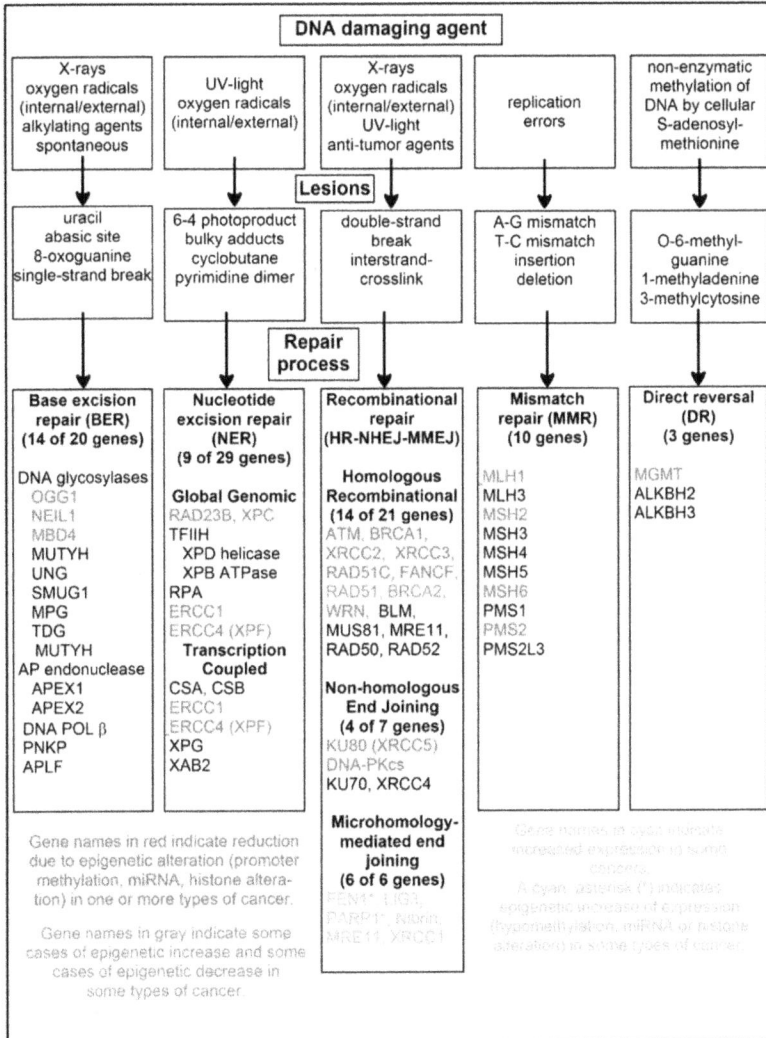

Figure 11.8. A chart displaying various DNA damaging agents (including ultraviolet and x-ray/gamma ray ionizing radiation) along with examples of the types of DNA lesions they induce and the repair processes used to repair the DNA damage. Among the pathways to repair DNA damage are: **BER, NER,** recombinational repair (including **HR, NHEJ,** and **MMEJ), DNA MMR,** and **direct reversal**. Not included in this chart is another form of direct reversal—photoreactivation. Photoreactivation can reverse pyrimidine dimer formation by using light energy, but it does not occur in humans. Also shown in the chart are some of the genes involved in these DNA repair pathways and which genes are epigenetically regulated to effect altered expression in various cancers. Note that of the three pathways available for repair of double-strand DNA breaks, only HR is high fidelity; both NHEJ and MMEJ are error-prone. This DNA damage, repair, alteration of repair in cancer image has been obtained by the author(s) from the Wikimedia website where it was made available under a CC BY-SA 4.0 licence. It is included within this book on that basis. It is attributed to Bernstein0275.

11.58 Molecular aspects of HR

HR can be divided into four key steps:

1. Resection
2. Strand exchange
3. Branch migration
4. Resolution.

The first step, resection, is a modification of the double-strand lesion site. Sections of DNA around the 5′ ends of the break are cut away in the 5′ to 3′ direction from the DSB site on both strands. This produces two single-stranded DNA ends with dangling 3′ tails. Next comes strand exchange, which involves pairing of the broken DNA end with the homologous region of its sister chromatid. This is followed by the overhanging 3′ ends of the 'broken' DNA molecule 'invading' the homologous DNA (the intact DNA molecule). This strand invasion process forms a DNA crossover or Holliday junction.

Strand invasion generates segments of heteroduplex DNA (which is double-stranded DNA made of strands from two different sister chromatids), providing a primer to initiate new DNA synthesis.

Following this comes branch migration when the Holliday junction is trans-located down the DNA by new DNA synthesis. This extends the region of heteroduplex away from the initial crossover site.

Finally, the Holliday junction is resolved by cleavage of the junction to form the two separate duplex DNA molecules again. The molecular details of the process are described below. In human cells (and other eukaryotic cells), the MRN complex is a key initiator of double-strand DNA break repair. The MRN complex is a multimer made of Mre11-Rad50-Nbs1. Although the MRN complex is the first step in HR, it is also the initiator of the NHEJ pathway. MRN detects the presence of DSBs by scanning the DNA molecule and binding to any free ends of the DNA fragments when it discovers a DSB. After binding at the break site, MRN then recruits multiple protein kinases to the site. This launch signal initiates the transduction pathways for DDR (DNA damage repair). The genes for the key kinase enzymes involved in this launch are ATM, ATR, and DNA-PKcs (DNA-dependent protein kinase catalytic subunit). The MRN complex then changes its 3D conformation, which better enables its NBS1 subunit to interact with the ATM dimer. These kinase pathways are extremely complicated and involve the phosphorylation of over 700 proteins. Among these are checkpoint kinases (such as Chk1 and Chk2) that promote cell cycle arrest, other kinases that promote chromatin remodeling, and still others responsible for double-strand DNA damage repair.

11.59 Resection

The Mre11 subunit of the MRN complex has a catalytic function during the resection phase. It has both endonuclease and exonuclease actions. These activities result in the 3′ overhangs on both DNA strands. CtIP proteins (carboxy-terminal binding protein-interacting protein) have a role in the activation of the G2/M checkpoint (G1 restriction point) that halts the progression into M phase. CtIP proteins may be recruited to the breakage site during S phase and G2 phase of the cell cycle, when sister chromatids can act as a template. Following phosphorylation by cyclin dependent kinases (Cdk's), these CtIP proteins may interact with the MRN complex that is bound to the damaged DNA.

11.60 Nucleoprotein filament formation

Resection processing of these DSBs by the MRN complex results in the formation of a 3′ overhang on each resulting strand. After the resection step, a slew of proteins join in, such as RPA (replication protein A), BRCA1, BRCA2, Rad51, Rad52, and Rad54. RPA binds to the newly created single-strand DNA 3′ overhangs. Facilitated by the actions of BRCA2, RPA is then replaced by Rad51, which is a key recombinase enzyme that searches for a homologous strand of DNA (that is, an exactly matching sister chromatid) and initiates exchange with a homologous sequence. It mediates the strand invasion process, which leads to the so-called Holliday junctions (named for the molecular biologist who first proposed the concept, Robin Holliday). Multiple subunits of Rad51 will bind to the site being repaired. The 3′ overhang with the Rad51 subunits bound to it forms the so-called nucleoprotein filament. The Rad51-loaded nucleoprotein filament starts the recombination events involving the intact (undamaged) homologous sequences on the sister chromatid. Such homologous sequences are called donor DNA in HR

11.61 Synaptic phase

The donor DNA template used for lesion repair is the sister chromatid. When the correct donor DNA is found, the nucleoprotein filament invades the double-stranded (duplex) DNA of that donor. This strand invasion leads to the displacement of the homologous DNA strand, forming a structure called the D-loop. Next, the displaced DNA strand is used as a template for repair and for DNA synthesis. This phase leads to the creation of a segment of DNA with a nucleotide sequence that precisely complements the homologous donor strand.

11.62 Post-synaptic phase

Finally, two Holliday junctions are produced, and these Holliday junctions may be resolved without crossing over (by Mus81-Mms4 endonuclease) or with crossing over (by presently unknown polymerase and ligase enzymes).

11.63 Mechanisms of NHEJ

NHEJ is mediated by Rad50, Xrs2, Mre11, Hdf1, Ku80, and Dnl4, as well as the **silencing factors** (**silent information regulators**) Sir2, Sir3, and Sir4, among other molecules. NHEJ plays a large role in illegitimate recombination (meaning that it makes mistakes). The **MRN complex** of HR is also involved in NHEJ at the very start. DNA-PKcs is also involved in the early stages of NHEJ, particularly in bridging the ends of the broken DNA molecule. A molecule called Ku then forms a complex with DNA-PKcs. Ku is actually a heterodimer consisting of Ku70 and Ku80. Ku may function as a docking site for other proteins involved in NHEJ.

In the next step, end processing, damaged or mismatched nucleotides are removed by nucleases and replaced by DNA polymerases (**Pol l** and **Pol m**). Of course, this step is unnecessary if the two ends are already compatible and ready for annealing. **Artemis** is a protein that may participate in end trimming during NHEJ. Finally, the two DNA ends are ligated together by the **DNA ligase IV complex**, which includes DNA ligase IV and the cofactor **XRCC4**. While the precise role of another protein, **XLF,** is still under study, it is known to interact with the XRCC4/DNA ligase IV complex and probably participates in the ligation step (perhaps by restoring the ligase (through re-adenylation) and allowing it to continue functioning. However, there are many other proteins involved in this complex process. The known enzymes involved in NHEJ include ATM, Ku70/80, DNA-PKcs, DNA ligase IV, XRCC4, XLF, Artemis, DNA polymerase μ, DNA polymerase λ, PNKP, aprataxin, APLF, BRCA1, BRCA2, and CYREN. Mutations in genes that participate in NHEJ lead to various disorders, including **ataxia-telangiectasia**, **Fanconi anemia**, and **hereditary breast and ovarian cancers** (*BRCA1/2* genes) as well as forms of **severe combined immunodeficiency**.

It is worth mentioning that in addition to its role in unreliably repairing radiation-induced double-strand DNA breaks, NHEJ is involved in a process called **V(D)J recombination**, which is important for generating diversity in making **antibodies** (in B lymphocytes) and **T cell receptors** (in T lymphocytes). Here, the tendency to make errors is invaluable since the immune system must adapt quickly to new pathogens, and this tendency towards errors actually helps create the needed novel antibody and T cell receptor diversity. The tendency for NHEJ to incorrectly duplicate the original DNA sequence is one of many means that the immune system uses to create the needed diversity.

11.64 Microhomology-mediated end joining

MMEJ, also known as **alternative end-joining** (**alt-NHEJ** or **a-EJ**), is one of the four known pathways for repairing double-strand DNA breaks. Thus, it joins homologous repair and NHEJ as means of dealing with DSBs. (But there is one last means of repairing double-strand breaks called single-strand annealing.) While HR is high fidelity and essentially error free, NHEJ is relatively error prone, and MMEJ is even more mutagenic.

The first step in MMEJ is a limited cleavage of nucleotides in the 5' to 3' direction on either side of the DSB. This process is called end resection. This end resection process is mediated by **MRE 11 nuclease** and generates 3' overhanging single-stranded tails, which reveal regions of microhomology. These single-stranded overhangs anneal at microhomologies between the two strands. (**Microhomologies** are short regions of complementarity, often 5–25 base pairs long, but MMEJ sometimes uses even shorter regions of microhomology.) For instance, a specialized form of MMEJ, called **polymerase theta-mediated end-joining**, is able to repair breaks using just 2 base pairs of homology. Most repairs carried out by alternative end-joining utilize microhomologies of 2–6 base pairs and use the enzyme DNA **polymerase theta (Polθ)**, encoded by the gene *POLQ*). As well as being the first step in MMEJ, this end resection process actively impedes NHEJ (by blocking Rad51, Ku70, and Ku80 recruitment) but promotes HR. It should be mentioned again that MMEJ or alt NHEJ is completely independent from classical (or 'canonical') NHEJ and does not rely on the NHEJ core factors such as Ku protein, DNA-PK, or DNA ligase IV.

Poly (ADP-ribose) polymerase 1 (PARP1) is required as another early step in MMEJ. After base pairing of the microhomology regions, an enzyme called flap structure-specific endonuclease 1 (FEN1) arrives to remove overhanging flaps. This is then followed by recruitment of XRCC1–LIG3 to the site for ligating the two DNA ends, thereby restoring double-stranded DNA.

Because the identification of a region of microhomology might be several base pairs away from the DSB, and because the annealing occurs at the site of the discovered microhomology, MMEJ is virtually always accompanied by a deletion. But given how tiny these required regions of microhomology are (just a few bases), and given how limited the nucleotide 'alphabet' is (just A, C, G, and T), in practice one does not have to go very far before encountering a series of bases that match and qualify as a 'microhomology'. Nevertheless, MMEJ is considered a highly mutagenic pathway for DNA repair. It is evident that MMEJ is frequently associated with subsequent chromosome abnormalities such as deletions, translocations, inversions, and other complex rearrangements.

In comparison, NHEJ directly reattaches both ends of the DSB and is (relatively) accurate, although small (usually less than a few nucleotides) insertions or deletions do occur occasionally. HR, in contrast, has high-fidelity since it uses the sister chromatid as a template for accurate repair of the DSB. MMEJ is distinguished from these other repair mechanisms by its use of random microhomologous sequences to align the broken strands. As mentioned, this results in frequent deletions and occasional insertions, which tend to be much larger than those generated by NHEJ.

In humans, HR actively suppresses the mutagenic MMEJ process (although ironically the first step of MMEJ actually promotes HR). Thus, when HR is available (e.g. in the right phase of the cell cycle), the cell avoids the mutation-producing MMEJ pathway. On the contrary, when cells are deficient in either classical NHEJ or HR, they will use MMEJ.

A clinically relevant observation is that approximately half of all ovarian cancers are deficient in HR. Perhaps to compensate for this deficiency, these and other HR-deficient tumors upregulate polymerase theta (Polθ), thereby increasing the use of MMEJ. Perhaps of therapeutic importance, since these tumors are hyper-reliant upon MMEJ, inhibition of polymerase theta results in substantial lethality of these abnormal cancer cells when combined with DNA-damaging chemotherapy or radiation therapy; such cancer cells are left with nearly no means of dealing with therapy-induced DNA damage. The concept of exploiting a known key mutation that causes reliance on a specific pathway and then medically inhibiting that pathway in cancer cells is called **synthetic lethality**. The use of PARP inhibitors in ovarian cancer is one example of synthetic lethality.

As mentioned repeatedly, MMEJ is a very mutagenic pathway. It would therefore seem that cells with increased MMEJ usage would have higher genomic instability and a predisposition towards transformation into cancer cells, but this has not yet been confirmed.

11.65 Single-strand annealing (SSA)

Another double strand DNA repair pathway has been identified that can serve as a backup for HR, NHEJ and MHEJ—single-strand annealing or SSA. But like MHEJ and NHEJ, it is also considered to be error prone. Despite its name, single-strand annealing is another double-strand DNA break repair mechanism. SSA is a DNA end-joining process that primarily operates on the multitude of repetitive DNA segments scattered throughout the genome. SSA uses the homologous repeats flanking a double-strand break to join the DNA ends and is quite error-prone. It often removes one entire repeat along with some DNA fragments between repeats. This leaves a sort of 'signature', announcing that 'SSA was here'. DNA deletions in cancer cells often have homology at breakpoint junctions; this telltale sign suggests the involvement of SSA in these cancer cells. This error prone repair pathway involves RAD52 and DNA ligase I, among many other enzymes. Some authorities consider single-strand annealing to be a subtype of homologous recombination that uses the repeated sequences in same strand of DNA as the template for repair.

References and further reading

[1] Griffin R J *et al* 2020 Understanding high-dose, ultra-high dose rate, and spatially fractionated radiation therapy *Int. J. Radiat. Oncol. Biol. Phys.* **107** 766–78

[2] El-Deiry W 1993 WAF1, a potential mediator of p53 tumor suppression *Cell* **75** 817–25

[3] Hernández Borrero L J and El-Deiry W S 2021 Tumor suppressor p53: Biology, signaling pathways, and therapeutic targeting *Biochim. Biophys. Acta, Rev. Cancer* **1876** 188556

[4] Dulić V *et al* 1994 p53-dependent inhibition of cyclin-dependent kinase activities in human fibroblasts during radiation-induced G1 arrest *Cell* **76** 1013–23

[5] Verma A, Halder K, Halder R, Yadav V K, Rawal P, Thakur R K, Mohd F, Sharma A and Chowdhury S 2008 Genome-wide computational and expression analyses reveal G-quad-ruplex DNA motifs as conserved cis-regulatory elements in human and related species *J. Med. Chem.* **51** 5641–9

[6] Han H and Hurley L H 2000 G-quadruplex DNA: a potential target for anti-cancer drug design *Trends Pharmacol. Sci.* **21** 136–42

[7] Wengner A M, Scholz A, Haendler B and Targeting D N A 2020 Damage response in prostate and breast cancer *Int. J. Mol. Sci.* **21** 8273

[8] Mladenov E, Fan X and Dueva R *et al* 2019 Radiation-dose-dependent functional synergisms between ATM, ATR and DNA-PKcs in checkpoint control and resection in G2-phase *Sci. Rep.* **9** 8255

[9] Smith J, Tho L M, Xu N and Gillespie D A 2010 The ATM-Chk2 and ATR-Chk1 pathways in DNA damage signaling and cancer *Adv. Cancer Res.* **108** 73–112

[10] Kligerman A D, Halperin E C, Erexson G L and Honoré G 1990 The persistence of lymphocytes with dicentric chromosomes following whole-body X irradiation of mice *Radiat. Res.* **124** 22–7

[11] Gong B, Radulovic M, Figueiredo-Pereira M E and Cardozo C 2016 The ubiquitin-proteasome system: potential therapeutic targets for Alzheimer's disease and spinal cord injury *Front. Mol. Neurosci.* **9** 4

[12] Plyusnina E N, Shaposhnikov M V and Moskalev A A 2011 Increase of Drosophila melanogaster lifespan due to D-GADD45 overexpression in the nervous system *Biogerontology* **12** 211–26

[13] Sullivan K D, Galbraith M D, Andrysik Z and Espinosa J M 2018 Mechanisms of transcriptional regulation by p53 *Cell Death Differ.* **25** 133–43

[14] Brooks C L and Gu W 2011 p53 regulation by ubiquitin *FEBS Lett.* **585** 2803–9

[15] Zhou L, Jiang Y and Luo Q *et al* 2019 Neddylation: a novel modulator of the tumor microenvironment *Mol. Cancer* **18** 77

[16] Lee C L, Blum J M and Kirsch D G 2013 Role of p53 in regulating tissue response to radiation by mechanisms independent of apoptosis *Transl. Cancer Res.* **2** 412–21

[17] Mijit M, Caracciolo V, Melillo A, Amicarelli F and Giordano A 2020 Role of p53 in the regulation of cellular senescence *Biomolecules* **10** 420

[18] Błasiak J 2021 Single-strand annealing in cancer *Int. J. Mol. Sci.* **22** 2167

Chapter 12

Radiation and respiration

12.1 Alternatives to carbon and water?

All life forms have three basic needs: a source of energy, a source of carbon and a solvent for their biochemistry. This is obvious when it comes to the familiar forms of life on Earth, but the same basic principles might apply to any life forms anywhere in the Universe with some variations on the theme. For instance, it is a good exercise to consider the possibilities of biological solvents besides water. Might liquid ammonia serve all the biological needs that water addresses? What about liquid hydrocarbons which might abound on Saturn's moon Titan? Students are encouraged to ponder the possibilities of potential life based on a non-polar solvent such as liquid methane. During such astrobiological deliberations, it is instructive to recall the basic chemistry of hydrogen bonding, which is so essential for life as we know it. Hydrogen bonds form only between hydrogen and three highly electronegative elements in the left upper region of the periodic table—nitrogen, oxygen and fluorine. Thus, if we assume that hydrogen bonding is likely to be important in the structure and function of extraterrestrial biological macromolecules, our imagination does have some constraints when it comes to solvents. Topics to consider when pondering alternative solvents for life include the temperatures and pressures at which the solvent in question remains liquid as well as the intrinsic aggressiveness of the solvent (e.g. protic solvents such as sulfuric acid that readily donate and accept protons from solutes vs chemically mild and non-reactive aprotic solvents such as liquid nitrogen or methane).

Similarly, one can scientifically speculate about the prospects and limitation of life based on elements aside from carbon. Carbon chemistry is so versatile that there is an entire field of chemistry devoted to the details of carbon-based molecules, their synthesis, interactions and nuances of covalent bonding—organic chemistry. Students who have taken courses in organic chemistry will recall that this field is not for the faint-hearted. Although theoretically restricted in scope, in practice, this branch of chemistry is vast and complicated. And this complexity is thanks to the enormous chemical flexibility and versatility of carbon. Do other elements in the same group as carbon on the periodic table (Group 14 or IVa) possess the same versatility? Silicon is a member of this group. But does it truly possess the same chemical creativity as carbon? Silicon does have the electron structure that allows it to be tetravalent. However, the bond angles created by silicon are different from those created by carbon and the bond lengths tend to be longer, thanks to silicon's larger covalent radius. Such subtle differences affect silicon's ability to create the myriad ring structures characteristic of organic chemistry. Additionally, the critically important double and triple bonding potential of carbon is not nearly as available through silicon.

How about sulfur or boron? After some analysis, many will come to agree with Carl Sagan's sentiments as a staunch 'carbon chauvinist'. As with alternative solvents for biological chemistry, students are encouraged to consider alternatives to carbon as the scaffold element on which life is based. Of course, when scientifically considering the possibilities, one must remain within the realistic constraints imposed by the natural abundance of these alternative elements in the Universe as well as the chemical bonding behavior exhibited by such elements. Curiously, the chemistry of certain competitors to carbon might not seem to have much potential under the familiar conditions of temperature and pressure at sea level on Earth, but when those conditions vary, newfound versatility emerges. For example, when the temperature and solvent in question is changed (e.g. using sulfuric acid or a liquid hydrocarbon cryosolvent instead of water), the chemical potential of silicon as a scaffold for life are improved.

12.2 Phototrophs and chemotrophs

So far, this discourse has been rather exotic and perhaps so seemingly remote from reality that it could be considered completely irrelevant. However, when we focus on the third key necessity of life, a source of energy, we see that the possible variations on a theme can (and indeed have) been explored in earnest by extant Earth life forms. The most familiar source of energy among primary providers on Earth is sunlight and organisms that obtain their energy from photons are called **phototrophs**. In other words, electromagnetic radiation is the ultimate source of energy for the base of our food chain. But an alternative nutritional biosphere exists that derives its energy not directly from photons but rather from the energy stored in the covalent bonds of chemical compounds. That is to say, higher on the food chain lie **chemotrophs**, who capitalize on the light energy harvested by plants and eat them as a source of energy. In the sections below, we shall explore these means of powering biology through the acquisition of energy from the environment. And we shall examine the possibility of yet another source of energy for biology—ionizing radiation. And whether that ionizing radiation is limited to Earth or could be exploited by extraterrestrial life is another fascinating subject.

12.3 Nutritional taxonomy

Familiar lifeforms can be classified by how they satisfy their nutritional needs, more specifically, by how they obtain their **carbon**, how they obtain their **energy** and how they obtain **reducing equivalents**. Organisms need a chemical scaffold for biochemistry, and as discussed above, the element used for biological chemistry on Earth is carbon. Thus, all organisms must continuously obtain carbon if they are going to grow and reproduce. Next, all of these processes require energy and so all biological entities must have a means of acquiring that needed energy. The choices here are sunlight or chemical bond energy. Finally, in order to build complex biological macromolecules and even simple structural necessities such as cellular membranes, life has chosen redox reactions as the means of driving these biosynthetic processes. Therefore, reducing equivalents must be available for cells to synthesize lipids, nucleic acids and complex carbohydrates (e.g. polysaccharides). In other words, this classification depends on what type of **electron donors** they use.

Another terminology digression is that the suffix '-trophism' alludes to nutrition and should not be conflated with a similar suffix, '-tropism', which alludes to movement. **Phototropism** is the tendency for plants to migrate towards a source of light, whereas **phototrophism** alludes to the use of light as a source of energy for nutritional needs.

Starting with carbon acquisition, the two basic pathways are **autotrophy** (making carbon compound from carbon dioxide) or **heterotrophy** (obtaining carbon by consuming other organic substances—that is, eating). Obtaining carbon and converting it into useful organic compounds through the incorporation of environmental inorganic carbon dioxide is known as **carbon fixation**. The next basic classification metric—obtaining energy—also offers two basic options: light (**phototrophy**) or chemical compounds (**chemotrophy**). The last nutritional classification parameter, reducing equivalent acquisition, can be divided again into two groups: **organotrophy** (acquisition of reducing power through organic compounds) or **lithotrophy** (use of inorganic chemical compounds as the source of reducing power).

To summarize, a means of metabolically classifying organisms divides everything into eight possible groups based on permutations of the six above-mentioned variables. By convention, these are named with the energy acquisition prefix first, the reducing equivalent source second and the source of carbon third. In other words, the sequence is photo-/chemo-, followed by organo-/litho- and ending with auto-/hetero-troph. The possibilities thus include:

1. **Photoorganoheterotrophs**
2. **Photoorganoautotrophs**
3. **Photolithohheterotrophs**
4. **Photolithoautotrophs**
5. **Chemoorganoheterotrophs**
6. **Chemoorganoautotrophs**
7. **Chemolithoheterotrophs**
8. **Chemolithoautotrophs.**

Although eight possible categories exist, that does not imply that all eight must be or are used in terrestrial biology. Nevertheless, the variability exhibited by life on Earth is astounding, and in fact, representative examples from all eight theoretical categories have been discovered. Humans (along with all animals and fungi) obtain the energy through chemistry rather than electromagnetic radiation (photons), obtain their reducing power through organic compounds rather than inorganic compounds and obtain their carbon by eating organic material instead of fixing inorganic carbon dioxide. In this way, humans, other animals and fungi are **chemoorganoheterotrophs**. Plants, algae and cyanobacteria are **photolithoautotrophs** as they extract energy from sunlight, they obtain reducing power through use of an inorganic molecule (water) as their electron donor, and they acquire their carbon through fixation of atmospheric CO_2. Of course, not all three prefixed must be used when describing nutritional metabolism of organisms and in practice, only two are typically explicitly mentioned. Plants are typically alluded to as **photoautotrophs** with the omission of the litho/organo category of what they use as an electron acceptor. **Chemoautotrophs** include several metabolically unusual organisms such as the methanogens, sulfur oxidizers and sulfur reducers, nitrifiers and thermoacido-philes. Medically important microbes are almost universally **chemoheterotrophs** that catabolize organic substances obtained from their hosts.

Chemolithoautotrophs, which by definition obtain energy from breaking chemical bonds, use inorganic electron donors for reducing equivalents and use carbon dioxide as their source of carbon are all microbes. Some belong to the Archaea, but most are bacteria. Generally speaking, organotrophic organisms are usually also heterotrophic, using organic compounds as sources of both electrons and carbon. Similarly, lithotrophic organisms are often also autotrophic, using inorganic electron donors and inorganic carbon (CO_2).

12.4 Energy metabolism and biological reducing equivalents

Despite the vast variety, the common chemical currency of cellular energy is **adenosine triphosphate (ATP)**. All organisms use ATP as their ultimate energy unit, indicating an underlying uniformity of all life on Earth. But there are various means of generating ATP. Regardless of the specifics, cells never oxidize reduced organic compounds all in one step, as in a fire. Rather, they slowly and steadily break these reduced compounds down in a stepwise fashion from one chemical form to another to capture the chemical energy in a stepwise fashion and ultimately convert it into ATP. This metabolic breakdown of complex molecules is called **catabolism**. Catabolism is contrasted to the opposite form of metabolism, the building of more complicated biomolecules, or **anabolism**.

Just as the universal energy commerce molecule is ATP, there is a limited range of organic molecules used in terrestrial biology for **reducing equivalents**. Among the molecules used to store the reducing power originally found in reduced chemicals are **nicotinamide dinucleotide (NADH)**, **nicotinamide dinucleotide phosphate (NADPH)** and **$FADH_2$**. As mentioned, metabolic processes that build other biological molecules are called anabolic pathways, whereas the breakdown of biological molecules (for example breakdown of glucose to extract the stored chemical energy) is called catabolism. Most anabolic pathways tend to use the reduced form of the molecule **NADPH** as the source of reducing power, although an important biochemical pathway that builds glucose (**gluconeogenesis**) uses the reduced form of a related molecule, **NADH**. Of relevance to the topic of radiation, NADPH provides the reducing equivalents for redox reactions involved in the regeneration of glutathione, which is a key molecule in neutralizing **reactive oxygen species** generated by ionizing radiation and other sources. Another key redox function of NADPH is through the *generation* of free radicals in certain immune cells for combating invading pathogens by the enzyme NADPH oxidase in a cellular process called the respiratory burst.

12.5 Fermentation

One of the simplest forms of catabolic processes is **fermentation**, which starts with a reduced carbon compound such as the monosaccharide carbohydrate **glucose** and moderately oxidizes it to two molecules of **pyruvate** through a pathway called **glycolysis** (also known as the **Embden-Meyerhof** or **Embden-Meyerhof-Parnas pathway**). Along the way, two ATP molecules are generated, along with two molecules of reduced NAD^+ (i.e. NADH). The process of making ATP this way is called **substrate-level phosphorylation** to distinguish it from other means of phosphorylation of adenosine diphosphate (ADP) into ATP such as oxidative phosphorylation and photophosphorylation. To keep the glycolytic pathway active, there must be a means of re-oxidizing that NADH back to NAD^+. These means constitute the various final phases of the different fermentation pathways. In humans, under anaerobic conditions when muscles demand energy very fast, for instance, the pyruvate is reduced to **lactic acid** while the NADH is re-oxidized into NAD^+. Other organisms such as yeasts employ **alcoholic fermentation**, which yields ethanol. Other fermentative pathways can yield hydrogen gas, butyrate and acetone.

Although glycolysis (the Embden-Meyerhof-Parnas pathway) is the most commonly used fermentation pathway and is the one that most familiar to students, alternative fermentative pathways exist. For example, the **Entner-Doudoroff pathway** also starts with glucose and ends with pyruvate but employs different intermediates and enzymatic catalysts. Technically, since the word 'glycolysis' literally means the breakdown of glucose, the Entner-Doudoroff pathway can also be called glycolysis. But in practice, the term is primarily reserved for the Embden-Meyerhof-Parnas pathway. Another pathway found in certain bacteria including *Escherichia coli* is the **pentose phosphate pathway** or **hexose monophosphate shunt**. This may operate simultaneously with the Embden-Meyerhof pathway, but in addition to being able to catabolize the six-carbon glucose molecule, this pathway provides a means of making various five-carbon monosaccharides such as ribose and ribulose 5-phosphate, which are essential in nucleotide metabolism and the Calvin cycle, respectively. The pentose phosphate pathway yields only one ATP molecule per glucose but manufactures NADPH. The details of the various steps and enzymes in these pathways are beyond the scope of this text, but interested students are encouraged to look up the details in standard biochemistry and microbiology textbooks.

12.6 Respiration

Regardless of the specifics, fermentation is a low-yield metabolic pathway. There is still a great deal of chemical energy remaining in the pyruvate molecule, and this energy can be tapped via **respiration**. Whether the respiration is aerobic or anaerobic, the common thread in both is the use of a **semi-permeable membrane** to create an **electrochemical gradient**. The electrochemical gradient for all forms of respiration uses hydrogen ions (that is, protons) as the ion species to concentrate on one side of the membrane. Thus, a proton gradient or pH gradient is created across a biological membrane. Recall that pH, by definition, reflects the concentration of protons:

$$\mathbf{pH = -\log[H+]}.$$

Hence, the membrane-based electrochemical gradient used in biological energy production is synonymous with 'pH gradient'. The electrical potential energy stored in the electrochemical gradient can be released upon discharge. This discharge is called **chemiosmosis** and refers to the natural movement of protons back down their electrochemical gradient across the membrane at specific sites. The electrical energy released is coupled with ATP synthesis and thereby converted into biochemically useful energy.

The membrane-based electrochemical gradient in respiration is created through the serial oxidation of specific reduced organic compounds that are made in metabolic processes. In terrestrial biology, the reduced chemical entities that are used to generate electrochemical gradients are NADH, NADPH and FADH$_2$. In the next section, we shall review the various types of respiration, the creation of a membrane-based electrochemical energy gradient, and how it is transduced into ATP.

12.7 Aerobic respiration and oxidative phosphorylation

Aerobic respiration is a form of cellular respiration that uses **molecular oxygen** (O_2) as the final electron acceptor. In **anaerobic respiration**, final electron acceptors other than molecular oxygen are used. Although oxygen is not the final electron acceptor in anaerobic respiration, the process still uses a membrane-associated electron transport chain and creates an electrochemical gradient across it. The energy contained in the electrochemical gradient can then be tapped for creation of ATP. This the definition of respiration. This is an important distinction from fermentation, which bears superficial similarities to anaerobic respiration in that it does not use oxygen, but importantly, it does not use a membrane.

Both forms of respiration start out with reduced organic compounds such as NADH, NADPH or FADH$_2$ as electron donors. These highly reduced chemical compounds are made elsewhere by glycolysis, by the **Krebs cycle** (also called the **tricarboxylic acid cycle**) or, in plant cells, the **non-cyclic photophosphorylation** process of photosynthesis. The 'goal' of respiration is to create an electrochemical potential gradient across a membrane that can be used as a biological battery. Upon discharge of the battery, electrical energy is converted into chemical energy—the universal biochemical energy commerce molecule, ATP.

The first step is the establishment of an electrochemical gradient (usually a proton gradient) across a membrane. This is achieved through a stepwise process of oxidation/reduction reactions of compounds along the membrane. Redox reactions are defined as the transfer of electrons from donors to acceptors, but in biochemistry, oxidation is often synonymous with dehydrogenation, which implies transfer of an electron and a proton. The electrons are always readily released and accepted by the electron carriers, but the protons are not. Protons that are not accepted migrate outward and accumulate along the outer lining of the membrane. Thus, a chemical concentration gradient of protons develops. In bacteria and archaea, this occurs on the outer cell membrane, which explains their negative intracellular charge. The result is that the cytoplasm of the prokaryotic cell is more alkaline and more negative, leading to both a chemical and electrical potential difference. This leads to a **proton motive force**, which can be used to do work. For **mitochondria** in eukaryotic cells, the situation is similar but more efficient. These organelles are double membraned. The protons accumulate in the intermembrane space between the inner and outer membranes. Mitochondrial inner membranes are highly folded into structures called **cristae**, which increase the surface area. This in turn increases the overall efficiency and yield of the desired end product—a proton motive force that can be tapped for conversion into chemical energy in the form of ATP.

The reduced chemical compounds (e.g. NADH) that are initially entered into the microbial or mitochondrial **electron transport chain** are oxidized by a series of membrane-bound proteins (i.e. **integral proteins**) with sequentially increasing reduction potentials. Electrons are shuttled from carrier to carrier, in order of their standard reduction potential, with the final electron acceptor being molecular oxygen in aerobic respiration or another chemical substance in anaerobic respiration.

Like a battery, the electrochemical potential so generated can be tapped. This 'battery' can be discharged, and discharge is linked to a proton motive force. The proton motive force converts electrical energy into chemical energy as protons move across the membrane and down the gradient (a process called chemiosmosis). The way this energy conversion is made is through an integral membrane protein called **ATP synthase**. As protons pass through the channel made by ATP synthase, the electrical current drives the synthesis of ATP from ADP and inorganic phosphate. Passing of protons through the channel of ATP synthase leads to a conformational change; relaxation is then coupled to the synthesis of ATP. The overall process is termed **oxidative phosphorylation**.

12.8 Anaerobic respiration

As mentioned, in aerobic respiration the final electron acceptor is molecular oxygen. Molecular oxygen is a strong oxidizing agent and thus, an excellent electron acceptor. Anaerobes instead use lower redox potential (less-oxidizing) substances as the terminal electron acceptor such as nitrate, fumarate, sulfate, elemental sulfur or other agents. These final electron acceptors have lower reduction potentials than O_2 and therefore release less energy per oxidized molecule. For that reason, anaerobic respiration is less efficient than aerobic respiration. In fact, examination of the **redox tower (electron tower)** (figure 12.1) shows that oxidation of glucose by molecular oxygen is among the highest redox couples on the entire electron tower, with an extremely positive standard electrode potential. Thinking astrobiologically, one would be hard pressed to come up with a better electron acceptor than molecular oxygen for exotic alien biochemistry. Students are encouraged to use their imagination!

While aerobic respiration is tough to beat, there are many contenders among the anaerobic respiration options right here on Earth. The metabolic diversity exhibited by microbes is astonishing with an astounding assortment of oxidants known to function as final electron acceptors. Theoretically, the best acceptors are those lower down on the redox tower when in their oxidized form (but of course, this depends on just how the redox tower is drawn—most positive at the top or most negative at the top!). Some common electron acceptors found in anaerobic respiration include nitrate (NO_3^{3-}), ferric iron (Fe^{3+}), sulfate (SO_4^{2-}), carbonate (CO_3^{2-}) or even certain organic compounds like fumarate, but the overall list is impressively diverse. Recall that the ability of an atom to attract electrons and serve as a biochemical electron acceptor is related to its **electronegativity**, the inherent tendency to attract electrons to itself when placed near an adjacent atom in a molecule. In practice, this is quantified by its E_0' **value**, the **standard reduction potential**. These standard reduction potentials can then be related to the free energy through:

$$\Delta G = -nFE_0'{}_{cell},$$

where

$$E_0'{}_{cell} = E_0'{}_{cathode} - E_0'{}_{anode}.$$

The more positive the standard reduction potential, the stronger its attraction of electrons from other atoms and the better it is as an electron acceptor. Among the biologically documented variations on the theme of anaerobic respiration, (in order of redox potential) are:

Perchlorate respiration
Iodate respiration
Iron reduction
Manganese reduction
Cobalt reduction
Uranium reduction
Nitrate reduction (denitrification)
Fumarate respiration
Sulfate respiration
Methanogenesis (carbon dioxide reduction)

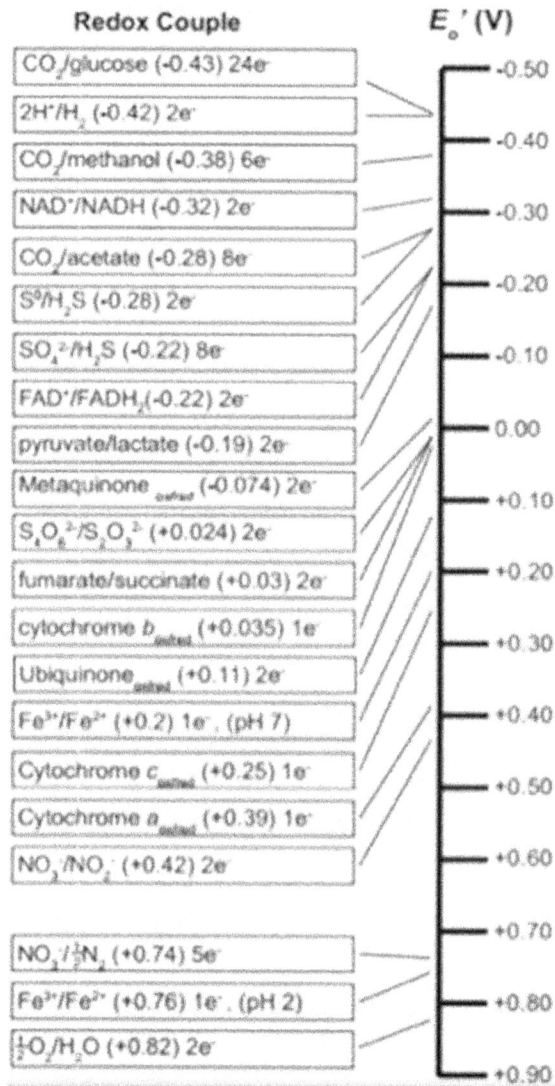

Figure 12.1. An 'electron tower' (redox tower). By convention the redox half reactions are written with the oxidized form of the compound on the left and the reduced form on the right. Compounds that make good electron donors are found near the top of the tower. Hence, glucose and hydrogen gas are good electron donors. Note that as electron donors, they are reduced and therefore found on the right-hand side of the paired half reactions. At the lower end of the tower are agents that are good electron acceptors, such as molecular oxygen and nitrite (NO_3^-). By convention, electron acceptors (that is, the oxidized form of the compound in a pair) are written on the left side of the redox pair and electron acceptors have a positive E_0' value (whereas electron donors have negative E_0' values). Available at: https://bio.libretexts.org/Courses/University_of_California_Davis/BIS_2A_(2018)%3A_Introductory_Biology_(Singer)/MASTER_RESOURCES/Reduction%2F%2FOxidation_Reactions%23. Reproduced from Biology LibreTexts (2023).

Sulfur respiration (sulfur reduction)
Acetogenesis (carbon dioxide reduction)
Dehalorespiration.

One of the most frequently used forms of anaerobic respiration is sulfate-reduction. Sulfate reducers employ sulfate (SO_4^{2-}) as their terminal electron acceptor. Some sulfate-reducers can use less oxidized forms of sulfur, including sulfite (SO_3^{2-}), or elemental sulfur as their electron acceptors. In so doing, they reduce oxidized sulfur to **hydrogen sulfide** (H_2S). In that they do not use oxygen but rather sulfate and do not make water via respiration but rather hydrogen sulfide, they are sometimes said to 'breathe' sulfate. Interestingly, some bacteria can reduce sulfate to biosynthesize needed sulfur-containing cellular components; this is known as **assimilatory sulfate reduction**. In contrast, sulfate-reducing bacteria reduce sulfate to obtain energy. They then expel the **hydrogen sulfide** as waste; this process is known as **dissimilatory sulfate reduction**. The end product, hydrogen sulfide, yields the familiar stench of rotten eggs. This pungent fingerprint left behind by sulfate reducers can sometimes be reeked in marshes, sewers, **meromictic lakes** (lakes that do not turn over and mix bottom water with surface water annually) and occasionally in well water. As hydrogen sulfide is poisonous in high doses (ironically because it poisons the electron transport chain in our aerobic respiratory process in a manner like how hydrogen cyanide does), it is wise to heed the natural biological warning of its offensive odor. Sulfate reduction as a means of respiration is an evolutionarily ancient process with traces dating back to around 3.5 billion years ago. As will be discussed below, certain bacteria such as the purple sulfur bacteria and green sulfur bacteria can use the product (H_2S) as an electron donor in anoxygenic photosynthesis. This mode of photosynthesis is also very ancient and predates the oxygen-producing variety that uses the ample molecule, H_2O, as its electron donor. Students of paleontology will recall the Permian extinction, which occurred around 250 million years ago and was the greatest of all the mass extinctions. Although the proximate cause of the extinction even was probably prolonged massive volcanic outgassing from the Siberian Traps, an aggravating contributing cause might have been the accompanying shift from an aerobic biosphere dominated by macrofauna (animals) back to an anaerobic biosphere dominated by sulfur metabolizing microbes. The anaerobic respirers, who were displaced by the oxygen-utilizing eukaryotes (including animals) after the global oxidizing caused by the profligate release of molecular oxygen by cyanobacteria (more on this topic below) nearly got their revenge at the Permian extinction.

It is important to recognize the similarities between aerobic and anaerobic respiration despite the very different final electron acceptors. Both forms of respiration can start with glycolysis, meaning the part beginning with glucose and ending with pyruvate. Also, in both forms, the pyruvate so formed can be sent to the Krebs cycle to make NADH (and some ATP via some substrate-level phosphorylation). Furthermore, oxidative phosphorylation is then used to generate most of the ATP in both categories. This means the use of an electron transport chain, the establishment of an electrochemical potential gradient and proton motive force and finally an ATP synthase that creates chemical energy units (ATP) out of the stored electrical energy. The only real difference is that the final electron acceptor may not always be oxygen.

Just how much ATP is generated in anaerobic respiration depends on the final electron acceptor. No matter which one we select, based on the redox tower, it will not be capable of making as much ATP as aerobic respiration can. On Earth, molecular oxygen appears to be the best possible electron acceptor. Use of an alternative electron acceptor pushes an organism higher up the electron tower. This decreases the energy difference between the electron donor and the acceptor, thereby reducing the amount of ATP possibly produced.

Another interesting observation is that the actual yield of oxidative phosphorylation, regardless of whether it is coupled to aerobic or anaerobic respiration, is not a simple stochiometric conversion. Some texts will say that each NADH molecule yields three ATP and each $FADH_2$ yields two ATP, but it is not really that simple and direct. If it were, one might expect a grand total of 38 ATP molecules per molecule of glucose oxidized. This would be:

> Two ATP from substrate-level oxidation in glycolysis
> Six ATP via oxidative phosphorylation from the two NADH molecules made in glycolysis
> Six ATP via oxidative phosphorylation from the two NADH molecules made in the preparatory step from pyruvate to acetyl CoA that enters the Krebs cycle
> Two ATP from the two GTP molecules made via substrate-level phosphorylation in the Krebs cycle
> 18 ATP via oxidative phosphorylation from the six NADH molecules made in the Krebs cycle
> Four ATP via oxidative phosphorylation from the two $FADH_2$ molecules made in the Krebs cycle
> 38 ATP molecules total

Prokaryotes could approach this theoretical maximum, but eukaryotes cannot since they conduct glycolysis in the cytoplasm while their electron transport chain and oxidative phosphorylation occurs in the mitochondria. And it takes energy (two ATP) to shuttle the NADPH molecules across the mitochondrial membrane. Furthermore, as one might suspect given all the conversion steps involved, the overall process is not perfectly efficient. For instance, there is some proton leakage across the electrochemical gradient after it is set up and the theoretical yield of three ATP per NAD and two ATP per $FADH_2$ is not achieved. A more realistic estimate is 2.5 ATP per NADH and 1.5 ATP per $FADH_2$. Thus, a more reasonable expectation is a total of about 27 ATP per molecule of glucose oxidized in eukaryotic cells.

12.9 Phototrophy, photophosphorylation and photosynthesis

As discussed above, plants, algae and cyanobacteria are well-known phototrophs—organisms that extract energy directly from sunlight or other available sources of photons. This contrasts them with chemotrophs, organisms that must reply on other sources of energy such as the organic compounds made by plants. But for plants to then take light energy and convert it into something biochemically useful requires additional steps. Among those important steps is **photophosphorylation**, the manufacture of chemical energy in the form of ATP from the harvested sunlight. This will now be described below.

Among the familiar macroscopic plants and animals, there are two general means of making an electrochemical gradient that is then used to power ATP synthesis. Oxidative phosphorylation as just discussed above is used in animals, including humans. In plants, another mechanism is photophosphorylation. Both processes set up an electrochemical potential gradient across a biological membrane and then harvest the energy stored in that potential gradient for chemical synthesis of the universal energy currency molecule, ATP. Plants can do this in the organelles called **chloroplasts**, while animals do this in the organelles called mitochondria (plant cells also contain mitochondria and can carry out oxidative phosphorylation as well).

In the way of terminology and taxonomy, **tracheophytes** (vascular plants including ferns and allies, gymnosperms and angiosperms), **bryophytes** (mosses and related groups), **algae** and **cyanobacteria** are the extant oxygen-producers on planet Earth through oxygenic photosynthesis. The subcellular organelles they all possess which carry out this photosynthesis are called **chloroplasts**.

The term **photosynthesis** is a confusing, all-encompassing term that alludes to *both* the conversion of photon energy into chemical energy (the so-called **light reactions**) as well as the conversion of CO_2 into organic compounds such as glucose (the **dark reactions**). A more appropriate synonym for the dark reactions is '**light-independent reactions**'. Adding to the confusion is the fact that there are several variations on the dark reaction theme of creating organic compounds from carbon dioxide (such as the **Calvin-Benson cycle**, **Hatch-Slack pathway** and **crassulacean acid metabolism**), which will not be covered in this text. For this reason and more, the catch-all name of 'photosynthesis' is potentially quite confusing. Nevertheless, it is here to stay.

Also in the way of terminology, **chemosynthesis** is another catch-all phrase that refers to the biological conversion of inorganic (or organic) carbon and nutrients into more complex organic matter using the chemical energy obtained through the oxidation of inorganic molecules (e.g. ferrous iron ions, hydrogen gas, hydrogen sulfide, etc) or organic molecules (e.g. methane). This contrasts chemosynthesis with photosynthesis, which uses photon energy to build up complicated biological molecules.

And although all modern land plants (embryophytes) employ the same '**z-scheme**' for photophosphorylation, there are many variations on the theme of making ATP and reducing power (NADPH) from sunlight, and these variations are employed extensively by different photosynthetic bacteria. They all use the same basic concept of setting up an electrochemical potential (a proton gradient) across a biological membrane using the energy from photons as the proton pump. The proton motive force can be discharged with the energy harvested to generate ATP. The ATP can then be used in the Calvin cycle or another dark reaction to make glucose and other carbohydrates. Some light reaction pathways lead to the creation of molecular

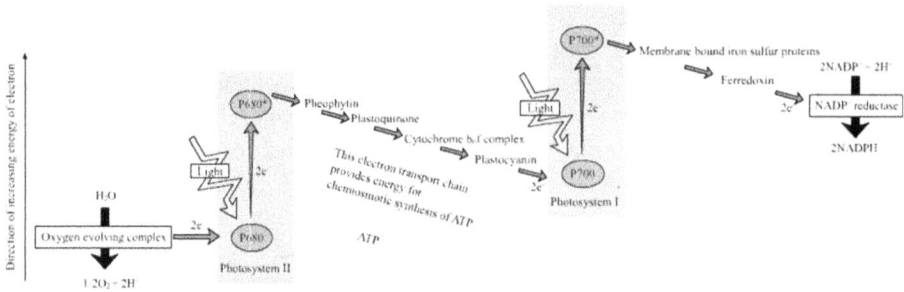

Figure 12.2. The so-called Z-scheme of oxygenic photosynthesis, illustrating the two separate 'photosystems', Photosystem II on the left and Photosystem I on the right. The passage of electrons from donors to acceptors along the electron chains converts the photon energy into chemical energy in the form of ATP—thus the name, photophosphorylation. Photosystem I uses cyclic photophosphorylation, meaning that the electron initially liberated from the P700 molecule is borrowed and eventually returned. In contrast, Photosystem I does not return the liberated electron and thus engages in non-cyclic photophosphorylation. Note that the first step, photolysis of water is what yields molecular oxygen and that this step only occurs in Photosystem II. Photosystem I is anoxygenic. This Z-scheme image has been obtained by the author from the Wikimedia website where it was made available under a CC BY-SA 3.0 license. It is included within this book on that basis. It is attributed to Bensaccount. From Biology LibreTexts: https://bio.libretexts.org/Bookshelves/Botany/ Botany_(Ha_Morrow_and_Algiers)/04%3A_Plant_Physiology_and_Regulation/4.01%3A_Photosynthesis_ and_Respiration/4.1.05%3A_The_Light-dependent_Reactions. Image by Jen Valenzuela (http://dropindol-phin.com/).

oxygen and are therefore lumped together as **oxygenic photosynthesis**. Note that oxygenic photosynthesis uses a certain type of chlorophyll molecule (**chlorophyll a**), which is different from the bacteriochlorophyll molecules used in the various forms of anoxygenic photosynthesis discussed below. Specifically, the particular pathway involved in making oxygen by **photolysis** of water is called **non-cyclic photo-phosphorylation**. Although this is more complicated than **cyclic photophosphoryla-tion**, we shall review the non-cyclic process first since that is the oxygen-generating one that happens to be used in familiar photosynthetic organisms.

    Cyanobacteria and all their descendants in eukaryotic organisms (that is, the **chloroplasts** in algae and plants) employ the **z-scheme** (figure 12.2). It is a two-stage process that when plotted out looks like a 'Z', thus the name. This two-step process begins with non-cyclic photophosphorylation. This starts out with the absorption of a photon by a certain type of chlorophyll a molecule called **P680**. The name P680 indicates that it absorbs photons of 680 nm. This initial step takes place in a part of the chloroplast called the **thylakoid membrane** (which is functionally analogous to the cristae of mitochondria when it comes to creation of a membrane potential gradient). P680 is part of a molecular **reaction center** called **Photosystem II**. Without going into the specific molecular details, P680 absorbs a photon and become ionized. The electron is passed onto other membrane-associated molecular electron acceptors in Photosystem II. The electron is subsequently passed from one electron acceptor to another, setting up a proton gradient along the thylakoid membrane, just as occurs in mitochondria. Oxidized (ionized) P680 is a very powerful oxidizing agent; it can even oxidize water. Thus, the electrons that are lost from P680 are replaced by electrons donated by water. In other words, during non-cyclic photophosphorylation, the electron donor is water. And when water is oxidized this way, the products are protons and molecular oxygen. Hence, the name **oxygenic photosynthesis**. Meanwhile the electrons that are passed along the electron transport chain lead to protons being transported from the chloroplast **stroma** into the **thylakoid lumen**. The thylakoid lumen thus becomes more positive and acidic, while the stroma becomes negative and alkaline. Eventually the electrons are passed onto $NADP^+$, which is reduced to NADPH. And this NADPH provides reducing power that can then be used in the Calvin cycle or other light-independent ('dark') reactions to fix inorganic carbon and make carbohydrates.

In the second part of the z-scheme, **cyclic photophosphorylation**, the special photon-absorbing chlorophyll a molecule in the reaction center is called **P700** since it absorbs photons of ~700 nm. P700 is located in a pigment complex called the **Photosystem I** reaction center. Absorption of a photon excites P700, which then donates an electron to a series of electron acceptors in another electron transport chain along the chloroplast stromal lamellae. As with the other electron transport chains discussed, a proton motive force is created across a membrane, and this gradient can be used to power ATP synthesis. The electrons are passed from one electron acceptor to another along the chain until it is returned to Photosystem I. Since the electron is returned to its original source, it is called cyclic photophosphorylation. Importantly, the cyclic photophosphorylation of Photosystem I, water is not an electron donor and oxygen is not liberated.

As mentioned above, green plants and cyanobacteria produce oxygen via oxygenic photophosphorylation, which is made in Photosystem II during non-cyclic photophosphorylation. The oxygen is manufactured because water is the electron donor. Evolution of such a mechanism was fabulous since water is so abundant. It guaranteed any cells that used it that they would never be at a loss for photosynthetic basic building blocks (provided they were not dangerously desiccated and had other things to worry about than maintaining photosynthesis!). But water is not the only possible electron donor in photosynthesis. There are many variations on the photophosphorylation theme—but when water is not the electron donor, oxygen is not the waste product. These different photosynthetic pathways are collectively called **anoxygenic photosynthesis**. As mentioned earlier, purple sulfur bacteria and green sulfur bacteria are common examples of anoxygenic photosynthesizers. Instead of H_2O as the electron donor, these organisms use H_2S. And instead of liberating oxygen gas, they produce granules of elemental sulfur. Microbes that participate in anoxygenic photosynthesis use pigments other than chlorophyll a (the green pigment in cyanobacteria, algae and plants). Instead, they employ bacteriochlorophylls. Bacteriochlorophylls are lettered a–g. Bacteriochlorophylls are molecules that are structurally related to the chlorophylls and also have a central magnesium coordinate ion. Like chlorophylls, bacteriochlorophylls are tetrapyrrole ring structures that can be classified as chlorins (although technically, bacteriochlorophylls a, b and g are partially hydrogenated bacteriochlorins, while bacteriochlorophylls c, d, e and f are true chemical chlorins). The bacteriochlorophylls of anoxygenic photosynthesis absorb wavelengths far longer than those absorbed by chlorophyll a. For example, bacteriochlorophyll b has a wavelength of maximum absorption at 790 nm. And anoxygenic photosynthesizers do not use a two-system Z-scheme—they only use a single photosystem, thereby restricting them to cyclic electron flow. Green sulfur bacteria have a special bacteriochlorophyll molecule called P840, whereas purple sulfur bacteria start their non-cyclic photophosphorylation process off with even longer wavelengths via their P870 bacteriochlorophyll molecules in their reaction centers. Incidentally, some phototrophic bacteria also use short wavelengths (<400 nm) as well.

For thoroughness, it should be mentioned that phototrophy comes in two basic forms: chlorophototrophy (which is the chlorophyll-based form discussed above) and the non-chlorophyll-using retinalotrophy, which uses another molecule (rhodopsin) as an alternative to chlorophyll. This rhodopsin-based phototrophy is perhaps the simplest and most phylogenetically ancient persistent form of photosynthesis. It relies on simple photon-induced conformational changes in rhodopsin molecules to pump electrons across a membrane and set up the ATP-generating electrochemical potential. It can be seen in certain archaea and Proteobacteria.

Upon reflection, one might have anticipated that Life would somehow and inevitably exploit a means of using the ample water (H_2O) and abundant sunlight to power biochemistry. Cyanobacteria were among the first to develop a highly efficient means of using simple sunlight and water to manufacture ATP. But in so doing, they polluted the atmosphere and hydrosphere with the then-toxic byproduct—molecular oxygen. This extremely reactive waste product was directly toxic to many of the extant forms of life when cyanobacteria first began exploiting oxygenic photophosphorylation for their own selfish gain. While pulling carbon dioxide from the atmosphere (and thereby potentially affecting the climate) and generating molecular oxygen (thereby poisoning the prokaryotic biosphere), cyanobacteria were the first mass polluters. And the pollution they profligately poured out most definitely had a major impact on the planet's biosphere. Of course, this impact was not all bad, since without the 'pollution' (i.e. oxygen) produced by the cyanobacteria, we and all forms of metazoan life (with the possible exception of the **Loricifera**), would not be here. Molecular oxygen, the key to the highly efficient aerobic respiration process, was necessary to take the next step in evolution from simple, single-celled life forms to large, mobile, multicellular and intelligent animals. It is amazing but perhaps not surprising that these two completely separate but unequivocally biochemically superior processes evolved on our biosphere, with the first one (manufacture of molecular oxygen) being an absolute prerequisite for the other (aerobic respiration).

And the conversion of an essentially anoxic planet with an anaerobic biosphere into one that was oxygen-replete and capable of capitalizing on the highly efficient aerobic respiration mechanism, is certainly relevant to our main text. The oxygenation of Earth's biosphere was accompanied by a new global radiochemistry and radiobiology—one that had become dramatically more affected by the indirect effects of radiation through reactive oxygen species.

12.10 Chemolithotrophy

By definition, chemotrophs obtain their energy through the oxidation of electron donors (i.e. reduced chemical compounds) found in their environments. This distinguishes them from phototrophs that obtain their energy from light. Recall that autotrophs are organisms that obtain their carbon through fixation of carbon dioxide, whereas heterotrophs must obtain their carbon by eating food made of organic substances. As introduced above, another nutritional distinction is sometimes made among heterotrophic chemotrophs that use organic compounds as their energy source (**chemo-organoheterotrophs**) and those that use inorganic compounds as their energy source (**chemolithoheterotrophs**). Students should be aware of a simpler use of similar terms that is also encountered in the literature in which autotrophs are called **lithotrophs**, while heterotrophs are **organotrophs**. The inconsistent use of terms in the literature can be very confusing (which is why so much introductory biology was reviewed in the early part of this chapter!). Here we will *not* make 'lithotrophy' entirely synonymous with autotrophy. Rather we will define chemolithotrophs (which literally means 'rock eaters') as organisms that can metabolically obtain energy from the oxidation of inorganic compounds. They happen to usually obtain their carbon from the fixation of carbon dioxide. Thus, chemolithotrophy describes a form of energy metabolism that can use the oxidation of reduced inorganic substances as a source of energy for cell biosynthesis and maintenance in the absence of light. This definition originally dates back to Rittenberg in 1969.

Using this definition, all chemolithotrophs are microorganisms. Although chemolithotrophs may be bacteria or archaea, there are no currently known macrofauna or macroflora capable of using inorganic compounds as electron sources. There are two major objectives to chemolithotrophy: the generation of energy in the form of ATP and the generation of reducing power in the form of NADH, NAHPH, etc. Chemolithotrophs might harvest and transduce energy from the oxidization of inorganic molecules through aerobic or anaerobic respiration. Many but not all lithoautotrophs are extremophiles.

Iron-oxidizing bacteria, which use ferrous iron (Fe^{2+}), constitute a major group of chemolithotrophs, as do the **sulfur bacteria**, which use reduced inorganic sulfur compounds (such as hydrogen sulfide, elemental sulfur, thiosulfate and sulfite) as their energy source. Other chemolithotrophs can derive energy from oxidation of other inorganic substances including other reduced divalent cations (e.g. Mn^{2+}), carbon monoxide, reduced nitrogen compounds and hydrogen gas. Aerobic oxidations of NH_3 (ammonia) to NO_2^- (nitrite) and of NO_2^- (nitrite) to NO_3^- (nitrate) are classic examples characteristic of (but not restricted to) chemolithotrophs such as *Nitrosomonas* and *Nitrobacter* species. Gram-negative and Gram-positive bacteria that oxidize molecular hydrogen gas (H_2) as a source of energy are called **Knallgas bacteria**.

Although all known chemolithotrophs are microbes, there are examples of mutualistic symbiotic relationships between chemolithotrophic bacteria and deep-sea vent tube worms of Phylum Annelida. The chemolithotrophs grow as prokaryotic symbionts. In this manner, these chemolithotrophic symbionts might serve as a living analogue to how modern plants acquired chloroplasts. Chloroplasts and other plant plastids are believed to be endosymbionts that were engulfed and permanently retained within cells of the ancestors of plants. It may be that the giant tube worms and other metazoan inhabitants of this completely **aphotic** marine environment are extant examples of macrofauna acquiring new and novel microbial organelles through endosymbiosis. Given the complete absence of light in this environment of extreme heat and pressure, the only source of energy is chemical, and the primary producers at the deep-sea vents are chemosynthetic bacteria and archaea. Amazingly, an entire ecosystem has evolved around these aphotic vents with chemolithoautotrophs at the base of the food chain.

12.11 'Radiotrophy'

A truly extraordinary Gram-positive rod belonging to the Phylum Firmicutes, *Candidatus Desulforudis audaxviator*, can be described as a motile, sporulating, sulfate-reducing, chemoautotrophic thermophile and obligate anaerobe that can fix its own nitrogen and carbon by using biosynthetic machinery it might have acquired via horizontal gene transfer from the Archaea. *Candidatus Desulforudis audaxviator* was discovered in a gold mine in South Africa, roughly 2.8 km (or 1.7 miles) deep. The name 'audax viator' translates to 'bold traveler' thanks to its discovery in this extreme environment. And it is unique in that it is the only species known to be completely alone in its isolated subterranean ecosystem. But these features are not what makes it so extraordinary. What is most remarkable about this organism (from the perspective of this book) is that it was the first bacterial species shown to obtain its energy via ionizing radiation.

This extremophilic species uses hydrogen as its source of reducing power and is thus a lithotroph, specifically a Knallgas bacterial chemolithotroph. Not surprisingly, given its extreme underground environment, it also employs anaerobic respiration as its means of making ATP. But what is most unusual about *Candidatus Desulforudis audaxviator* is that the primary source of its needed hydrogen is **radiolysis** of water via radioactive decay of uranium, thorium and potassium-40. In other words, this organism lives in a **radiolysis-powered ecosystem**.

Candidatus Desulforudis audaxviator is quite unusual in that it indirectly exploits the energy of gamma photons released by radioactive decay for its nutritional needs. Radiolysis of water molecules and minerals yields *in situ* H_2, formate and sulfate. Such radiolytically generated reduced organic and inorganic molecules then provide the needed energy and nourishment to sustain microbial metabolism in the ecosystem. Hydrogen gas and formic acid are the main electron donors in the lithotrophic microbial metabolism while sulfate reduction is the primary electron-accepting process in its anaerobic respiration. Much of the chemical energy available in the radiolytic products that are generated through uranium decay is harvested for biosynthesis, homeostasis and cell division, but some of it is devoted to repairing the inevitable DNA damage induced by ionizing radiation in their relatively radioactive environment.

The gamma rays from radioactive decay cleave water molecules (i.e. they cause radiolysis of H_2O) and minerals in rocks to generate, among other radiolytic products, sulfates and hydrogen peroxide. These substances in turn provide the bacteria with the energy they need to thrive. Based on these bizarre and remarkable findings, Dimitra Atri has speculated that alien life forms might possibly utilize similar mechanisms to survive in radiation-replete extraterrestrial environments. Contrary to earthbound expectations and biases, such life might oddly benefit from the absence of a thick convivial atmosphere or a strong planetary magnetic field. Because the alien life-sustaining radiation source in such circumstances could include galactic cosmic rays, the absence of a cosmic ray-absorbing atmosphere or a charged particle deflecting planetary magnetic field could actually be an advantage for the evolution of such exotic alien life.

Unlike terrestrial life forms, which can tap into the ample energy from sunlight on the surface of Earth, alien life forms that are dependent on galactic cosmic rays as their main source of energy would have to remain extremely tiny, just as deep-Earth *Candidatus Desulforudis audaxviator*. Atri has suggested that Mars is the best nearby candidate to host such cosmic ray-powered life based on the planet's surface composition (which is rich in silicates and other minerals) and the possibility of frozen subsurface water. These minerals and water molecules could be disintegrated and spallated by galactic cosmic rays to generate different energy-rich chemicals that could be used by Martian microbes—just as *Candidatus Desulforudis audaxviator* uses such chemicals produced by radioactivity in deep Earth. These observations, most ironically, make Mars and similar extraterrestrial bodies attractive potential abodes for truly exotic radiation-powered life.

12.12 Anaerobic respiration with radioactive electron acceptors

As discussed above, organisms can use molecular oxygen as an electron acceptor in the highly efficient aerobic respiration process, or they can use less efficient alternatives to molecular oxygen in the anaerobic respiration process. Among these alternatives to oxygen are oxidized states of iron, manganese, cobalt, nitrogen and sulfur—among several others. In the category of odd 'others' is uranium. Specifically, some microbes are known to use soluble U(VI) as an electron acceptor and reduce it to insoluble U(IV).

Among these unusual microorganisms is **Geobacter metallireducens**. *G. metallireducens* is an obligate anaerobe Gram-negative proteobacterium that can oxidize several organic compounds with Fe(III) (oxidized ferric iron) but can also use oxidized uranyl cations (U(VI)) as an alternative. *G. metallireducens* is the most effective known microorganism for this reduction of uranium, but others exist. For instance, **Shewanella oneidensis** is a facultative bacterium that can survive and proliferate in the absence or presence of oxygen. Organisms that are restricted to one choice or the other (aerobic or anaerobic environments) are called obligate. For example, we are obligate aerobes. Organisms that can survive under aerobic or anaerobic conditions and switch from aerobic respiration to fermentation accordingly are known as facultative anaerobes. In any case, *S. oneidensis* is a proteobacterium that was first discovered in Lake Oneida, NY, hence its name. Like *G. metallireducens*, it can use U(VI) (oxidized uranyl) cations in anaerobic respiration, thereby reducing these soluble uranyl ions to an insoluble form. *G. metallireducens* and *S. oneidensis* are iron-reducing bacteria that can also reduce oxidized uranium. In a similar fashion, *Desulfovibrio desulfuricans* is a sulfate-reducing bacterium that can reduce uranium as an alternative. Another bacterial species with the ability to reduce U(VI) is *Anaeromyxobacter dehalogenans*.

These species are remarkable for their versatility in the choice of oxidized electron acceptors, which includes uranium. But given the toxicity of heavy metals such as uranium, there is great interest in using these and similar microorganisms in remediation of hazardously contaminated environments. U(VI) is soluble and mobile, meaning that in sufficient concentrations, it can cause a problem through groundwater contamination. Microbial bioremediation can, in principle, convert the soluble and mobile form of uranium into an insoluble and immobile form, allowing its removal.

Further reading

[1] Petkowski J J, Bains W and Seager S 2020 On the potential of silicon as a building block for life *Life* **10** 84

[2] Atri D 2016 On the possibility of galactic cosmic ray-induced radiolysis-powered life in subsurface environments in the Universe *J. R. Soc. Interface* **13** 20160459

[3] Lin L H *et al* 2006 Long-term sustainability of a high-energy, low-diversity crustal biome *Science* **314** 479–82

[4] Terranova M L 2021 Radioactivity to rethink the Earth's energy balance *Glob. Chall.* **5** 2000094

[5] Atri D 2020 Investigating the biological potential of galactic cosmic ray-induced radiation-driven chemical disequilibrium in the Martian subsurface environment *Sci. Rep.* **10** 11646

[6] Hooper A B and DiSpirito A A 2013 Chemolithotrophy *Encyclopedia of Biological Chemistry* 2nd edn ed W J Lennarz and M Daniel Lane (New York: Academic) pp 486–92

IOP Publishing

Space Radiation
Astrophysical origins, radiobiological effects and implications for space travellers
James S Welsh

Chapter 13

Elementary zoology

Defining life itself has proven surprisingly challenging. From a standard dictionary, one will find descriptive entries like 'the quality that distinguishes vital and functional microorganisms, animals and plants which are made of organic matter from inanimate inorganic matter, including the capacity for growth, reproduction, functional activity, and continual change preceding death'. While such a definition may be descriptive of the living condition as opposed to death, it is cumbersome. NASA has provided a more succinct yet precise definition: 'a self-sustaining chemical system capable of undergoing Darwinian evolution'. While no definition of life is perfect, many agree that NASA's current definition is a practical, working definition that is the best we have—until someone comes up with something better.

Thanks to the natural radioactive decay of once-abundant primordial radioisotopes, background radiation on Earth is presently far lower than it once was. For example, the neptunium chain ($4n+1$ series) is now long extinct on Earth. And the parent radionuclide of the actinium series (the $4n+3$ series) begins with U-235, which only has a half-life of around 700 million years. Thus, over Earth's 4.6-billion-year lifetime, most of the radioactivity from this series has decayed away as well. Earth was once far 'hotter' in the radioactive sense than it is now, and life initially evolved in a much more intense radiation environment than we have presently. Thus, early organisms must have developed various means of coping with this increased radiation—and it would be surprising if life today had 'forgotten' how to cope with that radiation even though the background levels are much lower than they once were.

In upcoming chapters, we will be discussing inherent radioresistance in various organisms. To facilitate such discussion, a brief review of elementary zoology might be helpful. Organisms may be placed into three major Domains of life: the Bacteria, the Archaea and the Eukaryota. The first two Domains share the common feature of prokaryotic cells. The Domains may be divided into five or six Kingdoms: Animalia, Plantae, Fungi, Protista, Archaea and Bacteria (or Eubacteria). Some experts prefer to use only five kingdoms, lumping the archaea and bacteria together as Kingdom Monera. Below the kingdoms, smaller taxa (categories) follow the Linnean system including Class, Order, Family, Genus, and Species. Individual species are italicized and named using the binomial nomenclature approach of genus plus species with the genus name capitalized but the species name not (e.g. humans belong to the species *Homo sapiens*).

Unicellular 'animals' are commonly called protozoans and are lumped with unicellular plants and algae in Kingdom Protista. Green plants belong to Kingdom Plantae, which may or may not include fungi. Modern taxonomy typically separates fungi from plants and allocates them to their own Kingdom, Fungi. Kingdom Animalia, the multicellular animals, can be broken down into numerous phyla and other taxa. Animals are multicellular heterotrophs composed of eukaryotic cells that lack cellulose cell walls or photosynthetic pigments. While it might appear a bit anthropocentric to focus so heavily on animals and zoology, in fact there are far more known species of animals than in any of the other Kingdoms or Domains. In fact, there are more known insect species than all the species in all other Kingdoms and Domains combined. Note that while the modern custom is to not capitalize the various Linnean taxa, in this chapter an older convention of capitalizing them has been used.

13.1 Parazoans: sponges

Because sponges are more like simple amalgamations of different types of cells with different functions rather than organisms with their cells arranged into true tissues (cells organized into functional units), some treatments pull sponges (Phylum Porifera) out as the so-called Parazoa, leaving all other animals in subkingdom Eumetazoa (or simply the metazoans).

13.2 Metazoan animals

Metazoan animals can be classified according to various schemes. One such older method was based on body morphology wherein those with radial symmetry were lumped together as the **Radiata**, while those animals with bilateral symmetry constituted the **Bilateria**. Nowadays, Radiata is no longer considered a legitimate scientific grouping since the organisms exhibiting radial symmetry as adults are not all genetically (i.e. ancestrally) related. Furthermore, adult starfish or sea urchins might demonstrate radial symmetry, but their larval forms exhibit clear bilateral symmetry, so these echinoderms would be enigmatic in the Radiata/Bilateria classification scheme. In other words, the Radiata is **paraphyletic**, meaning it is a grouping that is not truly ancestrally related; this is to be contrasted with **monophyletic** groups who do share common ancestry. Modern taxonomy schemes aim to organize animals into groups that are ancestrally related, although getting it right is often very challenging.

One well-known means of categorizing animals is to break them down grossly into two groups—those with backbones (vertebrae) or those without vertebrae; these are the **vertebrates** and **invertebrates**, respectively. Since the vertebrates are typically much larger and more familiar to us, this broad classification is popular, but one must recall that the vertebrates all belong to only one phylum (Phylum Chordata), while there are approximately 30 other phyla that make up the invertebrates.

Classification *based on embryonic development* remains a legitimate means of organizing animals into monophyletic categories. For example, animals in which the **embryonic blastopore** becomes the anus are all called **deuterostomes**. (The blastopore is an opening in the blastula, which is an early hollow spherical embryonic stage.) Animals in which the embryonic blastopore becomes the mouth form another large taxon, the **protostomes**. Humans are vertebrates, vertebrates are chordates, and chordates are deuterostomes. As unlikely as it might seem, our closest relatives among the invertebrates are other deuterostomes, such as the echinoderms (like starfish and sea urchins). Deuterostomes are formally classified under **Superphylum Deuterostomia** and include Phylum Echinodermata, Phylum Hemichordata and Phylum Chordata.

The protostomes and deuterostomes also differ in some striking other aspects of development. In all animals, the sperm and egg unite to first form a **zygote**. The zygote then undergoes division to form cells called blastomeres. Cell division in early embryonic cells is called **cleavage**. After the zygote has undergone cleavage, a compact solid mass, the **morula**, is formed. Following additional cleavage, a **blastula** is created. Subsequent cell divisions follow various geometric patterns that are described as radial, spiral or rotational with respect to the primary axis of the embryonic body. While cleavage patterns were once used for classification, they are unreliable for creating truly phyletic groupings. For example, while mammals display rotational cleavage, so does the nematode worm, *Caenorhabditis elegans*. And whereas most mollusks display spiral cleavage, cephalopod mollusks (such as squid and octopus) exhibit a variant called meroblastic bilateral cleavage. Therefore, embryonic cleavage patterns can no longer be used as a legitimate means of categorizing animals into genetically related taxa. As an interesting aside, however, cleavage differs from ordinary mitosis in that cleavage increases the number of cells and nuclear mass without simultaneously increasing the cytoplasmic mass. In other words, upon each cell division during cleavage, the amount of cytoplasm in each daughter is only half as much as in the parent. In this manner, the N:C ratio (ratio of nuclear to cytoplasmic material) increases steadily during cleavage. This is not characteristic of mitotic cell division in dividing adult tissues.

Protostomes typically exhibit **determinate cleavage (mosaic cleavage)**, which means that the ultimate fate of individual cells in early embryonic development is set. Removal of a cell from the blastomere of a protostome might lead to an individual without an eye, for instance. Deuterostomes, on the other hand, exhibit **indeterminate cleavage (regulative cleavage)**, which means that the ultimate fate of individual cells in very early development are not pre-set, and if one cell was removed from or killed in a blastula, others can eventually compensate for the loss of that one individual cell during development in embryos displaying indeterminate cleavage. It will not lead to the absence of a particular structure in the adult animal. Additionally, each individual cell might be capable of developing into a whole organism if separated.

13.3 Endoderm, mesoderm and ectoderm

It turns out that both protostomes and deuterostomes have **three embryonic germ layers** or **primary embryonic layers** of cells. Thus, protostomes and deuterostomes are **triploblastic**. All metazoans are either triploblastic or **diploblastic** (with only two embryonic germ layers). The latter include the cnidarians and ctenophores, two groups in the older category of Radiata. And all the Bilateria are triploblastic metazoans, conferring a modicum of credibility to the old taxonomic classification scheme. The three germ layers, from inside to outside are the **endoderm**, **mesoderm** and **ectoderm**. In the adult organism, various organs can trace their origins to these embryonic germ layers. For instance, in humans, the liver (and nearly the entire gastrointestinal tract) is derived from the endoderm; the heart, skeletal muscles and kidneys are largely of mesodermal origin; and the skin, enamel and nervous system are primarily of ectodermal origin. Among the triploblastic metazoans (that is, the animals with three embryonic germ layers), some have a **body cavity** or **coelom**, while others do not. It turns out that the presence, absence and pattern of formation of these body cavities is a useful means of classifying animals.

13.4 The coelom

As just mentioned, animals can be classified according to how their internal body cavity (coelom) is created. Protostomes are **schizocoelous**, meaning that their coeloms are formed by a splitting of their middle body layer, the mesoderm. In contrast, the deuterostomes create their coeloms by a process called **enterocoely**. Enterocoelous animals generate their internal body cavities though outpouching of the embryonic gut (**archenteron**).

The triploblastic metazoans exclude sponges (poriferans), jellyfish and their kin (cnidarians) and comb jellies (Phylum Ctenophora). While flatworms (Platyhelminthes) and ribbon worms (Nemertea) have a mesoderm and are therefore triploblastic, they do not possess a clear body cavity in the adult and are called **acoelomate**. Roundworms (Nematoda) and **rotifers** (Rotifera) do possess a body cavity, but this coelom is better called a 'pseudocoelom' because it is not completely lined by mesoderm. All more complex metazoan animals have a 'true' coelom, that is a body cavity fully lined by mesoderm. Internal organs are harbored within these true coeloms and held in place by linings called mesenteries. In humans for instance, the coelom (or **peritoneal cavity**) is lined by a layer called the **peritoneum**. And extensions and folding of the peritoneum create **mesenteries** and **omenta** that line and hold the internal organ (**viscera**) in place.

Among those with true coeloms (the so-called **eucoelomates**), those that exhibit schizocoely (splitting of mesoderm to make the coelom) are the protostomes, and those exhibiting enterocoely (outpouching of the primitive gut or archenteron to form a mesoderm-lined coelom) are the deuterostomes.

A brief summary of Kingdom Animalia is as follows:

1. **Subkingdom Parazoa**: multicellular animals *without* genuine tissues or a digestive cavity:
    a. Phylum Porifera (sponges).

2. **Subkingdom Metazoa**: multicellular animals with cells organized into tissues and having a digestive cavity or tract:
    a. **Diploblastic metazoans** with radial symmetry but lacking a central nervous system:
        i. Phylum Cnidaria (Coelenterata)—corals, jellyfish, anemones and box jellies
        ii. Phylum Ctenophora—comb jellies
    b. **Triploblastic metazoans** with bilateral symmetry and a central nervous system:
        i. **Acoelomates** (metazoans without a body cavity)
            1. Phylum Platyhelminthes (flatworms)
            2. Phylum Nemertea (ribbon worms)
            3. Phylum Gnathostomulida
            4. Phylum Mesozoa
        ii. **Pseudocoelomates** (body cavity incompletely lined by mesoderm)
            1. **Phylum Rotifera** ('wheel animalcules')
            2. Phylum Loricifera
            3. Phylum Gastrotricha (gastrotrichs)
            4. Phylum Nematomorpha (horsehair worms)
            5. Phylum Nematoda (nematodes)
            6. Phylum Acanthocephala (spiny-headed worms)
            7. Phylum Kinorhyncha
        iii. **Eucoelomates** (animals with a 'true' body cavity lined by mesoderm)
            1. Phylum Bryozoa (moss animals)
            2. **Superphylum Protostomia**: eucoelomate metazoans where the blastopore becomes the mouth. Protostomes create their body cavities through schizocoely.
                a. **Phylum Tardigrada** (tardigrades or water bears)
                b. Phylum Brachiopoda (brachiopods or lampshells)
                c. Phylum Mollusca (Mollusks—clams, mussels, oysters, scallops; snails; cephalopods (octopus and squid))
                d. Phylum Annelida (segmented worms)
                e. Phylum Sipunculoidea (peanut worms)
                f. Phylum Arthropoda (insects, crustaceans, arachnids, millipedes, centipedes and horseshoe crabs).
            3. **Superphylum Deuterostomia**: eucoelomate metazoans where the blastopore becomes the anus. Deuterostomes create their body cavities through enterocoely.
                a. Phylum Chaetognatha (arrow worms)
                b. Phylum Echinodermata (starfish, sea urchins and sand dollars, sea cucumbers, brittle stars and serpent stars and sea lilies)
                c. Phylum Hemichordata (acorn worms)
                d. Phylum Chordata (the chordates).

Although taxonomists presently debate the specifics, the chordates have tradition-ally been divided into other categories including:

Subphylum Urochordata (tunicates)
Subphylum Cephalochordata (lancelets)
Subphylum Vertebrata (vertebrates).

The vertebrates are then traditionally classified into eight Linnean Classes:

1. Class Agnatha (jawless fishes)
2. Class Placoderms (extinct armored fishes)
3. Class Chondrichthyes (cartilaginous fishes—sharks, skates and rays)
4. Class Osteichthyes (bony fishes)
5. Class Amphibia (amphibians)
6. Class Reptilia (reptiles)
7. Class Aves (birds)
8. Class Mammalia (mammals).

Purists will rightfully criticize these traditional Linnean Classes (named after Carl Linnaeus, as set forth in his 1735 work, *Systema Naturae*), complaining that some are paraphyletic and not valid when it comes to categorizing animals based on whether they are genetically related. For example, reptiles, birds and mammals might better be classified according to modern **cladistical analysis** based initially on patterns of **temporal fenestration** (holes in the side of the head for jaw muscles to bulge into). Using this approach, various clades can be created:

1. Anapsida
2. Diapsida
3. Euryapsida
4. Synapsida.

This classification is based on the specific type of temporal fenestration. Such cladistic analysis has proven value in vertebrate paleontology for forming valid monophyletic classification. But based on temporal fenestration, vertebrate 'dynas-ties' can be formed such that those with the synapsid skulls are the Synapsida, whereas the Sauropsida includes vertebrates with the other skull fenestration patterns. In this fashion, Class Reptilia can be demonstrated as an **illegitimate taxon** that is better represented by the **Sauropsida** that includes both birds and reptiles under a single umbrella. That birds and reptiles are best catalogued together is illustrated by one definition of the **dinosaurs**—'the group of vertebrates consisting of the most recent common ancestor of *Triceratops* and modern birds (Neornithes) and all its descendants'. By this definition, modern birds (Class Aves in the Linnean taxonomic system) are really extant dinosaurs.

A curiosity among metazoans is **Phylum Loricifera**, which currently contains the only members that are capable of surviving in the complete absence of oxygen.

With only 40 known species, these tiny (typically $\leqslant 1$ mm) creatures were unknown to science before 1983. Some **benthic** loriciferans live on the seafloor >3000 meters below the surface of the Mediterranean Sea (in the L'Atalante basin) and have been found flourishing in completely oxygen-free habitats. The **halocline** (salinity gradient) causes a pronounced **pycnocline** (density gradient) in some parts of the lower Mediterranean Sea that precludes vertical mixing thanks to the absence of strong oceanic currents. Thus, these benthic regions meet the definition of 'extreme' in the sense that they are hyper-saline, high pressure, highly sulfidic (i.e. replete with toxic hydrogen sulfide), aphotic (without light), and anoxic. The fact that this environment remains totally devoid of oxygen made it most surprising to discover any metazoan animals thriving there. The exact mechanism whereby these animals can live without oxygen is uncertain, but it is presently believed that their secret to success might lie in specialized organelles called **hydrogenosomes**—modified mitochondria that can produce ATP even in the absence of molecular oxygen and release molecular hydrogen as a byproduct. These specialized organelles are found in certain unicellular anaerobic ciliates, trichomonad and fungi. Given their hypoxic environments and the fact that ionizing radiation magnifies the biological effectiveness of low-linear energy transfer radiation (such as gamma photons and beta electrons) through the **oxygen enhancement effect**, it would be interesting to learn more about the radiation sensitivity of these organisms.

With this basic background in zoology and taxonomy, we shall move onto radiation-resistant extremophiles in the next chapter.

Further reading

[1] Neves R C, Kristensen R M and Møbjerg N 2021 New records on the rich loriciferan fauna of Trezen ar Skoden (Roscoff, France): description of two new species of *Nanaloricus* and the new genus *Scutiloricus PLoS One* **16** e0250403

[2] Fang J 2010 Animals thrive without oxygen at sea bottom *Nature* **464** 825

[3] Danovaro R, Dell'Anno A and Pusceddu A *et al* 2010 The first metazoa living in permanently anoxic conditions *BMC Biol.* **8** 30

Chapter 14

Introduction to radiophiles

Across species, there is tremendous variability when it comes to tolerance of toxins. For instance, dogs and cats are not nearly as severely affected by the delta-hemotoxins in the venom of the Sydney funnel-web spider (*Atrax robustus*), which can cause dangerous complications in humans. Similarly, massive mighty lions can be felled by a cobra bite, while a ten-pound mongoose might be undeterred by their venomous bites thanks to protective mutations in their nicotinic acetylcholine receptors. Another example is provided by the commonly used topical agent for androgenetic alopecia (also called pattern hair loss), minoxidil. Although it is typically well-tolerated by people, even small single doses can lead to death or serious toxicity in cats since they uniquely lack an efficient enzymatic degradation system for this drug. If there can be such variability in tolerance to drugs, might there be a similar variability to radiation tolerance across species?

It is well known that there is significant variation in radiation tolerance of cells even within a single organism. For instance, slowly or non-dividing cells in adult organs such as muscle and bone are far more resistant than rapidly dividing cells in the skin, gastrointestinal tract or seminiferous tubules. Some cells that do not divide, such as lymphocytes, can also express significant radiosensitivity. In clinical radio-therapy, it is observed that some cancers (e.g. lymphomas, germinomas/seminomas, small cell lung cancers) are exquisitely radiosensitive and may apparently respond completely to even low to modest doses (meaning about 20–30 Gy given in daily fractions over a few weeks). It should be noted that radiation sensitivity and rapidity of response to radiation is not synonymous with curability—small cell lung cancer tends to be radiation-responsive but remains notoriously difficult to cure because of its tendency to recur distantly (that is, to spread metastatically).

All animals have limits when it comes to the tolerance of total body radiation exposure, although significant variability is observed. Such limits can be quantified by metrics such as the **LD50**. Another name for the LD50 is **median lethal dose**. To make the LD more meaningful, an additional number is typically added that signifies the timeframe during which the dose is likely to be lethal. For example, the LD50/50 represents the dose that would be lethal for 50% of the exposed population within 50 days. In practice, the LD50 is typically accompanied by numbers that correspond to familiar units of time such as 30 days or 60 days. According to the **United States Nuclear Regulatory Commission**, for adult humans, the **LD50/30** (the acute total body dose that would be fatal for 50% of the population within 30 d) is between 400 and 450 rem (4–4.5 Sv). For humans, this measure of lethality is complicated by the fact that medical intervention might mitigate the radiation exposure and avert death. For example, the human LD50/30 falls into the 4–4.5 Sv range for acute exposure with no medical intervention, whereas the LD50/30 might range up to over 8 Sv when intensive salvage medical intervention is provided. For animals, plants and microorganisms in the wild, such medical intervention is normally not provided, so the LD50/30 is a more consistent metric. Regardless, there is enormous variation across the biosphere in terms of tolerance to ionizing radiation. In this chapter, we shall review some of the most radiation-resistant known organisms and the mechanisms behind their radioresistance.

14.1 Radioresistant prokaryotes

Prokaryotic microorganisms (the archaea and bacteria, collectively known as Kingdom Monera) are generally more radiation resistant than eukaryotic organisms; some prokaryotes are by far the most radiation-resistant organisms known. The radiation-resistant prokaryotes may be described as non-spore-forming bacteria or archaea that can protect their cytosolic proteins from radiation-induced damage and can tolerate multiple DNA double-strand breaks after exposure to acute doses of ionizing radiation >1 kGy. The survival metric typically used in such microbial studies is the D10, which is a reduction in colony forming units down to 10% of the initial value.

14.2 Manganese, deinoxanthin and *Deinococcus radiodurans*

Among the most well-studied radioresistant bacteria is the Gram-positive, non-photosynthetic, obligate aerobic chemoorganoheterotrophic, non-sporulating, red-pigmented *D. radiodurans* with a D10 ≈ 15 kGy. The radiation resistance of *D. radiodurans* resides partly in cytosolic manganese-dependent molecular mechanisms that protect its proteins from various oxidative modifications (such as the introduction of carbonyl groups).

Like other radiation-resistant prokaryotes, *D. radiodurans* accumulates about 300-fold more manganese(II) than radiation-sensitive bacteria typically do. This has led to the belief that radiation resistance in this, and similar bacteria might lie not exclusively with DNA damage control and repair but also with protein preservation. Such protein protection seems to be conferred through operational antioxidant activity of high intracellular levels of manganese(II). The general concept of simultaneous protein preservation does make sense since the extensive radiation-induced DNA damage must somehow be repaired by protein enzymes that have survived the onslaught of high-dose radiation. And widespread double-strand DNA damage in *D. radiodurans* has indeed been observed following very-high-dose radiation exposure; its DNA is not impervious to radiation damage. But there is incredible capacity for DNA damage repair following irradiation. Experiments have demonstrated that *D. radiodurans* can withstand >10 kGy of ionizing radiation, a dose that causes approximately 100 double-strand DNA breaks per genome in this species.

Additionally, intracellular antioxidants play a pivotal role in the radioresistance of *D. radiodurans*. *D. radiodurans* has a robust reactive oxygen species (ROS) and free radical scavenging system. For example, *D. radiodurans* synthesizes a potent unique **ketocarotenoid** scavenger called **deinoxanthin**. Deinoxanthin has been shown to have higher scavenging activity than similar carotenoids such as β-carotene and xeaxanthin.

14.3 Gene redundancy in *D. radiodurans*

D. radiodurans achieves its extraordinary radiation resistance through an array of mechanisms. In addition to deinoxanthin, the highly efficient DNA repair machinery and the manganese-mediated ability to fend off enzyme damage, *D. radiodurans* possesses multiple copies of its genome. Unlike most bacteria, which have but one copy of their circular DNA, *D. radiodurans* is a **multigenomic** organism, which means that it carries multiple copies of the entirety of its genetic information (generally two to eight copies). We humans have two copies of our DNA in each somatic cell; human cells are diploid, while *D. radiodurans* cells are polyploid. But this polyploidy is exhibited in an unusual manner—each *D. radiodurans* genome is made up of multiple copies of two chromosomes, a **megaplasmid** and a smaller plasmid. (Plasmids are small, circular, extrachromosomal DNA molecules found in prokaryotes; megaplasmids are just larger versions of plasmids.) In contrast, in humans and other eukaryotes, the DNA is packaged in nuclear chromosomes. However, eukaryotic cells also contain some DNA within their mitochondria and plastids (in plants); but in many ways, this organelle DNA is more like prokaryotic DNA than like eukaryotic DNA.

14.4 DNA repair in *D. radiodurans*

The first step in repairing radiation-damaged bacterial chromosomes in *D. radiodurans* is **single-strand annealing**, which reconnects DNA fragments. The next step is through classic **homologous recombination**. As discussed elsewhere, homologous recombination is a high-fidelity repair mechanism that (unlike non-homologous end joining) typically does not introduce new mutations (illegitimate recombination). Therefore, even after massive doses that shatter the chromosome into many fragments, overwhelming and lethal accumulation of radiation-induced mutations is not normally observed in this microorganism.

14.5 *D. radiodurans* as an information storage unit

D. radioduran's uncanny ability to repair damaged DNA with impeccable fidelity encouraged researchers to explore its potential as an information storage device. They translated the lyrics of the song 'It's a Small World' into a nucleotide code based on the four nucleotides of bacterial DNA. They then constructed artificial

DNA strands containing various parts of the whole song and then inserted these DNA segments (about 150 bases each) into various bacterial species. After it multiplied for 100 generations, the embedded information for the song was later retrieved from the *D. radiodurans* specimens without distortion, suggesting a novel and robust means of storing data in the event of a nuclear disaster.

D. radiodurans was first discovered in 1956 by Arthur Anderson and colleagues at the Oregon Agricultural Experiment Station. The team was conducting experiments to verify that canned foods could be sterilized through high doses of gamma rays. Cans of meat were exposed to radiation doses that were expected to kill all forms of life. Despite the irradiation, in some instances the canned specimens spoiled, and *D. radiodurans* was isolated from these samples.

D. radiodurans is a **polyextremophile.** This species of bacteria is not only radiation resistant but can also withstand cold (making it a **cryophile** or **psychrophile**), desiccation (**xerophile**), low pH (**acidophile**) and the vacuum of space. It can withstand heat, making it a thermophile (but as explained below, it does not necessarily prefer extremely hot environments). Consistent with its heat-resistance, *D. radiodurans* has a relatively high GC content. (Recall that G≡C is more stable than A=T because it has three hydrogen bonds, whereas A=T has only two. This makes the melting temperature of GC-rich DNA higher than AT-rich DNA.)

14.6 'Radiophilic' versus radiation-tolerant organisms

As a note on nomenclature, *D. radiodurans* and other organisms that can tolerate and thrive in extreme environments are often called **extremophiles.** Several subtypes of extremophiles exist. For instance, a **thermophile** is an organism that displays optimal growth characteristics at high temperatures. Similarly, a **halophile** is an organism that displays optimal growth in very salty solutions. So, a true **acidophile** might be an organism that actually grows better when the pH is highly acidic (e.g. <3.0) and could be contrasted with an organism that simply displays acid tolerance.

Using this terminology, by analogy, a '**radiophile**' would be an organism that exhibits *optimal* growth in environments with high levels of ionizing radiation. Since most so-called radiophilic organisms do not normally encounter high-level radiation in their natural environments and do not truly exhibit better growth in highly radioactive environments, they might better be described as radiation-tolerant organisms (instead of 'radiophilic' organisms). The organisms that we typically call radiophilic might not truly exhibit *better or optimum* growth under conditions of high radiation. Rather, they simply can tolerate high levels of radiation. Nature does not normally provide truly high radiation environments to test this subtle difference. Their radiation resistance could be an **exaptation.** Exaptations are genetic traits that materialized as organisms evolved ways to cope with unrelated challenges in their environments. In developing solutions to one environmental challenge, they inadvertently also acquired resistance to radiation along the way. Naturally, this explanation leaves many unanswered questions.

14.7 Radiophiles as xerophiles

Because *D. radiodurans* is radiation tolerant but not necessarily radiophilic, the origins of its radiation resistance remain enigmatic. Natural selection for extreme radioresistance cannot be the explanation since no microbes are routinely exposed to intense ionizing radiation in the wild and probably never have during their evolutionary history. (A possible exception might be the environment around the natural nuclear reactors in Oklo, Gabon around 2 billion years ago.) Ionizing radiation probably never provided a selective pressure in the evolution of this organism.

The most likely explanation for the resistance to radiation seen in *D. radiodurans* (and in the other radiation-resistant organisms discussed below) is that, while evolving means of coping with the desiccation they occasionally faced in their actual environments, they developed exaptations that also happened to confer radiation resistance. The radiation resistance probably serendipitously came along with the evolution of drought resistance. In other words, radiophilicity came along with xerophilicity.

14.8 A Martian origin?

The xerophilic tendencies of this prokaryote are exhibited also in radioresistant metazoans such as certain **rotifers** and **tardigrades**. Nevertheless, the extraordinary resistance to ionizing radiation displayed by *D. radiodurans* led some scientists to once speculate that the bacterium has an extraterrestrial (specifically, Martian) origin. Unlike anywhere on Earth, Mars does indeed have a high-radiation environment and organisms that can survive or thrive under such conditions would have a selective advantage.

D. radiodurans would have benefitted from its radiation resistance on the surface of Mars and theoretically would have been subjected to selection pressures that refined its radiation resistance there. Then (according to the hypothesis) the hardy bacterium was incidentally transported to Earth on a meteorite. Despite the imaginative idea, scientific scrutiny does not bear witness to this tale. Detailed analysis of its genome and biochemistry makes this hypothesis unlikely, since aside from its extraordinary radiation resistance, 'Conan the Bacterium' is rather ordinary and earthlike at the genetic, metabolic and cellular level. One would expect a Martian microbe to be quite alien in its metabolic makeup. Despite its extraordinary capabilities, *D. radiodurans* is fairly mundane molecularly.

However, in a sense, maybe *most* organisms are in fact, truly radiophilic—in comparison to a zero ionizing radiation background, many organisms display better growth characteristics when grown under natural conditions of background radiation rather than an artificially lowered, zero radiation environment.

14.9 Other tough prokaryotes

D. radiodurans loosely translates in Greek to the 'terrible berry that withstands radiation'. Although touted as the world's toughest bacterium, *D. radiodurans* does have some competition when it comes to radiation resistance. It seems to be able to tolerate 5000 Gy with essentially no loss of viability and even acute doses of 15 000

Gy only reduces viability down to 37% or maybe 10%. Incidentally, the dose required to decrease survival down to 37% (that is, $1/e$) is often called the D_0, and the dose that diminishes colony forming units down to 10% is called the D10 or D_{10}. In comparison, an acute dose of 5000 Gy is sufficient to kill even the most radio-resistant metazoan animals, and only 5 Gy is likely to be lethal for a human. Bacteria are typically tougher than eukaryotes, and it takes 200–800 Gy to sterilize *Escherichia coli*.

14.9.1 Thermococcus gammatolerans

But some other prokaryotes are truly radiation resistant by any definition. Examples include the cyanobacteria *Chroococcidipsis*, actinomycetes species of genus *Rubrobacter*, and certain archaea. Among the archaea, *Thermococcus gammatolerans* stands out as a real challenger to *D. radiodurans*. It has been shown to survive doses up to 30 000 Gy and can sustain acute assaults of 5000 Gy without any obvious loss in viability. These archaea normally inhabit and thrive in the hot, high-pressure waters of deep-sea hydrothermal vents, making them thermophiles and barophiles (or piezophiles—organisms that thrive under pressure). *T. gammatolerans* is a true thermophile—it not only tolerates heat but prefers it. This species thrives in temperatures between 55 °C and 95 °C with an optimum development around 88 °C. But as with other radiotolerant species, it might just happen to coincidentally be radiation resistant rather than genuinely radiophilic.

T. gammatolerans possesses a unique version of **proliferating cell nuclear antigen** (**PCNA**), which plays a key role in DNA replication and repair. Upon investigation through x-ray crystallography, it appears that PCNA is a protein with a poorly conserved amino acid sequence across species, but with a highly conserved 3D structure. The PCNA from *T. gammatolerans* (called $PCNA_{Tg}$) has the highest percentage of charged amino acid moieties. Most of these charged amino acids are negatively charged, with the relative abundance of glutamate over twice that of aspartate. Also noted was a high number of salt bridges and a lack of cysteines and tryptophan. As explained in chapter 10, thanks to nuances in its structure, the $PCNA_{Tg}$ protein has an intrinsic resistance to ionizing radiation. And even though it appears that $PCNA_{Tg}$ has a highly evolved structure that has gone through much natural selection, it remains unclear whether this was actively selected for its radiation resistance or simply tagged along as a **spandrel** (a trait that tags along during selection for something else). As in other radiophilic organisms, the resistance to radiation might be an exaptation rather than something specifically selected for.

14.10 Life below background

The above-mentioned 'radiophiles' might not be truly radiophilic in the sense that they do not actually prefer highly radioactive environments; they simply exhibit an unusual ability to withstand high levels of radiation. But it is known that many organisms flourish better when there is *some* radiation in their environment (as opposed to *no* radiation). In other words, background radiation levels are preferred over an absence of radiation altogether. Whether or not normal terrestrial background radiation is the optimal amount of radiation is a subject of ongoing investigation. It remains possible that at least for some organisms, the ideal amount of radiation is more than present-day environmental background levels. Given that Earth was far more radioactive in the distant past, it seems plausible that some organisms were evolutionarily adapted to background levels that are beyond today's levels.

So, what happens when organisms are grown under conditions of very low or absent radiation by depleting the potassium-40 in their diets, sheltering them from cosmic rays, and shielding them from the natural terrestrial radiation from soil and rocks? Intriguing experiments have been conducted at the **Waste Isolation Pilot Plant**, which is heavily shielded and 605 meters below the surface and thus largely protected from cosmic rays and other forms of background radiation. In one experiment involving *D. radiodurans*, nine-fold reduction of ambient radiation dose-rate (from a normal 72.1 nGy h^{-1} down to 0.9 nGy h^{-1}) delayed entry into the exponential growth phase, resulting in a reduction in biomass. Some specific changes seen in the low radiation environment included downregulation of genes involved in the proper folding of proteins, including newly synthesized and denatured/misfolded proteins. Additional gene downregulation was seen for enzymes involved in the assimilation of nitrogen for amino acid synthesis, genes involved in the control of copper transport and others involved in homeostasis to prevent oxidative stress.

The results showed that *D. radiodurans* is sensitive to the absence of background levels of ionizing radiation (or levels of radiation that are 'too low'). Its transcriptional response is insufficient to maintain optimal growth under such circumstances. Therefore, it is indeed possible that *D. radiodurans* is in a sense a real radiophile that grows better when there is an ideal amount of radiation in its environment. Certainly, it suffers when there is little or no background radiation. This phenomenon has also been observed in other bacteria, in multicellular and unicellular eukaryotes and in mammalian cell cultures. For instance, V79 Chinese hamster cells grown at below background radiation levels showed significantly lower viability compared to those grown at control levels.

Suboptimal growth or deleterious effects have been observed during growth under very low ionizing radiation dose/dose-rate conditions for several cells and organisms. Curiously, there is a growing body of evidence suggesting that all cells, regardless of their taxonomic affiliations (meaning eukaryotes as well as archaea and bacteria) depend on a minimum amount of ionizing radiation for ideal control of fundamental functions such as gene regulation, homeostasis, DNA repair and growth. Perhaps there is a 'Goldilocks' range of radiation at which organisms optimally thrive. These curious findings seem to suggest that many or all organisms are 'radiophiles' in a strict sense of the word.

14.11 Radiation-resistant metazoans: bdelloid rotifers

Rotifers are tiny (50 μm–2 mm), bilaterally symmetrical, unsegmented, pseudocoelomate metazoans belonging to phylum Rotifera. Like all metazoan animals, rotifers lack cell walls, and they are heterotrophic (as opposed to being autotrophic), meaning they must eat to obtain their carbon needed for biosynthesis. And like other animals, they obtain their energy by consumption of reduced organic foodstuffs; that is, they are chemotrophs (as opposed to plants, which are phototrophs capable of generating biochemically useful energy via harvesting photons and converting them into energy-rich reduced chemical compounds). Rotifers are characterized by bilateral symmetry, and they have a body cavity that can be described as a **pseudocoelom** (as described in chapter 13). The pseudocoelom functions as a circulatory system of sorts and provides space for a **complete digestive tract** (i.e. a digestive tract that has both a mouth and anus). The pseudocoelom is a structure that is also found in the closely related Nematoda (the roundworms) among a few other phyla. Phylum Rotifera has approximately 2000 species, and they exist primarily as freshwater zooplankton.

Such small animal life is sometimes alluded to as **meiofauna**, and the rotifers might be the smallest of all animals. All rotifer species have a ciliated organ in the head region (meaning the head has tiny 'hairs' that beat to create a current). This ciliary organ is used for feeding and locomotion and is called the **corona** (which stems from the Latin for 'crown' and should not be confused with the upper solar atmosphere!). The characteristic coordinated motion of the cilia on the corona confers the rotifers with their common name, 'wheel animals' or 'wheel animalcules', because the moving cilia in the corona look somewhat like a rolling wheel; 'rotifer' loosely translates into 'wheel bearer' in Latin. The characteristic appearance of the corona varies from species to species.

Within phylum Rotifera is **class Bdelloidea**, the **bdelloid rotifers** (pronounced with a silent 'b'). There are over 450 species of bdelloid rotifers, and they typically only need a few drops of water to survive, so they can be found thriving on moss, on lichens and in the soil. They can enter varying states of dormancy (**cryptobiosis**) or 'suspended animation' when the environment grows challenging. For example, through desiccation-induced dormancy (**anhydrobiosis**), these microorganisms can survive when their normally aquatic environment dries up. Recently, scientists have unearthed—and revived—bdelloid rotifer specimens frozen in northeastern Siberian permafrost for an incredible estimated 24 000 years. That state of cold-tolerant dormancy can be called **cryobiosis**. (While 24 000 years is astonishing, it is not the cryobiosis world record—that presently belongs to a pair of nematode worms recovered from Pleistocene permafrost dating back 30 000–40 000 years).

Furthermore, these ancient bdelloid rotifers specimens (as well as modern forms) are **obligately parthenogenetic**. In other words, they are exclusively **asexual**; they clone themselves without the need for a sexual partner. It appears that these microscopic invertebrate species have existed for tens of millions of years without sex or meiosis.

Another oddity about bdelloids include the fact that they have relatively few **transposable genes**. But the observation that they can survive and reproduce after complete desiccation (at any stage of their life cycle) is an important finding when it comes to radiation resistance. It may be that concurrent resistance to dehydration is a common theme when it comes to organisms that are radiation resistant.

Rotifers are short-lived; their total lifespan has been recorded at 6–45 d. Additionally, rotifers are **eutelic**, meaning that adults of a given species all have exactly the same number of cells. Unsurprisingly, most eutelic organisms are microscopic. The average cell count in adult rotifer species is in the 900–1000 range. In addition to the rotifers, other eutelic animals include certain nematodes (including the parasitic *Ascaris* and the popular laboratory species, *Caenorhabditis elegans*, in which all males have 1033 cells), the parasitic (and taxonomically enigmatic) **dicyemida** (also known as **rhombozoans**), **larvaceans** (of Class Appendicularia of the Tunicata or Urochordata) and **tardigrades**.

Bdelloid rotifers are notoriously radiation resistant. Some rotifers can recover and resume reproduction after receiving radiation doses that basically shatter their genomes. In some bdelloid rotifers, radiation doses up to hundreds of gray (which introduces hundreds of double-strand DNA breaks) can be tolerated, and the damaged DNA can still be repaired. For instance, examination of the post-irradiated DNA in the bdelloid rotifer species, *Adineta vaga*, revealed that a dose of 560 Gy introduced approximately 500 double-strand DNA breaks, while 1000 Gy generated over 1000 double-stranded breaks.

And the radiation resistance is not limited to low-linear energy transfer (LET) radiation, it extends to space-like particle radiation as well. For example, even when their genomes were disarticulated into small DNA fragments by proton irradiation, some *A. vaga* specimens were able to correct the damage. Therefore, it appears that their ability to withstand radiation lies not with an ability to avoid DNA damage but instead with a remarkable ability to repair such damage. The radiation resistance in at least some species of bdelloid rotifers lies in their extraordinary ability to protect their high-fidelity DNA-repairing machinery from radiation-related oxidative damage. The enzymes needed for such repair are preserved despite high doses of radiation and remain ready for action. For instance, the DNA repair enzymes of *A. vaga* are far more resistant to radiation-induced **protein carbonylation** than is the more radiosensitive nematode *C. elegans*.

However, after gamma-irradiation of another bdelloid rotifer, *Brachionus koreanus*, the activity of the antioxidant enzyme, glutathione S-transferase, increases dose-dependently, suggesting another mechanism whereby this particular species can cope with the oxidative stresses associated with radiation. Nevertheless, gamma irradiation also increases the expression of genes and enzymes responsible for DNA repair such as p53. In this way, the radiation resistance of bdelloid rotifers may be thanks to a combination of effective antioxidant protection mechanisms along with remarkable double-strand DNA break repair. The antioxidant activity might not prevent all DNA double strand breaks, but it might protect and preserve the proteins required for repairing DNA damage.

Nevertheless, if the dose and quality of the radiation is sufficient, even bdelloid rotifers can undergo radiation-induced mutagenesis. A recent study used high-LET carbon and argon ion-beam irradiation to induce mutations that created rotifer mutants with larger loricas (shell-like structures). These mutants were being made as food for larval fish in aquaculture ('larviculture'). The study found that heavy-ion-beam irradiation was an efficient tool for mutagenesis of rotifers and the researchers identified three lines of created rotifers with larger lorica length and adequate population growth for fish larviculture. The radiation used was carbon ions

(1.62 GeV, LET = 23 keV μm^{-1}) at six irradiation doses of 100, 150, 200, 300, 400 and 600 Gy and argon ions (3.8 GeV, LET = 312 keV μm^{-1}) at six irradiation doses of 25, 50, 75, 100, 150 and 200 Gy generated at the RIBF (Radioactive Isotope Beam Factory at RIKEN, Japan). Researchers discovered that carbon ion beam radiation at a dose of 200 Gy was the optimal irradiation condition for the rotifer *B. plicatilis* and 50 Gy of argon-ion-beam irradiation was also effective. They produced the valuable TYC78 strain in this manner.

Like other organisms displaying exceptional radiation resistance, bdelloid rotifers have evolved a suite of adaptations for surviving dry spells. It is probable that some of these xerophilic adaptations have concomitantly contributed to the bdelloids extraordinary radioresistance.

Bdelloid rotifers are undoubtedly among the strangest of all animals. No male bdelloid rotifers have been observed. The females reproduce through **parthenogenesis**, meaning they produce offspring from unfertilized eggs. Fossil evidence suggests that these creatures have reproduced entirely asexually and have avoided sex for 25–80 million years. Despite this, the bdelloids have diversified into over 450 known species. One might wonder how such speciation can occur despite the disadvantage of not having sexual reproduction at their disposal.

Asexual reproduction as a way of life has its disadvantages—and hazards. Without the DNA reorganization that regularly accompanies sexual reproduction, asexual reproduction is considered a poor choice as a long-term strategy since it does not facilitate the genetic diversity needed to adeptly adapt to new environmental challenges. Thus, parasites and diseases may lead to a local epidemic that could in principle wipe out a genetically undiversified asexually-reproducing population. However, their choice of environments that are prone to dehydration allow bdelloids to survive under conditions that their enemies and competitors cannot. Furthermore, dry conditions facilitate lofting and relocation of desiccated bdelloids to new, pathogen-free environments. Thus, their ability to withstand desiccation and their predilection for such habitats may be a means of compensating for the absence of sexual reproduction.

However, one particularly fascinating possibility is that the radiation- and desiccation-induced process of splintering and rebuilding their genomes may provide the bdelloid rotifers with genetic benefits that offset asexual reproduction. When desiccation is severe, the DNA is fractured into multiple small segments. Upon rehydration, their DNA will be restored but it appears that the process of interspecies (and even inter-Kingdom) **horizontal gene transfer** may be possible. This means that foreign environmental DNA (eDNA) from other species (perhaps as distantly related as bacteria, fungi and plants) may be taken up and incorporated into the genomes of bdelloid rotifers. In this manner, up to 10% of the bdelloid genome may be of bacterial, fungal or plant origin.

Additionally, transposons or **jumping genes** can pose a particular hazard to organisms that do not reshuffle their genes regularly through sexual reproduction. Transposons are bits of DNA that jump about in the genome, moving into and out of the DNA and possibly causing gene malfunctions along the way. There are two main categories of transposons:

1. **Retrotransposons**, which move via an RNA intermediate that is copied back into DNA by reverse transcriptase.
2. **DNA transposons**, which move only as DNA via a 'cut-and-paste' transposition mechanism.

Unchecked intragenomic proliferation of retrotransposons may theoretically lead to the early extinction of lineages that abandon sexual reproduction. Nevertheless, bdelloid rotifers frequently repair their DNA and remove uninvited DNA sequences such as transposons. Despite their asexual way of life, bdelloids are devoid of transposons in their genomes. Perhaps this absence of accumulated transposons is a prerequisite for continued success as an asexual lineage.

14.12 The rotifer-B study

A colony of *A. vaga* bdelloids was sent to the International Space Station on the 19th SpaceX Dragon supply mission in December 2019. The goal of this study was to better elucidate the abilities of bdelloid rotifers to withstand radiation and other extreme conditions such as microgravity. The Rotifer B1 colony returned to Earth in January 2020. Results have not yet been published.

14.13 Radiation-resistant metazoans: water bears (tardigrades)

Like the bdelloid rotifers, tardigrades are extremophiles that exhibit tolerance to radiation as well as heat, cold, osmotic stress, pressure, drought and other environmental challenges. In fact, some tardigrades have been exposed to and survived the radiation, frigid temperatures and vacuum of space. Also like the bdelloids, tardigrades are eutelic, meaning that adults have a fixed, predetermined number of cells depending on the species. Furthermore, they are meiofauna, meaning they are tiny animals whose body length may reach 1 mm in the largest species (but typically are in the range of just 0.25–0.50 mm). Tardigrades are eight-legged creatures that may reproduce sexually or asexually through either parthenogenesis or through self-fertilization (thanks to their hermaphroditism).

There are approximately 1300 known species of tardigrades. They belong to phylum Tardigrada. Their precise taxonomic relationships are unknown but most likely they are related to the arthropods. Some zoologists place them in the superphylum Ecdysozoa along with the sister phyla of Arthropoda, Onychophora and Cycloneuralia (although there are studies suggesting a closer kinship to the nematodes). Some tardigrades live in freshwater or seawater, while others are terrestrial. Those that are terrestrial typically live in damp soil, sand, leaf litter, lichens or mosses. As they live in the moisture associated with such habitats, they are called **limnoterrestrial**.

Another feature shared by bdelloid rotifers and tardigrades is the capacity for horizontal gene transfer. However, only a small fraction of tardigrade genes (approximately 1.2%) seem to have been acquired from other species in this fashion.

Along with nematodes and rotifers, tardigrades are one of the three invertebrate groups where adaptations for total desiccation and freezing in the egg, juvenile and adult stages are commonly seen. Such adaptations allow these organisms to survive in extremely harsh environments. Tardigrades and other organisms may survive such severe ecological challenges by undergoing a transformation called **cryptobiosis**. This condition, in which the animal dries out and looks like a lifeless speck of dust, is called the **tun** state and this state of suspended animation is characterized by a metabolic rate as low as 0.01% of the active state.

Cryptobiosis caused by desiccation is known as **anhydrobiosis**, while cryptobiosis caused by freezing is called **cryobiosis**. Tardigrades can undergo both forms of cryptobiosis. In their anhydrobiotic state, their water content is reduced to around 1%–3% of the normal condition, and they can remain in this condition for a decade or so.

Several cryptobiotic invertebrates, including tardigrades and bdelloid rotifers, along with the brine shrimp crustacean *Artemia salina* (of Sea Monkey fame) and *Polypedilum vanderplanki* (a chironomid species of midge insects) have been found to be remarkably tolerant to radiation.

Tardigrades of the species *Paramacrobiotus areolatus* display a very high tolerance to ionizing radiation, with surprisingly similar dose responses in the desiccated (tun) and hydrated states. The LD50/24 h doses are usually reported to be between 5 and 6 kGy. Similar studies on *Milnesium tardigradum, Richtersius coronifer* and *Hypsibius exemplaris* using γ-rays have yielded similar results, with LD50 estimates between 3 and 5 kGy.

The observation of similar dose–responses in the tun and hydrated states seems to be a general feature across various water bear species. This is an unexpected finding. One might expect that, given the far higher indirect effects of radiation in the presence of water, the radiobiological effects should be augmented in the hydrated state. However, this does not appear to be borne out experimentally.

Similarly, the effects of low-LET versus high-LET radiation on tardigrades are paradoxical as well. In general, high-LET radiation is more effective for most biological endpoints. And the direct hits on DNA in the case of high-LET radiation usually implies that chemical radioprotectors (e.g. ROS and free radical scavengers) would be expected to be less effective with high-LET radiation. Despite these expectations, experimental exposures of tardigrades to high-LET radiation (including protons, alpha particles (helium ions) and iron ions) have shown that adult tardigrades are typically as tolerant to high-LET radiation as they are to low-LET radiation. Surprisingly, in some cases, the tardigrades were even more tolerant of high-LET than low-LET radiation. For instance, in a study using protons, the value of LD50/24 h was significantly but perplexingly higher at around 10 kGy in *R. coronifer*.

In another puzzling study using high-LET helium ions on desiccated and hydrated *M. tardigradum* specimens, at the higher doses the animals tended to survive better in the hydrated state compared to desiccated animals. Such interesting but confusing data is presently being analyzed.

Irrespective of the peculiarities, tardigrades have demonstrated an uncanny capacity for withstanding space radiation. And this is not just 'space-like radiation' but actual space radiation. Tardigrades were the first animals to survive unattenuated exposure to outer space. In 2007, tardigrades were taken into low Earth orbit on the **FOTON-M3 mission** as part of the BIOPAN astrobiology payload. Three projects were conducted under the FOTON-M3 mission. The **Tardigrade Resistance to Space Effects (TARSE)** Project was the first of these. TARSE aimed to analyze the impact of the space environment on DNA damage incurred in space on the eutardigrade *Paramacrobiotus richtersi*. In the TARSE project, both active and tun tardigrades were exposed to space radiation and microgravity. The specimens (both active and anhydrobiotic) displayed high survival rates.

The next project in the FOTON-M3 mission was **TARDIS (Tardigrada In Space)**. In this space experiment, eggs of *R. coronifer* and *M. tardigradum* were directly exposed to the vacuum of space, space particulate ionizing radiation and space ultraviolet radiation. No eggs hatched when exposed to all three assaults simultaneously, but exposure to space vacuum or cosmic radiation alone or in combination had no significant effect on hatchability.

Also, some *M. tardigradum* or *R. coronifer* tardigrade specimens were desiccated before the flight and then exposed to the extreme conditions of outer space, including the vacuum, while they orbited the Earth outside a spacecraft for 10 d. Upon their return to Earth, they were rehydrated. Approximately 68% of the animals had survived the extreme cold, vacuum and space radiation. Overall, the experiments proved that tardigrades can survive exposure to the space vacuum, but the addition of factors such as solar ultraviolet radiation, ionizing solar radiation and galactic cosmic rays significantly reduced their survival.

The third project from FOTON-M3 was the **RoTaRad (Rotifers, Tardigrades and Radiation)** experiment. In RoTaRad, the effects of space radiation on short-term and long-term survival were explored along with the preservation of fertility in these hardy metazoans.

Following this was the 2011 Endeavour mission with its Italian project **TARDIKISS (Tardigrades in Space)**. The goal of this project was to glean an understanding behind the repair mechanisms of DNA damage caused by space flight on the final flight of Space Shuttle Endeavour and onboard the International Space Station. This study showed that microgravity and cosmic radiation did not significantly decrease the survival rate of the tardigrade passengers. Perhaps not unexpectedly, the data demonstrated significant differences in ROS scavenging enzyme activity, total glutathione content and the fatty acid composition between the tardigrades sent into space and control animals on Earth.

Although it is extremely unlikely that there were any survivors, thanks to the crash landing of the 2019 Israeli lunar probe, Beresheet, that was carrying about 1000 tardigrades in tun as part of its payload, there might be some stranded tardigrades on the Moon.

The mechanisms behind the extraordinary radiation resistance in tardigrades is under study. One species of eutardigrade, *Ramazzottius varieornatus*, produces a novel, tardigrade-specific, protective protein called **Dsup** (for **damage suppressor protein**), which contributes to the organism's radiotolerance. Strong electrostatic attractions allow Dsup to tightly bind to histones in nucleosomes, and the high protein flexibility helps form a coat-like molecular aggregate, which affords protection of the DNA. The Dsup proteins of *R. varieornatus* and *H. exemplaris* seem to promote survival after irradiation by binding to nucleosomes and shielding chromosomal DNA from hydroxyl radicals. Cultured human cells that have been manipulated to produce Dsup protein exhibited 40% higher tolerance of x-ray radiation than ordinary human cells.

It was once believed that the ability of tardigrades to tolerate desiccation was due to high intracellular levels of the nonreducing disaccharide **trehalose** as a water replacement. This mechanism is not uncommon in other drought resistant organisms including certain nematodes, *Artemia*, and *P. vanderplanki* where intracellular levels reach 15%–20% of the dry weight in desiccated specimens. However, recent research has shown that, in both tardigrades and bdelloid rotifers, the capacity for trehalose synthesis is insufficient for desiccation tolerance and trehalose concentrations never exceed 3% of dry weight.

Another possible explanation lies with certain desiccation-induced **intrinsically disordered proteins (IDPs)**. Three of these IDPs have been found to be confined to tardigrades and are called **tardigrade-specific proteins (TDPs)**. TDPs may maintain the structural integrity of membranes by associating with the polar heads of the phospholipids in the membrane bilayers. This minimizes membrane chemical and structural damage upon rehydration.

In addition to resistance to x-rays and gamma rays, some water bears are very tolerant of ultraviolet radiation as well. For example, the Dsup protein of *R. varieornatus* bestows resistance to ultraviolet-C by upregulating the various DNA repair genes that correct the molecular damage caused by extreme ultraviolet. A recent experiment serendipitously demonstrated that a new species of *Paramacrobiotus* was much more resistant to ultraviolet (with a peak wavelength of 253 nm) than other tardigrades or ordinary bacteria and nematodes. At an ultraviolet dose of 1 kJ per square meter, the bacteria and roundworms were killed after just 5 min. This dose-rate began to demonstrate lethality to another species of tardigrade, *H. exemplaris*, by 15 min; almost all died after 24 h. But a strange, reddish-brown new species survived this dose easily. Even upon escalating the dose four-fold, nearly 60% of the reddish-brown bears lived longer than 30 days. Fluorescent pigments under the tardigrades' skin transformed the ultraviolet radiation into harmless blue light. To confirm that this was the mechanism, researchers extracted the fluorescent pigments and coated *H. exemplaris* tardigrades and *C. elegans* roundworms with the pigments. They then challenged these animals to the high-intensity ultraviolet again. Animals with the fluorescent coating survived the irradiation at nearly twice the rate of animals without the coating. The chemical makeup of this protective pigment is presently under investigation. But it appears that this fluorescent compound confers protection to *Paramacrobiotus* in a manner similar to ultraviolet-absorbing **mycosporine** amino acids in cyanobacteria, **hippo-sudoric acid** in hippopotamus skin and **melanin** in various other organisms. The *Paramacrobiotus* tardigrades may have evolved this protective fluorescence as a means of coping with the high doses of ultraviolet found in its native environment of southern India.

14.14 Radiophilic fungi

14.14.1 Radiotropism in melanotic fungi

Surprisingly, some of the areas with the very highest radiation levels at the Chernobyl Atomic Energy Station former nuclear power plant have been found to host lush growths of fungi. These fungi are not only radiation resistant but are actually thriving in the highly radioactive environment. Thus, in contrast to most of the organisms discussed previously that seem to have acquired radiation resistance as an exaptation that accompanied the evolution of characteristics enabling them to survive in their extreme environments, these fungi may be **true radiophiles**. Bolstering this assertion, in one experiment, it was shown that both beta and gamma radiation promotes directional growth of their hyphae (fungal equivalents of roots and branches) *towards* the source of ionizing radiation. Similar to plants that grow towards the light and exhibit phototropism, these fungi exhibited 'radio-tropism' (if there is such a word). Other work showed that some of these fungal strains grew three times as fast when exposed to radiation levels 500 times higher than normal background levels.

Closer inspection of such fungi revealed that many are **melanotic** or **melanized** strains. That is, they contain more melanin than their normal non-radiation-adapted counterparts. Melanized fungi have been found to inhabit some of the most extreme habitats on Earth, such as the damaged nuclear reactor at Chernobyl and the highlands of Antarctica, both of which happen to be high-radiation environments.

Cryptococcus neoformans is a well-known fungus that typically exists as a single-celled fungal form (i.e. a yeast) that is environmentally ubiquitous. It is often found growing in bird excrement and normally does not cause illness in humans. However, it is an **opportunistic pathogen**, meaning that given the opportunity (as in immuno-compromised individuals), it can cause serious disease such as **cryptococcal meningitis**. The *Cryptococcus* strains found flourishing at Chernobyl are melanized.

14.15 Melanin

Melanin is not a single chemical entity but rather a family of related conjugated organic polymers, based on $C_{18}H_{10}N_2O_4$ molecular subunits synthesized originally from the aromatic amino acid tyrosine. There are five basic types of melanin: eumelanin, pheomelanin, neuromelanin, allomelanin and pyomelanin.

The most common type is **eumelanin**, of which there are two subtypes: brown eumelanin and black eumelanin. Pheomelanin is said to be responsible for the color of red hair. Neuromelanin gives certain brain regions, such as the **locus coeruleus** and the **substantia nigra**, their distinctive dark colors. Students of medicine will recall that the substantia nigra (literally meaning 'the black substance') is a region in the midbrain that degenerates in patients with **Parkinson's disease**. The ink ejected by some cephalopod mollusks when threatened is made largely of a form of melanin. Certain foods and beverages such as tea, coffee, cocoa, chocolate and some mushrooms also contain melanin.

Melanization is not restricted to modern fungi—some fungal fossils appear to be melanized. Curiously, melanized fungal spores are relatively common in the sediment layers of the early Cretaceous period. This is believed to be a time in Earth's history when radiation levels were higher. This is because the geomagnetic field (and therefore the shielding against incoming cosmic radiation) was reduced for a prolonged period (the **Mesozoic dipole low**, coupled with the **Cretaceous normal superchron**). (Incidentally, a **superchron** is a prolonged period of geologic history during which geomagnetic polarity reversals were not going on. In recent geologic history, the north and south geomagnetic poles reverse every 200 000 years or so. Therefore, with the last reversal about 780 000 years ago (known as the **Brunhes-Matuyama reversal**) we are way past due. But this present interval does not qualify as a 'superchron'; true superchrons are far longer—over 10–20 million years by some definitions.)

Melanin plays a role in radiation protection from both ultraviolet and non-ultraviolet ionizing radiation. Experimental data shows that melanin may decrease the frequency of radiation-induced mutations in animal germ cells. The radiation **adaptive response** is also affected by melanin. The adaptive response is basically the phenomenon in which exposure to a low dose of radiation provides protection against a large dose applied shortly thereafter. In a study examining chromosomal aberrations following radiation exposure, it was shown that adding a small 'priming dose' of 0.2 Gy 4 h before irradiation with a relatively high dose of 1.5 Gy, diminished the amount of chromosomal damage to about half of what was seen with a single dose of 1.7 Gy. But if melanin was present prior to the irradiations, the chromosomal aberration level decreased by a factor of about two in both scenarios.

14.16 Radiosynthesis?

Melanized microfungi (such as *C. neoformans*) that can grow into and decompose, 'hot particles' (such as the highly radioactive graphite from the reactor itself) have been isolated from the Chernobyl Atomic Energy Station. This is astonishing enough, but it has also been discovered that some organisms growing in radioactive environments might actually be utilizing radioactivity as a source of metabolic energy.

In addition to *C. neoformans*, other melanized radiophilic fungal species including *Cladosporium sphaerospermum* and *Wangiella dermatitidis* have been shown to grow faster in an environment with hundreds of times higher radiation levels than normal background. Exposure of melanized strains of *C. neoformans* to such radiation increased melanin-mediated electron transfer nearly four-fold over baseline within just 40 min of exposure.

This type of unexpected radiophilic fungal biology was first noted in the 1950s in fungal colonies grown in Nevada nuclear test sites but was later also seen naturally in the highlands of Antarctica. The Antarctic highlands are harsh enough thanks to the frigid temperatures, but in addition, this is a relatively high radiation environment (approximately 500- to 1000-fold higher than at equatorial sea level; this amounts to a dose of 0.5–1.0 Gy per year). And melanotic fungi have been found thriving here. But given that most fungi (melanized or not), can normally withstand doses up to 1.7×10^4 Gy there is no obvious requirement for melanin as a radiation protector at the relatively modest doses and dose-rates in Antarctica. However, Dadachova *et al* have recognized an ability of melanin to absorb electromagnetic radiation and transduce this radiation energy into other forms of energy. An example is the conversion of potentially dangerous ultraviolet into harmless heat. This naturally raises the question of whether melanin might function not solely for protection but also in harvesting non-optical electromagnetic radiation for biological use.

Dadachova *et al* speculate that exposure to ionizing radiation could change the electronic properties of melanin and directly influence the growth of melanized microorganisms. They propose that, since biological pigments such as chlorophylls and carotenoids convert photonic energy into chemical energy during photosynthesis, and since melanins can absorb photons of all wavelengths, perhaps melanins play an active role in biosynthetic metabolism (rather than simply working as a radiation protector). In essence, they speculate that certain melanotic fungi could be autotrophic, employing an analogue of photosynthesis—'**radiosynthesis**'.

If correct, melanotic fungi might be **radiosynthetic organisms**. And as such, they can be called '**radiotrophic**', meaning that they obtain energy from ionizing radiation, and they synthesize their food via processes utilizing the energy derived from radiation. And if all this is true, perhaps tasty cultures of delicious radiosynthetic fungi can be 'farmed' in outer space during long missions, using the radiation energy of cosmic rays to provide their growth needs. Presently, such claims remain unsubstantiated. The only facts known for certain are that certain fungi can survive (and actually grow faster) in high radiation environments and such fungi tend to be melanized. Whether their melanin is contributing to their metabolic and nutritional needs or simply working as a radiation protector is presently uncertain. Samples of melanized fungi were sent to the International Space Station to determine their growth characteristics in the low-gravity and high-radiation space environment. Peer-reviewed results have not yet been published. Finally, questions remain regarding the palatability of such radiosynthetic fungi—and whether space travelers will really want to eat them!

Along the same lines, researchers are investigating the potential of melanin from radiation-resistant fungal strains as a cost-effective 'cosmic ray screen'. A set of experiments grew a 'lawn' made of a radiotrophic strain of *C. sphaerospermum*. The fungi seemed to grow quite well in the space-like environment, exhibiting a 21% growth advantage over normal radiation baseline rates. Radiation levels underneath a 1.7 mm 'shield' of fungus were 2.17% lower than unshielded controls. Projections based on linear attenuation coefficients indicated that a 21 cm thick layer of fungus could largely negate the annual radiation dose-equivalent on the Martian surface. This fungal shielding could be reduced to only about 9 cm if incorporated into an equimolar mixture of melanin and Martian regolith.

Whether this 'bio-shielding' has any advantages over conventional radiation shielding is undetermined. And whether astronauts and space colonizers would enjoy living under a 21 cm thick layer of fungus is another question altogether!

Further reading

[1] Suma H R, Prakash S and Eswarappa S M 2020 Naturally occurring fluorescence protects the eutardigrade *Paramacrobiotus* sp. from ultraviolet radiation *Biol. Lett.* **16** 20200391

[2] Jönsson K I 2019 Radiation tolerance in tardigrades: current knowledge and potential applications in medicine *Cancers (Basel)* **11** 1333

[3] Shmakova L, Malavin S, Iakovenko N, Vishnivetskaya T, Shain D, Plewka M and Rivkina E 2021 A living bdelloid rotifer from 24,000-year-old Arctic permafrost *Curr. Biol* **31** R712–3

[4] Shatilovich A V, Tchesunov A V and Neretina T V *et al* 2018 Viable nematodes from late pleistocene permafrost of the kolyma river lowland *Dokl. Biol. Sci.* **480** 100–2

[5] Tsuneizumi K *et al* 2021 Application of heavy-ion-beam irradiation to breeding large rotifer *Biosci. Biotechnol. Biochem.* **85** 703–13

[6] Gladyshev E and Meselson M 2008 Extreme resistance of bdelloid rotifers to ionizing radiation *Proc. Natl. Acad. Sci. USA* **105** 5139–44

[7] Krisko A, Leroy M, Radman M and Meselson M 2012 Extreme anti-oxidant protection against ionizing radiation in bdelloid rotifers *Proc. Natl Acad. Sci. USA* **109** 2354–7 http://jstor.org/stable/41477476

[8] Marín-Tovar Y, Serrano-Posada H, Díaz-Vilchis A and Rudiño-Piñera E 2022 PCNA from *Thermococcus gammatolerans*: a protein involved in chromosomal DNA metabolism intrinsically resistant at high levels of ionizing radiation *Proteins* **90** 1684–98

[9] Castillo H, Li X and Smith G B 2021 *Deinococcus radiodurans* UWO298 dependence on background radiation for optimal growth *Front. Genet.* **12** 644292

[10] Castillo H, Winder J and Smith G 2022 Chinese hamster V79 cells' dependence on background ionizing radiation for optimal growth *Radiat. Environ. Biophys.* **61** 49–57

[11] Sutou S 2022 Low-dose radiation effects *Curr. Opin. Toxicol.* **30** 100329

[12] Nowell R W, Almeida P, Wilson C G, Smith T P, Fontaneto D and Crisp A *et al* 2018 Comparative genomics of bdelloid rotifers: insights from desiccating and nondesiccating species *PLoS Biol.* **16** e2004830

[13] Sghaier H, Ghedira K, Benkahla A and Barkallah I 2008 Basal DNA repair machinery is subject to positive selection in ionizing-radiation-resistant bacteria *BMC Genomics* **9** 297

[14] Anderson A W, Nordan H C, Cain R F, Parrish G and Duggan D 1956 Studies on a radio-resistant micrococcus. I. Isolation, morphology, cultural characteristics, and resistance to gamma radiation *Food Technol.* **10** 575–7

[15] Zahradka K, Slade D, Bailone A, Sommer S, Averbeck D, Petranovic M, Lindner A B and Radman M 2006 Reassembly of shattered chromosomes in *Deinococcus radiodurans Nature* **443** 569–73

[16] Pavlov A K, Kalinin V L, Konstantinov A N, Shelegedin V N and Pavlov A A 2006 Was Earth ever infected by Martian biota? Clues from radioresistant bacteria (PDF) *Astrobiology* **6** 911–8

[17] Wong P C, Wong K K and Foote H 2003 Organic data memory using the DNA approach *Commun. ACM* **46** 95–8

[18] Zhdanova N N, Tugay T, Dighton J, Zheltonozhsky V and McDermott P 2004 Ionizing radiation attracts soil fungi *Mycol. Res.* **108** 1089–96

[19] Casadevall A, Cordero R J B, Bryan R, Nosanchuk J and Dadachova E 2017 Melanin, radiation, and energy transduction in fungi *Microbiol. Spectr.* **5** 2

[20] Mosse I, Kostrova L and Subbot S *et al* 2000 Melanin decreases clastogenic effects of ionizing radiation in human and mouse somatic cells and modifies the radioadaptive response *Radiat. Environ. Biophys.* **39** 47–52

[21] Hulot G and Gallet Y 2003 Do superchrons occur without any palaeomagnetic warning? *Earth Planet. Sci. Lett.* **210** 191–201

[22] Dadachova E, Bryan R A and Huang X *et al* 2007 Ionizing radiation changes the electronic properties of melanin and enhances the growth of melanized fungi *PLoS One* **2** e457

[23] Shunk G K, Gomez X R, Kern C and Nils J H A 2020 Growth of the radiotrophic fungus Cladosporium sphaerospermum aboard the international space station and effects of ionizing radiation https://doi.org/10.1101/2020.07.16.205534

IOP Publishing

Space Radiation
Astrophysical origins, radiobiological effects and implications for space travellers
James S Welsh

Chapter 15

Essentials of astrophysics

15.1 The Hertzsprung–Russell diagram

The Hertzsprung–Russell (HR) diagram (figure 15.1) is to astrophysics as the periodic table of elements is to chemistry. Named for the independent co-discoverers Ejnar Hertzsprung and Henry Norris Russell who built on prior work conducted by Annie Jump Cannon among others, an HR diagram plots stellar luminosity on the y-axis (vertical axis or ordinate) versus spectral class on the x-axis (horizontal axis or abscissa). For this reason, it is also called the temperature-luminosity diagram. Like the periodic table, the importance of the HR diagram is that careful examination of such plots uncovered previously unappreciated patterns and ultimately revealed important secrets about the life cycle of stars.

The HR diagram plots stellar temperature on the horizontal axis, going from hottest to coolest from left to right. The general surface temperature range is roughly 30 000 K on the left to 3000 K on the far right. The temperature decreases from left to right because Hertzsprung and Russel based their original diagrams on the **Harvard spectral sequence** O-B-A-F-G-K-M. These spectral classes were arranged from left to right in order of increasing wavelength (decreasing energy). Thus, the left-to-right sequence along the x-axis goes from blue to red. This original sequence has now been modified to also include suffixes from the **Morgan–Keenan luminosity classes** as described below. In this manner, our Sun, a yellow star of Harvard classification G, is more entirely described as a yellow dwarf star on the main sequence or a GV2 star.

In essence, **class O and B stars** are hot and blue with surface temperatures around 30 000 K and are found in the upper left corner of the HR diagram, while **M stars** are cool and red with temperatures around 3000 K and reside in the lower right aspect of the HR diagram.

The vertical axis depicts luminosity on a logarithmic scale, typically in units of solar units (meaning our Sun's luminosity, $L_{\odot} = 1$). Our Sun lands on a very specific region of the HR diagram corresponding to its spectral type (G2V) and its luminosity $L_{\odot} = 1$. Because luminosity increases going upward and temperature increases leftward, stars that are hotter and larger than our Sun are towards the

doi:10.1088/978-0-7503-5444-8ch15

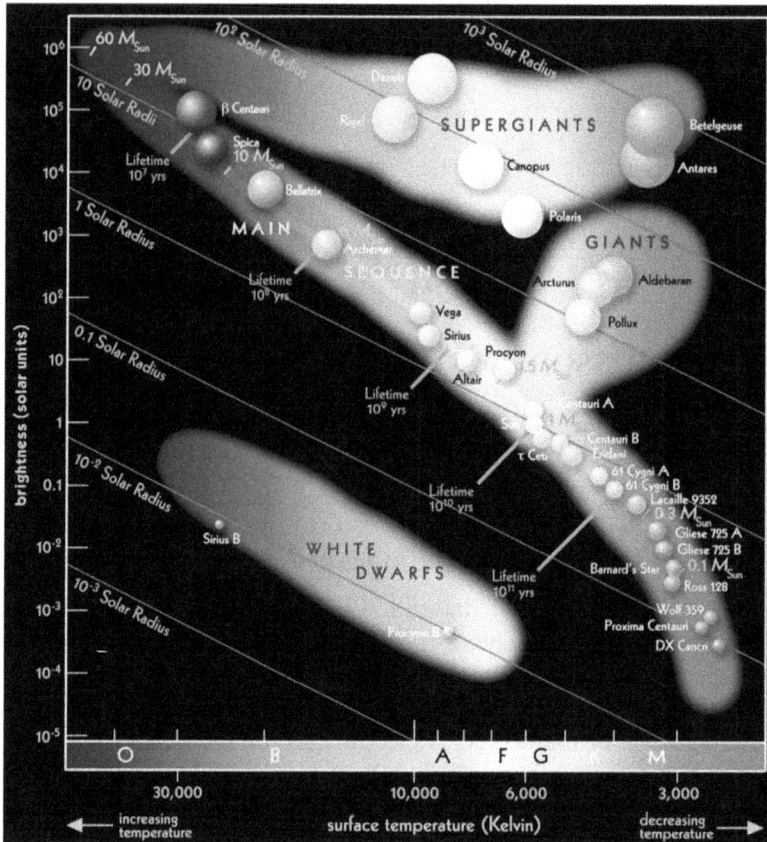

Figure 15.1. The **H–R diagram** courtesy of the European Southern Observatory. Brightness (luminosity) relative to the Sun is on the ordinate while surface temperature and spectral class are on the abscissa. Note that luminosity is on a logarithmic scale. Our Sun is a spectral class G star on the **main sequnce** (meaning it is burning hydrogen in its innermost core). Stars on the main sequence have their approximate lifespans listed. Bright, hot stars (blue giants) are in the upper left aspect of the H–R diagram and have relatively short lifetimes while dim, cool main sequence stars (red dwarfs) are in the lower right and have very long lifetimes. In about 5 billion years, the Sun will cease fusing hydrogen in its inner core and will move up and to the right as a red giant. Towards the end of that phase (actually two separate red giant phases, as discussed in chapter 16) it will blow off its outer layers to create a planetary nebula with a white dwarf remnant in the center. As a white dwarf, the Sun will eventually move towards the lower left aspect of the H–R diagram and then slowly slide to the right as it cools over time. Credit: ESO.

upper left relative to our Sun. Those that are cooler and smaller than our Sun will lie to our right and downward. This creates a band going from the upper left to lower right that is called the **main sequence**. Stars on the main sequence share the common (and defining) feature that they are all fusing hydrogen in their inner cores. Also, stars spend the vast majority of their fusion-generating lifetimes on the main sequence. For example, our Sun will spend approximately 10 billion years burning hydrogen on the main sequence but maybe only 1 billion years in its various, post-main sequence red giant phases.

15.2 Stellar classification—luminosity and spectral classes

Astronomers can assign stellar luminosity classes to these various groups. Using the standard **Morgan–Keenan luminosity nomenclature**, which assigns Roman numerals, the most luminous supergiants are Class I and the main sequence stars are Class V; other types of giants and subgiants comprise Classes II, III and IV. Luminosity class correlates with stellar radius, with luminosity class I stars being the largest and class V being the smallest. There is also a **Class VII for white dwarfs** and a **Class 0** for a category of stars called 'hypergiants'. The Harvard spectral types themselves can also have subdivisions. While class O stars are hot and blue and M stars are cool and red, there are levels within O, B, A, F, G, K and M. Hence, our Sun is most completely categorized as a G2V star. This tells us that it is a yellow dwarf, main sequence (hydrogen-fusing) star with a surface temperature of around 5800 K.

The position on the HR diagram of the various stars tells us valuable information about their sizes and distances. For example, if two stars have the same surface temperatures (based on their colors) and are therefore on the same place on the abscissa (x-axis) but exhibit very different luminosities (i.e. are at different heights on the y-axis or ordinate), they must either be different sizes or different distances from us.

15.3 Star sizes and luminosities

The luminosity of a star relates to its power (energy emitted per unit time). Power in turn is related to temperature by the Stefan–Boltzmann equation:

$$\text{Power (per unit area)} = \sigma T^4,$$

where the Stefan–Boltzmann constant, $\sigma = 5.7 \times 10^{-8} \text{W m}^{-3} \text{K}^{-4}$.

But the total luminosity of a star is more than just its power. It is power per unit area multiplied by total surface area. For a spherical star, this surface area is given by $4\pi r^2$. Thus, the **luminosity (L)** may be calculated as:

$$L = 4\pi r^2 \times \sigma T^4.$$

This can be rearranged to:

$$r = \sqrt{L/4\pi\sigma T^4}.$$

A star's color (spectral class) tells us its temperature, and this generally correlates well with its actual luminosity. By knowing both a star's observed luminosity and comparing it to its apparent luminosity, we can estimate the distance. Also, thanks to these relationships, if we measure a star's luminosity and know its distance, we can calculate its size.

15.4 Short-lived giants and eternal dwarfs

Large, hot stars are found in the upper left corner of the HR diagram. They are very luminous but have short lifespans. This is because of the incredible rate at which they burn their fuel. A 10 $M_\odot$ star (where $M_\odot$ represents one solar mass) residing on the main sequence will be about 10^4 times as luminous as the Sun. This means a 10 000-fold faster consumption of its endowment of fuel. So, although it has ten-fold more fuel to begin with it, it burns that fuel 10 000-fold faster and lives a net of 1/1000 as long as the Sun. So, instead of the Sun's 10 billion year lifetime, a 10 $M_\odot$ star would last only around 10 million years. (In reality, it can live a bit longer than this calculation since it would be able to utilize more of its core hydrogen than the Sun can.)

This explains why cool, dim red stars are so much more abundant than huge, hot blue stars—in addition to being lower in number to begin with, the majority of the most massive stars are long since dead and gone. If one thinks about this, it becomes obvious that the existence of massive, hot blue stars in the first place must mean that there are stellar nurseries out there that continuously replenish the rapidly dying giants. If we recall that the Milky Way rotates once about every 240 million years, newborn Class O or B stars will die long before they can complete even one lap around the galaxy.

At the other end of the spectrum, cool, dim, red K and M stars on the lower right-hand corner of the main sequence might have as little as 8% of the Sun's mass (0.08 $M_\odot$) and can potentially live tens of billions to even trillions of years. Given this longevity, it is not surprising that overall, far more stars reside on this lower right region of the HR diagram than on the high-mass upper left region.

In the way of nomenclature, Class M (and by some definitions, small Class K stars as well) are sometimes called **red dwarfs**. Red dwarfs are thus the smallest and coolest category of stars on the main sequence. Red dwarfs must be distinguished from **brown dwarfs**, which never make it onto the main sequence since they never attain the core temperature required for fusion of hydrogen (although brown dwarfs can fuse deuterium and sometimes lithium). Brown dwarfs remain to the right of the main sequence and never migrate onto it because they never attain the internal heat needed to undergo thermonuclear fusion of hydrogen into helium (which is the defining factor of a main sequence star).

15.5 A star's fate is predetermined

It all boils down to the fact that a star's birthweight (mass) is its most important feature. Stars, unlike biological entities, follow precise and unwavering natural histories that are essentially entirely predetermined by their initial mass. Mass dictates a star's luminosity because mass determines the rate of nuclear fusion within its core. More overlying mass means more gravitational inward force. To achieve hydrostatic equilibrium, greater outward pressure from fusion is needed to stave off gravity. This in turn, means a faster rate of fusion will ensue. Hence, larger stars are more luminous but have shorter lifespans.

Because a main sequence star's mass, surface temperature and luminosity are all closely related in a mathematically precise manner, one can get a good estimate of a star's mass simply by observing its spectral class. For instance, any random G2 star must have a mass and luminosity similar to our Sun's. In essence, a star's initial birthweight is a star's DNA—it determines everything about its future.

15.6 Red giants and white dwarfs

As covered in the chapter on the Sun, the Sun will someday run out of hydrogen fuel. When this happens, for a period of time it will no longer be capable of fusion in its innermost core. But it will still be able to undergo fusion in an outer shell of the core, where hydrogen remains plentiful. At that point, like all stars making the transition from hydrogen-burning in the inner core to hydrogen-burning in an outer core shell, the Sun will move off the main sequence and upwards and towards the right. This signifies its transition from a main sequence star into the red giant phase.

This rightward migration on the HR diagram is understandable since the surface temperature diminishes in a red giant, and its color is shifted to longer wavelengths (i.e. towards the red end of the spectrum). But the upward migration, which means its luminosity is increasing, is due to its growing size. Luminosity is a function of both surface temperature and surface area. And the surface area increases dramatically.

Although it is cooler, its monstrous size dominates the picture. A pattern that emerges is that stars increase in diameter as one moves from the lower left region of the HR diagram (where some white dwarfs reside) towards the upper right region (where the red giants and supergiants live). In fact, one can draw diagonal lines from upper left regions down towards the lower right that represent 'isometric' lines wherein the stars following this diagonal all have the same size. Importantly, stars do not fall randomly all over the HR diagram; they cluster into four main categories with very specific locations.

For instance, the main sequence extends from the top left region down towards the lower right. The main sequence once again represents stars that are fusing hydrogen into helium and is where most visible stars lie for the majority of their lives. Our Sun is presently on the main sequence. Stars in the far uppermost right-hand corner are the **red supergiants**. They are relatively cool but very bright because they are so large. Also, in the upper right region but just beneath the red supergiants are the **red giants**. Since they are so huge, red giants are much brighter than stars of similar spectral type on the main sequence. Our Sun will someday evolve into a red giant and live on this part of the HR diagram for a while. Red giants and red supergiants are not actively undergoing fusion in their innermost cores but are actively undergoing fusion in the shells around that often temporarily inactive innermost core. This is discussed further in the chapter on the Sun.

Below the main sequence, in the lower left regions of the HR diagram are some very hot but extremely small (and therefore not very luminous) stars. These are the **white dwarfs**. Ultimately, our Sun will become a white dwarf once it has ended its red giant stage. At this point, it will be comparable to the Earth in size but will be incredibly dense. As dense as they are, white dwarfs do not have sufficient mass or density to evolve into neutron stars or black holes (which are covered in other chapters). White dwarfs represent the end stage for many stars, including those around the mass of the Sun; they have run out of fuel and can no longer produce heat via thermonuclear fusion. After a long period of time, they fade away into **black dwarfs**.

15.7 Protostars and pre-main sequence stars

Stars form spontaneously when sufficient mass is gathered in one place. The gravitational force of attraction amalgamates the mass of a nebula (made of gravitationally unstable gas and dust) into a spherical **protostar.** The protostar will continue contracting into a smaller and smaller radius. As this radius decreases, the pressure increases and temperature in the innermost regions of this protostar heat up. The freefalling, progressively enlarging clumps and lumps gain speed and collide. The kinetic energy is converted into thermal energy, which opposes the inward pressure of gravity. As the temperature and pressure increases, the collapse slows, and the protostar evolves into a **pre-main sequence star**. It continues contracting but at a much slower pace, releasing about half of its gravitational potential energy in the form of photons, while the remainder is converted into heat. Eventually, if there is sufficient mass, nuclear fusion begins, and various elements are ignited. The nucleus with the lowest threshold for fusion is deuterium. After the special case of deuterium fusion (which occurs in brown dwarfs), if the temperature is above the threshold of approximately 10 million K, hydrogen fusion may occur.

Since hydrogen is by far the most abundant element in the Universe, once hydrogen fusion is initiated, the protostar will have ample fuel for long-term burning. It can now officially be called a 'star' and enters a prolonged period of relative stability. It continues in a stable state with a fairly constant temperature, luminosity and size. This period of stability due to hydrogen burning is called the **main sequence** phase. During this process of hydrogen fusion, the outward heat pressure from thermonuclear fusion balances the inward pressure of gravity and the main sequence star is in a state of hydrostatic equilibrium. But additionally, the main sequence star is in thermal equilibrium. This implies that the star maintains a constant temperature as the hydrogen is fused into helium at a fairly constant rate. If the equilibrium is somehow perturbed and the core gets too hot, fusion would slow down, and the outer parts of the star would expand and cool. This would decrease the rate of fusion in the core and the star would cool down. If on the other hand, the star's core somehow cooled, its outward pressure would diminish, and the star would contract under gravity. This heats the star up again and accelerates the fusion reaction rates, thereby reheating the star. In essence, the core is like a self-regulating thermostat.

As with most other features of stars, the duration of life on the main sequence is dictated by the star's initial endowment of mass. In stars, mass is just about everything. The ability to initiate hydrogen fusion depends on a protostar's mass; the duration of hydrogen fusion (i.e. longevity on the main sequence) is mass dependent. Whether or not a star can fuse elements heavier than hydrogen also depends on a star's initial mass. And the overall longevity of a star is heavily dependent on its mass (in an inverse fashion). Our Sun might spend approximately 10 billion years on the main sequence burning hydrogen into helium. Lighter stars will last far longer, with some potentially living a trillion years. Monstrous stars with masses far higher than the Sun might have lifetimes measured in only millions of years. Many generations of such stars have come and gone in the interval between the demise of the dinosaurs at the close of the Cretaceous Period 65 million years ago and today.

15.8 Stellar nucleosynthesis

There are several known pathways of generating new, heavier elements in the cores of stars. Thus, in a very real sense, stars can be considered 'element factories' and the processes that go on inside are genuine alchemy—new elements are created from other elemental starting blocks. This process, **stellar nucleosynthesis**, occurs only in the innermost and hottest region of the star—its core. Depending on the exact temperature and pressure in a star's core (which of course reflects its mass), the fusion might stop at the conversion of hydrogen into helium, or it could go beyond. (Deuterium fusion in **brown dwarfs** is a special case; protostars that cannot burn heavier elements than deuterium are not even considered 'real' stars by some.) Our Sun has sufficient mass to burn hydrogen for about 10 billion years and then will enter a red giant phase and other phases where helium burning can sustain it for a while longer. Upon finishing fusion of helium into carbon and oxygen, the Sun will reach the end of the line. Unable to generate enough heat to burn that accumulated carbon and oxygen, it will become a **white dwarf**. This white dwarf phase will last billions of years as it slowly fades away by blackbody radiation like a glowing ember and eventually becomes a **black dwarf**.

More massive stars can go beyond just helium fusion. They can take the 'ashes' of helium burning (carbon and oxygen) and fuse them into heavier elements such as neon and magnesium. If massive enough, they might next burn that neon and magnesium into even heavier elements such as silicon and sulfur. But this process is one of ever-diminishing gains. For instance, fusion of hydrogen into helium yields far more energy than fusion of silicon plus helium into sulfur. And this diminution in power means a correspondingly shorter duration of such burning. Hydrogen burning in the core of the Sun might last 10 billion years, whereas the fusion of chromium into iron might sustain a supergiant for only one day!

Anthropomorphically, it is as though the dying star is desperately trying to cling onto life by taking the ashes of previous fires and burning them to sustain itself for just a little bit longer. But although this strategy does work for a while, no matter how massive a star is, there is a hard-stop, ultimate end of the line. Fusion beyond iron and nickel fails to generate energy. It consumes rather than liberates energy. Rather than being an exothermic process, fusion of iron is endothermic. When fusion in the core begins to yield iron rather than a fusible element, it is all over. The star will be unable to generate sufficient heat and outward pressure to continue its battle against the inward crushing force of gravity. After millions of years of battle but balance, gravity wins in an instant. Within seconds, the core collapses and the ensuing implosion leads to an enormous explosion. This explosion is called a **supernova**. Recalling that in astronomy, any element heavier than helium is a 'metal', it is through the scattering of stellar 'ashes' during supernova explosions that our Universe is enriched with such metals. Thus, our Universe is endowed with far more diversity than just the hydrogen and helium (and minute traces of other light elements such as lithium) created in the Big Bang.

There are several distinct mechanisms whereby stars and supernovae create elements heavier than hydrogen. These mechanisms of nucleosynthesis include:

1. The proton–proton (p–p) chain
2. The carbon–nitrogen–oxygen (CNO) cycle
3. The triple alpha process
4. The alpha ladder
5. The r-process
6. The s-process.

15.9 The proton–proton chain

The proton–proton chain (p–p chain) is a sequence of thermonuclear reactions through which many stars, including our Sun, create helium out of hydrogen. The other means of making helium from hydrogen is the CNO cycle, which will be discussed below. Thus, because the defining feature of main sequence stars is their fusion of hydrogen into helium, the p–p chain and the CNO cycle are the two sources of fuel for main sequence stars.

The p–p chain has a threshold temperature of approximately 4 million K. It dominates in the Sun and in stars of similar masses or less, whereas the CNO cycle is predominant in stars with more than 1.3 solar masses.

In the p–p chain, four protons (hydrogen nuclei) are combined into one helium-4 nucleus. There is a slight mass difference between the reactants and products—the helium-4 nucleus is lighter than the combined mass of the four protons. Because of this mass difference, 0.7% of the original proton mass is converted into energy.

To initiate the p–p chain, a threshold temperature of about 4 million K is required. Above this threshold temperature, protons (that is, hydrogen nuclei) can take the first step in the p–p chain, fusion into deuterium. In this reaction, two protons fuse into a deuteron along with a positron for electric charge conservation and an electron neutrino for lepton number conservation.

$$^1H + {}^1H \rightarrow {}^2H + e^+ + \nu_e.$$

The positron immediately encounters an electron and promptly undergoes mutual annihilation into two gamma photons:

$$e^+ + e^- \rightarrow 2\gamma.$$

This mutual annihilation reaction generates 1.022 MeV. When both processes are combined, the energy yield or **Q value**, of step 1 is 1.442 MeV.

This first reaction is *the rate-limiting step in solar nucleosynthesis*. It is a slow reaction for several reasons:

1. First, the two protons, being positively charged, repel one another electro-statically. The energy barrier for fusion is approximately 1.6 MeV, but at a core temperature of approximately 15 million K, the kinetic energy of each proton is only about 2 keV—roughly 1000-fold less than what is required to overcome the electrostatic repulsion. It might seem like this is 'game over' right from the start, but the quantum **tunnel effect** allows the reaction to proceed. The energy needed is approximately 1000-fold more than what is available, so the protons must tunnel approximately 1000-fold more than the radius of a proton. While this is extremely unlikely, the tremendous pressures and temperatures in the core of the Sun mean it is not impossible. Two protons must come within about 1.5 fm for the strong force to take control and dominate over electrostatic repulsion.

2. Furthermore, once the two protons fuse, the product initially is helium-2, which is extremely unstable. It decays almost instantaneously back into two protons.

3. Nevertheless, a small fraction of these helium-2 nuclei manages to decay via a different reaction—one that leads to a deuterium (^{2}H) nucleus. It should also be noted that this step generates a neutrino, which indicates that the **weak force** is involved in this reaction. And, as with many weak interaction-mediated mechanisms, it moves along like a snail in comparison to other reactions that are primarily mediated through the strong force. Thus, this first reaction is the rate-limiting step in the p–p process.

Because of the three reasons listed above, it has been estimated that the odds of a typical proton in the core of the Sun undergoing fusion into deuterium might be less than once in a billion years despite nearly a billion collisions per second. Fortunately, there are a lot of protons in the Sun! Given the statement that the odds are under one in a billion, some readers might understandably doubt that fusion could truly proceed if the probability of interaction is that low. Students can work the math by recalling that the Sun is about 2×10^{30} kg, the core (where all the fusion occurs) constitutes about a third of the Sun's mass, the atomic mass of hydrogen is 1 g mole^{-1}, 1 mole contains Avogadro's number of atoms (6.02×10^{23}), and there are 31 556 952 s in a year. Thus, the concept of one event per billion years must be put into the context of something on the order of 10^{49} possible candidates for interaction. And this is without taking into consideration the temperature of around 15 million K and the number of collisions per second at such temperatures. Looked at from this angle, we need not worry that the Sun is not going to get the job done.

In the next step of the p–p chain, the deuterium formed in step 1 fuses with another hydrogen nucleus to create helium-3 along with a gamma photon:

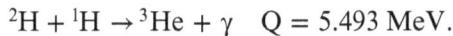

$$^2\text{H} + {}^1\text{H} \rightarrow {}^3\text{He} + \gamma \quad Q = 5.493\,\text{MeV}.$$

Being strictly mediated through the strong nuclear force, this reaction is not slowed by the involvement of the weak interaction and occurs quickly—each newly created deuteron lasts only around 1 s before encountering a proton and undergoing fusion into helium-3.

Upon creation of helium-3, there are theoretically four potential pathways to helium-4, all of which culminate in the creation of one helium-4 nucleus and two neutrinos. These paths are called the **p–p I**, **p–p II**, **p–p III** and **p–p IV** branches. The various branches differ significantly in the energies of the emitted neutrinos. In the Sun, it is the p–p I process that dominates.

15.10 The p–p I branch

This is the predominant p–p branch in our Sun. This reaction converts two helium-3 nuclei into helium-4 plus two protons. These two protons then feed back into the p–p chain via reaction 1. The reaction of the p–p I branch can be depicted as:

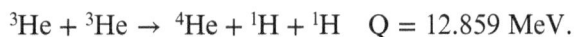

$$^3\text{He} + {}^3\text{He} \rightarrow {}^4\text{He} + {}^1\text{H} + {}^1\text{H} \quad Q = 12.859\,\text{MeV}.$$

The p–p I branch dominates when temperatures are relatively low, for instance in the range of 10–14 million K.

15.11 The p–p II branch

The p–p II branch is not as prevalent in the Sun but may be dominant in other stars where core temperatures range from 14 to 23 million K. The reactions involve beryllium and lithium intermediates and can be written as follows:

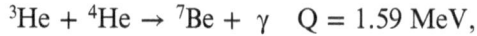

$$^3He + {}^4He \rightarrow {}^7Be + \gamma \quad Q = 1.59 \text{ MeV},$$

$$^7Be + e^- \rightarrow {}^7Li + \nu_e \quad Q = 0.861 \text{ MeV}(90\%) \text{ and } 0.383 \text{ MeV}(10\%),$$

$$^7Li + {}^1H \rightarrow {}^4He + {}^4He \quad Q = 17.35 \text{ MeV}.$$

While about 83.3% of all the helium is generated in our Sun is via the p–p I branch, the p–p II branch contributes about 16.68% to the total produced helium. The remaining 0.02% is made via the pp-III branch.

15.12 The p–p III branch

Although less frequent in the Sun, the p–p III branch is an important source of solar neutrinos and played an important role in solving the so-called **solar neutrino problem**. The neutrinos generated by the p–p III branch are the most energetic (ranging up to 14.06 MeV). This pathway uses a helium-4 nucleus as a catalyst and also involves beryllium and boron isotopes along the way:

$$^3He + {}^4He \rightarrow {}^7Be + \gamma,$$
$$^7Be + {}^1H \rightarrow {}^8B + \gamma,$$
$$^8B \rightarrow {}^8Be + e^+ + \nu_e,$$
$$^8Be \rightarrow {}^4H + {}^4H.$$

The overall energy released in the p–p III branch is 18.209 MeV with most of this going into the electron neutrinos generated in the third reaction.

The p–p III branch is not prominent in the Sun and only occurs at very high temperatures. Nevertheless, an appreciation for, and understanding of, this p–p III branch was crucial for solving the solar neutrino problem. The high-energy neutrinos produced were identifiable in experiments using tetrachloroethylene as a detection medium. Prior experiments could detect only lower-energy neutrinos and yielded perplexingly smaller numbers than predicted. This deficit in detected neutrinos was called the **solar neutrino problem**. This solar neutrino deficit was explained by the phenomenon of **neutrino flavor switching** in which the fusion-generated electron neutrinos created in the core of the Sun change flavor en route from the Sun to Earth during the 93 million mile (one Astronomical Unit or AU) trip. If the detectors were geared to detect only electron neutrinos, but the solar electron neutrinos transformed into muon neutrinos and tau neutrinos along the way, the detectors might provide a falsely lower overall number of detected neutrinos. And this is what happened during early experiments. Incidentally, flavor changing among neutrinos is evidence that the neutrinos have a small but non-zero mass, as opposed to being entirely massless as was once assumed.

The p–p III branch contributes only perhaps 0.02% of all the Sun's created helium, but the relative percentage of energy lost via 'useless' neutrinos (as opposed to light and heat energy that we can benefit from here on Earth) is far higher than with the p–p I or p–p II branches.

15.13 The p–p IV branch

This p–p IV reaction is theoretically possible but has not yet been observed. In this proposed p–p branch, helium-3 reacts with a proton to yield helium-4. The reaction should also create a high-energy neutrino.

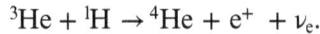

$$^{3}\text{He} + {}^{1}\text{H} \rightarrow {}^{4}\text{He} + e^{+} + \nu_{e}.$$

15.14 The proton–electron–proton reaction

As described above, the first step in the p–p chain is the fusion of two protons into deuterium:

$$^{1}\text{H} + {}^{1}\text{H} \rightarrow {}^{2}\text{H} + e^{+} + \nu_{e}.$$

It should be noted that this first step generates a neutrino, which is a tell-tale sign that the weak force was at work. And as with many weak interaction-mediated mechanisms, it moves along at a relatively pedestrian pace in comparison to other reactions that are primarily mediated through the strong force. The deuterium thus created can go into the next standard step, fusion of deuterium with another proton to make helium-3:

$$^{2}\text{H} + {}^{1}\text{H} \rightarrow {}^{3}\text{He} + \gamma.$$

However, there is an alternative means of making deuterium, namely the **proton–electron–proton (PEP)** reaction. This may be thought of as an **electron capture** reaction between a proton and an electron combined with fusion with another proton. As in all electron capture reactions, an electron neutrino is produced. The equation is given by:

$$^{1}\text{H} + e^{-} + {}^{1}\text{H} \rightarrow {}^{2}\text{H} + \nu_{e} .$$

Although both the PEP reaction and the first step in the p–p chain create deuterium and neutrinos, the neutrinos formed in the PEP reaction have greater energy (1.44 MeV compared to 0.42 MeV). These neutrinos have been detected experimentally, confirming that the PEP reaction does indeed happen in the Sun (albeit at a relatively low frequency). Perhaps only one in 400 deuterium nuclei created in the core are generated via the PEP reaction.

The overall equation of the p–p chain (when considering the electrons consumed by positron annihilation and ending with the p–p I branch) can be written:

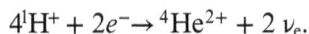

$$4{}^{1}\text{H}^{+} + 2e^{-} \rightarrow {}^{4}\text{He}^{2+} + 2\,\nu_{e}.$$

Note that the positrons and gamma photons made are omitted from this balanced equation, but the electrons consumed are included.

The overall p–p chain of six protons becoming one helium-4 with two regenerated protons, along with its intermediate steps of deuterium and helium-3 generation plus the creation of two positrons and two electron neutrinos, may be illustrated as shown in figure 15.2.

As with any energy-yielding fusion reaction, the mass of the end products (in this case helium-4) is slightly lower (about 0.7%) than the mass of the reactants (in this case four protons). The small mass difference leads to a significant energy yield via the Einstein relation, $E = mc^2$.

An interesting exercise is to compare the obviously enormous solar 'furnace' with the human body energy output (power, that is energy per unit time) on a per-kilogram basis. The Sun's power output or luminosity is approximately 4×10^{26} W, while the power generation by an average human body is roughly 100 W. But the Sun's mass is 2×10^{30} kg, whereas a human might weigh 80 kg. Upon doing the math, one might conclude that on a per-kilogram basis, in terms of heat generation, human muscle is 10 000 times more powerful than the Sun! So, how can the Sun emit so much energy for so long if it is so relatively 'dim'? The answer lies in the fact that human heat output is due to fast, short-lived chemical reactions that will burn out relatively quickly, whereas solar power is generated via nuclear fusion, which will last far longer thanks to the far higher output on an atom-per-atom basis. Nuclear fusion of hydrogen into helium will sustain the Sun for a total of approximately 10 billion years. Additionally, after its hydrogen fuel runs out, the Sun will temporarily be able to 'burn' its old 'ashes' to obtain more energy. For a while, the Sun will be able to use helium as a renewable energy source and fuse it into carbon. But the real secret lies in its immense mass. Despite the surprisingly low power of the Sun as a whole (on a per weight basis), its enormous size will sustain it for a long, long time.

15.15 The CNO cycle

As mentioned above, the p–p chain is but one means of making helium. The other mechanism is the **carbon–nitrogen–oxygen cycle**, or **CNO cycle**. The CNO cycle sometime goes by the name **Bethe–Weizsacker cycle** after two of the physicists who were independently instrumental in its formulation. The p–p chain is more efficient at the temperatures that prevail in our Sun's core and the CNO cycle plays a relatively minor role in the Sun (<2%). In smaller stars with cooler cores, the CNO cycle fails to operate at all. Additionally, as the name indicates, this pathway to helium requires carbon, nitrogen and oxygen as catalysts. Without adequate amounts of carbon, nitrogen and oxygen, this pathway is not available.

The CNO reaction has a higher threshold temperature than the p–p chain and becomes self-sustaining at about 15 million K. The Sun's core is estimated to be about 15.7 million K. At this temperature, only about 1.7% of the helium created can trace its origin to the CNO cycle. The likelihood of the CNO reactions increases with temperature and begins to dominate the p–p chain in stars with masses >1.3 $M_\odot$ and core temperatures $\geqslant$17 million K.

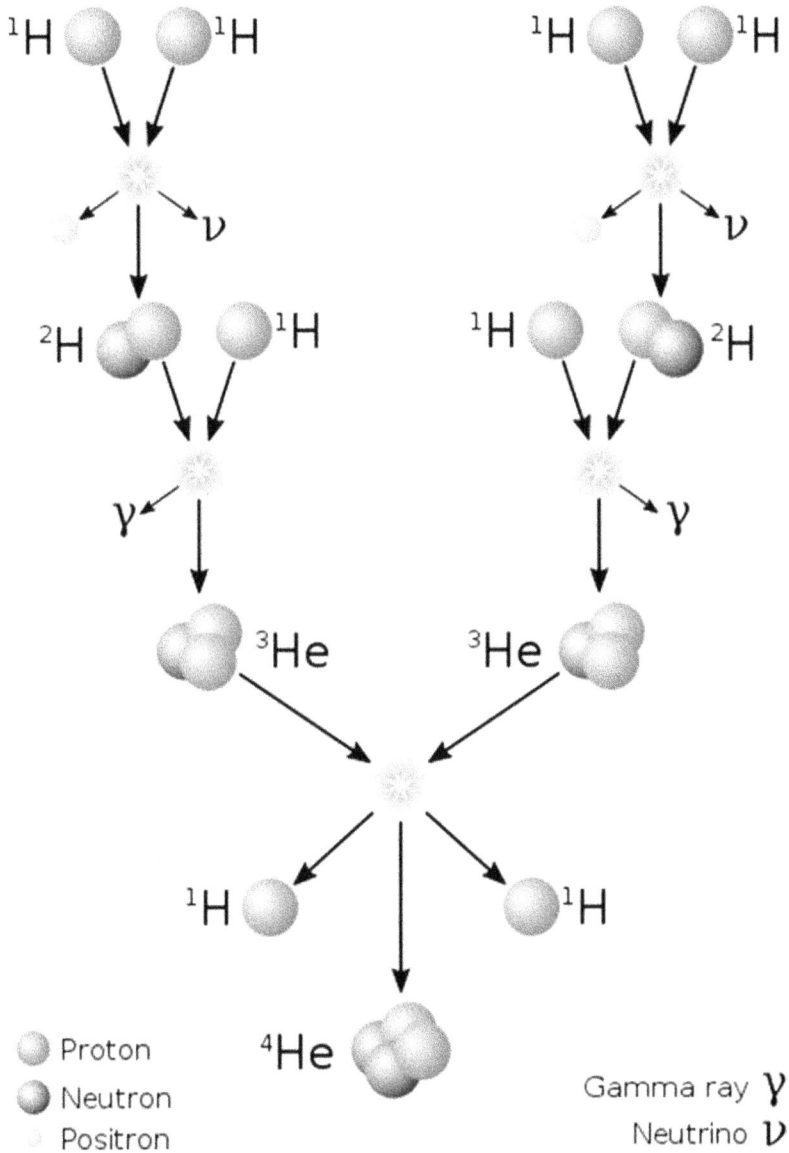

Figure 15.2. The proton–proton chain, ending with the most common pathway in the Sun, the p–p I pathway, which fuses two helium-3 nuclei into helium-4. This Fusion in the Sun has been obtained by the author from the Wikimedia website, where it is stated to have been released into the public domain. It is included within this book on that basis. It is attributed to Sarang https://commons.wikimedia.org/w/index.php?curid=51118538.

In the CNO cycle, four protons are fused into one helium-4 nucleus in a sequence of six individual reactions, using carbon, nitrogen and oxygen nuclei as catalysts. Since the first reaction consumes a carbon-12 nucleus, but the last reaction regenerates a carbon-12 nucleus, the sequence of reactions can be portrayed as a cycle, somewhat akin to the familiar Krebs cycle of biochemistry. In addition to the helium nucleus, three gamma photons, two positrons and two electron neutrinos are generated. As in the p–p chain, the produced positrons immediately encounter electrons and undergo mutual annihilation to yield two additional 511 keV gamma rays. When including the four annihilation gamma photons, the net reaction can be written:

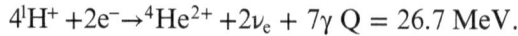

$$4^1\text{H}^+ + 2e^- \rightarrow {}^4\text{He}^{2+} + 2\nu_e + 7\gamma \quad Q = 26.7\,\text{MeV}.$$

The individual reactions involved in the CNO cycle may be portrayed as in the diagram (figure 15.3):

Although we often describe the CNO cycle as 'starting' with the fusion of hydrogen and carbon-12, given that it is closed loop, any of the six reactions can be used as a starting point. The rate-limiting step in the CNO cycle is the proton capture reaction by nitrogen-14.

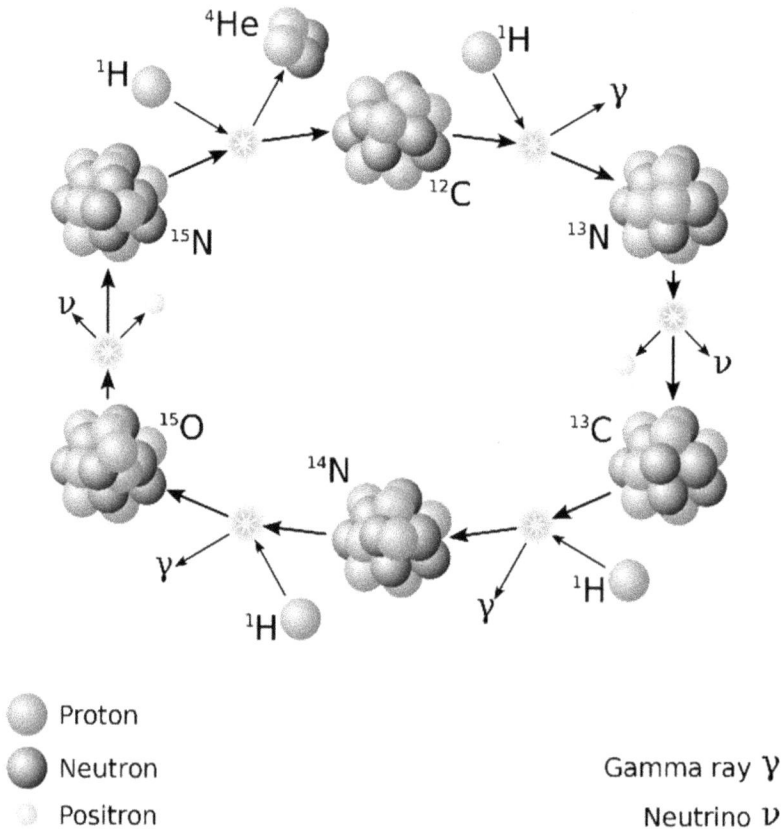

Figure 15.3. The carbon–nitrogen–oxygen or CNO cycle, in which four protons are fused into one helium-4 nucleus in a sequence of six individual reactions, using carbon, nitrogen and oxygen nuclei as 'catalysts'. In our Sun, the CNO cycle contributes approximately 1.7% of the helium-4 made via stellar nucleosynthesis. This CNO Cycle image has been obtained by the author from the Wikimedia website, where it is stated to have been released into the public domain. It is included within this book on that basis. It is attributed to Borb.

The reason for the higher temperature threshold for the CNO cycle compared to the p–p chain is the high coulombic barrier presented by fusion of hydrogen with carbon-12 (or equivalently the fusion of hydrogen with either carbon-13, nitrogen-14 or nitrogen-15, since they all have higher temperature thresholds than p–p fusion). Surprisingly, the confirmation of the CNO cycle in our Sun was experimentally confirmed only relatively recently in 2020. This discovery was made by the Borexino Collaboration, at the Borexino experimental facility located at the Italian National Institute for Nuclear Physics' Gran Sasso National Laboratory in the Apennine Mountains.

15.16 The Gamow factor

It should be pointed out that for any type of fusion reaction, there is a minimum (threshold) temperature as well as an optimal temperature. Since temperature is simply the average kinetic energy of the particles present (i.e. their speed), this means there is an ideal speed for fusion reactions. This ideal speed is called the **Gamow peak**, and the **Gamow window** is the range around that peak where the majority of fusion reactions occur. The probability for fusion is the **Gamow factor** and is related to two probability functions:

1. The odds of any two colliding nuclei having the appropriate speed
2. The odds of quantum tunneling through the coulombic barrier in order for the two positively charged entities to come close enough for the strong force to take control.

The Gamow factors differ for different fusion reactions of course, but generally speaking, the probability of tunneling increases with temperature. Competing with this is the fact that the density of fast, flying nuclei diminishes as the temperature increases. This leads to a characteristic Gamow peak in the probability curve.

15.17 The triple alpha process

When the Sun runs out of hydrogen in its inner core, it will have to resort to alternative means of sustaining itself. With the cessation of hydrogen fusion in the inner core, it will by definition leave the main sequence. Hydrogen fusion may still continue in the outer core or shell, and this signifies its transition into a red giant star. But eventually the temperature in the inner core may reach the threshold for fusion of the accumulated helium. The fusion of helium-4 is a more complicated procedure than hydrogen fusion into helium. Because alpha particles are helium-4 nuclei, the series of steps through which this occurs is known as the **triple alpha process**.

Part of the difficulty with fusing helium into heavier elements lies with the extreme stability of helium-4. As discussed elsewhere, helium is 'doubly magic' meaning that it has a magic number of both protons and neutrons. Helium and other doubly magic nuclides tend to be extraordinarily stable and less prone to undergo reactions that transmute them into other nuclides. Nevertheless, the energy binding curve proves that energy can be derived from fusion up to iron-56 is reached. So, the question arose about how the Sun and other stars might be able to circumvent the challenges of helium fusion.

Starting with the extremely stable helium-4 nucleus, one might wonder what could happen if the ultra-abundant hydrogen (i.e. a proton) were to try to fuse with it. The product would be lithium-5, a nucleus made of three protons and two neutrons. Lithium-5, however, is extremely unstable and decays with a half-life of only 3.7×10^{-22} s; it does not exist in nature. In fact, there are no stable nuclides with $A14 = 5$.

Similarly, one might wonder about fusion between two of the ample alpha particles (i.e. helium-4 nuclei). But given that helium-4 is doubly magic, it is likely to be more stable than its immediate neighbors and thus be unlikely to react readily. The fusion of two helium-4 nuclei would yield beryllium-8. And beryllium-8 is highly unstable with a half-life of around 10^{-16} s (8.19×10^{-17} s to be more precise). This presents a significant hurdle in nucleosynthesis and is called the **beryllium barrier**.

Under 'ordinary' circumstance, this beryllium barrier is quite formidable. It could not be surmounted under the conditions of the Big Bang, and small stars are unable to overcome it either. Our Sun does have the right stuff to clear this barrier, however.

When the Sun runs out of hydrogen fuel and leaves the main sequence, its core will contract and heat up (while its surface will balloon as a red giant). If a star's core density approaches several kilograms per milliliter and a temperature around 100 million K, fusion of helium does become possible. But we just said that beryllium-8 is highly unstable and there are no stable nuclei with five total nucleons. These realities impart serious constraints and obstacles. So how does one proceed beyond the beryllium barrier? The extreme densities and temperatures within the cores of stars allow a practical bypass of the beryllium barrier through a process called the **triple alpha reaction**. In this reaction, three alpha particles (helium-4 nuclei) essentially collide simultaneously and create a carbon-12 nucleus. The reaction is not exactly simultaneous but does require that the third alpha particle collides before the beryllium-8 nucleus can decay—meaning within 10^{-16} s.

The triple alpha process can thus be described as two separate steps:

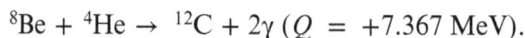

$$^{4}\text{He} + {}^{4}\text{He} \rightarrow {}^{8}\text{Be} \ (Q = -0.0918 \text{ MeV}),$$

$$^{8}\text{Be} + {}^{4}\text{He} \rightarrow {}^{12}\text{C} + 2\gamma \ (Q = +7.367 \text{ MeV}).$$

Again, the second reaction must occur within 10^{-16} s so that the unstable beryllium-8 nucleus does not have time to decay. The net energy release of the process is 7.275 MeV, and the entire triple alpha reaction may be represented as:

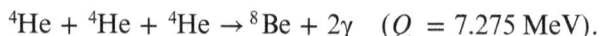

$$^{4}\text{He} + {}^{4}\text{He} + {}^{4}\text{He} \rightarrow {}^{8}\text{Be} + 2\gamma \quad (Q = 7.275 \text{ MeV}).$$

15.18 The Hoyle state

What seems to happen is that when the nearly simultaneous three-body collision of three alpha particles (helium-4 nuclei) occurs, the result is an excited state of carbon-12, the so-called **Hoyle state**, named after Fred Hoyle, who predicted its existence.

In a powerful display of **the anthropic principle** and ingenious reasoning, Hoyle asserted that since he and others exist (and were contemplating this problem), there simply *must* be a way that carbon-12 could be formed from helium. He postulated the existence of a highly excited state, a **resonance**, of carbon-12 (called ^{12}C*) that would be more likely to form from the collision of three alpha particles. The ^{12}C* would then decay into the normal, ground state of ^{12}C* with the emission of gamma photons. The existence of this fleeting ^{12}C* resonance thereby facilitates the reaction —and it enables the formation of carbon and all the heavier elements beyond the beryllium barrier. In the triple alpha reaction, three helium-4 nuclei react and create one carbon-12 nucleus plus three gamma photons (from the decay of beryllium-8 and the ^{12}C*)

15.19 The helium flash

As expected, the mass of a carbon-12 nucleus is slightly less than the mass of three helium-4 nuclei, and the mass defect is liberated as energy via $E = mc^2$. However, the energy released in this reaction is far lower than the energy released by conversion of hydrogen-1 into helium-4 through the p–p chain or the CNO cycle. But after spending much time as a red giant without active fusion going on in its innermost core, this sudden ignition of helium in the inner core sends a jolt through the star known as the **helium flash**. It is akin to a patient without a detectable pulse being jolted back to life with a defibrillator. The star's core has been revived and jumps back to life through fusion of helium-4 into carbon-12. However, this second lease on life is far shorter than the first go-around on the main sequence. While our Sun will continue burning hydrogen for about 10 billion years, helium burning will sustain it for far less time, perhaps just a 100 million years. This is not surprising since the mass deficit for carbon-12 versus three helium-4 nuclei is ten times smaller than for helium-4 versus four protons.

Once sufficient carbon-12 is available, another reaction becomes possible: the abundant helium-4 nuclei can bombard it and create oxygen-16 plus a gamma photon:

$$^4\text{He} + {}^{12}\text{C} \rightarrow {}^{16}\text{O} + \gamma \quad (Q = +7.162 \text{ MeV}).$$

Hence, the core of a star like the Sun, which first becomes filled with helium 'ashes', subsequently learns how to burn those ashes and subsequently becomes filled with carbon and oxygen ashes.

15.20 Nucleosynthesis in more massive stars

The lifespan of a star on the main sequence depends on two factors: how much fuel it has available to burn and how fast it burns that fuel. The first factor is simply how much mass it was born with. But the second factor is more complicated. More massive stars burn their fuel far faster than lighter stars. The luminosity of any given main sequence star is proportional to the rate at which it burns its fuel. But the luminosity is directly related to its mass cubed. So, while more massive stars have more fuel to burn, they burn through that fuel faster than small stars because one factor is directly proportional to the mass and the other factor is inversely proportional to the mass to the third power. Hence, the final outcome (the lifespan of a star on the main sequence) is inversely proportional to the square of its initial mass.

Depending on its mass, after a star has burned up all the hydrogen in its inner core into helium, it may or may not proceed to the next step. Our Sun is massive enough to someday take this next step in thermonuclear fusion (the triple alpha reaction) with a roughly 100 million K threshold temperature. But fusion of the resulting ashes (carbon and oxygen) requires far higher temperatures and pressures. Our Sun will never attain such temperatures and pressures. Thus, our star will end its life as a white dwarf surrounded by a planetary nebula (as described in the chapter on the Sun).

15.21 Stellar corpses and stellar nucleosynthesis

Stars can end their lives in several different fashions, some spectacular and others rather dull. But those spectacular explosive endings emit vast amounts of radiation along the way.

Atoms are electrically neutral thanks to a balance between the number of negatively charged orbiting electrons and the positive protons in the nucleus. Imbalances lead to ions (either anions if negatively charged or cations if positively charged). But when focusing exclusively on nuclei, the positive charges of protons lead to electrostatic repulsion via Coulomb's Law. To overcome this coulombic repulsion, the strong force must be significantly stronger than the electromagnetic force. And indeed, it is. Specifically, the strong force is 137 times stronger than the electromagnetic force. It must be kept in mind that this dominance only applies to very short distances on the order of one femtometer (also known as one Fermi). Thus, the electromagnetic force is 1/137 as strong as the strong nuclear force, and the value, 1/137, is a dimensionless quantity called α. It is also called the **fine structure constant** and is given by:

$$\alpha = 2\pi k e^2 / hc,$$

where k is Coulomb's constant, e is the charge of the electron, h is Planck's constant, and c is the speed of light.

Incidentally, the weak force is about 10^{-6} or one millionth the strength of the strong force, and gravity is about 10^{-39} times the strength of the strong force. Recent research has demonstrated that the fine structure constant, α, is not exactly 1/137 but rather is closer to 1/137.035 99 and has been measured to an accuracy of 11 digits.

In any case, for stellar nucleosynthesis to proceed, the coulombic repulsion of protons must be overcome. This can be achieved under the conditions found in the cores of stars where temperatures and pressures are astronomically high. For instance, the core of our Sun is about ten-fold denser than lead and approximately 15 million K, and these conditions permit the p–p chain and CNO cycles to proceed. Once the temperatures are sufficiently high, the particles will have sufficient kinetic energy to undergo collisions that exceed the energy threshold for getting the positively charged protons close enough for activation of the strong force—in other words, thermonuclear fusion can occur. (Technically, the requisite conditions are not fully met, but thanks to quantum tunneling, the processes can still proceed as discussed in the chapter on quantum mechanics.)

Protons and alpha particles will typically require speeds of thousands of kilometers per second, corresponding to temperatures of millions to billions of Kelvin. In thermonuclear fusion, the lightweight elements are 'burned', releasing heat and electromagnetic radiation and generating 'ashes'. In this regard, the ashes are heavier elements. But the thermonuclear furnaces of stellar interiors differ in one very important way from familiar, chemically mediated furnaces—while burning in an ordinary furnace via chemical means (i.e. oxidation of the coal, oil, etc) leads to ashes that cannot be further used, the 'ashes' of thermonuclear fusion can sometimes be used for further fusion reactions. One can thus 'burn' the ashes of old reactions to produce more energy if the conditions allow. For instance, fusion of hydrogen produces energy and helium ashes. In some stars, when the hydrogen runs out, the helium can be burned to create carbon and oxygen; and in some stars, when the helium runs out, the carbon and oxygen can be burned to generate neon, sodium, etc. This continuous use of old ashes as new fuels can only occur if the star in question has sufficient temperature to exceed the threshold for thermonuclear fusion of the new fuels. And this temperature can only be attained if the gravitational potential energy is sufficient to raise the temperature above the threshold. And the gravitational potential energy depends on the mass of the star. Again, it all boils down to the initial mass of a given star (and for that reason, one can consider the star's birthweight as its DNA blueprint—everything depends on that initial mass). The threshold temperatures for thermonuclear fusion of heavier and heavier elements is higher and higher. For example, **fusion of hydrogen can occur at a mere 14 million K**, but **fusion of helium requires 100 million K. Fusion of carbon requires 600 million K**, and **fusion of oxygen has a threshold of about 1.5 billion K.**

However, this new lease on life—the ability to burn old ashes to create new energy —only works up to iron. **Iron-56** is the end of the line as far as nuclear fusion energy goes; iron is the ultimate nuclear ash. To understand why this is so, we will briefly revisit the binding energy curve of the nuclides.

15.22 The nuclear binding curve

The competition between electrostatic repulsion of multiple protons versus the attractive strong force is manifested by the measured binding energy of the various nuclides. As discussed elsewhere, the curve demonstrates a large change in binding energy between hydrogen (given an arbitrary value of one since there are no other nucleons present aside from the single proton) and helium with its two protons. The difference is roughly seven units. As one increases in atomic number, the binding energy per nucleon continues to drop all the way down to $Z = 26$ (iron). Iron-56 represents the most efficiently packed (that is, most stable) nuclide; beyond this, the nuclides have increasing binding energy. In other words, iron is the trough or nadir on this curve. (Sometimes the plot is inverted such that binding energy increases, and iron-56 is at the peak of the curve rather than the bottom as is the case in figure 15.3.)

Regardless of how the curve is depicted, energy is not provided by fusion of elements beyond iron; fusion then consumes energy. As one works backwards (right to left) from the elements with very high Z such as uranium, the binding energy decreases from right to left as one approaches the trough of iron. What this means in the physical world is a tendency to release energy via fission of certain heavy nuclei such as uranium-235. Alternatively, such elements might gain further stability via alpha emission as they attempt to approach the ideal stability of iron-56. Given this reality, one might wonder how nature ever created elements heavier than iron in the first place? Among the mechanisms of nucleosynthesis beyond iron are:

1. The **alpha ladder**
2. The **r-process (rapid neutron capture)**
3. The **s-process (slow neutron capture)**.

Consistent with the nuclear binding energy curve (figure 15.4), these means of producing atoms heavier than iron-56 demand energy. Such additional energy might be supplied through supernovae and neutron star mergers.

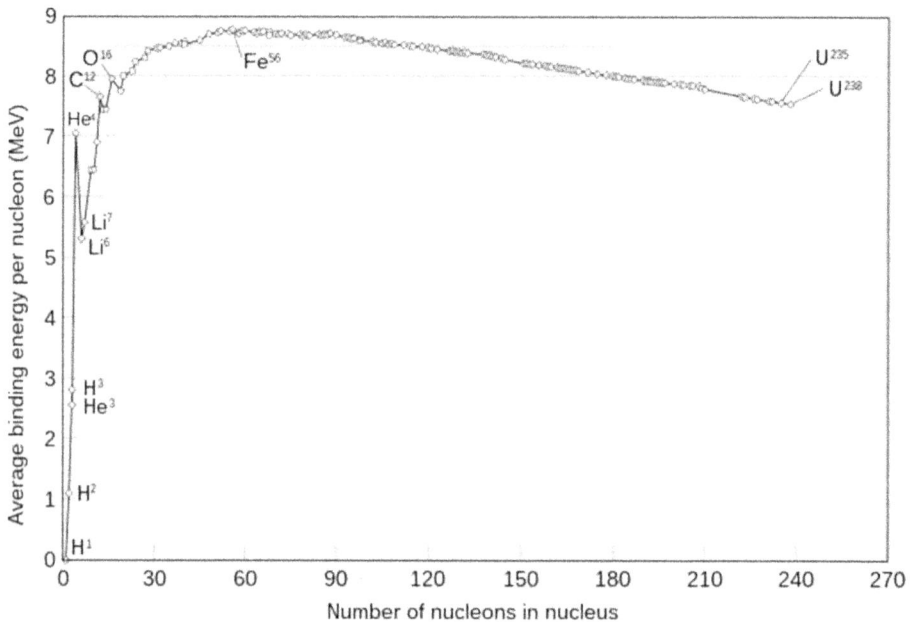

Figure 15.4. In this nuclear binding energy curve the average binding energy is depicted as increasing from left to right as one moves from hydrogen to around iron. This means that energy is released through fusion of these lighter nuclei until iron-56 is reached. Beyond that, fusion no longer releases energy, it consumes it. This Binding energy curve–common isotopes image has been obtained by the author from the Wikimedia website, where it is stated to have been released into the public domain. It is included within this book on that basis. It is attributed to Fastfission.

15.23 Release of the manufactured elements

If a star's core is the source of new elements via stellar nucleosynthesis, one might ask, how do such elements ever get out of the star? During most of a star's life, it burns hydrogen into helium as a main sequence star. And during that period, **convective dredge-up** brings material made internally up to the surface of a star. Once near the surface, such internally made materials (including heavier nuclei) can be blown into space via stellar winds, stellar flares and coronal mass ejections during the star's normal life on the main sequence. Or, towards the end of a star's life, material can be blown into space as a planetary nebula if the star ends as a white dwarf. However, if the star is even more massive, it might explode as a supernova, blasting its innards all over the place—and generating even more new nuclides in the process.

15.24 Supernovae

There are several different categories of supernovae. Although supernovae are historically classified based on **spectral characteristics** (*those without hydrogen lines being **Type I** and those with hydrogen lines being **Type II***), the underlying mechanisms behind supernova explosions might better be categorized as *exploding white dwarfs* or *collapsing supergiant stars* (*that is, **Type Ia supernovae** and everything else, respectively*).

The proposed mechanism behind Type Ia supernovae is that the two stars in a binary star system evolve at different paces such that the heavier one has already become a white dwarf while the other is just entering its red giant phase. If the red giant grows to where it encroaches on the white dwarf's gravitational field (that is, if it encroaches upon the neighbor's **Roche lobe**—the sphere or lobe in which the gravitational influence of one star begins to dominate over the influence of its partner's), it begins to spill mass onto the white dwarf. (Note that the white dwarf does not 'suck' material from the red giant, it is just that the red giant has grown so large that it is spilling over onto the gravitational realm of its white dwarf companion). Note also that this phenomenon of **Roche** lobe encroachment holds can occur regardless of whether the smaller companion is a white dwarf, neutron star or black hole. The smaller companion might have a more intense gravitational field thanks to its smaller radius, but it does not actively suck material from its companion. The red giant just invades the smaller companion's **Roche** lobe and mass gets transferred.

In any case, when a white dwarf gains sufficient mass from its red giant companion through this spillover process, it may eventually reach a certain threshold—the **Chandrasekhar limit**. If it grows beyond the Chandrasekhar limit (which is about 1.44 solar masses in theory, but in reality it depends on the elemental composition of the stellar remnant), it will detonate as a Type Ia supernova, exploding *completely* and leaving nothing behind except a huge cloud of gas and dust (a **shell-type supernova remnant**) that expands into the interstellar medium. Unlike other types of supernovae, this type does not leave behind a central stellar corpse in the form of a neutron star or black hole. Importantly, *all Type Ia supernovae explode with roughly the same energy production* (about 2×10^{44} J). This energy is sufficient to 'unbind' the star (that is, thoroughly overcome the gravitational attraction of all the matter that was once within the white dwarf). This amount of energy, being fairly uniform, allows astronomers to estimate the distances to these supernovae based on their observed apparent magnitudes (since their absolute magnitude is known to be just a shade brighter than -19). In other words, *Type Ia supernovae serve as important* **standard candles** *in astronomy* that allows a measurement of distances to remote galaxies. But for our present discussion, these supernovae not only create an abundance of elements near the bottom of the binding energy curve (i.e. near iron-56), but they also violently hurl these elements far out into interstellar space. In this fashion, other stars and planets (and possibly life forms), can use these new atoms for various purposes, such as new rounds of nucleosynthesis or for the neurocognitive activity associated with thinking about stellar evolution and nucleosynthesis.

Type Ia supernova may also come about through the collision of two white dwarfs (or theoretically, anything in between the extremes from a red giant to another white dwarf). The only prerequisite is that one partner is a white dwarf in a close binary system, and the nuclear detonation occurs on the white dwarf when it exceeds Chandrasekhar's limit.

There may be supernova oddities such as **Type Iax supernovae** in which the explosion is not potent enough to completely obliterate the white dwarf (as in a typical Type Ia supernova). The white dwarf's remains (called a **zombie star**) are then sent racing across the galaxy at phenomenal speeds thanks to the kinetic energy of the blast.

15.25 Supernova light curves

In addition to spectral features such as the presence or absence of hydrogen lines, another important way of describing supernova explosions is based on careful examination of their **light curves**—the plot of luminosity versus time right after the explosion. Type Ia supernovae display very characteristic light curves. Their luminosity is largely thanks to photons created through radioactive decay of **nickel-56.** This radionuclide decays first into cobalt-56, and this in turn decays into stable iron-56. The massive amount of newly generated nickel-56 is intensely radioactive and decays via positron (beta-plus) emission, with a half-life of only 6.077 d. And cobalt-56 is similarly radioactive with a half-life of 77.24 d and also decays via beta-plus emission. Of course, the positrons made during the decay of these two radioisotopes lead to a multitude of 511 keV gamma photons. The light curve of Type Ia supernovae reflects temporal characteristics consistent with the decay of these radioisotopes.

15.26 Core collapse supernovae

Moving onto the 'all other' category of supernova explosions, **Type Ib/c** and **Type II** supernovae share a common mechanism of massive star collapse. In the never-ending battle between inwardly directed gravity and outward pressure from thermonuclear fusion, some very massive stars reach a point where nuclear fusion is no longer able to withstand the relentless crush of gravity. This occurs when the inner core, after fusing one light element after another in nested layers within the core, finally reaches the end of the line—iron. Once iron is reached, the star can no longer generate heat through fusion. At this point, after millions of years of sustained nuclear fusion, things end in an instant. In basically one day (the day that iron starts forming), the inner core can no longer sufficiently generate outwardly directed, gravity-defying pressure, and gravity claims victory.

The core can collapse with one of two outcomes: it might collapse into a black hole with relatively little energy output (although a **gamma ray burst** can still occur as discussed elsewhere) or it might violently detonate the outer layers of the star in a supernova. Such supernova explosions create heavy elements beyond iron (all the way up to uranium) and blast them into space, contributing to the diverse contents of the interstellar medium. The supernova remnants in such cases typically are called **plerions (crab-type remnants)** with a **pulsar** stellar corpse in the middle. This distinguishes them from the Type Ia supernova explosions, which leave **shell-type remnants** with basically nothing in the center. (A quick aside on terminology: A plerion is a nebula that is within a supernova remnant and is powered by a central pulsar. Plerions glow in the radio, infrared, visible, x-ray and gamma ray parts of the electromagnetic spectrum.)

15.27 Pair-instability supernovae

Another unusual, proposed mechanism behind some supernova explosions is called **pair instability**. The concept is that in the cores of some supermassive blue giants, the temperatures might get so hot that the threshold for electron–positron pair production is exceeded (i.e. the kinetic energy exceeds the $E = mc^2$ threshold of 1.022 MeV; through Wein's law, this translates to around 11.8 billion K). Since the radiation pressure scales according to the fourth power of the temperature, in accordance with the Stefan–Boltzmann Law, at such phenomenal temperatures, the contribution made by gamma photons becomes a major fraction of the total pressure.

Once this process of pair production begins, the outward pressure formerly provided by radiation suddenly drops, and the core contracts. Recall that the gamma ray energy density from photons fleeing the core of a star exert a radiation pressure that aids in the battle against gravity. But particles such as electrons and positrons do not exert this same amount of pressure. The drop in pressure causes the core to quickly contract, which will then heat even further—leading to even more electron–positron pairs and an even greater decrease in pressure. This feedback loop can end in a runaway thermonuclear fusion reaction in which all the carbon and oxygen is detonated simultaneously and the whole star is blasted away in an instant. This is the so-called **pair-instability supernova model**. Pair-instability supernovae should only happen in stars with extraordinary masses—those with between 130 and 260 solar masses. Additionally, the parent star must be of low to moderate metallicity and not rotate too fast. This combination of low metallicity with extreme mass is what would be expected of Population III stars (the very earliest stars in the Universe). As in Type Ia supernovae, pair-instability supernovae should completely obliterate the parent star; no neutron star or black hole remnant is left behind.

Rather than a supernova, another consequence of the collapse of such super-massive stars might be a **hypernova explosion**—a special form of supernova in which material is ejected with extraordinary kinetic energy (over 10^{45} J, which is maybe an order of magnitude more than in typical supernova explosions) and with a luminosity over ten-fold higher than in a regular supernova. For this reason, hypernovae are sometimes called **superluminous supernovae** (although there are other forms of superluminous supernovae). Hypernovae might be generated by **collapsars** and theoretically also by pair-instability collapses. A proposed third mechanism behind hypernova explosions is the accretion of matter onto a neutron star that is part of a binary system. If the companion to that neutron star grows into a red supergiant and spills sufficient mass onto the neutron star, it might explode in a manner similar to a Type Ia supernova. But instead of detonation on a white dwarf, this is a scaled-up version with detonation on a neutron star. Hypernova explosions end in black holes and are a mechanism through which **long-duration gamma rays bursts** are made.

For stars with more than 250 or 260 solar masses, regardless of how little metal they contain, another mechanism comes into play. Following the core collapse caused by pair-production-induced loss of radiation pressure, the temperature might rise until yet another threshold is reached—**photonuclear disintegration** may occur in the most massive stars. This is a phenomenon that can occur with gamma ray photons that have enough energy to interact with and splinter atomic nuclei. Since photonuclear disintegration is an endothermic process, much of the energy released during the increased rate of fusion is consumed by these photodisintegration reactions, and the star might not explode as expected in a supernova or hypernova. Instead, the star collapses completely and directly into a black hole. This situation is sometimes called a **failed supernova**.

15.28 The Eddington limit

Returning briefly to the main sequence, some hypergiant stars in the upper left-hand corner of the HR diagram can be as much as 265 times the mass of our Sun (i.e. 265 $M_\odot$, which is around the estimated mass of the Wolf-Rayet star, R136a1). While still debated, it is believed at the present time that modern stars cannot exceed a certain mass based on the **Eddington mass limit** or **Eddington limit** (along with the observation that no known stars exceed this theoretical limit). Note that this is the limit for modern day young stars (that is, those that formed between 1 million and 1 billion years ago) with metal contents like our Sun. In other words, Population I stars. Very early stars that were relatively devoid of metals (i.e. the so-called Population II and Population III stars) could gain much more weight before reaching their Eddington limits.

The Eddington limit represents the maximum mass or luminosity of a star because of a balance point between outwardly directed radiation pressure and the inward gravitational attraction. Beyond this point, hydrostatic equilibrium cannot be sustained, and the overwhelmingly powerful radiation-driven stellar wind will blow away mass from its outer layers; the star will simply be too luminous to remain stable. (Incidentally, gamma ray bursts, novae and supernovae represent situations in which the Eddington limit is momentarily exceeded, causing intense intervals of acute mass loss.)

15.29 Supernova neutrinos

When gravity overpowers degeneracy pressure because the mass of the remnant is beyond Chandrasekhar's limit, the electrons are forced inward by gravity and combine with nuclear protons. The resulting nuclear reaction is one of **electron capture**, which was discussed in chapter 6. **Neutrinos** (specifically electron neutrinos) stem from electron capture whenever electrons combine with protons to create neutrons plus electron neutrinos. Beyond this, however, another (and more important) mechanism for neutrino production occurs during supernova explosions. A newborn neutron star might have a temperature approaching 100 billion K. At this enormous temperature, pair-production of neutrino-antineutrino pairs (of all three flavors) can occur. This process of neutrino production is called **thermal emission**. And the thermal emission process produces even more neutrinos than the electron capture process does.

The combined result is that within a brief 10 s blast, approximately 10% of the star's mass is converted into a burst of neutrinos. These neutrinos constitute the main output of a core-collapse supernova explosion—roughly 10^{46} J. This means that the immense blast of photons, the vast amount of kinetic energy associated with the expanding supernova remnant and all the ultra-energetic cosmic rays emanating from a supernova are just a mere side show of the main event—the neutrino spectacle. But there is yet another source of energy in a supernova blast. The freely infalling material of the collapsing star suddenly comes to an abrupt halt when it smashes into the surface of the small neutron star core that was just formed. Invoking Newton's Third Law, this sudden stop causes a rebound shock wave that helps blow off the outer layers of the star into interstellar space. All this energy contributes to one more important function—the creation of elements heavier than iron (and their dispersal).

15.30 Relative abundances of chemical elements

Nuclei from carbon to iron can be made in stellar interiors via fusion if the star is massive enough to carry on thermonuclear fusion. Recall what was mentioned earlier—fusion of hydrogen can commence at 15 million K, but fusion of helium requires 100 million K. Fusion of carbon requires 600 million K, but fusion of oxygen has a threshold of about 1.5 billion K. Certainly not all stars will have the mass needed to fuse elements all the way up to iron. For example, our Sun will never be able to synthesize elements heavier than carbon and oxygen. Nevertheless, across the universe, if one quantifies elemental abundances, the figures amount to around 74% hydrogen and roughly 24% helium. The remaining elements (known as 'metals' in astronomy) constitute only around 2%. And an important observation is that the relative abundances of these so-called metals is quite consistent across the Universe.

While chemists arrange elements in columns and rows of the periodic table according to their chemical characteristics, astrophysicists can derive useful information when arranging elements on a plot with relative abundance on the y-axis (ordinate) and atomic number (number of protons) on the x-axis (abscissa). The relative abundances on such a plot are given on a logarithmic scale. When arranged in this fashion, several interesting patterns emerge.

First, the relative abundances of hydrogen and helium are far more than the other elements. For quick reference, for every 10 billion hydrogen atoms there are about a million silicon atoms, 100 zinc atoms and one lead atom. The next characteristic of this curve is that there are several characteristic peaks and valleys along the way, on top of a general slow decline from hydrogen towards uranium. For instance, right after the hydrogen/helium peak, there is a very abrupt drop by a factor of nearly a billion to the so-called **light elements**: lithium, beryllium and boron. Lithium, beryllium and boron are sometimes called the **LiBeB elements** and are so rare because (for reasons explained elsewhere) they were not made in abundance during the Big Bang, nor are they made in large amounts during stellar nucleosynthesis.

Following the precipitous drop in abundance displayed by the light element trough is a high point associated with carbon, nitrogen and oxygen. Associated with these are neon, magnesium and their neighbors. This spike is called the **alpha-element peak**. The downward trend continues after the alpha peak until another prominent surge in abundance is encountered near $Z = 26$. This is the **iron peak** and is represented by iron and nickel. The gradual downward trend then resumes with a few more prominent peaks along the way. Specifically, there are three pairs of smaller bumps called **r-peaks** and **s-peaks** (with the r-peaks preceding the s-peaks). These r-peaks and s-peaks are covered a bit more later in this chapter. Another curiosity to note about this curve is a saw-toothed pattern displayed throughout the entire path from helium to uranium. This sawtooth pattern is thanks to a significant difference in cosmic abundance of elements with even atomic numbers compared to odd atomic numbers. The **Oddo-Harkins rule** states that for elements heavier than hydrogen, atoms with an even number of protons are more abundant than their odd-numbered neighbors. Nuclides with an even atomic number tend to be more stable, partly because paired protons can offset the excess unbalanced spin of an otherwise unpaired partner. The result is that even-numbered nuclei tend to be about 10- to 100-fold more abundant than their odd-numbered adjacent neighbors. Thus, magnesium-24 is about ten-fold more plentiful than sodium or aluminum, while sulfur-32 is roughly 100-fold more abundant than phosphorus or chlorine. An oddity is that oxygen-16 is only about ten times more abundant than its light odd-numbered neighbor nitrogen but over 10 000-fold more abundant than its heavier odd-numbered neighbor fluorine (figure 15.5).

Just as a proper understanding of the periodic table requires an understanding of quantum mechanics of electrons, a proper understanding of the natural abundances

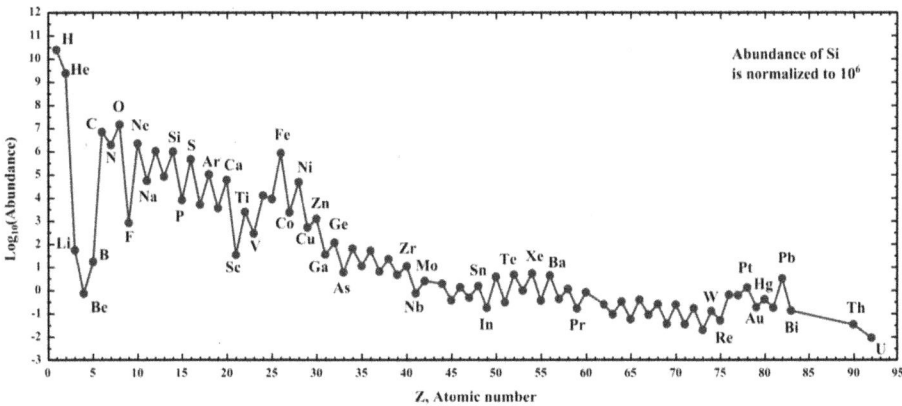

Figure 15.5. In this plot, the logarithm of the abundances of each element is on the vertical axis while the atomic number is on the horizontal axis. There is a sawtooth pattern superimposed on a gradual decline in overall abundance as one moves from left to right. The various peaks and valleys reflect the Oddo–Harkins rule and the numerous less-pronounced peaks may reflect neutron-capture-based nucleosynthesis through the r- and s-processes as explained in the text. This Solar System Abundances image has been obtained by the author from the Wikimedia website where it was made available under a CC BY-SA 3.0 license. It is included within this book on that basis. It is attributed to MHz`as.

curve requires an understanding of stellar and Big Bang nucleosynthesis. One will recall from discussions elsewhere that the first step in ether Big Bang or stellar nucleosynthesis is the formation of deuterium from two protons. But this reaction is slow thanks to being mediated by the weak force and because of the electrostatic repulsion of the two protons. Additionally, deuterium is relatively fragile and will break up at the temperatures under which fusion occurs unless it encounters a reactant sufficiently soon. In nucleosynthesis, this phenomenon is called **the deuterium bottleneck**. But under the right circumstances, the deuterium will fuse with another proton to create helium-3. Two helium-3 nuclei can collide to produce a highly stable helium-4 nucleus (along with two hydrogens).

Helium is extremely stable thanks to being '**doubly magic**'. In other words, it has a **magic number** of protons and a magic number of neutrons. When chemical elements possess a full outer electron shell (as in the noble gases), they tend to be chemically unreactive. Similarly, the stability conferred to nuclides with a magic number of protons or neutrons is thanks to full shells in their nuclear energy levels. Just as the noble gases have a complete octet in their outer electron shells and exhibit chemical stability, nuclides with a magic number of protons or neutrons have complete nuclear shells and tend to be happy just where they are. The magic numbers are:

2, 8, 20, 28, 50, 82 and 126.

A classic example of the stability conferred by possession of magic numbers is illustrated by tin with $Z = 50$. There are ten stable isotopes of tin, whereas its neighbors indium ($Z = 49$) and antimony ($Z = 51$) exhibit only two stable isotopes each.

With this background, one can better understand some underlying principles behind a Segrè plot of atomic number (Z, the number of protons) versus number of neutrons. Students might recall the observed trend of progressively increasing numbers of neutrons relative to protons as Z increases. A simple way of explaining this trend is that the escalating electrostatic repulsion of all those protons requires more 'neutron glue' to hold the nucleus together. Thus, there is a **zone of stability** or **valley of stability** where the ratio of neutrons to protons is optimal for any given atomic number. If a nuclide is too far from this optimal ratio, it tends to be radioactive and will undergo a decay that might promote better proximity to the zone of stability. An easy way to conceptualize this is that nuclides with a suboptimal ratio of neutrons to protons will be far from the ideal valley of nuclear stability and will want to 'roll down' into the valley via gain or loss of neutrons depending on exactly which side of the valley they start out on. The various modes of radioactive decay raise or decrease the relative number of neutrons to protons.

15.31 Restraints on nucleosynthesis

Among the lighter elements, hydrogen-3 (tritium) with its single proton and two neutrons, is unstable and decays with a half-life of 12.32 years. But some theoretical light nuclides simply do not exist or exhibit extreme instability. For instance, helium-2 (consisting of just two protons) is highly unstable thanks to the unmitigated repulsion of the two positively charged protons. Such constraints govern the rules of nucleosynthesis. A couple of classic examples come into focus during the creation of lithium and beryllium nuclei.

Given that helium-4 is doubly magic, it is very stable and generally reluctant to engage in nuclear reactions. Nevertheless, we know that various exergonic reactions involving helium-4 are possible and do occur. But the fusion of two helium-4 nuclei would yield beryllium-8. And as mentioned earlier, beryllium-8 is extraordinarily unstable with a half-life of less than 10^{-16} s. This hurdle in nucleosynthesis is called the beryllium barrier and basically put a halt to Big Bang nucleosynthesis. Similarly, if fusion between a hydrogen nucleus (i.e. a proton) and a helium-4 nucleus were attempted, the product would be lithium-5. However, like all elements with $A = 5$, lithium-5 is unstable and does not exist. Hence, the Big Bang was unable to synthesize elements heavier than helium, but the extraordinary conditions in the centers of stars provide unusual means of circumventing the beryllium barrier and the obstacle presented by the absence of any $A = 5$ stable nuclides. The solution in the cores of stars is the triple alpha reaction, wherein three helium-4 nuclei essentially fuse together simultaneously to create carbon-12.

15.32 The alpha process, the alpha ladder and silicon burning

Much of the mechanics of stellar nucleosynthesis was enunciated in a famous paper (100 pages long!) authored by Margaret Burbidge, Geoff Burbidge, William Fowler and Fred Hoyle (the 'B^{2}FH' paper). Their analysis showed that the extreme stability of the helium-4 nucleus, which was a problem in light element nucleosynthesis, becomes an asset in the synthesis of heavier nuclides. It turns out that at sufficiently high temperature and pressure, it becomes feasible to sequentially add helium-4 nuclei to other nuclei through various fusion reactions. For example, once carbon-12 is available via the triple alpha reaction, one can sequentially add other helium-4 nuclei to generate the other '**alpha elements**' such as oxygen-16, neon-20 and magnesium-24. The process of adding more helium-4 nuclei via fusion allows the creation of silicon-28, sulfur-32 and other even-numbered nuclei via what is called the **alpha process** or the **alpha ladder**. Creation of some even-numbered companions to those on the alpha ladder can be made when the temperatures reach about a billion Kelvin. At this temperature, fusion of heavier elements besides helium-4 are possible (e.g. fusion of carbon directly with carbon, or oxygen with oxygen). These reactions accompany the simple addition of helium-4 units to create heavier elements.

Nevertheless, as fusion reactions approach iron-56, the energy liberated becomes relatively smaller and smaller. Eventually, at iron-56, the ultimate nuclear ash, no longer is energy derived from fusion. Therefore, it is not surprising that there is a peak on the nuclide abundance curve around iron as these 'ashes' build up. Along with iron itself, other elements in **the iron peak** include chromium, manganese, iron, cobalt and nickel with Z ranging from 24 to 28. Iron does stand out prominently from the others as far as its abundance, however.

At some steps along the alpha ladder, the product nucleus may be unstable and undergo beta decay, and this beta decay might occur before another alpha particle is added. Since beta decay decreases Z by one, this is a means through which odd-numbered elements can be created.

A special subset of the alpha ladder occurs at the particularly high densities and temperatures found in the cores of the most massive stars (e.g. type O or B). **Silicon burning** is simply the alpha process starting with silicon ($Z = 14$), creating even-numbered nuclei up to chromium, iron and nickel. Silicon burning occurs in massive stars during their late **asymptotic giant branch stage**, after an oxygen-burning shell phase (figure 15.6). The temperature threshold for silicon burning is in the order of 2 billion K. It might last only 1 d or so. Once silicon-burning has run its very brief course, the massive star will then collapse and lead to a core-collapse supernova, leaving behind a neutron star or black hole. Incidentally, the alpha process of element manufacture also takes place in supernova explosions themselves.

15.33 Neutron capture reactions: the r-process and s-process

Further understanding of the mechanisms behind heavy element nucleosynthesis stems from **neutron capture reactions**. Free neutrons do not have to overcome coulombic repulsion and therefore have a lower energy threshold for fusion than do protons. Starting with iron-56, one can add neutrons to create iron-57, iron-58 and iron-59. But iron-59 is radioactive and undergoes beta decay to create cobalt-59. Cobalt-59 can then capture a neutron to make cobalt-60, which is also unstable and undergoes beta decay into nickel-60. Similar reactions can continue in this manner to create many elements heavier than iron. As one creates heavier and heavier nuclides this way, a sticking point is encountered around 50 neutrons. Adding more

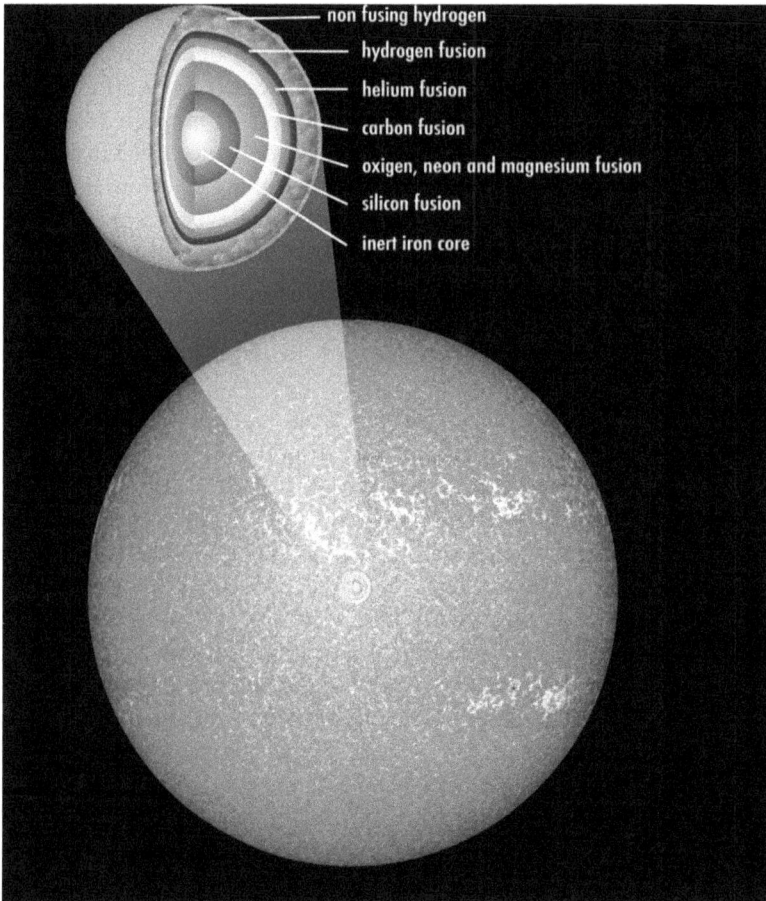

non fusing hydrogen
hydrogen fusion
helium fusion
carbon fusion
oxigen, neon and magnesium fusion
silicon fusion
inert iron core

Figure 15.6. Although our Sun is not massive enough to attain the internal pressures and temperatures needed to synthesize elements heavier than carbon and oxygen, very massive stars (>8 solar masses) can use the 'ashes' of prior fusion reactions to fuel additional energy-producing thermonuclear fusion. At temperatures near 1 billion K, carbon and oxygen nuclei can fuse into heavier elements such as magnesium, sodium, silicon and sulfur. The result is an onion-like core that has multiple concentric shells that are fusing different elements. If the star is sufficiently massive, core temperatures may reach the threshold for thermonuclear fusion of these products into the elements with the highest packing fractions and maximum nuclear binding energies, that is chromium, manganese, iron, cobalt and nickel. Nuclear fusion reactions leading to these nuclei are collectively called silicon burning; silicon burning lasts only 1 d or so. Iron-56 has the lowest mass per nucleon of all known nuclides and is thus the most efficiently bound nucleus. Fusion involving iron-56 requires rather than releases energy. When a massive star starts forming iron in its core, it can no longer sustain the needed outward pressure to stave off gravity, and it suddenly undergoes a catastrophic collapse. The massive star may explode as a supernova leaving behind a neutron star or black hole. This Layers of an evolved star image has been obtained by the author(s) from the Wikimedia website where it was made available under a CC BY-SA 4.0 license. It is included within this book on that basis. It is attributed to Pablo Carlos Budassi.

neutrons at this point tends to generate unstable nuclei that decay back to the $N = 50$ state. This makes sense when one recalls that 50 is a magic number, meaning nuclides with 50 neutrons have a full and stable nuclear shell. Thus, there is a slight overrepresentation of elements near zirconium-90, which has 50 neutrons.

When returning to the nuclide relative abundance plot, one can see small peaks around zirconium, barium and lead. These peaks correspond to the magic numbers of neutrons at 50, 86 and 126, respectively. In fact, lead-208 (the end product of the thorium chain) is doubly magic with 80 protons and 126 neutrons and thus is very stable.

Upon close scrutiny, these small peaks in the abundance plot are split and labeled 'r' and 's' for r-peaks and s-peaks. In each such pairs, the r-peaks come before the s-peaks (that is, the r-peak elements have lower atomic numbers than their partner s-peak elements). In this context, s stands for slow, and r stands for rapid; they describe the rate at which neutrons are captured.

During the r-process, different isotopes are created than during the s-process. The example of cobalt might be instructive in understanding the different mechanisms. If the addition of neutrons leads to an unstable radionuclide such as cobalt-60 with its half-life of 5.27 years, as long as the rate at which neutrons are added is relatively slow compared to this half-life, the cobalt can decay into stable nickel-60. But if the addition of neutrons is fast relative to this decay rate, the cobalt-60 can be converted into neutron-rich isotopes of cobalt-61, cobalt-62 or cobalt-63. Such very neutron-rich isotopes tend to be progressively more and more unstable as their neutron load is increased. For instance, while cobalt-60 has a half-life of 5.27 years, cobalt-61 has a half-life of about an hour and a half (1 h and 39 min), cobalt-62 has a half-life measured in minutes (1 min and 30 s), and cobalt-63 has a half-life measured in seconds (26.9 s). Thus, despite very rapid addition of neutrons in the r-process, there comes a point where addition of neutrons leads to radioisotopes with such short lifetimes that they will inevitably decay before further neutrons can be added. But the point is that the r-process and s-process can lead to different final elements. In the s-process, cobalt-60 may decay into its stable daughter product nickel-60, which may then captures a neutron to become nickel-61. In contrast, the r-process can possibly create cobalt-63, which decays into unstable nickel-63, which in turn decays into copper-63. Thus, one can see different end products of cobalt-60 neutron capture depending on the rate of capture. Working backwards, by examining the relative amounts of the various leftover final end products, one may deduce the relative contributions of the r-process and s-process during the formative phases.

Given the relative overabundance of neutrons in the products of r-process and s-process neutron capture, many of these elements are radioactive and decay back down towards the valley of stability, where the ratio of neutrons to protons is more favorable. Because of a tendency towards relative stability when the neutron numbers are magic (e.g. 50, 82, 126), these magic isotopes tend to accumulate relative to other isotopes. Nevertheless, if they are too far from the valley of stability, their magic cannot save them—they will be unstable and radioactive and will eventually decay into stable isotopes in or near the valley of stability. Such isotopes include selenium-80, tellurium-130 and platinum-132, respectively. Astute readers will note that these isotopes do not possess magic numbers of protons or neutrons themselves. However, their origins, which trace back to r-process nucleosynthesis, include a past in which magic numbers played a key role in their creation.

So, where do these free neutrons come from and where do these different neutron capture processes occur? The slow-process is known to occur over millennia in red giant stars (specifically during their **asymptotic giant branch** or AGB phase), where free neutrons are readily formed. Proof of this is found in the form of technetium in the spectra of such stars. Since technetium has no stable isotopes, its detection in the stellar spectrum of red supergiants is proof that this element was created there, *in situ*. The r-process in contrast is not associated with red supergiants but with supernova explosions. The core collapse of massive stars will lead to electron capture reactions that create neutrons (as well as neutrinos, as discussed above). This sudden abundance of free neutrons in a cataclysmic supernova explosion can drive elemental nucleosynthesis up the r-process ladder. After the brief but intense period of r-process nucleosynthesis, the resultant radionuclides can decay into stable isotopes over varying periods of time. The r-process also appears to occur during neutron star mergers, which may result in short-duration gamma ray bursts. Finally, the r-process may occur to a minor extent in thermonuclear weapon explosions. The elements einsteinium ($Z = 99$) and fermium ($Z = 100$) have been seen in nuclear weapon fallout and their presence there is consistent with the r-process.

15.34 Lithium, beryllium and boron: made through the x-process

The above discussions explain the abundances of carbon and the heavier elements through stellar nucleosynthesis. The most abundant elements, hydrogen and helium, are explained by Big Bang nucleosynthesis, which is covered in another chapter. But what about the light elements, lithium, beryllium, and boron? Some lithium-6 and lithium-7 was made in the Big Bang. Lithium is relatively fragile, however, and is destroyed via interactions with protons (i.e. hydrogen) at temperatures over 2.4 million K, which is easily attained in most stellar interiors. But beryllium and boron isotopes were not made through stellar or Big Bang nucleosynthesis. It is believed that these light elements (along with helium-3) were made mostly via **cosmic ray spallation**—the process in which heavier atoms (e.g. carbon, nitrogen and oxygen) are shattered into smaller fragments through cosmic ray collisions. Cosmic ray spallation is sometimes called **the x-process**. Such collisions can cause spallation right out in the open interstellar space, on asteroids, in Earth's atmosphere or on Earth's surface (the upper 10 m or so). Thus, these light elements are the so-called **cosmogenic nuclides**. The most likely culprits in cosmic ray spallation are ultra-fast protons given their relative abundance (87%–89%) in the galactic cosmic ray repertoire (with about 12% helium nuclei and 1% heavier nuclei). The interstellar medium is probably where most cosmogenic nuclide production occurs. Although cosmic ray spallation is the major source of helium-3, lithium, beryllium and boron, it also generates some hydrogen and helium-4, but the amounts of these nuclides produced this way compared to their background primordial abundances are negligible. Curiously, when one thinks about it, cosmic ray spallation reduces the average atomic weight of interstellar space.

Further reading

[1] Adelberger E G *et al* 2011 Solar fusion cross sections. II. The pp chain and CNO cycles *Rev. Modern Phys.* **83** 195

[2] Bellini G *et al* 2012 First evidence of pep solar neutrinos by direct detection in borexino *Phys. Rev. Lett.* **108** 051302

[3] The Borexino Collaboration 2020 Experimental evidence of neutrinos produced in the CNO fusion cycle in the Sun *Nature* **587** 577–82

[4] Morel L *et al* 2020 Determination of the fine-structure constant with an accuracy of 81 parts per trillion *Nature* **588** 61–5

[5] Xing Q F *et al* 2023 A metal-poor star with abundances from a pair-instability supernova *Nature* **618** 712–5

[6] Vennes S *et al* 2017 An unusual white dwarf star may be a surviving remnant of a subluminous type Ia supernova *Science* **357** 680–3

[7] Burbidge E M, Burbidge G R, Fowler W A and Hoyle F 1957 Synthesis of the elements in stars *Rev. Mod. Phys.* **29** 547

[8] Meneguzzi M, Audouze J and Reeves H 1971 The production of the elements Li, Be, B by galactic cosmic rays in space and its relation with stellar observations *Astron. Astrophys.* **15** 337–59

Chapter 16

The Sun

16.1 Natural history

If you were to serve as an invisible visitor to an alien world and wanted to gather as much information about the biological natural history of the inhabitants but only had one week in which to do it, what might be the best option? In an ideal situation, you might want to follow a few individuals along from birth to death but given that you are only staying for one week, that option is unavailable. But rather than follow any one alien individual along for that brief interval, a more reasonable strategy might be to examine a wide array of inhabitants of very different ages. In that manner you might learn about their conception, birth, infancy, childhood, pubertal phase, adolescence, adulthood, old age and death. In a similar fashion, by examining a wide array of stars at different stages in development, we can get a better understanding of stellar natural history. In this way, we have a fairly good idea of what our Sun was once like and what it will be like in the future.

One interesting observation is that the Sun is somewhat unusual in that unlike most stars in our Galaxy, it is a solitary star and not part of a binary system. Another (somewhat perplexing) observation is that, in absolute terms, although the Sun is less luminous than most of the brightest stars in the sky, this is an illusion—the Sun is actually more luminous than most stars in our region (up to an approximately 20 parsec radius). This is because stars of Morgan–Keenan spectral **type G**, like the Sun, are more massive and more luminous but less plentiful than the spectral **type K** and **M stars** (**red dwarfs**) that litter the skies (but are largely unseen because they are so dim). Nevertheless, we can consider the Sun to be fairly average in many ways, with an average mass and average luminosity. Nevertheless, being an 'average' star remains rather extraordinary by human standards.

16.2 Thermonuclear fusion in the Sun

The Sun burns prodigious amounts of hydrogen at an incredible rate. Every second over 600 megatons of hydrogen are consumed and then converted into helium through thermonuclear fusion. And in that process, about 4 megatons of mass are converted into energy through the **Einstein relation**, $E = mc^2$. By human standards, it is hard to comprehend how this rate could be maintained for any length of time. Yet, the Sun is just so massive that it can sustain this rate for about 10 billion years. And its rate of fuel burning is much slower than one might imagine.

Furthermore, the formation of deuterium from two protons depends on the **weak force**, which slows it down tremendously—the average proton might require around a billion years to take this initial step along the proton–proton chain (p–p chain) in fusion. But the Sun still manages to burn nearly 40 billion tons of hydrogen fuel per minute in its core. And it can do that every minute of its roughly 10 billion year life. How? Because it is big. (Big by human standards, that is. Other even larger stars might yawn a bored 'ho-hum' upon hearing this same old story.)

16.3 The Sun is a main sequence star

We know the Sun is presently in its prolonged adult phase, meaning it lies along the **main sequence** of the **Hertzsprung–Russell (H–R) diagram**. Like all main sequence stars, this *means it produces its energy via thermonuclear fusion of hydrogen in its inner core*. Before going on, a quick point about terminology is in order.

In any star, fusion is restricted to the inner region or **core**. But as we shall soon see, as stars evolve over their lifetimes, the regions where fusion occurs evolve too. The core can contain an innermost region that fuses different elements (e.g. helium into carbon) while an outer core region might continue fusing hydrogen into helium. Many texts call the innermost region the core and refer to the outer layers as '**shells**'. To minimize confusion, in this text we shall stick with the tradition of calling the region where fusion occurs the core as a whole. But we shall occasionally refer to an inner core and the surrounding shells of fusion as the outer core. This inner core and its surrounding shells occupy only a small volume within a star. In fact, in stars that have inner core, and outer cores or shells, relatively tiny proportions of their overall volumes are core. The outer regions of such stars (**red giants**) have expanded so much that, even though their cores have differentiated into various layers, the relative volume occupied by that fusion furnace is proportionately miniscule (figure 15.1 in chapter 15).

16.4 Anatomy and physiology

Spectral analysis of the outer layers can reveal the composition of the Sun. The Sun is overwhelmingly (over 90%) made of **hydrogen**. For every million protons (hydrogen nuclei), there are an estimated 85 000 helium atoms, 850 oxygen atoms, 400 carbons, 120 neons, 100 nitrogens and 47 iron atoms with just traces of all the other naturally occurring elements. Interestingly, helium was not known here on Earth until after it was discovered in the spectrum of our Sun. Thus, its name, which is derived from the Greek sun god, Helios.

Working our way out from the inside, the **core**, at around 15 million Kelvin is where the thermonuclear fusion furnace lies (figure 16.1). The core is the source of the Sun's energy. As reviewed in chapter 15, the core makes many gamma rays

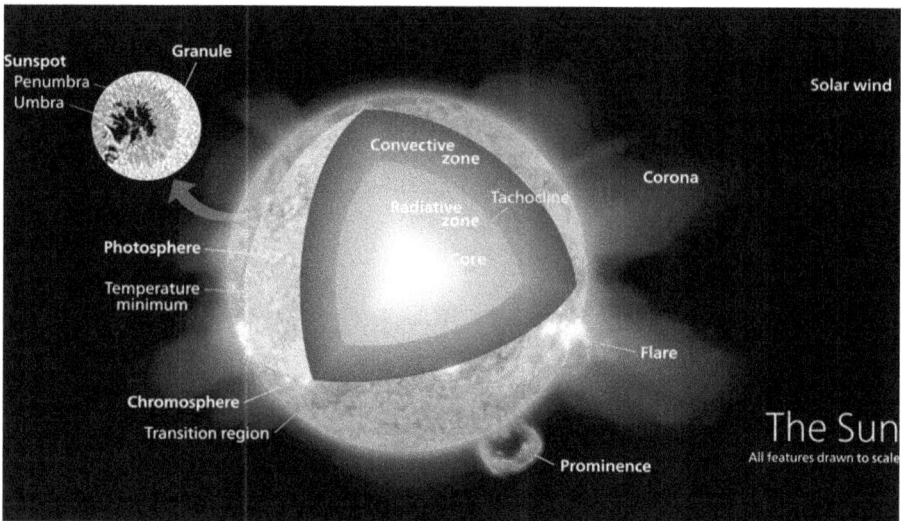

Figure 16.1. Solar anatomy. The Sun's core is where energy is created from matter through thermonuclear fusion reactions. The next layer out is the radiation zone or radiative zone. This is where energy moves slowly outward via radiation. Photons, although moving at the speed of light, do not travel far before becoming absorbed and re-emitted. A photon generated in the core may take over 170 000 years to radiate through this layer. Above the radiation layer lies the convection zone. Here, heat continues to move toward the surface but the transfer is through convection currents rather than radiation. The Sun's atmosphere consists of the photosphere, chromosphere and corona. The photosphere constitutes the visible surface of the Sun and is the lowest and coolest of the three layers with an effective temperature of about 5780 K. The chromosphere is shaped by magnetic field lines that restrain the underlying solar plasma. The ionized outermost layer, the corona, reaches temperatures over a million Kelvin and glows in x-ray and ultraviolet wavelengths. This Sun poster image has been obtained by the author from the Wikimedia website where it was made available under a CC BY-SA 3.0 licence. It is included within this book on that basis. It is attributed to Kelvinsong.

during thermonuclear fusion of hydrogen into helium. And these gamma ray photons enter the next layer up, which is the **radiation zone** (or **radiative zone**); above that is the **convection zone**. These regions are named after the way energy moves upward from the core to the surface. *In the radiation zone, energy is transferred via emitted photons.* But the plasma in the radiation zone is so electron dense that photons are an extremely inefficient means of moving that energy out. Photons will quickly encounter electrons and nuclei and get absorbed, only to be promptly re-emitted in a random direction. Energy moves through the radiation zone in a manner called **radiative diffusion**, which in this particular case, can be thought of as a combination of radiation and conduction.

The next layer up is the **convection zone** or **convective zone**. In this layer, energy is transported upward through convection, which is the familiar process in which heated blobs of matter grow less dense, rise upward and cool off as they release their heat and then sink back down again. Such giant cycles of moving solar material create **solar convection cells** within the convective zone. The bright speckles seen on the surface of the Sun are called **granules**. These granules are the visible evidence of underlying convection in action. Granules are simply transient pockets of extremely hot plasma that have risen to the surface, dropped off their heat energy, cooled off and will sink back down along their perimeter.

The Sun's visible surface itself is called the **photosphere**. This 'surface' is of course not solid but represents the interface where the underlying gas and plasma become opaque to visible light. Above the photosphere, the solar atmosphere seems transparent whereas below the photosphere, photons cannot fly freely but instead collide with electrons and bounce about as expected in opaque regions. The photosphere is well represented as a blackbody radiating at a temperature of **5778 K** (figure 16.1).

16.5 The Sun's atmosphere: the chromosphere and corona

Above the photosphere are the Sun's two atmospheric layers—the **chromosphere** and the **corona**. The lower atmospheric level, the chromosphere, is relatively thin but hot layer of plasma only around 10 000 km thick but about 10 000 K (on average). Some sources say the chromosphere is perhaps only 5000 km thick but has spicules that reach up to 10 000 km into the corona. Also, the temperature is not uniform but varies from around 6000 K at the photosphere boundary down to about 3800 K as one moves slightly higher up (at the **chromosphere temperature minimum**). But then the chromosphere paradoxically increases in temperature with altitude, increasing to 8000 K at the top. At the interface with the corona (the roughly 100 km transition region) temperatures rise rapidly.

Although it varies with height, in general the chromosphere is very rarefied at only 10^{-8} the density of our atmosphere at sea level, which makes it essentially invisible. Both the chromosphere and the corona are only evident during total solar eclipses. The chromosphere looks somewhat pinkish-red during an eclipse. This reddish hue comes from the many hydrogen atoms that are transitioning in electron energy down to the $n = 2$ energy levels. This transition corresponds to the **Balmer series**. This is what hydrogen atoms tend to do at these temperatures. A very prominent Balmer emission line in the corona is the Hα line at 656.3 nm, which represents the $n = 3$ to the $n = 2$ energy transition and lies in the red portion of the visible spectrum.

Surrounding the chromosphere is the **solar corona**, which is shaped by charged particles moving along magnetic field lines. As the Sun's magnetic fields change over time, so does the shape of the corona. But although the corona follows magnetic lines of force, it is always present, even when solar magnetic fields are minimal.

Like the chromosphere, the corona is hotter in some areas and cooler in others, but the odd thing is that it is generally extremely hot—over 2 million degrees in some parts. Its extremely high temperatures are betrayed by *the x-ray radiation emanating from the corona*. It is far thicker than the chromosphere and extends millions of kilometers into space. As one would expect based on its temperature, the corona contains highly ionized atoms moving at very high speeds. Ironically, although the temperature is very high, the density is so low (around 10^9 particles per ml) that, despite the temperature of individual particles reaching a couple of million degrees, the rate of heat loss of a 37 °C (98.6 °F) human body would greatly exceed the rate of heat gain. A person would paradoxically freeze to death in the 2 million degree solar corona, illustrating the difference between temperature and heat content. (Of course, this thought experiment conveniently ignores the heat emanating from the nearby 5778 K solar photosphere!)

16.6 Life history of our Sun

For approximately 10 billion years, our Sun will remain stable as a main sequence star (a G2V category star to be more specific). As with all main sequence stars, this means it will be productively obtaining large amounts of energy via thermonuclear fusion of hydrogen into helium. At roughly 4.6 billion years of age, the Sun is middle-aged.

However, the Sun will brighten significantly over the second half of its life. (In fact, it has already brightened about 30% since its infancy.) This phenomenon can be understood in terms of simple thermodynamics of ideal gases. As hydrogen undergoes fusion into helium, the number of discrete particles declines from four hydrogen nuclei into one helium nucleus. In other words, the **number density** (number of particles per unit volume) decreases sharply. From the **ideal gas law**, one might recall that pressure is proportional to temperature and number density:

$$P = kNT$$

where N is the number of particles per unit volume, and k is Boltzmann's constant.

Or in the more familiar form used in chemistry:

$$PV = nRT$$

where n is the number of moles, V is volume and R is the universal gas constant.

Because the number density in the core decreases as hydrogen is converted into helium, the volume will decrease. In order to maintain the outward pressure needed to fight the gravitational inward crush, the core temperature must rise. This increase in temperature with a reduction in volume results from a special example of the **virial theorem**, which relates the potential energy of a system to kinetic energy of a system of moving particles over time. In this particular case:

$$2\langle KE \rangle = n \langle PE \rangle \quad \textbf{Virial Theorem}$$

where the brackets signify average and $n = -1$.

In other words, the average kinetic energy is equal to negative one-half the average potential energy.

Note that (assuming uniform density and perfectly spherical shape) the gravitational potential energy caused by the mass of a star is:

$$PE = -3/5 \, GM^2/R$$

where M is the mass of the star and R is its radius.

Recall that the mean kinetic energy of an individual particle (for example, a hydrogen nucleus, that is a proton) is given by:

$$\langle KE \rangle = 3/2 k \langle T \rangle.$$

The virial theorem thus yields:

$$\langle T \rangle \approx GM^2/5kR.$$

Hence, the temperature increases as the radius decreases.

16.7 The endless battle between gravity and the outward forces

As an aside, from birth through death every star can be viewed as the product of a battle between gravity (which is trying to crush all the matter inward) and various forces that exerts outward pressure to counter that gravitational attraction. Sometimes the battle never really gets started and 'thermonuclear weapons' (i.e. nuclear fusion) are not necessary to combat gravity. Such is the case with planets and **brown dwarfs**, which are stars that never really get started (i.e. they never enter the main sequence of the H–R diagram). In other cases, a prolonged equilibrium is maintained as the outward pressure from thermonuclear fusion creates stability during the main sequence. Towards the end of a star's life, however, things can change drastically as gravity attempts to finish the fight. Our Sun will put up a mighty battle as it goes through a red giant phase when it runs out of hydrogen in its core and eventually switches from hydrogen fusion to helium fusion. But when it runs out of helium in the core, it will go ultimately die a relatively peaceful death as a white dwarf with a planetary nebula.

Other more massive stars will have more dramatic endings as they resort to carbon burning when they run out of helium. And they may then go through a series of different thermonuclear fusion options before running out of alternative fuels altogether. The end of the game is reached when iron (the 'ultimate nuclear ash', since it does not yield energy upon fusion) is created. Sensing weakness, gravity escalates the battle and squeezes the stellar remnant into a ball of neutrons, a **neutron star**. In some situations, gravity is ultimately and unequivocally victorious as it crushes the star into oblivion—a **singularity** with zero volume—when a **black hole** is formed. But as we shall see later, in other cases, the outward pressure does win the war and completely overcomes gravity, completely obliterating the entire star altogether, in certain types of supernovae.

16.8 The Sun's next few billion years

Returning to our Sun, because of the decreasing number density and consequent core contraction and temperature rise, the Sun's luminosity will increase over the next several billion years. Although the Sun is not due to run out of hydrogen fuel and resort to helium burning (and transitioning into an Earth-igniting **red giant**) for another 4 or 5 billion years, the increase in luminosity due to this change in number density is likely to cook Earth far sooner. In perhaps as soon as 1 billion years, Earth could become uninhabitably hot.

A billion years is not a terribly long time when put into perspective. Students of geology might realize that this is a relatively short interval of time considering the duration of the Archean Eon (circa 4 to 2.5 Ga) and the Proterozoic Eon (2.5 Ga to 543 Ma). In fact, part of the Proterozoic Eon goes by the nickname of the '**Boring**

Billion' or the dullest time in Earth's history. This billion-year span extended from 1.8 to 0.8 Ga (800 Ma) and was characterized by relatively stable tectonic activity, a stable climate and slow biological evolution. In stark contrast, Earth's next billion years are likely to be anything but boring. Unless something drastically changes such as a major diminution of the greenhouse effect (more on this later), thanks to increasing solar output, the Earth is likely to reach a tipping point of no return, putting an end to its habitable period and extinguishing its biosphere. (Despite the fact that the Sun itself will continue to burn for another 4 billion years after that before it becomes a red giant.)

16.9 The faint young Sun paradox

But in the distant past, the Earth had the opposite problem since the Sun was 30% less luminous than it is now. The **faint young Sun paradox** (also called the **dim young Sun** paradox) is the perplexing situation in which the fossil record shows signs of early life and a relatively warm Earth despite the Sun only being at 70% of its current output. The paradox is resolved by a relative abundance of greenhouse gases such as methane and carbon dioxide in Earth's atmosphere during this early phase. Given this past solution, one might imagine a future solution which reduces the abundance of greenhouse carbon dioxide from its present 400 parts per million to something lower and less likely to retain heat. Nevertheless, once the Sun's output increases sufficiently to start evaporating the oceans, water vapor (which is a very potent greenhouse gas) will accelerate the heating in a runaway positive feedback loop and Earth will likely suffer the same fate as Venus—a bone-dry, burning hot, biologically barren wasteland. And this is far before the Sun goes into its red giant phase.

16.10 Leaving the main sequence

When the Sun finally does run out of hydrogen in its core, by definition, it will leave the main sequence. On the H–R diagram it will be at a place called the **turnoff point**. It will have to find an alternative fuel source. Although hydrogen will still be quite ample everywhere outside the core, like a fission nuclear reactor, the only place nuclear reactions occur is in the core itself. And the previous temperature, which was adequate for hydrogen fusion, will not suffice for helium fusion. This is due to the increased electrostatic repulsion conferred by the two protons in helium (as opposed to just one proton apiece with hydrogen fusion) according to **Coulomb's Law**:

$$F = 4kq_1q_2/r^2$$

Where r is the distance between the two charges and q_1 and q_2 represents magnitude of the electric charges involved. In this case of two helium nuclei fusing, q_1 and q_2 are equal and represent the charge of a proton (one elementary charge unit, which equals $1.602\ 176\ 634 \times 10^{-19}$ coulombs).

Because of the increased electrostatic repulsion, the temperature will not initially be sufficient to move onto helium burning. With no immediate source of energy, the Sun will be 'out of balance' for the first time since it first ignited and went from a protostar to a genuine star around 10 billion years earlier. Gravity will begin to squeeze the core and force it to shrink. Curiously, while contraction is occurring in the core, the outermost layers of the Sun will expand as discussed below. But before then, this contraction will cause the Sun to move horizontally towards the right on the H–R diagram and will become a **subgiant**. (Movement along the H–R diagram for an individual star is called its **life track**.)

16.11 The Sun's red giant phase

Although fusion in the innermost core of the Sun will temporarily cease with depletion of hydrogen and insufficient heat for helium fusion, things will still be happening in the outer core or shells. The heat generated by the contraction of the helium-containing inner core will promote fusion in the outer hydrogen shell and increase the Sun's overall luminosity. This step is called **hydrogen shell burning** and is discussed further below. (This shell of hydrogen is just outside the helium inner core that has yet to ignite. It is undergoing fusion and therefore rightly can be defined as part of the core. But since it lies outside the innermost helium core, it is called a shell or outer core.) The increased hydrogen fusion will also affect the outer layers of the Sun, resulting in expansion. In other words, the contraction of the inner helium-containing inner core leads to a paradoxical expansion of the Sun as a whole.

This expansion of the outer layers of the Sun causes them to cool, shifting the color, as viewed from afar, from yellow to orange and then finally to a red color (thanks to **Wien's Law** and a spectral shift towards longer wavelengths). This phase of the Sun's development is its **red giant** stage. The Sun's outer expansion will continue, and its overall luminosity will correspondingly increase simply because of this size increase. This causes the Sun to move upward on its H–R diagram life track. Over perhaps a billion years, the Sun will progressively grow both in size and luminosity as a red giant. At the peak of its red giant phase luminosity, the Sun will be over a thousand times brighter than it is today and over a hundred times its present diameter. Estimates vary but at the time of this writing, most authorities feel the Sun will expand to around the orbit of Mercury at this point.

16.12 Hydrogen shell burning

During this prolonged period, the absence of the outward pressure from thermo-nuclear fusion in the innermost helium-containing core causes it to continually cool and contract under the compression of gravity. But this contraction releases gravitational potential energy, which in turn (according to the virial theorem) starts raising the core temperature. By this stage the core overall will have differentiated into a helium-containing inner core still incapable of fusion and an outer core ('shell') of hydrogen that can undergo fusion. When the temperature of this shell of hydrogen gets hot enough, fusion can begin in this surrounding shell—in other words, **hydrogen shell burning** commences. Hydrogen shell burning will actually proceed at a higher rate than present-day hydrogen burning in the inner core of today's Sun. It is this increased energy production that will push the Sun's superficial layers outward, causing it to balloon into a red giant. *Just as hydrogen fusion in the* *inner core is a distinguishing characteristic of the main sequence, a defining feature of the red giant stage is hydrogen shell burning.*

As hydrogen burning proceeds in the outer core ('shell') it contributes more helium to the inner core. In a sense, the outer core will be dumping its 'ashes' into the inner core. But since it takes four hydrogens to make one helium, by number density reduction, the shell shrinks. The compression makes it grow hotter, putting more pressure and temperature on the inert inner helium core. The increased heat accelerates the fusion rate in the hydrogen-burning shell, feeding more helium ash to the inner core, in a vicious cycle akin to a broken thermostat. The red giant continues to grow to the point where its surface gravity becomes quite weak and large volumes of mass can be blown off the red giant's surface in a strong stellar wind. This future red giant solar wind will eject far more matter than the Sun does today with its present solar wind, but the ejected matter will fly away at much slower speeds.

For stars lighter than the Sun, this might be the end of the line. **Degeneracy pressure** will halt the contraction of the red giant's core, and further fusion of the built-up nuclear ash will never occur. Such stars will end their lives as **helium white dwarfs**. Our Sun, however, has sufficient mass for one last hurrah and will be able to take another step in its lifelong battle against gravity. Being able to take this next step means our Sun will ultimately end up as a **carbon white dwarf**.

16.13 The Sun as a helium burning star

While the Sun is in its late red giant phase, the continued compression caused by gravity causes the inner core to shrink and heat even further. Eventually, the temperature will reach the **100-million-degree threshold for helium fusion**. At this point, the accumulated helium can begin to undergo fusion and the Sun will enter its next phase.

As the Sun becomes capable of burning its old ashes, it will become smaller and less luminous but hotter and more yellow than it was in the immediately preceding red giant phase. It will evolve into a **helium-burning star**. As is explained in further detail elsewhere, helium fuses into carbon via the so-called **triple-alpha process** in which three helium nuclei (alpha particles) fuse into a carbon nucleus. This carbon nucleus can fuse with another helium nucleus via alpha capture to create oxygen.

Of course, these fusion reactions that generate carbon and oxygen release energy; as with other fusion reactions, the mass of the product (in this case, carbon-12) is less than the combined mass of the reactants (three helium-4 nuclei) and that mass deficit is converted into energy via $E = mc^2$.

But this energy yield is far lower than the yield from hydrogen fusing into helium. As a result, although hydrogen-burning sustained the Sun along the main sequence for about 10 billion years, helium-burning will only sustain it for approximately 100 million years. Students of paleontology will recognize just what a brief amount of time this is, considering that various (non-avian) dinosaur species roamed the planet for nearly 180 million years (from around 240 Ma in the Triassic Period to the close of the Cretaceous Period 65 million years ago).

16.14 The helium flash

The transition from red giant to helium burning star has one exciting moment—the **helium flash**. This is caused by the sudden onset of helium burning, resulting in a dramatic rise in temperature. This in turn accelerates the fusion rate. Within a matter of seconds, over half of the accrued helium in the Sun's core will ignite and undergo fusion. This results in rapid expansion—and thus cooling—of the core, which then decreases the fusion rate. In its new equilibrium state, where it is fusing helium in its inner core and hydrogen in a surrounding shell, the total energy output is somewhat lower than it was during its peak productivity during the red giant phase. The outer layers contract and the surface temperature thus increases, changing the color (as viewed from afar) from red to yellow. Although it is now once again in a state of equilibrium with fusion taking place in its inner core, the Sun is off the main sequence, since by definition, only stars fusing *hydrogen* in their inner cores are on the main sequence. And unlike the main sequence phase, which lasted about 10 billion years or the red giant phase that lasted about a billion years, this helium-burning phase will only sustain the Sun for about 100 million years—about 1% of its lifetime on the main sequence.

Because helium-burning stars are smaller and hotter but less luminous than red giants, at the moment of their helium flash, the Sun will move down and to the left on its H–R diagram life track. And since they all fuse helium at roughly the same rate, *all helium-burning stars have roughly the same luminosity*. But they can have different surface temperatures (depending on how much of their mass they blew off in their red giant stellar winds). For this reason, helium-burning stars trace out a **horizontal branch** on the H–R diagram.

16.15 The Sun's second red giant phase

Once the Sun runs out of helium in its inner core, things will again change. Now, with an inner core consisting of a big pile of inert carbon (and oxygen) ashes, it will again fall out of balance and begin to shrink thanks to the crushing force of gravity. The inner carbon core will never attain the necessary heat for fusion, but the outer core shells can continue fusing their available fuels. As has been going on all along, there will be an outer core shell of hydrogen that will continue fusion. And there will also be a shell underneath that which is made of helium—the ashes of the outer shell hydrogen fusion. These helium ashes can ignite, and the Sun will become a **double shell-burning giant**. Recall that during its first red giant stage the Sun was a hydrogen shell-burning star; during this **second red giant stage**, it has become a double shell burning star. So, at this point the Sun has two distinct sets of fusion providing energy simultaneously—an inner shell burning helium via the triple alpha process and alpha capture, and an outer shell of hydrogen burning via the proton–proton chain (with a minor contribution from the CNO cycle). In between these two layers of the core lies another layer of inert helium ash that has not yet reached the threshold temperature for fusion. Meanwhile there is an inert inner core of carbon and oxygen ash accumulating at the very center. (Incidentally, in stars larger than the Sun, this onion-like configuration of multiple shells burning simultaneously, can be taken to its extreme with many layers fusing different elements in a desperate effort to stave off the crunching pressure of gravity.)

16.16 Solar radiation and thermal pulses

The furious rate of fusion in the Sun's double shell-burning phase will be so fast and hot that it will cause the superficial layers of the Sun to expand even farther than during its first red giant stage. In our Sun this new carbon and oxygen core will never reach the threshold temperature to initiate fusion. Unable to sustain its outward pressure-producing heat through thermonuclear fusion, it will begin to contract and heat up. Although it will never reach the threshold for carbon fusion in the inner core, the added heat will further accelerate the rate of hydrogen and helium fusion in the shells, releasing yet more energy. Then all this extra energy expands the outer regions of the Sun even further making the still larger, double shell burning red giant. Although most estimates are that it will at least engulf Venus, at this second red giant stage, it might extend all the way to Earth itself or beyond. On the H–R diagram, the Sun's life track will take another upwards turn as it moves onto this second red giant stage. But this frantic pace of fusion will burn out relatively quickly —maybe in only a few million years. And this double shell burning process will probably never attain a true equilibrium. Instead, the helium fusion will proceed in convulsions called **thermal pulses** every few thousand years or so.

Through the radiation pressure steadily exerted by the solar wind at this second red giant stage, much of the outer solar atmosphere will be blown away. In addition to this stable radiation pressure, the red giant will become unsteady and oscillate periodically in the sense that it will have occasional 'sneezes' or 'burps' (sternutation or eructation, respectively to use the medical terms) and blow off parts of its outer layers in a series of relatively non-violent eruptions. These discharges do not have to be too explosive since at this tremendous size, the outermost layers will be a vast distance away from the core and barely held in place at all by the Sun's gravity. In its old age, the red giant Sun will lose approximately 10%–20% of its mass through these solar 'hiccups' (singultation for medical and premedical students).

16.17 The Sun as a carbon star

Ignition of the accumulated carbon ashes through thermonuclear fusion would require a temperature of about 600 million degrees. And the Sun will never reach that temperature. **Degeneracy pressure** will halt the contraction caused by gravity, and without continued contraction the temperature will cease increasing. The Sun will have reached the end of its life as far as fusion goes. Thanks to its enormous size, it will have a tenuous grip on its outermost layers and increasing amounts of matter will fly away in the stellar wind. Meanwhile, powerful convection cells will dredge up carbon from the core below and bring it to the surface. Red giants with surfaces that are highly enriched in carbon this way are called **carbon stars**.

Carbon stars have relatively cool, slow stellar winds blowing into interstellar space. These winds, endowed with ample amounts of carbon, naturally cool down with distance. When the temperature falls to approximately 1000–2000 degrees, carbon atoms can start to cling together forming tiny solid particles of stardust. These carbon-rich grains of **interstellar dust** constitute a sort of carbon smog or 'pollution' in interstellar space. But this is most fortunate, since these carbon dust grains, emanating from carbon stars, are the basis for organic chemistry here on Earth (and in space). Without this interstellar pollution, terrestrial biology might not have materialized.

16.18 The Sun as a planetary nebula and white dwarf

After the Sun goes through it final phases of fusion, it will blow off its outermost layers, which form a **planetary nebula**. Often strikingly beautiful, planetary nebulae are typically ring-shaped shells of glowing, expanding ejected gas that are the last gasps of intermediate-mass stars like our Sun. Classic examples include the Ring Nebula, the Cat's Eye nebula, the Hourglass Nebula and the Helix Nebula (also known colloquially as the 'Eye of God') (see figure 32.1 in chapter 32).

After ejecting its outer layers, the only remnant of our Sun will be its naked but still very hot core. Having never attained the temperature needed to fuse carbon and oxygen into heavier elements, the Sun's core will have reached the end of the line. This endpoint is a **white dwarf**. Because it is so hot, *a young white dwarf emits ionizing radiation in the form of x-rays and ultraviolet*. The gas within a planetary nebulae also glows in visible light through fluorescence. The ultraviolet excites electrons and ionizes the various atoms in the ejecta. Upon recombination and de-excitation, these atoms emit visible light in the form of characteristically colored, visible emission lines.

The strong stellar winds of remnant cores within planetary nebulae also contain energetic charged particles. These charged particles crash into the ejected shells of matter and create shock waves that contribute to a diffuse x-ray glow that has been seen in some planetary nebula by the Chandra X-ray Observatory telescope.

After well under a million years, the Sun's planetary nebula will fade and disperse since the white dwarf stellar remnant will not be hot enough to continue irradiating the nebula. Our Sun will become a **carbon white dwarf** (to distinguish it from lighter **helium white dwarfs** but not to be confused with **carbon stars**, which are types of red giant stars as mentioned above). Some features of white dwarfs and other stellar corpses will be discussed in detail elsewhere.

Chapter 17

Cosmic rays

Cosmic rays constitute one of the most important forms of ionizing radiation in space. These energetic charged particles permeate the space environment and contribute to the population of the Van Allen radiation belts. In fact, the three primary source terms of ionizing radiation for astronauts presently include the Van Allen belts, solar radiation and extra-solar cosmic rays. Some cosmic rays can even penetrate our atmosphere and strike the surface. En route they can wreak havoc on satellite electronics and might possibly even cause trouble here on Earth in rare instances.

17.1 History

The original discovery of penetrating rays of extraterrestrial origin traces back to investigations of the electrical conductivity of air in the early 20th century. In 1912 Victor Hess made exploratory balloon ascensions to measure the intensity of ionization above the surface of Earth. Expecting the rates of ionization to diminish as he moved away from the radioactivity in Earth's crust, Hess was surprised to discover that the rate of discharge of his ion chambers increased when his instruments were carried aloft. In other words, the residual radiation actually was increasing as he ascended.

He attributed this phenomenon to 'a very penetrating radiation coming mainly from above and being most probably of extraterrestrial origin'. In the early 1920s Robert Milliken and colleagues at California Institute of Technology carried out a series of balloon flights which carried their instruments to heights of almost 16 km or 10 miles and quantified the radiation increase with altitude. They accompanied these high-altitude experiments with measurements made with ionization chamber electroscopes submerged to equal depths in lakes at different altitudes. These investigations conclusively showed that the mysterious radiation was far more penetrating than any gamma rays emitted by natural radioactive substances. They also firmly established the extraterrestrial origin of the radiation, leading Milliken to coin the name 'cosmic rays'. In 1928 Paul Dirac developed his relativistic version of the Schrödinger equation, which is now known as the Dirac equation. Solutions to the Dirac equation implied that every particle should have a corresponding antiparticle with the same mass but with opposite electrical charge. Upon advice from Millikan, Carl Anderson performed meticulous experiments on cosmic rays using a cloud chamber and discovered the first evidence of such antimatter in the form of the antielectron or positron in 1932. Later experiments using cosmic rays revealed another oddity, the muon.

Early cosmic ray research also demonstrated that cosmic ray intensities varied with latitude—specifically, they increased with latitude, peaking near the poles. The geomagnetic latitude effect is apparently due to the action of Earth's magnetic field on the charged cosmic ray particles. Although the geomagnetic field is weak, it extends out to great distances and hence may act on incoming charged particles for an appreciable time. Despite the weak magnetic field strength, the large radius of curvature is sufficient to cause lower-energy-charged particles to be completely deflected away from Earth. Higher-energy particles will be less deflected and may reach the Earth's atmosphere or surface. The minimum threshold energy required for vertical penetration at a geomagnetic latitude q is given by:

$$E_m = 15 \cos^2 q$$

where E_m is the minimum energy (in GeV) needed for vertical penetration of cosmic rays such that they reach our upper atmosphere at (geomagnetic) latitude q. On average, there is a steady flux of roughly 1500 particles per square meter per second of these so-called primary cosmic rays at the top layer of the atmosphere.

Early work showed the electric charge of the primaries was positive and magnetic rigidities strongly suggested protons. Most primary energies were around 6 GeV. The upper energy limit was not known at that time but was predicted to be in the EeV range (EeV stands for exa electron volts or 10^{18} eV). We now know that some primary cosmic rays do indeed possess kinetic energies of this magnitude. In any case, the upper energy limit was far too high to be accounted for by any known mutual mass annihilation processes. It seemed much more likely that particles were emitted with moderate energies and then accelerated to the energies observed here on Earth by non-uniform magnetic fields found in space.

17.2 The latitude effect and the east–west effect

Incoming particle directions were established by observations with 'cosmic ray telescopes' (series of Geiger Muller tubes connected in coincidence). The data showed that high-energy cosmic rays arrive at the Earth isotropically. This absence of directionality was assumed to be because the primaries had traveled enormous distances and in so doing encountered many irregular magnetic fields. Encountering a series of magnetic deflections will effectively erase any evidence regarding the point of origin. The observed isotropy strongly indicated that the primaries must originate from beyond the Solar System. Distances within our Solar System were too short for the known interplanetary magnetic fields to produce an appreciable deflection of particles with 10^5 GeV or more kinetic energy. But although high-energy cosmic rays rained in isotropically, cosmic rays of lower energies clearly demonstrated a preferred directionality.

Cosmic ray particles with less than a minimum threshold energy, face difficulties penetrating Earth's magnetic field. The end result is that more charged particles arrive from the west than the east because the cut-off energy for particles arriving from the west is lower than the cut-off threshold for particles arriving from the east. For instance, at the (geomagnetic) equator, the threshold for cosmic rays coming from the west and reaching our atmosphere is 10 GeV whereas for cosmic rays arriving from the east, the cut-off is 60 GeV. This 'east–west effect' is consistent with the idea that cosmic rays are mostly positively charged particles that are influenced by Earth's geomagnetic field.

Another phenomenon that demonstrates the charged particle nature of cosmic rays is the **latitude effect.** The latitude effect states that cosmic ray flux is lowest at the geomagnetic equator and increases with latitude. This is because the threshold for penetration of the geomagnetic field and reaching the atmosphere is highest at the geomagnetic equator. For example, the cut-off energy for vertically arriving cosmic rays at the equator is 15 GeV, whereas the cut-off energy at 50° latitude is only 2.7 GeV.

Thus 15 GeV is required for penetration at the magnetic equator but only 4 GeV is needed to breach the shielding and reach the upper atmosphere at 45° north latitude. It is important to remember that this alludes to geomagnetic rather than geographic latitude. Here on Earth there is about an 11 degree offset between the magnetic axis and the rotational axis. This discrepancy may be more substantial on other planets.

17.3 Modern understanding

Today, we have a far better understanding of these then-very mysterious charged particles from outer space. And study of cosmic rays broadens not only our perspective on this form of space radiation but also an understanding of the Universe since so-called galactic cosmic rays provide the only direct samples of matter originating from beyond our Solar System.

Primary cosmic rays are predominately positively charged protons or atomic nuclei along with some negatively charged electrons. As such their paths through space are scrambled by the multitude of magnetic fields permeating interstellar space. These constant magnetic deflections obscure their origins. However, some cosmic rays have such incredibly high energies that they are no longer easily deflected by the relatively weak magnetic fields of space. In recent years, cosmic rays with energies touching the exa electron volt (EeV) range have been detected.

Although determining the elemental representation among cosmic rays and their relative abundances is informative and relatively simple (since the different charges (that is, their atomic numbers) of individual cosmic ray particles give distinct identifying signals), an even more instructive characteristic would be the isotopic composition of these individual particles. Such work is more challenging and represents an ongoing effort. Such work will involve mass spectroscopy of individual cosmic ray particles.

17.4 Cosmic ray taxonomy

Cosmic rays can be categorized in several ways. First, there are two very broad categories of cosmic rays encountered in space based on where they come from. Those coming from the Sun can be called **solar cosmic rays** or **solar energetic particles (SEPs)**. These solar cosmic rays can be contrasted with cosmic rays coming from beyond the Solar System, which are called **galactic cosmic rays** or **GCR**s. Many discussions on the topic limit the term 'cosmic rays' only to the galactic cosmic ray flux and do not use the name solar cosmic rays at all. Here the general term, cosmic rays, shall include them both.

The next means of classifying cosmic rays addresses whether they are 'primary cosmic rays' or secondary cosmic rays'. Primary cosmic rays are unadulterated extremely energetic particles zipping about in outer space whereas secondary cosmic rays are those created in our atmosphere upon collisions between primary cosmic rays and nuclei of atmospheric atoms and molecules. The shattering of atomic nuclei by incoming projectiles like primary cosmic rays is known as spallation. Thus, while sailing through space, solar cosmic rays and galactic cosmic rays are **primary cosmic rays**. Upon smashing into nuclei in our atmosphere, a shower of **secondary cosmic rays** can be formed. In this fashion, primary cosmic rays are made of protons, helium nuclei, free electrons, heavy nuclei, gamma photons and a smattering of positrons and antiprotons. In contrast, secondary cosmic rays generated by interactions between primaries and our atmosphere include photons, electrons, neutrons, neutrinos, positrons, muons and pions. In fact, the last three on that list were first detected in cosmic rays rather than in the laboratory.

17.5 Galactic cosmic rays

The exact origin of primary cosmic rays is not known with certainty, but they clearly originate from beyond our Solar System. In fact, some galactic cosmic rays might originate from beyond our Milky Way, making the name 'galactic' too parochial. Supernovae are believed to be the primary source of galactic cosmic rays but distant active galactic nuclei may be another source. Galactic cosmic rays are mostly protons (hydrogen nuclei) and helium nuclei. By mass, they turn out to about 28% helium nuclei, which is not very dissimilar to the 24% primordial abundance of helium (generated in the Big Bang). At 28% by mass, this helium abundance translates to just under 9% abundance by particle numbers. About 1% of galactic cosmic rays are nuclei heavier than helium (the so-called metals).These heavier nuclei range in atomic number all the way up to $Z = 92$ (uranium). Curiously, the light elements (Li, Be and B) are grossly over-represented in the galactic cosmic ray flux compared to their abundance in the Universe as a whole. It is presently assumed that this is due to transmutation of heavier elements into these lighter elements in the interstellar medium space through spallation. Cosmic ray collisions in deep space with larger nuclei may lead to these smaller basic units. In a sense then, many of what what we call primary cosmic rays might be 'secondary' after all—but instead of being formed in our atmosphere, they were formed in the interstellar medium.

17.6 Origins of galactic cosmic rays

It is presently believed that cosmic rays are accelerated by blast waves from supernova explosions. Supernova remnants, which include expanding clouds of gas and dust and a potent magnetic field, persist for thousands of years. It is here that cosmic rays are believed to be accelerated. Random interactions with the remnant supernova magnetic field enable some charged particles to acquire large amounts of energy. Some might eventually gain sufficient energy and speed to escape the supernova remnant as cosmic rays. The 'escape velocity' depends on the size of the acceleration region within the remnant and its magnetic field strength.

It was Enrico Fermi who performed some of the early pioneering work on how 'magnetized clouds' might accelerate cosmic rays. Recent research involving NASA's Fermi Gamma-ray Space Telescope has expanded on this concept of **Fermi acceleration** and the latest data seems compatible with cosmic ray acceleration being caused by shock waves within supernova remnants. In principle, charged particles will bounce around within the magnetized supernova remnant and may occasionally cross the shock wave. They gain approximately 1% additional energy with each traversal. After numerous crossings, they will have acquired enormous amounts of energy and will be traveling at nearly the speed of light. At such energies and speeds some will finally escape the supernova remnant. When these liberated ultra-fast protons and electrons smack into nearby giant molecular clouds, gamma rays should be generated. The gamma rays may be made by different mechanisms. Energetic protons can strike other protons and create neutral pions (π^0). Neutral pions have a rest mass of 135 Mev c^{-2} and in 98.8% of the time, they decay into pairs of gamma photons on the order of 8.5×10^{-17} s. In contrast to this mechanism of photon production, if energetic electrons encounter the electromagnetic field near the nucleus of an atom in the giant molecular cloud, they can yield bremsstrahlung gamma photons.

The Fermi Gamma-ray Space Telescope focused on two supernova remnants—IC 433, which is about 5000 light years away in the constellation Gemini, and W44, which is about 10 000 light years away in the constellation Aquila—and analyzed gamma ray photons emanating from their adjacent giant molecular clouds.

Proton-produced photons (originating from neutral pion decay) should be slightly higher in energy than photons made by electrons. At the lower end of the scale, the energy spectra should differ slightly with more electron-produced photons than proton-produced photons. The Fermi data were indeed consistent with proton acceleration and neutral pion production. This evidence seems to confirm the concept that supernova remnants are the source of many cosmic ray protons through the process of Fermi acceleration. Importantly, unlike cosmic rays, gamma rays are not affected by interstellar magnetic fields, which means they will follow straight lines that can be traced back to their sources. Finding the predicted gamma photons from molecular clouds near these supernova remnants provides solid evidence for the origin of some cosmic rays within supernova remnants via this mechanism.

But the theory on the origins of cosmic rays needs to be further refined. Although it now seems likely that at least some cosmic rays can be created through this Fermi acceleration mechanism, it remains unclear just what fraction are made this way. And since cosmic rays accelerated this way in supernova remnants can only reach a certain maximum energy, how the highest energy cosmic rays have been formed is still a mystery. Some cosmic rays have been found to have far more energy than what can be generated via this supernova remnant acceleration method. Presently, the mechanisms of production and the origins of these ultra-high-energy (UHE) cosmic rays remains a mystery. Some possibilities include origins from beyond the Milky Way Galaxy stemming from gamma ray bursts, or quasars and other active galactic nuclei. Other more speculative ideas include an etiology in presently uncharted new physics beyond the Standard Model such as dark matter, super-strings, strongly interacting neutrinos or even topological defects in the fabric of the Universe.

17.7 Galactic cosmic ray composition

High-energy cosmic rays race across the Galaxy at close to the speed of light. They are composed primarily of atomic nuclei, especially hydrogen (which is to say they are mostly protons). Nevertheless, all naturally found atomic nuclei from hydrogen to uranium are represented in the galactic cosmic ray repertoire. A commonly cited breakdown (by numbers) is roughly 89% protons, 9% helium nuclei and 1% heavy atomic nuclei (i.e. the 'metals' of astronomers) and 1% bare electrons. Although one might argue semantically that the helium nuclei could be called alpha particles and the electrons could be called beta particles, in this text we shall continue to call them helium nuclei and electrons to emphasize the vast energy differences. Recall that alpha particles from radioactive decay average about 5 MeV, with an energy range of 2–9 MeV. Beta particles range from about 5.7 keV (average from tritium) to around 3.3 MeV (bismuth-214). In contrast, the energy of some cosmic rays has been recorded at over 100 EeV (EeV stands for exa electron volts or 10^{18} eV). Thus, these cosmic ray particle energies are many orders of magnitude higher than anything produced naturally via radioactive decay (or artificially in a particle accelerator for that matter) and because of that we will not make them synonymous with alpha particles and beta particles in this chapter.

Differences in the relative abundances of nuclei among the cosmic rays inform us somewhat about their origins and trip across the Galaxy. Although the metals make up only 1% of all primary cosmic rays, among this 1% are some extremely scarce elements and isotopes. When analyzing the relative abundances of the elements in cosmic rays, silicon (Z = 14) is the standard to which all other abundances are compared. Compared to Solar System abundances, galactic cosmic rays contain relatively far less hydrogen and helium. This bias remains unexplained but might be due to the fact that it is more difficult to accelerate protons and alpha particles to the very high energies of cosmic rays seen in heavier nuclei. Additionally, the elements between cobalt and iron are more well represented in galactic cosmic rays than in the Solar System overall. Finally, a remarkable difference is observed in the relative abundances of the light elements, lithium, beryllium and boron. These elements are quite rare in the Solar System (and the Universe at large) but are relatively plentiful in galactic cosmic rays.

The accepted explanation for the abundance of lithium, beryllium and boron among galactic cosmic rays is that these light nuclei are fragments of heavier cosmic ray elements, especially carbon, nitrogen and oxygen, that have undergone high-impact collisions in interstellar space. In other words, lithium, beryllium and boron are cosmic ray **spallation products**. However, lithium can be made by other means. These include a small amount produced in the Big Bang (via Big Bang nucleosynthesis), a larger fraction made in the asymptotic giant branch phase of certain stars and another source being supernova explosions. However, these explanations apply to lithium-7. The second most abundant natural, stable isotope, lithium-6, was probably made exclusively through cosmic ray collisions. Cosmic ray spallation is also responsible for the anomalous abundances of scandium, titanium, vanadium and manganese in galactic cosmic rays but in this case, iron and nickel nuclei probably crashed into interstellar matter and then shattered into these particular elements.

Similarly, the inventory of galactic cosmic ray elements between silicon and iron has been supplemented by fragments of shattered heavier galactic cosmic rays, thereby increasing the relative representation of elements in this range when compared to the Solar System as a whole. Confusingly, these fragments are sometimes called 'secondary cosmic rays'. In this book, we shall reserve the name secondary cosmic rays for those produced in our atmosphere from interactions between primary cosmic rays and atmospheric gas molecules or atoms. The knock-on or knock-out products that continue onward in the atmosphere are secondary cosmic rays. Nevertheless, it is helpful to distinguish primary galactic cosmic rays (that have just emerged from supernova remnants or whatever other sources exist) from the galactic cosmic rays that contain the remains of collisions between these true primaries with interstellar matter. In this text we shall allude to the galactic cosmic rays that are fresh from the supernova as 'primary galactic cosmic rays' and those that have resulted from interactions and spallation reactions as 'secondaries'.

17.8 The age of galactic cosmic rays

From the number of spallation products in galactic cosmic rays (the secondaries), such as the light elements and those between silicon and iron, coupled with the probability of interactions that will yield them, one can calculate the amount of material that cosmic rays have journeyed through. The probabilities of shattering interactions (the collision cross sections) can be measured by experiments conducted at particle accelerator colliders. The more matter the original cosmic rays traverse, the more likely they will be to be shattered into smaller cosmic rays. Naturally, the longer they have been traveling, the more matter they have transversed. And, the more matter they have traversed, the higher the probability of creating secondaries. Thus, the longer they have been around, the higher the concentration of secondaries.

Because of these relationships, one may look at the relative abundances of secondaries and get an estimate of the age of these cosmic rays. Based on all this, if one assumes that the cosmic rays have started and stayed within the Milky Way Galaxy, the ratio of two known and measurable quantities (the volume of matter traversed (which is proportional to the time the cosmic ray has been traveling) divided by the average density of this matter) should yield the age of these cosmic rays. The average density of interstellar space in the Milky Way is about one atom per milliliter. Such a calculation gives an age of around 2 million years for the average cosmic ray. But an alternative means of estimating the age of cosmic rays is through radioactive decay measurements. Several galactic cosmic ray species are radioactive, including beryllium-10, aluminum-26, chlorine-36, and iron-60. These are what we will call 'secondaries' in this chapter, meaning they were formed from collisions in outer space but not in our atmosphere ('secondary cosmic rays' for our purposes). After creation, these radioactive cosmic rays will begin to decay just as all radionuclides do. By quantitating the fraction that reaches us here on Earth, we can estimate their age. This method yields a different age for the cosmic rays—approximately 10 million years. It is believed that this older figure is more accurate because cosmic rays do not reside solely in the 'high density' regions of one atom per milliliter. Rather, they spend much of their time in lower-density regions such as the galactic halo and rebound back and forth across the galactic disc several times. Of course, special relativity plays a significant role in these calculations given that galactic cosmic rays are the fastest bits of matter known to man, with some reaching speeds over 99.9999% the speed of light. Incidentally, at such extreme energies and speeds, if it were possible to create an Earth-destroying mini- or micro-black hole through ultra-relativistic particle collisions, we likely would have seen this already, as super-energetic cosmic rays can and do regularly crash into our atmosphere's atomic nuclei. Since such collisions have well over a million times the energy of any collisions passible at the Large Hadron Collider (LHC) at CERN, the fears propagated by the popular media of mad scientists destroying the world through particle accelerator experiments are totally unfounded.

17.9 Galactic cosmic ray energies

Among the most intriguing and amazing aspects of cosmic rays is the extraordinary energy possesed by some of these particles. Some cosmic ray energies have been measured at values in the hundreds of PeV with the record presently standing at an astounding 3×10^{20} eV or 300 EeV. This particle (given the moniker of the Oh-My-God or OMG particle) defies the commonly imposed restriction on the quantum world and reaches into the realm of everyday experience—such a particle has an energy comparable to a well-hit tennis serve or a 60 mph baseball! This energy is around 40 million times the maximum energy capacity of even the most powerful manmade particle accelerators such as the LHC. Of course, most galactic cosmic rays do not have nearly this much energy and the abundance versus energy curve shows a progressive dropoff with increasing energy. Most galactic cosmic rays that reach Earth have an energy just above 1 GeV per nucleon. A plot of cosmic ray flux versus energy yields a progressively decreasing linear curve with slight bends in the curve known as the 'knee' and 'ankle' regions at very high energies. It is presently believed that the very highest energy cosmic rays actually originate from beyond our galaxies, making the name 'galactic cosmic rays' a bit too limited! Furthermore, at energies approaching the so-called Greisen–Zatsepin–Kuzmin limit (GZK limit), ultra-energetic cosmic rays should interact with the cosmic microwave background. These interactions could reduce the comic ray energy to below the GZK limit. Therefore, although ultra-energetic cosmic rays appear to originate from beyond the Milky Way, there is a speed limit imposed on such extragalactic particles.

17.9.1 HZE ions

Primary cosmic rays made of metal nuclei are sometimes called HZE ions. HZE stands for *h*igh atomic number (*Z*) and *e*nergy. Although they are alluded to as ions, they are entirely bare atomic nuclei with no orbiting electrons. This means the electric charge of these HZE ions (and all galactic cosmic rays) equals the atomic number of the nucleus involved. HZEs are mostly galactic cosmic rays. However, a minority of them are of solar origin, being fired out during strong solar storms and solar flares. Nevertheless, solar HZEs are typically far lower in energy than galactic HZEs (maybe MeV energies vs GeV energies, respectively). But from a radio-biological perspective, HZEs have very **high linear energy transfer** (LET) and therefore potentially are **high relative biological effectiveness** radiation. In other words, HZEs are very densely ionizing in biological tissue. Some of the prominent HZE ions in space include carbon, oxygen, magnesium, silicon and iron nuclei. The radiobiology of HZEs is an ongoing topic of research regarding future space travel. Although HZEs constitute only about 1% of primary cosmic rays encountered by astronauts, it is possible that their radiobiological contribution is equal to the vast number of energetic protons that space travelers will encounter. HZE ions such as iron-56 nuclei with energies measured in GeV per nucleon are biologically relevant because they are both highly ionizing (i.e. they possess high LET) and they are highly penetrating. This contrasts them with energetic protons (which are not very ionizing except near the very ends of their tracks) and alpha particles from radioactive decay (which are highly ionizing but short in range—typically measured in tens of microns, meaning just a few cell diameters). HZEs like iron-56 galactic cosmic rays combine the worst of both worlds.

17.10 Solar cosmic rays and primary cosmic rays

Primary cosmic rays are those above our atmosphere. They can be solar in origin (meaning they are solar cosmic rays or solar energetic particles) or they can arrive from beyond our Solar System (meaning they are galactic cosmic rays). In either case, they can interact with atoms and molecules in our atmosphere and generate additional radiation. That is to say, primary cosmic rays engender secondary cosmic rays upon smashing into the atmosphere.

The International Space Station (ISS) orbits the globe every 90 min at a speed of about 17 500 mph (28 000 km h^{-1}) and an average altitude of 248 miles (400 km).

This is well above Earth's atmosphere (which 'officially' ends at around 50 miles or 80 km per NASA and the United States Air Force as this altitude is their definition of the Kármán line) or more scientifically, at around 76 miles or 122 km where 'atmospheric interface' occurs for space vehicles returning to Earth. Therefore, the cosmic rays encountered by residents of the ISS could be considered primary cosmic rays.

At under 1% of all primary cosmic rays, high-energy gamma photons represent a minor contingent of the primary cosmic ray flux. These photons can exceed 50 MeV in energy and are apparently of both galactic and extragalactic origins. The Fermi Observatory has identified sources from both diffuse and discrete sources within the galactic plane of the Milky Way as well as discrete point sources from beyond our Galaxy. The precise nature of these sources presently remains unclear but sources emitting gamma radiation with photon energies between 100 GeV and 100 TeV are called very-high-energy gamma ray sources, while those with photon energies above 0.1 PeV are known as UHE gamma ray sources. Very recently several gamma ray photons with energies exceeding 1 PeV (with one recording an energy of 1.4 PeV) were detected. The sources of such gamma rays have been nicknamed Pevatrons. The Fermi Gamma-ray Space Telescope documented that nearly 60% of the very highest energy gamma photons sources are blazars although the famous Crab Nebula in our own galaxy is also a PeVatron (figure 17.1).

Somewhat surprisingly, the Sun is also a source of some high-energy gamma rays. Some of these photons can exceed 100 MeV (even reaching the GeV range) and appear to originate in solar flares and coronal mass ejections. The Sun can theoretically be a source of secondary photons as well when UHE galactic cosmic rays collide with the Sun's atmosphere and generate photons in the high TeV range that might then head our way. In a similar manner, the Moon has been found to shine in gamma rays with energies exceeding 20 MeV thanks to collisions with energetic primary cosmic rays.

Primary cosmic rays also contain antimatter. Although less than 1% of primary cosmic rays are antiprotons, they tend to have higher energies than their normal matter proton counterparts. While most of the protons come in around 1 GeV (with perhaps peak intensity at around 300 MeV), primary antiprotons tend to peak at 2 GeV. This hints at a different fundamental (but presently unknown) etiology.

17.11 Secondary cosmic rays

Primary galactic cosmic rays can reach amazing energies. Some have been recorded with energies up to 10^{20} eV as described elsewhere. When particles of such energies strike nuclei of molecules and atoms in our atmosphere, the nuclei are shattered and lead to a so-called '**air shower**' of secondary particles as the energetic shattered remains then go on to shatter other nuclei and so on. These extensive air showers greatly amplify the number of radiation particles. The number of particles created is proportional to the incoming primary cosmic ray energy. In some situations, the number of secondary particles can reach into the billions. These secondary cosmic ray particles rain down from the sky within a cone of about 1° of the incoming

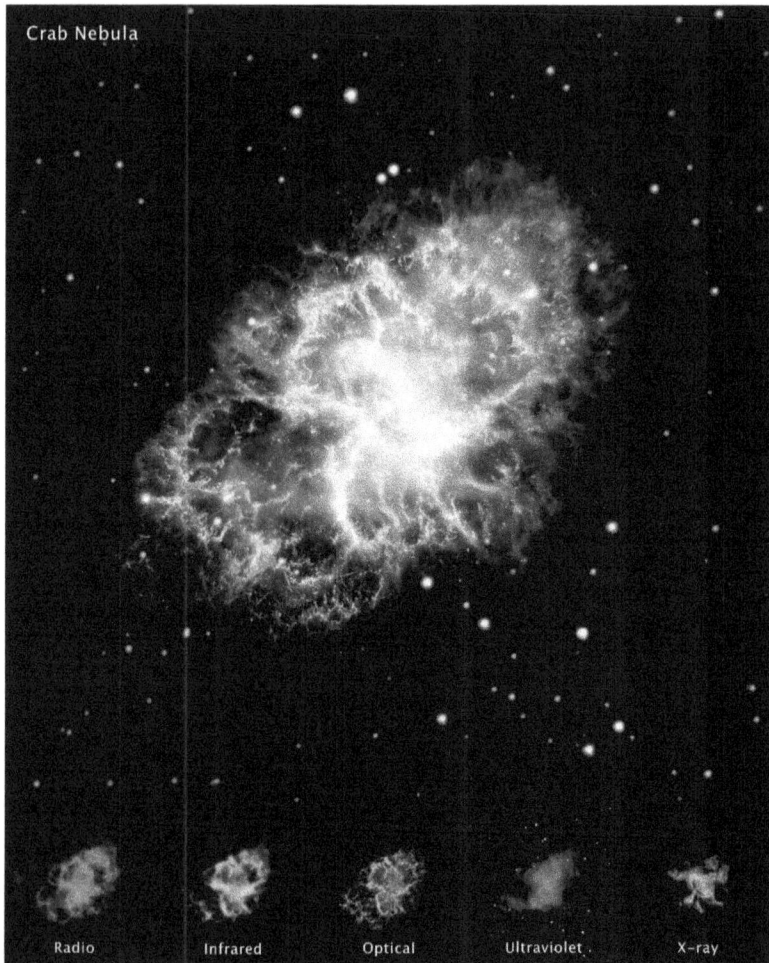

Figure 17.1. A composite view of the Crab Nebula, the remnant from a supernova explosion recorded in 1054 AD. After exploding, the former massive star left a neutron star corpse that generates an intense magnetic field and releases radiation that energizes the nebula. The Crab Nebula is the brightest known 'pulsar wind nebula' meaning it is being blown outward by an ultra-relativistic electron–positron wind emanating from the central neutron star (the Crab Pulsar). The supernova was around 6500 light years away. If the star had detonated much closer, it would have heavily irradiated Earth and the consequences on the biosphere might have been significant. It is presently believed that galactic cosmic rays likely originate from supernova explosions. This image combines data from five different telescopes: the Karl G. Jansky Very Large Array (VLA) provided the radio wavelengths in red; the Spitzer Space Telescope provided the infrared data in yellow; the Hubble Space Telescope provided the visible light in green; XMM-Newton provided the ultraviolet frequencies in blue; and the Chandra X-ray Observatory provided the x-ray photons in purple. The Crab Nebula is also one of the brightest and most energetic gamma ray sources in the sky, with energies exceeding 1 PeV. The detection of such extreme energy photons was made through Cherenkov detectors that pick up atmospheric secondaries (air showers) made by the gamma rays. These phenomenal photon energies qualify the Crab Nebula as one of a dozen recently discovered 'PeVatrons', natural astrophysical accelerators that generate photons with unimaginable energies. Credit: NASA, ESA, G Dubner (IAFE, CONICET-University of Buenos Aires) *et al*, A Loll *et al*, T Temim *et al*, F Seward *et al*, VLA/NRAO/AUI/NSF, Chandra/CXC, Spitzer/JPL-Caltech, XMM-Newton/ESA and Hubble/STScI.

17-11

primaries. The inbound primary cosmic ray does not need to possess extraordinary energy in order to generate a cascade. Any primary with more than a few hundred MeV per nucleon can generate an extensive air shower. The production of these secondary showers is sensitive to atmospheric density, which in turn is a function of temperature and pressure.

Secondary cosmic rays include a wide array of different forms of radiation such as gamma photons; pions and kaons and other mesons; nucleons (neutrons and protons); and leptons such as electrons, positrons, muons and neutrinos. The neutrinos which tend to sail right through the remaining atmosphere and indeed through the entire Earth without interacting or depositing dose. Aside from neutrinos, the muons are the particles most likely to reach the surface at sea level and these do deposit dose. But even traveling at nearly the speed of light, a muon with a lifetime of 2.2 microseconds (msec) created in our upper troposphere should never reach the surface. Recall that the troposphere is between 5 and 9 miles (8 and 14 km) thick depending on latitude; it is thickest at the equator and thinnest at the poles. In any case, even if one assumes muons are speeding along at the full speed of light, the math shows they can only traverse a distance of less than a half mile before decaying:

$$vt_0 = (2.994 \times 10^8 \text{ m s}^{-1}) \times (2.2 \times 10^{-6} \text{ s}) = 659 \text{ m}.$$

However, most muons are born at altitudes of 6000 m or more. As students of physics will recall, the fact that muons can reach the surface is an excellent demonstration of special relativity. As a simple illustration, if we assume a realistic muon velocity of $0.998c$, special relativity reveals:

$$t = \gamma t_0 \quad \text{where } \gamma = \frac{1}{\sqrt{1 - \beta^2}} \text{ and } \beta = v/c$$

$$t = (2.2 \times 10^{-6} \text{ s})/[1 - (v^2/c^2)]^{\frac{1}{2}} = 31.6 \text{ μs}.$$

Thus, from our frame of reference, muons live about 14 times longer than expected. Although its resting lifetime t_0 is only 2.2 μs from its resting frame of reference, when moving at this speed, it appears to have a lifetime of nearly 32 s from our frame of reference. The distance it can thus travel is given by:

$$d = vt_0 = (2.994 \times 10^8 \text{ m s}^{-1}) \times (32 \times 10^{-6} \text{ s}) = 9581 \text{ m}.$$

Thus, one can see that when traveling along at such impressive speeds, these muons (from our frame of reference) have far longer lifetimes and some can therefore reach the surface even if they were born in the upper troposphere.

Returning to the generation of secondary cosmic rays, among the first spallation products of collisions between energetic incoming primaries and nuclei in the atmosphere are pions and nucleons. The neutral pions decay very quickly (lifetime ~85 attoseconds or 8.5×10^{-17} s) into a pair of energetic gamma photons. Since the mass of the neutral pion is ~135 MeV, the two photons generated by neutral pion decay are about 67.5 MeV each. In rare instances, the neutral pion might decay also into an electron–positron pair along with a photon, the so-called **Dalitz decay**. Even less frequently both photons can undergo internal photon conversion into electron–positron pairs such that the neutral pion decays into two electrons and two positrons (the **double-Dalitz decay**). These are examples of the **chiral anomaly**, a prediction of quantum chromodynamics. Neutral pions, like all mesons, are bosons composed of quark–antiquark pairs. Neutral pions are specifically composed of up–antiup pairs or down–antidown pairs. The chiral anomaly arises from a phenomenon in which the neutral pion's valence quarks transiently transition into different flavors, which then undergo self-annihilation and pair production. Importantly, the lifetime of the neutral pion is consistent with six, and only six, flavors of quarks.

Charged pions possess slightly higher masses (139.57 MeV) than neutral pions (134.98 MeV) and rather than decaying into gamma photons, they decay into muons (which are always electrically charged) and neutrinos. More specifically, for conservation of lepton number and matter–antimatter, a positive pion will decay into a positively charged antimuon and a muon neutrino whereas a negative pion will decay into a negatively charged muon (which is normal matter) plus a muon antineutrino. Students and readers will recognize that the presence of a neutrino signifies the weak force in action. And as such, the reaction will proceed far slower than the decay of neutral pions (which is mediated purely by the electromagnetic force and is measured in attoseconds). Thanks to the involvement of the weak force, charged pions decay with lifetimes measured in nanoseconds. More precisely, charged pions have a mean lifetime of 26.033 ns (2.6033×10^{-8} s).

17.11.1 More about muons

The muons created via charged pion decay then continue downward and eventually shower Earth's surface with a flux of roughly 1 muon per square centimeter per minute. This contributes about half of our natural annual background radiation dose.

As with any charged particle moving through matter, muons lose energy through ionization of the matter they traverse; this energy loss is proportional to the length and density of matter they pass through. Thus, any given medium through which charged particles pass can be characterized by its density (in g cm^{-3}) times the distance traveled (in cm). This is called the **interaction length.** For cosmic ray muons, this is proportional to the density of the atmosphere (in g cm^{-3}) times the length of the atmosphere they penetrate (in cm). As with protons passing through human tissue, the interaction length of muons in the atmosphere has units of g cm^{-2}. Secondary cosmic ray muons lose energy at a fairly constant rate as they sail through the atmosphere—approximately 2 MeV g^{-1} cm^{-2}. On average, a column of air extending from sea level to the top of the atmosphere weighs 14.7 pounds per square inch or 1.03 kg cm^{-2}. Thus, we can assume an atmospheric mass of about 1000 g cm^{-2}. If we assume an atmospheric vertical depth (total interaction length) of 1000 g cm^{-2}, muons will lose roughly 2 GeV through ionization before reaching the Earth's surface. Since the average energy of muons at sea level is around 4 GeV, this means they must have had about 6 GeV at their moment of creation. Some of the more energetic muons can penetrate Earth's crust for tens of meters, perhaps up to 50 meters or so, indicating far higher energies than this average 4 GeV figure. But some rare muons were observed in the Soudan 2 detector, which was nearly 2 km underground. For a muon to penetrate to that depth through solid rock, the minimum kinetic energy needed at Earth's surface must be $\geqslant$800 GeV; the primary cosmic ray energy to create such muons in the atmosphere must have been around 10 TeV.

As mentioned above, muons are unstable fundamental leptons with a mass of 105.66 MeV c^{-2} or roughly 207 times the mass of an electron. They decay after a mean lifetime of 2.2 msec. The decay is 'slow' (by subatomic particle standards) because it is mediated via the weak force and because the mass difference between the muon and its decay products is small (providing relatively few degrees of freedom for decay). Negative muons normally decay into three particles following standard conservation laws: an electron (with the same electrical charge as the parent muon), a muon neutrino (for conservation of total muon number) and an electron antineutrino (for conservation of electron number and balance of matter–antimatter):

$$\mu^- \rightarrow e^- + \nu_e + \nu_\mu.$$

The positive muon would decay according to the same basic principles, but the products would be a positron, a muon antineutrino and an electron neutrino.

17.12 Other secondary cosmic rays

Another type of meson created when primary cosmic rays strike atmospheric atomic nuclei are positive and negative kaons or K mesons. Kaons are like pions in that they are bosons and mesons made of up and down quarks (or their antiquarks). In addition, however, kaons contain strange quarks or antiquarks. Kaons quickly decay (in about 12 ns) into muons, pions, electrons/positrons and their associated neutrinos/antineutrinos, further contributing to secondary cosmic ray air shower.

In addition to the mesons and leptons in the secondary cosmic ray cascade, high-energy photons are present. Some of these originate from the decay of neutral pions. Others can be made via bremsstrahlung when energetic electrons and muons encounter nuclei of air atoms. If sufficiently energetic, these gamma rays can in turn lead to electron–positron pair production or triplet production when they encounter other atoms in the atmosphere.

As mentioned, muons are the predominant secondary cosmic ray particles penetrating the full thickness of the atmosphere and reaching the surface. It is estimated that for every 10 000 muons making it to the surface, there are an accompanying 200 primary cosmic rays, 20 high-energy (>1 GeV) electrons and 4 pions.

In each cascade overall, there can be as many as 100 000 low-energy electrons created. However, in contrast to the above-mentioned rarer, high-energy or 'hard component' electrons, these 'soft component' electrons tend to be absorbed relatively quickly and are unlikely to reach the surface. Of course, if the shower is created at low enough altitude or if it is initiated with sufficiently high energy, electrons can be an important component radiation reaching the surface.

Finally, in addition to the above-mentioned mesonic, leptonic and photonic components, there is an important nucleonic component. Nucleons are protons and neutrons. These are both baryons, meaning they are composed of three quarks of different colors. Specifically, the nucleons are made only of first-generation quarks, that is, up and down quarks. More precisely still, protons are composed of two up and one down valance quarks whereas neutrons are composed of two down and one up quarks. The three quarks must be of different colors to maintain color neutrality. Although the search is on for proton decay (because some grand unification theories predict it), this has not yet been observed. The current data place constraints on proton decay such that the lifetime must be at least 1.67×10^{34} years. For all intents and purposes, the proton is stable. In contrast, free neutrons decay with a mean lifetime (τ) of 879.6 s or 14 min and 39.6 s. This translates to a half-life ($t_{1/2}$) of 611 s or 10 min and 11 s by $\tau = 1.44\ t_{1/2}$. An interesting branch of physics at this time centers on the ongoing disagreement between so-called 'bottle' versus 'beam' methods for estimating the mean lifetimes of free neutrons since the two modalities come up with slightly different results (as also covered in chapter 6). The discrepancy was previously attributed to simple technical differences between the two methods but after years of painstakingly careful work, the discrepancy remains, hinting at some new physics. In fact, at the time of this writing, the most recent and meticulous bottle method data have increased rather than decreased the discord between the bottle versus beam techniques, which presently stands at >10 s. This recent version of the bottle method came up with a mean lifetime of 877.75 s while the latest beam method figure is 887.7 s. Without going into details, the bottle methods aim to trap neutrons and measure the neutrons that remain in that magnetogravitational trap or bottle. In contrast, beam methods measure the protons appearing within a neutron beam.

Secondary cosmic ray neutrons may cause electronic malfunctions by clanging into semiconductors and memory chips within computers and other electronic equipment. The effects of space radiation on electronics are covered a bit further in the chapter on Jupiter (chapter 23).

17.13 Applications of cosmic rays

Incoming primary cosmic rays initiate nucleonic-muon-electromagnetic cascades in the Earth's atmosphere. Neutrons are among the nucleonic component of secondary cosmic ray cascades. Neutrons can readily generate new radionuclides which can be quantified. Among these cosmogenic radionuclides are beryllium-7 ($t_{1/2}$ of $\approx$53 d), sodium-22 (2.6 yr), carbon-14 (5730 yr), chlorine-36 (3×10^5 yr) and beryllium-10 (1.4×10^6 yr).

The most familiar of these cosmogenic radionuclides is radiocarbon or carbon-14. Through the ^{14}N(n,p)^{14}C reaction, the most abundant gas in our atmosphere (molecular nitrogen, which constitutes approximately 78% of our atmosphere by volume), interacts with free neutrons to generate the cosmogenic radionuclide, carbon-14:

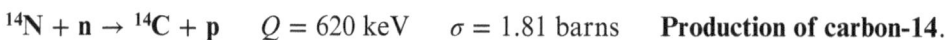

$$^{14}\mathbf{N} + \mathbf{n} \rightarrow {}^{14}\mathbf{C} + \mathbf{p} \qquad Q = 620\ \text{keV} \qquad \sigma = 1.81\ \text{barns} \qquad \textbf{Production of carbon-14.}$$

Most of this carbon-14 is generated in the upper troposphere and lower stratosphere at altitudes from 9 to 15 km (5.6–9.3 miles or around 30 000–49 000 feet). Thus, during a typical commercial jet flight, passengers might spend time in the atmospheric carbon-14 production zone. Additionally, more carbon-14 is made at higher latitudes, nearer the geomagnetic poles. The neutrons involved in this reaction are typically in the thermal neutron energy range, making the $^{14}N(n,p)^{14}C$ reaction a neutron capture reaction. The cross section for production of carbon-14 from the $^{14}N(n,p)^{14}C$ reaction is 1.81 barns and the abundance of nitrogen gas is so high that for all intents and purposes, this is the only reaction that matters. But for completeness, carbon-14 can also be created out of carbon-13 and oxygen-17 by $^{13}C(n,\gamma)^{14}C$ and $^{17}O(n,\alpha)^{14}C$ reactions, respectively. Another even rarer source of carbon-14 is through 'cluster decay' of heavy radioisotopes of radium (radium-223, radium-224 and radium-226). Note that these last two sources of carbon-14 are radiogenic but not cosmogenic.

In this way, carbon-14 makes up one of the three naturally existing isotopes of carbon along with stable carbon-12 (about 99% of all carbon atoms on Earth) and carbon-13 (about 1%). The trace amounts of carbon-14 constitute around 1–1.5 atoms per trillion (that is, 1–1.5 in 10^{12} carbon atoms). Carbon-14 is radioactive with a $t_{1/2}$ of 5730 years and decays through beta-minus decay as follows:

$$^{14}C \rightarrow {}^{14}N + e^- + \bar{\nu}_e \quad t_{1/2} = 5730 \text{ years} \quad \textbf{Decay of carbon-14.}$$

The overall reaction releases an energy of 156 keV. This energy is shared between the beta electron and the electron antineutrino such that the maximum beta energy is 156 keV but the average is 49 keV. This is consistent with most beta decays in which the average beta energy is roughly one third the maximum. One can also use the rule of thumb that the range of electrons in human tissue in cm is about half the energy in MeV. Thus, one half of 0.156 MeV yields 0.078 cm or 0.78 mm as the maximum range, while most of the beta particles go only half of 0.049 MeV or 0.026 cm or 0.26 mm. These beta decays contribute roughly 1 mrem yr^{-1} or 0.01 mSv of ionizing radiation dose to each person. This can be compared to the 39 mrem yr^{-1} (0.39 mSv yr^{-1}) dose due to decay of potassium-40.

The range of these beta particles in air is roughly 22 cm.

Newly produced carbon-14 rapidly reacts with atmospheric oxygen to yield radioactive carbon monoxide, which is then further oxidized to radioactive carbon dioxide. This radioactive version of carbon dioxide is rapidly taken up into the food chain via photoautotrophs ('photosynthesizers') at the base of the food chain. Carbon-14 is then distributed fairly evenly throughout the biosphere. In this manner, the organic compounds in all living organisms possess roughly one atom of carbon-14 per trillion carbon atoms. Upon death however, biological materials cease to take up new radiocarbon into their organic material and the carbon-14 in the organic matter begins to decay with its characteristic $t_{1/2}$. In this fashion, organic matter can be dated to the time it ceased taking up new radiocarbon, which is a surrogate for when the organism died. Because of the limitations of any radiometric dating methods, the radiocarbon dating technique is only useful for organic objects going back around 10 half-lives or about 57 000 years.

Of course, the radiocarbon dating method is only useful for organic materials and cannot be used to directly date stone or metallic artifacts. But if such materials are found adjacent to wooden artifacts or in a wooden coffin or near some charcoal ash, the organic materials can be dated to provide some insight. On the other hand, in the realm of radiocarbon dating the definition of the term 'organic' might have to be expanded beyond its conventional restrictions and include non-biological carbon compounds as well. For instance, speleothems (rock formations in caves such as stalactites and stalagmites) made of calcite and aragonite (which are mineral polymorphs of calcium carbonate) can possibly be dated by radiocarbon methods. In practice however, geological dating often has requirements far beyond the practical age limit of radiocarbon dating.

It is assumed that the uptake of radiocarbon into the biosphere is constant and uniform but in fact, just as with carbon-12 versus carbon-13, biological **isotopic fractionation** dictates that the lighter isotopes are preferentially taken up. Nevertheless, the approximations are good enough to provide reasonably accurate dates through radiometric analysis. Furthermore, for radiocarbon dating, the production of carbon-14 is assumed to be constant, and the atmospheric concentration is assumed to be stable. Neither of these assumptions is totally accurate. Hence, carbon-14 dating can provide reasonable estimates, but high precision should not be assumed.

Primary cosmic ray flux can vary depending on factors such as modulation by the solar wind. On a short-term basis, the 11 year sunspot cycle (Schwabe cycle) imposes a slight variation in galactic cosmic rays that reach our atmosphere and generate free neutrons. Generally speaking, more solar activity means fewer galactic cosmic rays striking the atmosphere and creating the neutrons that lead to carbon-14. This is known as the **Forbush effect**. Thus, thanks to such Forbush decreases, increased sunspot activity (in principle) transiently slows down carbon-14 production.

On the other hand, grand solar minima (which are prolonged periods (several decades) during which there are fewer than normal numbers of sunspots) might allow increased galactic cosmic ray flux in to strike our atmosphere and generate more secondary cosmic rays including neutrons. So, during the Maunder minimum for instance, there might have been more carbon-14 production than at other times in Earth's history such as the present. However, although it is generally thought that galactic cosmic ray primaries are the ultimate source of the neutrons that make carbon-14, sufficiently energetic solar primary cosmic rays also have this potential. In fact, solar proton events might be responsible for two natural spikes in carbon-14 that seemed to occur during 774–75 and 993–94 AD. Data from ancient tree rings (that is, the field of dendrochronology) indicate a sharp enrichment in atmospheric carbon-14 in AD 775 amounting to an increase of about 12‰ (parts per mil or parts per thousand) in 1 year (about 20 times larger than changes attributable to standard solar modulation). The relative abundance of atmospheric carbon-14 might trace its origin to transient high fluxes of solar protons with energies above 100 MeV. Corroborating the carbon-14 evidence is an increase in other cosmogenic radionuclides (beryllium-10 and chlorine-36) in Arctic and Antarctic ice cores. The data for the larger of these two events (774–75) suggest that the solar particle event was over fivefold stronger than any recorded event in recent history and even more potent than the 1859 Carrington Event. The authors argued in favor of a solar particle event over a gamma ray burst (GRB) because the relative ratios of the three cosmogenic radionuclides studied (carbon-14, beryllium-10 and chlorine-36) were inconsistent with a GRB origin since a GRB would be unlikely to yield as much beryllium-10 as was detected. (The secondary neutron energy needed for beryllium-10 production via spallation of oxygen is beyond the energy of a GRB.) A similar large spike in cosmogenic radionuclides (known as a Miyake event), suggestive of another extreme solar proton event occurred around 2600 years ago in 660 BC. Recent cosmogenic radionuclide data have revealed additional evidence of Miyake events in 7176 and 5259 BC. Curiously, the extreme event of 7176 BC left its signature during what appears to have been a solar minimum. This calls into question the concept of solar energetic particles being the etiology for these cosmogenic radionuclide abundances, or it might represent a rare energetic solar event that occurred out of sync with the solar cycle. Recently, an even more powerful Miyake event was discovered that occurred approximately 143 000 years ago. The solar storm responsible for this phenomenon was perhaps ten-fold more powerful than the Carrington event of 1959.

This is a very active scientific discipline at the moment. Curious readers might wish to be acquainted with common metrics for quantifying the strength of solar particle events including the F_{30} or fluence of particles above 30 MeV (the integrated flux of particles with kinetic energy above 30 MeV per unit area). Another useful parameter is the spectral hardness, or the proportion of high-energy protons (>200 MeV) compared to lower-energy protons (>30 MeV). Some solar proton events can possess sufficient flux of high-energy protons to be recorded by ground-based instruments and are called **ground level enhancements** (**GLEs**). At the moment, GLE number 5 of February 1956 (GLE no.5-1956-02-23) with an F_{30} of 1.42×10^9 protons cm^{-2}, is the largest hard event detected by ground-based methods. Somewhat concerningly, the 774–75 event (the largest solar energetic particle event described so far) had an estimated F_{30} almost one order of magnitude larger than GLE no.5-1956-02-23.

Unnatural spikes in radiocarbon have also occurred in recent times. For instance, above-ground nuclear bomb tests between 1955 and 1980 blasted several tons of carbon-14 into the atmosphere. This led to a near doubling of atmospheric carbon-14 with peak levels in 1964 for the northern hemisphere and 1966 for the southern hemisphere. Over subsequent decades, this bomb carbon pulse has subsided as the artificially introduced radiocarbon has spread throughout the carbon cycle reservoirs. Another manmade source of carbon-14 that has had a smaller effect than nuclear bomb testing is fossil fuels burning. Since the 19th century large amounts of coal and oil have been burned as fuel or for heat. Since these fossil carbon sources contain essentially no carbon-14, the combustion product, carbon dioxide, similarly has virtually no carbon-14. Therefore, the carbon-14 concentration in the atmosphere since the dawn of the Industrial Age has been diluted. Dating organic material from the 20th century will thus yield an apparent age older than its true age thanks to this dilution. This phenomenon is sometimes called the **Suess effect** honoring Hans Suess who first reported it in 1955.

Aside from well-known tritium and carbon-14, there are many other radio-isotopes which are created by interactions between cosmic rays and terrestrial matter. A table of such cosmogenic radionuclides is below (table 17.1).

Table 17.1. Adapted from the *European Atlas of Natural Radiation* (Chapter 8, Cosmic Radiation and Cosmogenic Radionuclides) with data from the NuDat online database maintained by Brookhaven National Laboratory (available at https://remon.jrc.ec.europa.eu/About/Atlas-of-Natural-Radiation/Download-page and https://www.nndc.bnl.gov/ensdf/).

Radioisotope	$t_{1/2}$	Decay radiation	Target nuclides	Air (troposphere)	Rain water	Ocean water
^{10}Be	1 600 000 yr	β	N, O	—	—	2×10^{-8}
^{26}Al	716 000 yr	γ, β^+	Ar	—	—	2×10^{-10}
^{36}Cl	300 000 yr	β	Ar	—	—	1×10^{-5}
^{81}Kr	229 000 yr	K x-rays	Kr	—	—	—
^{14}C	5730 yr	β	N, O	—	—	5×10^{-3}
^{32}Si	172 yr	β	Ar	—	—	4×10^{-7}
^{39}Ar	269 yr	β	Ar	—	—	6×10^{-8}
^{3}H	12.33 yr	β	N, O	1.2×10^{-3}	—	7×10^{-4}
^{22}Na	2.60 yr	β^+	Ar	1×10^{-6}	2.8×10^{-4}	—
^{35}S	87.51 d	β	Ar	1.3×10^{-4}	7.7–107×10^{-3}	—
^{7}Be	53.29 d	γ	N, O	0.01	0.66	—
^{37}Ar	35.0 d	K x-rays	Ar	3.5×10^{-5}	—	—
^{33}P	25.3 d	β	Ar	1.3×10^{-3}	—	—
^{32}P	14.26 d	β	Ar	2.3×10^{-4}	—	—
^{28}Mg	20.91 h	β	Ar	—	—	—
^{24}Na	14.96 h	β	Ar	—	3.0–5.9×10^{-3}	—
^{38}S	2.84 h	β	Ar	—	6.6–21.8×10^{-2}	—
^{31}Si	2.62 h	β	Ar	—	—	—
^{18}F	1.83 h	β^+	Ar	—	—	—
^{39}Cl	55.6 m	β	Ar	—	1.7–8.3×10^{-1}	—
^{38}Cl	37.24 m	β	Ar	—	1.5–25×10^{-1}	—
^{34m}Cl	32.0 m	β^+	Ar	—	—	—

Further reading

[1] Mekhaldi F *et al* 2015 Multiradionuclide evidence for the solar origin of the cosmic-ray events of AD 774/5 and 993/4 *Nat. Commun.* **6** 8611

[2] Schimmerling W 5 February 2011 *The Space Radiation Environment: An Introduction.* The Health Risks of Extraterrestrial Environments. https://three.jsc.nasa.gov/concepts/SpaceRadiationEnviron.pdf

[3] Miyake F, Masuda K and Nakamura T 2013 Another rapid event in the carbon-14 content of tree rings *Nat. Commun.* **4** 1748

[4] O'Hare P *et al* 2019 Multiradionuclide evidence for an extreme solar proton event around 2,610 B.P. (~660 BC) *Proc. Natl Acad. Sci. USA* **116** 5961–6

[5] Brehm N *et al* 2022 Tree-rings reveal two strong solar proton events in 7176 and 5259 BCE *Nat. Commun.* **13** 1196

[6] Paleari C I *et al* 2022 Cosmogenic radionuclides reveal an extreme solar particle storm near a solar minimum 9125 years BP *Nat. Commun.* **13** 214

[7] National Research Council (US) 1999 Executive Summary *Evaluation of Guidelines for Exposures to Technologically Enhanced Naturally Occurring Radioactive Materials* (Washington, DC: National Academies Press (US)) 1–15

[8] UCNτ Collaboration 2021 Improved neutron lifetime measurement with UCNt *Phys. Rev. Lett.* **127** 162501

[9] Bard E, Miramont C and Capano M *et al* 2023 A radiocarbon spike at 14 300 cal yr BP in subfossil trees provides the impulse response function of the global carbon cycle during the Late Glacial *Phil. Trans. R. Soc. A* **381** 20220206

Chapter 18

Planet Earth

18.1 The differentiated interior of Earth

The interior of the Earth is divided into several layers based on density, chemical composition and seismological properties. Basically, these layers are the **crust**, **mantle** and **core** from outermost to innermost. More specifically, the crust can be divided into a lighter continental **crust** and denser **oceanic crust**. In general, continental crust tends to be composed of **granitic** rock made of **felsic minerals** whereas oceanic crust tends to be composed of denser **basaltic** rocks made of **mafic** minerals. Felsic minerals tend to be less dense and are composed of aluminum, calcium, sodium and potassium silicates (e.g. feldspars) and silica (quartz). In contrast, mafic minerals are denser and are chemically composed chiefly of iron and magnesium silicates (i.e. ferromagnesian silicates) (figure 18.1).

Seismological waves can be detected from earthquakes, volcanic eruptions and even large nuclear bomb detonations. The crust and the mantle generate very distinct seismological signals and the boundary between them is called the **Mohorovicic discontinuity** (or 'Moho'). Seismic wave velocities increase below the Moho because of the increased density of the mantle compared to the crust.

Based on the theory of **plate tectonics** and specific definitions, the superficial layers of the Earth can be divided into 15–20 **lithospheric plates**. However, the seven major tectonic plates cover 95% of Earth's surface. The **lithosphere** is the crust plus the uppermost mantle. The lithosphere is on average about 100 km or 62 miles thick. It rests upon a softer layer of mantle called the asthenosphere. These lithospheric plates, driven by convection in the mantle, move about on the relatively softer, plastic asthenosphere. The pace is pedestrian by most standards—maybe a few centimeters per year, which is comparable to the 2.5–3.7 mm per month average rate of human fingernail growth. Incidentally, 'tectonics' simply refers to large-scale activity on a planet's surface that results in obvious structures. Mars and Venus have (or had) tectonic activity that created large volcanoes, chasms and mountains. But

doi:10.1088/978-0-7503-5444-8ch18

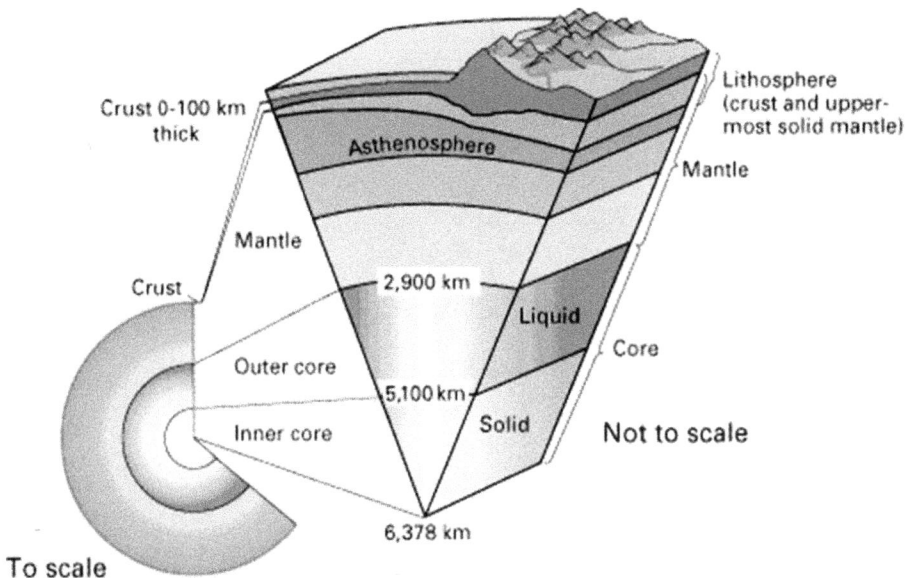

Figure 18.1. A diagram of Earth's interior showing the solid metallic inner core, the liquid metallic outer core that generates our geomagnetic field, the silicate mantle and the superficial lithosphere. The lithosphere is composed of the crust (either the thinner oceanic crust or the thicker continental crust) and the uppermost mantle. The lithosphere rests upon a softer layer of mantle called the asthenosphere. Lithospheric plates glide very slowly (averaging a few centimeters per year) over the asthenosphere in the process of plate tectonics. Credit: pubs.usgs.gov.

only planet Earth has the special lateral movement of lithospheric plates in the fashion of plate tectonics.

New oceanic crust (oceanic lithosphere) is produced at **oceanic spreading centers** such as the well-known **Mid-Atlantic Ridge**. Basaltic igneous rock is produced at these spreading centers and pushes the oceanic plates apart. Thus, the oceanic crust is continuously growing. Given that the Earth as a whole is not expanding, this new surface material must mean crust elsewhere is either being consumed or piling up (figure 18.2).

18.2 Plate boundaries

Plate boundaries can be described as **divergent, convergent** or **transform**. **Divergent plate boundaries** are sites where magma rises from the mantle at spreading centers such as the mid-oceanic ridges. It might be worth mentioning that the lava pouring out of these spreading centers does not prove that the underlying mantle was liquid to begin with. The divergence depressurizes the extremely hot underlying mantle and this allows a phase change from solid to liquid. **Convergent plate boundaries** are where plates collide. Such collisions can cause buckling, leading to the formation of mountain chains (orogeny). Alternatively, the denser plate (usually basaltic oceanic) might bend downward while the lighter plate (usually granitic continental) bends upward. These sites are known as **subduction zones** and are marked by deep oceanic

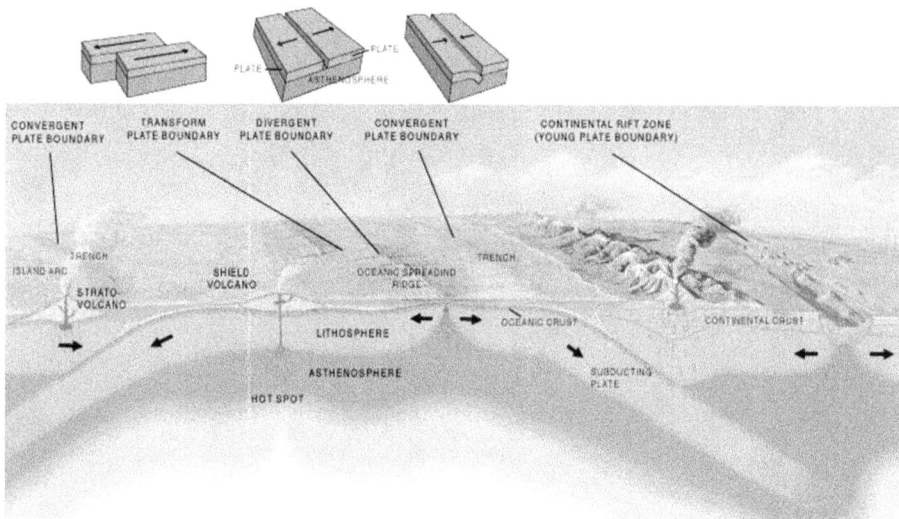

Figure 18.2. Oceanic spreading centers are the source of new oceanic crust which moves away on both sides and eventually undergoes subduction at deep ocean trenches. The collisions between oceanic crust (technically, oceanic lithospheric plates) and continental crust (continental lithospheric plates) lead to volcanic activity and earthquakes. Spreading centers may also occur on continental plates, leading to rift zones. An example of this is the Great Rift Valley in Eastern Africa. This Tectonic plate boundaries image has been obtained by the author(s) from the Wikimedia website, where it is stated to have been released into the public domain. It is included within this book on that basis. It is attributed to Jose F. Vigil. USGS.

trenches. A well-studied subduction zone, the **Mariana trench**, exists under the western Pacific. The **Challenger Depth** in the Mariana trench is the deepest known spot in any of our oceans at 35 876 feet deep (10 935 m or 6.94 miles). Volcanic activity and earthquakes are commonly encountered near convergent plate boundaries. The Pacific **Ring of Fire** is a large chain of volcanoes parallel to the convergent plate boundaries in the Pacific Ocean.

At convergent plate boundaries, the denser crust is forced down into the mantle where it is subjected to friction and great heat. This causes it to begin to melt. But the melt may not be complete. Partial melting of rocks follows **Bowen's reaction series**. It turns out that the minerals that melt at the lowest temperature happen to be felsic minerals such as quartz, muscovite, biotite, orthoclase feldspar and sodium-rich plasioclase feldspar (albite). These light-colored felsic minerals also happen to be lighter in weight than their mafic mineral counterparts. The net result is that partial melting at convergent plate boundaries creates new and lighter continental crust, which is granitic (or andesitic) and 'floats' on the older, denser basaltic crust. It is believed that this mechanism of partial melting and creation of lighter continental crust at convergent plate boundaries is what created the continents on today's Earth. Meanwhile the denser subducted plate continues to sink deeper into the mantle until, after an average of 200 million years, it reaches the mantle–core interface. The subducted rock is not dense enough to penetrate the core. At such great depths, the temperature is enormous, and this hot mantle will tend to rise if given a chance.

Because superficial mantle is moving away from itself at divergent plate boundaries (oceanic spreading centers), this deep, hot mantle will move upwards to fill these voids. In this manner, **mantle convection** occurs.

Thanks to mantle convection-driven plate tectonics, the surface of Earth has very little truly ancient material on it. The recycling process of destruction of crust at subduction zones, coupled with new crust formation at spreading centers, means that the oceanic crust anywhere is generally no more than a mere 180 million years old. And although the continental crust does not undergo the same cycle of subduction and destruction, it is subjected to physical and chemical weathering, which constantly erodes the old surface away. Hence, unlike the Moon or Mercury where many surface rocks can be billions of years old, finding pristine ancient rock on Earth is far more difficult. This is one of the reasons why studying the other terrestrial planets and our Moon is so important—they can teach us things about Earth's past that have long since been erased from Earth's history book.

When two plates slide past each other rather than crashing head-on, a **transform plate boundary** is formed. One of the most famous transform plate boundaries is the **San Andreas fault**. Earthquakes are common along such fault lines. In contrast to convergent and divergent boundaries, while old crust is cracked and broken at transform faults, new crust is not created and old crust is not destroyed.

18.3 The supercontinent cycle and Wilson cycle

All this jostling and migration of plates has an obvious consequence: Alfred Wegener's initial 1912 hypothesis of 'continental drift' was essentially correct. But if one traces the plate movements backwards over hundreds of millions of years, an inescapable conclusion is that at one time, all the continents must have been linked together in a supercontinent. The last supercontinent was **Pangea**. Pangea first amassed in the Paleozoic Era around 330 million years ago (during the Carboniferous Period) and began disassembling in the Mesozoic Era perhaps 215 million years ago (near the end of the Triassic Period). Pangea was an equatorial supercontinent, which had implications for climate. It is suspected that the geographic location of Pangea was instrumental in setting the stage for the **Permian extinction** around 250 million years ago. The Permian extinction was the most severe of the five major mass extinction events in Earth's history.

Unlike today's oceans with the **global conveyor system** actively circulating the world's ocean waters, when Pangea was in place, the oceans were arranged into just two superoceans, the **Panthalassa** and **Paleotethys** Oceans. The thermostat effect of the global conveyer system was not as effective with this arrangement and once the **Siberian Traps** discharged sufficient volcanic greenhouse gas into the atmosphere, the stage was set for a runaway positive feedback loop with ever-increasing temperatures. This is believed to be the root cause of the Permian extinction, which wiped out around 95% of all species then living on the planet.

Pangea was the most recent supercontinent but certainly will not be the last. And there were several supercontinents before Pangea. Some of these include the relatively short-lived **Pannotia** (650–500 million years ago or **mega-annum, Ma**) in the late Proterozoic Eon, **Rodinia** (also in the Proterozoic Eon, spanning about 1.1 billion to 700 million years ago) and perhaps a couple of others such as Columbia (Nuna) and **Kenorland**. The study of supercontinents is an evolving scientific discipline with a fair amount of disagreement presently. Some contend that before Kenorland (2.7–2.1 **giga-annum (Ga)** in the late **Archean Eon**), there were two other supercontinents, **Ur** and **Vaalbara**, but the evidence is debated.

The assembly and breakup of supercontinents over millions or billions of years is known as the **supercontinent cycle**. This is related to but not synonymous with the **Wilson cycle** (named after J Tuzo Wilson), which is the opening and closing of oceans as ocean basins grow and shrink due to the relative dominance of mid-oceanic ridges (spreading centers or divergent boundaries) versus subduction zones (convergent boundaries). A classic example is the present expansion of the Atlantic Ocean as the **Mid-Atlantic Ridge** continues to create new oceanic crust. But this present opening (that is, the Atlantic Ocean), was preceded by the closure of the previous ocean (the **Iapetus Ocean**) in the same region when the supercontinent Pangea formed.

18.4 The asthenosphere and mesosphere

The mantle is made mostly of silicate solid rock. But the mantle is not chemically uniform. The mantle is differentiated, meaning that it has segregated itself internally such that denser minerals are deeper and lighter minerals are more superficial. In the deep mantle, **mafic** and **ultra-mafic** silicate-poor ferromagnesian minerals, such as olivine dominate (leading to rocks such as basalt, gabbro, peridotite and komatiite). But as one moves upwards within the mantle, the mineral content changes from mafic to more **felsic** composition. But since the felsic minerals have lower melting points, as one moves more superficially, the mantle becomes mushy; it is not quite as solid as it is deep down. This relatively softer, plastic layer in the upper mantle is called the **asthenosphere** and extends roughly from about 100 km (60 miles) to about 700 km (450 miles) below the surface. It is on the softer asthenosphere that the lithospheric plates slide about, allowing the phenomenon of plate tectonics (colloquially called continental drift) to exist.

Below the asthenosphere is a more rigid region of mantle made mostly of mafic rocks and minerals called the **mesosphere** (or **lower mantle**). It extends from just below the asthenosphere down to the core–mantle boundary around 2900 km down. This makes the mesosphere the subpart of the Earth with the greatest volume. In fact, the mantle as a whole makes up about 84% of the Earth's volume.

18.5 Seismology and the core

Below the mantle is Earth's core, and the layer between the mantle and core is called the **Gutenberg discontinuity** or **core-mantle boundary** (sometimes abbreviated as **CMB**). (Note: the name 'Gutenberg discontinuity' is confusingly also given to another seismic wave transition zone, namely the lithosphere–asthenosphere boundary below oceanic tectonic plates. And 'CMB' also stands for cosmic microwave background.) **Primary seismic waves (P-waves)** slow down here while **secondary seismic waves (S-waves)** disappear entirely.

P-waves are longitudinal waves whereas S-waves are transverse. Of these two types of body waves, the P-waves move faster: they arrive first at seismic station detectors hence the name 'primary'. P-waves are also called **seismic pressure waves** while S-waves are also called **seismic shear waves**. Both are examples of so-called **body waves**, which travel through the body of the Earth (distinguishing them from slower moving surface waves).

S-waves cannot transmit through liquids, so this is evidence that the outer core is liquid. Additional seismological data indicate that the core is composed primarily of iron and nickel metal. Since nickel is denser, it may be the primary component of the inner core. Nickel also has a higher melting point than iron. Hence, given the particular temperature and pressure in the core (about 6000 K and 3.5 mega bars), it is likely that the inner core is solid and largely nickel whereas the outer core is mostly molten iron. The outer core is about 2300 km thick and extends down to around 3400 km; the inner core begins around 5150 km (3200 miles) below the surface and has a radius of about 1220 km or 760 miles.

This layering of the core into a solid inner core and molten metal outer core allows rotational decoupling between the rest of the planet. This means the liquid outer core could presently be spinning faster than the rest of the Earth and the solid inner core could be rotating at yet another rate. This differential spin velocity, coupled with convection and turbulence in the liquid conducting metal (which contains ions and allows currents) generates a magnetic field. The geomagnetic field in turn extends far beyond the core and surface of Earth. It extends into space to form the **magnetosphere**, which helps protect the planet from incoming charged particulate radiation. Without such magnetic shielding, the solar wind might have eroded Earth's atmosphere away in a manner like what has apparently happened to Mars.

18.6 The carbonate–silicate cycle: Earth's thermostat

Speaking of comparative planetology, it is worth emphasizing again that Earth is the only planet in our Solar System known to display true plate tectonics (as opposed to simple volcanoes and such). But since plate tectonics serves as a powerful global thermostat, this means that only Earth has this sort of climatological negative feedback system to keep its temperature convivial to life. On Venus for example, the constant volcanic outgassing has led to colossal carbon dioxide buildup in the atmosphere. On Venus, this accumulation of atmospheric carbon dioxide is a one-way street; the thermostat is broken. But on Earth, we have the **carbonate–silicate cycle**. And a key part of this cycle is the plate tectonics-driven burial of massive amounts of carbonate rocks and minerals at subduction zones. This in essence takes greenhouse gas carbon dioxide from the atmosphere and buries it for millions of years as carbonate rock in the mantle. So, unlike Venus with its 96%–97% carbon dioxide atmosphere (which incidentally is similar in terms of carbon dioxide percentage to Mars, which is 95%–96%) Earth's atmosphere is only 0.040% (400 parts per million) carbon dioxide. A truly remarkable difference! And it should be mentioned that Venus and Mars do not inherently possess more carbon than Earth; both planets were conferred with the around same quantity of carbon at the time of their conceptions. But vast atmospheric carbon dioxide difference is largely thanks to the unique phenomenon of plate tectonics and the carbonate–silicate cycle.

The biosphere also plays an important role in sequestering carbon dioxide; botanical activity banks a good deal of carbon dioxide in the form of wood, peat and coal.

In brief, the carbonate–silicate cycle starts by taking carbon dioxide from the atmosphere and dissolving it in water to make a dilute solution of **carbonic acid** (H_2CO_3) in clouds. The carbonic acid then precipitates out of the sky as rain, and through chemical weathering, this carbonic acid solution slowly dissolves silicate rocks and minerals.

The dissolution of silicate rocks and minerals then dumps cations (including calcium) and anions (including bicarbonate) into lakes, streams and rivers that eventually lead to the oceans. It should be mentioned that ocean water is not simply concentrated river water; the chemical composition is quite different. River water has relatively more calcium and bicarbonate. So, what has happened to that calcium and bicarbonate in the ocean? Life, in the form of animals, protists and algae that create **calcareous (calcium carbonate) skeletons**, takes the calcium and bicarbonate, and converts them into massive coral reefs, countless clamshells and thick layers of ocean sediment (called calcareous ooze) that eventually become limestone. The famous White Cliffs of Dover are massive outcroppings of limestone that were once ocean sediments left behind by calcareous coccolithophore plankton. Basically, these and other limestone deposits are disinterred graveyards of calcareous ocean organisms. (Of course, limestone and other carbonate rocks can be created by non-biological mechanisms as well.) But after the corals, clams and coccolithophores have died and left their calcium carbonate skeletons on the ocean floor, plate tectonics can bury this calcium carbonate at subduction zones, effectively taking it out of circulation for millions of years. In this manner, Earth (uniquely in all the Solar System) has a built-in thermostat that helps prevent wild fluctuations in climate by removing greenhouse gas carbon dioxide and sequestering it in the form of limestone or other carbonates for very long timescales.

18.7 The Urey reaction

The reaction between atmospheric carbon dioxide (in the form of carbonic acid) with calcium silicate minerals to create carbonates is called the **Urey reaction** after Harold Urey who first proposed the reaction in 1952. An example of the Urey reaction is when water (which on Earth is a dilute solution of carbonic acid) reacts with the calcium silicate mineral wollastonite to yield calcium carbonate and silica. This is a classic chemical weathering reaction. It is important in the carbonate–silicate cycle since it releases both dissolved calcium carbonate and silica into river water. Over time, in the oceans these building blocks eventually form limestone and chert, respectively.

18.8 Snowballs and Hothouses—failures of Earth's thermostat

Eventually the buried carbon can return to the atmosphere as carbon dioxide when the carbonate rocks and minerals are heated, melted, boiled and erupted through volcanism. In fact, it was through massive volcanic activity that Earth probably escaped from the few **Snowball Earth** scenarios during which it seemed like the planet was forever doomed to exist as a giant frozen ball of ice floating around the Sun.

(If it were not for greenhouse gases, Earth's average temperature would probably hover around zero Fahrenheit or −18 °C which is about 33 °C colder than its present average of 15 °C.) Immense releases of greenhouse carbon dioxide gas through volcanoes broke Earth free from the grasp of these snowball Earth episodes and re-established climatic equilibrium. But it might have been a paucity of greenhouse gas carbon dioxide in the atmosphere in the first place that allowed the pendulum to swing so far to the frigid side. This subject is covered in greater detail in chapter 29.

On the other extreme, it is suspected that the worst mass extinction of all time, the **Permian extinction** event, which occurred roughly 250 million years ago, was caused primarily by unbridled greenhouse gas belching by the **Siberian Traps**. This led to a **'Hothouse Earth'** and a series of chain reactions (positive feedback loops) that made a bad situation worse. Eventually around 95% of all species perished during this 'mother of all extinctions'. Mass extinctions are discussed in further detail in chapter 30.

18.9 Venus's runaway greenhouse

The runaway greenhouse effect on Venus was caused by its thick atmospheric blanket of ~96% carbon dioxide. This is why Venus has an average surface temperature of 867 °F or 464 °C—it is *not* primarily because of its relative proximity to the Sun. (In fact, without the greenhouse gases, Venus's surface temperature might be 700 °F (390 °C) cooler than it would be without a greenhouse effect. Students might enjoy calculating very rough estimates of planetary surface temperatures without greenhouse effects by making use of the Stefan-Boltzmann equation,

$$S = e\sigma T^4$$

Where S is the average amount of sunlight absorbed per unit surface area, σ is Stefan's constant, T is the temperature and e is the emissivity (albedo) of the planet in question and setting the energy coming in (from the Sun) equal to the energy lost. However, one should factor in the very high albedo (reflectivity) of Venus's cloud cover, which is in the range of 75%–85%. Because of this, the amount of sunlight reaching the surface of Venus is actually less than the amount reaching Earth's surface despite the difference in proximity to the Sun. When accounting for this, some have estimated an ice-cold surface temperature on Venus of −50 °F or −45 °C! However, this can be criticized as a 'chicken versus egg' dilemma since Venus would probably not have its thick reflective sulfuric acid cloud cover without its astronomical temperatures.

18.10 Plate tectonics and geomagnetism

Plate tectonics might also play a role in our cosmic ray-deflecting geomagnetic field. To generate a magnetic field that partially shields us from charged particulate radiation, the core must have a **geodynamo** in the form of circulating, conducting liquid metal. But to maintain this geomagnetic field over the past 4 billion years or so, the Earth would have had to cool its core by about 3000 °C during that interval. But the core can only cool as quickly as the mantle cools. Thanks to mantle convection and plate tectonics, the mantle has cooled faster than it would have if there were no such convection cycles. This requires a delicate balance, for if the outer core cools too much and freezes solid, it will no longer sustain a geodynamo and, like on Mars, the magnetic field will cease to exist.

However, there have been some discrepancies in recent modeling of the Earth's early interior temperature with recent work suggesting that the core could not have cooled in accordance with the classical theory. An alternative mechanism has been proposed, which asserts that continuous mantle deformation due to the Moon's gravitational tidal effects could promote motion within the liquid outer core sufficient to generate the geomagnetic field. Furthermore, the inner core was probably not always solid. In fact, it might have only cooled and solidified about a billion years back, with debated dates ranging from 565 million to 2 billion years ago. This is a hot topic if you'll excuse the pun. But once the inner core did solidify, the strength of the geodynamo increased dramatically, as did the geomagnetic field strength on the surface.

18.11 The Goldschmidt classification of elements

Despite the high temperatures, the immense pressure on the inner core causes it to convert to the solid state. Along with the nickel and iron, it is possible that some other heavy metals might be present in the inner core such as gold, silver, platinum, tungsten and palladium. This conclusion is based partly on the densities of these elements but also on the chemical tendency of elements to 'follow the leader' and flow along with the molten iron that sank to the center. The **Goldschmidt geochemical classification of elements** groups the elements according to where they prefer to partition geologically. For example, the **siderophiles** ('iron lovers') tend to readily dissolve in iron (either in the molten phase or in solid solution) and are therefore inclined to follow iron to the core. The Goldschmidt classification has five different groups:
1. Aerophiles
2. Lithophiles
3. Chalcophiles
4. Siderophiles.

The **aerophiles** tend to become gases or liquids and partition into the atmosphere or hydrosphere. **Lithophiles** are 'rock-loving' elements that tend to combine readily with oxygen and form silicate minerals that in turn, form rocks. The lithophiles, like the aerophiles, did not sink into the core during early Earth differentiation. The lithophiles instead largely remained in the lithosphere and mantle. **Chalcophiles** are the 'ore-formers' and also tend to remain on or near the surface because they tend to combine with sulfur or other chalcogens, but usually not oxygen. Chalcophiles form sulfides that are generally denser than the silicates formed by lithophiles and therefore the chalcophile elements have positioned themselves below the lithophiles during differentiation. Thus, the superficial crust is relatively depleted in chalcophiles relative to lithophile elements. The siderophiles theoretically followed the molten iron downwards as it sank to the center. However, several elements that are considered highly chalcophilic also have highly siderophilic tendencies (such as Au, Cu, Te, Ru, Rh, Pd, Os, Ir, and Pt).

18.12 Sources of Earth's internal heat

The Earth is internally differentiated. This internal stratification stems from reshuffling during its early history when it was in a molten state. The Earth is believed to have formed through a series of accretions first involving small dust particles, then moderate-sized clumps of rock and ice, and then finally large boulders, asteroids and planetesimals. This primitive agglomeration was all held together (and attracting ever more material) through progressively increasing gravitational attraction.

This process of accretion through collisions converted gravitational potential energy into kinetic energy and then ultimately into heat energy. Thus, accretion itself generated intense heat that contributed to the melting; the heat generated by this process is called accretion heat. The resulting hot, semi-liquid surface might have made the burgeoning planetesimals 'stickier', thereby allowing more rocky matter to glom on rather than rebound away. Furthermore, as the protoplanet grew larger, it was better able to retain its heat thanks to the square-cube law. This simply restates the obvious fact that the surface area (which can dissipate heat back into space) increases as the square of the radius of a sphere ($A = 4\pi r^2$) while the volume (which promotes retention of heat) goes according to the cube of the radius ($V = 4/3\pi r^2$). Basically, the larger the protoplanet became, the more effectively it retained heat.

But the liquid protoplanet would then begin to assort its internal structure based on density. And this internal rearrangement process led to further **frictional heating** as the denser material sank downward and less dense substances floated upward. Alternatively, this reassortment process could be interpreted as further **conversion of gravitational potential energy into heat** energy as the denser metallic matter moved downwards towards the core. And as parts of the protoplanet began to congeal, additional heat was produced thanks to the **heat of fusion** released during the liquid–solid phase changes. Finally, and perhaps most importantly, there was a tremendous amount of energy released due to radioactive decay of primordial radionuclides. This radioactive heating is known as **radiogenic heat**. Along with gravitational attraction, these various heat sources converted the cold, amorphous dust cloud (primordial solar nebula) into the roiling and boiling molten proto-Earth.

18.13 Radiogenic heating

In order of decreasing total energy released (when including the entire decay chain) the main sources of radiogenic heat (i.e. the radioactive heat-producing elements) are:

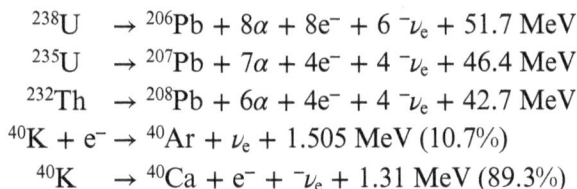

$$^{238}U \rightarrow {}^{206}Pb + 8\alpha + 8e^- + 6\,{}^-\nu_e + 51.7\,\text{MeV}$$
$$^{235}U \rightarrow {}^{207}Pb + 7\alpha + 4e^- + 4\,{}^-\nu_e + 46.4\,\text{MeV}$$
$$^{232}Th \rightarrow {}^{208}Pb + 6\alpha + 4e^- + 4\,{}^-\nu_e + 42.7\,\text{MeV}$$
$$^{40}K + e^- \rightarrow {}^{40}Ar + \nu_e + 1.505\,\text{MeV}\,(10.7\%)$$
$$^{40}K \rightarrow {}^{40}Ca + e^- + {}^-\nu_e + 1.31\,\text{MeV}\,(89.3\%)$$

Other relatively abundant primordial radioisotopes with long half-lives and activities above 1 millibecquerel per kilogram of Earth's crust that could continue to cause heating include: samarium-147 (which decays into neodymium-143 with a

Table 18.1. Adapted from National Research Council (US) Committee on Evaluation of EPA Guidelines for Exposure to Naturally Occurring Radioactive Materials. Evaluation of Guidelines for Exposures to Technologically Enhanced Naturally Occurring Radioactive Materials. Washington DC: National Academies Press (US); 1999. 2, Natural Radioactivity and Radiation. Available at: https://www.ncbi.nlm.nih.gov/books/NBK230654/.

Radionuclide	Half-life in yr	Major radiations	Typical crustal concentration, Bq kg^{-1}
^{40}K	1.28×10^9	β, γ	630
^{50}V	1.4×10^{17}	γ	2×10^{-5}
^{87}Rb	4.75×10^{10}	β	70
^{113}Cd	9×10^{15}	β	$<2 \times 10^{-6}$
^{115}In	6×10^{14}	β	2×10^{-5}
^{123}Te	1.24×10^{13}	X-rays	2×10^{-7}
^{138}La	1.05×10^{11}	β, γ	2×10^{-2}
^{142}Ce	$>5 \times 10^{16}$	β	$>1 \times 10^{-5}$
^{144}Nd	2.29×10^{15}	α	3×10^{-4}
^{147}Sm	1.06×10^{11}	α	0.7
^{152}Gd	1.08×10^{14}	α	7×10^{-6}
^{174}Hf	2.0×10^{15}	α	2×10^{-7}
^{176}Lu	3.73×10^{10}	β, γ	0.04
^{187}Re	4.3×10^{10}	β	1×10^{-3}
^{190}Pt	6.5×10^{11}	α	7×10^{-8}

103 billion year half-life), rubidium-87 (which decays into strontium-87 with a 48.8 billion year half-life) and lutetium-176 (which decays into −176 with a 37.3 billion year half-life). A more comprehensive table of non-chain-forming primordial radioisotopes is below (table 18.1).

Given the half-lives of uranium-235 (704 million years), potassium-40 (1.25 billion years), uranium-238 (4.5 billion years) and thorium-232 (14 billion years) one can see that there must have been far more radiation and radiogenic heat when Earth first formed about 4.567 billion years ago due to the relatively high abundance of these primordial radioisotopes compared to today. However, one must also consider the numerous 'short'-lived radioisotopes that no longer even exist since they have long since decayed away. For example, neptunium-237 has a half-life of only 2.14 million years. It, and the entire **neptunium series (4n + 1 series)** which ends in thallium-205, are now extinct in nature. But it was once actively contributing radiogenic heat to the early Earth and assisting in the melting process. The total amount of energy released from decay of neptunium-237 into thallium-205 (including the energy lost to neutrinos) is 50.0 MeV which puts this decay chain near the top of the list above, second only to the **uranium series (4n + 2 series)**. And the fact that it releases all that energy in a much shorter time frame means increased heating effectiveness.

And there were several other primordial radionuclides at Earth's inception that have long since expired because of their relatively short half-lives. Important among them was aluminum-26. This radioisotope is produced in supernova explosions. Evidence left behind in chondrite meteorites suggests that aluminum-26 was rather abundant during the creation of the Solar System. Specifically, there is an anomalous amount of the daughter product, magnesium-26, within the feldspar minerals inside these very ancient chondrite meteorites (that are believed to be among the very oldest objects in our Solar System). The implication is that if aluminum-26 was abundant in chondrite meteorites, it was probably abundant throughout the early Solar System, including the newborn Earth. But since aluminum-26 decays into magnesium-26 with a half-life of only about 720 000 years, by around 10 million years, this radioisotope was essentially extinct. Nevertheless, this was sufficient time for it to contribute meaningfully to the early radioactive heating of the newly forming planets.

Incidentally, aluminum-26 can decay through either emission of positrons or through electron capture. It primarily decays through positron emission but approximately 15% of decay is via electron capture. In either situation, the resultant magnesium is in an excited state that releases further energy in the form of gamma rays (mostly 1.81 MeV). Thanks to its short half-life, there is no leftover primordial aluminum-26 in the Earth's interior today, but it is still produced today as a cosmogenic radionuclide. Additionally, some aluminum-26 was artificially manufactured by nuclear weapons testing during the 1960s.

Does radioactivity contribute much to Earth's present internal heat? Although both the remnant accretion heat and radiogenic heat have both waned significantly from the birth of the Earth, the presence of volcanoes, plate tectonic activity and our geomagnetic field reveal that there is still plenty of heat deep inside Earth's interior. In fact, it is estimated that approximately 47 TW of heat emanate from the interior today. Most estimates put the relative contribution of radiogenic heat at about 50%, with the remainder being leftover accretion heat. This estimate is supported by the **KamLAND Collaboration study of geoneutrinos**, which revealed that the combined radioactive decay of uranium-238 and thorium-232 contribute about 20 TW, while the decay of potassium-40 contributes an additional 4 TW.

18.14 The georeactor theory

In 1992 J Marvin Herndon proposed the concept of a 'georeactor'—a natural nuclear fission reactor as a source of heat in the cores of the giant planets Jupiter, Saturn, Uranus and Neptune. From there, it is a short jump to the concept of a similar fission reactor in the Earth's core. Subsequently, Herndon and Hollenbach further refined the theory and modeled the Earth's interior with radically different ideas. Rather than a 3450 km radius liquid iron outer core and a 1220 km radius solid iron–nickel inner core, they proposed a multi-layered core with a 3400 km radius liquid sulfur–iron outer core, a 1200 km radius nickel–silicide inner core, then below that a 6 km radius 'subshell' of decay and fission products and finally a 4 km radius uranium plus actinides 'subcore'. According to this model, there may be hundreds or thousands of natural nuclear reactors formed from super-critical concentrations of uranium radioisotopes in Earth's core. They argue that since uranium is among the very densest natural elements, large volumes of this heavy metal sank by gravity into the core during the differentiation process.

They then modeled a georeactor that would function as a **fast neutron breeder reactor**. Once a critical mass was gathered in a sufficiently compact space, the natural reactor would fission uranium-235, which produces neutrons that fission other uranium-235 nuclei in a standard self-sustaining chain reaction. But the free neutrons could also be absorbed by uranium-238 nuclei and form plutonium-239, another fissile nuclide. This process, known as 'breeding', can significantly extend the lifetime of a nuclear reactor. Purported evidence for the georeactor exists in the form of ratios of helium-3:helium-4 in basaltic lava bubbling up from extreme depths reaching down near the core. Helium-3 is a cosmogenic nuclide (meaning it can be produced by cosmic ray collisions with other atoms) but deep within the Earth, helium-3 is a primordial isotope that was present in a preset amount after the Earth condensed from the primordial solar nebula. The same is true for helium-4 but in addition, alpha decay contributes to its abundance. Measurements of the helium-3:helium-4 ratio in lava extruded at geological **hotspots** such as Hawaii and Iceland, are consistent with predictions for a georeactor and up to 34-fold different from the ratio of helium isotopes expected by cosmic ray interactions.

Furthermore, they contend that a georeactor can explain the tendency of Earth's magnetic field to fluctuate in strength and sporadically flip its north–south orientation. Such changes could be expected when georeactors turn on and off periodically. And natural georeactors could operate on an on/off cycle because fission products (which slow down the reactivity) are typically less dense than the fissile uranium and plutonium in the georeactor and they might literally float away from the active site in the core. Fission products are 'waste' in a nuclear reactor and they can slow down fission. Xenon-135 and samarium-149 for example have very large cross section for neutron capture and could gobble up neutrons that would otherwise sustain the fission of uranium and plutonium. Such fission products are called **neutron poisons** since they impede the reactivity of a reactor. But once the lightweight 'waste' has drifted off, the reactor would then resume activity to full swing.

While the idea of a deep-Earth georeactor is a fascinating idea, it remains unproven at this time. One argument against the concept is that although uranium is extremely dense in the metallic form, chemically it is a Goldschmidt chalcophile and thus forms chemical compounds that preferentially partition into the crust and mantle rather than the core. However, ongoing research on neutrinos emanating from Earth's interior (the **Borexino geoneutrino measurement**) is putting stringent constraints on the power output, or even existence, of natural internal reactors. Present data suggest that if they exist at all, such georeactor output is limited to no more than 3 TW.

18.15 Natural reactors

Natural uranium is 0.7202% uranium-235 and 99.2745% uranium-238. A trace amount (about 0.0054%) is uranium-234, an intermediate in the decay chain of uranium-238. These figures are highly precise. Therefore, in 1972 when a facility in France determined that the fractional content of its uranium-235 was only 0.7171%, there was an initial concern that theft for weapons production had occurred.

Was the batch intentionally manipulated for diversion of the fissile isotope? Although a deficit of only 0.0031% is tiny, when dealing with material capable of creating an 'atomic bomb', a formal investigation is mandatory. The uranium in question was mined in the Oklo Mine in Gabon, Africa. Further investigation of uranium ore from this mine revealed some samples with deficits as high as 0.440%. Once theft was ruled out, the only logical explanation for the depleted uranium was that sometime in the distant past, the uranium at the Oklo Mine must have gone critical and fissioned away some of its uranium-235!

It now appears that 16 natural uranium fission reactors 'came to life' completely unaided by man about 1.5 billion years ago when the uranium-235 concentration was about 3%. These natural reactors like modern nuclear reactors hosted self-sustaining, nuclear chain reactions that generated a great deal of power. To get the reactor going, a moderator would have been required. This was supplied by geological and geographical shifts that allowed groundwater to bath the under-ground uranium deposit. Ordinary water can serve as a moderator, which slows neutrons down to the energies that are optimal for capture by uranium-235. Thus, when water was available, the reactors fired up on their own, cranking out power at a rate around 100 kW. They seemed to have operated in 'pulse mode' meaning they would become active, heat up for a while and then turn off and cool down for a bit. What was actually happening was, the water moderator would seep in, allow the reactors to ignite and then things would heat up so much that the water would boil away causing the reactor to dry up and temporarily shut down. Then, when things cooled off enough, the water would return and refill the reactor region, and the cycle would start anew. The operating interval was determined to be about 2.5 h. And this went on for a few hundred thousand years. Evidence supporting this theory is present in the form of fission products remaining at the site, embedded in the ore along with the original uranium itself. The fact that these fission products were still there on-site 1.5 billion years later suggests that they did not 'contaminate' the region through widespread dispersal of waste products. In all, about 11 907 pounds of waste were made but because the reactor functioned as a breeder, about 3307 pounds of plutonium-239 were 'bred' on the premises. It would be most interesting to learn about any radiophilic extremeophiles that might have been able to survive and thrive under such truly harsh conditions.

18.16 Kuroda's principles

These observations confirmed the validity of three hypothetical necessary conditions that would allow the existence of a natural nuclear reactor. These postulates, which were first proposed in 1956 by Paul Kuroda are now known as **Kuroda's principles.**

1. The first requirement addresses the critical mass of natural uranium needed. In order for the released fission-generated (2 MeV mean energy) neutrons to be absorbed without escaping and not producing fission, the uranium deposit must be at least 66 cm in width.
2. Next, the fraction of uranium-235 in the uranium ore deposit must exceed 3%. At the time of the Oklo reactors, this requirement was met.

3. Also, there must be a neutron moderator such as water present. This slows down the fast neutrons to an energy where they are more prone to induce fission in other uranium-235 nuclei. This requirement was also intermittently met at Oklo as water flowed in and served the purpose of moderation but subsequently boiled off, shutting the reactors down in a perpetual cycle.
4. Finally, the local environment had to contain a limited amount of neutron poisons (absorbers) such as boron and lithium, which in excess, would scram the reactor.

18.17 Formation of uranium ore deposits

Although it was not one of Kuroda's basic requirements, another curious possible prerequisite was the evolution of an oxygen-rich atmosphere. And, as discussed below, this is because large deposits of uranium ore probably did not exist until we had ample amounts of free molecular oxygen.

Thus, natural reactors such as Oklo probably did not fire up before 2.4 billion years ago, despite the even higher abundance of uranium-235 during those days. The so-called **Great Oxidation Event (GOE)** took place during the early part of the Proterozoic Eon (that is, during the Paleoproterozoic Era) between the interval between 2.4 and 2.1 Ga, as discussed in chapter 28. This changed our atmosphere from a slightly reducing atmosphere, virtually devoid of oxygen, into one that was rich in oxygen at levels nearly 10% of today's concentrations. When free oxygen is plentiful, uranium becomes oxidized and in oxidized states, uranium is water soluble, often as uranium cations paired with carbonate cations. In contrast, under anaerobic conditions, uranium is less water soluble and precipitates out of solution readily, typically as uraninite.

Granitic magmas—those containing more felsic rocks and minerals—are more enriched in uranium than andesitic or basaltic magmas. Such magmas erupt from volcanoes very explosively because of the higher silica content and greater viscosity. The granitic magma then falls back to the surface as ash fall or **volcanic tuff** (solidified ash). Water then encounters these tuff deposits and dissolves the contained uranium. The dissolved uranium then heads off in the groundwater. In reducing anoxic waters, the uranium then precipitates out of solution, creating a uranium deposit. Thus, the formation of large uranium ore deposits requires:
1. a source of uranium,
2. oxidizing conditions that allow solubility of the uranium in water and
3. reducing (anoxic) conditions in a stagnant pool of water that encourage precipitation and concentration of the uranium.

Based on this model large, concentrated uranium ore deposits were probably not possible before the GOE. And therefore, natural reactors such as Oklo probably did not exist before then.

18.17.1 Atmosphere

It can be debated as to whether Earth's atmosphere is a simple mixture of different gas molecules or legitimately can be considered a gaseous solution. Compositionally, (by volume) it is 78.084% diatomic molecular nitrogen (N_2), 20.946% diatomic molecular oxygen (O_2) and about 0.934% argon. It is believed that much of this argon stems from the decay of radioactive potassium-40 over the last 4.567 billion years of Earth's existence. (Potassium-40 has a half-life of 1.25 billion years and presently makes up about 0.012% (120 parts per million) of today's natural potassium). The remaining 0.043% (about 400 parts per million) is composed of the important greenhouse gas, carbon dioxide, but this figure has increased roughly twofold over the last 120 years. The balance of Earth's atmosphere is various trace gases including neon, helium, methane, krypton and hydrogen in order of abundance.

Not mentioned above is water vapor, which has a highly variable concentration. It can range from nearly zero in cold, dry regions to around 5% in warm, humid conditions. The amount of water vapor that can 'dissolve' in air is a strong function of temperature. At low temperatures, the water vapor precipitates out but at higher temperatures, air can hold far more water vapor. (Or alternatively, water is more soluble in warm air than in cold air.) Curiously, this phenomenon makes water vapor the only greenhouse gas whose concentration dramatically increases as the temperature increases. As the atmosphere warms, water vapor content increases and can further contribute to warming. Other greenhouse gases of importance include carbon dioxide, methane, nitrous oxide and manmade fluorinated gases (such as hydrofluorocarbons, perfluorocarbons, sulfur hexafluoride). These fluorinated gases have occasionally been used as substitutes for ozone-depleting chlorofluorocarbons (which themselves have greenhouse gas potential).

The amount of water vapor in the atmosphere at a given temperature is reflected by the **relative humidity** ϕ. The relative humidity is the ratio of how much water vapor is actually in the air to how much water vapor the air could maximally contain at a given temperature. It can be expressed mathematically in terms of partial pressures where P_{H2O} is the observed partial pressure of water and P^*_{H2O} is the maximum vapor pressure possible at the temperature and ambient pressure in question (sometimes described as the equilibrium vapor pressure):

$$\phi = (P_{H_2O})/(P^*_{H_2O})$$

Some representative maximum water vapor capacity values (or equilibrium vapor pressures) which correspond to 100% relative humidities in g m^{-3} at different temperatures at sea level are:

0 °C	4.8
5 °C	6.8
10 °C	9.4
15 °C	12.8
20 °C	17.3
30 °C	30.4
40 °C	51.1

Such figures are non-linear but can be calculated by the **Clausius–Clapeyron equation** which demonstrates an increase in the atmospheric water vapor carrying capacity of about 7% for every 1 °C (1.8 °F) increase in temperature:

$$P* = P_o^* \, e^\wedge -\{L_v/\mathfrak{R}_v \, (1/T_o - 1/T)\}$$

where $P*$ is the water vapor pressure, P_o^* is the water vapor pressure at 0 °C which is 0.6113 kPa, $\mathfrak{R}_v$ is the water–vapor gas constant (461 J K^{-1} kg^{-1}), T_0 is 0 °C or 273.15 K and L_v is the latent heat of vaporization, 2.5 × 106 J kg^{-1}. In this form $L_v/\mathfrak{R}_v = 5423$ K. Students are reminded that the temperature must be in Kelvin to use the Clausius–Clapeyron equation in this form.

Another common form of the Clausius–Clapeyron equation is:

$$\ln (P_1/P)_2 = \Delta H_{\text{vap}}/R(1/T_2 - 1/T_1)$$

which can be derived by integration of the empirically observed equation:

$$P \, \alpha \, e^{-(\Delta H_{\text{vap}}/RT)}$$

Where P_1 and P_2 are the water vapor pressures at temperature T_1 and T_1, ΔH_{vap} is the enthalpy of vaporization and R is the gas constant (8.3145 J mol^{-1} K^{-1}).

Further reading

[1] Hollenbach D F and Herndon J M 2001 Deep-earth reactor: nuclear fission, helium, and the geomagnetic field *Proc. Natl Acad. Sci. USA* **98** 11085–90

[2] Herndon J M 1993 Feasibility of a nuclear fission reactor at the center of the Earth as energy source for the geomagnetic field *J. Geomagn. Geoelectr.* **45** 423–7

[3] Bellini G *et al* 2010 Observation of geo-neutrinos *Phys. Lett.* B **687** 299–304

[4] The KamLAND Collaboration 2011 Partial radiogenic heat model for Earth revealed by geoneutrino measurements *Nat. Geosci.* **4** 647–51

[5] National Research Council (US) 1999 Executive Summary *Evaluation of Guidelines for Exposures to Technologically Enhanced Naturally Occurring Radioactive Materials* (Washington, DC: National Academies Press (US)) 1–15

[6] Andrault D, Monteux J, Le Bars M and Samuel H 2016 The deep Earth may not be cooling down *Earth Planet. Sci. Lett.* **443** 195–203

IOP Publishing

Space Radiation
Astrophysical origins, radiobiological effects and implications for space travellers
James S Welsh

Chapter 19

Space weather

Succinctly, space weather describes the variations in the space environment between the Sun and the Earth. But space weather focuses on the various phenomena that impact systems and technologies in orbit, in interplanetary space and on Earth. In a broader sense, space weather is caused by variations in the Sun's heliosphere, solar activity and solar wind, as well as Earth's magnetosphere, ionosphere and thermosphere.

The **National Oceanic and Atmospheric Administration (NOAA)** has a branch called the **Space Weather Prediction Center (SWPC)**, which serves as an official source for space weather forecasts and information, as well as how such events may affect Earth. The SWPC depends on the **Geostationary Operational Environmental Satellite Series (GOES)-R Extreme Ultraviolet and X-ray Irradiance Sensors** to issue warning about large x-ray solar flares.

NOAA has developed scales that describe the environmental disturbances for three categories of space weather events:
1. Geomagnetic storms
2. Solar radiation storms
3. Radio blackouts.

The existence of space weather was somewhat surprising given that the Sun is quite steady and nearly constant when it comes to optical and near-infrared radiation. Nevertheless, there is considerable variability in the Sun's output of extreme ultraviolet, x-rays and radio waves during solar storms. During such storms (which are accelerated by coronal mass ejections (CMEs) and solar flares), the solar wind speed and plasma density increase. High-energy **solar energetic particles (SEPs)** in the form of **solar particle events (SPEs)** become more abundant.

Space weather can affect the performance and reliability of space-based and ground-based electronic technological systems. There is a concern that serious space weather may endanger human health and safety, especially for astronauts and

doi:10.1088/978-0-7503-5444-8ch19

airplane crew. Around the time of the Apollo Mission, a large SPE occurred (August 1972), which might have endangered astronauts had they been in space at the time.

19.1 Space physics

Space physics (**solar–terrestrial physics** or **space–plasma physics**) is the study of plasmas in the outer atmosphere and within the Solar System at large. In an even broader interpretation, it can encompass the study of plasmas beyond our own Solar System, such as those around other stars and exoplanets, as well as in interstellar and intergalactic space. In this broad interpretation, space physics encompasses several relevant related topics, such as **heliophysics** (which is an integrated investigation of the physics of the Sun and its environment). Space physics might thereby also include study of the solar magnetosphere (**heliosphere**), solar wind, planetary magnetospheres and ionospheres, auroras, cosmic rays and synchrotron radiation as well as the interplay between these various factors. Thus, space weather is a fundamental part of the study of space physics. Space weather, and therefore space physics, clearly has implications for modern daily living, including the proper functioning of electronics here and in space, as well as accurate operations of weather and communications satellites. Damage to the electrical power grid and accompanying blackouts are possible if space weather becomes too severe. Ironically, as a society, the world today is probably more vulnerable than ever to the incapacitating consequences of a major geomagnetic storm.

From a 'down-to-Earth' perspective, there are three main categories of space weather storms:
1. Radio blackouts
2. Space radiation storms
3. Geomagnetic storms.

19.2 Radio blackouts

Extensive radio blackouts may result in a complete loss of high-frequency (HF) contact on the entire sunlit side of the Earth. HF radiocommunication is in the range of 3–30 MHz. Very-high-frequency (VHF, which is in the 30–300 MHz range) radio communication may be affected as well.

Such radio blackouts incapacitate HF radio communications with mariners and aviators in the affected sector. Space radiation-induced radio blackouts may last for several hours. According to the **NOAA Solar Radiation Storm Scale**, there are five categories of space radiation-caused radio blackouts labeled **R1–5** for minor, moderate, strong, severe and extreme. These categories of radio blackouts correspond to solar flare x-ray brightness, which can be measured using the **GOES X-ray Brightness Scale**. The correlation is such that a minor (R1) radio blackout is caused by an x-ray class of M1, a moderate R2 radio blackout is caused by an M5 x-ray event, R3 (strong) is caused by an X1 event, R4 (severe) is caused by an X10 event and R5 (extreme) is caused by an X20 event. Solar flare categories are covered in more detail in section 19.10.

Radio blackouts are the result of solar bursts of extreme ultraviolet and x-ray radiation caused by solar flares. This electromagnetic radiation leads to ionizations that then result in increased electron densities. The enhanced electron densities then cause losses of radio wave energy and the ability to use radio as a form of communication. Since they are caused by x-rays and ultraviolet radiation, which travels at the speed of light, radio blackouts are the first consequence of space weather storms. Given the 8 min it takes for such radiation to reach the Earth after leaving the Sun, advanced warnings are not available.

19.3 Solar radiation storms

Large magnetic eruptions, which are often accompanied by solar flares and CMEs, energize and accelerate protons and other charged particles to very high velocities. Upon attaining such energies, the fastest of these excited protons can traverse the 150 million km Earth–Sun distance in just tens of minutes (recall that gamma photons and x-rays move at the speed of light and reach Earth in about 8 min). When they reach Earth, the fast-moving protons penetrate the magnetosphere that shields Earth from lower energy-charged particles. Once inside the magnetosphere, the particles are guided down the magnetic field lines and penetrate the atmosphere near the north and south poles.

The start of a solar radiation storm is defined as the time when the flux of protons at energies $\geqslant$10 MeV equals or exceeds 10 **proton flux units (pfu)** (where 1 pfu = 1 particle cm^{-2} s^{-1} ster^{-1}). The end of a solar radiation storm is defined as the last time when the flux of $\geqslant$10 MeV protons is measured at or above 10 pfu. This definition allows multiple injections from flares and interplanetary shocks to be encompassed by a single solar radiation storm. A solar radiation storm can persist for time periods ranging from hours to days. Solar storms are categorized by NOAA using a logarithmic five-level system known as the **S-scale**:

S1	Minor	(Flux of >10 MeV particles threshold = 10)
S2	Moderate	(Flux of >10 MeV particles threshold = 100)
S3	Strong	(Flux of >10 MeV particles threshold = 1000)
S4	Severe	(Flux of >10 MeV particles threshold = 10 000)
S5	Extreme	(Flux of >10 MeV particles >10^5)

Solar storms may have consequences for human health and satellites. Among those potentially affected by the high radiation hazard of solar radiation storms are astronauts engaging in extra-vehicular activities. Conceivably there could be biological and health consequences to passengers and crew in high-flying aircraft at high latitudes at the time of a severe or extreme solar storm. Satellites that lie beyond the protective confines of our atmosphere are vulnerable to single event upsets caused by solar radiation storms. Solar protons may affect operations by transient and relatively mild issues such as the creation of electronic noise that interferes with star tracking, or it may be so severe that is can damage electronics and render a satellite useless. Space radiation damage to electronic technology is covered further in chapter 23. Other effects include degradation of solar panel efficiency, memory impacts that can cause loss of control and introduction of serious noise in image data. Solar radiation storms can cause complete blackout of HF radio communications possible through the polar regions, and position errors make navigation operations extremely difficult. The solar protons may penetrate the magnetosphere and follow magnetic lines of force down to the atmosphere in the polar regions. The abundance of such solar protons in the atmosphere ionizes the D-layer of the ionosphere. This D-layer ionization prevents HF radio waves from reaching the higher E, F1 and F2 layers that are normally involved in reflecting the radio signals back to Earth. This type of solar radiation storm-related radio blackout is called a **polar cap absorption (PCA) event**. PCA events can affect HF radio communications in the polar regions for days. The severity of PCA events correlates with the NOAA S-scale.

19.4 Geomagnetic storms, the ring current and the disturbance storm time

Geomagnetic storms are another example of space weather. These are transient disturbances of the Earth's magnetosphere caused by intense solar activity such as CMEs or penetrating solar winds. Geomagnetic storms result from variations in the solar wind that cause changes in the currents, plasmas and fields in our magnetosphere. Large solar storms are associated with CMEs. CMEs can contain billions of tons of energetic plasma that is fired from the surface of the Sun. This plasma may take several days to reach Earth where it arrives with its embedded magnetic field. In rare instances, the kinetic energy of the CME may be so high that it reaches Earth in as short as 18 hours, which is still far slower than the fastest protons from the strongest solar flares, which can reach Earth in well under one hour. The Carrington Event, which is discussed further below, was one such high-speed event. The solar wind conditions that most effectively produce geomagnetic storms are sustained periods of high-speed solar wind (meaning persisting for several hours) along with a southward solar wind magnetic field (which is the opposite direction of Earth's magnetic field) on the day side of the magnetosphere. This set of conditions is most efficient at transferring energy from the solar wind into Earth's magnetosphere and creating geomagnetic storms.

The solar wind typically races along at velocities between 250 and 750 km s^{-1} and is **supersonic** (meaning that it travels faster than the associated **magnetosonic** or **magnetoacoustic waves**). In some situations, the solar wind might reach speeds up to 1000 km s^{-1}. Consistent with the **Biot–Savart law**, which simply states that a current induces a magnetic field, as a stream of rapidly moving charged particles (i.e. a current), the solar wind carries with it an induced magnetic field. This is called the **interplanetary magnetic field**.

There are several ways of quantifying geomagnetic storms including the commonly used five-tiered NOAA Geomagnetic Storm Scale (**G-Scale**) which is described below.

In simple terms, a geomagnetic storm can be defined as a change in the **disturbance storm time (Dst) index**. The Dst measures the **ring current** strength of electrons and protons around the Earth at a given time. During geomagnetically quiet times, the Dst is between +20 and −20 nano-Tesla (nT). A more negative Dst value (as during geomagnetic storms) is associated with a decrease in geomagnetic field strength. For example, during a geomagnetic superstorm, the Dst falls below −250 nT.

Based on Dst indices, geomagnetic storm strength may be classified as:

1. Moderate (minimum Dst −50 nT to −100 nT)
2. Intense (minimum Dst −100 nT to −250 nT)
3. Superstorm (Dst < −250 nT).

The Earth's electrical ring current system lies in the equatorial plane and circulates clockwise around the Earth (when viewed from the north; that is, it is a westward current). It may be thought of as a band at an altitude of 3–8 R_E (Earth radii) above the Earth, meaning about 18 000–48 000 km up (although some sources say 10 000–60 000 km). This ring current is the result of protons and other positively charged ions moving westward while electrons drift to the east within the Van Allen belts.

During geomagnetic storms, the number of charged particles in the ring current increases. This decreases the geomagnetic field at Earth's surface. Such decreases are detected at sea level by ground magnetometers and are described by the so-called **Sym-H index**. It was previously supposed that the ring current was not consistently present and only existed during geomagnetic storms. But it is now evident (based on findings from the Van Allen Probes **Radiation Belt Storm Probes Ion Composition Experiment** that a persistent baseline high-energy proton current exists, and this is accompanied by a lower-energy proton current injected during geomagnetic storms. Joining the protons in the ring during geomagnetic storms are oxygen ions (O^+). This compositional change can affect the decay process of the ring current through charge exchange and wave-particle scattering losses.

In addition to the Dst and Sym-H indexes, there is another unit that measures geomagnetic storms called the **planetary geomagnetic disturbance index (Kp)**. Geomagnetic storms produce electrical currents in the magnetosphere that follow the magnetic field. These field-aligned currents connect to intense electrical currents in the auroral ionosphere. These auroral currents are called **auroral electrojets** and they can spawn large magnetic disturbances of their own. Collectively, these high-altitude currents and the magnetic changes they produce at the surface are incorporated into the Kp. The Kp in turn is the basis for the **G-Scale**, which is one of NOAA's three Space Weather Scales.

At this moment in 2024, our Sun is nearing the end of its 11-year Schwabe cycle (specifically, Solar Cycle 25). At that point, solar activity will peak—it will enter solar maximum. As part of its natural solar magnetic cycle, the solar magnetic field will flip (i.e. the north and south magnetic poles of the idealized dipole will exchange positions) and solar activity will subsequently calm down for a few years. That is, it will enter a solar minimum. The sequela of a solar maximum is a visible increase in the number of sunspots and a surge in solar activity, including more solar flares and CMEs. In other words, space weather worsens.

19.5 The G-Scale

As for radio blackouts and solar radiation storms, the USA NOAA Space Weather Scale includes a five-category **G-Scale** for geomagnetic storms. The Carrington Event was a category G5 ('extreme') storm on the G-Scale and had a Kp of 9 as the physical measure on this G-Scale. Based on the G-Scale, geomagnetic storms range from:

1. G1 (mild or minor; Kp = 5)
2. G2 (moderate; Kp = 6)
3. G3 (strong; Kp = 7)
4. G4 (severe; Kp = 8)
5. G5 (extreme; Kp = 9).

Given our far higher reliance on electronics and electricity today, a Carrington Event category G5 storm could potentially have catastrophic effects on modern society through damage to electrical power grids, destroyed communication satellites and severe disruption of radio communications. The consequent electrical blackouts could require significant time for restitution.

The last extreme (G5) geomagnetic storm was the Carrington Event in 1859 but it is estimated that such events might recur on Earth every 200–500 years or so. This estimate is based on the observation that the required solar phenomena for an extreme geomagnetic storm seem to occur at a rate of about four per solar cycle (meaning 4 days per 11-year cycle). Of course, the causative CME must have the right geometry, meaning it must be pointed straight at Earth to cause local space weather. Less severe solar storms may happen twice a century. A severe solar storm in 1989 caused a power outage in Quebec, Canada and significantly affected some of the USA power grid; it is considered the most significant storm of the Space Age. In 2022, 38 SpaceX satellites were disabled by a geomagnetic storm soon after launch.

19.6 CMEs and coronal holes

During CMEs, billions of tons of charged particles blast from the Sun's corona. The average amount of mass ejected during CMEs might be about 1.6×10^{12} kg. The velocities of the charged particles may range from 250 to 3000 km s^{-1} Thus, it might take a day or so for a CME to travel the 93 million miles (1 astronomical unit) to Earth. If sufficiently strong, solar flares and CMEs can disrupt electronics in space and here on the surface and could potentially seriously disrupt the electrical grid, thereby blacking out electrical power to many regions. Present technology cannot predict the potential impact of a CME until it strikes satellites stationed around a million miles from Earth. Depending on the velocity, this provides only 20 min to an hour for a CME to reach us after the satellites have been hit.

Curiously, the magnetic orientation of a CME influences how strongly it will interact with Earth's geomagnetic field and potentially provoke problems. However, when a CME erupts, scientists are unable to immediately ascertain its magnetic orientation. Only upon arrival at one of the orbiting satellites or space stations can we accurately predict the possible impacts to our atmosphere and surface.

Solar-derived space weather does not only originate from solar flares and CMEs. **Coronal holes**, which are cooler spots in the solar corona, are sources of intense solar winds. If a coronal hole aligns with Earth as the Sun rotates, it can send strong solar winds in our direction. Hence such phenomena are called **co-rotating interaction regions** (**CIRs**). The resulting fast solar winds may have effects on Earth and warp our magnetic fields, leading to geomagnetic storms. CIRs may result in geomagnetic storms that are not quite as intense as CME-related geomagnetic storms but may be more durable because they deposit more overall energy in Earth's magnetic field over a longer time.

19.7 The Carrington Event

The largest known geomagnetic storm was the **Carrington Event**, which occurred from 1 to 3 September 1859, during Solar Cycle 10. In addition to intense auroras that were visible at relatively low latitudes, this geomagnetic storm seriously disrupted much of the USA telegraph network, and started fires in telegraph stations and even electrically shocked some telegraph operators. The Carrington Event was associated with a particularly bright solar flare seen on 1 September 1859 by Richard Carrington and Richard Hodgson and happens to be the first formally recorded observation of a solar flare. Carrington was examining a unique patch of sunspots that were extraordinary in size when the flare caught his attention. Modern estimates of the solar flare's temperature are in the range of 50 million Kelvin, implying that it was a powerful x-class flare (i.e. one rich in x-ray and gamma ray emissions).

This solar flare was an indication of unusually strong solar activity at the time, and the Carrington Event itself was probably caused by a CME striking our magnetosphere. There was an unusually large aurora on 28 August 1959 that was seen as far south as the equator and was visible directly overhead in Cuba. This was apparently associated with a precursor CME to the main event, and might have paved the path, so to speak, by depleting the geomagnetic reserves. A much larger CME then struck us and compressed the magnetosphere from its normal 60 000 km sunward extent down to just a few thousand kilometers (or maybe down even further into the upper atmosphere). The Van Allen radiation belts were transiently eradicated, and their stored radiation contents expelled into the atmosphere. The released electrons and protons contributed to the bright red auroras.

The Carrington Event solar flare and CMEs energized solar protons to >30 MeV. Unlike most CMEs that take several days to arrive at Earth, the Carrington Event CME probably reached Earth in as short as 17–18 h. Near the geomagnetic poles (where the magnetic field offers the least protection) these high-energy protons could penetrate the atmosphere down to an altitude of 50 km (which is at the stratopause or the top of the stratosphere). One consequence of such radiation dose deposition was perhaps a 5% reduction in stratospheric ozone according to calculations by Brian C Thomas. Such ozone depletion in the stratosphere might have taken up to 4 years for full recovery. The most powerful protons (>1 GeV) could have sparked nuclear reactions with atmospheric nitrogen and oxygen atoms. These nuclear reactions released neutrons. With modern detectors, such showers of neutrons would be evident as **ground level events** (or **ground level enhancements**) but such sophisticated measuring systems were not in place in 1859. Nevertheless, the radiation would have caused ionization of the atmospheric nitrogen and oxygen molecules, leading to nitrogen oxides, including nitrates. Ice-core data from glaciers in Greenland and Antarctica document marked spikes in the concentration of trapped nitrate gases at the time of the event. Such increased nitrate concentrations correspond to increased incoming solar particles. The **nitrate anomaly of 1859** was allegedly the largest one recorded over the last 500 years. Its magnitude was roughly equal to the sum of all major events recorded over the past 4 decades (although during the last few years, there have been a few 'severe' geomagnetic storms (category G4) in Solar Cycle 25). It should be mentioned however that some studies have challenged the assertion that high levels of nitrate were truly related to the Carrington SEP event. In one study, most ice cores did not confirm a nitrate peak corresponding to 1859 and the one core in Greenland that did might be better explained by biomass burning rather than space weather.

19.8 Beryllium-10 as a space weather surrogate

Isotope analyses of ancient ice cores have revealed evidence of other significant space weather events in the distant past. Specifically, increases and decreases in the beta-emitting radioisotope, beryllium-10, offer clues about space weather incidents. Beryllium-10 has a half-life of 1.4 million years and is a cosmogenic radionuclide made by cosmic ray spallation of atmospheric nitrogen and oxygen nuclei. Various spallation paths may lead to beryllium-10, such as $14N(n, p,\alpha)^{10}Be$. Over the last several centuries, sunspot numbers (inversely) correlate well with beryllium-10 found in ice cores. This is consistent with the concept that increased solar activity fends off galactic cosmic rays (GCRs), and diminished cosmic ray influx means lower beryllium-10 concentrations. Conversely, increased cosmic ray activity increases the amount of beryllium-10.

Like beryllium-10, carbon-14 is made in the atmosphere by cosmic rays and when cosmic ray influx increases, atmospheric and biospheric carbon-14 concentrations increase. Hence, another surrogate for space weather is the carbon-14 found in tree rings. That is, **dendrochronology** (i.e. tree ring analyses) may reveal past space weather events. When tree rings confirm an unusual outbreak of space weather, the example is called a **Miyake event**. One notable Miyake event occurred around 774 or 775 and so is also termed **the 774–775 carbon-14 spike**. It is the largest and most rapid rise in carbon-14 yet recorded in the tree ring data. Dendrochronology has confirmed that an unusually large bolus of carbon-14 was dumped into the biosphere at this time and this was essentially all injected into the atmosphere instantly. Old Japanese cedar tree ring data showed an increase in carbon-14 of about 1.2% in tree rings from this time. Other analyses of ancient wood from Finland, Germany, New Zealand, Russia and the United States of America have confirmed the increased carbon-14 and proved that this was a worldwide event.

Correlating with the 774–775 dendrochronological data is an increase in beryllium-10 in Antarctic ice cores. The amount of beryllium-10 and carbon-14 by some estimates might imply a space weather event some 80-fold more powerful than the Carrington Event. But whether this Miyake event was due to an SPE like the Carrington Event, or some other kind of space weather phenomenon is presently debated. Alternative explanations include a sporadic **solar superflare**, gamma ray bursts or supernova-related cosmic ray showers. Superflares, during which as much as 10^4 times as much energy is released as a typical solar flare, are not expected from our Sun. Younger stars are more prone to superflare phenomena. Nevertheless, stars like our Sun have been seen to produce superflares in rare instances and thus it is possible that the Sun has let loose some such superflares. These alternative suggestions have arisen because of a doubtful correlation between the timing of the 774–775 event and solar maximum.

Perplexingly however, increased solar activity should repel cosmic rays and therefore decrease, rather than increase, carbon-14 and beryllium-10. In addition to the 774–775 carbon-14 spike, several other Miyake events have been recorded by proxy evidence in the tree rings and glacial ice. As chronicled in the radionuclide records, rivalling the 774–775 event was a similarly pronounced Miyake event in 7176 BCE. Recent data suggest that about 14 300 years ago an even more potent Miyake event occurred. This might have been associated with a solar storm that was 2–10 times as powerful as the Carrington event.

19.9 Cannibalistic CMEs

Space weather can be exciting and exotic. For instance, it was recently observed that a relatively inconspicuous sunspot (AR3370) suddenly erupted and fired a CME in our direction but the following day another CME was also aimed at Earth stemming from a somewhat larger sunspot (AR3363). That second CME was larger and was travelling much faster. It overtook the smaller one and formed a '**cannibal CME**' in which a bigger and quicker CME consumes a smaller and slower one on the way to Earth. Such combination CMEs may precipitate generous geomagnetic storms. Geomagnetic storms, whether initiated by CMEs or solar flares are the sources of the aurora. The aurora borealis and aurora australis are atmospheric phenomena that are covered in greater depth in other chapters. They are caused by radiation-induced ionizations of atmospheric gases when solar particle activity is high.

19.10 Solar flare categories

Solar flares are classified according to their brightness in the soft x-ray wavelengths from 1 to 8 Å (that is, 10–80 nm). Solar flares come in five classes. **X-class** flares are major x-ray flares; they may trigger planet-wide radio blackouts and cause prolonged radiation storms. **M-class** flares are medium-sized; they may produce brief radio blackouts that particularly affect polar regions. Minor radiation storms sometimes follow an M-class flare. **C-class** flares are the smallest; typically, they have no or few noticeable consequences. Originally, only classes X, M and C existed but as instruments grew more sensitive, two other categories were added (A and B), which are even lower in their soft x-ray output.

Solar flares may be quantified in peak intensity, I (measured in power per unit area or W m^{-2}) of the x-rays with wavelengths between 1 and 8 Å.

X:	$I \geqslant 10^{-4}$ W m^{-2}
M:	$10^{-5} \leqslant I < 10^{-4}$ W m^{-2}
C:	$10^{-6} \leqslant I < 10^{-5}$ W m^{-2}
B:	$10^{-7} \leqslant I < 10^{-6}$ W m^{-2}
A:	$I < 10^{-7}$ W m^{-2}

Each x-ray flare category has nine subdivisions, labeled from 1 to 9. For example, X1–X9, M1–M9 or C1–C9. Like the Richter and Fujita scales in seismology and meteorology, respectively, these are based on a logarithmic scale. (For instance, M1 is ten times stronger than C1; X1 is ten times stronger than M1; etc.) X-class flares with a peak flux exceeding 10^{-4} W m^{-2} may be assigned a numerical suffix equal to or greater than 10 (figure 19.1).

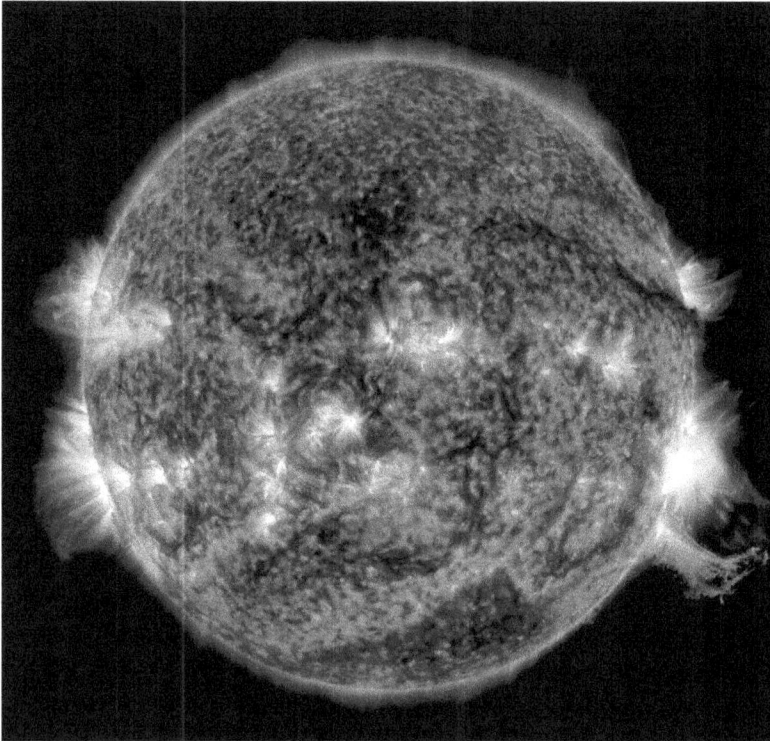

Figure 19.1. An image provided by NASA of an X1.2 class **solar flare** at the 4 o'clock position that occurred on 5 January 2023. The image was captured by NASA's **Solar Dynamics Observatory** and shows a blend of high-energy electromagnetic radiation from the 171 Å (17.1 nm) and 304 Å (30.4 nm) wavelengths. This would be in the extreme ultraviolet radiation range if one defines x-rays as beginning at wavelengths of 10 nm or 100 Å. Our atmosphere is largely opaque to wavelengths shorter than 200 nm because oxygen allotropes absorb ultraviolet with these wavelengths (https://www.nasa.gov/directorates/esdmd/hhp/space-radiation/). Credit: NASA.

19.11 Earth's atmosphere

Earth's atmosphere may be divided into regions that decrease in temperature as a function of altitude and regions that increase in temperature with height. Using this classification, our atmosphere comprises the:

1. Troposphere
2. Stratosphere
3. Mesosphere
4. Thermosphere
5. Exosphere.

For review, the **troposphere** is the lowest part of our atmosphere, and it grows colder with altitude. The average temperature at the surface is around 60 °F or 15 °C but at the top of the troposphere, the temperature falls to around −75 °F or −60 °C. The troposphere extends up to the **tropopause** which lies between 17 km at the equator and under 9 km near the polar regions. The troposphere may be estimated to average about 10 km or 6.2 miles (about 33 000 feet) in thickness at middle latitudes. Atmospheric pressure at the bottom of the troposphere is about 1 bar but decreases to about a fifth of that at the top. Thus, the barometric pressure is about 200 millibars at the tropopause. This means that about 80% of the atmosphere's mass is in the troposphere (figure 19.2).

Above the tropopause lies the **stratosphere**, which in stark contrast to the troposphere, increases in temperature with altitude. Average temperatures at the bottom and top of the stratosphere are about −75 °F (−60 °C) and 32 °F (0 °C), respectively. This curious aspect explains why the stratosphere is stratified—since warm, less dense air rises while colder, denser air sinks, a parcel of air at the bottom of the stratosphere will not rise. It will remain layered rather than turbulently rising and causing the mixing that is characteristic of the weather-containing troposphere. The stratosphere stretches up to the **stratopause**, which averages around 50 km or 31 miles above ground. The barometric pressure decreases from about 200 millibars at the tropopause to 1 millibar at the stratopause. Many commercial airplanes fly in the lower stratosphere, which puts them above most of the weather and associated turbulence. The stratosphere contains our ultraviolet-filtering **ozone layer** at heights between 15–35 km (9–22 miles) above sea level.

Above the stratopause, the **mesosphere** begins. Like the troposphere, temperatures fall with increasing altitude in the mesosphere. The mesosphere extends up to the **mesopause**, which is the coldest part of Earth's atmosphere. Temperatures at the mesopause are as low as −143 °C. Barometric pressures start at about 0.001 bar at the base of the mesopause (the stratopause) and fall to only 10^{-6} bar at the top (the mesopause). In general, the densities and pressures in the mesosphere are too thin for balloons but too thick for satellites to safely orbit in. The mesosphere is where most infalling meteors burn up and is also the home of some unusual meteorological phenomena such as red sprites and blue jets (mesospheric lightning effects).

The atmospheric layer above the 80 km (50 mile) high mesopause is the **thermosphere**. Like the stratosphere, the thermosphere temperature increases with

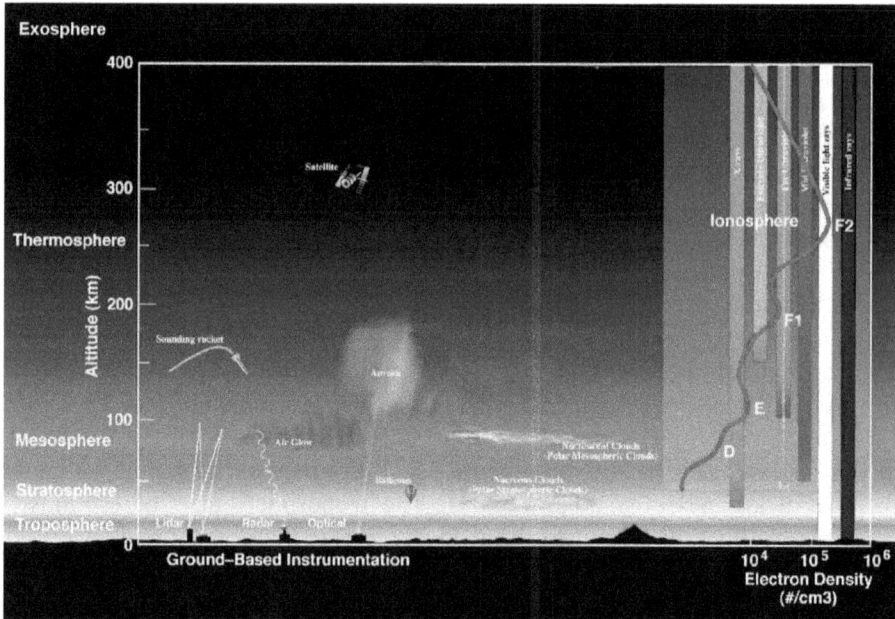

Figure 19.2. Earth's atmosphere as defined by the change in temperature with altitude. The majority of the atmosphere is in the troposphere, which decreases in temperature with altitude up to the tropopause. Above this, the stratosphere increases with temperature until the stratopause is reached at around 50 km. The lower to mid-stratosphere is the site of our ozone layer. The mesosphere then resumes the pattern of temperature dropping with altitude. It is the layer where meteors tend to burn up and the home of the highest known clouds —noctilucent clouds. Temperatures continue declining with altitude until the coldest part of the atmosphere, the mesopause, is reached. Above the mesopause, averaging around 85 km above sea level, the thermosphere begins. Like the stratosphere, the thermosphere increases in temperature as altitude increases. Finally, above the thermosphere lies the gossamer exosphere where the existing atoms and molecules no longer behave like classical gases because of the extremely low density. The Karman line, which serves as an arbitrary line that defines the start of outer space is indicated at 100 km above ground level. Other means of defining atmospheric layers exist such as the degree of atmospheric gas mixing. This breaks the atmosphere into the homosphere where gases mix homogenously and the heterosphere where gases no longer mix evenly but rather are stratified by mass. The homosphere and heterosphere are separated by the turbopause. Another way of classifying atmospheric layers is to divide them into a lower neutral atmosphere and an ionosphere. The ionosphere stretches from about 50 miles or 80 km up to around 400 miles or 640 km. It is ionized and electrified by extreme ultraviolet radiation from the Sun. The ionosphere waxes and wanes with solar input and changes significantly from day to night. The ionosphere reacts to both terrestrial weather below and solar storms from above, creating a complex space weather system of its own. The ionosphere constitutes the inner edge of the magnetosphere and plays an important role in radiocommunications. Figure 19.2a. Credit: NASA/Goddard. Figure 19.2b. This Atmosphere layers image has been obtained by the author(s) from the Wikimedia website, where it is stated to have been released into the public domain. It is included within this book on that basis. It is attributed to NOAA & User:Mysid.

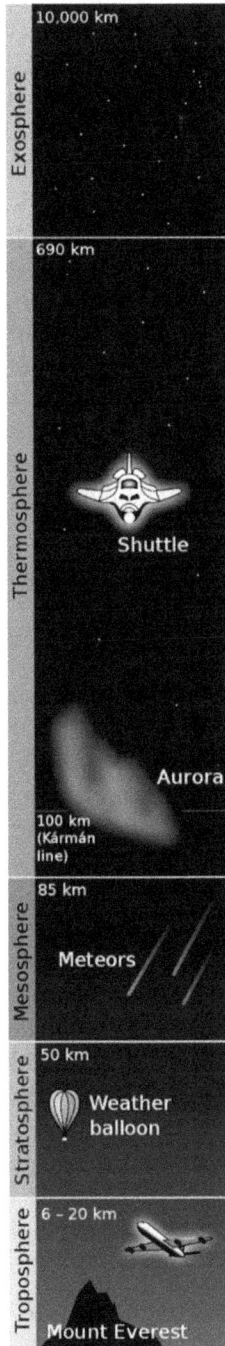

Figure 19.2. (Continued.)

altitude. Solar ultraviolet radiation causes photoionization of gas molecules, and the absorption of solar radiation increases the temperatures of individual gas molecules, atoms and ions. But despite temperatures of 2000 °C or higher, the thermosphere has a very low heat content due to the paucity of particles. Recall that atmospheric temperature is simply a reflection of the average kinetic energy of the gas or plasma particles while heat is a measure of the total energy in that parcel of air. If the air contains very few particles, it has very little heat, even if the temperatures of the individual particles are very high. In the very sparsely populated thermosphere, speeding air molecules may have to fly over a kilometer before crashing into another gas molecule. Nevertheless, this rate of collisions is still sufficient for the thermosphere to behave like a gas (in contract to the exosphere as described below). The thermosphere is where most auroras form. Also, as will be discussed below, the International Space Station (ISS) orbits within the thermosphere at an altitude of around 330–420 km or 205–260 miles. Many satellites and scientific instruments including the Hubble Space Telescope orbit in the thermosphere as well. The Hubble orbits around 545 km or 339 miles above the surface, which means it is in the upper thermosphere or lower exosphere. These orbits are in the range of what is called **low Earth orbit (LEO)**. The thermosphere extends to the **thermopause**, which separates the thermosphere from the exosphere and varies very significantly in altitude over time, season and solar activity and lies between 500 and 1000 km (311–621 miles) above sea level. The thermopause is also sometimes called the **exobase**.

Above the thermopause or exobase is the **exosphere**. The location of the bottom of the exosphere is variable (between 500 and 1000 km) and the top is ill defined. Most say that it stretches to around 10 000 km or 6200 miles beyond Earth's surface. Some argue that the edge of Earth's atmosphere is really the upper boundary of the thermosphere and does not include the exosphere. This is because particles in the exosphere do not behave like a gas. Despite being gravitationally bound to the planet, they are so rarified that they do not regularly collide; they simply encircle the Earth on ballistic trajectories. Earth's exosphere is mostly hydrogen and helium, with some heavier molecules and atoms flying about near its base. Small planets like Mercury and several moons, including our own, have exospheres despite not really having atmospheres made of particles behaving like gases. Numerous satellites orbit in the exosphere, which is still part of LEO (roughly 200–2000 km above sea level). Among these orbiting satellites is the **Iridium satellite constellation**, which is involved in mobile phone communications and consists of 66 satellites at an altitude of 780 km (484 miles). **Medium Earth orbit (MEO)** also overlaps the exosphere and begins at a higher altitude of around 2000 km or 1200 miles. MEO extends out to geosynchronous orbit which lies at an altitude of 35 786 km or 22 236 miles.

19.12 The homosphere, heterosphere and turbopause

As an alternative to using temperature changes with altitude as a means of describing layers of the atmosphere, another way is by describing the degree of mixing of gas molecules. Using this defining feature, the **homosphere** is the lower level and is where atoms and molecules collide incessantly and remain aloft due to these constant collisions. In the homosphere, these gas particles are thoroughly mixed by turbulent air motions. The layer lying above the homosphere is the **heterosphere**, where turbulence does not rule but rather, simple diffusion is the main means of gas particle interaction. In the heterosphere, the atmospheric atoms and molecules no longer remain mixed; they become layered according to atomic/molecular weight. Between the homosphere and heterosphere is the **turbopause**, which lies roughly 100 km or 65 miles above the surface. The turbopause thus lies at roughly the same level as the **mesopause** (the boundary between the mesosphere and the thermosphere) which is at about 80–90 km in altitude. Incidentally, the boundary defining the edge of space—the **Kármán line**—lies at roughly the same altitude as the mesopause and turbopause. But even this line is not perfectly constant. As defined by the Fédération Aéronautique Internationale the Kármán line lies at an altitude of 100 km (62 miles) above mean sea level, roughly matching the turbopause. But NASA and the United States Armed Forces set the Kármán line at a lower altitude of 50 miles or 80 km high, which more closely approximates the mesopause. The Kármán line is supposed to represent the imaginary boundary between Earth and outer space. From a more practical perspective, it is a line that separates the practicality of aeronautically propelled aircraft from the need for rocket-powered spacecraft. Above the Kármán line, aerodynamic lift can no longer keep an aircraft aloft. Thus, it serves as a surrogate for regulatory authority by the United States Federal Aviation Administration below the line or NASA above the line. The Kármán line is technically defined only for Earth, but curiously, despite the far more rarefied atmosphere of Mars, if one were to calculate the altitude of the Kármán line on Mars, it would lie at the same altitude of around 50 miles or 80 km. (But on Venus, it would lie at 160 miles or 250 km above the surface.)

Incidentally, at altitudes over 19 km (63 000 feet) the atmospheric pressure falls so low that water will boil at the normal temperature of the human body (98.6 °F or 37 °C). Not unexpectedly, dire health consequences or death will ensue quickly at such altitudes (and of course, at far lower altitudes as well) if the aircraft cabin is not adequately pressurized; pressurized oxygen masks cannot sustain pilots at this altitude. The pressure at this altitude is about 0.0618 atm (61.808 millibars or 6.3 kPa). This is called the **Armstrong pressure** and the altitude is called the **Armstrong limit**. It is named after Harry George Armstrong MD, the Air Force General who first recognized this limiting challenge of ultra-high aircraft flight.

19.13 Low Earth Orbit or LEO

Most of our orbiting artificial satellites are in LEO, with an altitude below about one third of Earth's 6400 km or 4000 mile radius (1 R_e). Thus, LEO refers to a band of space below an altitude of 2000 km (1200 mi). Satellites in the lower levels of LEO lie in the exosphere or thermosphere and suffer from relatively fast orbital decay. They require either periodic re-boosting to maintain a stable orbit or the launching of replacement satellites when old ones re-enter.

All manned space stations have orbited within LEO, well beneath the inner Van Allen belt. For instance, the ISS is in an LEO at an altitude of about 400 km (~250 mi) to 420 km (~260 mi) above Earth's surface while the Tiangong Space Station orbits between 340 and 450 km (210–280 miles). As such, these space stations require re-boosting a few times a year to fend off orbital decay. The Skylab launched in May 1973 and orbited in LEO at a height between 434 and 442 km (270–275 miles), making 15.4 orbits per day. It was before the time when the Space Shuttle could re-boost it (since the Space Shuttle was unavailable until 1981). Thus, the Skylab's orbit decayed, and it disintegrated in the atmosphere with pieces landing in the Pacific Ocean and Western Australia on 11 July 1979.

LEO overlaps with the exosphere and the thermosphere. Students and other readers are cautioned that there is a great deal of overlap between these variously defined atmospheric layers. For example, the exact altitude of the thermopause (top of the thermosphere) varies by the time of day, solar flux, season, etc, and can range between 500 and 1000 km (310 and 620 miles), which means the thermosphere overlaps with the ionosphere (which ranges from about 48 km (30 miles) to 965 km (600 miles)), as well as the heterosphere (the atmospheric region where molecular diffusion dominates over turbulence and the atmosphere is no longer well mixed; in the heterosphere, air is layered according to chemical composition and atomic weight). A portion of the magnetosphere dips below the thermopause and into the thermosphere as well. Thus, one can see that these differently defined atmospheric layers have substantial overlap.

19.14 MEO and the GPS

Under normal conditions, the **Global Positioning System (GPS)** (the USA part of the **Global Navigation Satellite System**), can provide positional information with an accuracy of a meter or less. They orbit in MEO at an altitude of approximately 20 200 km (12 550 miles) and each of the 32 satellites, arranged in 6 orbital planes, circle the globe twice a day. Sets of satellites that work together as a system, such as the GPS, are sometimes called **satellite constellations**.

MEO lies between 2000 and 35 786 km (1243 and 22 236 miles) above sea level.

The boundary between LEO and MEO (2000 km or 1243 miles) is arbitrary, whereas the boundary between MEO and **high Earth orbit** (**HEO**) is more objective. MEO extends up to the altitude where a geosynchronous orbit is possible.

At the altitude of 35 786 km (22 236 miles) above sea level, satellites attain a **geosynchronous orbit**, in which a satellite moving in a circular orbit takes 24 h to circle the Earth, which matches the period of Earth's own rotation (or more precisely, 23 h 56 min and 4 s). A special case of geosynchronous orbit is the **geostationary orbit**. A satellite in a geostationary orbit appears to remain in the same exact position in the sky relative to observers on the surface. Satellites in a geostationary orbit travel in a circular geosynchronous orbit *directly over Earth's equator with zero inclination and eccentricity*. Thus, the upper boundary of MEO is not arbitrary but instead corresponds to the altitude where geosynchronous and geostationary orbits are possible; objects in orbits beyond this are in high Earth orbit or HEO. All satellites in MEO have a period under 24 h, with the minimum orbital period of about 2 h for a circular orbit at the lowest MEO altitude.

Aside from geosynchronous orbits, another important orbit in MEO is the **semi-synchronous orbit**. A satellite in semi-synchronous orbit hovers at a consistent altitude of approximately 20 200 km (12 550 mi) and has an orbital period of 12 h. It will pass over the same two spots over the equator every day. This consistent and predictable semi-synchronous orbit in MEO is what is used by the GPS constellation.

19.15 The Van Allen belts

The Van Allen belts are a collection of highly energetic charged particles, gathered in place by Earth's magnetic field. Other planets with strong magnetic fields have similar planetary radiation belts. Planetary radiation belts are thus regions in space with amplified amounts of energetic electrons and protons surrounding a planet. Around our Earth, the radiation belts are called the Van Allen belts. Historically, the discovery of the Van Allen belts in 1958 by the Explorer 1 satellite is considered by many to be the scientific event that ushered in the 'Space Age'. The Van Allen belts can be described as giant donuts or toroids filled with energetic charged particles derived from outer space or the upper atmosphere. They are in the same plane as the equator and are confined to a volume which extends about 65° north and south of the celestial equator. That is, there are weak spots or low-radiation regions in the belts near the magnetic poles. Thus, they are the Van Allen radiation 'belts' rather than a Van Allen sphere of radiation (which would make escape from Earth challenging). Low radiation 'holes' in the donuts are important for astronauts as they travel beyond LEO and onwards to the Moon, Mars or elsewhere. As mentioned elsewhere, the outer Van Allen belt contains many relativistic, highly energetic electrons (>5 MeV) whereas the inner belt contains many high-energy protons. The protons in the inner belt have kinetic energies ranging from about 100 keV to over 400 MeV (figure 19.3).

Importantly, the MEO band (at altitudes of 2000–35786 km) significantly overlaps with the two **Van Allen radiation belts** (640–58 000 km). As a rough rule of thumb, the inner Van Allen belt is centered about 1.6 Earth radii (R_e) away whereas the outer Van Allen belt is centered at about 4 R_e (and recall that 1 R_e is about 6400 km or 4000 miles). But the size and shapes of the two belts can change over time and they may even merge on occasions. They can even separate into three discrete belts in rare situations. But generally, the inner belt stretches from 400 to 6000 miles above Earth's surface and the outer belt stretches from 8400 to 36 000 miles above Earth's surface. Thus, overall Earth's two main Van Allen belts extend from an altitude of about 640–58 000 km (400–36 000 mi) above the surface. Given these altitudes, not only do the Van Allen belts overlap with MEO, but they also overlap with HEO as well as the ionosphere, thermosphere/exosphere and heterosphere. The Van Allen belts can wax and wane in response to incoming solar activity, sometimes swelling up sufficiently to expose satellites in LEO to significant radiation.

On the other hand, all objects circling in MEO are not necessarily in harm's way from radiation. While it is true that any spacecraft travelling beyond LEO clearly enters the radiation zones of the Van Allen belts, there is a lower radiation region

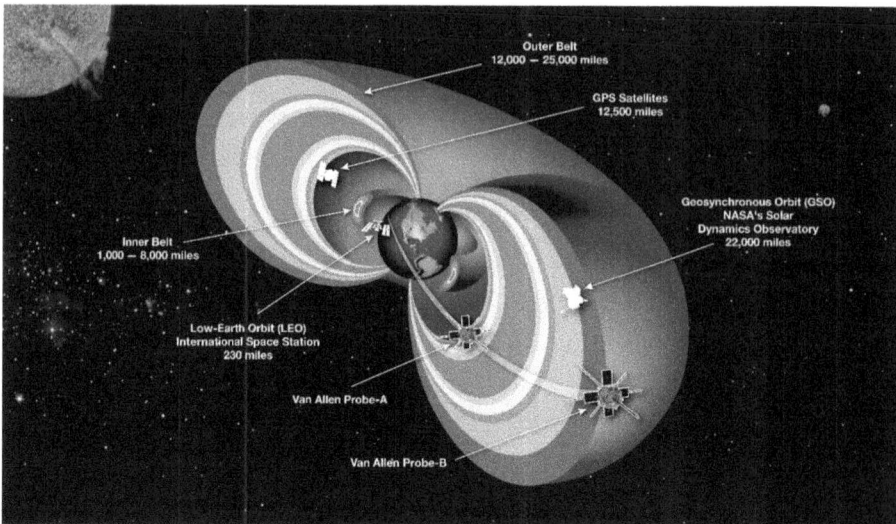

Figure 19.3. A cutaway diagram of the Earth's radiation belts (the Van Allen belts) showing some typical distances of the inner and outer belts under ordinary conditions (the distances vary depending on space weather) along with the altitudes of the ISS in LEO, the global positioning satellites (GPS constellation) in MEO and the relative location of geosynchronous orbit. NASA's Van Allen Probes (formerly known as the **Radiation Belt Storm Probes**), which were part of the Living With a Star program, provided unprecedented detail about the radiation environment. Credit: NASA.

between the inner and outer Van Allen belts that is sometimes referred to as the safe zone. This safe zone lies roughly between two and four Earth radii in altitude over the equator. Nevertheless, satellites in MEO may be perturbed by solar radiation pressure; radiation is the dominant non-gravitational perturbing force for objects orbiting in MEO. Satellites flying in MEO may experience radiation-induced damage to their electronic systems without special shielding (figure 19.4).

The Van Allen belts lie along the inner region of Earth's magnetic field or magnetosphere. Most of the particles that form the Van Allen belts originate from the solar wind with some other particles being captured cosmic rays. For the most part, the belts trap and thus are composed of energetic electrons and protons. Other atomic nuclei, such as alpha particles (helium nuclei) are less prevalent but nevertheless, there are some exotic species in the Van Allen belts including captured galactic cosmic rays (i.e. heavy nuclei) and antiprotons. These antiprotons are probably generated by interactions between incoming energetic cosmic rays and atoms in the atmosphere. The kinetic energies of these antiprotons are in the range between 60 and 750 MeV.

As mentioned, under ordinary conditions, there are two distinct Van Allen belts —an inner belt and an outer belt. There is a radiation void or safe zone between the two belts. What keeps the belts apart and why there is this low radiation vacancy in between them has been a mystery. NASA's **Van Allen Probes** which were part of the **Living With a Star program** provided data that the inner edge of the outer Van Allen belt is quite sharply demarcated. Recalling that the outer Van Allen belt is largely

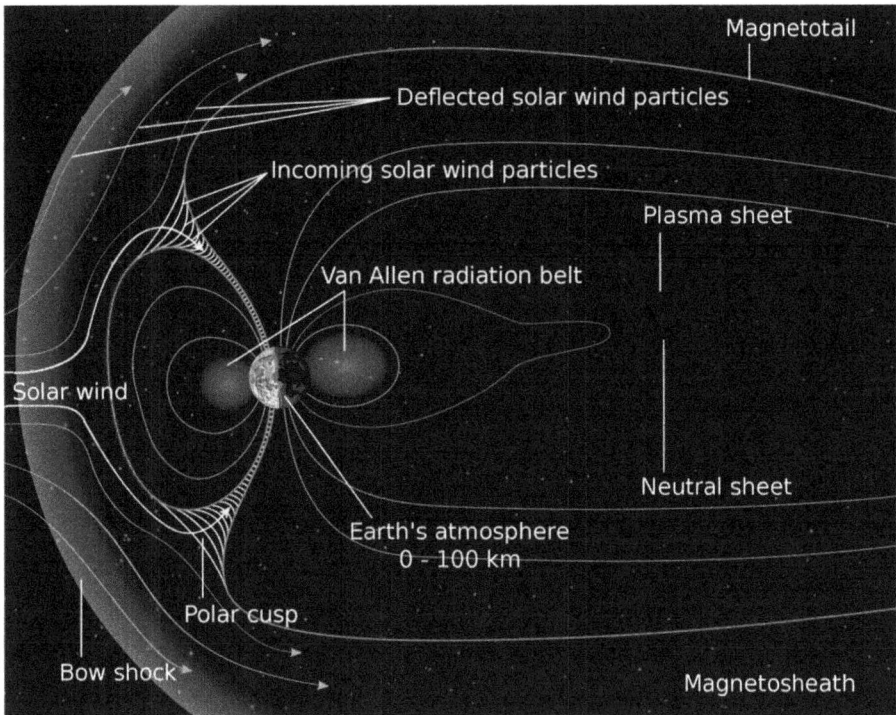

Figure 19.4. An illustration of Earth's magnetosphere and Van Allen radiation belts. The charged particles in the solar wind may be captured and accelerated by magnetic field and accumulate in the Van Allen belts like a storage ring. This structure of the magnetosphere image has been obtained by the author(s) from the Wikimedia website, where it is stated to have been released into the public domain. It is included within this book on that basis. Original: NASA. Vector: Aaron Kaase, Medium69.

electrons, it turns out that the electrons ordinarily cannot penetrate this obstacle (the inner edge of the outer belt). Hence, very-high-energy electrons can only come to within a certain distance from Earth; they are effectively blocked by this barrier. The presence of other charged particles appears to be the mechanism behind the barrier.

In addition to the Van Allen belts, a giant 'cloud' of relatively cool, charged particles occupies the Earth's exosphere and uppermost thermosphere beginning at an altitude of about 600 miles (966 km) and extending partway into the inner Van Allen belt. This cloud is called the **plasmasphere**. Charged particles at the outer edge of the plasmasphere cause particles in the outer Van Allen radiation belt to scatter, thereby removing them from the belt. This may seem surprising given that the outer radiation belt electrons are extremely energetic (>5 MeV) and move at great velocities. But they travel in immense rings around Earth and the Van Allen Probes have found that despite their enormous energies and overall velocities, in the vector normal to Earth's surface (that is, in direction directly toward Earth), these energetic electrons have very low velocity. That is, they are moving tangentially and have relatively little vertical drift towards Earth. This slow and weak downward drift is easily rebuffed by the scattering force of the plasmasphere.

However, when space weather is severe (e.g. during a powerful CME or solar flare) and vast droves of charged particles are jetted our way, the plasmasphere can be overpowered and electrons from the outer Van Allen belt may be shoved into the normally vacant slot between the belts. In addition, an exceptionally strong solar wind can cause the plasmasphere's boundary to be blown inward. Severe space weather can essentially erode the outer plasmasphere, push its boundaries inward and allow electrons from the Van Allen belts to move inward as well.

19.16 Astronauts and the Van Allen belts

Spacecraft orbiting or travelling beyond LEO may enter the high radiation zones of the Van Allen belts. A region of reduced radiation between the inner and outer Van Allen belts lies roughly at two to four Earth radii and is sometimes referred to as the 'safe zone'. Beyond the Van Allen belts, space travelers and their electronic equipment face additional radiation hazards from GCRs and SPEs.

Solar cells, integrated circuits and sensors can be damaged by this radiation. Geomagnetic storms occasionally damage electronic components on spacecraft. Miniaturization and digitization of electronics and logic circuits have made satellites more vulnerable to radiation, as the total electric charge in these circuits is now small enough so as to be comparable with the charge of incoming ions. Thus, electronics on spacecrafts and satellites must be 'hardened' against radiation to operate reliably. Hardening may be achieved through many means aside from physically toughening the equipment. As mentioned in another chapter, **triple mode redundancy** may be implemented in which case, one or several chips will be required to perform the same calculation in triplicate. Then a 'vote' is taken to ascertain the most likely answer. This and similar methods can reduce radiation-induced errors and disruptions. As an example of another radiation safety measure, the Hubble Space Telescope, among other satellites, will often turn their sensors off when passing through regions of intense radiation.

Nevertheless, when passing through high radiation regions such as the Van Allen belts, significant radiation exposure becomes inevitable. For instance, a satellite shielded by 3 mm of aluminum orbiting in a highly elliptical orbit (200 by 20 000 miles or 320 by 32 190 km perigee and apogee respectively) passing through the radiation belts will receive about 2500 cGy or rads (25 Gy) per year. (For comparison, a full-body dose of 5 Gy or 5 Sv is likely to be lethal to a human.) The majority of the radiation will be received while passing through the inner belt. The Apollo missions marked the first time when humans traveled through the Van Allen belts. As this was an anticipated hazard, the Apollo spacecrafts planned for this by minimizing the amount of time spent flying through the belts. Thus, the astronauts received low radiation doses from the Van Allen belts thanks to the brief period of time spent flying through them. The Apollo astronauts' overall dose was predominately due to energetic solar particles encountered outside of Earth's magnetic field. The total radiation received by the astronauts varied from mission to mission but was measured to be between 0.16 and 1.14 cGy or rads (1.6–11.4 mGy). Note that this was much less than the presently established standard of 5000 mrem (5 rem or 50 mSv) per year set by the United States Nuclear Regulatory Commission for people who work with radiation. In comparison, a CT scan of the chest, abdomen and pelvis might confer a radiation dose of around 7–15 mSv or 7–15 mGy (0.7–1.5 cGy). The upper end of this range exceeds the highest estimates of the radiation dose received by the Apollo astronauts. And the dose from a PET/CT scan for cancer detection might be as high as 23–32 mSv or 2.3 to 3.2 cGy, which is far higher still.

19.17 The ionosphere, solar flares and radio communications

The ionosphere is an electrically charged (ionized) layer of the upper atmosphere that ranges roughly from about 50–965 km (30–600 miles) but becomes most distinct and important as an entity above 80 km or roughly 50 miles above sea level. Thus, it can include parts of the mesosphere, thermosphere and exosphere. The uncharged atmosphere below the ionosphere may be alluded to as the **neutrosphere** or neutral atmosphere.

The ionosphere is created primarily by ionization caused by ultraviolet radiation, but of course higher-energy solar x-rays and gamma rays can contribute to the ionization. The density of the ionosphere is so low that ionization can lead to free electrons that persist for a prolonged period before being captured by cations and undergoing **recombination** back into atoms. Since the ionosphere is partly ionized and contains positive ions and electrons, it is by definition a plasma. And key ionospheric plasma parameters are electron density, ion composition, and electron/ion temperature.

This plasma is quite relevant to us since much of our modern technology is affected by the ionosphere or even directly exploited for communications. The ionosphere plays a major role in communications. For example, GPS signals travel through the ionosphere and alterations of the density and composition of the ionosphere may affect these signals. More directly, certain radio communications work by bouncing back and forth off the ionosphere to reach their destinations. Again, changes in the ionosphere can significantly disrupt these radio communications signals. **Skywave propagation** is a type of radio communication that relies on the reflection of radio waves off the ionosphere and back to Earth.

Because the ionization process is caused by short-wavelength (high-energy) photons, and because the solar output of high-energy photons varies with the solar cycle, the more active the Sun is, the greater the degree of ionization. Thus, ionization is far higher during the day but the ionosphere thins out at night as ions and electrons recombine back into neutral atoms and molecules, thereby improving AM radio reception at night. The ionization in the ionosphere also varies over the 11-year Schwabe cycle. Thanks to the inclination of Earth's spin axis, there is a slight seasonal variation as well such that ionization is higher in the summer months.

The ionosphere is divided into layers which, from bottom to top are:
1. The D layer
2. The E layer (Kennelly–Heaviside layer)
3. The F layer (Appleton–Barnett layer).

Because they are made by solar activity, these layers of ionization differ between day and night. The **D layer** is largely present only during the day. The D layer lies between 48 km (30 miles) and 90 km (56 miles) above sea level. The ionization in the D layer is largely due to the ionization of nitric oxide by ultraviolet, although x-rays made during solar flares may ionize nitrogen and oxygen molecules and at times also contribute to the D layer. At night, the D layer practically vanishes; only a trace remains thanks to cosmic ray activity. On the other hand, during intense solar proton events, the D layer ionization can be magnified significantly, especially in the polar regions. These unusual events are called **PCA events**.

The next layer up is the **E layer** or **Kennelly–Heaviside layer**, which extends from 90 to 150 km (56–93 miles). The ionization here is mostly because of soft x-rays and far ultraviolet radiation on diatomic molecular oxygen. As in the D layer, the ionization here is mostly molecular (resulting from the ionization of neutral nitrogen and oxygen molecules). But in contrast to the D layer, the ionization of the E layer persists through the night (albeit markedly diminished). Historically, it was the E layer of the ionosphere region which was responsible for the reflections involved in Guglielmo Marconi's original wireless transatlantic radio communication in 1902.

The **F layer** or **Appleton–Barnett layer** is the outermost region of the ionosphere. It spans from about 150 km to >500 km (93 miles to beyond 310 miles). Due to intense ionization, this layer has the highest concentration of free electrons and electron density; the F layer is the most important in terms of radio communications. Most of the ionization here is thanks to extreme ultraviolet radiation (10–100 nm wavelengths) ionizing atomic oxygen. Although there is still a trace of an E layer, at night the F layer is the most significant ionosphere layer. At night, the F layer ordinarily consists of just one layer (**F2**) but during the daytime, a smaller inner layer (**F1**) often appears. It is the more highly ionized F2 layer that plays the biggest role in **skywave** radio propagation and **shortwave** (HF) radio communications. Some authorities allude to only the F2 layer as the Appleton layer.

Above the F layer is the **topside ionosphere**, which extends up to the **plasmasphere**. This portion of the ionosphere contains less oxygen, but more hydrogen and helium ions are present. The uppermost ionosphere contains significant amounts of ionized hydrogen (that is, protons and free electrons); for this reason, the outermost ionosphere is sometimes called the **protonosphere**.

Airglow is the visible glow of the upper atmosphere visible from space. It can occur when positively charged ions in the ionosphere recombine with free electrons to recreate neutral atoms and molecules. Airglow may also occur during the initial ionization and excitation processes. In both circumstances, photons are emitted during the relaxation process and give rise to the airglow phenomenon.

Because the ionosphere is composed of charged particles, it is highly responsive to changes in the magnetic and electric environment around it. Thus, space weather (like solar-charged particle events) may strongly affect our ionosphere.

Ordinary but severe regular weather (i.e. not space weather), such as hurricanes and large thunderstorm systems, can create pressure waves that affect the ionosphere. Incidentally, **tropical cyclones** are powerful, low-pressure, circular storms that originate over warm tropical ocean waters with winds exceeding 119 km (74 miles) per hour. They are called **hurricanes** if they occur in the North Atlantic or Eastern Pacific and are called **typhoons** when they occur in the Western Pacific or Northern Indian Ocean. Cyclones bring bad, wet and windy weather whereas **anticyclones** generally bring dry, fair weather. Anticyclones are high-pressure systems that rotate in the opposite direction as their cyclone counterparts would in the same hemisphere. For instance, a cyclone in the Northern Hemisphere will rotate counterclockwise while an anticyclone in the Northern Hemisphere spins clockwise. Incidentally, tropical cyclones and severe thunderstorms may be associated with extreme lightning and such lightning has been found to spawn gamma rays and x-rays. Thus, airline passengers and flight crew are exposed to additional natural background radiation dose in the skies thanks to the radiation generated by lightning in our troposphere during severe thunderstorms.

The **Global-scale Observations of the Limb and Disk (GOLD)** instrument and the **Ionospheric Connection Explorer (ICON)** spacecraft provide comprehensive observations of our ionosphere. ICON is in LEO near the equator at around 360 miles (~580 km) altitude whereas GOLD is in geostationary orbit at 22 000 miles (35 406 km) altitude over the Western Hemisphere. Also in geostationary orbit is the GPS.

GPS radio signals travel from a satellite in MEO to a receiver on the ground, passing through the ionosphere. The charged plasma of the ionosphere refracts the radio signal similar to the way a lens bends the path of visible light. In the absence of significant space weather, GPSs can effectively compensate for the ionospherically induced distortion. This is accomplished using a model of the ionosphere, called the **Klobuchar model**, which calculates and cancels the positioning error introduced by the ionosphere when single-frequency GPS receivers are used. However, when the ionosphere is disturbed by severe space weather, the Klobuchar model is no longer sufficiently accurate, and the receivers are unable to calculate a precise and accurate position.

19.18 Total electron count

Geomagnetic storms and other space weather can lead to disturbances in the ionosphere. The injected energy and electrical currents going into the ionosphere by geomagnetic storms augment and intensify the ionosphere. The result is an increase in the height-integrated sum of ionospheric electrons, which is measured by the **Total Electron Count (TEC)**.

The TEC is the sum of all electrons present along a path between a radio transmitter and receiver. Radio waves are affected by the presence of electrons. Thus, the more electrons in the path of the radio wave, the more the radio signal will be affected. For ground to satellite communication and satellite navigation, TEC is a good parameter for monitoring and estimating possible space weather impacts.

TEC is measured in electrons per square meter. By convention, 1 TEC unit (TECU) = 10^{16} electrons m^{-2}. Vertical TEC values in Earth's ionosphere typically range from a few to several hundred TECU.

19.19 Sunspot number

The solar cycle is most readily ascertained by counting sunspots at a given time. Solar maximum ('solar max') is the time when the Sun is most active in terms of solar flares, prominences, CMEs and the strength of the solar wind. It is also the time when sunspots are most abundant. But counting sunspots might seem to be a whimsically subjective endeavor that could vary substantially from place to place, time to time and from telescope to telescope. Therefore, some objectivity has been introduced with a metric called the **sunspot number**.

There are two official sunspot numbers in common use. One is the daily '**Boulder Sunspot Number**', which is computed by the NOAA Space Environment Center using a formula devised by Rudolph Wolf in 1848:

$$R = k(10g + s)$$

where R is the **sunspot number**, g is the number of sunspot groups on the solar disk, s is the total number of individual spots in all the groups and k is a variable scaling factor (usually <1) that accounts for observing conditions and telescope type.

Scientists combine data from multiple observatories—each with their own k factor—to arrive at a daily sunspot number.

The Boulder sunspot number is usually about 25% higher than the second official sunspot index, the **International Sunspot Number**. Both the Boulder and the International numbers are calculated from the same basic formula, but they incorporate data from different observatories. As a rule, if one divides either of the two official sunspot numbers by 15, one will obtain the approximate number of individual sunspots visible on the solar disk (counted by projecting the Sun's image on a paper plate with a home telescope for example).

19.20 Space weather and the thermosphere

Space weather definitely affects our atmosphere, and it may indirectly affect our weather and climate. For example, the thermosphere shrinks and cools off during solar minimum and it then expands and warms up during solar maximum. The thermosphere is the atmospheric layer in between the mesosphere and the exosphere. Thus, it begins at an altitude of about 80 km (50 miles) above the Earth's surface and, at times, it may reach an altitude of ~750 km (or 466 miles). Importantly, when the thermosphere reheats and swells during significant solar activity, this expands the overall extent of Earth's atmosphere. This atmospheric expansion can lead to increased aerodynamic drag on satellites in LEO. Parenthetically, there is no official definition of LEO, but it usually pertains to orbits between 200–2000 km (or 124–1240 miles) above Earth. Expansion of the thermosphere, leading to drag may shorten satellite lifespans and cause them to tumble down early. The thermospheric expansion may be measured using the '**Thermosphere Climate Index**' (**TCI**). This system indicates how much heat energy is being emitted into space from carbon dioxide (CO_2) and nitrogen oxide (NO) molecules in the thermosphere. Basically, it is a 60-day average that provides a measure of the temperature of the top of the atmosphere (in watts). During solar maximum, the TCI is high and during solar minimum, the TCI is low.

The TCI is based on measurements from the **Sounding of the Atmosphere using Broadband Emission Radiometry (SABER) instrument** which is onboard **NASA's Thermosphere, Ionosphere, Mesosphere, Energetics and Dynamics (TIMED) spacecraft**, which launched in 2001.

For those interested in statistics, the present records are:

$$\text{Maximum TCI: } 49.4 \times 10^{10} \text{ W Hot (10/1957)}$$

$$\text{Minimum TCI: } 2.05 \times 10^{10} \text{ W Cold (02/2009)}$$

Recall that the **ionosphere** is the region of the atmosphere that contains a high concentration of ions and free electrons, which enables it to reflect radio waves. It (usually) lies above the mesosphere and extends from about 50 to 600 miles (80–1000 km) above Earth's surface. Note that the thermosphere and ionosphere overlap. This is not surprising since they are layers of the atmosphere defined by different parameters.

19.21 Forbush decreases

There is a yin–yang relationship between solar activity and incoming galactic cosmic rays (GCRs). During heightened solar activity the solar wind grows stronger, and more solar flares and CMEs blast off the Sun; all of this repels GCRs. Thus, in a loose sense, intense solar winds blow galactic cosmic rays out of the inner solar system. On top of the solar wind, periodic CMEs also clear cosmic rays, leading to noticeable reductions in cosmic ray counts called 'Forbush decreases'.

To qualify as a Forbush decrease, the reduction in GCRs must be at least 10% on the surface, but other definitions range from about 3% to 20%. But even larger decreases of >30% have been recorded aboard the ISS. As discussed below, Forbush decreases have been indirectly linked to various adverse health effects. The mechanism behind the putative association is presently unclear.

GCRs can alter the electrochemistry of the atmosphere of Earth's upper atmosphere and might possibly affect our climate through the **Svensmark effect** (as discussed elsewhere). The Svensmark hypothesis basically states that diminished solar activity means diminished ability to fend off incoming GCRs; since cosmic rays facilitate cloud formation, the climate cools thanks to this increased cloud cover. Hence, some speculate that the Little Ice Age might have been mechanistically connected to the Maunder Minimum and other grand solar minima during that time.

Additionally, GCRs may trigger lightning and they can penetrate commercial airplanes. According to at least one study, aircraft crew members tend to have higher rates of cancer than the general population. However, there are many confounding variables in such studies such as irregular sleep habits and chemical contaminants among other cancer risk factors. Of course, for astronauts beyond our atmosphere, the intensity of cosmic rays is magnified further. It has been suggested that excessive cosmic ray exposure poses a health hazard to astronauts and frequent polar air travelers. Therefore, understanding how solar activity and cosmic rays are interrelated is important.

And as mentioned, heavy lightning, as seen during severe thunderstorms, can generate low atmospheric x-rays and gamma rays that contribute to our natural background radiation budgets. If it is true that our climate and weather may be affected by a chronic deficit of solar wind and activity through the Svensmark effect, then severe thunderstorms—and the associated dose from lighting-generated gamma rays—might increase during solar quiet periods. Conversely, through Forbush decreases, increased solar activity might clear the skies of cosmic ray-generated clouds and thereby decrease thunderstorm and lightning-linked gamma rays. Much of this remains hypothetical at the time of this writing but is a fascinating aspect of space weather and its impact on our regular weather and our natural background radiation doses.

19.21.1 Neutron counters

When cosmic rays strike our atmosphere, through spallation, they create a spray of secondary particles that rain down on Earth. Among these particles or secondary cosmic rays, are neutrons. As discussed elsewhere, these neutrons can interact with the most abundant atoms in our atmosphere (nitrogen-14) and convert them into carbon-14 atoms through $^{14}N(n,p)^{14}C$ reactions. The newly formed radiocarbon can become part of carbon dioxide molecules which then are taken up by plants, algae and cyanobacteria via photosynthesis and get incorporated into the biosphere. Thus, all living organisms are in a state of equilibrium wherein uptake of carbon-14 is balanced by losses in a dynamic steady state. But upon death of any organism, the uptake of carbon-14 ceases and radioactive decay then dictates the amount of radiocarbon present in the remains. Thus, radiocarbon dating allows one to calculate the time of death of the biological organism that left behind the organic matter being examined.

Nevertheless, some cosmic ray-manufactured neutrons can make it all the way down to Earth's surface. The **Sodankyla Geophysical Observatory** in Oulu, Finland, has been measuring secondary cosmic ray neutrons since 1964. Thanks to this and similarly collected data, we have a good grasp of how solar activity modulates the (presumably) steady incoming stream of GCRs. Although it is assumed in radiocarbon dating that the rate of carbon-14 production in the atmosphere (and therefore the amount of radioactive carbon dioxide that gets into the biosphere) is constant, there is some variability. This variability is most pronounced during unusual space weather phenomena such as Miyake events and Forbush decreases.

19.22 The Chapman–Ferraro distance

The **Chapman–Ferraro distance** of a given planet with a magnetic field is that distance at which the planet's magnetosphere precisely balances the solar wind pressure. Beyond this distance, space radiation is unmitigated by either a planet's atmosphere or by magnetic deflection. Mars has no magnetic field to boast of and has a very thin atmosphere. Therefore, it is no surprise that radiation levels on the surface of Mars are far different from those on Earth.

19.23 High natural background radiation regions

Based on data collected by the **Radiation Assessment Detector Instrument** on the Mars Science Laboratory, for a 180-day transit to Mars or a 500-day stay on Mars, the estimated amount of radiation would be a bit over 300 mSv. This was deemed unacceptably hazardous by some when the data first arrived. Recent analyses however suggest that the average doses on Mars may not be incompatible with survival or even long-term health. Given the apparently normal health, longevity and cancer rates of people residing in high natural background radiation regions in certain areas such as **Ramsar**, Iran; Karunagappally in the State of **Kerala**, India; some regions of **Yangjiang** County (Guangdong province) in China; the Minas Gerais State and Guarapari in Brazil, doses on the surface of Mars may not be prohibitive. Unshielded Martian doses are not extremely different from such natural high background regions here on Earth. For example, there are some regions in Ramsar where natural background radiation reaches 260 mSv (26 rem) per year. This is comparable to the estimated 240–300 mSv per year on the surface of Mars. Note that the International Commission on Radiological Protection (ICRP)-recommended limit for occupational radiation exposure is 50 mSv per year and a lifetime total of no more than 1000 mSv or 1 Sv. One inhabited domicile in Ramsar was found to have an annual effective dose of 131 mSv (13.1 rem) from exposure and an annual internal committed dose of 72 mSv (7.2 rem). The health consequences of residing in such regions is an area of active study. Most studies have not demonstrated any increased rates of cancer, premature aging or early childhood deaths in areas with elevated natural background radiation.

While proper shielding will reduce the dose, nevertheless, in interplanetary space, with the complete absence of an attenuating atmosphere and no magnetosphere, powerful and unpredictable SPEs could be more dangerous than on Mars itself; such acute doses might even be fatal. It is believed that the Solar Storm of August 1972 could have been life-threatening had astronauts been in transit between Earth and the Moon at that moment. As that unusually strong solar storm was very near the time of Apollo 17 (7–9 December 1972), a radiological disaster was barely averted.

Incidentally, the radiobiological effects of space radiation are still largely uncertain. It is known that equal doses of different types of radiation yield different biological outcomes. These differences are quantified by the relative biological effectiveness (RBE) of the radiation in question. High linear energy transfer (LET) radiation such as fast neutrons, low- to moderate-energy protons, alpha particles and high-Z, high-LET galactic cosmic rays (HZEs) are generally considered high-RBE radiation types but students and readers must remember that RBE is not just a function of radiation type ('quality') but also total dose, dose rate, fractional dose (if the total dose is broken up into separate fractions) and most importantly the biological endpoint in question. RBE is the ratio of doses needed to achieve the same biological endpoint (e.g. cancer cell killing, induction of cataracts, etc) using one type of radiation versus a standard (typically 250 kVp x-rays or cobalt-60 gamma rays). That is:

RBE = (dose of a standard radiation type to produce a given biological effect)/
(dose of the radiation in question to produce the same biological effect)

or

$$RBE = D_X / D_R$$

where D_X is the dose needed in 250 kVp x-rays and D_R is the dose needed of the radiation in question.

Another way of thinking of this is that the RBE is the ratio of biological effectiveness of one type of radiation in achieving a given outcome compared to a standard form of radiation given the same amount of absorbed energy.

RBE may vary for the same particle type, depending on factors such as energy, dose rate, target organ and other factors, most notably LET. For example, there is a huge difference between the high-energy protons used in proton beam radiation therapy and the low-energy protons set in motion by neutron radiation exposure. Generally, for radiation-induced cell death, RBE reaches a maximum at a particle LET of $100 \text{ keV } \mu m^{-1}$. Beyond that LET, the RBE diminishes. But this is not because the higher LET radiation is not as lethal; it is simply because at higher LET values the total energy deposited by very high LET radiation is also very high. RBE is defined as the ratio of doses (energy deposited in a given mass), and very high LET radiation has no choice but to deposit inordinate amounts of energy, which increases the deposited dose (which is the denominator in the defining equation). In essence, very high LET radiation is 'overkill' and this reduces RBE because of its definition as a ratio of two doses. (One can deposit an indefinite amount of energy in a cell but that cell can only be killed once.)

The amount of radiation deposited per unit track length by high-LET radiation may be very high but the RBE, if using cell death as the biological endpoint, decreases with increasing LET. The extra energy above and beyond that needed to kill the cell just serves to decrease the ratio that defines RBE. But the dead cell is still dead. No more, no less.

19.24 Weighting factors

In radiation protection calculations, different particle types are assigned different **radiation weighting factors** (W_R, formerly called **quality factor**, Q) that are supposed to represent an average of calculated relative biological effectiveness or RBE values for the particle in question. To identify the relative biological risk of a certain type and dose of radiation, the physical dose (in Gy) is multiplied by W_R to estimate the so-called **equivalent dose** in units of Sieverts (Sv). For the most part, radiation weighting factors (and tissue weighting factors as discussed below) are used for the calculation of stochastic effects (e.g. cancer induction). Presently, different regulatory bodies (e.g. the United States Nuclear Regulatory Commission and the ICRP) use slightly different numerical values for W_R when dealing with neutrons. For gamma photons, x-rays, muons and electrons W_R is taken to equal 1; for protons and charged pions $W_R = 2$; and for alpha particles, fission products and heavy charged ions $W_R = 20$. For neutrons however, the ICRP uses complicated formulas that vary with neutron energies. As an example, for neutrons with kinetic energies between 1 and 50 MeV, the formula is:

$$W_R = 5.0 + \{17e^{-[ln(2E)]2/6}\}.$$

As mentioned, RBE is a function of LET, which in turn is a function of energy. And absorbed dose is a function of LET as well since LET is the equivalent of dE/dx. Given the complexities associated with estimating proper values for W_R, some radiation scientists recommend reporting physical absorbed dose in Gy rather than the calculated (estimated) biological dose in Sv.

Further complicating the estimates of radiobiological risks, for radiation protection purposes, the human body is divided into several sets of different tissues, and each tissue is assigned a **tissue weighting factor** (W_T), which is intended to reflect its estimated sensitivity (in this case, the predicted risk of developing cancer). Thus, one multiplies the physical absorbed radiation dose by a radiation weighting factor to obtain the equivalent dose and then multiplies the equivalent dose by a tissue weighting factor to finally obtain the **effective dose**.

The **effective dose** (E) is a summation over radiation type and tissue type using the various radiation W_R and W_R values:

$$E = \sum W_T \times H$$

where D is dose, $H = W_R \times D$ and W_T is obtained from tabulated data.

Some representative examples of W_T are:

$W_T = 0.12$ for lung, stomach, colon and red bone marrow (hematopoietic marrow that manufactures blood cells).

$W_T = 0.05$ for adrenals, pancreas, kidney, muscle, esophagus, breast, bladder, brain, liver and thyroid.

$W_T = 0.01$ for skin and bone surfaces.

19.25 The interplanetary radiation environment

As mentioned repeatedly, the fluence of HZEs in the interplanetary space of our Solar System varies inversely with the solar cycle. Typical dose rates range from 50 to 100 mGy yr^{-1} during solar maximum, to 150–300 mGy yr^{-1} at solar minimum. In contrast to these steady and predictable backgrounds, the fluence and frequency of sporadic SPEs are highly erratic. But such SPEs tend to occur far more frequently during solar maximum and may yield dangerously high dose rates reaching up to 1400–2837 mGy h^{-1}. Depending on the duration and specific dose rates, unprotected astronauts receiving such total body doses could be at risk of developing acute radiation syndrome.

The shielding requirements for HZE GCR particles are quite different from the shielding requirements for solar protons. For instance, even if spacecraft shielding were effective at reducing radiation from SPEs, the spallation byproducts made by GCRs as they crash into shielding material atoms may lead to a significant dose. Additionally, the kinetic energies of some GCRs may reach into the thousands or even millions of GeV whereas solar protons are mainly in the MeV range. Thus, spacecraft shielding must take into account the extraordinary kinetic energies of GCRs in interplanetary space.

Assuming present standards of shielding and normal background fluences, during space travel beyond LEO, on average every nucleus in every cell within an astronaut's body would be traversed by a proton or a delta ray electron every few days. Every nucleus in every cell would be traversed by an HZE GCR every few months. Despite their relative infrequency, HZEs are believed to contribute substantially to the total biological dose received by astronauts venturing beyond LEO. This is partly because of the high LET (dE/dx) and associated RBE of these particles and partly because these HZEs are just so energetic that they can penetrate most shielding. Although thicker shielding could provide protection, this option is limited by practicality in terms of mass and volume requirements.

Present versions of lightweight aluminum shielding greater than 20–30 g cm^{-2} would only reduce the GCR effective dose rate by about 25% or less, and an equivalent amount of polyethylene would only decrease the GCR dose by about 35%. Although this degree of shielding has been achieved on the ISS in LEO, a similar degree of shielding is impractical for interplanetary explorations due to the bulk of such shielding and the present limitations of launch lifts.

The spacecrafts of the Apollo Mission were the only ones that transported astronauts beyond LEO. The average shielding of the Apollo command modules was only 6.15 g cm^{-2}. Thus, these spacecrafts were only effectively shielded against solar protons with energies of $\leqslant$75 MeV. However, while SPEs consist mainly of protons with energies in the tens of MeV, they may occasionally contain protons reaching kinetic energies reaching the GeV range and the HZEs of GCRs routinely possess energies in this range and beyond. Current spacecraft shielding cannot effectively block such energetic particles. And of course, the dose rates associated with intense SPEs are orders of magnitude higher than baseline rates. One particularly powerful particle event in October 1989 would have delivered dose rates as high as 1454 mGy h^{-1} to an astronaut traveling through interplanetary space. (Recall that the average (effective) daily dose for astronauts onboard the ISS is approximately 0.5–1.0 mSv or an absorbed dose of about 0.282 mGy d^{-1}.) Additionally, 10%–15% of the protons in this particular SPE had energies exceeding 100 MeV.

In addition to the potential for causing acute radiation syndrome (which is covered in chapter 23), high acute SPE doses may subsequently cause chronic and degenerative effects such as ocular damage (e.g. cataracts), respiratory and digestive diseases, neurological impairment and damage to the microvasculature in years to come. Of course, at such high doses, cancer is another possible late complication. Although these effects usually have a long latency period (meaning they may not appear for months or many years after exposure) and do not pose an immediate risk to crew health or their ability to carry out the mission, such late effects of acute, high-dose exposures remain important when planning missions.

19.26 Limitations of terrestrial analogs

Accurate estimates of the biological impact of space radiation are extremely challenging and the use of terrestrial radiation sources in lab animals is a poor surrogate. The molecular damage caused by space radiation is quite different from the damage associated with the artificial laboratory irradiations that are occasionally used as a surrogate in space radiobiology studies. First, high LET charged particle radiation, including GCRs and SEPs, primarily causes **direct radiation effects** wherein the molecular biological effects result from interactions between the particle and the struck molecules. As charged particles continuously lose energy through interactions, each energy loss event may result in damage to the biological tissue being traversed. However, because of limited resources, terrestrial analogs often use low-LET radiation such as gamma rays that cause **indirect radiation effects**. Low-LET radiation causes the formation of free radicals, free electrons and reactive oxygen/nitrogen species that ultimately cause the biological damage. It is difficult to derive meaningful conclusions from lab experiments using very different types of radiation like this.

Additionally, the terrestrial experiments often use highly inbred and immuno-logically compromised lab rodents. These animal models may not even be able to exhibit the same immunologically mediated responses to space radiation that humans do, making them questionable surrogates for healthy human astronauts exposed to space radiation. Of course, this scientific criticism must be considered in the context of reality—actual experiments in space with suitable surrogate animals are prohibitively expensive, and there are very few facilities here at home that can deliver truly space-like radiation for experiments.

It has been suggested that close study of humans residing in high natural background regions could be very informative, especially if the high natural background radiation is high-LET. Regions in Ramsar, Iran provide high back-ground radiation levels that reach an astounding 260 mSv yr^{-1} and much of this radiation is from alpha emitters, implying that it is high-LET and high-RBE, much like space radiation is. This effective dose rate is actually quite comparable to the estimated *effective* dose rate of 0.7 mSv d^{-1} that might be expected on the Martian surface (recalling that the *absorbed* dose on Mars is only 76 mGy yr^{-1} based on data from RAD on the Curiosity Mars rover).

It is important to keep track of whether reports are providing absorbed radiation doses or effective doses. In terms of equivalent and effective dose, since much of the radiation there is due to alpha emitters, some extreme areas in Ramsar have effective dose rates up to an astonishing 20 mSv h^{-1} at a level about 1 m above ground. Although most people do not live in such extremely high background dose regions, approximately 2000 people do live in a relatively high background region. In the extreme, some people might theoretically be receiving annual radiation exposures nearly an order of magnitude higher than the 20–50 mSv yr^{-1} maximum recommended limits for radiation workers. Studies thus far have not demonstrated any negative major health consequen-ces to such high and high-LET chronic radiation exposure. In fact, some studies have shown considerably fewer chromosomal aberrations in lymphocytes of individuals from Ramsar after a challenge dose of 1.5 Gy compared to lymphocytes from people from typical background regions similarly exposed to a dose of 1.5 Gy. This could indicate a chronic **adaptive response** thanks to the radiation exposure, or it could be a genetic phenomenon in which the people of Ramsar, after many generations of living there have evolved an inherent enhanced ability to cope with ionizing radiation. This is an area of active study since if it is the latter phenomenon, it would imply that astronauts embarking on long-term, high-dose missions might best come from regions of high natural background radiation because people inhabiting such regions might be naturally radiation resistant. If, on the other hand, the former phenomenon is the explanation, it would imply that almost anyone who can manifest an adaptive response could safely embark on a long-term mission, provided the total dose and dose rates are not so high that they abrogate the induced adaptive response.

19.27 Airplanes and ground level events

Solar proton events increase radiation in airplanes flying at high altitudes. Although the risks per flight are low, flight crews may be exposed repeatedly. Depending on the dose rate, total dose and the individual crew members' biological responses to radiation (in terms of efficiency of radiation-induced DNA damage repair), such chronic exposure could have consequences. While it is generally expected that such consequences would be uniformly negative, radiobiological phenomena such as the adaptive response alternatively suggest that the consequences could be paradoxically positive. Such surprising results have been seen in naval nuclear shipyard workers and residents in cobalt-60 contaminated buildings in Taiwan. But until the specific biological details are better understood (along with a confirmation that such adaptive responses to low doses of radiation are universal), the monitoring of solar proton events by satellite instruments is important. Such monitoring allows airplane radiation exposure estimates, which may lead to flight path and altitude adjustments to avoid excessive radiation dose. For example, high flights over or near the North Pole intrinsically expose passengers and crew members to more radiation than flights over lower latitudes; during an SPE, flights might aim to avoid these flight patterns.

Passengers and crew airborne at the time of an extreme event (a G5, Carrington-like event) would be exposed to radiation dose up to 20 mSv or 2 rems, which is about twice the dose from a CT scan of the chest, abdomen and pelvis and the same as a PET/CT scan, and about two-thirds the dose of a multi-phase CT of the abdomen and pelvis. Using conventional (linear no-threshold (LNT)-based) models, such levels compute to an increased cancer risk of 1 in 1000 for each person exposed. Such calculations must be considered in the context of the lifetime risk of non-skin cancer, which is about 1 in 3 (if one includes the baseline rates of non-melanoma skin cancers, the risk of developing cancer during one's lifetime rises even higher). And as mentioned, the risk model, the LNT model, is probably invalid and grossly overestimates risks at these low doses.

Ground level enhancements (GLEs) (also known as **ground level events**), occur when an SPE contains protons or other particles with sufficient energy to overcome the magnetic repulsion of our magnetosphere and penetrate the atmosphere. These energetic particles can be recorded at ground level in the form of increased secondary neutrons. Since 1942, when records first began, an average of about one GLE is detected per year. They tend to cluster around solar maximum. The increased radiation can have effects at ground level, as in the Carrington Event. Such GLEs do increase radiation dosage to people but they are not expected to increase the risk of cancer. Even the ground level radiation from a G5 geomagnetic storm would be dwarfed by the background doses received annually by residents of naturally high background regions such as in parts of Ramsar and Kerala.

Aside from cancer, there may be other health effects of space weather. The fields of clinical **cosmobiology** and **heliobiology** investigate the possible links between space weather (such as Forbush decreases associated with increased solar activity) and human health. Some studies suggest a link between cosmic ray-associated neutron dose at the surface of Earth and rates of sudden cardiac death. Other recent studies indicate that heart rate variability is statistically significantly affected by variations in cosmic ray intensity or geomagnetic activity. Of course, such claims require further verification and an elucidation of the underlying mechanisms.

19.28 Heliobiology and clinical cosmobiology

The true impact of inclement space weather on human—and animal—health is presently unknown. It has been suggested that solar activity and the consequent geomagnetic storms may be linked to whale beachings. Animals that rely on geomagnetism for navigation (such as birds and honeybees) might also be affected by unusually powerful geomagnetic storms.

Earth's thick atmosphere and its protective magnetosphere confer radiation protection to humans and animals at ground level, but astronauts may be subject to significant doses of radiation during solar storms. Beyond the attenuating atmosphere and repelling geomagnetic field, high intensity charged particle radiation may reach astronauts unabated and lead to health consequences. As mentioned, HZEs can cause DNA and chromosome damage, which, if extensive and unrepaired, may lead to cancer and other health problems. Very high acute doses (e.g. >10 Gy) can quickly be fatal through acute radiation syndrome.

Nevertheless, the radiobiology of space radiation is a topic of great interest but incomplete understanding. It was believed that SPEs containing solar protons with energies >30 MeV are particularly hazardous, but one must remember that as energy goes up, the LET and RBE go down. Thus, while highly energetic solar protons have greater penetration, they have less biological effectiveness. And the overall biological impact is the product of the total dose times the radiobiological effectiveness. Hence, low-energy protons, which have higher LET and RBE are more hazardous particle for particle and dose for dose (but they are less penetrating and may not even be able to pierce the metal hull of a spacecraft).

It should be emphasized that crew living on the ISS, while above the protection of Earth's atmosphere, are still partially protected from the full impact of the space radiation environment by Earth's magnetic field, since the magnetosphere deflects charged solar particles and extends far beyond the average 250 mile (400 km) height of the ISS. The magnetosphere (on the solar side) extends about 6–10 Earth radii or about 24000–40 000 miles out (38 600 to 65 000 km away).

19.29 Dose on the ISS and other space missions

Measurements from personal dosimeters on astronauts staying on the ISS indicate that they receive a wide range of radiation from day to day. This variability is partly due to the steep inclination of the ISS's orbit (51.6°) and its variable altitude (it orbits at altitudes between 370 and 460 km)

The dose also is a function of space weather, altitude, latitude, the phase of the solar cycle, the number of spacewalks performed, location within the ISS itself (due to shielding differences) and the level of solar activity. For example, at an altitude of 500 km (slightly above where the ISS usually is), the average daily dose (absorbed dose) would be around 50 millirads (5×10^{-4} Gy). Also, when the orbit takes the ISS over the vicinity of the **South Atlantic Anomaly**, radiation dose rate may be 1000 times higher. As a generalization, astronauts aboard the ISS receive an average around 80 mSv for a 6-month stay during solar maximum and an average of about twice that (160 mSv) for a 6-month stay during solar minimum. Thus, average radiation exposure is higher during solar minimum. But although a lower average background can be expected during solar maximum, solar maximum is also the period during which powerful SPEs are more likely to occur, punctuating the radiation doses during this otherwise lower average. Using these rough figures, one might say that astronauts might be expected to receive an average effective radiation dose of 0.5–1 mSv of radiation per day (0.2–0.5 mGy absorbed dose) with significant variability.

This daily dose is about the same as someone would get from natural sources on Earth over about 50–100 days (assuming a rate of 1 millirem or 0.01 mSv d^{-1}, as in much of the United States of America). An easy, albeit imperfect, way of remembering this is that since many places in the United States of America average about 1 mrem d^{-1} of natural background radiation, it takes about 100 days on Earth to receive a day's worth of the high end of the range of roughly 1 mSv d^{-1} dose rate on the ISS (since 100 mrem equals 1 millisievert). Recent, high-precision dose measurements made over 3 years by the Japanese Kibo module installed on the ISS as part of an environmental monitoring study for the Tanpopo mission measured radiation doses in two particular areas within the ISS. The study showed mean annual dose rates were 231±5 mGy yr^{-1} at the exposure facility and 82 ± 1 mGy yr^{-1} at the pressurized module.

Additional radiation inside the ISS, and other space stations such as Skylab and Mir, can be created when the fast, heavy ions in cosmic rays collide with the aluminum hull, spraying showers of secondary particles into the living quarters.

Keeping in mind that the dose in deep space is higher than the dose on the Space Shuttle or ISS (or on the surface of Mars), below are some representative radiation doses during various space missions:

Mission	Radiation dose
Space Shuttle Mission 41-C (8 d orbiting the Earth at 460 km)	5.59 mSv
Apollo 14 (9 d mission to the Moon)	11.4 mSv
Skylab 4 (87 d mission orbiting the Earth at 473 km)	178 mSv
ISS Mission (up to 6 months orbiting Earth at 353 km)	160 mSv
Estimated Mars mission (1 yr)	660 mSv
Estimated Mars mission (3 yr)	1200 mSv

Adapted from NASA Space Faring: The Radiation Challenge radiationchallenge.pdf (nasa.gov).

Below is some tabulated data on radiation doses during the Apollo Missions that spent time on the surface of the Moon. Students are always advised to note the difference between physical absorbed dose reported in gray or rads versus calculated effective doses given in sieverts or rem.

Mission	Total duration	Lunar surface duration	Average radiation dose
Apollo 11	08 d, 03 h, 13 min	21 h, 38 min	0.18 centigray
Apollo 12	10 d, 4 h, 31 min	31 h, 31 min	0.58 centigray
Apollo 14	09 d, 01 min	33 h 31 min	1.14 centigray
Apollo 15	10 d, 01 h, 11 min	66 h, 54 min	0.30 centigray
Apollo 16	11 d, 01 h 51 min	71 h, 2 min	0.51 centigray
Apollo 17	12 d, 13 h, 51 min	74 h, 59 min	0.55 centigray

Adapted from NASA Space Faring: The Radiation Challenge radiationchallenge.pdf (nasa.gov).

Further reading

[1] Dobrzynski L, Fornalski K W and Feinendegen L E 2015 Cancer mortality among people living in areas with various levels of natural background radiation *Dose Response* **13**

[2] Papailiou M, Ioannidou S and Tezari A *et al* 2023 Space weather phenomena on heart rate: a study in the Greek region *Int. J. Biometeorol.* **67** 37–45

[3] Stoupel E 2019 50 Years in research on space weather effects on human health (Clinical Cosmobiology) *EC Cardiol.* **6** 470–8

[4] Unger S 2019 The impact of space weather on human health *Biomed. J. Sci. Tech. Res.* **22** 16442–3

[5] Wolff E W, Bigler M, Curran M A J, Dibb J E, Frey M M, Legrand M and McConnell J R 2012 The Carrington event not observed in most ice core nitrate records *Geophys. Res. Lett.* **39** L08503

[6] Abbasi S, Mortazavi S A R and Mortazavi S M J 2019 Martian residents: mass media and ramsar high background radiation areas *J. Biomed. Phys. Eng* **9** 483–6

[7] Ghiassi-nejad M, Mortazavi S M, Cameron J R, Niroomand-rad A and Karam P A 2002 Very high background radiation areas of Ramsar, Iran: preliminary biological studies *Health Phys.* **82** 87–93

[8] Welsh J S, Bevelacqua J J and Mortazavi S M J 2022 Ramsar, Iran, as a natural radiobiological surrogate for Mars *Health Phys.* **122** 508–12

[9] Kodaira S, Naito M, Uchihori Y, Hashimoto H, Yano H and Yamagishi A 2021 Space radiation dosimetry at the exposure facility of the International Space Station for the Tanpopo Mission *Astrobiology* **21** 1473–8

[10] Bard E, Miramont C and Capano M *et al* 2023 A radiocarbon spike at 14 300 cal yr BP in subfossil trees provides the impulse response function of the global carbon cycle during the Late Glacial *Phil. Trans. R. Soc. A.* **381** 20220206

[11] Ghiassi-nejad M, Mortazavi S M, Cameron J R, Niroomand-rad A and Karam P A 2002 Very high background radiation areas of Ramsar, Iran: preliminary biological studies *Health Phys.* **82** 87–93

Chapter 20

Mercury and Venus

20.1 Mercury

Mercury is both the smallest planet and the one closest to the Sun. Following Kepler's Third Law, Mercury is also the swiftest planet in terms of linear velocity and the time it takes to orbit the Sun. One Mercurian year is the shortest in the Solar System at 88 days. Because it is so close to the Sun, tidal friction caused by gravity has slowed its rotation. The result is that unlike Earth, which rotates once every 24 h, Mercury takes nearly 59 Earth days to spin once about its axis (with respect to rest of the Universe). This long day, coupled with the short 88 day year, makes for some very interesting patterns that the Sun traces across the Mercury daytime sky. Kepler's Second Law dictates that planets move faster when they are at perihelion. And given Mercury's rather eccentric orbit, there is considerable difference in orbital velocity between perihelion and aphelion. At perihelion, the orbital velocity exceeds the rotation rate and the Sun will temporarily move 'backwards' in its daytime trajectory.

Additionally, the combination of the long period of rotation and the short period of revolution makes the apparent 'day' (meaning sunrise to next sunrise or a 'solar day') far longer than the time it takes to rotate once around its axis relative to the rest of the Universe (the 88 day sidereal day). The end result is that, from the perspective of someone sitting on the surface of Mercury, a solar day (or diurnal day)—sunrise to next sunrise—is 176 days. This is the second longest day among the planets, with Venus taking top honors at 243 Earth days for a sidereal rotation.

This slow rate of rotation is thanks to its proximity to the Sun, whose immense gravity has caused a slowdown in rotation rate because of tidal friction. For a while it was believed that Mercury was tidally locked (like our Moon is) meaning that its rate of rotation was the same as is rate of revolution and only one side faced the Sun. However, Mercury actually rotates a bit faster than predicted and a day does not equal a year on Mercury. Instead of being locked in a 1:1 ratio of day to year, radar Doppler data have demonstrated a 3:2 spin–orbit resonance. Its nearly 59 day sidereal day is repeated three times for every two 88 day sideral years. While not quite as extreme as a 1:1 resonance, a 3:2 resonance reflects significant tidal friction.

The surface of Mercury is heavily pockmarked with craters. This indicates that it surface is very old. In fact, data from the MESSENGER Mission suggest that some of Mercury's surface dates back 4.1 billion years. (**MESSENGER** stands for **MErcury Surface, Space ENvironment, GEochemistry, and Ranging**.) This differs from Earth where the oceanic crust is constantly being recycled. The oldest oceanic crust on Earth is rarely older than 180 million years thanks to subduction, with a few pockets of older oceanic crust in places that have been spared from subduction such as the **Herodotus Ridge** in the Mediterranean Sea (the former **Tethys Ocean**). Also, on Earth, the continental crust is constantly subjected to weathering, which wears down the rocks and other surface features including mountains and craters. Things do not last very long here on the surface of our dynamic Earth.

As one might imagine, Mercury is extremely hot. But Mercury's extremely gossamer atmosphere cannot retain heat effectively. The atmosphere is exceedingly thin at about 1–2 picobar pressure (vs Earth's standard sea level pressure of 1013.2 mbar), which is similar to the Earth's atmosphere at an altitude of 50 km. This thin atmosphere results in a day–night temperature difference of over 1000 °F. Thus, while daytime temperatures in the scorching Sun reach around 700 K (427 °C or 800 °F), at night the temperature plummets to a frigid 100 K (−179 °C or −290 °F). Mercury has a very eccentric elliptical orbit meaning there is a great deal of variation between perihelion and aphelion. In fact, unlike Earth, where our seasons are caused by our 23.5° axial tilt which causes the Northern Hemisphere to be closer in summers and further in winters (with the opposite situations for the Southern Hemisphere), the seasons on Mercury are due to its relative proximity at perihelion and distance at aphelion. Mercury's axial tilt or obliquity is nearly zero (0.027°, which beats Jupiter's second place finish at 3.1°). Mercury also has the greatest eccentricity of all the planets. Although its semimajor axis is just under 0.4 astronomical units (AU) (0.3871 to be precise), it has an eccentricity of 0.206, which means that during its closest approach it is only 0.307 AU from the Sun, while at aphelion it is 0.467 AU away.

Perhaps due to a catastrophic collision that blasted away much of its crust and mantle early in its history, Mercury is nearly all core, making it a dense planet. Its dense metallic core occupies roughly 85% of its radius whereas its less dense crust plus mantle are only about 400 km thick in total. In contrast, Earth's mantle represents about 2900 km of its 6378 km total thickness. Despite being nearly all core, curiously Mercury is the second densest planet in the Solar System—Earth ranks number one. Mercury's average density is 5.240 g cc^{-1} whereas Earth's average density is 5.513 g cc^{-1}.

Mercury's core might be partly molten just as ours is, and this gives rise to the planetary magnetic field that was detected by a **Mariner 10** flyby and confirmed by the magnetometer on MESSENGER. This magnetic field is relatively weak however, at about only 1.1% of Earth's geomagnetic field strength. Despite its weakness, it is still strong enough to deflect some of the solar wind. Furthermore, the magnetic field of Mercury interacts with the magnetic field of the solar wind to create a Mercurian magnetosphere. MESSENGER revealed that on occasions, these interactions induce intense magnetic 'tornadoes' that channel the energetic solar wind plasma right down to the surface. The twisted magnetic flux tubes, known as **flux transfer events**, create openings in Mercury's already weak magnetic shield. Through magnetic reconnection, the solar wind enters through these flux tubes (which may be up to 800 km in diameter or nearly a third of Mercury's radius) and penetrates all the way down to the surface. These energetic ions may blast atoms into space through a process called sputtering. Such phenomena can occur on Earth as well but fortunately our atmosphere serves as a backup protector when it comes to incoming charged particulate radiation. Overall, MESSENGER showed that the reconnection rate is tenfold higher than on Earth. But sputtering might explain a longstanding mystery—why does Mercury still have an atmosphere at all? When the high-energy ions of the solar wind come crashing onto the surface of Mercury, they may launch atoms into the sky and replenish Mercury's atmosphere.

20.2 Venus

The second closest planet to the Sun is also our nearest neighbor. Venus's semimajor axis is about 0.72 AU compared to Mars' 1.5 AU (Earth's distance is defined as 1 AU). At 5.24 g cc^{-1}, its density is not very dissimilar to Earth ($\sim$5.5); Venus is the third densest planet, behind Earth and Mercury. Given that Venus is about 95% of Earth's diameter, in many ways, it is a twin sister to us.

At one time, it was believed that Venus might be a lush tropical garden of paradise. But data from the Soviet Union's Venera mission showed quite the contrary. Rather than a verdant tropical garden, Venus is searing hot and shrouded in inhospitable **clouds of sulfuric acid**. In fact, it is these clouds which confer such a high **albedo** to Venus. Albedo is a measure of reflectivity; the brightness of Venus is thanks to its high reflectivity of incoming sunlight. In fact, Venus has the highest albedo of all planets in the Solar System. Anyone who has looked into the pre-dawn or early evening sky has probably noticed the brilliant 'morning star' or 'evening star', which is really planet Venus.

Unlike the Moon, Mercury and Mars, which barely have any atmosphere at all, Venus was long known to host a thick atmosphere, which is why it initially was hoped that Venus could be an abode for life as we know it. But upon closer scrutiny, that atmosphere was not convivial at all. First, the atmospheric pressure is a whopping 92 bars—92 times the surface pressure on Earth. Second, the atmosphere is mostly carbon dioxide, which is a potent greenhouse gas; there is no free oxygen to speak of. Thanks to this thick, planet-encompassing atmosphere of greenhouse gas, the surface of Venus is about 480 °C everywhere and all the time, hot enough to melt lead (the melting point of lead is 450 °C or 850 °F). As on Earth, the lowest layer of the Venerean or Venusian atmosphere is the troposphere. But unlike Earth's troposphere which averages about 13 km or about 8 miles in height, the troposphere on Venus is 65 km (40 miles) tall. This dense troposphere then transitions directly into a thinner mesosphere, and then into a gossamer exosphere where the atoms and molecules no longer behave like typical gases since collisions are so rare. Unlike

Earth, there is no stratosphere or thermosphere with their characteristic temperature inversions. The exosphere on Venus begins at an altitude of 220–350 km (137–217 miles) as opposed to Earth's exosphere which starts at a height of about 600 km (373 miles). Ironically, Venus's dense, crushing atmosphere is actually about 50% shorter in overall height than Earth's multi-layered atmosphere.

As discussed elsewhere, based on its proximity to the Sun coupled with its high albedo thanks to its highly reflective clouds, Venus's equilibrium surface temperature should be roughly −45 °C or −50 °F. Being off by nearly 900 °F indicates something is seriously amiss! Thanks to a 96% carbon dioxide atmosphere (4% nitrogen), the greenhouse effect is so intense that Venus's equilibrium temperature is far, far hotter than one would predict based solely on solar proximity and albedo.

Earth has a strong greenhouse effect going on as well, which is good since without it our surface temperature would probably be about 60 °F colder than it presently is: −0 °F or −18 °C as opposed to 60 °F or 16 °C. Earth has plants that take carbon dioxide out of the atmosphere and convert it into biomass. And as discussed elsewhere, Earth has a very effective long-term thermostat (the carbonate–silicate cycle) that recycles greenhouse carbon dioxide over millions of years. Venus does not have such a thermostat—it is a one-way switch that is always on 'high'. Thus, Venus has experienced a runaway greenhouse effect. This is because the geological phenomenon called plated tectonics is not operational on Venus. Through subduction of carbonate rocks such as limestone (which are created from atmospheric carbon dioxide and rock cations through chemical weathering), plate tectonics buries and sequesters these carbonates in a separate carbon reservoir for many years. Thus, far less carbon accumulates in Earth's atmosphere as CO_2 compared to Venus. Unlike Earth, which has seven major plates and several minor plates that move about relative to each other and participate in **subduction**, Venus has just one thick crust without individual motile plates. Nevertheless, radar data reveal that this thick crust has two relatively elevated regions colloquially called 'continents'. These continents or highlands cover about a fifth of the Venerean surface.

Venus also has mountains, valleys, plains and craters among other surface features. There are over 160 known Venerean volcanoes with a diameter over 100 km (62 miles). Some are similar to terrestrial shield volcanoes here at home. But there are several unique versions of volcanoes on Venus. One example is the orb web-shaped **arachnoid volcano** type. There have been nearly 100 known examples of arachnoid volcanoes discovered. They tend to be quite wide at 40–200 km (25–125 miles) across. Even broader than the arachnoid volcanoes are the huge **pancake volcanoes** of Venus. As the name suggests, these are very flat, circular outpourings of lava. Another unusual type of Venerean volcano are the **coronae volcanoes** which are probably formed when crust is pushed up by the upwelling magma but then crumbles back down after eruption to create concentric circles around the base. Coronae have been observed on the moon Miranda, which orbits Uranus, but arachnoid and pancake volcanoes are unique to Venus.

The largest known crater on Venus is called Mead and is about 280 km or 174 miles in diameter (which makes it comparable to Earth's largest known crater (Vredefort crater in South Africa) and it was probably formed about 2 billion years ago and originally measured about 300 km or 186 miles across. There are no craters less than 5 km or 3 miles in diameter on Venus, probably because the meteorites that would leave such scars burned up in the thick Venerean atmosphere on the way in. For reference, the famous Barringer Meteor Crater in Arizona is only about 1.3 km (0.8 miles) in diameter. Compared to Mercury, Venus has fewer impact craters, indicating that its surface is younger than Mercury's (or the Moon for that matter). Based on the crater count, it is estimated that the surface of Venus is no more than 200 million to 1 billion years old anywhere. This is because of massive planet-wide 'resurfacing events' that erased all previous evidence of impacts. During these global resurfacing affairs the entire Venerean crust collapsed into the mantle and was replaced by a new surface of magma; this may have happened more than once on Venus.

There are some active volcanoes that imbue the atmospheric with greenhouse gas carbon dioxide and provide the sulfur dioxide for the high clouds of sulfuric acid. There is precipitation ('rain') on Venus in the form of sulfuric acid from these high clouds. The surface temperature is so sweltering however that this sulfuric acid precipitation evaporates before reaching the ground. These sulfuric acid clouds are about 50–70 km (31–43 miles) above the surface but the sulfuric acid rain evaporates at an altitude around 20–30 km (about 12–19 miles) above the ground. Curiously, 'brimstone' is an old word for sulfur, so coupled with the 850 °F temperatures, Venus certainly does not meet the expectations of a Garden of Eden but thanks to 'fire and brimstone', it appears to qualify as quite the opposite! (Of course, with no free oxygen, real fire is not present on Venus despite the sufficiently high temperatures.) On Earth, when rain or snow evaporates or sublimates before reaching the surface (in hot, dry deserts for example) the phenomenon is called meteorological **virga**.

Although Venus is our twin sister in some ways, it is extremely different from Earth in several other ways beside the climate. For one, Venus is unique in all the Solar System in that it rotates the 'wrong' way. Viewed from above (celestial north), all the planets revolve around the Sun in a counterclockwise fashion, and as one would expect from angular momentum, they also rotate about their axes in a counterclockwise manner too. Venus is a notable exception to the rule in that it spins about its axis in a clockwise manner. This can be described as **retrograde rotation**. Of course, this is the same as saying that its rotational axis is off by about 180° (177.4° to be precise). This results in the Sun rising in the west and setting in the east. But such a day on Venus takes 243 Earth days. This is the slowest rotation among the planets in our Solar System. Curiously, a year on Venus (the time of a complete orbit around the Sun) is only 225 days, so a sidereal day takes longer than a year on Venus. While no other planet displays this extreme an anomaly, Uranus takes second honors as it rotates 'on its side'. In other words, its axial tilt is approximately 90° (more precisely, 97.77°).

As mentioned, Venus's density of 5.42 g cc^{-1} (or equivalently, 5440 kg m^{-3}) is similar to Earth's, indicating that it is composed of silicate rocks and an iron core. While Venus's iron core is probably comparable to Earth's in relative size, there is no planetary magnetic field, indicating that the core is perhaps no longer molten.

Or if it is molten, it is spinning sluggishly and not generating the required dynamo for producing a global magnetic field. Therefore, it might seem that, when it comes to space radiation protection, Venus is outstanding when it comes to atmospheric attenuation but fails when it comes to inherent magnetic deflection of charged particles. But this perspective might be too simplistic.

Venus's thick atmosphere does interact with the incoming solar electromagnetic radiation to create an ionosphere—an atmospheric layer laden with charged particles. The thermal pressure of the Venerean ionosphere pushes back against the magnetic pressure of the incoming solar wind. The solar magnetic field lines drape around the back of the planet as the plasma piles up on the solar side. Thus, an induced magnetosphere forms around Venus with a bow shock on the solar side and a long, trailing extension (magnetotail) on the anti-Sun side.

When the solar wind (and its associated magnetic field) crashes into this ionosphere, it heaps up akin to a traffic jam after a crash on the highway. The resulting magnetic barrier inhibits penetration of the solar wind deep into the Venerean atmosphere. In this fashion, the solar wind does not easily erode the atmosphere of Venus despite its absence of an inherent planetary magnetic field.

In November 2011, a powerful coronal mass ejection struck Venus while the ESA's Venus Express was in in the vicinity and showed that Venus's ionosphere and induced magnetosphere reacted spectacularly to the solar storm. Ionospheric plasma density increased threefold while the magnetic barrier increased its strength. The Sun-side bow shock was compressed and widened while the magnetotail waved and bent on the anti-Sun side. Although the induced magnetosphere provides a magnetic barrier that offers some protection from the incessant solar wind, large coronal mass ejections do drive away ions from the ionosphere leading to some atmospheric loss.

Data from Venus Express have demonstrated that the 'holes' in Venus's ionosphere (first detected by the NASA Pioneer Venus Orbiter in 1978) are not really simple holes but instead, long drawn-out channels or cylinders of relatively low ion density on the anti-Sun side of Venus. These tubes of low ion density form not only during solar maximum but also during solar minimum (which was an unexpected finding based on the Pioneer Venus Orbiter data).

Further reading

[1] Slavin J A *et al* 2008 Mercury's magnetosphere after MESSENGER's first flyby *Science* **321** 85–9
[2] Slavin J A *et al* 2009 MESSENGER observations of magnetic reconnection in Mercury's magnetosphere *Science* **324** 606–10
[3] Collinson G A *et al* 2014 The extension of ionospheric holes into the tail of Venus *J. Geophys. Res. Space Phys.* **119** 6940–53
[4] Qi Xu *et al* 2019 Observations of the Venus dramatic response to an extremely strong interplanetary coronal mass ejection *Astrophys. J* **876** 84

IOP Publishing

Space Radiation
Astrophysical origins, radiobiological effects and implications for space travellers
James S Welsh

Chapter 21

Mars

Mars is the most extensively studied planet in our Solar System. Nicknamed the Red Planet, Mars' reddish hue is due to rust—iron oxide minerals. Although Venus is at times closer to us than Mars ever gets, Mars has a far more hospitable climate and a higher probability of recently hosting plentiful surface water—and therefore, life. At an average distance from the Sun of about 1.5 astronomical units (AU), the fourth and final terrestrial planet in our Solar System is the Red Planet. Mars is the most extensively studied planet in our Solar System. Nicknamed the Red Planet, Mars' reddish hue is due to rust—iron oxide minerals. Although Venus is at times closer to us than Mars ever gets, Mars has a most hospitable climate and a higher probability of recently hosting plentiful surface water—and therefore, life. And Mars has been the subject of speculation for quite some time. In 1877, astronomer Giovanni Schiaparelli described Martian 'canali' which in his native Italian meant channels. Such channels could be of completely natural origin, but the word was mistranslated into English as canals. Since on Earth, canals are man-made, it was misinterpreted that the canali of Schiaparelli were Martian-made. Percival Lowell, using a more sophisticated telescope in a private observatory in Flagstaff Arizona, expanded the concept of canals on Mars as he described a network of interlacing canals across the planet in his 1906 book, *Mars and Its Canals*. As depicted in H G Wells' 1898 *The War of the Worlds*, prevailing sentiment around the time of the turn of the 19th century was that there was technologically advanced, intelligent life on Mars—and it might not necessarily be friendly. This perception persisted until 1965, when the **Mariner 4** spacecraft (launched 28 November 1964) obtained detailed photographs of the Martian surface. Nevertheless, Mars remains the most likely planet for human exploration and potential colonization thanks to its many similarities to Earth and the relatively hospitable environment.

Unlike Venus's extremely slow retrograde rotation, Mars' rotation rate is remarkably similar to our own. One day on Mars (called a **sol**) is 24 h, 39 min and 35 s. Thus, human colonizers would probably not have too much difficulty with their circadian rhythm. But one Martian year is nearly twice as long as an Earth year at 687 Earth days (669.6 sols) so the seasons are prolonged. And the obliquity or axial tilt (which confers the seasons on Earth) is also quite similar—23.5° on Earth and 25° on Mars. However, while axial tilt is the reason for our seasons here, Mars' orbit is far more eccentric (in fact, more eccentric than any planet aside from Mercury). At perihelion Mars is 1.3814 AU from the Sun but 1.666 AU at aphelion. Thus, on Mars the seasonal variation is due to a combination of both the effects of axial tilt and distance from the Sun.

21.1 The Martian atmosphere and its weather

While Mars does have an atmosphere, the atmospheric pressure is extremely low. In contrast to Venus's 92 bars, Mars (at about 2 millibars) has an atmospheric pressure well under 1% of our sea level pressure. Pressure suits will be absolutely mandatory for any manned Martian mission. Interestingly and in contrast to popular depictions, the windstorms on Mars would probably not pose much of a problem. Despite gale-force gusts up to 70 km h^{-1} (38 miles h^{-1}), the rarefied air on Mars would be unlikely to blow people over or blow heavy equipment away. Nevertheless, these winds can and do cause odd Martian dust storms. The **Viking** orbiters were the first to photograph dust devils, and the **Curiosity Mars rover** encountered several during its prolonged mission. The solar-powered **Spirit rover** was temporarily incapacitated until a fortuitous dust devil in 2005 blew the occluding dust off the solar panels. Similarly, the **Opportunity rover** was temporarily offline due to obscuration of its solar panels until another presumed dust devil blew it clean and operational again. Of interest, Curiosity, being powered by a radioisotope thermoelectric generator (RTG), has been immune to the obfuscation caused by dust and has greatly exceeded its lifetime expectations. It is still operational at the time of this writing in 2024, well over a decade after its launch in November 2011. Originally designed to operate for at least 1 Martian year (687 Earth days), Curiosity's plutonium-powered RTG has exceeded expectations and is still driving the rover along today.

Larger than dust devils are Martian dust storms, which were first observed by the **Mariner 9** probe in 1971. These occur roughly annually (mostly during Martian perihelion, when there is ~40% more **insolation** (incident solar radiation)) and can last several weeks. And they can be quite large as well, often covering continent-sized patches of Mars. Not infrequently (maybe a few times per Martian decade), such dust storms go global—monstrous, planet-wide dust storms that may last up to 6 months. These global dust storms can host wind speeds up to 60 miles h^{-1} (~96.6 km h^{-1}). Again, thanks to the low atmospheric pressure on Mars, these winds do not have much physical punch despite what Hollywood might suggest. Nevertheless, dust storms can lower ground temperature and raise atmospheric temperature, promote water loss and affect solar-powered Mars rovers. Spirit and Opportunity, which both landed on Mars in 2004 and were temporarily out of commission for several weeks during a strong storm in 2007. The Martian planetary dust storm of 2018 finally put an end to the Opportunity mission.

21.2 Geology of Mars

Mars hosts a number of interesting geological features including the largest known volcano in the Solar System, **Olympus Mons**, and a colossal canyon that if superimposed on North America would span the width of the continental United States of America, called the **Valles Marineris**. Olympus Mons is a shield volcano that stands over 72 000 feet tall (13.6 miles or 21.0 km) based on data from the **Mars Orbiter Laser Altimeter**. To get a perspective on just how tall this is, one might think about commercial jet airplanes cruising at 30 000 feet or the fact that Mount Everest is 29 032 feet or 8.85 km tall. Hence, Olympus Mons is around two and a half times as tall as Mount Everest. Olympus Mons is just outside the northwest perimeter of highlands called the **Tharsis Bulge**, which dominates Mars' western hemisphere. Three large volcanoes (Arsia Mons, Pavonis Mons and Ascraeus Mons) collectively called the **Tharsis Montes** punctuate the Tharsis Bulge region. Reaching from Tharsis out to the east is the Valles Marineris. This canyon then stretches eastward for nearly a quarter of the planet's circumference.

At around 6792 km or 4220 miles across, Mars is approximately half the diameter of Earth but only about one tenth the mass. Thus, the average density of Mars is significantly lower than Earth at just over 3.9 g cc^{-1} (compared to Earth's 5.5). In fact, with this density value, Mars finishes last among the terrestrial planets. However, a planet's overall size affects its density because the larger a planet (or moon) is, the more compressed it will be thanks to its own gravity. This allows a fairer comparison based on 'uncompressed density'. Based on this metric, the uncompressed densities of Mars and Earth are not very different at 4.0 versus 4.2 g cc^{-1}, respectively. This shows that Mars and Earth are made of similar materials. Just as on Earth, Mars has a crust and mantle made of silicate rocks and a denser iron core. But unlike Earth (and just as on Venus), the Martian crust does not appear to have sliding plates. Instead, there is but one thick crust on Mars without the possibility of genuine plate tectonics. Regardless of the absence of plate tectonics, based on data from the NASA InSight lander, recent seismic studies of Martian earthquakes ('Mars-quakes'?) suggest that there could still be molten magma beneath the crust. This molten magma may be capable of limited volcanic activity and reshaping of the Martian surface to this day.

Mars is home to some amazing canyons (such as the **Valles Marineras**) and the largest known volcano in the Solar System, **Olympus Mons**. The origin of Valles Marineras is probably due to increased geological activity early in Mars' history. Although, Mars presently is and perhaps always has been devoid of true plate tectonics, other types of tectonic activities were not precluded, and the Valles Marineras and other Martian canyons represent ruptures in the crust from long ago. Without weathering or plate tectonics, such geological surface scars can remain intact for billions of years on Mars. Although most of Mars is geologically quiescent, the **InSight lander** has identified some intriguing action in the region called **Cerberus Fossae**. Certain geographic features called **graben** and **horsts** have been observed here but up until recently it was unclear whether these depressions and elevations were truly due to seismic activity. InSight has confirmed the presence of seismic activity here in the form of Martian earthquakes, which hints of still molten magma down below.

Until recently, it was believed that Mars' core was not molten and spinning, since there is no strong global magnetic field at this time. However, recent data from NASA's InSight spacecraft have confirmed that Mars' core is still partially molten. This is an area of active investigation. Although there does not appear to be a strong planetary magnetic field at this time, the Martian surface does have some magnetized rocks, implying that a functional dynamo and prominent planetary magnetic field were in place at one time in the distant past. Since craters can help determine the age of a planet's or moon's surface, crater counting in magnetized Martian crustal regions hints that Mars' magnetic dynamo died out around 4 billion years back. This ongoing research is most interesting since it demonstrates that the absence of a geomagnetic field does not guarantee that the planet's core is not at least partially molten and that the presence of a still molten core does not guarantee the presence of a planetary magnetic field.

21.3 The moons of Mars

Unlike Mercury and Venus, Mars does have moons. But its two tiny moons, **Phobos** and **Deimos**, are irregularly shaped and rather small at just 22 and 12 km across (14 and 7 miles), respectively. They are most likely captured asteroids. Phobos is the closer one and orbits Mars at only 6000 km (3700 mi) above the surface. Relatively speaking, this is closer to its home planet than any other moon in the Solar System. In fact, the orbit of Phobos is around three times as close to Mars as the global positioning satellites are to Earth. Thanks to this proximity, Phobos zips around Mars in just 8 h, which is quicker than Mars' rotation rate of just under 25 h. This means that observers of Mars would see this moon rise in the west and set in the east only about 4 h later—and this would ordinarily happen twice a day. And thanks to the alignment of Phobos's orbit with Mars' plane of rotation around the Sun, there is a 30 second solar eclipse each and every time Phobos passes in front of the Sun (about twice daily). As one might expect from its close orbit, Phobos is subjected to substantial tidal friction and this tidal friction is gradually decreasing Phobos's orbital speed—and distance to Mars—over time. Therefore, some time in the next 50 million years or so, Phobos will plunge below Mars' Roche limit and come crashing down or get torn asunder into a ring system like Saturn's. In fact, one theory is that Phobos participates in a repeating cycle in which it comes too close and gets pieces ripped off which form rings that move inward and the remaining remnant of Phobos then moves outward. Then the rings eventually crash onto Mars and the cycle repeats, as the new and smaller Phobos begins another inward spiral. This could explain the small, but otherwise difficult to explain, 2° orbital inclination of Deimos. And according to this model, Deimos, despite outward appearances, is not a captured asteroid but instead would have coalesced along with Mars. This model also helps explain the observed 3:1 mean-motion resonance between these lightweights, Phobos and Deimos.

Deimos, being smaller and almost fourfold farther away, experiences less tidal friction than Phobos. Rather than spiraling inward over time, Deimos is gradually migrating outward. Thus, instead of someday plummeting out of the sky and smashing into the surface (and/or creating Martian rings), Deimos will eventually step beyond Mars' gravitational grasp and become unfettered.

21.4 The Martian seasons

Mars has some fascinating and peculiar polar ice caps that wax and wane with the seasons. Both have frozen water ice all year long but settling atop these water ice caps are large seasonal dry ice (carbon dioxide) condensations that arise each winter. The amount of water ice in these polar caps may be quite substantial, perhaps rivaling the amount in the Greenland Ice Sheet. In addition to the plentiful frozen water at the poles presently, there is permafrost just below the surface in non-polar regions. If taken from this sequestered subsurface locale and spread out evenly across the surface, the estimated volume of ice would be enough to cover the surface with a thick glacier well over 30 m thick. Approximately a quarter of all the Martian atmospheric carbon dioxide precipitates out as winter 'snowfall' creating a large glacial carbon dioxide cap on top of the water ice of the poles. The carbon dioxide sublimates each summer, leaving behind just the frozen water ice caps. Incidentally, much like Venus and vastly different from Earth, Mars' atmosphere is about 95% carbon dioxide. The balance is mostly molecular nitrogen and argon.

Our seasons are essentially entirely because of our axial tilt. In contrast, the marked eccentricity of Mars' orbit makes a substantial contribution to the seasons independent of the axial tilt. And Mars' axial tilt is not nearly as stable as ours. While there are some important but fairly regular variations in Earth's axial obliquity that are dictated by the Milankovitch cycles and influence our climate, Mars' obliquity is far more prone to whimsical and drastic changes. For instance, over the last 5 million years, Mars' axial tilt has ranged from as low as 15° to as much as 35°. On a longer timescale, the obliquity might have varied as much as from near 0° to up to 70°. The reason for the vast difference in axial stability between Earth and Mars is mostly because of our huge, stabilizing Moon. Our Moon helps stabilize our overall angular momentum, which in turn helps prevent large shifts in axial obliquity. Mars' feeble Phobos and Deimos do little for stabilizing Mars' wobbling axial tilt. Furthermore, Mars' relative proximity to Jupiter imposes additional gravitational effects that can change Mars' obliquity.

21.5 Surface water on Mars

There is geomorphological evidence of standing and flowing water in the distant past in the form of ancient river valleys and waterfalls. And they are not exactly rare, at over 40 000 at last count. Some, like the 2400 km (1500 miles) long **Kasai Valles**, are quite impressive. But the extinct Martian waterfalls (**cataracts**) are even more amazing. At least one cataract in Kasai Valles has a drop of around 500 m (1600 feet) with a lateral extension of more than 100 km (62 miles). For perspective, this is about ten times taller and a hundred-fold wider than Niagara Falls. There are also fan-shaped river deltas where fast-flowing rivers drained into deep, slow-moving lakes. Some pebbles and gravel show signs of erosion in what were probably once stream beds under rapidly flowing waters. Based on the size and shapes of these tumbled river rocks, the water was perhaps knee-deep and moving at about a meter per second (3 feet s^{-1}).

Further evidence of ancient abundant water on Mars is found in sedimentary rocks, which form when sediments settle or precipitate out of aqueous solutions and accumulate on the bottom of lakes and ponds. The Curiosity rover spotted such layers of suspected sedimentary rock at Gale Crater. Another bit of geomorphological supporting evidence for prehistoric Martian oceans comes from the Mars Global Surveyor, which showed extremely flat lowlands over a vast expanse of the northern hemisphere. On Earth, we talk of Kansas being 'as flat as a pancake' (it is actually far flatter than your average pancake when properly scaled) but Gale Crater easily wins the comparison. The only places on Earth comparably flat are the abyssal plains found at the bottom of oceans. This hints of an ancient ocean in the northern hemisphere of Mars.

A different sort of evidence supporting the concept of ample water in the distant past comes from Martian mineralogy. **Hydrated minerals** incorporate the atoms of water into their crystal structures and can only form in the presence of liquid water. Especially in the equatorial areas, hydrated minerals are plentiful on Mars. Jarosite —a potassium and iron hydrous sulfate that is formed on Earth in ore deposits and acidic, oxidizing aqueous environments—was the first hydrated mineral detected on Mars (by the Opportunity rover, using **Mössbauer spectroscopy**). The **Compact Reconnaissance Imaging Spectrometer for Mars (CRISM)** instrument on the **Mars Reconnaissance Orbiter** subsequently showed the western side of an elongated pit depression in the eastern Noctis Labyrinthus valley. Additionally, the Mars Reconnaissance Orbiter CRISM instrument found further confirmation of phyllosilicates, a class of hydrated minerals first identified on Mars by the **OMEGA (Observatoire pour la Mineralogie, L'Eau, les Glaces et l'Activitié)** instrument.

21.6 Aqueous minerals and the missing carbonate problem

A similar class of Martian minerals called **aqueous minerals** can only form through chemical reactions in water, even though the mineral products themselves do not contain water. **Grey hematite** is an aqueous mineral that only forms in hot springs and pools of standing water on Earth.

The **Mars Global Surveyor** identified 'signatures' of grey hematite (α-Fe$_2$O$_3$) in **Meridiani Planum**, further suggesting the presence of water on Mars in ancient times. In particular, Martian hematite spherules or 'blueberries' have been found in abundance at Meridiani Planus, Eagle Crater and Victoria Crater. Opportunity unequivocally found such 3–6 mm diameter blueberries at Meridiani Planum, which indicates that salty, acidic, liquid water once covered this area. Another category of aqueous minerals, carbonate minerals (such as calcite, magnesite and siderite), been spotted at **Huygens Crater** and **Nili Fossae**. Although carbonates have been found, they are not quite as abundant as one might expect given the calculated amount of water. Solutions to this so-called **missing carbonate problem** are:

(1) Perhaps water was not as plentiful as initially thought.
(2) Perhaps the carbonate silicate cycle did not operate the way it does on Earth and predictions are therefore inaccurate.
(3) Perhaps the carbon dioxide was lost from the atmosphere before it could form carbonates.
(4) Perhaps the Martian water was too acidic.

This last (conjectural) point is interesting. On Earth, formation of carbonate minerals is favored by a slightly alkaline pH. Shallow ocean water has a pH of around 8.1, which readily permits carbonate formation. But it is possible that Martian water was too acidic—and instead of carbonates forming, oxalates were produced. The Mars Science Laboratory, Phoenix and Viking showed signs of oxalates on Mars.

Another possibility is that a once-thicker carbon dioxide-dominated atmosphere was gradually eroded through **sputtering**, a process whereby the solar wind degrades and strips away the upper atmosphere. NASA's **Mars Atmosphere and Volatile Evolution (MAVEN) mission** has shown that about 100 g of atoms and particles are still being stripped from Mars' outer atmosphere every second via this process. Thus, sputtering is probably the major instigator of atmospheric loss currently. Sputtering favors loss of light carbon (carbon-12) and thus over time, the ratio of heavier carbon-13 to carbon-12 increases. But data indicate that the carbon-13 to carbon-12 ratio is far higher than could be accounted for through sputtering alone. Sputtering does slightly favor the loss of carbon-12, compared to carbon-13, but this effect is small. The Curiosity measurement shows that today's Martian atmosphere is far more enriched in carbon-13—in proportion to carbon-12—than it should be as a result of sputtering alone, so a different process must also be involved. It has been proposed that another mechanism, photodissociation, also was at work. Energetic photons from the Sun first break up CO_2 into CO and atomic oxygen; next, the CO is dissociated into carbon and oxygen. The carbon-12 so generated is more likely to escape Mars' atmosphere than the heavier isotope. In essence, particulate plus electromagnetic radiation provided a one-two punch that favored the egress of light carbon-12. This explains not only the mysterious carbon-13 to carbon-12 ratio but also explains the relative paucity of carbonate rocks and minerals. At one time (about 3.8 billion years ago), the Martian atmosphere was perhaps nearly as thick as present-day Earth's is at close to 1 bar. And the temperatures on the Martian surface would have been conducive to liquid water and the associated chemistry. But through photodissociation, this thick carbon dioxide atmosphere was eaten away, the associated greenhouse effect disappeared and liquid water evaporated or froze. And the carbon dioxide that was responsible for the initially favorable conditions was literally blown away rather than being preserved in carbonates as initially predicted.

Regardless of the missing carbonate problem and the perplexing heavy to light carbon ratio, the Curiosity Mars rover has confirmed the presence of frozen water in samples from **Gale Crater**. Along with the suspicious mineralogy, this hints that Gale Crater might have once been a large lake.

21.7 Martian meteorites and Mars' ancient water

Rare among meteorites are those that have been blasted off the surface of Mars in the distant past and subsequently landed here on Earth. Such Martian meteorites include **shergottites**, **nakhlites** and **chassignites** (the so-called 'SNC' group of Martian stony meteorites). Some relatively young nakhlites indicate the presence of Martian water as recent as 620 million years ago. Recalling that the Phanerozoic Eon began with the Paleozoic Era and Cambrian Period about 543 million years ago, one is tempted to speculate about the possibility of life in such Martian waters. The fascination with water on Mars is not only for life support of future astronauts and colonizers, but also because on Earth where we find water, we find life. Throughout the Solar System, the best bet for biology is wherever there is or was liquid water. Hence, Mars might have been an abode to biological entities at one time. As the climate and overall environment of Mars changed for the worse, Martian life might have gone extinct entirely—or maybe though hitching a ride on meteorites, it found its way here and seeded another hospitable world.

21.8 The Martian deuterium:hydrogen ratio

Martian water is somewhat different from our water. The isotopic ratio of deuterium (hydrogen-2) to hydrogen (hydrogen-1) or **D/H ratio** in water from the two planets is quite different—Mars has far more deuterium than one would expect. Water from Martian polar ice caps is around 7–8 times as rich in deuterium than our ocean water is. Clues to why this is and how it happened can be gleaned by careful comparisons of the D/H ratios in Martian meteorites of different ages. Long ago, the D/H ratio was lower, suggesting that Mars has selectively lost light hydrogen (hydrogen-1) compared to heavy hydrogen (hydrogen-2). Reconstruction of the past history suggests that Mars lost slightly over half its water in the first 400 million years of its 4.6 billion year history and then lost the balance over the next 4.2 billion years.

Loss of Martian hydrogen is likely due to interactions between the upper atmosphere and the solar wind, which dissociates water molecules into constituent atoms and also provides sufficient kinetic energy to liberate such atoms and particles from the atmosphere forever. But just as lighter carbon-12 is more likely to escape than carbon-13, lighter hydrogen more readily reaches escape velocity this way than deuterium. Thus, the D/H ratio gradually increases over time and this is reflected in the isotopic analysis of residual water on Mars.

21.9 Martian geological chronology

Because of the size difference, smaller Mars probably coalesced as a protoplanet from the protoplanetary disc slightly before Earth. One estimate is that Mars is about 100 million years older than Earth. And because it is smaller, its surface to volume ratio allowed it to cool far faster than Earth. Therefore, liquid lakes and oceans probably appeared on Mars much sooner than here.

Earth's geological history is divided into several paleontological eons, eras and periods. Mars' history is similarly expressed as three main periods:

1. The **Noachian Period**, from Mars' inception to about 3.5 billion years ago. Noachian age surfaces are scarred by many large impact craters. There was probably extensive flooding late in the Noachian. The period is named after Noachis Terra.
2. The **Hesperian Period**, from 3.5 to between 3.3 and 2.9 billion years ago. This period is marked by what appear to be vast lava plains. Named after Hesperia Planum.
3. The **Amazonian Period**, from between 3.3 and 2.9 billion years ago to the present. Amazonian period surfaces have relatively few meteor craters. This period is named after Amazonis Planitia.

The huge shield volcano Olympus Mons probably formed during the Hesperian Period and erupted well into the Amazonian Period. In contrast, the Valles Marineris probably started forming around 4 billion years ago, in the Noachian period.

21.10 What happened to Mars' atmosphere and cordial climate?

Obviously, today's Mars is very different from the wet and warm world we have been discussing above. In addition to being a barren and dry world, Mars is currently ice cold. This is partly due to its distance from the Sun but more importantly it is because it does not have a thick, insulating atmosphere. Despite being about 95% greenhouse gas carbon dioxide, Mars' atmosphere is simply too thin to retain much heat. But this begs the question—what happened to Mars' once lush atmosphere, abundant water and warm climate?

The most likely scenario is that being small and having an unfavorable surface area to volume ratio, Mars lost its internal heat early on and its core no longer sustained a magnetic field. With no intrinsic magnetic field, Mars lost its protective magneto-sphere. In a domino effect, this allowed the solar wind to gradually erode the once ample and insulating atmosphere. This atmospheric erosion was probably accelerated by the fact that in its younger days, the Sun probably had an even more intense solar wind along with greater ultraviolet output. (Although its overall output was considerably lower thanks to the faint young Sun paradox that is discussed elsewhere.)

During the early Noachian Period, Mars might have been more volcanically active and thus the atmosphere was probably thicker and contained more carbon dioxide in absolute terms. Additionally, this intense volcanism might have exhaled a lot of methane as well (which is a far more potent greenhouse gas than carbon dioxide).

If this were true, it should be evident from carbonate deposits in ancient rocks. However, there appears to be a deficit in carbonates on Mars and this is known as the missing carbonate problem. It could be that the carbonate–silicate cycle operates differently from the way it does on Earth (e.g. because of a lack of plate tectonics) or there might be other explanations for the missing carbonate problem as discussed above. But one possible explanation for the missing carbonate problem could be that the carbon dioxide was simply lost from the atmosphere before it could create carbonates. If substantial amounts of carbon dioxide were lost, it could lead to a global cooling. This in turn would permit freezing out of carbon dioxide at the polar caps, which would further pull carbon dioxide from the atmosphere and exaggerate the cooling. In this way, a runaway inverse greenhouse effect might have occurred leading to the opposite situation from Venus.

21.11 Radiological evidence of water on Mars

Radiological evidence of extant Martian water was provided by the Mars Odyssey, which counted neutrons emanating from the Martian surface. With little atmospheric attenuation or magnetic deflection, primary cosmic rays regularly crash right into the Martian surface. There, ejected neutrons are the products of such unmitigated cosmic ray collisions with atoms in the Martian crust. These are called **albedo neutrons**. But there appears to be a deficit in albedo neutron count in some areas. Since hydrogen is the most efficient absorber of neutrons, and hydrogen is abundant in water (H_2O), the inference is that there is still subsurface water in these places. The **Dynamic Albedo Neutron instrument (DAN)** on the Mars Science Laboratory's Curiosity rover is a Russian Federal Space Agency project that fires 14.1 MeV neutrons into the Martian crust and detects the reflected neutrons. Since neutrons interact most strongly (via simple elastic collisions) with similarly sized protons (i.e. hydrogen nuclei), the presence of water, hydroxyl groups or other forms of hydrogen can be detected by the change in reflected neutrons. DAN can measure the amount of hydrogen in the soil below and function in this active mode (generating its own neutrons and shooting them into the ground) or can passively rely on galactic cosmic rays to make the neutrons beneath the surface.

21.12 Martian chlorates and perchlorates

Any life on Mars will have to be shielded from the intense radiation on the surface and the reactive chemical species generated by that radiation. This was inferred by an analysis the shergottite Martian meteorite, **Elephant Moraine 79001** (or simply **EETA79001**). This meteorite was probably ejected and launched from Mars around 600 thousand years ago but it is the remnant of Martian rocks going back about 180 million years. It contains chlorate, perchlorate and nitrate anions in sufficiently high concentrations to suggest that these anions were and probably still are widespread on Mars. These reactive species would damage any organic compounds they come in contact with. Therefore, any bio-organic molecules would need to be safely sequestered well under the surface to survive the onslaught of space radiation. Perchlorate contains chlorine in its highest oxidation state (+7) and is a strong oxidizing agent (although not as strong as chlorite or hypochlorite). But intense radiation can turn already-hazardous chlorate and perchlorate anions into even more hazardous, highly reactive oxychlorine species. Furthermore, iron oxides and hydrogen peroxide, which are both abundant on the Martian surface, act in synergy with irradiated perchlorate products to provide over tenfold increased bactericidal efficacy. Hence, the surface radiation chemistry could be quite challenging for astronauts and future colonizers, as well as for plants or any other familiar life forms. On the plus side however, the abundant chlorates might provide a good source of much-needed oxygen. Moreover, it is known that some unusual micro-organisms (both bacteria and archaea) are capable of growth via perchlorate reduction. In this relatively strange form of metabolism, perchlorate reductase and chlorate dismutase are enzymes that reduce perchlorate into chloride anions and molecular oxygen. The fact that terrestrial organisms can exploit perchlorate in this fashion leads one to wonder about exotic microbial life forms that might similarly metabolize perchlorate on planet Mars.

21.13 Modern day Martian radiation

Martian atmospheric pressure, at around 1% of Earth's, is much too low to allow liquid water to survive on the surface. Today, any superficial liquid water will quickly freeze or evaporate away (or sublimate after freezing). But for the first billion years or so, Mars may have hosted a global magnetic field that led to a magnetosphere, which protected its atmosphere enough to remain dense and adequately pressurized for liquid water to flow and accumulate in abundance. Additionally, Mars probably once had a far thicker, carbon dioxide-rich atmosphere that provided warmth through a greenhouse effect. The thick Martian atmosphere and existent magnetosphere endowed early Mars with a far different radiological environment than it has today.

Today however, Mars' thin atmosphere affords only a relatively small shielding effect against cosmic rays and provides very modest protection against electro-magnetic radiation from space. Without an ozone layer like what we have in our stratosphere, solar ultraviolet is relatively unattenuated, albeit reduced thanks to the inverse square law. And there is no present-day Martian magnetosphere to ward off charged particulate solar and galactic cosmic rays. The consequence is an average natural radiation level on Mars that is about 40–50 times the average on Earth. Initial data from NASA's special instrument called the Martian Radiation Experiment (or MARIE) on the 2001 Mars Odyssey spacecraft showed that the radiation levels on Mars were nearly 2.5 times higher than what astronauts experience on the International Space Station—0.22 mGy or 0.022 cGy per day. This would translate to 8 cGy (rads) per year. But MARIE also detected two solar proton events during which the radiation levels reached nearly 2 cGy in a single day, and a few other solar proton events that provided around 0.1 cGy. Of course these figures would have to be multiplied by the appropriate radiation weighting factor, wR (quality factor) to obtain the equivalent dose. Further data from the Radiation Assessment Detector (RAD) on the Mars Science Laboratory's Curiosity rover at Gale Crater provided similar galactic cosmic ray doses of 0.210 mGy (0.021 cGy) per day. Combined with the other sources of radiation, this amounts to around 24–30 cGy (rads) per year. This absorbed dose can be compared to the average annual absorbed or equivalent dose on Earth (excluding medical radiation) of roughly 0.365 cGy (or 3.65 mSv or 365 mrem). Including radiation exposure from medical procedures and diagnostic tests, the average radiation dose in the United States of America is now 6.2 mSv y^{-1}.

However, there are some regions on Earth with far more radiation than average. Among these high natural background radiation regions is Ramsar, Iran where a few inhabitants are exposed up to 260 mSv (26 cGy or rads) annually. People have lived here for countless generations over thousands of years without obvious medical consequences such as inordinate rates of cancer or grossly diminished lifespans. Of relevance, and despite being over five times the amount of radiation that radiation workers in the United States of America are permitted, this annual amount of radiation is quite comparable to what Martian settlers might receive without intensive shielding. Furthermore, as much of the radiation exposure in Ramsar is due to alpha particles, the radiation quality in both Ramsar and on Mars is comparable in linear energy transfer and relative biological effectiveness.

Of note, the calculated total amount of radiation absorbed by travelers to and from Mars must include not only the dose absorbed on the surface of Mars but also the considerable radiation exposure during the traversal of interplanetary space. Data from the **Radiation Assessment Detector (RAD)** instrument mounted on the Mars Science Laboratory's Curiosity rover suggested that even the shortest, energy-efficient (**Hohmann transfer orbit**) round trip would result in radiation doses around 660 mSv (0.66 Sv) for a 180-day trip to Mars and a 180-day return trip from Mars. This would have to be added to the 0.64 mSv/day on the Mars surface itself. While some scientists were very concerned about this predicted dose, it is helpful to keep in mind that some patients with hematologic malignancies are given total body irradiation doses well over twice this amount (e.g. 1.5 Gy given in ten fractions spread over 5 weeks). Nevertheless, NASA has documented punctuated doses of relatively intense radiation from solar particle events that could pose significant health problems. For instance, in September 2017, a strong solar particle event was seen from orbit by instruments on the MAVEN mission as well as the European Space Agency's **Mars Express orbiter**. This solar storm sparked an aurora on Mars over 25 times brighter than any previously seen aurora by the MAVEN orbiter. On the surface, the RAD instrument confirmed an ionizing radiation dose more than double what had ever previously been recorded by RAD since its landing in 2012. These high readings persisted for over 2 days. The peak radiation dose rate only persisted around 8 hours but during that time interval the dose rate approached 600 microgray per day (0.6 mGy per day). Again these absorbed dose figures must be multiplied by a weighting factor, wR to obtain the equivalent dose figure. But it may be useful to recall that only sufficiently energetic protons can penetrate the Martian atmosphere. Such protons will likely be in the range of greater than 100 MeV and therefore no different from the protons used in clinical radiation therapy. Surprisingly, this powerful particle event (a coronal mass ejection) occurred during a part of the solar cycle that is typically quiescent (past the solar maximum and en route to solar minimum).

Further reading

[1] Welsh J S, Bevelacqua J J and Mortazavi S M J 2022 Ramsar, Iran, as a natural radiobiological surrogate for Mars *Health Phys.* **122** 508–12

[2] Janiak M K, Pocięgiel M and Welsh J S 2021 Time to rejuvenate ultra-low dose whole-body radiotherapy of cancer *Crit. Rev. Oncol./Hematol.* **160** 103286

[3] Block A M, Silva S R and Welsh J S 2017 Low-dose total body irradiation: an overlooked cancer immunotherapy technique *J. Radiat. Oncol.* **6** 109–15

[4] Stähler S C, Mittelholz A and Perrin C *et al* 2022 Tectonics of Cerberus Fossae unveiled by marsquakes *Nat. Astron.* **6** 1376–86

[5] Ćuk M, Minton D A, Pouplin J L L and Wishard C 2020 Evidence for a past Martian ring from the orbital inclination of Deimos *Astrophys. J. Lett.* **896** L28

[6] Hu R, Kass D and Ehlmann B *et al* 2015 Tracing the fate of carbon and the atmospheric evolution of Mars *Nat. Commun.* **6** 10003

[7] Clement M S, Kaib N A, Raymond S N and Walsh K J 2018 Mars' growth stunted by an early giant planet instability *Icarus* **311** 340–56

[8] Wadsworth J and Cockell C S 2017 Perchlorates on Mars enhance the bacteriocidal effects of UV light *Sci. Rep.* **7** 4662

[9] Hassler D M, Zeitlin C and Wimmer-Schweingruber R F *et al* 2014 Mars' surface radiation environment measured with the Mars Science Laboratory's Curiosity rover *Science* **343** 1244797

IOP Publishing

Space Radiation
Astrophysical origins, radiobiological effects and implications for space travellers
James S Welsh

Chapter 22

Asteroids

Between Mars and Jupiter lies the largest asteroid, **Ceres**. First discovered in 1801 by Giusepe Piazi, Ceres was long thought to be a planet. However, Ceres now meets the definition of a **dwarf planet**. It is round and it is not orbiting another planet. But Ceres has not cleared out its own orbit entirely, and thus does not meet all three modern defining characteristics. To qualify as a 'planet', a celestial body must:

1. Primarily orbit the Sun (as opposed to orbiting another planet, which would mean the body is a moon).
2. Be sufficiently large enough to attain a spherical shape.
3. Have cleared its orbiting path around the Sun of smaller bodies and companions.

And while it does primarily orbit the Sun and it is not orbiting another planet, Ceres does have companions—millions of them. And this disqualifies Ceres (or any other large object in this orbiting path around the Sun) from being called a planet. But Ceres is the largest member of the **main asteroid belt**, which lies between Mars and Jupiter. Nevertheless, Ceres does qualify as a dwarf planet and is the only dwarf planet within Neptune's orbit. Note that aside from the requirement of being large enough to be round under its own gravity, size does not matter. For instance, Pluto and Ganymede (a dwarf planet and a moon, respectively) are both larger than planet Mercury. Nevertheless, of these three only Mercury meets all three requirements and is a genuine planet by the present definition.

Additionally, Ceres hosts the closest known **cryovolcano** to the Sun. Cryovolcanoes erupt 'lava' made not of molten magma rock, but instead of lightweight volatile chemicals such as water, ammonia and methane.

NASA's Dawn spacecraft approached Ceres in 2015 for some close-up photographs. Analysis of its size, shape and gravitational effects has led to the conclusion that it is partially differentiated into a mantle/core made of silicate rocks and water ice with a relatively thick crust that makes up a surprising 30% of its total volume.

Recall that Earth's crust is but a thin veneer over its mantle that makes up less than 1% of the planet's total volume. By some estimates, Ceres could have even more total water that Earth does.

Ceres just happens to be the largest and first-discovered asteroid but there are literally millions of others. These range in size from as small as a meter wide to the dwarf planet Ceres, with its mean diameter of 939 km. Most of these reside in the main asteroid belt. Behind Ceres, the three other largest asteroids in the main belt are **Vesta**, **Pallas** and **Hygiea**. One curious observation is that although there are millions of asteroids out there, the calculated *combined mass of all of those in the main asteroid belt amounts to less than 4% of the mass of the Moon*. In fact, Ceres alone makes up roughly a third of the entire asteroid belt's mass; along with Vesta, Pallas and Hygiea these four confer approximately half of the total mass.

Bode's law predicts that there should have been another planet located at 2.8 astronomical units (AU) based on the Bode's law formula: $[(0, 3, 6, 12, 24, \ldots) + 4] \div 10 = n$ AU from the Sun. The formula worked reasonably well for Mercury, Venus, Earth and Mars but there was no planet seen at the predicted 2.8 AU. Thus, a search was initiated in 1800 which culminated in Piazza's 1801 discovery of Ceres, which has a semi-major axis of 2.77 AU. Subsequently many other asteroids have been discovered in this giant torus between Mars and Jupiter, which is now called the main asteroid belt. This distinguishes asteroids that reside here from other categories of asteroids such as the **near-Earth asteroids (NEAs)** and **trojan asteroids**. There is another group of rocky/icy objects mainly in the space between Saturn and Neptune (i.e. in the vicinity of Uranus) called **centaurs**. These objects seem to have characteristics of both asteroids and comets; thus, they are named after the mythical man–horse mixtures.

Being relatively tiny, the orbits of asteroids are easily perturbated. This can set them on collision courses with other planets—including Earth. They share this feature with comets. But incoming comets tend to originate in the outer Solar System, where they were formed and normally reside. Asteroids on the other hand, reside in the inner Solar System and were probably formed there. Additionally, asteroids tend to be denser and made of carbon, silicates or metal whereas comets may be aptly described as ice-rich 'dirty snowballs'. And although comets can certainly come shooting in from afar and smack into Earth or other planets (as did Comet Shoemaker–Levy 9, when it hit Jupiter in July 1994), the vast majority (>99%) of near-Earth objects are asteroids.

22.1 Trojan asteroids

The trojan asteroids do not reside in the main asteroid belt but instead occupy the same orbit as Jupiter, around 5.2 AU from the Sun. Jupiter's gravity causes these asteroids to collect in small bundles before and after Jupiter in its path around the Sun. These stable locations are gravitational balance points between Jupiter's gravity and the Sun's gravity called **Lagrange points**. Viewed from the Sun's perspective, the trojan asteroids are settled about 60° east and west of Jupiter. These gravitationally balanced spots are sometimes called the **L4** and **L5** Lagrange points. There are other gravitationally-balanced Lagrange points at different orbital locations called L1, L2 and L3 Lagrange points. More and more trojans are being discovered, and the estimated total (of those >1 km diameter) could be as much as a million. They are not as confined to the ecliptic plane as the main belt asteroids are and have orbital inclinations as high as 40°. This allows far more asteroids to be packed into these tight locations than if they were all confined to the plane. Other planets also have Lagrange points, and a 300 m wide asteroid has been discovered traveling ahead of Earth at its L4 Lagrange spot. The recently launched James Webb Space Telescope is parked at Earth's L2 Lagrange point.

22.2 Asteroid classification

In addition to classification based on location and orbital characteristics, asteroids may be categorized based on their color and reflectivity or albedo. These visible superficial spectral features reflect their presumed compositions (which can be further deduced from their densities). Thus, historically we have:

1. C-type asteroids—carbonaceous
2. S-type asteroids—siliceous
3. M-type asteroids—metallic
4. V-type asteroids—'vestoids' with a spectral type similar to Vesta

The C-type asteroids are dark and have low albedo; they are the most abundant asteroids, comprising around 75% of known asteroids. Because they are dark, they are hard to spot—and therefore there could be many more than we currently know of. Containing a lot of carbon, they have relatively low densities at about 1.7 g cc^{-1} on average.

The S-type asteroids constitute another 17%. They have spectral characteristics indicative of a stony (that is, silicate-based) mineralogy. They have higher densities than C-type asteroids, with an average of 3.0 g cc^{-1}.

The M-type asteroids have the highest albedos, especially for microwave or radio waves (**radar albedo**). In fact, M-type asteroids have the highest radar albedos among all the common Solar System objects. Although no known M-type asteroids have densities consistent with pure iron–nickel, they tend to have higher densities than the other types, suggesting that they do indeed contain more metal than the others do. It is believed that M-type asteroids are fractured pieces from a planetesimal's prior metallic core. If this is true, study of M-type asteroids (such as (16) Psyche) could provide invaluable information about the cores of planets (including Earth's) that we will probably never be able to directly sample.

There is another category of asteroids, the V-type asteroids, which have a spectral type similar to the largest example, Vesta. These are the so-called vestoids. Vestoids comprise only about 5% of all known asteroids. These V-type asteroids contain more of the mineral pyroxene than S-types do. Since pyroxenes as a rule do not contain K$^+$ cations, the vestoids are probably not particularly rich in radioactive postasium-40. Curiously, the visible light spectrum of Vesta itself, is similar to the spectra of meteorites called the **HED (howardite–eucrite–diogenite) achondrite meteorites**, suggesting that these meteorites are fragments of Vesta that found their way here.

While such categorization is convenient, of course many asteroids have over-lapping compositional characteristics. For instance, some M-type asteroids have spectra consistent with the presence of hydrated silicates (as evidenced by their absorption characteristics at 3 µm). Furthermore, as observations have gotten more comprehensive, other spectral type classifications have emerged such as the Tholen, SMASS and Bu-DeMeo classification schemes.

There is an organization of asteroid types based on location. The main asteroid belt is wide and the types of asteroids within in varies from place to place. At the innermost edge of the main belt are the S-type asteroids. At the outer edge of the main belt, the asteroids are mainly C-type asteroids. The C-type asteroids account for around 80% of outer belt asteroids ($\sim$3.5 AU from the Sun) whereas inner asteroids ($\sim$2 AU) are only 40% C-type. Scattered throughout the asteroid belt are M-type (metallic) asteroids. These M-type asteroids are believed to be the frag-mented remains of the inner cores of planetesimals or larger asteroids that have been shattered apart.

One fascinating fact is that samples of asteroids may be found here on Earth in the form of meteorites. While some rare meteorites trace their origins to the Moon or Mars, the vast majority are believed to stem from asteroids. And some specific meteorites are linked to specific asteroids. For instance, the so-called **howardite–eucrite–diogenite (HED) meteorites** are believed to be fragments of the asteroid Vesta. Although HED meteorites constitute 60% of all achondrites, they only comprise 5% of all discovered meteorites overall.

22.2.1 Meteorites

As an aside on meteorites, there are several classification schemes but the most popular system breaks them down into three groups:

1. Stony meteorites
2. Iron meteorites
3. Stony-iron meteorites.

Most meteorites are categorized as **stony meteorites** which are either **chondrites** (if they contain millimeter-sized, visible grains called **chondrules**) or **achondrites** if they do not contain chondrules. Roughly 86% of meteorites are chondrites. The chondrules that give chondrites their name are remnants of molten droplets in the protoplanetary disk that were flash-melted by the early Sun during formation of the Solar System.

Among the chondrites are a special group known as **carbonaceous chondrites** that are so interesting because of the organic compounds they can contain. Other types of chondrites include **enstatite chondrites** and **ordinary chondrites**. Of these, the ordinary chondrites are the most abundant. Chondrites are considered the oldest types of meteorites and were formed at the very inception of the Solar System. They are largely leftover materials from the main asteroid belt that failed to congeal into asteroids or planets. They are the building blocks of the Solar System.

Over time, these chondrite building blocks coalesced into larger bodies, eventually forming planetesimals. Upon reaching a diameter of over 200 km, the heat released by radioactivity of these growing bodies (coupled with the heat released by powerful collisions) encouraged melting. Then these planetesimals would undergo differentiation with a dense metallic interior with a lighter surrounding mantle and crust. Massive collisions between these newly formed planetesimals blasted pieces into space. Achondrite meteorites are believed to be the remnants of the crust and mantles of these planetesimals, whereas iron meteorites are shattered remains of the inner cores. Rare stony-iron meteorites such as pallasites represent remains of the interface region between the mantles and cores.

Around 8% of all meteorites are achondrite stony meteorites. These include the HED group derived from Vesta. Other achondrites trace their ancestry back to the Moon or Mars.

Approximately 5% of meteorites are **iron meteorites** or 'irons', that contain iron–nickel minerals such as **kamacite** and **taenite**. Depending on the relative proportions of these minerals, irons can be classified structurally as **hexahedrites, octahedrites** or **ataxites** (in order of increasing taenite content). Incidentally, kamacite and taenite are minerals that are not normally found on Earth. Upon acid etching, the octahedrites demonstrate the classic **Widmanstätten patterns**.

At only about 1%, the rarest of all meteorites are the **stony-irons** or siderolites, which contain characteristics of both. These are believed to be remnants of the core–mantle interface of planetesimals that were smashed by monstrous collisions and hurled into space. Examples of stony-irons include **mesosiderites** and **pallasites**. Of note, the pallasites are not named for the large asteroid Pallas, but rather after the great 18th century naturalist, Peter Simon Pallas.

22.3 NEAs

While comets can approach close to Earth and pose a threat, over 99% of all near-Earth objects are asteroids—the so-called NEAs. By definition, **NEAs** are those that come within 1.3 AU of the Sun. Then, within that group, there are other asteroids called **potentially hazardous asteroids (PHAs)**. PHAs are the closest asteroids, coming within about 8 million km (5 million miles) of Earth. Additionally, they are large enough to pass through Earth's atmosphere and cause significant damage. Technically, to qualify as a PHA, an asteroid must:

1. Approach within 0.05 AU (7.5 million km or 4.6 million miles) of Earth. This is roughly 20 times the Earth–Moon distance of 385 000 km or 239 000 miles.
2. Be larger than 120 m in diameter. Smaller asteroids will likely burn up in the atmosphere before impact.

NEAs are classified into four subgroups based on their orbits:

1. **Armors**—orbits completely beyond Earth's orbit but closer than Mars' orbit.
2. **Apollos**—orbits mostly outside of Earth's orbit but occasionally crossing it.

3. **Atens**—orbits mostly inside Earth's orbit but occasionally crossing it.
4. **Atiras**—orbits completely inside Earth's orbit.

Thanks to their orbits which may intersect Earth's orbit, Apollos and Atens have the greatest risk of a direct hit on Earth.

According to calculations, after only a few million years, any NEA should be ejected from our vicinity. Nevertheless, there are still tens of thousands of such objects, indicating that they must be regularly replenished from somewhere. Most of these come from the main asteroid belt. Jupiter's gravity disrupts the paths of asteroids and makes their orbits more eccentric and inclined to the ecliptic. Such inclined and eccentric orbits are more prone to collisions with one another, thereby setting them way off course. Thus, occasional internal collisions agitate the orbits of these main belt asteroids, and they may subsequently become NEAs. But Jupiter's gravity is also the reason why these millions of asteroids never coalesced into a planet in the first place. Astronomers initially wondered why the numerous scattered asteroids were a failed planet that never conglomerated into a single entity. The gravitational kick from Jupiter received by the asteroids tends to make them too energetic to ever stick together sufficiently to amalgamate into a genuine planet in this part of the Solar System. Jupiter's gravity is also responsible for some observed gaps in the asteroid belt called the **Kirkwood gaps**. Gravitational perturbations become amplified during certain orbital resonances (for instance a 3:1 orbital resonance where an asteroid would make three orbits for every one of Jupiter's). Asteroids in such resonance regions become hurled out of these particular orbits, leaving the Kirkwood gaps.

22.4 Asteroid strikes

Earth has been clobbered by some hefty ones in the past. The most famous one is probably the dinosaur-killer that hit the Yucatan Peninsula around 65 million years ago, ending the Cretaceous Period. It was an estimated 10 km (6 miles) in diameter and left a large impact crater at Chicxulub, at the northern end of the Yucatan. Even larger than this one is a 160–300 km (110–200 mile) diameter crater at Vredefort in South Africa; thanks to erosion the exact size is now difficult to accurately estimate. This is the scar of a massive impact that occurred around 2 billion years back during the Proterozoic Eon. The asteroid that made this crater was probably at least 10 km wide, with recent estimates as large as 25 km (16 miles) and moving as swiftly as 25 km s^{-1} (56 000 mph).

22.5 Asteroid mining and associated hazards

Besides the morbid fascination with asteroids because of their potential for mass extinctions and the end of humanity, there is another reason why people are so interested—precious resources. NASA is planning to send a spacecraft to the 225 km diameter, M-type, main belt asteroid called (16) Psyche also called 16 Psyche or just Psyche. The spacecraft is eponymously named 'Psyche'. This will be the first such expedition to a mostly metallic celestial body (as opposed to an icy or silicate body).

Based on its high density ($\sim$4 g cc^{-1}), (16) Psyche is the largest M-type asteroid yet discovered and is predicted to be perhaps as high as 60% metal. If that metal (probably mostly iron/nickel) could be harvested and safely returned in bulk, it could meet the world's needs indefinitely. Also, there has been some speculation that this asteroid contains a sizeable amount of gold, although this is yet unconfirmed. Nevertheless, there is ongoing discussion, both serious and speculative, about someday regularly mining asteroids for their resources. And although a seemingly odd choice, there is a fair amount of internet traffic on the topic of colonization of asteroids. Perhaps if harvesting resources from asteroids becomes a reality, short-term colonization will be needed.

In any case, if humans were to embark on missions to and from the asteroids, the radiation environment must be taken into account. Microgravity will be an obstacle, along with the frigid cold and absence of oxygen. But since none of the asteroids have an atmosphere to speak of or are likely to possess a substantial dynamo-generated magnetosphere, people and instruments traveling to the asteroids will be subjected to the full impact of interplanetary radiation. However, the presence of some magnetism is not impossible, especially in those that are largely metallic such as (16) Psyche. The Psyche spacecraft will host a magnetometer to search for remnant magnetism.

If (16) Psyche is a remnant of a planetesimal core that once generated its own magnetic field through a dynamo, then it is possible that some vestige of this primordial magnetism may still remain detectable (just as on the Moon). NASA's Psyche spacecraft will be searching for such residual magnetism. But to our main point, such relic magnetism would be unlikely to be strong enough to ward off incoming charged particle radiation from outside our Solar System or from the Sun itself. Thus, radiation shielding and evasive maneuvers will be important when working on or residing on asteroids. While it may be possible in theory to seek shelter underground, some asteroids might be more like large rubble heaps than solid ground, making the formation of such shelters difficult (or maybe easier, depending on the technology available).

22.6 Radioactivity of asteroids

While asteroids are not as a rule very radioactive since they do not contain more than their fair share of primordial radionuclides, they will (like any celestial body) contain some long-lived radioactive potassium, thorium and uranium atoms. Despite their present unimpressive radiological status, the data indicate that at one time, at least some of these asteroids were far more radioactive than they presently are.

Meteorites, especially chondrites with abundant calcium–aluminum inclusions (CAIs) such as the Allende meteorite contain a relative excess of Mg-26 in its CAIs. This stable isotope is the decay product of Al-26, a radionuclide with a half-life of 717 000 years that decays via positron emission or electron capture with a total decay energy of 4 MeV (i.e. Q = 4.0 MeV). Of course, decay through electron capture leads to gamma rays or conversion electrons, as well as characteristic x-rays and Auger electrons (whereas positron emission leads to two annihilation gamma photons of 511 keV when the positron encounters an ordinary electron). Iodine-129 was the first primordial radionuclide discovered in meteorites thanks to an over-abundance of its decay product xenon-129. Although iodine-129 has a long half-life (15.7 million years or roughly 16 Ma) from the perspective of nuclear fission products from reactors, this is very short-lived from the perspective of the origin of the Solar System. Hence, there is no remaining iodine-129 at this point, nearly 4.6 billion years after the inception of the Solar System. Nevertheless, early in the history of asteroids, iodine-129, aluminum-26, and other primordial radioisotopes were abundant, and could have provided a significant source of heat. Besides iodine-129 and aluminum-26, many other short-lived primordial radioisotopes exist including ^{7}Be ($t_{1/2} = 53$ days), ^{10}Be ($t_{1/2} = 1.5$ Ma), ^{36}Cl ($t_{1/2} = 0.3$ Ma), ^{41}Ca ($t_{1/2} = 0.1$ Ma), ^{53}Mn ($t_{1/2} = 3.7$ Ma) and ^{60}Fe ($t_{1/2} = 1.5$ Ma). The relative abundance of these primordial radionuclides at the birth of the asteroid can be calculated from the amount of stable daughter products left behind.

Analysis of the Jbilet Winselwan meteorite, a 6 kg carbonaceous chondrite that probably originated as a piece of a C-type asteroid, showed that it contains a carbonate mineral—calcite—that typically requires the presence of liquid water for formation. This led to the question of how liquid water might have ever existed on a small, cold asteroid so far from the Sun. Using radiometric dating, it appears that the calcite formed just 2.6 million years after the Solar System formed. This implies that at that point in time, the parent asteroid was warm enough for liquid water to exist on the asteroid. Furthermore, some of the meteorite minerals appear to have thermally decomposed (thermal metamorphism), indicating that the temperature must have gotten quite hot—perhaps over 300 °C. Of the various possible heat sources, radioactive decay is the most likely explanation for this thermal meta-morphism. Experiments using aluminum-26 have shown that this radioisotope represents a plausible heat source for thermal metamorphism in this scenario. It appears feasible that some sections of a small asteroid could be heated up to temperatures beyond 300 °C through the decay of abundant aluminum-26.

Further reading

[1] King S D, Bland M T, Marchi S, Raymond C A, Russell C T, Scully J E C and Sizemore H G 2022 Ceres' broad-scale surface geomorphology largely due to asymmetric internal convection *AGU Adv.* **3**

[2] Fujiya W *et al* 2022 Hydrothermal activities on C-complex asteroids induced by radioactivity *Astrophys. J. Lett.* **924** L16

[3] Allen N H, Nakajima M, Wünnemann K, Helhoski S and Trail D 2022 A revision of the formation conditions of the Vredefort crater *J. Geophys. Res.: Planets* **127**

Chapter 23

Jupiter and the effects of space radiation on electronics

Beyond the main asteroid belt lie the outer planets. Named after the Roman god Jove (also called Jupiter), these so-called **Jovian planets** are quite different from the inner or **terrestrial planets** in many ways. For one, these planets are giants. Even the smallest of the outer planets, Neptune, is 3.9 times the diameter of Earth. Jupiter is by far the largest planet in our Solar System and is about 11.2 times the diameter of Earth. Jupiter's volume is over 1321 times that of Earth's. But this leads to another distinction between the outer and inner planets. Despite being over 1300 times the volume of Earth, Jupiter is only ~318 times as massive as Earth. This indicates a remarkable difference in density. Unlike Earth's average density of roughly 5.5 g/cc, Jupiter's density is about 1.33 g/cc, making it under a quarter as dense. This density difference underscores another major difference between the small, inner terrestrial planets and the giants of the outer Solar System—their compositions. Because of their low densities and compositions, Jupiter and Saturn are called **gas giants** whereas Uranus and Neptune are **ice giants**. And to address the main focus of this text, the radiation environments around the Jovian giants are also quite different from the terrestrial planets. For instance, all four Jovian planets have extensive radiation belts but of the four terrestrial planets, only Earth has a radiation belt. NASA's **Juno mission Jupiter Energetic Particle Detector Instrument (JEDI)** has confirmed that the radiation belt around Jupiter is somewhat like our Van Allen belts but is millions of times more intense.

23.1 Characteristics of Jupiter

Jupiter is enormous. As mentioned, Jupiter's diameter is over 11 times that of Earth's and is therefore over 1300 as voluminous. In terms of diameter, 'Jupiter is to Earth as the Sun is to Jupiter' with roughly tenfold relative differences.

Its mass is nearly 320 times as great as Earth's mass. In fact, Jupiter's mass is over two and a half times the mass of all the other planets and asteroids combined. Despite its mass, it spins at an impressive rate; its sidereal rotation period is 9.92 h. This rapid rotation leads to a centrifugal effect that changes its shape from spherical to a widened **oblate spheroid** with a distinctly broader equatorial radius than it north–south radius (71 492 km vs 66 854 km, respectively). The rapid rotation rate causes some extreme weather phenomena including the famous Great Red Spot, a hurricane larger than planet Earth which has been going on for the last several centuries and will be discussed further in this chapter.

Jupiter orbits the Sun at a distance of about 5.2 astronomical units (AU) (which can be remembered through Bode's mnemonic: (0, 3, 6, 12, 24, *48*... + 4)/10 = 5.2). In accordance with Kepler's Third Law, at this distance Jupiter is traveling quite slowly and takes about 12 Earth years to complete 1 Jovian year.

The Voyager 1 space probe discovered a faint ring around Jupiter in 1979, making Jupiter the closest planet to the Sun with a ring system. Jupiter's ring system includes, from inside to outside, an inner halo ring, a main ring and two gossamer rings (the Amalthea gossamer ring and the Thebe gossamer ring).

23.2 Chemical composition of Jupiter

The elemental composition of Jupiter is far more reflective of the Universe as a whole and the primordial solar nebula from which the Solar System condensed than the Earth is. The atmospheric chemical composition is about 89% hydrogen, 10% helium, 0.3% methane (CH_4) and 0.03% ammonia (NH_3). Jupiter's atmosphere is rather dry. Based on the Galileo mission it was believed that, like the Sun, Jupiter's atmosphere had very little water. However, recent results from NASA's Juno mission suggest that at the equator, water makes up about 0.25% of the molecules in Jupiter's atmosphere (which is almost three times that of the Sun). Somewhat surprisingly, Jupiter's atmosphere contains relatively more hydrogen deuteride (molecular hydrogen in which one hydrogen is hydrogen-1 and the other is hydrogen-2) at about 0.003%. The interior of Jupiter is also largely diatomic molecular hydrogen but thanks to the immense pressure at depths within Jupiter, this hydrogen transforms into metallic hydrogen which consists of a solid lattice of protons and delocalized electrons. Metallic hydrogen behaves very differently from molecular hydrogen as it is a good conductor of heat and electricity. The rapid rotation of this metallic hydrogen is what probably confers the strong magnetic field to Jupiter (which is about 10 times stronger than Earth's).

23.3 Jupiter's structure

Being a gigantic ball of gas, it is hard to assign a surface to Jupiter. Of course, one definition might focus on its visual characteristics such that the surface is defined as when the gas becomes opaque to visible light. Another definition of its surface is the depth within Jupiter's atmosphere where the pressure equals that on the surface of Earth, that is around 1 bar. If we choose that definition for Jupiter's surface, the gravitational attraction is much stronger than on the surface of Earth. Instead of the familiar gravitational acceleration of 9.8 m s^{-2}, the gravitational acceleration on Jupiter would be about 25 m s^{-2}.

Among the most striking visual features of Jupiter are the colorful bands moving west to east and vice versa on its surface. While these rapidly moving clouds are mostly white hydrogen and helium, they contain a smattering of other elements that confer the brown and reddish hues. Ammonia clouds are largely white, superficial horizontal bands. Clouds of water can be found in deeper layers. The brownish-red colors in other bands signify the presence of sulfur. Because of Jupiter's amazingly fast rotational speed, these clouds stretch out into elongated, lively bands.

Underneath Jupiter's layer of metallic hydrogen is what is likely an icy layer, which surrounds a relatively small rocky core although details remain uncertain. Some recent data from Juno have been quite surprising and suggest that the core might not be as solid as previously believed but occupies a substantial fraction of the planet's total volume with a radius estimated to be 30%–50% of Jupiter's planetary radius. This 'fuzzy' core contains heavy elements and has an estimated mass between 7 and 25 times the Earth's mass. One hypothesis to explain this oddity is that Jupiter was struck by a massive object that stirred up its insides, mixing some lighter hydrogen and helium with the material in the original dense core.

23.4 The Great Red Spot and weather on Jupiter

A probe dropped by the Galileo spacecraft provided evidence of Jovian winds in excess of 400 mph. Comet Shoemaker–Levy 9 crashed into Jupiter in 1994 but through careful tracking of its stirred up molecular fingerprints over the years, astronomers from the Laboratoire d'Astrophysique de Bordeaux were able to calculate phenomenal wind speeds of over 900 mph ($\sim$400 m s^{-1}) in the polar regions of Jupiter. While not as fast as the fastest rifle shots, this is definitely in Superman's league since it does exceed the velocity of some speeding bullets.

The Great Red Spot is an enormous storm located in Jupiter's southern hemisphere yet rotating in a counterclockwise fashion. This distinguishes this enormous hurricane from those here on Earth which would rotate in the opposite direction. For instance, hurricanes that hit the southeastern United States of America rotate in the counterclockwise direction since they are in the northern hemisphere while tropical typhoons in the southern hemisphere spin in a clockwise fashion. These terrestrial storms (hurricanes and typhoons) are just special names for tropical cyclones. And on Earth, these cyclones are low-pressure systems. The Great Red Spot is colloquially called a 'hurricane' but technically, because it is a high-pressure rather than low-pressure system, it is an **anticyclone**. As such, it spins in the opposite direction from the familiar cyclones here at home. The most intense winds of the Great Red Spot are along the outer edge of its oval shape, reaching up to 250 mph (400 km h^{-1}). Data from the Juno mission show that the Great Red Spot is not only wide but incredibly deep—about 350 km (217 miles) in height. This makes it roughly 20 times as tall as a major hurricane on Earth.

Although there is some debate about whether it was truly the same storm, the Great Red Spot was first noted perhaps as long ago as 1665 by Giovanni Cassini. If it is the same storm, then the Great Red Spot hurricane was raging when Issac Newton was a student at Trinity College and 20 years before the birth of Johan Sebastian Bach! But modern observations show that the spot is shrinking. At the present rate of shrinkage, it could be gone in another 70 years.

At 16 350 km (10 160 miles) wide, the Great Red Spot is about three times the diameter of Earth and thus could easily engulf the entire Earth. But although this is the largest and most famous storm on Jupiter, there are many more. Multiple, white ovals representing smaller storms, are seen scattered about the surface. On occasion, these storms have been observed merging into even larger ones.

Jupiter itself is still slowly contracting after all these years. And as the gas contracts and undergoes phase transitions, a great deal of heat is released. This internally generated heat drives upward convection currents. These upward convection currents, coupled with the rapid rotation of the surface are the driving forces behind Jupiter's complex weather. In addition to the Earth-based observations and the Hubble Space Telescope, numerous data regarding Jupiter have been gleaned from the Pioneer missions, the two Voyager expeditions, Galileo, Cassini and of course Juno.

23.5 Jupiter's moons

Orbiting Jupiter are over 75 moons. Some of these are amazing little worlds—and not so little at that, since Ganymede is unequivocally larger than Mercury while Callisto is 99% of the diameter of Mercury, and even our gigantic Moon is smaller than three of the four **Galilean moons** around Jupiter. These Galilean moons are **Io**, **Europa**, **Ganymede** and **Callisto** in order of distance from Jupiter. In size order from smallest to largest, they are: Europa, Io, Callisto and Ganymede. Nevertheless, while Europa is slightly smaller than our Moon, it is still bigger than dwarf planet Pluto, which was considered for many years to be a genuine planet. Incidentally, there was a great debate surrounding the 1610 discovery of these Jovian satellites, including allegations of plagiarism. After a formal analysis of the historical records, it was concluded that Galileo Galilei published his discovery one day earlier than Simon Mayr (Latinized to Simon Marius). Nevertheless, the names we use today for the four large moons of Jupiter were given by Marius.

Io is said to be the most geologically active body in our Solar System (although Earth gives it a good run for the money). Its numerous active volcanoes spray sulfur-containing compounds into space, contributing to the unusual energetic ion species found in Jupiter's radiation belts. Io's intense geological activity is due to the constant tugging of Jupiter's immense gravity while Io conducts its eccentric, 2 day orbit. Jupiter's gravitational tidal forces lead to incessant squeezing and stretching (like kneading dough) and generate frictional heat internally. This internal heat is then manifested geologically as volcanoes. Io orbits Jupiter within its incredibly intense radiation belts. The massive radiation dose leaves a visible mark on Io's polar regions, which are noticeably darker than other areas. Given today's technology, it seems unlikely that any manned missions to Io will ever be possible. And even unmanned missions will have to have specially radiation-hardened electronics to avoid quick inactivation or even destruction. Hardening of electronic devices is covered later in this chapter.

23.5.1 Europa

Europa's relatively smooth, icy surface is geologically speaking relatively young since it lacks many impact craters. (Young in this context means in the 40–100 million year old range.) Old craters could not persist since the shifting surface ice would eventually erase the, evidence. Europa's ice surface floats atop a deep subsurface ocean and this ice behaves in some ways like our crust. In fact, some of that surface ice has been observed to break up and become subsumed into the ocean below. In a sense, this is just like how plate tectonics operates on Earth. If one accepts this type of subduction and generation of new ice in the resulting gaps as 'plate tectonics', then Europa is the only place in the Solar System aside from Earth to exhibit plate tectonics. Curiously, with two to three times the amount of water as on Earth, and with much of this water in a liquid form (albeit below the frozen surface), one must naturally wonder if life is teeming about in that massive Europan ocean. Several key elements of biological chemistry—carbon, hydrogen, nitrogen, oxygen—are found in abundance on Europa. Furthermore, Europa appears to offer a very stable environment over vast expanses of geological time. Climatological shifts are buffered by Jupiter's gravity which provides a steady source of heat. This contrasts Europa with Mars, whose highly variable axial tilt puts it at risk of dramatic climatological changes, or Venus, whose runaway greenhouse effect has made its surface a Hadean nightmare. It seems like Europa's subsurface ocean has probably been consistently liquid for billions of years. This might have allowed organic synthesis to proceed unimpeded for extensive periods of time, possibly culminating in biological entities. For life to evolve and persist, a reliable energy source is mandatory. Unlike cold and barren Mars, Europa probably has the equivalent of numerous geothermal vents at the bottom of its seas. These deep-sea vents are probably churning up nutrients from the rocky mantle below, just as they do on Earth. Although at Europa's distance no sunlight can penetrate the thick ice and deep ocean, these hydrothermal vents have perhaps been a dependable source of heat for billions of years, thereby possibly allowing a chemosynthetic ecosystem to evolve. Thus, Europa is arguably the most fascinating—and promising—extraterrestrial world in our Solar System from an astrobiological perspective.

Nevertheless, one must also factor radiation into the equation. On Earth, our geomagnetic field wards off much of the charged particulate radiation and our ample atmosphere absorbs energetic electromagnetic radiation and further protects us from charged particles (although it does generate secondary cosmic rays that vastly outnumber the primaries). Europa has neither a protective magnetic field nor a thick atmosphere. Nevertheless, its thick layer of ice and the underlying liquid water effectively shield the deep sea from radiation. The total depth of water is estimated at 100 km (62 miles) but just how thick the surface really is remains debated. Best estimates for the thickness of the ice layer are in the range of 15–25 km or 10–15 miles. Hopefully the planned NASA **Europa Clipper** and the ESA **Jupiter Icy Moons Explorer (JUICE)** will provide more definite answers.

23.5.2 Ganymede

Historically, Ganymede was the first celestial body observed orbiting another entity. This observation strongly supported the Sun-centered Copernican model (which replaced the older, geocentric Ptolemaic model). Ganymede is the next farthest out of the four Galilean moons, after Io and Europa. It orbits Jupiter in a 1:2:4 orbital resonance with the other Galilean moons Europa and Io, respectively. This is tantamount to two consecutive 2:1 mean motion resonances; the resultant 4:2:1 period ratio among the three bodies involved is known as a **Laplace resonance**.

Larger than the planet Mercury (but not nearly as massive), Ganymede has an odd mixture of features including a magnetic field but virtually no atmosphere. What little atmosphere it does hold onto appears to contain oxygen in atomic, diatomic molecular and ozone forms along with some hydrogen. It appears to be made of roughly equal amounts of silicate rocky material and ice, along with a small central liquid metal core. Like Europa, Ganymede probably has a vast internal ocean beneath its icy surface.

Like Io and Europa, Ganymede is still within Jupiter's radiation belts. While far less than on the surface of Europa, the radiation level on Ganymede is still prohibitively high at an estimated 50–80 mSv d^{-1}. Total body exposure to unshielded radiation doses of this magnitude can have health consequences (such as lymphopenia and thrombocytopenia—a drop in circulating lymphocytes and platelets in the blood, respectively) after just a week or two. Nevertheless, it is worth recalling that definitive total body radiation therapy for hematologic malignancies has been safely and effectively given at higher doses still—at 10–15 cGy per fraction given until the cumulative dose over about 5 weeks reaches 150 cGy (1.5 Gy). Curiously, Ganymede is the only moon in our Solar System known to have its own magnetic field. As such, Ganymede also has a small radiation belt of its own. Ganymede's radiation belt contains electrons with energies ranging from 30 keV to over 1 MeV along with protons with energies ranging from 30 keV to 2 MeV. Energetic oxygen and sufur ions are also present in Ganymede's radiation belts.

23.5.3 Callisto

Callisto is the farthest out of the four Galilean moons and is the second largest after Ganymede. It is the third largest moon in the Solar System after Ganymede and Saturn's Titan. Despite its size (99% the diameter of planet Mercury), it does not appear to be internally differentiated. Unlike the other three Galilean moons, it is not in an orbital resonance and its orbit is not very eccentric. Thus, it is not subject to significant tidal heating like the others. Nevertheless, it is tidally locked, meaning that the same hemisphere always faces Jupiter.

Like Europa and Ganymede (as well as Saturn's moons Enceladus, Dione and Titan and possibly Neptune's Triton), it appears that Callisto has a large subsurface salty ocean. Since on Earth, anywhere we find liquid water we find life, the possibility of extraterrestrial life in this subglacial liquid water ocean is intriguing. But the odds of microbiological activity in the Callistoan ocean are less than in the Europan ocean since Europa is extensively heated through tidal friction whereas Callisto is largely dependent on residual radioactive isotopes to heat its subglacial ocean. Furthermore, the Europan ocean is likely in direct contact with nutrient-rich silicate minerals on its floor, whereas the Callistoan ocean might not be.

Callisto does seem to have a very thin atmosphere composed primarily of carbon dioxide with some diatomic molecular oxygen. This thin atmosphere does appear to host an intense ionosphere. And Callisto is the only one of the Galilean satellites that is outside Jupiter's radiation belt. The relatively low radiation levels make Callisto the most suitable of the four Galilean moons for manned exploration and possible future moon bases. Best present estimates for the radiation exposure on the surface of Callisto are around 0.1 mSv d^{-1} (or 10 mrem d^{-1}). Since natural background radiation in the United States of America is roughly 1 mrem d^{-1}, the radiation on Callisto is about ten times as high as here.

23.6 Radiation around Europa and acute radiation syndromes

Given its extensive, and probably nutrient-rich subsurface ocean, Europa may very well be the most astrobiologically intriguing body in the Solar System. But just how much radiation is there in the vicinity of Europa and the other Galilean moons? Given the intense and extensive radiation surrounding Jupiter, manned missions to the Galilean satellites including Europa would certainly be radiobiologically prohibitive. In fact, if the estimated dose rates are correct, a human would not survive the radiation dose from just a single day on the surface of Europa. Presently we do not have any actual measurements of the radiation on the surface of Europa, but estimates are around 5.4 Gy per day. The dose on the surface of Io is even worse since Io is right smack in the middle of Jupiter's intense radiation belts. Daily doses up to 36 Gy d^{-1} might be absorbed on Io's surface. Given that a potentially fatal total body dose to a human is around 2.5–5 Gy, a daily dose of 36 Gy is certainly a showstopper. (But then again, so are the volcanoes!)

According to the CDC, and as defined by the LD50/60 (which is a dose that will kill 50% of people exposed within 60 days without treatment) potentially fatal **hematopoietic acute radiation syndrome (H-ARS**, sometimes simply called the **bone marrow syndrome**) may occur at total body acute doses of 2.5–5 Gy if no treatment is rendered; with treatment the LD50/60 for H-ARS increases to around 6–7 Gy. The hematopoietic syndrome, if untreated, may result in bone marrow failure and potential death a few weeks to months after the acute exposure due to an inability to fight off infection thanks to neutropenia (loss of neutrophils) and bleeding thanks to thrombocytopenia (loss of platelets). There are other forms of acute radiation syndrome (**ARS**) aside from H-ARS such as the **acute gastrointestinal (GI) syndrome** and the **acute central nervous system (CNS) syndrome**. Technically, to qualify as an ARS, several conditions must apply: the dose must be acute (that is, all within minutes rather than spread out over many hours, days or longer), it must be to the total body (or very close to it), and it must be sufficiently large (i.e. over 0.7 Gy).

Additionally, according to most definitions, the radiation must be penetrating; it cannot be confined to the skin surface as might be the case with radiation from pure beta emitters or electron beam radiation exposure. However, there is an **acute radiation cutaneous (or skin) syndrome** with a threshold dose of about 10 Gy. The latter is an example of what is known as **cutaneous radiation injury** or **local radiation injury**.

Acute total body doses beyond 10 Gy (but as low as 6 Gy) may lead to the **GI ARS** (often just called the GI syndrome). This typically leads to death within 2 weeks due to dehydration, electrolyte imbalances and infection. The **CNS syndrome** (also called the **neurovascular** or **cardiovascular (CV) syndrome**) will occur at acute total body doses over 50 Gy but may appear at total body doses as low as 10 Gy. This form of ARS is invariably fatal, and death typically ensues within a few days. At 36 Gy d^{-1}, the radiation on the surface of Europa would quickly overwhelm any unshielded astronauts through these various acute radiation syndromes.

While these various manifestations of acute radiation exposure are given different names, they overlap significantly and may be better described as clinical **subsyndromes** that occur in phases over timeframes of hours to weeks following exposure, as the various manifestations appear. Thus, some authorities refer to the above-mentioned syndromes as **subsyndromes of ARS**. Thus, there are the hematopoietic, GI and neurovascular subsyndromes of ARS. All these subsyndromes display a common clinical pattern of response following the radiation exposure:

1. The prodromal phase
2. The latent phase
3. Manifest illness
4. Recovery or death.

The **prodromal phase** is most often expressed as GI symptoms such as nausea, vomiting, diarrhea and anorexia. This may be accompanied by fever. The prodromal phase is triggered by inflammation that is caused by the release of **proinflammatory cytokines** or mediators released from the injured tissues. The classic mediator is **serotonin** produced from the intestines. Following the prodromal phase is the **latent phase**. Depending on the total dose, this latent phase may last several weeks, or it may be as short as just a few hours. The higher the radiation dose, the shorter the latent phase. During the latent phase, the acute inflammation of the prodromal phase subsides, and the irradiated patient feels better. Unfortunately, in heavily irradiated individuals, the latent phase is just a temporary reprieve and the next phase, **manifest illness** inevitably ensues. This phase of ARS is provoked by the damaged tissues and organs. When vulnerable tissues and organs have suffered significant radiation injury, they manifest the damage through the specific subsyndromes. For instance, if the GI system is heavily damaged by total body doses of say, 8 gray, the (presently) irreversible GI subsyndrome may be the displayed manifest illness. As with all ARSs, the GI subsyndrome follows a prodromal and latent phase. The GI subsyndrome manifest illness phase usually appears a couple of weeks after the exposure and is reminiscent of the prodromal phase—nausea, vomiting and diarrhea may return. In addition to these symptoms however, dehydration, severe electrolyte imbalances and infection (due to breach of the GI system lining) also manifest themselves. These problems are the causes of death in those with the GI subsyndrome. The LD50/60 for the GI subsyndrome is 6–7 Gy even with modern supportive care. The hematopoietic subsyndrome manifest illness phase appears later than the GI or CNS subsyndromes; its latent phase is prolonged and it may not be evident for several weeks following the irradiation. Importantly, the hematopoietic subsyndrome is potentially reversible through intensive medical care. Patients with this particular manifest illness are vulnerable to infections and bleeding. Thus, proper wound care and isolation to minimize risks of infection or judicious use of antibiotics may be indicated. Also, platelet transfusions may reduce bleeding as the thrombocytopenia worsens. Mature red blood cells (erythrocytes) circulating in the bloodstream typically live 100–120 days and have no nuclei; they are not very sensitive to radiation. On the other hand, the erythrocyte precursors (erythroblasts) in bone marrow are. Some patients with hematopoietic subsyndrome may exhibit signs and symptoms related to decreased red blood cell counts (i.e. anemia). Transfusions may be of substantial help with such symptoms. Use of cytokines to promote platelet and erythrocyte production (such as thrombopoietin and erythropoietin) have been suggested but have not yet been shown to be more clinically useful than transfusions of platelets and red blood cells. And there is some concern that the use of such agents might be leukemogenic (meaning it might lead to leukemia in the future), but this has not been confirmed. Cytokines to promote restoration of the immune system (white blood cells) such as the colony-stimulating factors filgrastim, pegfilgrastim and sargramostim may be helpful in warding off infection in those with hematopoietic subsyndrome. Finally, bone marrow transplant (hematopoietic stem cell transplant) may be considered in severe cases although to date, this approach does not have clinical proof of efficacy.

Sadly, for those who have been exposed to radiation doses high enough to produce the manifest illness of the CNS subsyndrome (CV or neurovascular subsyndrome), there is no treatment aside from supportive comfort care. All such heavily irradiated individuals will die in a matter of days. The prodromal phase usually appears within minutes after the irradiation and tends to be severe. The latent period is brief and typically lasts only up to 6 h. Multiple organ failure leads to death within the next few days in most such cases.

Being within Jupiter's radiation belts, Europa's surface is constantly being bombarded by blasts of radiation from Jupiter. Analysis of data from the Galileo and Voyager 1 spacecrafts demonstrates that the highest radiation levels are concentrated around the equator, with less radiation near the poles. (Note that this distribution of space radiation is the opposite of the pattern on Earth's surface.)

Even if there is life deep beneath the Europan sea, evidence of such life may never be found on the surface. The intense radiation will break chemical bonds and eventually degrade and erase all traces of complex organic chemistry that floats to the surface and becomes exposed. Hence, missions to Europa likely will not find any traces of life or even intact complex biological molecules by simply scanning the superficial ice.

But the thick coat of ice will filter out even these massive radiation doses after a certain depth. The depth of penetration for either electromagnetic or particulate radiation is directly proportional to the energy. So, the question arises as to how far down the robots on lander missions must drill or dig into that ice to find any preserved evidence of life. Modeling of the Europan radiation environment and its ice layer has led to the conclusion that it might not have to dig too deep. The radiation intensity will likely be sufficiently diminished to allow preservation of biosignatures by 10–20 cm even in the high-radiation zones. In the middle- and high-latitude polar regions, where the radiation is less intense, one might need only drill 1 cm deep to find preserved biosignatures frozen in the ice.

23.7 Does Europa glow in the dark?

In addition to Europa's astrobiological importance, there could be an aesthetically appealing '**moon glow**' on Europa emanating from the surface thanks to the perpetual irradiation. As Jupiter's radiation incessantly blasts Europa's surface with electrons and other particles, these particles may cause Europa to glow in the dark through fluorescence. Experimentally blasting ice that was designed to mimic Europa's frozen surface with radiation like what is expected around Jupiter caused the ice to glow greenly. This ice glow was most intense at around wavelengths of 525 nm and was enhanced by **epsomites** (magnesium sulfate) which should be found in abundance in Europa's ice. On the other hand, carbonates and plain old sodium chloride (which is also likely to be plentiful on Europa) strongly quenched the ice glow. We would never be able to see such ice glow from our vantage point here on Earth or with our space telescopes; one would only see the eerie glow from the dark (antisolar) side. However, the planned Europa Clipper could possibly detect this predicted fluorescence. If so, analysis of the detected glow should reveal much about the chemical composition of Europa's ice. And if the Europa Clipper does confirm the existence of such ice glow, other moons in high-radiation zones (such as Io and Ganymede) might be expected to exhibit similar phenomena (figure 23.1).

23.8 Jupiter's magnetosphere

Jupiter is 'radio loud'. In fact, Jupiter emits more radio noise than any other object in the Solar System besides the Sun. It was radio noise that provided the first hints that Jupiter has a strong magnetic field. Jupiter emits radio waves in a host of different wavelengths. Decimetric radiation (wavelengths between 3 cm and 3 m) is synchrotron radiation that originates from relativistic electrons moving in a magnetic field. The intense radio noise thus provided clear evidence of a large magnetic field around Jupiter.

We now know that Jupiter and the other giant planets have powerful planetary magnetic fields that lead to hazardous radiation belts. Jupiter has the strongest magnetic field among all the planets in our Solar System. Earth's geomagnetic field

Figure 23.1. An artist's illustration of Jupiter's moon Europa as viewed from the side facing away from the Sun. Europa is within Jupiter's gigantic radiation belt, meaning its surface is relentlessly blasted with high-energy particles. Bombarment with 10–25 MeV electrons might make Europa glow in the dark, that is fluoresce. The color and intensity of the radiation-induced ice glow would depend on the incoming radiation and the chemical identity of the targeted material. Thus, spectral analysis might reveal details about the chemical composition of Europa's surface. Credits: NASA/JPL-Caltech.

is generated by electrical currents in circulating liquid iron in the outer core. Similarly, Jupiter's planetary magnetic field is generated by electrical currents in its outer core, which is composed of liquid metallic hydrogen. The result is a magnetic moment approximately 18 000 times stronger than Earth's (dipole moment of 1.55×10^{20} T m^3 versus 7.91×10^{15} T m^3). Curiously, although all four of the giant planets host radiation belts, of the four terrestrial planets, only Earth has a radiation belt.

The vast area around Jupiter where its magnetic field affects its regional environment is its **magnetosphere**. A magnetosphere may be conceptualized as the volume in space around a planet (or moon or star) wherein that body's magnetic field affects the movement of charged particles. Jupiter is surrounded by an enormous magnetosphere, which has over a million times the volume of Earth's magnetosphere. Jupiter's magnetosphere is among the largest objects in the Solar System; if it were visible, it would be larger than a full moon (6° versus 0.5° in angular diameter). The Jovian magnetosphere averages about 5.3 million km (3.3 million miles) wide, which is nearly 15 times the width of the Sun.

Contributing to the size of Jupiter's magnetosphere is a far feebler solar wind pressure at this great distance from the Sun—only about 4% of what we experience here at Earth. This means there is less compression of Jupiter's magnetosphere by the incoming solar wind. In fact, by volume Jupiter's magnetosphere is the second largest structure in the whole Solar System, second only to the Sun's magnetosphere —the heliosphere. (Note—one definition of the heliosphere may include not just the solar magnetosphere but also its '**astrosphere**' or stellar wind bubble that includes all the gas blown into the interstellar space by the solar wind.) The Jovian magneto-sphere extends far beyond its four large Galilean moons Io, Europa, Ganymede and Calisto. On the sunward side, Jupiter's magnetosphere begins diverting solar wind protons nearly 3 million km before they reach Jupiter. On the anti-Sun side, the Jovian magnetosphere sculpts the solar wind even beyond the orbit of Saturn.

23.9 The Jovian radiation belts

As with Earth's magnetosphere which forms the **Van Allen belts**, charged particles become trapped by Jupiter's magnetosphere and form the **Jovian radiation belts**. Charged particles (mostly electrons, protons, alpha particles and oxygen and sulfur nuclei) become ensnared and boosted to high energies in the Jovian magnetosphere and form huge radiation belts surrounding Jupiter. The captured charged particles are accelerated to enormous energies by Jupiter's strong magnetic field, ultimately reaching energies that qualify them as ionizing radiation. These trapped energetic charged particles then constitute Jupiter's incredibly intense radiation belts. These Jovian radiation belts are similar to our Van Allen belts but are many millions of times more intense. Generally speaking, trapped charged particles in the Jovian radiation belts are about ten times more energetic than those in our Van Allen belts. In addition, the ions are several orders of magnitude more abundant in the radiation belts of Jupiter. Thus, in comparison to our radiobiologically formidable Van Allen belts, the radiation belts of Jupiter would pose an absolutely hazardous radiological environment for astronauts and unmanned space probes.

Jupiter's inner radiation belts contain energetic electrons (>70 MeV and possibly even >100 MeV). These energetic electrons emit intense synchrotron radiation while weaving along Jupiter's curved magnetic lines of force. (Recall that a change in direction is an acceleration, and an accelerating charged particle emits electro-magnetic radiation; in this case the electromagnetic radiation is called synchrotron radiation.) This synchrotron radiation falls in the radio range and can be detected with radio telescopes. Jupiter's radiation belts are complex. For instance, radio telescope images of the synchrotron emission show two distinct populations of high energy electrons (1–100 MeV) in the inner belts with different pitch angles. In addition to electrons in the inner belts, Jupiter's radiation belts host protons with energies in excess of 1 GeV. Furthermore, the Jovian radiation belts (unlike those of Earth and Saturn) also contain a high flux of oxygen, sodium and sulfur ions.

23.10 Source of ions in the Jovian radiation belts

Jupiter's magnetic field is so strong that even ultra-relativistic, 100 GeV protons (over 50 times higher in energy than those in Earth's Van Allen belts) can become trapped in the radiation belts. The protons within Jupiter's high-intensity radiation belts originate from a wide array of sources. Among these sources of ions are those migrating upwards from Jupiter's ionosphere along with others being captured from the solar wind. Another source of radiation belt protons is radioactive decay of the free neutrons produced by **spallation**. (Spallation here alludes to collisions between primary cosmic rays and atoms in Jupiter's upper atmosphere; the collisions shatter these atoms into smaller components, including neutrons.) These secondary neutrons may move up from their production sites in the atmosphere and undergo β-decay into charged protons and electrons where they encounter the force of Jupiter's magnetism and become trapped in and add to the planet's radiation belts. This process is sometimes called **Cosmic Ray Albedo Neutron Decay (CRAND)**. CRAND feeds the radiation belts of Jupiter (as well as those on Earth and Saturn) with protons of energies between several MeV and up to the proton-trapping limit of each planet (which is in the GeV range around Jupiter).

But these ion sources are not very exotic; they are like what happens here in Earth's Van Allen belts. In addition to these mundane sources however, Jupiter has significant input from a highly unusual source. Some of Jupiter's radiation stems from the most volcanically active body in our Solar System—its innermost Galilean moon, Io. Io spews roughly a ton of gaseous sulfur dioxide into space every second. The SO_2 molecules are then ionized and further fragmented into individual sulfur and oxygen atoms, ions, electrons and nuclei. Consequently, the densest part of the Jovian magnetosphere is the **Io torus**, which is located at 5.5–8 Jovian radii (1 Jovian radius is 71 492 km or 44 423 mi). The resultant electrons and nuclei generate mega-ampere currents between Io and Jupiter's ionosphere. These large-scale currents, along with **Alfvenic waves**, produce magnificent **aurorae** around the poles of Jupiter. In addition to dwarfing our ephemeral aurora borealis and aurora australis, the Jovian aurorae are virtually permanent. Additionally, they exhibit a broad spectrum of electromagnetic radiation ranging from radio waves up to 3 keV x-rays. These potent Jovian aurorae probably heat the polar atmosphere, and this polar heat might spread across the entire Jovian atmosphere, making it significantly warmer than otherwise expected.

Thus, the high-atomic number/high-energy ions (or HZEs) circling Jupiter (such as oxygen and sulfur nuclei), originate primarily from its moons through volcanism or **particle sputtering**. Sputtering is simply the process whereby incident energetic ions collide with atoms (for instance on the surfaces of Io and Europa) and eject these atoms, provided the incident projectile has more kinetic energy than the bonding energy of the intact atom. Besides oxygen and sulfur, other ions including sodium, magnesium, carbon and iron have also been observed but in lower abundances. (Sputtering is sometimes confused with spallation but sputtering simply refers to the ejection of particles from an irradiated surface whereas spallation refers to the shattering of nuclei by radiation. Sputtering does not change the chemical identity of the ejected elements whereas spallation can.)

Another intriguing phenomenon is that as Io passes through Jupiter's magnetic field, it may become highly electrically charged and polarized—negative on one side and positive on the other. This polarization might create a potential difference of as much as 400 000 volts. Thus, when Io arrives at certain locations along its orbit around Jupiter, electrical discharge may occur. This may result in an electric current of up to 5 million amperes flowing between Jupiter's ionosphere and Io. Such discharges produce immense bursts of detectable radio waves.

Much of the energy driving Jupiter's magnetosphere is derived from the rotational energy of the giant planet. Jupiter's rapid rotation along with material from Io's volcanoes collectively collaborate in the Jovian magnetic field's effort to push against the relatively weak solar wind at this distance. The net result is a magnetosphere of enormous proportions. And within this gigantic system, the Jovian radiation belts grow into one of the most radiologically hazardous regions of our Solar System. The trapped charged particles have extraordinary fluxes and may reach energies characteristic of those found in galactic cosmic rays.

Another distinction of Jupiter's magnetosphere is that it contains a multitude of moons as well as a ring system. These moons and rings regularly orbit within Jupiter's magnetosphere. (In contrast, our Moon resides within Earth's magnetosphere for only 2–4 days per month.) These interlopers refine the Jovian radiation belts by absorbing energetic particles, thereby obstructing their transport and energization. Of course, this means that these moons and rings are themselves extreme radiation environments, which makes expeditions to these regions quite foreboding. Furthermore, some of these moons (such as Io) directly serve as plasma (i.e. ion) sources, which makes the Jovian radiation belts most unusual. The robust rotational energy, coupled with the effects of the rings and moons and the exotic sources of additional ions, have an extensive effect on the Jovian radiation belt's structure and dynamics.

One interesting observation is that as ion energy increases, especially in the vicinity of Jupiter's moon **Amalthea**, the radiation belt's ion composition transitions from primarily sulfur ions to oxygen ions. This appears to be due to a local source of energetic (>50 MeV per nucleon) oxygen ions. Amalthea is one of four tiny moons (mean radius only 84 km) within the orbit of Io, the closest of the large Galilean moons. Despite its puny size, Amalthea boasts the title of 'reddest object in the Solar System'. Furthermore, it seems to give off more heat than it receives from the Sun. While tidal friction due to proximity to Jupiter could be one source of this heat, another possibility arises from the fact that as it orbits within Jupiter's magnetic field, electric currents are induced in its core.

But as far as the source of the oxygen ions in the Jovian radiation belt in the vicinity of Amalthea, it appears that the magnetosphere could be accelerating sulfur nuclei that subsequently strike—and spall—the Jovian ring system, thereby creating oxygen nuclei. Another possibility is stochastic oxygen heating by **low-frequency plasma waves**, also called **Alfven waves**. (Recall that Alfven waves push electrons caught in Earth's magnetic field north and south along magnetic lines of force, accelerating them to speeds as high as 45 million mph (72 million km h^{-1}), at which speed they strike and excite nitrogen and oxygen atoms in the upper atmosphere, to generate our local auroras.) Regardless of the precise mechanism, the resulting ultra-energetic oxygen ion products put Jupiter's radiation belt in the same league as the galactic cosmic ray-generating astrophysical particle accelerators.

Another previously unrecognized radiation zone around Jupiter has been discovered by NASA's Juno **JEDI**. This high-radiation region is located at equatorial latitudes just above the Jovian atmosphere. Thus, it is relatively close to Jupiter's surface and well within the other radiation belts—and even within the rings of Jupiter. The ions within this inner radiation belt have energies up to hundreds of keV and atomic numbers up to sulfur ($Z = 16$). However, there is perplexing paucity of energetic electrons in this innermost radiation belt. And the source of the ions here remains a mystery. Albedo neutron decay only provides electrons and protons; therefore, this heavy ion-rich radiation belt cannot be supplied exclusively through the CRAND process. A proposed source of the positively charged heavier ions is electron stripping of neutral, energetic atoms from the high atmosphere. This mechanism could explain the relative rarity of electrons in this radiation belt.

The Juno mission continues to provide exciting new information about Jupiter's radiation environment. Juno's **Stellar Reference Unit (SRU)** star camera provided evidence of energetic heavy ions (i.e. $Z > 1$) within high latitudes of **Jupiter's relativistic electron radiation belt**. This radiation belt was already known to contain electrons moving at nearly the speed of light, but heavy ions were not previously observed at high-latitude locations within this electron belt. SRUs are high-resolution cameras whose primary purpose is to compute the spacecraft's precise orientation based on stellar observations. However, the SRU on board the Juno spacecraft is one of the most heavily shielded instruments, with six times more radiation protection than the other components in its well-shielded radiation vault. For ions to reach this detector, they must possess extraordinary energy. The ions that penetrated this radiation shielding must have had over tenfold more energy than the typical energic protons and over 100-fold more energy than the energetic electrons in this area. Thus, they must have possessed energies well beyond 100 MeV per nucleon, with some exceeding 1 GeV. With energies measured in the GeV range, these ions are the most energetic charged particles discovered yet by Juno. The ion species probably ranged in atomic number up to $Z = 8$ (that is, oxygen nuclei) but possibly up to sulfur ($Z = 16$). Such radiation species are not only a major radiobiological hazard that would make manned missions impossible but are also capable of damaging spacecraft electronics and causing computer errors.

23.11 Single-event upsets and single-event effects in space electronics

Most of the radiation dose in shielded electronic components comes from energetic protons. These energetic protons contribute to so-called **single-event upsets (SEUs)**. During SEUs, a brief electrical discharge might lead to random glitches, memory flips or maybe to '**latch-ups**' (which are localized short circuits). An SEU can occur when an energetic particle travels through and strikes a susceptible node in an electronic device such as a microprocessor, semiconductor memory or transistor. This then induces uninvited electrical signals within the electronic device. These signals are the result of the free electrical charge liberated by ionization in a node of a combinational logic element (a memory 'bit').

The SEU caused by the particle strike results in an error in the electronic device's operation. This is an example of what is called a **soft error**. This 'soft error' designation is because an SEU is not considered permanently damaging. Another example of a non-permanent, nondestructive, soft error is a **single-event transient (SET)**. SETs occur when the electric charge from a radiation-induced ionization event discharges and leads to a spurious signal that travels through the electronic circuit. In essence, this is the result of electrostatic discharge (ESD).

Transient soft errors affect electronic equipment functionality, but the damage is not permanent. This contrasts soft errors with other classes of potentially permanent and potentially destructive radiation-induced electronic damage. These more serious errors are called **hard errors**. Hard errors include, among others, **single-event latch-up (SEL)**, **single-event gate rupture**, **single-event snapbacks** and **single-event burnout**. These are examples of hard errors stemming from the effects of radiation on electronic devices called **single-event effects (SEEs)**. They are called 'single-event' because they are caused by a single, energetic particle. Despite being caused by just a single particle track, these hard errors may be permanent and destructive.

In SELs, the circuit becomes short-circuited and stays shorted until the device is power-cycled. This radiation-triggered short circuiting is known as **latch-up**. In an SEL, damagingly high operating current (which exceeds device specifications) may ensue, causing the part to fail irreversibly. Thus, SELs are classified as potentially irreparable hard errors.

23.12 Total ionizing dose and radiation hardening of electronic devices

SEUs occur often when spacecrafts pass through the Earth's Van Allen belts. This is especially true in the northern and southern **auroral zones** (that is, at latitudes >60° north or south) and over the **South Atlantic Anomaly**. SEUs occur on almost all low-Earth orbit satellite systems. Awareness of this fact dates back as far as 1958 when Explorer 1 discovered the Van Allen belts.

More important than an SEU is the '**Total Ionizing Dose**' (TID). High cumulative TID leads to a gradual degradation of device functionality over time, as chronic radiation exposure results in a buildup of internal defects within a component. Unlike biological systems, electronic units do not undergo repair. (Or at least the present generation of space electronic equipment is not yet capable of such self-repair.)

Radiation-hardened products are typically tested against susceptibility to soft errors such as SEUs as well as one or more other tests, including TID, enhanced low dose rate effects, neutron and proton displacement damage and single-event effects (SEEs).

Just as an interesting aside, most so-called 'Single-Event Upset Facilities' are radiotherapy medical centers. In these facilities, electronic parts are tested both while they are running and while they are turned off. In addition to testing the differential sensitivity of the components while on or off, this testing evaluates the robustness (redundancy) of the entire integrated system. When a component is found to be vulnerable, then electronic radiation countermeasures are considered. Such countermeasures may include additional physical shielding, replacement with more radiation-resistant materials or remedial modifications in the software such as sophisticated error detection systems. Redundant elements at the circuit level are another way to harden electronic devices against radiation. Sometimes, the aptly named 'triple mode redundancy' approach will be implemented. In this case, a single bit may be replaced with three bits with separate 'voting logic' for each bit. One or several chips will be required to perform the same calculation in triplicate. Then a 'vote' is taken to ascertain the most likely answer. This method can reduce radiation-caused errors and disruptions, but a disadvantage of triple modular redundancy is that it increases the area of a chip design fivefold, so it is reserved for smaller designs.

23.13 Other sources of space radiation-induced electronics damage

It is not just the high-energy heavy ions that can do damage to sensitive electronics. Even low-energy electrons can lead to spacecraft **surface charging**, potentially leading to electrostatic discharge (ESD) and degradation of thermal properties. High-energy electrons may lead to **dielectric charge buildup** and damaging electric arcing. Dielectric charge buildup is a well-known cause of problems for orbiting satellites. And of course, high-energy electrons can deposit dose directly to electronic equipment as well as generate bremsstrahlung gamma rays that further deliver dose.

Although protons and electrons cause damage to a few devices through direct ionization, energetic protons and heavier charged particles may interact indirectly with devices to cause damage. This is achieved through particle interactions with atomic nuclei in electronic components. These inelastic nuclear interactions are another source of upset events. Hence, high-energy proton flux is an important factor in SEU rates. Furthermore, through **non-ionizing energy loss**, relatively low-energy protons (<10 MeV) can damage **charge-coupled devices (CCDs)**, which are integrated circuits containing arrays of coupled capacitors. CCDs are very commonly used in digital imaging (which of course makes them critically important in space missions).

While all of this is relevant to spacecrafts and satellites orbiting Earth (including our formidable Van Allen belts), the radiological concerns are magnified immensely in the vicinity of Jupiter. The ESA **Jovian Specification Model (JOSE)** has worked out a detailed radiation model for the **JUICE** spacecraft. This model helped in planning orbital trajectories that reduce radiation exposure and to determine the susceptibility of electronic components to the expected radiation. Thanks to the forecasts from JOSE, JUICE will fly by Callisto 21 times and will end up in orbit around Ganymede, but it will only fly by Europa twice because of the radiation hazards. Despite only two Europa fly-bys, JUICE is expected to sustain a full one-third of its radiation allowance in the vicinity of Europa. JUICE's most radiation sensitive electronics are secluded within radiation vaults whose carbon-fiber walls are lined with lead. Some components have additional localized shielding with aluminum or tantalum. A great deal has been learned from the JOSE model and JUICE is expected to provide a great deal of more data about the actual radiation environment around Jupiter.

Further reading

[1] Becker H N *et al* 2021 High latitude zones of GeV heavy ions at the inner edge of Jupiter's relativistic electron belt *J. Geophys. Res.: Planets* **126**

[2] Roussos E, Allanson O and André N *et al* 2021 The in-situ exploration of Jupiter's radiation belts *Exp. Astron.* **54** 745–89

[3] O'Donoghue J, Moore L and Bhakyapaibul T *et al* 2021 Global upper-atmospheric heating on Jupiter by the polar aurorae *Nature* **596** 54–7

[4] Roussos E, Cohen C, Kollmann P, Pinto M, Krupp N, Gonçalves P and Dialynas K 2022 A source of very energetic oxygen located in Jupiter's inner radiation belts *Sci. Adv.* **8** eabm4234

[5] Kollmann P *et al* 2017 A heavy ion and proton radiation belt inside of Jupiter's rings *Geophys. Res. Lett.* **44** 5259–68

[6] Liu S F, Hori Y and Müller S *et al* 2019 The formation of Jupiter's diluted core by a giant impact *Nature* **572** 355–7

[7] Nordheim T A, Hand K P and Paranicas C 2018 Preservation of potential biosignatures in the shallow subsurface of Europa *Nat. Astron.* **2** 673–9

[8] Gudipati M S, Henderson B L and Bateman F B 2021 Laboratory predictions for the night-side surface ice glow of Europa *Nat. Astron.* **5** 276–82

[9] Kollmann P, Clark G and Paranicas C *et al* 2022 Ganymede's radiation cavity and radiation belts *Geophys. Res. Lett.* **49** e2022GL098474

IOP Publishing

Space Radiation
Astrophysical origins, radiobiological effects and implications for space travellers
James S Welsh

Chapter 24

Saturn

Saturn is the sixth planet from the Sun and has a semi-major axis of around 9.6 astronomical units (AU) from the Sun. At perihelion it is 9 AU from the Sun while at aphelion it is 10.1 AU out. This can be recalled through Bode's mnemonic: [(0, 3, 6, 12, 24, 48, *96*...) + 4] ÷ 10 = 10 AU; thus, one can remember that Saturn is roughly 10 AU from the Sun (recalling that Bode's rule includes the asteroid belt at $n = 24$). In many ways Saturn is like the other gas giant planet Jupiter, although it is far less massive. In Earth radii, Jupiter measures 11.2 R_E compared to Saturn's 9.4 R_E, meaning Saturn is 84% as wide as Jupiter. But at 95.2 M_E (Earth masses) versus 318 M_E, Saturn is only 29.9% as massive. Recall that if one were to compare the mass of Jupiter to the sum of all the masses of the remaining planets, Jupiter would be about 2.5 times as massive as all the others combined. This implies that Saturn is far less dense than Jupiter. With an average density of only 0.7 g cc^{-1}, Saturn is the least dense of all the planets. Recalling that water has a density of 1.0 g cc^{-1}, this means that if we had a big enough bathtub, Saturn would float in it!

Saturn has an axial tilt (rotational obliquity) that is comparable to Earth's: 27° compared to Earth's 23.5°. This means that we see Saturn's rings from different perspectives over time. Saturn rotates rapidly—a sidereal day on Saturn takes only 10 h and 33 min. On the other hand, in accordance with Kepler's Third Law, it is revolving around the Sun very slowly. A Saturnian year takes the equivalent of 29.46 Earth years.

24.1 Saturn's rings

Undoubtedly, the most distinctive feature of Saturn is its magnificent system of rings. While the other Jovian giants also possess rings, none can compare with Saturn's. Saturn's rings are named alphabetically in chronological order of discovery. The dense, bright **A, B and C rings** (from outside to inside) were the first three to be discovered and make up the main rings of Saturn. The remaining rings are lower

doi:10.1088/978-0-7503-5444-8ch24

in density and made of smaller particles. The innermost ring is the D ring. There are three fainter outer rings—the E, F and G rings. The E ring is the outermost of the 'classic' rings. The rings are composed of about 99.9% water ice, which has a very high albedo, accounting for their brilliance. In comparison, the rings of the other Jovian planets are dark and dusty. For instance, Jupiter's rings are composed primarily of silicate material, probably fragments of the inner moons. Voyager 2, the only spacecraft to visit the two ice giants, Uranus and Neptune, did not carry a mass spectrometer. Thus, the composition of these rings is known with less certainty. The dark color of these rings suggests they are composed of (or coated with) carbon soot. Also, as will be discussed later, the intense radiation in the vicinity of the rings around Uranus could have degraded and darkened any frozen methane on the rings and moon surfaces. For now, suffice it to say that the rings of each planet have their own specific characteristics, with Saturn's being by far the most extensive and visually stunning.

Beyond the classic set of Saturnian rings lies the **Phoebe ring**. Much farther out than any of the other rings, the Phoebe ring is not aligned with the planet's axial equator and other remaining rings. Instead, the Phoebe ring is aligned with the plane of the Solar System (the ecliptic) and thus is inclined 27° away from the other rings. Extremely faint and rarefied (at around ten dust particles per cubic kilometer), the Phoebe ring is only detectable in the infrared and was discovered with NASA's Infrared Spitzer Telescope in 2009. Its origin is probably material ejected by meteorite impacts on Saturn's moon Phoebe. Phoebe is an **irregular moon**, meaning that its orbit displays anomalies such as high eccentricity or steep inclination. In the case of Phoebe, the irregularity is even more pronounced—its orbit is **retrograde**, meaning that is revolves about Saturn the 'wrong' way. Phoebe is not alone in its retrograde revolution, however. There are now over three dozen catalogued companions with retrograde orbits. Except for Phoebe (a female Greek Titan), these retrograde irregular moons are named for Norse gods, since the number of Greco-Roman gods, Olympians and Titans was running short. Hence the retrograde Saturnian satellites are called the **Norse group** of irregular moons. Phoebe is probably a captured **centaur** that originated somewhere in the Kuiper belt. And just as Phoebe revolves around Saturn in a retrograde fashion, so does the Phoebe ring.

Another interesting feature of the Saturn ring system is that the **E ring** is fed by **cryovolcanoes on Enceladus** that are incessantly spewing out the micrometer-sized ice and dust particles that constitute the E ring.

Within the main A, B, C ring system are a few noticeable gaps. The largest such gap is the **Cassini division**, which separates the A ring from the inner neighboring B ring. This gap appears to be thanks to a 2:1 **mean motion resonance** with the moon Mimas. Mimas orbits once for every two times that a hypothetical moon in the Cassini division would orbit. This orbital resonance produces gravitational effects that have cleared out the debris in the Cassini division. Another gap called the **Encke gap** is found within the A ring itself. The Encke gap is caused by the tiny moon **Pan** embedded right in the A ring at this orbital distance. Pan is puny, with a mean radius of about 14 km or 8.8 miles, making it less that the full length of Chicago.

Incidentally, Pan is within Saturn's **Roche limit**—a distance within which a parent body's gravitationally induced tidal forces exceed the cohesive gravitational forces holding an orbiting body together. Therefore, one might expect that Pan should suffer the same fate as the rest of the ring system and disintegrate. Such disintegration is believed to be the root cause of all planetary rings. The general approximation for the Roche limit is:

$$d = r_M^3 \sqrt{2 \frac{\rho_M}{\rho m}}$$

where d is the Roche limit, r_M is the radius of the parent planet (Saturn in this case), ρ_M is the density of the parent planet and ρ_m is the density of the orbiting moon.

This can be expressed equivalently as:

$$d = r_M^3 \sqrt{2 \frac{M}{m}}$$

where M is the mass of the parent planet and m is the Moon's mass.

When the parent and moon are made of similar materials, the theoretical Roche limit is around 2.44 times the planet's radius.

A better approximation considers the orbiting moon's deformation as it becomes an oblate spheroid under tidal stress of the parent planet. But this analysis applies to gravitationally held piles of rubble or regolith. In real situations, the inherent cohesive force (tensile strength) must also be considered. Since Pan is a big block of ice with greater tensile strength than a lump of gravitationally held gravel, it can withstand the stresses of being within Saturn's Roche limit.

Other tiny moons besides Pan are within the Roche limit and affect the rings. For example, the F ring is very narrow and was believed to be shaped by two 'shepherd moons', **Prometheus** (on the inside) and **Pandora** (on the outside) that herd the particles into the narrow F ring. Recent data from Cassini suggest that Prometheus alone might account for the size and shape of the F ring, however.

The most likely explanation for Saturn's ring system is that a weakly held together, $\geqslant 500$ km (300 miles) wide icy moon came within Saturn's Roche limit and broke apart. But if these rings formed at the same time as Saturn around 4.56 Ga, they should have vanished long ago thanks to internal collisions between particles that dissipate the rings. Saturn's rings are probably only between 10 and 100 million years old, suggesting that Jurassic dinosaurs looking up at the night sky would not have seen around Saturn! By the same logic, Saturn's rings will not last forever. The constant collisions within the rings dissipate orbiting energy and result in particles falling out of the sky. Based on estimates from Voyager and Cassini, at the present rate of 'ring rain', the entire ring system might disappear over the next 300 million years—and perhaps much sooner than that. Some estimates predict ring depletion within 10–100 million years.

24.2 Helium rain

As a gas giant, Saturn has a similar structure and chemical composition to Jupiter. Presently it is believed that Saturn has a relatively small metallic core, surrounded by a rocky outer core, surrounded by a layer of ice, surrounded by a large layer of metallic hydrogen and helium and finally a superficial layer of molecular hydrogen and helium. Above this opaque 'surface' is Saturn's transparent gaseous atmosphere.

But the internal anatomy could be more complicated and interesting than originally believed. Saturn's atmosphere is relatively poor in helium, compared to Jupiter. So, where is Saturn's helium? A hint comes from the observations that Saturn radiates about twice as much energy into space as it receives from the Sun; it is not in thermal equilibrium. This radiation is primarily in the infrared wavelengths (mainly from 20 to 100 μm). This suggests that Saturn possesses an active, internal source of heat.

One possible source of this energy stems from helium precipitating out of solution. At deep depths below Saturn's surface, the extreme pressure will force liquid helium and molecular hydrogen into metallic states. But there is a pressure and temperature range where the liquid molecular hydrogen will transition into metallic form while the liquid helium does not yet convert into the metallic state. When this happens, the resulting two elements no longer homogenously mix. The two immiscible elements separate and the helium precipitates out as denser 'rain-drops'. As these helium droplets rain through the metallic phase of hydrogen and descend into deeper levels, potential energy is converted into the kinetic energy of helium droplet motion. Friction fights this motion and creates heat, which is carried upward via convection. This heat energy is then radiated into space as the detected infrared. The phase transitions may also contribute additional heat.

Helium rain might also affect the radiation environment around Saturn. As the liquid helium rains out of the mixture, it forms a layer that inhibits convection and the planetary magnetic dynamo. Thus, Saturn's unusual internal anatomy and bizarre 'physiology' could have a drastic influence on the overall shape of the magnetic field. And of course, the magnetic field will in turn influence the morphology and intensity of the surrounding radiation belts.

24.3 Saturn's weather

Saturn's atmosphere is like Jupiter's with bands, spots and other weird weather phenomena. These storms, vortices, spots and horizontal bands have some color to them, but none are as stark or durable as Jupiter's Great Red Spot. The different colors are due to varying amounts of ammonia, methane and other chemical compounds in the background of hydrogen and helium. Given Saturn's size and rapid rotation, it is not surprising that there is extreme weather on Saturn. But unlike Jupiter with its contracting and hot interior, there is less vertical convection originating from Saturn's interior to drive the complex weather seen on Jupiter.

Nevertheless, the convection that does exist, coupled with the relatively weak solar irradiation and rapid rotation of the gas giant lead to winds near Saturn's equator reaching nearly 1770 km h^{-1} (1100 mph). And since Saturn is inclined at 27° it has seasons. And since a year (time for a complete orbit) on Saturn is about 29 Earth years, this means that Saturn's seasons last around seven and a half years each.

24.4 Saturn's magnetosphere

The earliest evidence of an intrinsic planetary magnetic field around Saturn was found in the form of weak radio emissions (around 1 MHz) coming from Saturn. Discovered conclusively in 1979 by the Pioneer 11 spacecraft, Saturn's magnetosphere is now recognized as the second largest of any planetary magnetosphere in the Solar System. As with any planetary magnetosphere, Saturn's internal magnetic field deflects the solar wind away from its surface, preventing it from directly striking its atmosphere.

However, Saturn's magnetosphere is composed of plasma that is very different from that of the solar wind. Saturn's magnetosphere contains plasmas derived from both Saturn itself and its moons. And as with Earth's magnetosphere, the boundary separating the solar wind's plasma from the plasma within Saturn's magnetosphere is called the **magnetopause**. This represents the region where a balance occurs between the dynamic pressure of the solar wind (ρv^2) and the magnetic field pressure ($B^2/2\mu_o$).

The magnetopause distance from Saturn's center at the **subsolar point** (the sunny side) varies widely from 16 to 27 R_s, with an average distance of about 22 R_s. (An R_s is the equatorial radius of Saturn: 1 R_s = 60 330 km.) The magnetopause's position at any given time depends on the pressure from the solar wind, which in turn depends on solar activity. Staying on the Sun's side, in front of the magnetopause, at a distance of about 27 R_s from the planet, lies the **bow shock**. As with other magnetospheric bow shocks, this is a wake-like disturbance in the solar wind created by the collision between the incoming solar wind and the magnetosphere. Thus, the bow shock represents a discontinuity in the medium through which the solar wind is moving and its speed changes from supersonic to subsonic. And as with Earth's and other planetary magnetospheric bow shocks, it blunts and flattens the dayside of the magnetopause.

The region between the bow shock and a planet's magnetopause is called the **magnetosheath**. The slowing of the solar wind speed results in a conversion of kinetic energy into thermal energy, resulting in a hot but subsonic turbulent plasma in the magnetosheath.

On the opposite side of the planet (the anti-Sun, antisolar or night side), the solar wind stretches Saturn's magnetic field lines into a long, trailing **magnetotail**. Saturn's magnetotail consists of two lobes (a north and south lobe wherein the magnetic flux is pointed in opposite directions) separated by a thin layer of plasma called the **tail current sheet** or **plasma sheet**. Like Earth's magnetotail, Saturn's magnetotail functions like a channel through which solar plasma enters the inner regions of the magnetosphere, including the radiation belts. But in addition to providing an entry, like Jupiter's magnetotail, the magnetotail of Saturn is also an egress through which the internal magnetospheric plasma can leave Saturn's magnetosphere altogether. Importantly, plasma moving from the magnetotail to the inner magnetosphere becomes heated and contributes to Saturn's radiation belts.

24.5 Enceladus as a source of ions

Just as Jupiter's moon Io supplies ions to the Jovian magnetosphere and radiation belts, a main source of ions for Saturn's magnetosphere and radiation belt is Enceladus, which sprays water ice from cryogeysers near its south pole. This source heavily influences the plasma composition of the Saturnian magnetosphere and radiation belts. Data from the Cassini spacecraft show that the chemical composition of the plasma in Saturn's inner magnetosphere is mostly water group cations such as hydronium ions (H_3O^+), H_2O^+, O^+, OH^+, HO_2^+ and O_2^+. In addition, nitrogen ions (N^+) and of course, protons are also present. The main source of the water-derived ions is Enceladus. Enceladus is situated right within Saturn's radiation belts and contributes ions to the belt and shapes the belt by serving as a radiation barrier.

Like some of the Jovian moons and some other moons of Saturn, Enceladus appears to be covered with a crust of ice which covers a saltwater ocean beneath it. From this subsurface ocean, Enceladus squirts out an astonishing 300–600 kg s^{-1} of water vapor through its cryogeysers near its south pole. The ejected water (and the resulting ions and radicals) forms a thick molecular torus around Enceladus at ~4 R_s, with densities up to 10 000 molecules per milliliter. Of the 300–600 kg s^{-1} shooting through Enceladus's cryogeysers, over 100 kg s^{-1} eventually becomes ionized and added to the plasma of the Saturnian magnetosphere. Further water group ions originate from other icy moons around Saturn as well as its rings. In the outer aspects of Saturn's magnetosphere, the dominant ions are protons; these originate either from the solar wind or Saturn's ionosphere.

24.6 Saturn's magnetic field

Saturn's magnetic field is probably produced the same way that Jupiter's is—through a dynamo generated by swirling liquid metallic hydrogen. However, due to the significant differences in mass and density, the liquid metallic hydrogen on Saturn is relatively much deeper beneath Saturn's surface than is the case for Jupiter. This makes the magnetic field strength on its surface far weaker than on the surface or surrounding space of Jupiter. Moreover, the magnetic moment of Saturn that makes its magnetic field is far weaker than Jupiter's.

Unlike Earth, where the rotational axis is offset by 9.41 from the magnetic dipole axis, Saturn's magnetic dipole is strictly aligned with its rotational axis. The term for this tight alignment is **axisymmetric**. The magnetic field strength at Saturn's equator is about 21 µT (0.21 G). This is only about two-thirds of the magnetic field strength at Earth's equator (~31 µT or 0.31 G). This means that Saturn's magnetic field is slightly weaker than Earth's at their respective surfaces; however, its **magnetic moment** is about 580 times larger when scaled to similar size. Saturn has a dipole magnetic moment of about 4.6×10^{18} T m^3. Jupiter still has the strongest planetary magnetic moment at 1.55×10^{20} T m^3. This is about 2×10^4 times larger than Earth's magnetic moment. Of note, like Earth, Jupiter's magnetic dipole axis is not axisymmetric—it is offset from its rotational axis (about 10°). In contrast, the tilt of Saturn's dipole magnetic moment is inclined less than 1° to the rotation axis. It may be worthwhile to state again that planetary magnetic fields are imperfect dipoles— while they may be modeled as dipoles, the north and south magnetic poles of a planet are rarely truly antipodal.

(Just as a curious digression, Uranus has two anomalous tilts—both its rotation axis and its magnetic dipole axis are out of harmony with the other planets in our Solar System. Uranus basically spins on its side—its rotational axis is tilted 97.5° to a line normal to its orbital plane. And the magnetic dipole axis of Uranus is angled 58.6° from the rotational axis. In contrast, Neptune has a more typical obliquity at 28.3° which is not very different from Earth's 23.5°. However, the tilt of Neptune's magnetic moment is 47°—less than that of Uranus but much larger than those of the Earth, Jupiter and Saturn. Uranus's magnetic dipole moment of 3.9×10^{17} T m^3 is about 50 times that of Earth's magnetic moment. Neptune has a magnetic moment of 2.2×10^{17} T m^3 or just over 25 times Earth's magnetic moment.)

As another aside, students of physics might recognize that the units used to express the magnetic moments of these planets (T m^3) are not truly dimensionally correct for magnetic moment. The proper unit for magnetic moment is A m^2. Alternatively, this is the equivalent of J T^{-1} since 1 A m^2 = 1 J T^{-1}.

24.7 Aurorae on Saturn

As on Earth, Saturn has brilliant aurorae encircling its poles. Because Saturn's magnetic field is axisymmetric, these aurorae directly encircle the geographic poles. However, the Saturnian aurorae arise from interactions of the solar wind, not with nitrogen and oxygen molecules are here on Earth, but rather with the dominant molecule in Saturn's upper atmosphere, hydrogen. And compared to Earth, the particles leading to Saturnian aurorae are far more energetic. This accounts for the detected shine in the far-ultraviolet that was first spotted by Pioneer 11 in 1979 and by Voyager 1 and Voyager 2 during flybys. NASA's Hubble Space Telescope can also detect ultraviolet emissions and captured additional magnificent images of Saturn's aurorae in 1997 with its Space Telescope Imaging Spectrograph. Finally, Cassini's Grand Finale allowed high-resolution close-up views of these aurorae as it circled the poles before crashing into Saturn itself on 15 September 2017. Since Earth's atmosphere is largely opaque to ultraviolet, good views of Saturn's aurora had to wait until space telescopes and orbiters could see them in ultraviolet. Curiously, compared to Jupiter and Earth, there appears to be a paucity of auroral x-ray emission from Saturn's poles (figure 24.1).

Saturn's aurorae also glow in infrared wavelengths. Tracking of these infrared aurorae by the Keck Observatory Near Infrared Echelle Spectrograph has solved a puzzling problem about conflicting length estimates of the Saturnian day. As plasma becomes entangled and spirals along the magnetic lines of force to and from the poles, synchrotron radiation is emitted in the radio range. But in the polar regions,

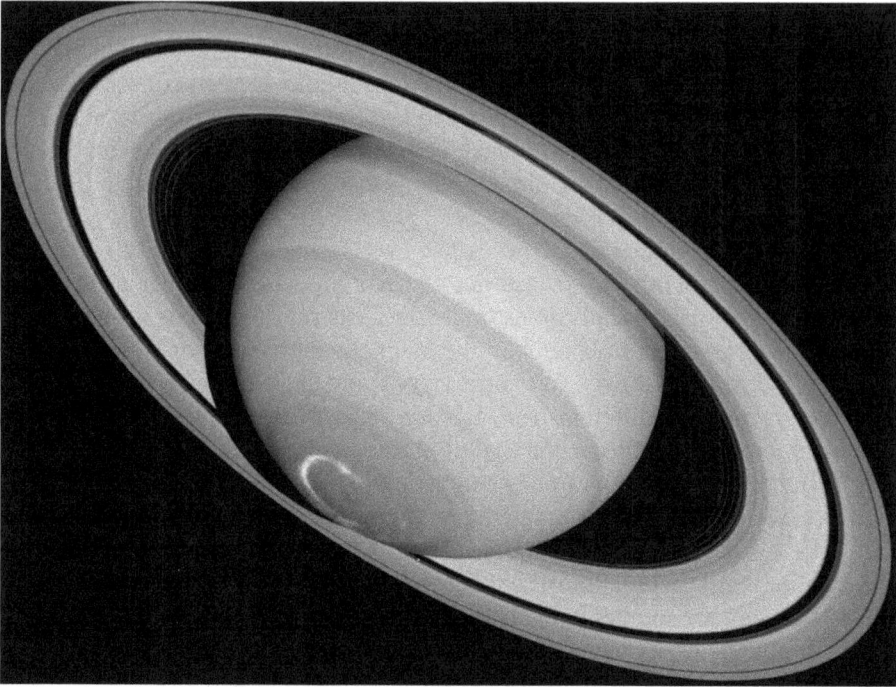

Figure 24.1. A view of Saturn taken by the Advanced Camera for Surveys on 22 March 2004 from NASA's Hubble Space Telescope. Around the south pole is an image of an ultraviolet aurora (false-color enhancement) captured by the Imaging Spectrograph superimposed on the optical image obtained by the Hubble. Unlike the colorful aurorae on Earth, the aurorae on Saturn only shine in the ultraviolet part of the electromagnetic spectrum. NASA, ESA, J Clarke (Boston University) and Z Levay (STScI).

high-speed winds (over 10 000 km h^{-1}) lead to spinning vortices that affect the magnetic field lines. The result is that measurements of Saturn's rotation made by monitoring the magnetic field (via radio waves in the polar regions) produce different results from measurements obtained by examining the same phenomenon in lower latitudes. And as the wind speed changes and the vortices are altered, one might obtain different estimates of the length of Saturn's day from time to time. Thanks to an understanding of this confounding factor, the length of a day on Saturn has been better constrained. It is now estimated to be 10 h, 33 min and 38 s.

These aurorae become brighter and move closer to the poles as solar activity increases. Saturn's aurorae probably contribute to heating of Saturn's upper atmosphere near the poles.

24.8 The moons of Saturn

Saturn has seven spherical major moons—Mimas, Enceladus, Tethys, Dione, Iapetus, Rhea and Titan in size order of smallest to largest. In order from inside out they are: Mimas, Enceladus, Tethys, Dione, Rhea, Titan and Iapetus. But as of this time, there are a total of 83 known moons around Saturn. Most of these are regular moons, meaning that they follow antegrade orbits (going in the same direction as Saturn's rotation), have fairly circular orbits and orbit in a plane near Saturn's equator but there are several irregular moons as well. Rather than being named after figures from Roman mythology, the irregular moons with steeply inclined orbits are named after Inuit gods, while another group with less orbital inclination are named after Gaelic gods. Another group of irregular moons, those with retrograde orbits, are named after Norse giants.

The moons of Saturn are made mostly of various ices including water ice, frozen ammonia and solid-state methane. Titan is by far the largest of Saturn's moons. In fact, if all the moons were gathered together for a weigh-in, Titan alone would make up about 96% of the total mass. Behind Jupiter's giant Ganymede, Titan (at 5150 km or 3200 miles diameter) is the second largest moon in our Solar System. Coincidentally, Titan is about the same distance from Saturn as Ganymede is from Jupiter. But unlike Jupiter's Galilean moons, which are all sizeable, Titan has no companions in its size league. The second largest moon in the Saturnian system is Rhea at only 1582 km or 950 miles diameter. Rhea is only around half the size of Jupiter's smallest Galilean moon Europa. And Saturn's smallest round major moon Mimas is only 400 km or 248 miles across.

Moons are generally not very dense and no moons in the Solar System are as dense as even the least-dense terrestrial planet, Mars (3.93 g cc^{-1}). Io is the densest moon in the Solar System (at 3.53 g cc^{-1}), with our Moon gathering second-place honors (at 3.346 g cc^{-1}). Titan is the densest of Saturn's moons at 1.88 g cc^{-1}. This is comparable to Jupiter's other large moons such as Ganymede and Callisto. But this modest density is higher than any of the other Saturnian moons, indicating that they have less rock and more ice. Some of Saturn's moons have a density lower than liquid or frozen water (around 1 g cc^{-1}), suggesting that they are porous with sponge-like interiors. (For reference, recall that Earth is the densest planet in the Solar System at 5.513 g cc^{-1}.)

Titan is the only moon in the Solar System with a significant atmosphere. While no other moons anywhere in our Solar System possess ample atmospheres, Titan makes up for the deficiencies of its fellow moons in spades—Titan's atmosphere is even thicker than ours. The atmospheric pressure at the surface of Titan is about 1.5 bars, 50% higher than at our surface. And even though Titan is far less massive than Earth, its overall atmospheric mass is about 20% more than Earth's atmosphere. The overall density of Titan's atmosphere is fourfold more than Earth's. But because of the reduced gravity on Titan, its atmosphere extends far higher relatively —and absolutely—than ours. For instance, Titan's mesosphere extends to nearly 600 km (373 miles), which is about tenfold taller than our mesosphere. And Titan's thermosphere reaches nearly 1200 km or 746 miles. (For reference, our mesopause is about 80 km above the surface and our thermopause averages around 600 km above the surface, although there is a great deal of variability.) One reason for the persistence of Titan's thick atmosphere is the relatively weak solar wind at this long distance from the Sun. If one could withstand the frigid temperature and ignore the absence of oxygen, one could possibly fly around with a set of artificial wings on Titan. And thanks to the density of Titan's atmosphere and the reduced gravity, if one's wings fell off mid-flight, the landing might not be too harsh since terminal velocity is only around a tenth of what it would be here.

Like Earth, Titan's atmosphere is mostly nitrogen. Ours is 78% nitrogen, while Titan's is around 95% nitrogen. However, Earth has about 21% oxygen whereas Titan has no free oxygen. But it does have around 5% methane. The surface temperature on Titan is around −180 °C (−290 °F), which is not cold enough to precipitate out liquid nitrogen (the boiling point of nitrogen is near −196 °C (−320 °F), but liquid methane can exist on the surface and it feeds Titan's atmosphere through vapor pressure. (Methane has a boiling point of roughly −162 °C (−259 °F) at 1 bar pressure.)

Titan was explored in depth by the **Huygens probe**, which was named after Christiaan Huygens who discovered Titan in 1655. The Huygens probe landed on Titan on 14 January 2005 after separating from the combined **Cassini–Huygens spacecraft** on 25 December 2004 (after which Cassini continued as an orbiter). Huygens revealed what appears to be Titanian **methane-based hydrology** with large lakes, streams and even **methane rain**. In many ways, Titan is fairly earthlike as it possesses dunes, mountains, plains and volcanoes. But the 'sand dunes' are not real sand but solid hydrocarbons; and the volcanoes are cryovolcanoes that shoot out ice, ammonia and chilled water rather than molten magma. And of course, the liquid lakes of hydrocarbon methane and ethane are a bit 'alien'. Nevertheless, the abundance of alkanes suggests the possibility of more complex organic chemistry on Titan.

But in addition to the raw materials for organic chemistry, there is probably a large reservoir of liquid water below the surface of Titan, just as on some of the moons of Jupiter. Data from Cassini suggest that the subsurface water lies about 100 km (62 miles) beneath the surface and has a relatively high density. This high density indicates that the subsurface ocean is probably quite salty, perhaps rivaling the Dead Sea (which unlike the ocean's average salinity of about 3.5%, approaches 40%). Presently, it is believed that the solutes are largely sodium, potassium and sulfate ions. With many of the necessary ingredients for the basic building blocks of life, one might speculate about alien biochemistry on Titan. Whether some sort of exotic extremophilic life can survive in such super-salty water remains to be seen. And of course, the question of ultra-exotic life forms (using not water as a solvent but liquid hydrocarbons), thriving in lakes of methane and ethane, is an even more imaginative thought experiment.

As far as suitability for human colonization, Titan has several advantages over even the Moon and Mars. Since neither the Moon nor Mars has a magnetosphere (nor a robust atmosphere), galactic cosmic rays will come raining down to strike their surfaces. But while Titan similarly has no magnetic shield, its thick atmosphere would ward off primary cosmic rays. Additionally, Saturn's magnetosphere will deflect many incoming galactic cosmic rays. And unlike Jupiter's Io, Europa and Ganymede, which are within Jupiter's radiation belts, Titan lies beyond the radiation belts of Saturn. Nevertheless, some energetic cosmic rays will strike Titan's atmosphere and generate secondary cosmic rays. And the galactic cosmic rays influence the **isotopologues** of various nitrile species. Isotopologues are molecules that differ only in their isotopic content, such as light water (H_2O) and heavy water (HDO and D_2O). Data from the **Atacama Large Millimeter/ Submillimeter Array** or **ALMA** show that the nitrogen-14/nitrogen-15 isotopic ratio in various nitriles changes with altitude in Titan's atmosphere. Acetonitrile (CH_3CN) in Titan's lower atmosphere is relatively enriched in nitrogen-14. This has been attributed to the fact that wavelengths of ultraviolet radiation that can break the triple covalent bonds between $C\equiv N$ when the nitrogen is nitrogen-14 versus nitrogen-15 are preferentially absorbed in Titan's upper atmosphere. However, in the lower atmosphere where photodissociation does not occur as readily (thanks to ultraviolet attenuation), acetonitrile production can still occur thanks to cosmic ray-induced dissociation of N_2. Thus, acetonitrile made from the products of cosmic ray products at these lower altitudes will have a different nitrogen-14/nitrogen-15 isotopic ratio from acetonitrile in the upper atmosphere.

24.9 Other moons of Saturn

Saturn has a host of other interesting moons, including **Iapetus** with its one dark hemisphere and one light hemisphere. The vastly different reflectivity of these two sides gives Iapetus its nickname of the yin-yang moon; it is characterized by the dark side albedo of around 0.04 compared to the bright side albedo of nearly 0.6. The dark side is called **Cassini Regio** and appears to be covered with dust derived from Phoebe and another tiny moon, Ymir. Iapetus is relatively distant from Saturn at nearly 60 R_s, which makes it the farthest from Saturn of the seven major moons. Titan is the next most distal to Saturn and has been discussed above.

Rhea is the second largest of the major moons and is the closest major moon to Titan, orbiting about halfway between Saturn and Titan. Its diameter is about 1529 km or 950 miles across. Curiously but controversially, Voyager 1 and the Cassini mission revealed a pattern of Saturnian magnetospheric depletion and electron absorption in the vicinity of Rhea. This has been interpreted as a mini-ring system of particles around the moon.

Moving inward, **Dione** is the next most proximal of the seven Saturnian major moons and is about 1130 km or 700 miles in diameter and is slightly rockier than Rhea (1/3 rock on Dione versus 1/4 rock on Rhea, with the rest mostly ice). Dione might have two tiny companions in its orbital path around Saturn at its Lagrange points in a manner akin to Jupiter's Trojan asteroids.

Moving in further, the next main moon is Tethys, which is 1060 km or 660 miles in diameter. **Tethys** is slightly closer to Saturn than Dione but much less dense. In fact, Dione is the least dense of the seven major moons—and is the least dense of all the spherical moons in the Solar System. At only 0.984 g cc^{-1}, Tethys is less dense than water, suggesting a sponge-like interior. Like Dione, Tethys too has mini-moon companions at its Lagrange points.

Enceladus is the next moon we encounter moving inward. At only 504 km or 313 miles in diameter, Enceladus is only about half the diameter of Tethys. Despite its small size, Enceladus wins a few first-place prizes including the title of most reflective body in the Solar System. With an albedo of 0.99, Enceladus is most impressive, given that the albedo of freshly fallen snow is around 0.9 and that of aluminum foil might reach 0.95. This high albedo causes the moon to be extremely cold; the surface temperature of Enceladus is around −200 °C or −330 °F

Some areas on Enceladus have very few craters, indicating that this surface must be relatively young. One such region is the so-called **tiger stripes**, which are calculated to be under 100 million years old and maybe as young as 4 million years. The four nearly parallel tiger stripes are **fault lines** associated with cryovolcanic activity. More conclusive evidence of cryovolcanic activity has been documented near Enceladus's southern pole as jets of ice and water (containing organic molecules) have been seen shooting into space. These jets are shooting out of fault lines that are warmer than other surface areas. The erupted ice, gas and aqueous solution are forcibly fired out at approximately 800 mph or 1287 km h^{-1} with a volume of nearly 250 l s^{-1}. This ice is actively replenishing Saturn's E ring and Enceladus is integrally embedded in this ring. The vast majority of the cryovolcanic material does not end up in the E ring however since it falls back to the surface as highly reflective ice and snow. In addition to water, other gases include methane and carbon dioxide and maybe some ammonia, carbon monoxide and nitrogen.

The ice and liquid water spewing from Enceladus's surface suggests that there must be a liquid ocean beneath the frozen surface, which might encircle the entire globe. Estimates of the depth of this ocean are as much as 30 km (18 miles). For reference, the deepest point below Earth's oceans is **Challenger Deep** in the southern Mariana Trench at roughly 10.9 km or 6.78 miles. While this subglacial ocean is believed to be rich in salty solutes such as sodium, potassium and sulfate ions, there was doubt about its phosphorus content. Given the central role of adenosine triphosphate as the primary biochemical energy currency and the key role of nucleic acids in molecular biology, an ocean devoid of phosphate could probably not sustain life as we known it. However, recent modelling of weathering of Enceladus's seafloor indicates that orthophosphate (HPO_4^{2-}) concentrations could be very similar to those of Earth's oceans.

In comparison to Rhea and Dione, Enceladus is relatively rocky at about 50:50 rock and ice. This high silicate fraction is second only to Titan among the major moons of Saturn. It is possible that the rocky interior explains the liquid ocean and the ongoing cryovolcanic activity—radioactive decay of isotopes in these rocks could be releasing heat that keeps Enceladus's interior warm enough for liquid water to exist. Alternatively, or supplementally, gravitational forces due to the 2:1 mean motion resonance of Enceladus with Dione might be contributing some tidal heating. While the organic chemistry in the subsurface sea of Enceladus might be astrobiologically intriguing, *Enceladus is one of the moons within Saturn's radiation main belt*, which presents serious challenges to any future manned exploration.

The innermost of the seven, round main moons is **Mimas**. At only 386 km or 240 miles wide, Mimas is near the theoretical limit of a gravitationally created spherical celestial body. As mentioned earlier, Mimas is the probable cause of the largest gap in Saturn's ring system, the **Cassini division**. Mean motion resonance between Mimas and material that would lie within the Cassini division leads to instability that precludes population of this gap between the A and B rings. And like Enceladus (and the minor moons **Janus** and **Epimetheus**), *Mimas lies within Saturn's main* *radiation belt.* Thus, like Jupiter's Europa, *Mimas holds the distinction of being the most heavily irradiated (icy) moon in its system.* (Although Io is even more heavily irradiated than Europa, it is a hot, volcanic world rather than an ice moon like Europa.) But the high-energy electron flux is far lower on Mimas than on Europa— about 40-fold lower. As Mimas moves around Saturn, its leading hemisphere takes the brunt of high-energy electrons (>1 MeV) whereas the trailing hemisphere is bombarded mainly by electrons <1 MeV. Mimas creates a corridor within Saturn's radiation belt that is relatively bereft of high-energy protons, probably because of absorption of slowly diffusing **cosmic ray albedo neutron decay (CRAND) protons** by Mimas.

24.10 Saturn's radiation belts

Inside the magnetosphere are Saturn's radiation belts, which house energetic charged particles capable of ionization. This distinguishes the radiation belts from the rest of the magnetosphere, which also contains charged particles (that is, plasma) but the plasma of the rest of the magnetosphere generally does not possess sufficient energy to qualify as what many would call 'radiation'.

The Saturnian magnetosphere and radiation belts are far less intense than the Jovian belts and do not emit much microwave or radio wave electromagnetic radiation. In contrast to the **decimetric radio emissions** that gave away Jupiter's magnetosphere and radiation belts, Saturn does not produce sufficient decimetric radio signals for detection here on Earth.

Nevertheless, Saturn's proton radiation belts are enormous: they stretch from the planet's innermost ring to the orbit of Saturn's moon Tethys. And while some of the energetic particles trace their origins to the solar wind, other ions populating Saturn's belts arise from Enceladus and Saturn's atmosphere.

The electrons in the main belt probably stem from the solar wind or Saturn's outer magnetosphere. Such electrons get transported via diffusion and are heated adiabatically. In contrast to these electrons however, the energetic proton populations have two distinctly different origins. The first population (mostly with moderate energies <10 MeV) has the same origin as the electrons. The second population of protons is more energetic with a maximum flux of ~20 MeV. These protons stem from spallation of solids in the Saturnian system caused by galactic cosmic rays. This spallation produces free neutrons, which decay in minutes via the weak force into protons, electrons and electron antineutrinos. This is known as the **CRAND** process (for Cosmic Ray Albedo Neutron Decay).

The extent of the radiation belts of Saturn exceeds 285 000 km into space. While this is not as far into space as our Moon is from Earth, it does encompass Saturn's rings (except the Phoebe ring) and some of Saturn's moon (including the relatively large moons Janus, Epimetheus, Mimas and Enceladus). This contrasts with the situation on Earth, where our Moon lies far beyond our magnetosphere and Van Allen radiation belts. (The Moon is about 384 400 km or 238 800 miles away from us while the Van Allen belts only reach as far as 58 000 km or 36 000 miles.) The presence of moons within the Saturnian radiation belts dramatically affects the overall configuration. For example, these moons serve as dense radiation barriers to protons. And the absorption of these protons stymies migration of energetic protons from their site of origin, thereby introducing low-exposure gaps into the belts. And while the moons affect the radiation distribution, the radiation itself also has a substantial effect on the moons. The surfaces of the ice-clad inner moons of Saturn are constantly battered and worn down through **radiation weathering**. The radiation incessantly sputters water, oxygen and various ionized water products from the moon's surfaces.

One consequence of having moons within the radiation belts is the formation of areas within the belts that are completely isolated from each other. This is different from Earth's Van Allen belts where particles might migrate within a belt and external particles can enter and contribute to the radiation belts. Furthermore, Saturn's main rings inhibit the inward migration of trapped charged particles as well. Hence, energetic particles in the outer radiation belts cannot directly reach and populate the low-altitude radiation region beneath the rings. The source of the radiation in these low-altitude regions was thus once a mystery. Potential sources of radiation here include CRAND and **multiple-charge exchange processes** described below.

24.11 Saturn's radioactive rings?

The innermost region of Saturn's magnetosphere (near the rings) is largely devoid of particulate radiation since the numerous electrons, protons and positively charged ions are absorbed by the rings. This makes one wonder if the rings of Saturn have become radioactive through various nuclear activation processes. Conceivably, inelastic nuclear interactions between energetic protons and nuclei in the rings (such as the $^{16}O(p,\alpha)^{13}N$ or $^{16}O(p,pn)^{15}O$ nuclear reaction channels) could possibly lead to short-lived radioisotopes via proton activation. And oxygen-16 is quite abundant in Saturn's icy rings.

Sufficiently energetic galactic cosmic rays (with energies over 20 GeV for the A ring or 72 GeV for the D ring) can overcome the diverting effects of Saturn's magnetic field and reach these rings and produce secondary radiation. Radiation measurements from Pioneer 11 and Cassini have led to GEANT (Geometry ANd Tracking)-based models on the generated gamma photons, electrons, protons, neutrons and pions from these cosmic ray collisions with ice clumps in the rings.

Saturn's rings do shine in soft x-rays. Oxygen Kα line x-ray emissions have been seen emanating from the rings. These x-rays are likely from fluorescent scattering of solar x-rays. Thus, Saturn's rings join other bodies in the Solar System (including Venus, Moon, Mars) that shine in soft x-rays thanks to scattering of solar x-ray radiation. Saturn itself emits x-rays that are probably scattered from the Sun. NASA's Chandra X-ray Observatory has documented x-ray flares from Saturn's lower latitudes following powerful solar flares, indicating that Saturn does somehow reflect explosive activity from the Sun. The exact mechanism behind Saturn's impressive ability to reflect solar x-rays is still puzzling.

The main Saturnian radiation belt region ranges between 2.3 and 3.5 R_s, which is in between the outer edge of the A ring and the inner edge of the Enceladus gas torus. This main belt mainly consists mainly of energetic protons and relativistic electrons with energies ranging from hundreds of keV to tens of MeV. In the distance beyond 3.5 R_s but less than 6 R_s these protons and electrons are heavily absorbed by neutral particles—abundant, cold, uncharged gas atoms and molecules. **Energetic neutrals** can be created by collisions between energetic positively charged ions with atmospheric neutral particles when a **charge exchange process** occurs. An electron is transferred from the cold neutral particle onto the energetic positive ion and that ion becomes an energetic neutral. In such charge exchange processes, soft x-rays may be emitted through the so-called **charge exchange x-ray emission process**. Such high-energy neutrals are no longer guided by the magnetic field and may be lost from the magnetosphere, migrate deeper into the atmosphere, or engage in another charge exchange process with another ion. Finally, in the region beyond 6 R_s slightly less energetic radiation sometimes reappears in the form of particles with energies in the hundreds of keV.

While Tethys (located at just under 5 R_s from Saturn) is normally located just outside the perimeter of Saturn's permanent radiation belts, sometimes solar energetic particles colliding with Saturn's magnetosphere can generate a transient new radiation belt just beyond the orbit of Tethys. Curiously however, unlike what happens to Earth's Van Allen belts, the protons from these strong coronal mass ejections and flares do not affect Saturn's inner radiation belts. It appears that Tethys's absorption of energetic solar protons isolates the inner radiation belts from such solar activity. But increased solar activity does affect Saturn's radiation belts in a different way as described below.

24.12 The Van Allen belts versus the Saturnian radiation belts

Our Van Allen radiation belts have two primary proton sources—the solar wind and galactic cosmic rays. The outer Van Allen belt is largely populated by electrons originating in the solar wind. These solar wind electrons enter the magnetotail on the anti-solar side ('downwind') and eventually wind up in the outer Van Allen belt. Some galactic cosmic rays are energetic enough to reach our atmosphere where they spall atmospheric molecules and atoms, leading to secondary free neutrons—so-called **albedo neutrons**. Through CRAND, these neutrons decay into protons that can find their way into the inner Van Allen belt. Hence, in contrast to the outer belt, which is largely populated by energetic electrons, the inner Van Allen belt is mainly protons.

Solar activity plays a crucial role in modulating galactic cosmic rays since the solar wind wards off these incoming energetic particles to an extent. The *decrease* in incoming galactic cosmic rays thanks to stronger solar magnetic fields associated with increased solar wind, sunspot activity, solar flares and coronal mass ejections during solar maximum is known as the **Forbush effect**. Thanks to the Forbush effect, more galactic cosmic rays reach Saturn's magnetosphere during solar minimum when there are fewer sunspots and a weaker solar wind.

On Saturn, the Sun's activity also affects the radiation belts in this way, as well as through a very different manner. The solar wind is far weaker at Saturn's distance of nearly 10 AU away. Nevertheless, solar extreme ultraviolet can heat Saturn's atmosphere, resulting in exacerbation of the already turbulent winds caused by the planet's rapid rotation. These winds affect the ionosphere, which is linked to the magnetosphere through the planetary magnetic field. The result is that during intense solar activity, the protons in the radiation belts are spread out more than usual. This radial spread encourages them to encounter some of Saturn's moons and get absorbed. The overall intensity of the radiation belts thereby diminishes substantially in a manner that directly reflects solar extreme ultraviolet radiation emissions.

This discovery was thanks to the unanticipated longevity of Cassini and its **Magnetospheric Imaging Instrument (MIMI); Low-Energy Magnetospheric Measurements System** data were recorded over a period of time that included a complete solar cycle. The Cassini mission was active in space for almost total 20 years despite being planned for only 4 years in the vicinity of Saturn from 2004 to 2008 (after its launch on 15 October 1997). Cassini's 'Grand Finale' ended on 15 September 2017 in a planned crash into Saturn itself to avoid the risk of contaminating any of Saturn's astrobiologically intriguing moons with terrestrial microbes if an unplanned crash landing occurred after running out of fuel. Cassini's longevity has been attributed to robustness (i.e. redundancy) in its mechanical and electronic hardware coupled with its three reliable plutonium-238 powered **radioisotope thermoelectric generators (RTGs)**. Plutonium-238 is an alpha emitter with a half-life of 87.7 years (in contrast with the more familiar fissile isotope, plutonium-239, that can be used in nuclear reactors and nuclear weapons and has a 24 110 year half-life). These 33 kg (73 lb) RTGs were composed of plutonium dioxide. Similarly designed RTGs powered the Galileo, Ulysses and New Horizons space probes.

The possibility of an inner radiation belt located in the region between Saturn itself and its main rings had been suspected based on computer simulations. But it was not until the Grand Finale of the Cassini mission that direct measurements were obtained from this region using Cassini's MIMI. We now know that Saturn has a second radiation belt located just inside the innermost D Ring. This inner radiation belt is bounded by Saturn's atmosphere at its inner edge and by the D73 ringlet (a component of the D ring located at 1.22 Saturn radii (1 R_s = 60 268 km) at its outer limits). Thus, the inner radiation belt spans the equatorial plane at distances between 1.03 and 1.22 R_s. Another ringlet (D68, at 1.12 R_s) splits the energetic charged particles of this inner radiation belt into two groups. This radiation belt probably consists of energetic protons and electrons formed via the CRAND process along with some energetic ions from the main radiation belt. Radiation in this inner belt can be quite energetic. It is dominated by energetic protons with energies ranging from 25 MeV up to the GeV range. Radiation losses from this belt are likely dominated by energy deposition and scattering of charged particles by dust and atmospheric neutrals.

Further reading

[1] O'Donoghue J, Moore L, Connerney J, Melin H, Stallard T, Miller S and Baines K H 2018 Observations of the chemical and thermal response of 'ring rain' on Saturn's ionosphere *Icarus* **322** 251–60

[2] Chowdhury M N, Stallard T S, Baines K H, Provan G, Melin H and Hunt G J *et al* 2022 Saturn's weather-driven aurorae modulate oscillations in the magnetic field and radio emissions *Geophys. Res. Lett.* **49** e2021GL096492

[3] Crida A, Charnoz S and Hsu H W *et al* 2019 Are Saturn's rings actually young? *Nat. Astron.* **3** 967–70

[4] Roussos E *et al* 2018 A radiation belt of energetic protons located between Saturn and its rings *Science* **362** eaat1962

[5] Krimigis S M *et al* 2005 Dynamics of Saturn's magnetosphere from MIMI during Cassini's orbital insertion *Science* **307** 1270–3

[6] Yan C and Stanley S 2021 Recipe for a Saturn-like dynamo *AGU Adv.* **2** e2020AV000318

[7] Hao J, Glein C R, Huang F, Yee N, Catling D C, Postberg F, Hillier J K and Hazen R M 2022 Abundant phosphorus expected for possible life in Enceladus's ocean *Proc. Natl Acad. Sci. USA* **119** e2201388119

[8] Iino T, Sagawa H and Tsukagoshi T 2020 14N/15N Isotopic Ratio in CH3CN of Titan's Atmosphere Measured with ALMA *Astrophys. J.* **890** 95

IOP Publishing

Space Radiation
Astrophysical origins, radiobiological effects and implications for space travellers
James S Welsh

Chapter 25

Uranus

25.1 The gas giants versus the ice giants

Jupiter, Saturn, Uranus and Neptune have much in common:

1. They are located in the outer Solar System, which distinguishes them from the inner terrestrial planets.
2. They are huge compared to the terrestrial planets.
3. They are compositionally different from the rocky terrestrial planets.

For these reasons, the four outer planets are typically lumped together and called the **Jovian giant planets** or sometime the **gas giants**. However, Jupiter and Saturn are different from Uranus and Neptune in a few ways. First, Jupiter and Saturn are both far larger than either Uranus or Neptune. Jupiter's radius is 11.2 Earth radii (R_E) and Saturn's is 9.4 R_E; in comparison Uranus is 4.0 R_E while Neptune is 3.9 R_E.

Secondly, although all four are very different from the terrestrial planets, the compositions of Jupiter and Saturn are relatively enriched in the lightest elements, hydrogen and helium. Thus, Jupiter and Saturn qualify as genuine 'gas giants'. In contrast, Uranus and Neptune contain relatively more 'ice' than the other two; roughly 80% of the mass of Uranus and Neptune is a dense slurry of water, ammonia and methane in various phases. Therefore, Uranus and Neptune are sometimes lumped together as the '**ice giants**' to distinguish them from the two gas giants. 'Ice' in this context includes not only solid and liquid H_2O but also other solid and liquid volatile molecules such as ammonia, methane and ethane. And although they are called ice giants, they are probably largely liquid. Hence, the terms gas giants and ice giants do not imply states of matter as much as they reflect chemical composition.

Jupiter and Saturn contain a great deal of molecular hydrogen and helium, which both remain in the gaseous state for a large fraction of the planet's volume. Although Jupiter's surface temperature might hover around −110 °C and Saturn is

doi:10.1088/978-0-7503-5444-8ch25

even colder around −140 °C, the boiling point of diatomic molecular hydrogen is around −253 °C and that of helium is around −269 °C, suggesting that hydrogen and helium will remain in the gas phase on the surfaces of these gas giants. However, the pressure is far greater on the surfaces of these giant planets than here on Earth, so one cannot simply translate the boiling points at our surface pressure (1 bar) to the situation on the giant planets; one must analyze the phase diagrams. And naturally, the pressure increases dramatically with depth. And as the pressure increases, so does the temperature.

The result is that the gas giants contain molecular hydrogen and helium in the gas phase but at greater depths and pressures in these planets, molecular hydrogen will exist in the liquid phase and can even convert into an unusual metallic hydrogen form. It is the rotation of this liquid hydrogen metal that generates the magnetic fields of Jupiter and Saturn. But liquid hydrogen probably cannot form in the interiors of the ice giants.

25.2 Superionic water

On Uranus and Neptune, the amount of hydrogen and helium is relatively and absolutely reduced compared to Jupiter and Saturn. Additionally, the pressure in the interior is far less than inside the gas giants. Therefore, there is a relative paucity of liquid metallic hydrogen to generate planetary magnetic fields. On the other hand, the amount of water, ammonia and methane is relatively increased on Uranus and Neptune. And at great depths and pressures in these planets, H_2O can take on metallic features. This very strange form of H_2O is called **superionic water** or **ice XVIII**. Basically, in superionic water the individual H_2O molecules break down and form an interconnected solid lattice in which the oxygen exists as fixed O^{2-} ions in which a sea of H^+ ions (that is, protons) swim freely. Such superionic H_2O is predicted to exist in conditions like the deep interior of Uranus and Neptune where the pressure can exceed the required 100 gigapascals (1 megabar) and temperatures can exceed 2000 K. Superionic H_2O has been created under extreme conditions in the laboratory.

In fact, the temperature near the core of Uranus probably exceeds 5000 K. While this is incredibly hot, the core of Uranus is probably slightly cooler than that of Earth and certainly much colder than the estimated 24 000 K in Jupiter's interior. Uranus does not generate as much internal heat as Jupiter and Saturn do. It exhibits a low **thermal flux**. In fact, Voyager 2 found that Uranus radiates about as much heat as it receives from the Sun, which conflicts with the situation on Jupiter and Saturn (which radiate about twice as much heat as they receive from the Sun). For that matter, Uranus's low thermal flux is far lower than that of Neptune, which radiates 2.61 times as much heat as it takes in from the Sun. Just why Uranus has such a low thermal flux remains unknown.

Nevertheless, conditions are suitable for the formation of superionic H_2O, and it is probably this unusual state of metallic water that generates the magnetic fields of Uranus and Neptune. The depth at which water assumes this metallic character is deeper in the ice giants than the depths at which hydrogen assumes its metallic character in Jupiter and Saturn. For this reason, the magnetic fields of the gas giants are stronger than those of the ice giants. But thanks to their spinning sources of 'metal', both Uranus and Neptune do have planetary magnetic fields and associated radiation belts.

25.2.1 Uranus

In comparison to Jupiter and Saturn, which have diameters approximately 11.2 and 9.4 times greater than Earth's, Uranus is only approximately four times as large as Earth. Uranus has a mass around 14.5 times that of Earth's mass. Because this is far less than the masses of Jupiter and Saturn, Uranus (and Neptune) are insufficiently massive to compress molecular hydrogen into metallic hydrogen. Uranus has a good deal of hydrogen and helium, but the overall proportions of these elements are less than those on Jupiter and Saturn. On the other hand, the relative abundance of water, ammonia and methane is enriched. Methane tends to reflect blue and green wavelengths of light, giving Uranus its faint blue-green color. Data from the Gemini Telescopes in Hawaii indicate the presence of hydrogen sulfide (H_2S) in the upper cloud cover of Uranus. Since 'natural gas' for cooking is doped with a trace of hydrogen sulfide to confer a characteristic stench to the otherwise odorless methane, Uranus probably reeks of the same smell. Of course, without any oxygen, we need not worry that the planet's atmosphere will go up in blazes.

Uranus is located about 19.8 astronomical units (AU) from the Sun, which can be roughly recalled through Bode's mnemonic: [(0, 3, 6, 12, 24, 48, 96, *192*...) + 4] ÷ 10 = 19.6 AU. In compliance with Kepler's Laws, this long distance from the Sun implies a very slow orbital pace. It takes around 84 earth years to take one orbit around the Sun.

One of the strangest things about Uranus is that unlike any of the other planets, it basically rotates on it side—its axial tilt is 98° relative to its plane of revolution around the Sun. This sideways spin is unique among the planets (although since Venus essentially spins 'backwards' with its 177° axial tilt, still takes top honors in the weird rotation category). It is likely that a monstrous collision with another celestial body early in its history is responsible for this odd axial tilt. Regardless of its rotational obliquity, Uranus still rotates rapidly—1 Uranian day takes only 17.2 Earth hours.

Despite the very odd axial tilt and the relatively short day, the weather on Uranus is not all that bad (although the steady winds are fierce!). Images from Voyager 2 showed comparatively little atmospheric activity relative to the ferocious storms seen on Jupiter and Saturn. Other images from the Hubble Space Telescope and ground-based telescopes (using adaptive optics to reduce atmospheric distortion) have revealed some storms and bands but overall, the weather is calmer than on Jupiter or Saturn. But the relative absence of isolated major storms is not synonymous with stationary doldrums—the wind speeds in Uranus's jet streams are a blistering 560 miles per hour or about 900 km h^{-1}. These jet streams are largely confined to three regions—a westward equatorial band and two eastward bands near the polar regions. These wind speeds are faster than those on Jupiter but not quite as swift as those on Saturn.

As mentioned above, Uranus has a remarkably low thermal flux, meaning that it does not radiate nearly as much heat as Jupiter, Saturn and Neptune. Thus, Uranus is extremely cold—in fact, a tropospheric temperature of −224.2 °C (−371.5° F or 49 K) was estimated from Voyager data, making this the coldest part of any planet in the Solar System.

Moving from the atmosphere to the deeper interior, the relative abundances of hydrogen and helium decrease while the fraction composed of the 'icy molecules' increases. The depth at which these molecules begin to dominate the light gases is around 7000 km or 4350 miles. At this depth, which is approximately a third of the way down, the pressure approaches 100 000 bars while the temperature reaches 1725 °C. Under such extreme conditions, water molecules disintegrate and become positively charged hydrogen ions and negatively charged hydroxyl ions. At greater depths yet, the pressures exceed 1.5 megabars and the temperature reaches 3725 °C. This allows water to assume the superionic state or ice XVIII, wherein H_2O behave like a metal. These unusual phases of water are good conductors of electricity. Motions in these ionic and superionic layers generate electrical currents that in turn provide the dynamo that leads to Uranus's magnetic field.

As one goes deeper, there is a solid ice and rock core that cannot spin and contribute to a magnetic field. Therefore, the Uranian magnetic field is generated in much more superficial layers of the planet than in Earth and the Jovian gas giants. And although the ice giants Uranus and Neptune are huge, they are not huge enough to compress molecular hydrogen into metallic hydrogen which can generate magnetism.

25.3 Moons around Uranus

At this time, Uranus has a moon count of 27. But given how small these moons are and how relatively unexplored Uranus is, there could be more. The largest moon is Titania. Titania has four large companions, which constitute the five major moons of Uranus. In descending size order these are: Titania, Oberon, Umbriel, Ariel and Miranda. From closest to farthest from Uranus, the order is: Miranda, Ariel, Umbriel, Titania and Oberon. None of these major moons of Uranus are even close to our own Moon in size.

The two largest, Titania and Oberon, are less than half the diameter of the Moon; Ariel and Umbriel are only around a third of the Moon's diameter. Miranda is the closest to Uranus and smallest with a mean diameter of only about 290 miles (470 km). Miranda has some curious geographical features including craters, canyons, striations and fault lines.

Although small, these five major moons are large enough to attain spherical shapes due to gravity. The remainder of the 27 Uranian moons are smaller and non-spherical. Overall, Uranus has 18 regular moons with fairly circular, prograde orbits and 9 irregular moons that are on the perimeter and follow highly elliptical, inclined or retrograde orbits. The celestial mechanics is consistent with the 18 regular moons being co-formed with Uranus from the primordial accretion disc whereas the outer nine irregular moons were probably captured Centaurs or Kuiper belt objects.

25.4 Rings of Uranus

Uranus's rings were discovered incidentally in 1977 when James Elliot, Edward Dunham and Jessica Mink were taking advantage of a stellar occultation by Uranus to examine its atmosphere. Upon analyzing data from the Kuiper Airborne Observatory, they noticed that the star faded from view five times before and after full occultation by Uranus itself. They concluded that there must be a ring system around Uranus with at least five components. In 1986, these suspected rings and more were confirmed by Voyager 2 and additional rings have been discovered by the Hubble Space Telescope. The grand total currently is 13 rings. However, these rings are thin and dark with relatively wide separation. They are more like Jupiter's rings than Saturn's spectacular system. Although the particles in Uranus's rings can be as big as houses (but some of Saturn's are as well), they are far less reflective than Saturn's ice-rich rings. The composition remains unknown but their darkness might reflect a mixture of rocky material and ice that is relatively rich in organic material. The overall configuration of the ring system is consistent with a very young age (perhaps less than 1 million years old) or the presence of presently unidentified additional shepherd moons (aside from the known moons Cordelia and Ophelia) that stabilize the rings, as is the situation with Saturn.

As a homework problem, students may want to contemplate where the rings will be around a planet like Uranus, which is basically tipped on its side.

25.5 The Uranian magnetic field

The geomagnetic dynamo created via the deep ionic and superionic water layers in Uranus leads to a planetary magnetic field, but this field is quite unusual. Unlike Earth's geomagnetic field, which can be approximated as a giant dipole magnet, the field on Uranus is not a simple dipole. Rather, it is multipolar. This means that instead of a north and south magnetic pole, there are several sites on the surface of Uranus that are areas where magnetic lines of force are perpendicularly piercing the planet. Navigating by compass would be complicated!

Initially, it was assumed that Uranus's anomalous and extreme axial tilt coupled with its highly inclined magnetic axis was the explanation for the peculiar multipolar magnetic field. However, Voyager later found that Neptune also exhibits this bizarre multipolar magnetism. Yet Neptune does not display the same degree of extreme axial and magnetic tilt. (Neptune has an axial obliquity that is slightly greater than Earth's—28.3° versus 21.5°.) Therefore, the peculiar multipolar nature of the ice giant's magnetic field is probably more related to the mechanism of production (via spinning ionic and superionic water) rather than the axial tilt.

Radio emissions created by trapped charged particles were detected by Voyager 2, providing the first conclusive indication that Uranus possesses a magnetosphere. But not only does a planetary magnetic field exist around Uranus, it is highly skewed with its axis tilted at nearly a 60° angle to Uranus's anomalous rotational axis. On Earth, by comparison, the rotational and magnetic axes are only offset by about 11.5°. (Recall that Jupiter's rotational and magnetic axes are offset by around 9.6° while Saturn, being axisymmetric, has an offset of less than 1°.)

Furthermore, if one models the Uranian magnetic field as a dipole the magnetic axis is also off-center; it does not go through the center of the planet. The magnetic center is displaced by almost 8000 km or 5000 miles from the center of Uranus— approximately 31% of the radius of Uranus. It is displaced mainly towards the north pole.

The intensity of the magnetic field at Uranus's surface is about 0.23 gauss, roughly comparable to that on the surface of Earth's (about a half a gauss, ranging from 0.25 to 0.65 gauss). Recall that 1 gauss = 10^{-4} Tesla. However, the magnetic moment associated with the field is far more powerful on Uranus—its magnetic dipole moment of 3.9×10^{17} Tm^3 is about 50 times the terrestrial moment. But the magnetic field intensity on the surface varies substantially from point to point on the surface of Uranus because of the large offset from center.

The size of a planet's magnetosphere depends not only on the intrinsic strength of the magnetic field but also on the solar wind in the planet's vicinity. As the solar wind pressure is much lower at the great distance of Uranus, the magnetosphere is larger than it would be if the planet were in closer to the Sun. The result is a large magnetosphere despite the relatively weak magnetic field. On the upwind (solar) side, it extends about 18 Uranian radii to the magnetopause and bow shock.

And as with the planetary magnetospheres of Mercury, Earth, Jupiter and Saturn, there is a long magnetotail extending millions of miles behind Uranus on the side facing away from the Sun. This magnetotail reaches over 10 million kilometers or 6.2 million miles beyond Uranus. However, the steep inclination of Uranus's magnetic axis coupled with the extreme obliquity of Uranus's rotation (essentially on its side) causes the magnetic field lines to follow a highly unusual path. Within the extensive cylinder of Uranus's magnetotail, the field lines twist in a complicated helical fashion.

In general, planetary magnetospheres fend off the solar wind, channeling it around the planet into its trailing magnetotail on the distal side. However, this occurs only when the solar wind and the planet's magnetosphere are oriented in the same direction. Because of Uranus's bizarre and shifting magnetic fields, there are orientations that allow solar wind particles to penetrate via magnetic reconnection. On Earth on occasion, magnetic reconnection occurs near the north and south magnetic poles, leading to intense aurorae. But it is predicted that such intense aurorae might occur daily (every 17.24 h) on Uranus through periodic magnetic reconnections. These intermittent reconnections play a key role in alternating the global magnetosphere between an 'open' and 'closed' configuration. Voyager 2 only made a singly flyby as it passed Uranus in 1986, so we presently do not have documentation of these predicted aurorae.

25.6 X-rays from Uranus

Just as with other planets, Uranus reflects solar x-rays. Data from 2002 and 2017 obtained from the Chandra X-ray Telescope provided clear evidence of such scattered solar x-rays (albedo x-rays). Such x-rays are modest in energy, ranging from 0.6 to 1.1 keV. But in addition to the reflected solar x-rays, there was an observed unusual burst of x-rays in which x-ray brightness increase fourfold over the course of a day. The source of these x-rays remains uncertain, but one possibility is the Uranian ring system. As highly energetic charged particles (such as galactic cosmic rays) bombard these rings, fluorescent x-rays and gamma rays may be generated. Similar processes occur with Saturn's rings. Alternatively, since Uranus has a higher concentration of energetic electrons in its radiation belts than Saturn, perhaps these high-energy electrons could be striking the rings to create the observed x-rays through bremsstrahlung. Another possible source is unusual Uranian auroral activity. Aurorae around Jupiter and Saturn are known to generate high-energy ultraviolet photons and occasional x-rays.

25.7 Radiation belts of Uranus

Voyager 2 identified planetary radiation belts around Uranus with an intensity similar to those of Saturn, although they differ markedly in ion composition. Unlike Jupiter's and Saturn's belts that contain ions derived from the rings and moons within the radiation belts, the radiation belts of Uranus are dominated by protons (hydrogen ions) and electrons. There is no evidence of alpha particles (helium nuclei) which would originate from the solar wind and no evidence of heavier ions which would originate from sputtering on the moons around Uranus.

Regardless of composition, the radiation in Uranus's radiation belts is intense enough that it would quickly (that is, within 100 000 years) blacken any methane present in the Uranian rings or the icy surfaces of its inner moons. Such irradiation may have contributed to the darkened surfaces observed on Uranus's moons and ring particles. The main Uranian moons orbit *within* the radiation belts and absorb some of the trapped protons. Protons, and all charged particles, follow magnetic lines of force. Those protons following force lines that intersect a moon's orbit will be absorbed by the moon, thereby causing gaps in the radiation belts in these orbits.

However, some of the 1986 data from the Voyager 2 flyby remain enigmatic. The data indicate that Uranus' electron radiation belt is more intense than expected, and near the theoretical limit (the **Kennel–Petschek limit**). In contrast, the proton radiation is far less intense and the energy of these protons is lower than expected. One possible explanation for this is the asymmetry of Uranus's magnetic field. Models indicate that the larger the charged particle's **gyroradius (Larmor radius** or **cyclotron radius**: mv/qB), the greater the impact of the asymmetry. This means that energetic protons with greater than 100 keV are more affected and wind up getting disproportionately depleted through trajectory changes that cause them to eventually be absorbed by the rings, moons or atmosphere. This might account for the relatively weak and low-energy proton radiation belt but intact and energetic electron belt.

Further reading

[1] Millot M, Hamel S, Rygg J R, Celliers P M, Collins G W, Coppari Federica, Fratanduono D E, Jeanloz R, Swift D C and Eggert J H 2018 Experimental evidence for superionic water ice using shock compression *Nat. Phys.* **14** 297–302

[2] Masters A, Ioannou C and Rayns N 2022 Does Uranus' asymmetric magnetic field produce a relatively weak proton radiation belt? *Geophys. Res. Lett.* **49** e2022GL100921

[3] Cao X and Paty C 2017 Diurnal and seasonal variability of Uranus's magnetosphere *J. Geophys. Res. Space Phys.* **122** 6318–31

[4] Dunn W R, Ness J-U, Lamy L, Tremblay G R, Branduardi-Raymont G, Snios B, Kraft R P, Yao Z and Wibisono A D 2021 A low signal detection of x-rays from Uranus *J. Geophys. Res.: Space Phys.* **126** e2020JA028739

IOP Publishing

Space Radiation
Astrophysical origins, radiobiological effects and implications for space travellers
James S Welsh

Chapter 26

Neptune

Discovered in 1846, Neptune is the only planet not visible to the naked eye. It was discovered after John Couch Adams and Urbain Le Verrier noticed anomalies in Uranus's orbit consistent with perturbations caused by a large body in an outer orbit. Johann Galle then explored the region predicted to harbor the missing mass and found planet Neptune. It is located 30.1 astronomical units (AU) from the Sun, making it too faint to see without a telescope. At this distance, light travelling around 186 000 miles per second or nearly 300 000 km s^{-1} still takes about 4 h to reach us.

Neptune has an axial tilt not very dissimilar from Earth's (28.3° versus our 23.436°). Like the other Jovian giants, it rotates rapidly, and a day takes only 16 h and 36 min. But because of its great distance from the Sun, it moves slowly and orbits once every 165 Earth years.

With a radius of about 24 764 km or 15 470 miles, Neptune is approximately four times the size of Earth. Neptune is slightly smaller than Uranus (which has a radius of roughly 25 362 km or 15 759 miles). But despite Uranus being roughly 2% wider, Neptune is more massive. And this mass discrepancy is not insignificant—Neptune is roughly 17% more massive. This indicates that Neptune is substantially denser. In fact, Neptune is the densest of all four Jovian giant planets. The proportion of rocky material and solid ice is thus higher inside Neptune than Uranus or the gas giants, Jupiter and Saturn.

Much of what we know about Neptune comes from a single flyby in 1989 by Voyager 2 while it was whizzing by at 62 000 mph (100 000 km h^{-1}).

26.1 Neptune's atmosphere

Neptune has an atmosphere similar to Uranus's, accounting for their generally similar colors. Neptune's atmosphere is approximately 79% nitrogen (very similar to Earth). But instead of 21% oxygen, the balance of Neptune's atmosphere is about 18% helium and 3% methane. The slightly higher concentration of methane on

doi:10.1088/978-0-7503-5444-8ch26

Neptune is probably what confers its deeper blue hue compared to Uranus (or perhaps there is another unidentified chemical component that gives Neptune the blues). Icy methane clouds hover high in Neptune's atmosphere. Both planets exhibit stable fast jet stream winds in their upper atmospheres with distinct zones: a westward zone at the equatorial regions and two eastward wind zones (near each pole).

26.2 The weather on Neptune

But while both ice giant planets are roughly the same size at about four times the width of Earth, Neptune has wild weather while Uranus is relatively calm (despite the fast but stable upper atmospheric jet streams). This is surprising because weather typically is dictated by temperature changes and since Neptune is much farther from the Sun, it would be expected to be colder, calmer and less turbulent. But despite relatively less insolation on Neptune, its atmosphere is nonetheless more dynamic.

Neptune has enormous hurricanes reminiscent of Jupiter's Great Red Spot. These storms manifest as dark spots with associated white methane clouds. A large oval storm called the **Great Dark Spot of 1989** was visible in the southern hemisphere for a while but has since dissipated. The storm was larger than the entire Earth. Like the Great Red Spot of Jupiter, this was a high-pressure **anticyclone**. Other gigantic storms come and go periodically.

Neptune's weather is enigmatic. Neptune is about the same size as Uranus and its rotation rate is similar (albeit it does not spin on its side like Uranus does) so wild weather on one but not the other is unexpected. Present theories to account for the discrepancies relate to the relative heating of each planet's atmosphere by the planet's respective interiors. Like Jupiter and Saturn, Neptune generates significant internal heat. Recall that in contrast, Uranus has a cool interior and does not radiate significant heat (i.e. it is in low thermal flux). Neptune's relatively high thermal flux might be providing heat that energizes its troposphere, which in turn leads to the powerful storms. Thanks to the heat emanating from its interior, Neptune's atmosphere is about the same temperature as Uranus's atmosphere; this is despite being so much farther from the Sun than Uranus (approximately 50% father). At its distance, Neptune receives less than half of the insolation (solar energy) than Uranus (roughly $1/e$ as much) so a similar atmospheric temperature is most surprising.

26.3 Neptune's rings

Like the other three Jovian planets, Neptune has a ring system. These rings were long suspected based on the unanticipated discovery of rings around Uranus in 1977 thanks to a fortuitous occultation. But it was not until the 1989 flyby of Voyager 2 that these rings were confirmed. The images from Voyager 2 revealed three narrow rings and two relatively broad rings, for a total of five.

Like Uranus's rings (but markedly different from Saturn's), those around Neptune are relatively dark. However, unlike the relatively large chunks that make up Uranus's rings, the rings of Neptune are composed mostly of microscopic (micron-sized) particles. In this regard, Neptune's rings are somewhat similar to those around Jupiter. The small-sized grains in Neptune's rings were initially deduced by the absence of detectability via Voyager's radio signals, which indicates that the particles are no larger than 1 cm. Further evidence of their tiny sizes was provided by the fact that the rings were difficult to see from reflected sunlight but easier to spot when looking back towards the Sun, consistent with dust-sized particles. The precise chemical composition remains uncertain, but it is hypothesized that the darkened color is the result of irradiation of methane ice crystals.

26.4 Neptune's moons

Neptune presently is known to host 14 moons. Of these, however, only one—Triton —is large enough to attain a round shape. In the grand tally, Neptune finishes last in the round moon competition with Saturn in first at seven, Uranus in second at five and Jupiter in third with four. Triton, with a 1353 km mean radius, is a pretty large moon and ranks seventh in the Solar System overall (behind Ganymede, Titan, Callisto, Io, Moon and Europa in size order). This means that although Saturn has seven round moons, Neptune's solitary round moon Triton is larger than six of Saturn's seven.

Triton is only about half the diameter of Jupiter's Ganymede or Saturn's Titan. However, it is denser than Titan suggesting that it contains a larger proportion of rock and a lower proportion of ice. Neither Triton nor Titan is nearly as dense as the Moon is though, which is lacking in water ice but makes up for this deficiency in lightweight rocks.

But Triton is unique among all the Solar System's large round moons in that it revolves around its parent planet the 'wrong' way—Triton orbits Neptune in a retrograde fashion, that is, in a clockwise direction (when viewed from above) which is opposite of the spin of Neptune (and the direction of revolution of all the planets around the Sun).

Furthermore, Triton exhibits a highly inclined orbit (approximately 23°) relative to Neptune's rotational equator. Thus, Triton is an irregular moon and probably did not form concurrently with its parent, Neptune, from the protoplanetary accretion disc. More likely, Triton was captured by Neptune early in the Solar System's history.

In addition to a set of moons, Neptune also has a small population (22 at last count) of Trojan asteroids at the stable Lagrangian points, L4 (ahead) and L5 (behind).

26.5 The Neptunian magnetosphere

Voyager 2 discovered a strong and complex intrinsic magnetic field around Neptune with an equally complex magnetosphere and magnetotail. The bow shock wave in the solar wind was detected at 34.9 Neptune radii (R_N), (where 1 R_N = 24 764 km). While the magnetic field has a dipole nature, it also has a pronounced quadrupolar character as well. However, the dipole of Neptune is oriented in the opposite direction of Earth's. A compass that points north on Earth would point south on Neptune.

The magnetic field strength at the equatorial surface amounts to about 0.14 gauss (14 μT), which is slightly less than the field strength at the surface of Earth (roughly 0.5 gauss or 50 μT with a broad range from 25 to 65 gauss). On the other hand, the underlying driving dipole magnetic moment is 2.2×10^{17} T m^3 or nearly 27 times the terrestrial magnetic moment. But like Uranus, Neptune has a complex magnetic field that may be modelled as a dipole but also has strong non-dipole components, including a prominent quadrupolar quality. Also, much like with Uranus but even more extreme, the magnetic axis is markedly displaced from the central axis of rotation. Uranus's magnetic field axis is offset by about 31% of the planet's radius and is inclined by 59 degrees with respect to the rotation axis; Neptune's magnetic field axis is offset by about 55% of the planet's radius and is inclined by 47 degrees with respect to the rotation axis. Such offset tilted magnetic dipoles are sometimes abbreviated as OTD.

Much of this news about Neptune was initially surprising since it was previously assumed that Uranus's strange magnetic field was largely due to the fact that Uranus rotates on its side and has a magnetic field axis that is steeply inclined to the rotational axis. It now appears that the odd and asymmetrical magnetospheres of the two ice giants share much in common. So, this commonality cannot be attributed to their rotational axes. As described below, the similarities between the two planetary magnetic fields are probably due more to their common mode of generation by spinning superionic water in relatively superficial layers of their planetary interiors.

Neptune has a fairly typical obliquity of 28.3° but the tilt angle of its magnetic moment is 47° off from its rotational axis. While this misalignment is smaller than that of Uranus (nearly 60° discrepancy), it is still rather oblique and much more pronounced than the dipole tilts of Earth, Jupiter and Saturn. Because of the misalignment, with each roughly 16 h rotation, the magnetosphere experiences tempestuous fluctuations. This, coupled with the similar underlying generating mechanism (spinning ionic fluids at relatively shallow depths compared to the liquid iron in our deep outer core) probably account for the bizarre and highly asymmetrical magnetospheres of both Neptune and Uranus.

Neptune's planetary magnetic field is most likely generated in a shell of liquid at intermediate depths in the interior rather than deep in the core as with Earth. And as with Uranus, the magnetic field of Neptune is assumed to be due to an internal reservoir of rotating ionic liquid. Given that Neptune and Uranus have roughly the same compositions, but Neptune is more massive and denser than Uranus, the transition depth to the strange ionic (and superionic) phases of water occurs at shallower depths on Neptune. It seems that at a depth of roughly 5000 km or 3000 miles below the 'surface', the pressure reaches 250 000 bars and the temperature exceeds 2000°, which permits a transition to the ionic and superionic phases. Rotation of this conducting fluid creates the internal dynamo responsible for Neptune's magnetic field. As with other planets that have permanent magnetic fields, Neptune is surrounded by a planetary radiation belt.

26.6 Neptune's radiation belts and the L parameter

The **Voyager 2 Cosmic Ray System** detected significant fluxes of high-energy (>1 MeV) trapped electrons and protons in the magnetosphere of Neptune, thereby confirming the presence of a radiation belt. The measured radiation intensities decreased closer to the planet thanks to particle absorption by moons and rings and the radiation intensity was maximum near a **magnetic L-shell** of seven.

The **L-shell**, **L-value** or **McIlwain L-parameter** (named after Carl E McIlwain) is a method of organizing and representing magnetically trapped particles in planetary magnetic fields. The L-parameter is a metric that describes particles in a particular set of planetary magnetic field lines. In particular, the L-value describes the set of magnetic field lines which cross a planet's magnetic equator at a number of planetary radii equal to the L-value in question. For example, if $L = 7$, we are describing the set of the magnetic field lines which cross the magnetic equator seven planetary radii from the center of the planet. Hence, in the simplest sense, the L-value represents the radius of a field line at the magnetic equator. Under normal circumstances, L is the maximal radial distance of a magnetic field line from Earth's center.

The L-shell parameter is helpful in describing the shape and structure of the magnetic field lines around a planet, but it only works when the planet in question may be adequately modeled as a dipole like Earth. In such cases, the equation for L is:

$$R = L \cos^2 [\lambda]$$

where R is the radial distance (in planetary radii) to a point on the magnetic field line, λ is its geomagnetic latitude and L is the L-shell of interest.

If $R = 1$, we are on the surface of the Earth and the value of λ represents the **foot point** of the closed field line at the surface of the Earth. This value is the so-called **invariant latitude**. Solving for λ, the invariant latitude, at $R = 1$ (that is, on the Earth's surface) we get:

$$\lambda = \cos^{-1} (R/L)^{1/2}.$$

For each value of L there is a unique invariant latitude. For example, if $R = 1$ (meaning we are at the Earth's surface), and $L = 3$, one may calculate a value of $\lambda = 54.7°$. One can see from the equations that as one approaches the magnetic poles (90° from the magnetic equator) L falls to zero.

More thoroughly, standard radiation belt models describe the trapped particle fluxes in the **McIlwain magnetic coordinate system (B,L)**, where B is the magnetic field intensity at the point of measurement and L is the drift shell of a trapped particle mirroring at this point. Alternatively, one may think of the coordinates as uniquely defining a location based on the intensity of the magnetic field at the mirror point and the L-parameter. Hence, the McIlwain coordinate system is based on the magnitude of the magnetic field and adiabatic invariants.

On Earth, during the most severe geomagnetic storms, such as the **Carrington Event** of 1859 (estimated to be category X45 ±5 in terms of soft x-ray class), trapped electrons moved to as close as $L\sim1.41$ and some trapped protons may have moved to

$L{\sim}1.36$. The Van Allen radiation belts generally correspond to $L = 1.2{-}3$ for the inner belt and $L = 4{-}6$ for the outer belt. That is, the altitudes above the surface of Earth at the equator for the inner Van Allen belt is roughly 0.2–2.0 Earth radii while the outer Van Allen belt lies at 3–5 Earth radii. Note that L is measured from Earth's center, not from the surface. The Earth's plasmapause is typically located around $L = 5$.

As previously discussed, the Jovian magnetic field is the strongest planetary field in our Solar System. Its powerful magnetic field traps electrons with energies exceeding 500 MeV, creating high-intensity and high-energy radiation belts. The characteristic L-shell of Jupiter is $L = 6$, where the electron distribution undergoes a marked radiation hardening (increase of electron energy). Other important L-shells around Jupiter are those of $L = 20{-}50$, where the electron energy decreases to the very-high-frequency realm and the magnetosphere merges with the solar wind. Because some of Jupiter's trapped electrons are so energetic, they can diffuse across different L-shells more readily than trapped electrons in Earth's magnetic field can.

In the region of the inner moons of Neptune, the radiation belts have a complicated structure. The presence of these moons imposes severe constraints on the magnetic field geometry of the Neptunian inner magnetosphere. Electron phase-space densities have a positive radial gradient, which implies that electrons diffuse inwards from a source in the outer Neptunian magnetosphere. Electron spectra harden in the region of peak flux with energies ranging from 1 to 5 MeV. Protons in the Neptunian magnetosphere and radiation belts have significantly lower fluxes than electrons do, and they exhibit large anisotropies. The Neptunian radiation belts resemble the Uranian radiation belts to the extent allowed by the different locations of their respective moons and rings, which uniquely shape the flux densities around each planet.

Putting the radiation belts of the Solar System in perspective, Earth's Van Allen radiation belts are very strong for the size of the parent planet, Earth. The Van Allen belts are comparable to the radiation belts around Saturn. The ice giants, Uranus and Neptune, have relatively weaker magnetic fields and hence, weaker radiation belt intensities. Jupiter is not surprisingly the king when it comes to radiation and its radiation belts are very intense and energetic thanks to the magnetic moment of Jupiter, which is about 2×10^4 as large as Earth's. Mercury, Mars and Venus have very weak or virtually absent magnetic fields and therefore have no radiation belts.

Further reading

[1] Garrett H and Evans R 2017 *The JPL Neptune Radiation Model (NMOD)* (Pasadena, CA: California Institute of Technology) JPL Publication 17-2 https://ntrs.nasa.gov/api/citations/20170006886/downloads/20170006886.pdf

[2] Garrett H B, Jun I and de Soria-Santacruz M 2017 *Conf. Paper at Applied Space Environments Conf. (ASEC) Modeling trapped radiation—a comparative study of the terrestrial, Jovian, Saturnian, Uranian, and Neptunian radiation belts (Huntsville, AL, 15-19 May 2017)* https://asec2017.exordo.com/files/papers/12/final_draft/IRB_Talk3_V3.pdf

IOP Publishing

Space Radiation
Astrophysical origins, radiobiological effects and implications for space travellers
James S Welsh

Chapter 27

Radiation belts beyond our Solar System

27.1 Earth's Van Allen radiation belts

Earth harbors two more or less permanent radiation belts arranged like two giant concentric donuts circling the magnetic equator, with Earth in the donut hole. The radiation distribution diminishes with latitude and the belts are practically absent over the magnetic poles. These toruses are filled with highly energetic charged particles. Technically therefore, the material in our Van Allen belts is in the state of matter called plasma (that is, an effectively neutral but electrically conducting state of matter made of electrons, ions and bare atomic nuclei that generally behaves like gas or liquid but also contains electromagnetic fields that affect its behavior). The Van Allen belts are arranged such that the inner belt is where most of the highly energetic (>30 MeV) protons reside. Some protons in the inner belt may have over 700 MeV of kinetic energy. This is made possible by the relatively stronger geomagnetic field in the region of the inner belt compared to the outer belt. The peak flux of protons in the inner belt is about 20 000 particles per second across a spherical surface area of 1 square cm. The protons of the outer Van Allen belt generally have much lower energies than those residing in the inner belt.

In contrast to the heavy concentration of high-energy protons in the inner belt, the outer belt contains the most energetic electrons. The relativistic electrons in the outer belt may possess kinetic energies reaching up to several hundred MeV. The Sun is the source of most of these energetic particles in the outer belt. This differs from the inner belt's supply of particles, which is largely derived from interactions between galactic cosmic rays and Earth's atmosphere. This inner cadre of protons includes many that were created through the decay of free neutrons that were made by spallation of atmospheric atoms by cosmic rays, the so-called cosmic ray albedo neutron decay (CRAND) process.

The outer belt mainly contains charged particles of solar origin, including alpha particles (helium nuclei) from the solar wind. (Recall that the solar wind consists of about 5% helium nuclei by number density, about 10% by charge density and about

doi:10.1088/978-0-7503-5444-8ch27

20% by mass density.) In comparison, the inner belt is virtually devoid of helium ions, suggesting that its radiation is not derived from a solar source.

27.2 The magnetic mirror effect

These captured and cosmic ray-generated charged particles 'spiral' back and forth along Earth's magnetic lines of force (more correctly, they follow helical paths). As the charged particles approach either the north or the south magnetic poles, the increased strength of the geomagnetic field repels them; they get reflected back towards the equatorial latitudes. Thus, the particles bounce to and fro between the magnetic polar regions because of this so-called **magnetic mirror effect**. This mirroring typically occurs at altitudes of 300 miles (500 km).

Given sufficient time, these particles eventually collide with atoms in the thin upper atmosphere and are thereby removed from the belt. Meanwhile, new particles from the Sun and cosmic rays replenish them. However, the residence time of charged particles in the Van Allen belts is surprisingly long, averaging around 10 years. This long lifetime allows large numbers of particles to accumulate and accounts for the high particle flux in these radiation belts.

27.3 The waxing and waning Van Allen belts

Both belts wax and wane with space weather but the inner belt is generally more stable than the outer one. The highest radiation intensity in the inner Van Allen belt is centered at a height of about 1800 miles or 3000 km above the equator and normally spans an altitude between 620 mi (~1000 km) and 7500 mi (12 000 km) above the Earth's surface. In other words, the inner Van Allen belt typically extends from 0.2 to 2 Earth radii (that is, **L-values** of 1.2–3) under ordinary circumstances. It does dip considerably closer to Earth over the South Atlantic in the area called the **South Atlantic Anomaly (SAA)**.

Nevertheless, the inner belt does swell from time to time and this expansion may bring hazardous radiation near sensitive items in low Earth orbit, including the International Space Station (ISS) and many communications satellites.

It should not be misconstrued that the two belts are entirely independent entities; they are best thought of as two regions of maximum particle density that gradually overlap but have a region of lower density between them. Having said this, the inner edge of the outer belt does serve as a powerful barrier to very-high-energy electrons; relativistic electrons essentially cannot penetrate beneath this inner edge of the outer belt. Furthermore, the gap between the inner and outer belts varies from time to time, becoming wider during intervals of low geomagnetic activity. It may be that this slot widens significantly or that the two belts practically disappear altogether when the geomagnetic field temporarily collapses during the transition period when the magnetic poles are reversing.

Although the outer radiation belt expands and shrinks substantially, typically spans from around 8400 to 36 000 miles (approximately 13 520–58 000 km) above the surface. The highest intensity of radiation in the outer belt is usually found between about 9000–12 000 miles (roughly 14 480–19 312 km) above Earth's surface.

27.4 The SAA

The SAA is an area over eastern South America and the South Atlantic Ocean where the inner Van Allen belt lies lower than elsewhere; over the SAA, the Van Allen belt might descend to an altitude of only 120 miles or 200 km. This is significantly below the orbit of the ISS (which averages around 250 miles or 400 km above the surface) or the Hubble Space Telescope which is about 340 miles or 550 km up. Thus, sensitive spacecrafts and satellites select orbits that avoid or minimize transits through the SAA. The ISS, which orbits at an inclination of 51.6°, has special shielding and radiation-hardened instruments to cope with the extra radiation encountered in the SAA. The Hubble Space Telescope shuts down its operations when traversing the SAA. Nevertheless, the SAA does present problems. The Skylab's Apollo Telescope Mount, which was able to observe the Sun in x-rays, ultraviolet and visible light, suffered radiation-induced solar flare false alarms when in the SAA. Passage through the SAA might have contributed to the failure of some Globalstar satellites in 2007. Furthermore, the advanced x-ray astronomy observatory called Hitomi (also known as ASTRO-H and NeXT for New X-ray Telescope) of the Japan Aerospace Exploration Agency (JAXA) had tremendous potential but lost contact only about a month after its launch in 2016 and eventually tumbled out of orbit. It is suspected that this was because of radiation-engendered damage suffered while passing through the SAA.

The SAA traces its origin to the fact that the Earth's magnetic dipole axis (in addition to being about 11° off the spin axis) does not pass through Earth's center; that is, it is not concentric or antipodal. The point where the rotational and magnetic axes intersect is not at the center of the Earth, but rather is around 280–310 miles (450–500 km) from the center.

The SAA results from this non-concentric orientation of the magnetic axis and represents the point on the Earth where the magnetic axis is farthest from the surface. Thus, the magnetic field strength is weakest over the SAA and the Van Allen belts approach Earth most closely over this region. Conversely, the Van Allen belts are normally farthest from Earth over an area in the northern Pacific Ocean.

Another possible contributor to the SAA is a large deposit of dense rock some 1800 miles (2900 km) below the African continent known as the **African large low-shear-velocity province (LLSVP)**. LLSVPs are large regions where seismic shear waves (S-waves or secondary waves) move considerably slower, indicating a dramatic change in chemical and physical properties. LLSVPs are located deep in the mantle, near the core–mantle boundary. Two such LLSVPs have been discovered: the African and Pacific LLSVPs. They are named for the geographic regions overlying them at the surface. The African LLSVP (nicknamed Tuzo) covers most of the southern African continent and extends into the eastern South Atlantic. It is believed that it might be somehow affecting the magnetic field strength in the overlying Earth, thereby contributing to the existence of the SAA.

27.5 Antiprotons and the transient third Van Allen belt

During spells of severe space weather, the density and energy of the charged particles within the Van Allen belts increase, and the belts grow significantly in size.

Such enhanced radiation may pose a danger to astronauts and susceptible spacecraft technology. In fact, during very intense solar activity (such as a large coronal mass ejection), a transient third radiation belt containing ultra-relativistic charged particles may arise. This third belt occupies the space in the gap between the inner and outer belts and appears to be due to a splitting of the outer belt. NASA's **Van Allen Probes** documented the persistence of this third belt or radiation zone for about four weeks in 2013.

The **Payload for Antimatter Matter Exploration and Light-nuclei Astrophysics (PAMELA)** telescope, which was launched in June 2006 from Kazakhstan, detected antiprotons in the inner Van Allen belt when traveling through the SAA in 2011. These antiprotons are probably produced by the interactions between galactic cosmic rays and the upper atmosphere, specifically when antineutrons escape upwards and decay into antiprotons and positrons. Based on this proposed mechanism, Saturn may be expected to harbor a particularly antiproton-rich radiation belt thanks to collisions between galactic cosmic rays and the ring system. The kinetic energy of these rare antiprotons (only 28 were actually identified) has been recorded in the range from 60–750 MeV.

27.6 Space weather and the Van Allen belts: whistlers and electron rain

'Whistlers', also called **whistler waves, whistler-mode waves**, or **whistling atmospheric electromagnetic waves** are radio waves in the **very-low-frequency** category. Their frequencies range between 300 Hz to 30 kHz but the most common frequencies typically lie between 3 kHz to 5 kHz. Although they are electromagnetic waves, their frequencies coincide with audio frequencies. With appropriate technology, whistler waves can thus be converted into audible sounds that are perceived as descending tones which last for a few seconds. Hence, the name whistler waves. Whistlers may repeat at regular intervals (usually of several seconds). The repeated whistler waves get progressively longer and fainter over time. Detailed studies of whistler waves documented the existence of Earth's **plasmasphere**, which is a region of cool (i.e. low energy) plasma within the lowest portions of the magnetosphere. The plasmasphere begins above the ionosphere and extends upwards and co-rotates with the Earth. It normally extends to a height of around four Earth radii (R_E) (that is, about 16 000 miles or 26 000 km); this outer limit is called the **plasmapause**. Beyond the plasmapause, electron concentrations fall off rapidly (by about an order of magnitude) and their movements differ substantially. Under conditions of low solar activity, the plasmapause is higher up and may reach 8 R_E but during times of very intense solar activity, the plasmasphere is compressed and the plasmapause may be only 3 R_E above ground. Of course, in contrast to this cool plasma in the plasmasphere, the hot plasma (i.e. high-energy ions and electrons) in the magnetosphere constitute the regions we call the radiation belts. Thus, the Van Allen radiation belts partly overlap with the plasmasphere.

Whistler waves are usually produced by powerful lightning strikes where the resulting impulse propagates along Earth's north–south magnetic field lines in the ionosphere. (The ionosphere is that region within our atmosphere where the concentration of ions is high enough to affect radio wave propagation. Although it varies significantly with space weather, the ionosphere typically begins at an altitude around 50 km (30 miles) above sea level.) As the ionosphere is filled with plasma, whistler waves are thus electromagnetic waves that move through plasma. (They should not be confused with the related plasma waves called **Alfven waves**, which are magnetodynamic-hydrodynamic wave-like variations or oscillations of the ions in a plasma, and which propagate along the magnetic field lines within plasmas.) The whistler waves then travel north–south along 'ducts' in the ionosphere (regions of increased ionization along magnetic field lines) from one hemisphere to another. Eventually they rebound back towards the equator when they reach the corresponding geomagnetic latitude in the opposite hemisphere.

Higher up, electrons in the Van Allen belts similarly travel along magnetic lines, bouncing between the north and south magnetic poles. Under certain conditions, such as geomagnetic storms, intense whistler waves may be generated within the outer Van Allen belt and the gap region. These whistlers (particularly a subtype called chorus waves) energize and accelerate resident electrons to relativistic velocities and alter their pitch angles (the angle between electron linear velocity and magnetic field direction). The result is that the electrons' pathways become distorted, and energetic electrons may emerge from the Van Allen belts. When this happens, relativistic electrons precipitate into the Earth's upper atmosphere, creating heavy '**electron rain**' in the polar regions. This relatively intense **whistler wave-induced electron precipitation**, which was observed by the **Time History of Events and Macroscale Interactions during Substorms** (**THEMIS**) spacecraft in equatorial latitudes and by the **Electron Losses and Fields Investigation** (**ELFIN**) satellite on a low-altitude polar orbit, was only recently recognized and was previously unappreciated. This extremely energetic electron precipitation is from the outer Van Allen belt and heats up the outer atmosphere and alters its chemical properties. It contributes to the aurorae but could possibly threaten the integrity of satellites and spacecrafts and might even pose a health hazard to astronauts caught in the rain.

The Earth's inner radiation belt was discovered in 1958 by James Van Allen through an analysis of data from a cosmic ray detector (basically, a Geiger counter) on NASA's very first mission, the **Explorer 1** spacecraft. The outer radiation belt was discovered shortly afterward using data from **Explorer IV** and **Pioneer 3**, which were both launched in 1958. The Van Allen belts (and all planetary radiation belts) are part of their parent planet's magnetic environment, that is, their inner magneto-spheres. Planetary magnetic fields steer electrically charged particles and dictate how they move and where they will reside. Hence, every planet in our Solar System with a strong magnetic field has a detectable radiation belt and the specific features of that radiation belt depend on the physical, atmospheric and magnetic features of the parent planet. In many ways, the radiation belt around a planet is like a fingerprint that reveals aspects of the planet's magnetic characteristics and the space radiation milieu in the vicinity of that planet.

27.7 Radiation belts around exoplanets?

Radiation belts are the regions of a planetary magnetosphere where highly energetic charged particles, such as electrons, protons and heavier ions, are abundant and trapped, thereby creating a more or less permanent high-radiation zone. All the Jovian planets, as well as Earth, have radiation belts. Thus, with the exception of Mercury (maybe), perceptible permanent planetary radiation belts exist around every planet in our Solar System that has a sufficiently strong magnetic field to create a stable magnetosphere. Even Jupiter's moon Ganymede, which is the only moon with a magnetic field, has its own radiation belt. This begs the question, do **exoplanets** (planets beyond our own Solar System) similarly harbor radiation belts? Basic principles of planetary science would suggest the answer should be a clear yes. And although our Sun does not have a permanent radiation belt because of its fluctuating magnetic field that reverses its dipole orientation every 11 years, other stars and brown dwarfs that have more magnetic stability might be expected to possess radiation belts as well.

At the time of this writing, exoplanet surveys, most notably through the **Kepler Mission**, have led to the discovery of 5502 confirmed exoplanets in 4096 planetary systems with 928 systems having more than one planet. And there are a few thousand additional candidates awaiting confirmation. Of these exoplanets, a surprising 1673 have been identified as rocky planets that are several times the mass of Earth and up to eight times the diameter—hence their designation as '**Super-Earths**'. Thus, nearly a third of all identified exoplanets fall into the Super-Earth category.

27.8 Super-Earths

A Super-Earth is a type of exoplanet with a mass higher than Earth's, but substantially below those of the Solar System's ice giants, Uranus and Neptune, which are about 14.5 and 17 times Earth's mass, respectively. By convention, they are over twice the size of Earth and up to ten times its mass. In the mass range of three to ten times Earth's mass, there may be a vast array of planetary compositions including 'water worlds', 'snowball planets' or planets like Neptune that are largely composed of dense gas, liquid and ice. Hence, the name 'Mini-Neptune' or 'Sub-Neptune' is sometimes used synonymously with Super-Earth for larger exoplanets in this size range (but technically such Mini-Neptunes should be more icy and less rocky than Super-Earths). Super-Earths were a bit of a surprise since nothing of the kind exists in our Solar System. In addition to these Super-Earths or Mini-Neptunes, a surprising number of other oddities not seen in our Solar System have been discovered including **Sub-Earths, Hot Jupiters, Hot Neptunes and Super-Neptunes**.

The various types of exoplanets may be categorized as **gas giants**, **Neptunian planets**, **Super-Earths** and **terrestrial planets**. Along with Hot Jupiters (which are gas giants with very short orbital periods—under 10 days, with some even shorter than one earth-day—meaning they are unusually close to their parent stars), Super-Earths appear to be the most common type of exoplanets. Of course, this is based on the detection capabilities we presently have at our disposal.

The unanticipated discovery of Super-Earths has led to a host of unanticipated questions, including questions about magnetospheres and radiation belts around

Super-Earths. For example, if a rocky planet is very massive, will its interior be too dense to support a liquid state of iron in its core? Since our geomagnetic field is produced by rotating liquid iron in the outer core, the absence of an internal reservoir of spinning liquid metal might preclude the possibility of a magnetic field around Super-Earths. Based on their size and densities, the interiors of Super-Earths might consist of undifferentiated, coreless mantles. Or if they do have a core, it might be entirely solid and unable to sustain a magnetic field. Also, the geophysical peculiarities of Super-Earths casts doubt on the likelihood of plate tectonic on such planets. Given how conducive plate tectonics is to life on Earth, one must wonder about the overall conviviality of Super-Earths, even if they are located in the **habitable zone** of the planetary system (where water would exist as a liquid on the surface).

The possibility of Super-Earths not having magnetospheres has raised a good deal of interest about the significance of our own magnetosphere. As mentioned many times, our magnetosphere shields us from incoming charged particles. But of course, our magnetosphere cannot protect us from all incoming cosmic rays and it does nothing against uncharged radiation such as electromagnetic radiation. On the other hand, a Super-Earth would likely host a far thicker atmosphere than our Earth does. And such an atmosphere would be a better screen against electromagnetic radiation. But without a magnetic field, primary galactic cosmic rays would strike the atmosphere unabated and generate more secondary cosmic rays. The likelihood of such secondaries reaching the surface would then depend on the thickness and density of the Super-Earth's atmosphere and the energy of the particles in question. Such interesting questions will pose exciting challenges for computer modelers.

But perhaps the initial calculations and predictions about Super-Earths being devoid of magnetospheres and radiation belts were not correct. For example, recent data suggest that iron metal in the outer core of Super-Earths four to six times more massive than Earth would solidify much more slowly than inside a smaller planet. Thus, the initial expectation that the high pressures would lead to rapid solidification of the core and loss of a magnetic field might need to be revised. Perhaps Super-Earth magnetospheres last longer than initially believed. Kraus *et al* used 16 high-powered lasers at the National Ignition Facility at Lawrence Livermore National Laboratory to generate conditions like those expected inside the cores of Super-Earths. They employed *in situ* x-ray diffraction to see whether the iron in the sample was in a liquid or solid state under such circumstances. The results showed that even under such extreme conditions, molten iron can crystallize similarly to what occurs at the base of Earth's outer core. Thus, an earthlike partly liquid metallic interior might exist under the extraordinary heat and gravitational pressures inside some Super-Earths. And even at pressures up to 1000 gigapascals (or about three times the pressure of Earth's inner core) iron in the inner regions of Super-Earths might still behave in a familiar fashion (that is, remain liquid) and generate durable geo-dynamos that can continue to shield alien life against cosmic radiation.

Moreover, we have noted that other planets in our own Solar System rely on different mechanisms to power their geomagnetic dynamos. For instance, rather than liquid metallic iron, other planets in our Solar System use spinning metallic hydrogen or superionic water (ice XVIII) to generate their planetary magnetic fields. Thus, given the likely internal compositions of Super-Earths, investigators have explored the possibility that such planets might still harbor magnetic fields despite their cores being presumably solid iron. Perhaps liquid iron is not the only way a Super-Earth might generate a geodynamo. It has been speculated that magnesium oxides and silicates, which are solid on and inside our Earth, may exist in a liquid metallic form at the pressures and temperatures found in Super-Earths. If true, spinning reservoirs of liquid metallic magnesium silicates might generate magnetic fields from the mantles of Super-Earths.

Magnesium silicates are abundant in our planet's mantle. Throughout most of Earth's mantle the stable phase of magnesium silicate is in the form of perovskites. Perovskites are named after Russian mineralogist Lev A Perovski, have a chemical formula of **ABX₃** and have the crystal **perovskite structure.** In the perovskite chemical formula of **ABX₃**, A and B are cations and X is an anion. In most perovskites A is typically Mg^{2+}, Fe^{2+} or Ca^{2+}; B is usually Si, Fe^{2+} or Ca^{2+}; and X is usually oxygen. Thus, many perovskites have the general chemical composition represented by $(Mg,Fe)(Fe,Al,Si)O_3$. The magnesium end-member of the silicate perovskite series is **bridgmanite**. Bridgmanite is thus a dense, high-pressure polymorph of magnesium silicate with the perovskite structure. It should be stated that the perovskite structure is named after the mineral **perovskite**, which was the first mineral found to have this specific structure. Confusingly, perovskite the mineral has the perovskite structure but does not adhere to the general perovskite chemical formula of $(Mg,Fe)(Fe,Al,Si)O_3$. Rather, the mineral perovskite has the chemical formula of $CaTiO_3$.

Models hint that the lower mantle consists mostly of this mineral, which means that about half our planet by volume is bridgmanite. Thus, the perovskite mineral bridgmanite is probably the most abundant mineral in the interior of Earth. Despite its abundance on planet Earth, it was only named in 2014 after it was isolated and identified in a meteorite. Bridgmanite is probably also dominant in the interior of rocky exoplanets, including Super-Earths. Curiously, although rocky inside Earth, bridgmanite may be a liquid metal at the pressures and temperatures found in Super-Earths. In principle, while in its liquid phase, bridgmanite might be able to generate a magnetic field in the mantles of Super-Earths.

The pressure within a Super-Earth mantle could exceed 1400 GPa as its mass reaches 10 Earth masses (M_E). Using the Sandia National Laboratories' Z Pulsed Power Facility, Fei and colleagues observed unprecedented high melting temperatures for bridgmanite—9430 K at 500 GPa. This somewhat surprising finding has bearing on the prospects of driving a geomagnetic dynamo in Super-Earths using liquid bridgmanite. It also provides an important constraint on the accretion heat required to melt the mantle of a Super-Earth in the protoplanet stage.

27.9 Radiation belts around ultracool dwarfs

All of the planets in our Solar System with strong, large-scale, persistent magnetic fields (including Earth, Jupiter, Saturn, Uranus and Neptune) have measurable radiation belts. However, no radiation belt has been clearly seen outside of our Solar System until very recently. In 2023, radiation belts have been indirectly detected around the brown dwarf, **LSR J1835+3259**. It was discovered using the **High Sensitivity Array** of 39 radio dishes spanning from the United States of America to Germany

Very low-mass stars such as M-class red dwarfs and brown dwarfs are sometimes collectively called **ultracool dwarfs**. Ultra-cool dwarfs are thus stellar or sub-stellar objects of spectral class M that have effective surface temperatures below 2700 K. For a while, some ultracool dwarfs have been known to produce planet-like radio emissions, which would make one suspicious of an earthlike (or Jupiter-like) magnetosphere and radiation belt. Such radio emissions have included those consistent with intermittent aurorae originating from presumed large-scale magneto-spheric currents.

Following the detection of bursts of radio emission from an M9 ultracool dwarf (**LP 944-20**) in 2001, active searches began at the **Arecibo Observatory** in Puerto Rico and the **Very Large Array** (also called the **Karl G Jansky Very Large Array**) in New Mexico to search for other radio signals suggestive of magnetospheres and radiation belts. The first suspected extrasolar aurorae were detected in July 2015 by the Very Large Array. These auroral phenomena occurred in the atmosphere of **LSR J1835+3259** and were detected by analysis of emitted radio waves. It is estimated that these probable aurorae were about 1 million times brighter than those on Earth.

27.10 LSR J1835+3259

Extremely energetic electrons and electrically charged particles become magnetically trapped and accelerated in torus-shaped regions wrapped around Jupiter's equator that constitute its radiation belts. These persistent zones of radiation contain relativistic particles with energies up to tens of MeV. Their geometric extent can range further than ten times the planet's radius. Recall that Jupiter is **radio loud** thanks to its powerful magnetosphere and radiation belts. The loud bursts of radio noise emanating from the Jovian radiation belts have been studied by radio astronomers since the 1960s.

Located only 20 light years away in the constellation Lyra, the ultracool dwarf **LSR J1835+3259** (more specifically, a brown dwarf) is very close to the size of Jupiter with a radius of 1.07 ± 0.05 Jupiter radii. Although it is about the same size as Jupiter, it is many times more massive at 77.28 ± 10.34 Jupiter masses and is thus far denser.

LSR J1835+3259 is of spectral type M8.5. Ultracool dwarfs are within the span of hydrogen-burning low-mass stars (which have made it to the main sequence) and massive brown dwarfs (which have failed to make it onto the main sequence though hydrogen burning). Ultra-cool dwarfs collectively comprise roughly 15% of the Sun's stellar neighborhood. One of the best-known examples is the red dwarf **TRAPPIST-1**. Incidentally, the name TRAPPIST-1 comes from the Transiting Planets and Planetesimals Small Telescope (TRAPPIST) at La Silla Observatory in Chile. TRAPPIST-1 has an interesting and impressive planetary system, consisting of seven Earth-like (terrestrial) exoplanets. This makes the TRAPPIST-1 planetary system of great astrobiological interest.

The TRAPPIST-1 terrestrial exoplanets are about 8% less dense than they would be if they had the same elemental makeup as planet Earth. And the orbits of all seven planets would easily fit within the orbit of Mercury around our Sun. However, the fact that TRAPPIST-1 is a cool red dwarf means that, despite the proximity of the seven exoplanets to its parent sun, some of these exoplanets (planets d, e, f and g) reside in the stellar system's habitable zone where water would be a liquid on the surface based on the calculated temperatures. (Of course, when one is considering this question properly, the surface gravity and atmospheric pressure should be factored in along with the temperature in order to determine the true probability of liquid water on the surface of such planets.)

Returning to the ultracool dwarf LSR J1835+3259, high-resolution radio imaging at 8.4 GHz outlined a double-lobed, axisymmetrical structure that is shaped like the Jovian radiation belts. Up to 18 stellar radii separate the two lobes. Also, these radiation belts have been found to be stably present across three observations spanning more than 1 year, which is a prerequisite for qualifying as a real 'radiation belt'. One may estimate electron energies based on synchrotron emissions, because the trapped accelerating electrons emit most of their power near the critical frequency, $\nu_{\text{crit}} \approx (3/2)\gamma^2 \nu_c \sin\alpha$ for pitch angle α. For a $\nu_{\text{crit}} \approx 8.4$ GHz, electrons with nearly perpendicular pitch angles have $\gamma \approx 30$. This is consistent with a cloud of electrons confined by the magnetic dipole of LSR J1835+3259 with energies around 15 MeV, which is comparable to the energies seen in Jupiter's radiation belts. Overall, however, this extrasolar radiation belt is almost 10 million times more intense than Jupiter's, which is itself millions of times more intense than Earth's Van Allen belts.

Further reading

[1] Kraus R G, Hemley R J and Ali S J et al 2022 Measuring the melting curve of iron at super-earth core conditions *Science* **375** 202–5

[2] Fei Y, Seagle C T and Townsend J P et al 2021 Melting and density of MgSiO$_3$ determined by shock compression of bridgmanite to 1254GPa *Nat. Commun.* **12** 876

[3] Kao M M, Mioduszewski A J and Villadsen J et al 2023 Resolved imaging confirms a radiation belt around an ultracool dwarf *Nature* **619** 272–5

[4] Hare V J, Tarduno J A, Huffman T, Watkeys M, Thebe P C, Manyanga M, Bono R K and Cottrell R D 2018 New archeomagnetic directional records from Iron Age southern Africa (ca. 425–1550 CE) and implications for the South Atlantic Anomaly *Geophys. Res. Lett.* **45** 1361–9

[5] Zhang X-J et al 2022 Superfast precipitation of energetic electrons in the radiation belts of the Earth *Nat. Commun.* **13** 1611

Chapter 28

Radiation and paleontology

Unicellular prokaryotic life first emerged somewhere around 3.5 billion years ago, and perhaps as early as 3.8 Ga, based on carbon-12:carbon-13 ratios found in archaic sedimentary rocks. Biological processes tend to prefer the lighter isotope of carbon; thus cells and tissues are relatively enriched in carbon-12. This is reflected in the $\delta^{13}C$ ('delta C thirteen'), which is simply a means of quantifying the carbon-12: carbon-13 ratio in a specimen. Biological specimens tend to have lower $\delta^{13}C$ values than specimens created non-biologically. The formula for the $\delta^{13}C$ is:

$$\delta^{13}C = \left(\frac{\left(\frac{^{13}C}{^{12}C}\right)_{\text{sample}}}{\left(\frac{^{13}C}{^{12}C}\right)_{\text{standard}}} - 1 \right) \times 1000.$$

Evidence of ancient life persists today as fossil stromatolites. Stromatolites are lithified deposits that have been left behind as microbes created slimy mats on the seafloor that bound sediments into layers upon layers. More minerals precipitated into these layers, eventually leading to petrified, stratified, sedimentary rock structures that we can observe today, billions of years after the organisms that created them have died. Some stromatolites date back nearly 3.5 Ga. Recent evidence of early life around ancient submarine hydrothermal vents on what is now Canada date back to over 3.77 Ga and may be as old as 4.28 billion years old.

28.1 Earth's eons

Based largely on paleontological chronological milestones, Earth's geologic history is divided into four major eons. These are the:

1. Hadean Eon: 4.56–4.03 Ga
2. Archean Eon: 4.03–2.5 Ga
3. Proterozoic Eon: 2.5 Ga–543 Ma
4. Phanerozoic Eon: 543 Ma–present.

As the name indicates, the Hadean was originally considered a time of hell on Earth—a period of intense volcanism with lakes of lava and frequent asteroid collisions that melted and resurfaced any areas that attempted to settle down and become stable. However, recent data suggest that oceans of water might have formed earlier than previously believed. In fact, the discovery of zircon crystals in the Jack Hills region of Australia dating back to 4.4 Ga suggests that oceans formed a mere 150 million years after Earth was born, meaning that the hellish period was far shorter than first assumed. Such zircon crystals form in the presence of water. Specifically, zircons with a relatively high $\delta^{18}O$ (that is a high ratio of heavier oxygen-18 to lighter oxygen-16) serve as an indication that such zircons formed under cool, wet conditions—not the conditions previously held to be the case. But while oceans might have formed during the Hadean, no rocks from this eon have survived. The oldest known rocks now date back to 4.03 Ga, marking the end of the Hadean and the start of the Archean.

28.2 The Archean Eon

The Archean Eon contains four eras:
1. Eoarchean Era: 4.03–3.6 Ga
2. Paleoarchean Era: 3.6–3.2 Ga
3. Mesoarchean Era: 3.2–2.8 Ga
4. Neoarchean Era: 2.8–2.5 Ga.

It is from the Eoarchean Era that the first **banded iron formations (BIFs)** date. **BIFs** are often massive, stripped sedimentary rock deposits created by precipitation of the iron oxides, hematite and magnetite, in layers between iron-poor chert. These BIFs are of great importance in paleontology since they *indicate the presence of free oxygen* in the local environment. BIFs will be discussed further in the context of **the Great Oxidation Event (GOE)**, which took place in the Proterozoic Eon.

Somewhere during the Archean Eon, maybe around 3.5 Ga, which falls in the **Paleoarchean Era**, large-scale rock structures called **cratons** formed and persisted. These thick, granitic, igneous cratons are over twice as thick as the surrounding lithosphere, which probably accounts for their durability. They are composed of the lighter—and therefore more buoyant—igneous products of partial melting. Being buoyant relative to basaltic oceanic crust, these cratons, made of **crystalline basement rock**, will tend to grow over time as more material from partial melts is added to their perimeters. *Cratons form the stable interiors of continents* such as North America, Africa and Australia. They harbor the oldest rocks on the surface of Earth. Also, it is from the Paleoarchean Era that the *earliest direct microfossil evidence of life* is found. These microfossils were found at the Apex Chert in Western Australia and are dated at 3.465 Ga.

During the **Mesoarchean Era**, the cratons grew and the first real continents were constructed. Genuine plate tectonics was in full force at this point, replacing the previous vertical, plume-driven tectonic activity. Somewhere around 3.1 Ga the first supercontinent **Ur** may have been assembled.

The last era of the Archean was the **Neoarchean**, a 300 million year interval between 2.8 and 2.5 billion years ago. It is at the close of the Neoarchean that perhaps the most important change in Earth's atmosphere took place—*the rise of oxygen* thanks to the cumulative effects of oxygenic photosynthesis by cyanobacteria. This event is arguably the most significant change in the history of the entire biosphere. It marks the close of the Archean Eon and the start of the next eon, the Proterozoic.

28.3 The Proterozoic Eon and the faint young Sun paradox

The name 'proterozoic' stems from ancient Greek words for 'early life'. The Proterozoic Eon is divided into three somewhat arbitrary eras that do not perfectly reflect the geological and biological events of the time. These eras are the:
1. Paleoproterozoic Era: 2.5–1.6 Ga
2. Mesoproterozoic Era: 1.6–1.0 Ga
3. Neoproterozoic Era: 1.0 Ga—543 Ma.

Before the Proterozoic, Earth's atmosphere was probably rich in methane. This abundance of methane might explain the '**faint young Sun paradox**'. The young Sun was only about 70% as luminous as today. With the reduced solar output, it is expected that the early Earth would have been an uninhabited frozen rock covered with ice. Nevertheless, the geologic record shows clear evidence of liquid water and primitive life early in Earth's history. Methane is a potent greenhouse gas—27.9 times as potent as carbon dioxide based on weight. But since the molecular weight of carbon dioxide is 44 g mol^{-1} compared to 16 g mol^{-1} for methane, one has to be careful about simple statements regarding how much more potent methane is than carbon dioxide as a greenhouse gas unless it is specified about whether we are speaking about per molecule or per mass. And as an aside, in today's atmosphere methane is short lived thanks to oxidation by molecular oxygen into carbon dioxide. This was not the case on the ancient anoxic Earth.

This atmospheric methane probably came from methanogenic microbes. Several species of prokaryotic bacteria and archaea produce methane as a waste product of their metabolism, just as cyanobacteria produce molecular oxygen as waste material. Thus, early life probably helped keep the planet alive through production of greenhouse methane that compensated for an underperforming Sun.

28.4 The GOE

But by the dawn of the Proterozoic, the waste products of cyanobacteria and their irresponsible use of **oxygenic photosynthesis** began to profoundly affect the atmosphere and the biosphere. For millennia, reduced ferrous iron (Fe^{2+}) remained in solution in the primordial oceans. But as molecular oxygen accumulated, it began oxidizing this iron in massive quantities into oxidized ferric iron (Fe^{3+}), which was insoluble and precipitated out as **BIFs** of red **hematite** and black **magnetite**. The rate of BIF reached a peak around 2.5 billion years ago and then markedly slowed after 1.85 Ga. Thus, when the pool of reduced iron ran out and was no longer available to soak up the oxygen, molecular oxygen began accumulating, first in the oceans and then in the atmosphere. This dramatic environmental change was the **GOE**. It was during the **Paleoproterozoic Era** that the GOE took place.

This release of free oxygen into the environment might have had a very major impact on the climate at the time. The previously abundant atmospheric methane became oxidized into carbon dioxide and water, causing the formerly pink skies to turn blue—but eliminating the greenhouse effect that was keeping Earth warm. Thus, temperatures plummeted during the Paleoproterozoic. Evidence suggests that from 2.4 to 2.1 billion years ago, Earth experienced global glaciation (called the **Huronian Glaciation**). **Tillites**, which are rocks typically deposited by glacial activity, were found as far south as the equator during this part of the Paleoproterozoic. This suggests that Earth was essentially frozen from top to bottom—a genuine **Snowball Earth** situation.

Although it remains uncertain, the *first eukaryotic cells* probably emerged early in the Paleoproterozoic, probably somewhere around 2 billion years ago.

Following the 900 million years of the Paleoproterozoic came the 600 million year span (from 1.6 to 1.0 billion years ago) called the **Mesoproterozoic Era**. It was during the Mesoproterozoic that another supercontinent (**Columbia**, also called **Nuna** or **Hudsonland**) broke up. Stromatolites made by photosynthetic cyanobacteria reached peak production during this time, thus the '**age of stromatolites**' falls under this era.

From 1 billion years ago to 543 million years ago was the **Neoproterozoic Era**. This 457 million year interval was a time full of geological, climatological and biological activity. Another supercontinent (**Rodinia**) formed early in the Neoproterozoic or late Mesoproterozoic and then broke up about 900 million years ago. Stromatolite reefs, which had peaked during the Mesoproterozoic, began decreasing in abundance. The *first multicellular animals* appear in the fossil record during this time.

Also during the Neoproterozoic Era, Earth experienced the **Cryogenian Period** from 720–635 Ma. Several super glaciations occurred during the Cryogenian. These were not the first Snowball Earth episodes, since the Huronian glaciation took place in the Paleoproterozoic. This first global glaciation (the Huronian) was probably due largely to the depletion of atmospheric methane because of the rising levels of atmospheric molecular oxygen (which oxidized and destroyed the methane). This loss of greenhouse methane placed Earth at the mercy of a still somewhat faint young Sun and a planetwide ice age ensued. But the next Snowball Earth episodes (in the Neoproterozoic (Cryogenian)) were probably due more to the loss of carbon dioxide and the breakup of the supercontinent Rodinia.

Rodinia began breaking apart around 830 million years ago. Such fragmentation led to extensive waterways between the separating continents. Since this was occurring in equatorial latitudes, the setup was ripe for intense intercontinental evaporation and severe storms and rain over these newly formed continents. This is the recipe for climate-cooling **chemical weathering** of silicate rocks. And around 723 million years back, a large outpouring of basaltic lava occurred; since basalt is particularly vulnerable to chemical weathering, this accelerated the overall cooling process.

Atmospheric carbon dioxide gets dissolved in rainwater and forms a dilute solution of the weak acid, **carbonic acid**. This carbonic acid solution dissolves silicate rocks through chemical weathering. The products include various cations, silica and bicarbonate anions which wash into the sea. The bicarbonate and carbonate anions then form carbonate minerals on the seafloor. Some of these carbonates are formed by life (in the form of calcareous organisms such as corals, coccolithophores, clams, mussels, etc). Thus, there was a massive pulldown of carbon dioxide out of the atmosphere by the **Urey reaction** of chemical weathering. The carbonates are then sequestered in rocks like limestone and dolomite (dolostone) for millions of years, thereby depleting the amount of carbon dioxide in the atmosphere. Without as much of this greenhouse gas, the Earth was susceptible to global cooling.

With the cooling, more snow fell. This in turn led to large glaciers since, thanks to drifting continents, there was now some land over the polar latitudes (which facilitate glacier formation). These glaciers and snow cover led to increased albedo, which in turn further cooled the planet. This positive feedback loop eventually reached a tipping point as progressively more solar light was reflected than absorbed. The planet was then in the grips of a worldwide ice age.

This time was the Cryogenian Period of the Neoproterozoic Era and spanned 720–635 Ma. The first global glaciation of the Cryogenian was the **Sturtian glaciation**. The Sturtian snowball lasted for about 70 million years. Then, after a brief reprieve, another global glaciation occurred called the **Marinoan** from 650 to 632 Ma.

28.5 Melting the snowball

How did Earth ever break the grip of these Snowball Earth episodes? The most plausible explanation is large-scale volcanic activity. The lava from these volcanoes is not what melted the ice; rather it was the erupted carbon dioxide gas that restored the thermostat. Since the rocks were mostly covered by snow and ice, the released CO_2 did not participate in the Urey reaction but instead accumulated in the atmosphere. It is estimated that carbon dioxide levels might have reached as high as an amazing 10%. (Recall that at present the atmospheric CO_2 level is about 400 parts per million or 0.04%.) This led to extremely warm temperatures (maybe more than 120 °F or 49 °C). Evidence supporting this dramatic shift in climate from frigid cold to sweltering heat comes in the form of '**cap carbonates**'. Cap carbonates are large deposits of dolomite or other carbonate rocks and minerals overlying ancient lithified glacial deposits (**glaciogenic strata**). Such caps form in warm, shallow, saturated waters (as off the shores of southern Florida and the Bahamas). The interpretation is that glaciers gave way to warm waters in a relatively short timeframe geologically speaking.

Incidentally, there were other early and extensive glaciations before these world-wide ice ages (that is, other than the Huronian of the Paleoproterozoic and the Sturtian and Marinoan of the Neoproterozoic). One of these was the **Pongola**

glaciation, which occurred around 2.9 Ga. Evidence supporting the concept of extensive glaciation includes tillites, diamictites and coarse breccias consistent with glacial activities over the South African craton. There was evidence of ice extending to a paleolatitude (based on the magnetic field recorded in the rocks) as far from the equator as 48°. Unlike the other mass glaciations, this earlier glaciation was likely not triggered by the evolution of photosynthetic cyanobacteria, however. With less energy emanating from the Sun, models predicted that it would have been inevitable that a Snowball Earth was in the forecast. But no **erratics** (**dropstones**, or boulders picked up and carried by glaciers and then dropped in odd locations) have been found. The Pongola glaciation might have paradoxically been caused by an excess of greenhouse gases. Methane accumulation in the ozone-empty stratosphere may have led to organic chemical reactions leading to long saturated hydrocarbons, giving the sky a hazy orange color. If sufficiently dense, this orange haze might have blocked out sunlight and caused a global cooling. But it appears that the extent of cooling was not enough to cause a full-blown Snowball Earth.

28.6 The rise of oxygen

We have discussed some of the global sequelae of the dramatic change in the atmospheric chemistry from an anoxic, reducing atmosphere into an oxygenated, oxidizing atmosphere. In addition to the climatic consequences, the biological consequences were also extremely severe. Essentially all the organisms in existence prior to the GOE were anaerobic bacteria and archaea. These obligate anaerobic microbes were very sensitive to the newly introduced environmental toxin, oxygen. To this day, their descendants have been markedly reduced in numbers and can now only survive in unpleasant, oxygen-depleted locales such as marshes and the colorectal region of the digestive tract. Although less well celebrated than the mass extinctions that killed macroscopic life forms, the introduction of free oxygen into the biosphere was the cause of what might have been the most profound mass extinction of all time. The flip side of this devastating impact on anaerobic microbes was that the availability of molecular oxygen now allowed the evolution of large, multicellular organisms such as animals. When it comes to generation of the universal biological commerce molecule, adenosine triphosphate (ATP), aerobic respiration is far more efficient than any of the previously available fermentation pathways or the anaerobic respiration methods.

So, where exactly did all this oxygen come from? As discussed in greater detail elsewhere, the standard answer is that it was the byproduct of a particular subtype of photosynthesis, namely oxygenic photosynthesis (and specifically **photosystem II** of the so-called **Z-scheme** of oxygenic photosynthesis). Unlike their other photosynthetic microbial counterparts, these new photosynthetic microbes produced oxygen as a waste product. (Before the cyanobacteria developed oxygenic photosynthesis, other bacteria used different methods of photosynthesis that do not generate oxygen. These bacteria (whose descendants are represented today by the purple bacteria, green sulfur bacteria, green non-sulfur bacteria and heliobacteria) had invented **anoxygenic photosynthetic approaches** perhaps 3.5 billion years ago.)

Present evidence suggests that cyanobacteria first appeared somewhere around 2.7 billion years ago. Regardless of the exact date, these oxygen-producing photo-synthesizers were restricted to relatively deeper waters than they are today because of the unfiltered solar radiation at the time. And they certainly could not exploit the direct sunlight on land at that time because of the intense, molecule-destroying ultraviolet radiation. No biological organisms were likely to be able to withstand that intense, direct ultraviolet on land back then. It was only when adequate atmospheric diatomic molecular oxygen accumulated in the atmosphere that our present life-preserving stratospheric ozone layer materialized. This mass production and accumulation of ultraviolet-blocking ozone (triatomic molecular oxygen) probably did not occur until well after the GOE.

28.7 Non-photosynthetic oxygenic chemotrophs

In addition to the cyanobacterial source of life-giving oxygen (or 'climate-destroying and mass extinction-provoking oxygen', from an anaerobic prokaryotic perspective!), there are at least four other microbiological sources of molecular oxygen. One oxygen-generating metabolic reaction exploits the abundant energy in hydrogen peroxide (H_2O_2), which is produced in trace amounts high in the atmosphere by reactions with radiation. But before the rise of molecular oxygen, there was even less hydrogen peroxide available than there is today, so this biochemical pathway probably did not contribute any significant amounts of oxygen to the primitive Archean ecosystem.

Generally speaking, these primitive microorganisms utilize non-photosynthetic chemotrophic metabolic pathways that take advantage of oxidants with a higher (more positive) or comparable redox potential than the 1.23-volt O_2/H_2O redox couple. Some examples of redox couples that meet this requirement include:

1. Hypochlorite (ClO^-)
   a. ClO^-/Cl^- $E_0' = +1.31V$
2. Chlorite (ClO^{-2})
   a. ClO^{-2}/ClO^- $E_0' = +1.28V$
   b. ClO^{-2}/Cl^- $E_0' = +1.08V$
3. Nitrous oxide (N_2O)
   a. N_2O/N_2 $E_0' = +1.36V$
4. Nitric oxide (NO)
   a. NO/N_2O $E_0' = +1.18V$
   b. NO/N_2 $E_0' = +1.27V$

The first group of **non-photosynthetic oxygenic chemotrophs** discovered were **perchlorate- and chlorate-reducing (respiring) bacteria**. These organisms can aptly be described as using **oxygenic chemosynthesis**. Today there are relatively few microbial species that *generate oxygen in the absence of sunlight* as part of their alternative metabolic processes, and most of these reduce nitrogen oxides. Molecular oxygen and molecular nitrogen are byproducts of nitrate and nitrite reduction reactions; the oxygen and energy released are used to oxidize methane, ammonia or other reduced compounds. They then make the stored chemical energy available for biochemical use in the form of ATP. One important species that is capable of oxygenic chemosynthesis is *Nitrosopumilus maritimus*. This and other marine **ammonia-oxidizing archaea** play an important role in the nitrogen cycle. The **nitrogen cycle** is the ecological cycle in which free molecular nitrogen is first 'fixed' in the form of ammonia by nitrogen-fixing bacteria. This now-bioavailable nitrogen then gets further incorporated in the biosphere in the form of amino acids that make up proteins and nucleotides that make up nucleic acids. Waste products of organisms (or death of these organisms) release the nitrogen in these macromolecules back to the envirnoment in the form of ammonia. The ammonia is then oxidized into nitrites by bacterial species such as *Nitrosomonas* and the nitrites are converted into nitrates by bacteria such as *Nitrobacter*. Finally, the nitrates are reduced back into molecular nitrogen to complete the cycle.

Along with the *Nitrosomonas species, N. maritimus* is one of the other species capable of oxidizing ammonia into nitrite for chemosynthesis. In so doing, it couples the oxidation of ammonia with the reduction of oxygen to form the nitrite. This is fine in today's world, where there is ample oxygen for this step. But researchers have explored the situation in which oxygen is not plentiful. It turns out that this species can generate its own oxygen to meet the needs of this reaction. It produces diatomic molecular oxygen (O_2) through the simultaneous reduction of nitrate into nitrous oxide (N_2O) plus diatomic molecular nitrogen (N_2). In other words, *N. maritimus* couples molecular oxygen production to the production of gaseous molecular nitrogen.

N. maritimus is extremely abundant in our oceans—it is estimated that up to one in every five microbes in a sample of seawater is one of these archaea. It is possible that the abundance of these archaea is due to the fact that they can carry out the oxidation of ammonia in the presence of oxygen—or in its absence—thanks to its ability to produce the needed oxygen inherently. This raises the question of whether similar oxygenic chemosynthesis could have contributed to the 'rusting' of the oceans during the GOE of the Paleoproterozoic. Maybe it was not entirely the fault of the profligate, polluting cyanobacteria after all!

28.8 Early eukaryotes

Exactly when the first unicellular eukaryotic cells evolved is unsettled but fossil **rhodophytes** (red algae) may date back to 1.9 Ga. Radiolarians, which are also unicellular and play an important role in the carbonate–silicate cycle thanks to their calcareous tests (microscopic shells) first show up in the fossil record around 1.2–1.3 Ga, in the Mesoproterozoic. Also, during this part of the Mesoproterozoic, the first fossil unicellular green algae (**chlorophytes**) have been found.

At 1.2 billion years ago the first evidence of **multicellularity** is found in the form of certain red algae species that have formed leaf-like colonies of single cells. Also, such multicellular organisms soon began taking advantage of working together as a colony. **Holdfasts** in the fossil red algae species *Bangiomorpha pubescens* are specialized cells that anchor the rest of the 'plant' to the seafloor. (The word plant is used colloquially here since algae are not true plants in the taxonomic sense.) Green algae do share a common ancestry with the **Embryophyta** or 'land plants' which includes all extant organisms we call plants today. This common ancestry is the justification for lumping green algae and embryophytes together as the **clade Viridiplantae**. Red algae remain separate from the Viridiplantae based on various biochemical and molecular differences. In any case, this fossil red algal species is held as among the earliest evidence of eukaryotic multicellularity yet identified.

The first eukaryotic 'animals' (protists) were probably **choanoflagellates**. Although no fossil choanoflagellates date back to the Proterozoic, biological molecular clock analysis suggests their appearance in the Neoproterozoic. These single-celled organisms possess flagella and closely resemble cells found in the walls of sponges called **choanocytes**. It is believed that sponges (phylum Porifera) are the most primitive multicellular animals. Sponges do not have true tissues; rather, they can almost be considered as simple collections of various differentiated, specialized cell types working together in a cooperative manner. Among these cells are choanocytes that beat their flagella and create a current that funnels water into the central cavity of the sponge (spongocoel) so that food may be filtered out of that water. Sponges thus represent an agglomeration of cells with various functions that work cooperatively for the benefit of the organism as a whole.

Molecular clocks use molecular biological techniques for ascertaining dates of phylogenetic separation. A classic example is the amino acid sequence (the so-called primary structure) of the hemoglobin protein molecule. There is no difference in either the alpha or beta chains of human chimpanzees. On the other hand, there are substantial differences between human hemoglobin molecules and those of other primates such as macaque monkeys. And there are greater differences still between humans and other non-primate mammals. The distinctions grow wider still when human hemoglobin is compared to that of non-placental mammals such as kangaroos. The molecular dissimilarities between human hemoglobin and that of birds, amphibians and fish grow progressively wider. It is estimated that small changes in the hemoglobin primary structure occur once every few million years or so. From this, one could conclude that humans are closely related to chimpanzees and branched away from a common ancestor a few million years ago, whereas humans diverged from the other species many tens or hundreds of million years back. But one must look at the entire panoply of molecules available for analysis (rather than just hemoglobin) since the pace of molecular evolution varies from one molecule to another. And since amino acid sequences can be identical despite changes in the DNA (silent point mutations), examination of DNA sequences can lead to more instructive information than protein sequences. Finally, such clocks are only as good as their calibration. Such calibration in turn is dependent on the quality and completeness of the fossil record being used for such calibration.

The rise in oxygen to around 1% of modern levels occurred in the Mesoproterozoic, around 1.2 billion years ago and this theoretically should have spurred on an upsurge in biological evolution. Nevertheless, this is not borne out in the fossil record. It was not until 630 million years ago that complex multicellular organisms appeared. This leads to the question of, what took so long? One explanation is that the oxygen level was simply still too low to allow evolution of complicated multicellular life forms to emerge. Another explanation however is that the influx of oxygen led to complex chemical changes in our oceans, leading to harmful excess sulfides.

Oxygen can react with the sulfide mineral pyrite (FeS_2). Oxidation of this iron sulfide mineral yields sulfate (SO_4^{2-}), which washes into the oceans. There, **sulfate-reducing bacteria** use it for their metabolic needs and convert it into **hydrogen sulfide gas (H_2S)**. This gas is highly toxic to aerobic organisms (including people) by poisoning the electron transport chain of aerobic respiration. Hydrogen sulfide also binds and sequesters key enzyme co-factor metals such as copper and molybdenum thereby inhibiting the evolution of certain enzymatic pathways. (Incidentally, two forms of the enzyme **superoxide dismutase**, which plays an important role in disarming some of the reactive oxygen species (ROS) generated via ionizing radiation, are copper-containing enzymes.) In any case, it appears that the rise of oxygen might have also led to an increase in oceanic hydrogen sulfide, and this **sulfidic** ocean (or Canfield ocean) might have held back biological evolution for a prolonged period. Another possible explanation for the delay is that the effects of ionizing radiation are augmented by the presence of oxygen.

28.9 Oxygen and radiation on early Earth

While the rise in oxygen drastically increased the biological impact of ionizing radiation thanks to the **oxygen enhancement effect**, it is important to remember that oxygen is extremely reactive outside of radiation chemistry as well. As discussed in other chapters, diatomic molecular oxygen is a **free radical**, meaning that it has unpaired valance electrons. Technically it is a **diradical** species in its ground state—**triplet oxygen**. Being a radical makes it quite chemically reactive, as any combustion reaction can attest to. But diatomic molecular oxygen can exist in an excited state, **singlet oxygen**, that is even more reactive. Also, oxygen exists as another biochemically hazardous **allotrope—triatomic molecular oxygen**, or ozone. And there are a host of **reactive oxygen species (ROS)** that are biochemically hazardous such as various peroxides, superoxides and the hydroxyl free radical. Such ROS are a natural hazard of working with oxygen. For instance, the production of ATP through the electron transport chain of oxidative phosphorylation occasionally yields superoxide as a byproduct.

Incidentally, these ROS may or may not be free radicals. For instance, hydrogen peroxide is an example of an ROS that does not happen to be a radical. It is the production of these ROS in oxygenated water that confers an oxygen enhancement ratio (OER) to sparsely ionizing radiation such as electrons, gamma rays and x-rays. But recall that these ROS are naturally present in small amounts even in the absence of ionizing radiation, simply through an oxygenated environment, and especially thanks to aerobic respiration. Therefore, the rise of free oxygen mandated the evolution of various biochemical mechanisms to cope with oxygen and its reactive relatives—and by association, ionizing radiation. In other words, the appearance of oxygen in the environment meant that new means of coping with ionizing radiation must have materialized.

Examples of these detoxifying mechanisms include the enzymes **catalase** and **superoxide dismutase**. And the evolution of these mechanisms has proven extremely useful in helping cells cope with radiation (which multiplies the concentration of these naturally encountered oxygen species). While **extremophiles** such as *Deinococcus radiodurans*, the **bdelloid rotifers** and water bears (**tardigrades**) have taken the ability to deal with massive doses of radiation to the limit, ordinary cells also have effective defense mechanisms in place. (Exceptions might be found in people with rare autosomal recessive genetic disorders—such as ataxia telangiectasia, xeroderma pigmentosum, Bloom syndrome, Fanconi anemia, Nijmegen breakage syndrome, Cockayne syndrome and Werner syndrome among others—that make patients exquisitely sensitive to ionizing radiation.)

28.10 Radiobiological consequences of the GOE

The GOE probably significantly increased the biological consequences of radiation exposure. Prior to this time, there was a higher absolute amount of ionizing radiation in the environment thanks to greater abundances of natural primordial radionuclides that had not yet decayed away. However, the radiobiological effectiveness of this radiation was reduced in that ancient anaerobic environment compared to today. This is because the presence of oxygen greatly enhances the biological impact of some types of radiation.

This generalization holds true for radiation species with **low linear energy transfer** (**LET**) but not for radiation species with **high LET**. Recall that LET is the deposition of energy per unit path length of the radiation as it traverses matter. Examples of low-LET radiation include high-energy photons and electrons (as in the gamma rays and beta particles from radioactive decay).

The phenomenon of increased radiobiological effectiveness of radiation in the presence of oxygen is known as the oxygen enhancement ratio or **OER**. The OER is mathematically defined as the ratio of the radiation dose needed to achieve a given biological effect in the absence of free oxygen compared to the dose needed to achieve the same biological effect in the presence of oxygen.

$$OER = \frac{\text{Radiation dose in absence of oxygen}}{\text{Radiation dose in presence of oxygen}} \quad \textbf{Oxygen enhancement ratio}$$

For low-LET radiation such as gamma and x-ray photons and electrons, the OER is said to be around three or four. But it is important to recognize that the OER depends on the specific biological endpoint in question. In radiation therapy for cancer, the relevant biological endpoint is the killing of malignant cells. Thus, while the OER of energetic photons and electrons for the killing of cancer cells might be three or four, it could be different for alternative biological endpoints (such as induction of cancer in an organism, production of genetic mutations or the formation of cataracts). Furthermore, for any given biological endpoint, the OER may change with fractionation of a total radiation dose and with fraction size, among other factors.

Technically, indirectly ionizing radiation such as gamma and x-ray photons does not truly possess a definable LET since it either moves through matter unaltered (aside from directional changes that do not deposit dose) or it vanishes altogether after it interacts through the photoelectric effect or pair production (or liberates an electron and changes directions in the Compton effect). It is actually the electrons liberated by these processes that have measurable LETs. Nevertheless, it is common practice to describe gamma and x-rays as 'low-LET radiation species', and it is understood that it is the secondary electrons or positrons that do the ionizing when we describe high-energy photons as a form of 'low-LET radiation'. The same principle holds when we talk about neutrons—these indirectly ionizing particles set protons loose that do the ionizing; in this case the radiation is high-LET.

In contrast to low-LET radiation, high-LET radiation (such as alpha particles or very energetic galactic cosmic rays) is not as dependent on oxygen for its biological effectiveness. Thus, the OER for alpha radiation is close to unity when it comes to cancer cell inactivation. The presence or absence of oxygen makes little difference. This represents an advantage in the radiotherapy clinic since many tumors contain hypoxic cores, and areas within such tumors are therefore relatively resistant to low-LET photons and electron beam radiation therapy. To ensure sterilization of such hypoxic tumor cores, higher doses of low-LET radiation might be necessary or techniques must be applied such as dividing the radiation dose into daily 'fractions' spread over several days or weeks. (Fractionation allows a phenomenon known as **reoxygenation** to occur as the oxygen-rich perimeter of the tumor is selectively killed with each fraction, allowing the previously hypoxic tumor core to become more and more exposed to oxygen over the course of treatment.) Additionally, fractionation allows another phenomenon known as **reassortment** to come into play. Cells, including malignant cells, are more sensitive to ionizing radiation at certain phases of the cell cycle such as mitosis (M phase). Fractionation of a course of radiation therapy allows rapidly dividing cancer cells that were not killed during the first fractions to continue their cell cycles and be killed during subsequent fractions when they enter more sensitive phases. In other words, fractionation permits killing of cancer cells through reassortment of phases of their cell cycles and 'catching' them on other days when they are in a phase that is more sensitive to the daily dose of radiation.

In any case, before the GOE in the Paleoproterozoic, gamma rays and beta electrons had lower radiobiological impact on life forms. But after oxygen became abundant, these low-LET radiation species became far more biologically potent. Therefore, students and researchers who contemplate the radiation environment of early Earth must take into account not only the physics of radioactive decay (e.g. the half-lives of primordial radionuclides to estimate the relatively increased abundance and physical dose conferred by such naturally occurring radiation in the distant past) but also the radiobiologically relevant environmental conditions (such as a threefold increase in biological effectiveness in an aerobic environment vs the anaerobic environment before the GOE).

Thanks to the 1.25 billion year half-life of potassium-40, the absolute absorbed radiation dose from primordial potassium-40 has decreased roughly eightfold over the last 4 billion years (from 5.5 mGy yr^{-1} to 0.7 mGy yr^{-1}, assuming a steady intracellular potassium concentration of 140–150 millimoles per liter over this timespan). This assumption might not be perfectly valid given the large variability presently seen in extant organisms, with *Escherichia coli* intracellular concentrations varying between 30 and 300 mM and human intracellular concentrations being about 140 mM while extracellular concentrations (e.g. blood serum) are 3.5 to 5.3 mM. Nevertheless, it is a very reasonable assumption and approximation. (Note that **plasma** potassium concentrations tend to be 0.5 mM lower than **serum** values because serum measurements require additional time for blood clotting and cells continuously release their internalized potassium during this clotting process. Plasma is the liquid fraction of blood that has had blood cells and platelets removed; serum is plasma without the clotting factor proteins.)

When taking the relative abundances of primordial thorium-232, uranium-235 and uranium-238 in the distant past into consideration, the annual doses from geologic sources have decreased from $1.6 \, \text{mGy yr}^{-1}$ 4 billion years ago to $0.66 \, \text{mGy yr}^{-1}$ today. Hence, the overall dose from both beta and gamma sources combined has fallen from around $7.0 \, \text{mGy yr}^{-1}$ to $1.35 \, \text{mGy yr}^{-1}$ over the last 4 Ga.

But as mentioned, the physical absorbed radiation dose must be modified by the ambient oxygen levels at the time. It is possible that shortly after the GOE, radiogenic mutation rates increased drastically thanks to the threefold to fourfold increase in biological effectiveness of gamma and beta radiation in a newly aerobic environment. This hypothesis might initially be dismissed or contradicted by the fossil record (recalling that the '**Boring Billion**' spans from 1.8 Ga to 800 Ma). And the Cryogenian period (720–635 Ga) also occurred during this post-GOE interval, which dramatically slowed down evolution, nearly grinding biology to a complete stop. Nevertheless, after the great thaw, biological evolution took off with a vengeance, reaching an extravagant pinnacle during the **Cambrian explosion** at the dawn of the Phanerozoic Eon ('phanerozoic' is derived from the ancient Greek words for 'visible life' signifying the change from hidden fossils to abundant and obvious life beginning with the dawn of this eon). It is tempting to speculate that the combination of sufficient oxygen for efficient aerobic respiration along with the increased potential for radiogenic mutations thanks to this elevated oxygen concentration teamed up to allow the amazing explosion of life during the Cambrian Period.

Another logical conclusion of all of this is that *life originated in a far more radioactive world*. As such, it evolved and adapted during times when both the physical absorbed doses and the radiobiologically effective doses of radiation were far higher than today. The implication is that life has had ample time to develop effective defense mechanisms against low doses of ionizing radiation. The redundant molecular mechanisms activated during exposure to low doses of radiation seem to bear witness to this hypothesis.

28.11 Oxygen and the natural reactors

It appears plausible that the large infusion of oxygen into the environment resulted in a proliferation of natural nuclear reactors around the globe. This is consistent with the observation that volcanically produced **uraninite** began disappearing from the geological record after the GOE. Although stable in anoxic water, uraninite readily dissolves in oxygenated water. As the dissolved uranium eventually precipitated out and concentrated elsewhere under the right conditions, many (perhaps millions?) of these natural reactors might have materialized given the much greater proportion of uranium-235 billions of years back.To date however, the only known natural nuclear reactors are those discovered at Oklo, Gabon.

28.12 Did radiation trigger the evolution of aerobic respiration?

After reintroduction of oxygenated waters into these ore deposits as moderators, the reactors went critical and fired up, providing a double whammy of intense radiation and oxygen into certain areas of the biosphere. As this was going on somewhere around 2 billion years ago, the only organisms in existence at this time were microbes. And since eukaryotic cells might not yet have hit the scene, these microbes were probably all prokaryotic. These prokaryotes were quite sensitive to the new toxin, oxygen, that was polluting their environments. They needed to evolve and adapt to the changing environment or face extinction. It is probable that one of the first **mass extinctions** did indeed occur at this time thanks to the immense introduction of oxygen. And since oxygen enhances the biological effects of beta and gamma radiation, it seems plausible that *radiation encouraged the evolution of oxygen tolerance* at this time. And if one takes the logical argument further, the newly increased biological effectiveness of radiation might have encouraged the evolution of microbes that not only tolerated oxygen but actually put it to good use through the creation of aerobic respiration.

Perhaps extremophilic microorganisms such as ancient ancestors of *D. radiodurans* and highly radiation-resistant cyanobacterial species first evolved under circumstances surrounding natural reactors. Whether the evolution of increased radiation resistance led to increased oxygen tolerance—and ultimately the exploitation of energy-rich oxygenic respiration—remains unknown.

It is known that some cyanobacteria have been observed thriving in the high-level radiation regions near the Chernobyl and Fukushima reactors. Desiccation-tolerant cyanobacteria of genus *Chroococcidiopsis* have been shown to survive irradiation doses as high as 15 000 Gy. While the idea that plentiful natural reactors spurred on all sorts of evolution is a fascinating hypothesis, to date, the only fossil reactor discovered is the single example at Oklo, Gabon, suggesting that such natural nuclear reactors were not as abundant as this model predicts.

28.13 A new era in Earth's history

The close of the Cryogenian (and the fertilizing nutrients stirred up and dumped into the oceans by the melting glaciers) coupled with the globally warm temperatures (caused by a volcanically induced influx of greenhouse carbon dioxide) led to an enormous bloom of cyanobacteria (and perhaps eukaryotic algae). This poured more oxygen into the seas and atmosphere and set the stage for the evolution of large aerobic animals. The last of the four great geochronological eons is the **Phanerozoic Eon**, which began with the **Cambrian Period** of the **Paleozoic Era** about 543 million years ago. Although multicellular life began proliferating and evolving well before the Cambrian (for instance, the **Ediacaran fauna** in the late Neoproterozoic, after the last Snowball Earth episode), the so-called **Cambrian explosion** is probably the single most dramatic proliferation of multicellular (metazoan) animal life in Earth's history.

Thanks to the far greater efficiency of aerobic respiration, abundant oxygen allowed the development of larger bodies with differentiated cells possessing specialized functions—the appearance of the first **biological tissues**. The ample amounts of oxygen now allowed multicellularity to develop since the needed oxygen could diffuse from one tissue to the next across cells. Prior to this time, aerobic cells had to remain in contact with oxygen-containing water in order to function. With sufficient oxygen for diffusion, multicellularity and larger bodies became possible. And of course, the next step in evolution in this vein (pardon the pun) was the development of dedicated respiratory and circulatory organs and systems.

The Phanerozoic Eon is comprised of three eras:

1. Paleozoic Era: 543 Ma—250 Ma
2. Mesozoic Era: 250 Ma—65 Ma
3. Cenozoic Era: 65 Ma—present.

These eras are then broken down into multiple periods as such:
1. The Paleozoic Era:

    1. Cambrian Period: 543–485 Ma
    2. Ordovician Period: 485–444 Ma
    3. Silurian Period: 444–419 Ma
    4. Devonian Period: 419–359 Ma
    5. Carboniferous Period: 359–299 Ma
    6. Permian Period: 299–252 Ma.

2. The Mesozoic Era:

    1. Triassic Period: 252–23 Ma
    2. Jurassic Period: 23–145 Ma
    3. Cretaceous Period: 224–65 Ma.

3. The Cenozoic Era:

    1. Paleogene Period: 65–23 Ma

        a. Paleocene Epoch: 66–56 Ma
        b. Eocene Epoch: 56–34 Ma
        c. Oligocene Epoch: 34–23 Ma.
    2. Neogene Period: 23–2.58 Ma

        a. Miocene Epoch: 23–5.33 Ma
        b. Pliocene Epoch: 5.33–2.58 Ma
    3. Quaternary Period: 2.58 Ma—present
        a. Pleistocene Epoch: 2.58 Ma—11 700 years ago
        b. Holocene Epoch: 11 700 years ago—present.

28.14 The appearance of hard body parts

It was during the Cambrian Period that mineralized shells and other exoskeletons became abundant in the fossil record. And since hard body parts are far more readily preserved than soft ones, it is not surprising that our knowledge of paleobiology expands greatly with the Cambrian Period. Thanks to the practically limitless supply of free, ionized calcium in the prehistoric oceans, marine life forms evolved calcium carbonate shells before their freshwater cousins in lakes and rivers. These readily preserved exoskeletons in the fossil record documented a profound phenomenon raging at the time. The evolution of tough exoskeletons—and hard voracious mouth parts—instigated an ever-escalating war between the increasingly powerful predators and the commensurately well-defended prey species. This phenomenon is sometime called the **Red Queen hypothesis**.

Parenthetically, the origins of strong skeletons remain an area of passionate debate and speculation but based on the fact that these Paleozoic seas were saturated with calcium cations and bicarbonate anions, it could be that skeletons were just a means of getting rid of excessive calcium. Perhaps it was simply bioenergetically favorable to dispose of these surplus ions by precipitating them together in the form of calcium carbonate. And these biomineralization byproducts just happened to have beneficial other uses (such as defensive shields and predatory weapons).

28.15 The Cambrian Period

Among the ferocious denizens of the Cambrian seas was the top predator, *Anomalocaris*. As a 6 foot shrimp that was at the top of the food chain, 'Anomalocaris' (derived from the Greek words for 'anomalous shrimp') is aptly named indeed. Also in great abundance during the Cambrian were other arthropods known as trilobites. But although the arthropods were dominant, they were accompanied by various other hard-bodied invertebrates such as corals, brachiopods, clams and other mollusks. Much of what we know about the Cambrian explosion is thanks to the **Burgess Shale**, a deposit in western Canada with a wide array of previously unknown specimens. The Burgess Shale is an example of what is called a **Lagerstätten**, which is a sedimentary rock deposit that contains abundant and diverse fossils with exceptional preservation.

Almost all extant phyla of animals trace their origins to the Cambrian Period. The earliest known **chordate** fossil is *Pikaia gracilens*, a cephalochordate dating back to 505 Ma. Hemichordates known as graptolites also appeared in the Cambrian. The earliest known fossil vertebrate, *Haikouichthys*, also dates back to the Cambrian Period at 518 Ma. The Cambrian explosion (sometimes colloquially called the **Biological Big Bang**) is perhaps the most pronounced example of animal **adaptive radiation** in history. For this reason, the Cambrian explosion is also known as the **Cambrian radiation**. This interval of exceptional animal diversification began around 538 Ma and lasted for 13–25 million years.

Note that in contrast to the main topic of this book, the use of the word 'radiation' here alludes not to invisible energy in the form of photons or particles but rather to a rapid evolution of life forms to fill available ecological niches. Adaptive radiation is a synonym for **divergent evolution**, in which previously closely related species gradually become more and more dissimilar as they specialize for the new niches they occupy. Divergent evolution contrasts with **convergent evolution** in which unrelated organisms adopt similar phenotypes (body forms) when they occupy and adapt to their particular ecological niches. A classic example is the similarity in body morphology between the sea-living dolphins, ichthyosaurs and fish. Birds, bats and pterosaurs provide another example of such convergent evolution.

28.16 The Ordovician Period

The Ordovician Period was when the phylum of animals known as the **Bryozoa** appeared. The bryozoans were the last major phylum of animals to appear on the fossil record. Although not quite as profound as the Cambrian explosion, there was another great radiation of marine life during this period. In fact, this expansion of biodiversity is sometimes called the **Ordovician radiation** or the **Great Ordovician Biodiversification Event**. The number of genera of marine animals increased roughly fourfold during the Ordovician radiation. Arthropods continued to do well as trilobites flourished and evolved into a vast variety of different species. Their daunting spines might have been helpful in defending them against the new top predator of the Ordovician, the **nautiloid orthocones**. The nautiloid orthocones were huge molluscan cephalopods with long straight shells (in contrast to today's shell-less big cephalopods such as squids and octopuses). But not all giant nautiloids were orthocones—although those of the Ordovician were largely straight, later species in the Silurian and Devonian Periods had curved and classic coiled spiral shells. Tabulate corals and rugose corals also appeared and diversified during the Ordovician, forming the world's *first true tropical coral reefs*. Although they might have evolved even earlier, the first fossil **land plants** or **embryophytes** appear in Ordovician rocks. The Ordovician was a time of more obvious continental breakup as Rodinia (which was fragmenting since 750 Ma) had broken into Gondwanaland, Laurentia, Baltica and Avalonia.

The Ordovician Period ended abruptly in the first of the world's five major mass extinction events. This topic will be revisited later in the chapter.

28.17 The Silurian Period

During the next Paleozoic period, the Silurian, the arthropods returned with a vengeance to top predator positions. The Silurian is sometimes called the 'Age of Eurypterids' thanks to the rise of the eurypterids or fierce sea scorpions. Some of these Silurian sea monsters such as *Jaekelopterus* and *Pterygotus* reached nearly 10 feet (~3 m) in length! Eurypterids are closely related to today's scorpions although they are not direct ancestors. Genuine giant scorpions such as *Brontoscorpio* were among the first animals to walk the land although it is presently believed that they were largely aquatic and spent relatively little time on land. *Brontoscorpio* lived during the late Silurian or early Devonian but the earliest true scorpions (e.g. *Dolichophonus*) date to Silurian times. Also during the Silurian, the first vascular land plants (**tracheophytes**) show up in the fossil record. These vascular plants contained xylem and phloem for transport of water and sugars respectively to other parts of the plant and were able to adequately withstand the desiccation associated with life on land.

28.18 The Devonian Period

Following the Silurian came the Devonian Period, sometimes called the 'Age of Fishes'. Unlike most of their predecessors, fish of this period had powerful jaws which were derived from their two most rostral (front) gill arches. While jaws had actually evolved earlier, and some sharks were present in the Ordovician and Silurian, it was during the Devonian Period that fish came to the fore. Sharks had grown more numerous and bizarre during the Devonian, but it was not the sharks who owned the top predator spot. Another group of fish-celled **placoderms** were the most ferocious predators of the day. Some, such as **arthrodire** species of *Dunkleosteus* reached 29 feet long (8.8 m) and had an enormous gape thanks to a dorsal joint between their armored heads and bodies that supplemented their lower jaws (mandibles) in opening wide. Based on biomechanics, it is believed that *Dunkleosteus* might have had one of the most forceful bites of all time. Giving *Dunkleosteus* some competition as most fearsome fish of the Devonian was *Dinichthys*, which also grew to a length of about 30 feet or 9 m. Of this total, more than 10 feet or 3 m consisted of an armored head shield, which was hinged in the neck region (as was characteristic of all arthrodires). This allowed the upper jaw to be raised in relation to the lower, something rarely seen in any other vertebrates. The name 'arthrodire' is derived from the Greek words for jointed neck. Another group of primitive fish called the acanthodians was also abundant in the Devonian, and the bony fish (class Osteichthyes) diversified later in the Devonian (although they first appeared in the fossil record earlier). These were the ancestors to today's modern fish—and to the tetrapods that eventually moved onto the land later in the period as amphibians.

Joining the nautiloid cephalopod mollusks in the Devonian were well-armored **ammonoid** cephalopods. Curiously, considering their superficially similar shelled exteriors, these ancient ammonoids were more closely related to modern octopuses, squids and cuttlefish than to their neighboring nautiloids (which include the familiar extant species *Nautilus*). The seas were not the only site of evolutionary action during the Devonian—the first significant adaptive radiation of land life took place during the Devonian Period. The earliest known amphibians date to the Devonian, as do the first winged insects.

There was also another major mass extinction event during the late Devonian, leading to the demise of many species of trilobites and chordate relatives called graptolites. This was the second of the 'Big Five' major mass extinctions.

28.19 The Carboniferous Period

The next Paleozoic Period was the Carboniferous. The name basically means 'coal-bearing' since so many Carboniferous strata contain coal because this was a time of marked plant proliferation. (Coal is largely compacted fossilized plant material, particularly **peat**—the leftover remains of dead plants in wetlands.) **Lycopsids**, which today are relatively small and inconspicuous '**club mosses**', grew up to 40 m (130 feet) tall in the Carboniferous and included giant members such as *Lepidodendron* and *Sigillaria*. Similarly, modern sphenopsids (horsetails) are relatively small and inconspicuous today but back in the Carboniferous, forms such as *Calamites* were dominant flora and probably contributed to much of the present day's coal supply.

All this plant growth meant that a great deal of photosynthesis was going on and record-high levels of atmospheric oxygen were attained. Unlike today's 21%, Carboniferous air contained approximately 35% oxygen. The high oxygen levels might have made the environment more susceptible to fire. And more relevantly, it might have also meant that there was greater sensitivity to radiation during the Carboniferous. Alternatively, the increased oxygen might have forced the evolution of biochemical mechanisms to adapt to the higher biological impact of the radiation already in existence.

But what is clear that the increased oxygen levels allowed arthropods (which do not have sophisticated respiratory and circulatory systems like we do) to grow to enormous proportions. *Meganura* was a dragonfly with a 70 cm or 28-inch wingspan and *Arthropleura*, an ancient millipede-like creature, grew to 2.5 m or 8 feet 2 inches long. *Pulmonoscorpius* was a genus of giant Carboniferous land scorpions, with some species exceeding 28 inches or 70 cm.

Also during the Carboniferous was a proliferation of land vertebrates (the tetrapods or four-footed vertebrates) after their first appearance in the Devonian. Among the tetrapods radiating during this period were large amphibians such as the temnospondyls, embolomeres and reptiliomorphs (anthracosaurs). *Eogyrinus* reached about 15 feet (5 m) in length but was lighter than *Archegosuarus*, which reached about 10 feet (3 m) but probably weighed a few hundred pounds. Another big one from the Carboniferous was *Megalocephalus*, which reached about 6 feet (2 m) and probably weighed up to 75 lb.

Diplocaulus was a meter-long late-Carboniferous salamander-like lepospondyl amphibian with a distinct boomerang-shaped head. Thanks to these sometimes-huge amphibians, the Carboniferous is sometimes called the '**Age of Amphibians**'.

Also during the Carboniferous, the first **amniotes** appeared in the fossil record. Amniotes are the clade of organisms that have **amniotic eggs** or derivatives of them such as **placentas**. This watertight egg (also called the **cleidoic egg**) permitted the ancestors of modern reptiles, birds and mammals to reproduce on land and colonize terrestrial habitats that were previously off limits to them. Amniotes were probably descended from a group of amphibians called the **anthracosaurs**. The amniotes include all tetrapods except amphibians. The distinction between early reptiles and advanced amphibians was not so much anatomic but instead was ecological—in contrast to the amphibians who were still tied to water for their reproduction, the reptiles, being amniotes with their watertight eggs, were capable of reproducing in regions remote from lakes and ponds. Since Pangea was largely arid land, this ability was a tremendous advantage.

28.20 Sauropsids, synapsids and archosaurs

Amphibians continued to flourish in the next period of the Paleozoic, the **Permian Period**, with species such as the batrachomorph temnospondyl known as

Prionosuchus, a huge crocodile-like prehistoric aquatic amphibian (possibly) 9 m long. On land and in swamps, **Eryops** was a 5 foot long, 200 lb predator. But it was the reptiles who really began radiating during the Permian. Among the two diverging paths taken were the **archosaurs** and the **therapsids**. At this point it may be instructive to discuss a bit about relevant vertebrate cladistics.

Early in the evolution of the amniotes, these vertebrates diverged into two great phylogenetic groups or clades. (Phylogenetic groups are those that are genetically related.) These two clades were:

1. Sauropsida (the sauropsids)
2. Synapsida (the synapsids).

All members of these two great clades of vertebrates are *tetrapods*. Also in common, they all share the amniotic egg or modern derivatives of it (that is, they are *amniotes*). Nevertheless, the two groups diverged significantly *with the sauropsids leading to modern reptiles and birds* (as well as the extinct dinosaurs) whereas *the synapsids led to modern mammals* (and their extinct relatives such as the 'mammal-like reptiles' (the therapsids)) and their ancestors, the pelycosaurs. One can distinguish these two great clades by their skulls—specifically the **temporal fenestrae** (holes in the side of the skull). The synapsids have a temporal fenestra that is located lower on the side of the head (that is beneath the **postorbital bone**). The sauropsids have varying types of temporal fenestrae including the **anapsid** skull with no evident holes; the **euryapsid** skull (also called **parapsid**) with the hole high on the side of the skull (above the postorbital); and the **diapsid** skull with two temporal fenestrae (one above and one below the postorbital).

As mentioned, the synapsid skull is characteristic of mammals and their ancestors. The anapsid skull is seen in extinct early reptiles and in modern turtles. (But the absence of temporal fenestrae in turtles might not be because they are related to the primitive reptiles but that they lost their holes somewhere during their course of evolution.) The euryapsid skulls were found in ancient extinct marine reptiles including the ichthyosaurs, pliosaurs and plesiosaurs. The diapsid skulls are found in extant birds and reptiles as well as the extinct dinosaurs and pterosaurs.

The pterosaurs were (often huge) flying reptiles of the Mesozoic. Thanks to their temporal fenestration pattern, they were diapsids. But in addition, they were **archosaurs**—a group of diapsid, sauropsid, amniote tetrapods with an extra opening in front of the eye socket. That is, an **antorbital fenestra** between the maxilla and lacrimal bones. Furthermore, the teeth of archosaurs (if present) were in sockets—the so-called **thecodont dentition** pattern. Archosaurs include living birds and crocodilians, as well as extinct dinosaurs and pterosaurs. But despite having diapsid skulls, the modern snakes and lizards however are not archosaurs. Neither were the giant extinct marine mosasaurs.

Curiously, the marine mosasaurs of the Mesozoic (late Cretaceous) such as *Mosasaurus hoffmannii* (which might have reached roughly 56 feet or 17 m) and *Tylosaurus proriger* (which might have been as long as 52 feet or 15.8 m) were not euryapsids like the sea-dwelling pliosaurs, plesiosaurs and ichthyosaurs. Instead, they appear to be more closely related to modern monitor lizards like the Komodo dragon. So, they were not close relatives of these other marine reptiles (the euryapsid sauropsids), nor were they close kins of the archosaur sauropsids such as dinosaurs or crocodiles. Nonetheless, the mosasaurs were categorized as diapsid sauropsids— but not archosaurs.

With this minor digression, we can now return to the Permian Period with proper context.

28.21 The Permian Period

The final period of the Paleozoic Era was the Permian. Early in the Permian or late in the Carboniferous, the continents again amassed into a supercontinent, Pangea. The rest of the globe was occupied by a super-ocean, the Panthalassic Ocean, with a smaller gulf-like sea to the east called the Paleo-Tethys Sea. Much like central Australia, the interior regions of Pangea were probably rather arid.

The Permian was a time of great diversification of the amniotes, with the development of the ancestral groups to modern mammals, turtles, snakes and lizards and the archosaurs. The early Permian terrestrial fauna was dominated by big, bulky amphibians and by early **pelycosaurs**. The pelycosaurs were early synapsid reptiles that often had a stunning *sail along their backs*. These sails might have been an experiment in primitive thermoregulation through selective solar energy absorption (by angling the sail into or away from direct sunlight), thereby paving the way to genuine inherent thermoregulation in their mammalian descendants. Among the classic examples of the large pelycosaurs was the fierce large predator, *Dimetrodon*. Another large pelycosaur was the herbivore called *Edaphosaurus*.

At this time, the fossil record shows that plant eating was relatively late in arriving. Vertebrates had lived on land for millions of years, yet herbivores were missing. One possible explanation for this delay is that digestion of plant material is more biologically challenging. Meat is far more calorically dense, meaning that one must eat more vegetable matter in order to obtain the same amount of calories. Modern herbivores, such as cows and other ruminants, recruit cellulose-degrading bacteria to their assistance in their multi-chambered foreguts. The microbes in the bovine rumen break down cellulose in grass and other 'low-quality', high-fiber foods to generate acetic (ethanoic), propionic (propanoic) and butyric (butanoic) acids. These **short-chain fatty acids** provide up to 70% of the cow's main calories. The rumen microbiota also produces B vitamins, vitamin K and certain amino acids for bovine nourishment. However, such microbes do not perform as well in the cold. They must be kept at a fairly warm and constant temperature for optimal function. Hence, it may be speculated that herbivory evolved relatively late because of the prerequisite for a warm internal environment. One theoretical solution to this need is gigantothermy, in which animals retain heat thanks to the square-cube law (heat retention is dictated by volume, which is proportional to linear dimensions cubed, whereas heat loss is dictated by surface areas, which is proportional to linear dimensions squared). But animals hatched from eggs can never be very large when young; they would therefore have a period during which they would not be able to digest vegetable matter well until they grew substantially. And while the *Edaphosaurus* was big, it was not huge; gigantothermy was not available to this species. So, was the sail the solution to the challenges of vegetarianism?

By the Middle Permian, the pelycosaurs led to the early '**mammal-like reptiles**' or **therapsids**. Unlike most reptiles, but much like mammals, therapsids had **heterodont teeth** (meaning teeth of different shapes and sizes) that were differentiated into incisors, canines and molars. Therapsids were also more biomechanically fit for life on land, with their legs more vertically positioned beneath their bodies (rather than sprawling out to the sides) and with more symmetrical feet that were better suited to forward motion with an axis directed forward instead of laterally).

In the Late Permian, more advanced and quite imposing advanced therapsids hit the scene. Among them were the **gorgonopsians** and **dicynodonts**. 'Dicynodont' means 'two dog toothed' from ancient Greek, indicating the two large canine tusks in their otherwise typically edentulous beaked jaws. These dicynodonts were the only non-mammalian animals to have true tusks. As they were herbivorous, the tucks might have been used in self-defense. The dicynodonts also developed a **secondary palate**, which allowed them to simultaneously chew and breath—a significant advantage for herbivores who needed to eat a great deal. Rather than adopting herbivory, the gorgonopsians evolved into ferocious predators with jaw gapes possibly exceeding 90°. Some, like *Inostrancevia*, reached an estimated 11 feet (3.5 m) and 660 lb (300 kg). Like saber-toothed cats, and all modern mammals, the gorgonopsians had seven cervical vertebrae. *Anteosaurus* was another large, probably predatory therapsid of the Late Permian. It might have stretched to over 20 feet long with a weight over 1 ton. Later in the Permian Period, the **cynodont** therapsids emerged. These were the creatures that *eventually evolved into mammals*. Like the dicynodonts, the cynodonts had a hard palate that allowed them to chew and breathe simultaneously. They also had modifications to **the tarsus** (ankle bones, including the astragalus and calcaneus) not found in other reptiles that better enabled their locomotion.

But also during the late Permian, another group of vertebrates appeared that would seriously rival the therapsids—the archosaurs. Unlike the therapsids, which were synapsids with the characteristic synapsid skulls, the archosaurs were sauropsids with diapsid skulls. (Of course, there were numerous differences between these two very different clades of animals, but these skull differences are among the most obvious.) The archosaurs were worthy competitors to the rapidly evolving therapsids, but it looked like the therapsids did hold an edge over them. But then the largest of all mass extinctions came and nearly wiped them all off the face of the Earth. The fact that the archosaurs rebounded first and expanded into the pterosaurs, crocodiles and dinosaurs that would dominate the world's ecosystems for nearly 200 million years might not have had so much to do with outcompeting the therapsids; it might have been plain luck. Some (notably Stephen Jay Gould) have speculated that if we rewound the tape of the history of life and let it freely replay, human intelligence might have never graced the face of the Earth. But given that some modern mammals such as cetaceans and primates possess substantial intelligence, it does appear that there has been some selective pressure that has driven animal intelligence, at least in recent times. This makes one wonder what the world would look like today (intellectually) if the Permian extinction had not intervened and set the therapsids and mammals back so many millions of years.

28.22 The Mesozoic Era

The Mesozoic Era is the next era of the Phanerozoic Eon. It began after the Permian extinction around 250 million years ago and is sandwiched in between the preceding Paleozoic and subsequent Cenozoic Eras. Thus, its name which means 'middle life'. It contains three periods, the Triassic, Jurassic and Cretaceous in chronological order from oldest to most recent. The Mesozoic Era spanned the time from about 250–65 Ma. After roughly 185 million years, the Mesozoic Era ended around 65 Ma with the fifth and final major mass extinction event. This is thought to have been caused by an asteroid strike and will be discussed elsewhere.

The early Mesozoic Era flora was dominated by ferns, cycads, ginkgophytes, bennettitaleans and other now rare or extinct plants. Modern gymnosperms (such as conifers) first presented in the fossil record during the early Triassic Period. The Late Jurassic was characterized by climate stability and this widespread ecology can be aptly described as 'tropical'. But the vast Jurassic forests were not filled with what we would consider tropical plants. Instead, they were full of gymnosperms—ginkgoes, giant cycads and conifers. **Angiosperms** or flowering plants had to wait until later in the Mesozoic to make their appearance. But in the Cretaceous Period, early angiosperms appeared and began to diversify. Gradually, the angiosperms assumed a dominant position among the land flora.

Some Mesozoic fish were quite fascinating—and very impressive. The whale shark (*Rhincodon typus*) is the largest living fish today, but the Jurassic *Leedsichthys* is thought to be the largest fish that ever lived, with an estimated maximum length of 72 feet or 22 m. Previous estimates of the size of this Jurassic fish were as high as 98 feet. Like the whale shark, *Leedsichthys* was a filter feeder. But *Leedsichthys* was a bony fish (osteichthyes) in contrast to the whale shark, which is a cartilaginous fish (chondrichthyes). Leedsichthys lived during the Middle to Late Jurassic around 165 Ma.

The Mesozoic is often called **the age of reptiles**. However, as discussed in the section on cladistics, we now know that the group commonly called 'reptiles' is polyphyletic, and therefore taxonomically illegitimate. Regardless of the pedantic systematics, there were many extinct Mesozoic creatures we colloquially call reptiles such as the enormous flying pterosaurs, monstrous marine-dwelling mosasaurs, ichthyosaurs and plesiosaurs and of course the dinosaurs. Of the Mesozoic fauna, the dinosaurs attract the most attention—largely because they were so large.

28.23 The dinosaurs

Dinosaurs may be formally defined as 'the clade of vertebrates consisting of *Triceratops*, modern birds, their most recent common ancestors and all their descendants'. Dinosaurs roamed the Earth during nearly the entire Mesozoic Era (from the Triassic through the Cretaceous Period). They diverged from their archosaur relatives early in the Triassic, maybe as soon as 20 million years after the Permian extinction, around 270 Ma. Although dinosaurs appeared in the Triassic, they did not really diversify and dominate the land flora until the Jurassic. But once they gained the upper hand, they held it as the dominant terrestrial animals for around 150 million years.

There was a major mass extinction event at the end of the Triassic Period (one of the 'Big Five'). But the early dinosaurs survived. And they subsequently flourished and filled the newly vacated ecological niches in another adaptive radiation event. The early dinosaurs such as *Eoraptor* were not terribly impressive (despite the name 'dinosaur', which means 'terrible lizard') weighing maybe up to 10 kg or 22 lb—and probably smaller than that. It had relatively long hind limbs and dinosaur-specific ankle anatomy that made it well suited for agility in an upright posture.

Anatomically, practically all dinosaurs share a certain skeletal feature—a hip socket with a hole in it (technically described as **perforated acetabulum**). This perforation or hole (also termed a fenestra) in the acetabulum is unique among all the tetrapods. This unique hip was accompanied by anatomical rearrangements of the hips and shoulder regions (that is, the pelvic girdle and pectoral girdle respectively) which better allowed dinosaur legs to be positioned directly under their bodies (unlike the sprawling body habitus of today's salamanders, lizards and crocodilians). In other words, the legs of dinosaurs were erect (vertically aligned) and oriented perpendicular to their body axis. This arrangement was more conducive to a larger (taller) body size in a strictly terrestrial animal (in contrast to the relatively flat, sprawling, aquatic crocodilians where water buoyancy helped compensate for the anatomical disadvantages), and set the stage for (big) things to come.

Taxonomically, dinosaurs were diapsid archosaurs with a perforated acetabulum who held their limbs erect and directly beneath their bodies. As mentioned, dinosaurs are unique among all tetrapods in having the fenestrated acetabulum (although it now appears that not every dinosaur species had this feature). Nevertheless, the perforated acetabulum is a **synapomorphy**—a distinguishing feature possessed by essentially all members of the group in question, but is not seen in other animals. Such synapomorphies serve as defining characteristics in cladistics.

In dinosaur cladistics, the Dinosauria can be divided into two main branches based on their hip anatomy. These are the so-called 'bird-hipped' dinosaurs (**ornithischians**) and the 'lizard-hipped' dinosaurs (**saurischians**). Ironically, although modern birds are now considered to be surviving dinosaurs, they are descended from the saurischians rather than from the ornithischians. Furthermore, since dinosaurs were considered 'reptiles' but birds (class Aves) are extant dinosaurs, it becomes evident that 'class Reptilia' is a polyphyletic, and therefore illegitimate, taxonomic grouping unless birds are included as reptiles too.

The pubis bone in saurischian dinosaurs is directed towards the head (cranially or rostrally) such that, combined with the dorsally directed ileum and the caudally directed (towards the tail) ischium, the saurischian hip is **tri-radiate** (with bones oriented in three different directions). In stark contrast, the pubis of ornithischians is directed posteriorly, and is parallel to the ischium, creating a bidirectional rather than tridirectional overall hip structure. Since *all ornithischians were herbivorous* but some adopted a bipedal stance, it is speculated that this shifting of the pubis was an anatomical adaptation that allowed these bipedal herbivores to maintain better balance while their big, vegetable-digesting guts enlarged further to accommodate chambers for symbiotic bacteria that aid digestion. Such an arrangement is not needed in quadrupedal animals such as modern cows or sauropod dinosaurs. The bipedal saurischians (theropods) were carnivores and therefore had much smaller guts (since meat is more readily digested than vegetable matter), thus there was no evolutionary selective pressure in theropods for restructuring of the pubic bone. (One must remember however that birds, who evolved from saurischians, also developed this posteriorly directed pubis. Therefore, there must have been additional selective pressures for the pubis rearrangement besides the upright posture of vegetarian dinosaurs.)

The saurischians included three main groups:

Saurischia:
1. Sauropodomorpha
    a. Prosauropoda
    b. Sauropods
2. Theropoda
    a. Birds (class Aves).

Prosauropods were odd, early dinosaurs that were facultative bipeds, meaning that they could walk on all fours in a quadrupedal fashion, or they could stand on two legs in a bipedal manner. They were not restricted to one or the other. They were herbivorous and lived during the Triassic and early Jurassic. Their name suggests that they were precursors or ancestors to the sauropods, but this older view is no longer widely accepted; they were probably a completely independent group. The sauropods were the familiar long-necked giants of the Mesozoic. Sauropods were strict herbivores whose heyday was in the Late Jurassic. As a group, the sauropods were one of the most successful animal groups of all time in terms of durability, as they remained prominent for 100 million years or so, ranging from the Late Triassic to the Late Cretaceous. Well-known sauropods include *Brontosaurus, Diplodocus, Apatosaurus, Barosaurus* and *Brachiosaurus*. The last surviving sauropods were the **titanosaurs**, some of whom lasted until the extinction event at the close of the Cretaceous Period. In addition to being the last of the sauropods, some titanosaurs were also the biggest. Noteworthy giants included *Patagotitan*, which reached an estimated 121 feet long (37 m) and weighed perhaps 76 tons (152 000 lb or 69 000 kg) and the similarly sized *Argentinosaurus*.

The last group of saurischians were the theropods. These were bipedal and were the only carnivorous, predatory dinosaurs. (Although some theropods later evolved into omnivores and herbivores.) Theropods first appeared in the Late Triassic, about 232 Ma. They diversified in the Jurassic and successfully persisted until the very end of the Cretaceous. Among the popular theropods were the giant carnivores such as the Jurassic *Allosaurus fragilis* and the Late Cretaceous *Giganotosaurus carolinii*, which rivaled the famous contemporaneous Cretaceous therapod, *Tyrannosaurus rex* for top honors of 'largest known terrestrial carnivore'. *Spinosaurus* might have been even larger but present evidence suggests this species lived an aquatic or semiaquatic lifestyle like crocodilians. Another imposing theropod of the Late Cretaceous was *Carcharodontosaurus saharicus*, whose genus name reflects its sharp, shark-like teeth (the great white shark is *Carcharodon carcharias*).

Although they were all herbivores, the ornithischians were more varied than the saurischians in body types and comprised five major subgroups.

Ornithischia:
1. Ornithopods
    a. Iguanodontids
    b. Hadrosaurs
2. Pachycephalosaurs
3. Ceratopsians
4. Stegosaurs
5. Ankylosaurs.

The first two groups—the ornithopods and pachycephaosaurs—were bipedal. The remaining three groups were well-armed quadrupeds. Some stegosaurs, such as *Stegosaurus*, had a set of four fearsome spikes on the end of their tails called **'thagomizers'**. (The name comes from a Gary Larson 'Far Side' cartoon in which a caveman points to a picture of a *Stegosaurus* tail and names the spiked structure in honor of 'the late Thag Simmons'.)

The ankylosaurs instead had a large, heavy, bony club at the end of their tails. Ankylosaurs were also typically very heavily armored over their entire bodies except their undersides. Both stegosaurs and ankylosaurs had osteoderms (bony plates, scales or similar structures in their dermis) and were related to each other. They are thus collectively grouped as **thyreophorans** ('armor bearers').

The **ceratopsians** or horned dinosaurs often had a protective nuchal (neck) shield as well as large, intimidating horns. The most well-known ceratopsian dinosaur was probably *Triceratops* (specifically, *Triceratops horridus*). Sharing bony elaborate structures on their heads, were the domed dinosaurs or **pachycephalosaurs**. Their thick bony domes might have been used in headbutting potential predators and maybe in interspecific rivalries, like how present-day rams do. However, their neck structures do not seem as well suited to full-speed, head-on collisions as rams' are. Despite being bipedal, the pachycephalosaurs were closely related to the ceratopsians and are lumped together in the tribe **marginocephalia**.

Among the **ornithopods** were the **iguanodontids**, such as *Iguanodon* (one of the very first dinosaurs to be discovered) and the **hadrosaurs** (duck-billed dinosaurs). Some hadrosaurs such as *Parasaurolophus* had bony head ornaments, giving them their nickname of 'crested dinosaurs'. Some of these crests grew into elaborate, hollowed-out structures which, because they were connected to the nasal passages, may have served to make loud audible signals. The internal structure of some of these crests suggests that when air was blown through them, they may have sounded like a brass instrument.

Like all archosaurs, dinosaurs had diapsid skulls and had a prefrontal fenestra. But not all the other amazing Mesozoic reptiles were dinosaurs or even archosaurs.

For instance, the marine **mosasaurs** were diapsids but were not true archosaurs; they were more closely related to modern monitor lizards than crocodilians or dinosaurs. Taxonomically speaking, mosasaurs were **lepidosaurs** rather than archosaurs. Some mosasaurs grew to enormous sizes such as the Late Cretaceous *Tylosaurus* (which might have reached 14 m or 45 feet) and the 11–17 m long (36–56 feet) *M. hoffmannii*, also of the Late Cretaceous.

The other great marine reptiles of the Mesozoic (ichthyosaurs and plesiosaurs) were not even diapsids, in that their skulls did not have two fenestrae on each side; they had what are called **euryapsid** (also called parapsid) skulls. However, recent research suggests that the euryapsid skull is really a diapsid skull that has secondarily lost one of the two fenestrae, specifically, the lower one (below the postorbital bone). This means they were neither traditional diapsids nor were they archosaurs. But they certainly were not dinosaurs. The **plesiosaurs** or **sauropterygians** came in two broad forms, the long-necked plesiosaurs (such as *Plesiosaurus*) and the more robust, larger-skulled, short-necked **pliosaurs** such as the Middle-Late Jurassic *Liopleurodon* (which might have reached 10 m or 33 feet in length but was probably smaller) and *Kronosaurus* (which lived during the Early Cretaceous and was even larger and might have reached 9–11 m (30–36 feet)).

Another prominent group of marine Mesozoic reptiles were the **ichthyosaurs**. In a classic example of **convergent evolution**, the ichthyosaurs were remarkably dolphin-like (and of course, dolphins are remarkably shark-like or fish-like). Thus, despite extremely different genotypes, the phenotypes or morphological appearances of ichthyosaurs, dolphins and sharks are surprisingly similar. Ichthyosaurs swam the Mesozoic seas from the Early Triassic, relatively soon after the Permian extinction, to around 90 Ma in the Late Cretaceous.

Blue light penetrates deeper into the ocean than red wavelengths and some ichthyosaurs seem to be very well adapted for seeing, even in deep waters. One example is the appropriately named *Opthalmosaurus* who had an orbital (eye socket) diameter nearly a foot in diameter, with a bony (osteosclerotic) rim around the eye that might have helped the animal cope with the immense pressures encountered during deep dives. Some Triassic species of ichthyosaurs such as *Shonisaurus* were immense and might have attained lengths of about 69 feet or 21 m.

28.24 The Pterosaurs

The marine monsters were not the only magnificent Mesozoic non-dinosaur reptiles. Like dinosaurs, the flying **pterosaurs** were diapsid archosaurs. Yet the pterosaurs were not true dinosaurs, as they did not possess the defining anatomy of a fenestrated acetabulum. Pterosaurs existed in two main groups—the earlier, tailed **rhamphorhynchids** and the later, often ornately headed **pterodactyls**. Whether these two groups are truly natural taxa or simple but artificial constructs is debated. Pterosaurs ranged in size from the tiny, 10 inch wing-spanned *Nemicolopterus* to the nearly 40 foot (~12 m) wide *Hatzegopteryx* and *Quetzalcoatlus*, both of the Late Cretaceous. Over time, the mean wingspan of the pterosaurs progressively increased, consistent with a general observation of size increasing over time called **Cope's rule**.

The pterosaurs were the first vertebrates to fly, arising in the Late Triassic and thereby beating their cousins, the birds, whose ancestors (such as *Archaeopteryx*) first appear in the fossil record in the Late Jurassic.

28.25 Phytosaurs

Another fascinating group of Mesozoic reptiles were the **phytosaurs**, large Triassic archosaurs that were remarkably similar in overall appearance to crocodilians, although they are only distant cousins genetically. This represents another classic case of convergent evolution. If one examines their skeletons closely, differences are apparent in ankle structure, body armor (including a more heavily protected belly in phytosaurs through dense abdominal ribs or **gastralia**), absence of a hard palate in phytosaurs and most conspicuously, a difference in the location of the nostrils—crocodilians have their nostrils at the tip of their snouts while phytosaurs had their nostrils on the top of their skulls, located near their eyes.

28.26 Mesozoic crocodiles

Speaking of Mesozoic crocodilians, the largest crocodile of them all was perhaps the early Cretaceous crocodilian, *Sarcosuchus imperator*. This species might have exceeded 40 feet (12 m) in length and weighed an estimated 8000 kg (17 600 lb), however other recent estimates provide a lower limit on its size at only 31 feet or 9.5 m. *Sarcosuchus* was not alone as a giant Mesozoic crocodilian, however. It had rivals in size such as the Late Cretaceous *Deinosuchus*. At the time of this writing, *Deinosuchus rugosus* is believed to be the largest crocodilian of all time, with a probable maximum length of 39 feet or 12 m. Its bite was something to be reckoned with as well. Among extant animals, the saltwater crocodile (*Crocodylus porosus*) presently has the strongest documented bite force, with a maximum force of 16 414 newtons (3690 lb of force). The estimated bite force of *Deinosuchus* easily exceeds this and has been calculated to be between 18 000 N (4047 lb) and 102 803 N (23 111 lb). Incidentally, crocodiles were one of the few large animals to survive the mass extinction at the end of the Cretaceous Period. During the Miocene Epoch of the Cenozoic Era, other giant crocodilians such as *Rhamphosuchus* and *Purassurus* lived.

Thus, one can see that there was a great array of different vertebrates that can colloquially be called reptiles, justifying the nickname, Age of Reptiles for the Mesozoic Era. But all this came to a rather abrupt end at the close of the Cretaceous. The **K–Pg extinction event** (using the abbreviation symbols for Cretaceous–Paleogene) was the last of the Big Five major mass extinctions.

28.27 The Cenozoic Era

The next and last of the three eras of the Phanerozoic Eon was the Cenozoic. The Cenozoic Ear had three periods, the Paleogene, the Neogene and the Quaternary. Up until recently, the Paleogene and Neogene were often lumped together as the **Tertiary Period**, but that terminology has been updated. These periods are divided into **epochs**. The **Paleocene**, **Eocene** and **Oligocene Epochs** comprise the Paleogene Period. The **Miocene** and **Pliocene Epochs** make up the Neogene Period. And the

Quaternary Period contains the **Pleistocene** and **Holocene Epochs**. We are presently in the Cenozoic Era, specifically in the Holocene Epoch of the Quaternary Period.

The Cenozoic began about 65 Ma, following the mass extinction event that ended the dinosaurs' domination. This event was previously called the **K–T extinction event**, with K being the abbreviation for Cretaceous and T being the abbreviation for Tertiary. But as mentioned, the name Tertiary has been abandoned in favor of the use of the Paleogene and Neogene Periods instead. Thus, what was formerly the K–T extinction is now the **K–Pg extinction event**.

The Mesozoic was understandably called the Age of Reptiles, but the Cenozoic deserves its nickname, the **Age of Mammals**. Relatively few large animals survived the K–Pg. The crocodilians were the only large predatory reptiles to make the cut; sea turtles also made it through. Among the dinosaurs, only the birds made it through. The mammals who did survive into the Paleocene were relatively small and unimposing. But over the epochs this would change, with mammals radiating into the various ecological niches left vacant by the dinosaurs. Cope's rule was followed by horses, who started out as the small (maybe a foot tall at the withers) 'dawn horse' (*Eohippus*) in the Eocene, and species progressively grew larger over time in the form of *Orohippus, Epihippus, Mesohippus, Miohippus, Parahippus, Merychippus, Pliohippus, Dinohippus* and today's *Equus*.

Another example of progressively increased body size over time was seen with whale evolution. Whales (order **Cetacea**) evolved from mammals belonging to order **Artiodactyla**. Today's artiodactyls include hippopotamuses, pigs, camels and **ruminants** (which are **ungulates** or hoofed mammals with a specialized foregut that can ferment plant-based foods and extract additional nutrients from them and includes goats, giraffes, deer and bovids such as buffalo, bison and cattle). The artiodactyls share the synapomorphy of an ostensibly even number of toes. This contrasts artiodactyls from the ungulate (hoofed) mammals with an odd number of toes such as the single-toed horses and the three-toed tapirs (at least on their rear legs) and rhinoceroses; these odd-toed ungulates are the **perrisodactyls**. The 'missing' toes in perissodactyls may be present as vestigial structures or may be grossly absent. The body weight in artiodactyls is distributed equally between the two toes (typically the 3rd and 4th digits)—an arrangement called **paraxonic** feet. The body weight in perissodactyls is distributed over the middle (3rd) digit, which is called the **mesaxonic** foot arrangement. Whales, dolphins and porpoises (order Cetacea) evolved from the artiodactyls that are most closely related to today's hippopotamus. Technically, this implies that order Artiodactyla is an illegitimate (paraphyletic) group—unless one includes the cetaceans as artiodactyls.

Pakicetus was a small amphibious artiodactyl mammal from the Eocene Epoch (around 50 Ma) but had an ear structure consistent with cetacean ancestry. *Ambulocetus* appeared slightly afterwards around 47–48 Ma and grew to about 9 feet long; it still had hind legs but was clearly comfortable in water. With its dolphin-like body, *Dorudon*, at 16 feet long was certainly not small, but was not huge either (for a whale). *Basilosaurus* (initially thought to be a giant lizard, accounting for its decidedly non-whale-like name) was a top predator that lived 40–34 million years ago and was indeed huge at 40–60 feet long. *Squalodon* was the first whale species known to have echolocation capabilities and lived 34–18 million years back. *Aetiocetus*, from around 25 million years ago, was a link between the toothed whales (**odontoceti**) and the baleen whales (**mysiceti**) as it had both baleen and a full set of teeth. Speaking of teeth (excluding tusks), no animal in history had larger teeth than the Miocene predatory whale *Livyatan melvillei*. Its ferocious teeth even exceeded the size of those on *T. rex*. Incidentally, in sheer size its teeth also outstrip those of *Megalodon*, the largest shark in history. It is interesting that these two sea monsters lived at the same time since it makes one wonder how a showdown between these two behemoths of the prehistoric seas would have gone down.

Although *Livyatan* was probably a lot more formidable than a modern sperm whale overall, it was no larger. And for that matter, no prehistoric fish, whale, shark or dinosaur can match the size of the present-day blue whale, *Balaenoptera musculus*. No animal in history ever matched the size of the blue whale. And as far as plants go, the largest tree alive now is the General Sherman giant sequoia (*Sequoiadendron giganteum*). And the Lindsay Creek tree (a coastal redwood, *Sequoia sempervirens*) that fell in 1905, might have had twice the trunk volume of General Sherman. Based on the samples from the fossil record, it is likely that these trees are larger than any prehistoric tree in Earth's past. Therefore, although we tend to think of the days of the dinosaurs as the time when the largest creatures lived, the fact is that the 'Age of Gigantism' might be right now.

The largest land animal at this time is the African bush elephant (*Loxodonta africana*). But over the Cenozoic Era, there have been many (about 160) different proboscidean species. Most of these elephant relatives lived during the Miocene Epoch. It appears that proboscideans originated in African and soon spread to every continent except Australia and Antarctica. The **gomphotheres**, with their blade-like 'shovel tusks' crossed the Bering Straits in the Miocene and various species migrated all the way to South America. The **mastodons** (from Greek words meaning 'breast teeth' because of their peculiar tooth shapes) are also called **mammutids** (because they belong to family **Mammutidae**) and had long curved tusks and long red-brown hair. They crossed the land bridge at the Bering Straits during the Pliocene Epoch about 3.5 Ma. Some species of mastodons persisted into the Pleistocene up to just 10 400 years ago in central North America. The third group of proboscideans to reach North America were the **mammoths**. Very confusingly, despite being called mammoths, they belong to family **Elephantidae** (not Mammutidae) yet the genus name is *Mammuthus*. Mammoths of various sorts arrived in North America about 2 million years back. They were the last group of proboscideans to evolve and were the last to reach North America migrating as far south as Central America.

The woolly mammoth (*Mammuthus primigenius*) was a cold-adapted species covered in thick, long, dark hair that lived on the tundra steppes of North America, Asia and Europe. Somewhat surprisingly, this was the largest biome on Earth during the Pleistocene Epoch. Frozen woolly mammoth carcasses have been found that were as recent as 5500 years old. Larger than the wooly mammoth was the Columbian mammoth (*Mammuthus columbi*), which reached 4 m (13 feet) at the shoulders and weighed around 10 tons (22 000 lb). The Columbian mammoth resided in southern North America and overlapped with early indigenous people. The Columbian mammoth went extinct at the end of the Pleistocene around 11 500 years ago. As massive as the Columbian mammoth was, it was not the largest mammoth species.

The two largest mammoth species were the southern mammoth (*Mammuthus meridionalis*), which lived in northern Eurasia around 2 million years ago, and its descendant, the steppe mammoth (*Mammuthus trogontherii*) which likely originated in northern China around 1.6 million years ago and subsequently spread across much of central Eurasia. The largest documented steppe mammoth specimen had an estimated shoulder height of 4.5 m (14 feet 9 inches) and weighed approximately 15.4 tons. As huge as these mammoths were, they were not the largest of the extinct proboscideans. That honor goes to *Palaeoloxodon namadicus*, the Asian straight-tusked elephant. This non-mammoth elephant species lived in India and other parts of Asia and went extinct in the late Pleistocene. By some estimates, it might have attained a size of 5.2 m (17.1 feet) in shoulder height and a weight of roughly 22 000 kg or maybe 24.3 tons (48 600 lb). Whether this was the largest land mammal in history is debated. The prehistoric, gargantuan, long-necked, hornless rhinoceros species, *Paraceratherium transouralicum* has been estimated as large as 30 000 kg (or 33 tons or 66 000 lb) but these early claims are now considered overestimates based on the incompleteness of fossil specimens. *Paraceratherium* was formerly known as *Indricotherium* and more likely was in the range of 33 000 to 44 000 lb. It lived during the early to late Oligocene Epoch (34–23 million years ago) in what was then the equivalent of Central Asia and eastern Eurasia.

Although elephants and mammoths are known for their enormous size, there were several dwarf species over the ages. Some dwarf species of stegodonts probably arouse through insular dwarfism (diminution of body size over time seen in animal species confined to an island). Some southeast Asian island species were as small as 400 kg as adults. Ironically, some different stegodont species were also among the largest of all proboscideans. Even smaller than the dwarf stegodonts was *Paleoloxodon falconer*, the Maltese or Sicilian pygmy elephant. An adult female specimen measured 80 cm (2 feet 7.5 inches) at the shoulder and weighed about 168 kg (370 lb).

The pygmy mammoth, *Mammuthus exilis* was on average, 1.72 m (5.6 feet) tall at the shoulders and 760 kg (1680 lb) in weight. Although much larger than the smaller pygmy elephants, it was lilliputian in comparison to its immediate ancestor, the 22 000 lb Columbian mammoth. These pygmy mammoths represent another example of **insular dwarfism**, as a population of Columbian mammoths was stranded on the Channel Islands off the coast of California. The pygmy mammoth, *M. exilis*, should not be confused with the little mammoths of Wrangel Island and Saint Paul Island, which was a race of the woolly mammoth (*M. primigenius*) and died out around 3000–4000 years ago. These relatively small island mammoths were the last of the mammoths. It was previously believed that these were pygmy species that shrank through insular dwarfism, but recent data refute that perspective.

28.28 Susceptibility to cancer and radiation among elephants

Recent studies have shown that modern elephants (the Asian elephant, *Elephas maximus* and the African bush elephant (*Loxodonta africana*)) are relatively cancer resistant, and the putative mechanism is a multitude of copies (40 alleles) of the **tumor suppressor gene**, *TP53*. This contrasts with the human diploid condition that has just two copies (two alleles—one maternal and one paternal in origin). The *TP53* gene codes for the critical **p53 protein**, which is nicknamed the 'guardian of the genome' because of its crucial role in protecting the integrity of nuclear DNA. Among the numerous functions of p53 are:

1. Halting of the cell cycle in late G1 phase upon DNA damage by radiation or other causes.
2. Initiation of DNA damage repair pathways.
3. Promotion of **apoptosis** (programmed cell death) if the DNA damage is too extensive for repair.

Many malignant tumor cells have **loss of function** in the *TP53* gene (confusingly called '**loss of heterozygosity**'), which means that both the maternal and the paternal copies of this critical gene have been compromised. The net result is that cells harboring such mutations will be less capable of repairing damage and may proceed with cell division despite their damaged DNA (and this process promotes further accumulation of potentially deleterious mutations). And cells with unrepaired DNA damage will not undergo apoptosis and take themselves out of circulation if they have defective or absent p53 protein. As covered in chapter 11, people who are born with a germline mutation in one of their two *TP53* genes (alleles) have the **Li–Fraumeni syndrome**. Li–Fraumeni syndrome predisposes people to cancer; such individuals have around an 80% chance of developing cancer at some time during their lives.

The redundancy of *TP53* in elephant cells may account for the qualitatively and quantitatively different response of elephant cells to radiation exposure. Elephant cells generally undergo higher rates of apoptotic death than equivalent human cells, thereby prohibiting potentially cancer-causing DNA damage from propagating. However, the multiple copies of TP53 are **retrogenes** or **pseudogenes**, meaning they are copies of the original gene inserted into the elephant genome by reverse transcriptase activity—and thus might not perform like fully functional genes. The functional capacity of these elephant *TP53* retrogenes is being studied presently and their real role in radiation resistance is being evaluated. However, a curious finding is that the preserved mammoth cells found in the tundra permafrost, like present-day elephants, also contained DNA with multiple copies (14 in this case) of the *TP53* gene. This contrasts with the standard single diploid set in fossil manatees and hyraxes (which are believed to be the closest relatives of order Proboscidea). This would argue that the evolution of large body size is indeed linked to the increased copy number of this key cancer- and radiation-resistance gene. Supporting this observation is the identification of a single canonical *TP53* gene plus an estimated *TP53* retrogene copy number in the Asian elephant genome of 12–17;

14 in the extinct but relatively recent Columbian and woolly mammoth genomes; but only 3–8 in the far older (50 000–130 000 year old) genome of the American mastodon.

Another gene with multiple copies in the genome of giant proboscideans is leukemia-inhibiting factor (*LIF*). Like p53, the protein made by this gene is a promoter of apoptosis in response to DNA damage by radiation or other causes. And at least one of the several copies of this gene (*LIF6*) has been shown to be functional and actively involved in induction of apoptosis in elephant cells. Thus, *LIF6* is probably among the genes involved in making elephant cells radiation sensitive (meaning more susceptible to radiation-induced apoptosis) and protecting these large beasts from getting cancer. Interestingly, the *LIF6* gene is transcriptionally upregulated by increased levels of p53 protein in response to DNA damage due to radiation or other causes. It should be stated that although the individual elephant cells appear to be more sensitive to radiation, this does not mean that elephants are more likely to develop cancer in response to radiation exposure; in fact, it is the opposite. The paradox is explained by the fact that the elevated amounts of p53 cause radiation-damaged cells to be either effectively repaired or simply taken out of commission quickly through apoptosis, thereby prohibiting their propagation as potentially malignant cells.

28.29 Hominoids, hominids and hominins

No discussion about the Cenozoic—the Age of Mammals—would be complete without briefly touching the topic of human and hominid evolution (even in a book on space radiation). To put humans (*Homo sapiens*) into context taxonomically, first they are **chordates**. That means that at some point in their lives they possessed the following:
1. Pharyngeal gill slits.
2. A single, hollow dorsal nerve cord (as opposed to the solid ventral nerve cord of annelid earthworms or the paired lateral nerve cords of flatworms (phylum Platyhelminthes) for instance).
3. A notocord (an embryonic structure that directs development of the vertebral column and eventually becomes part of the intervertebral discs in adult humans).
4. A post-anal extension (tail). Most invertebrates with a complete gastrointestinal system have the anus at the caudal end of their bodies; chordates are different in this regard.

These are the classic **synapomorphies** that serve to define the phylum Chordata. But in addition to being chordates, humans are vertebrates. As vertebrates, they also have the following defining synapomorphies:

1. An endoskeleton made of genuine bone (hydroxyapatite and collagen).
2. A vertebral column (segmented backbone) located on the dorsal (posterior) side of the body.

3. A brain and central nervous system. (Invertebrate chordates do not have an identifiable brain.)

Among the vertebrates are several taxonomic classes, including the Osteichthyes (bony fishes), Chondrichthyes (cartilaginous fishes), amphibians, reptiles, birds and mammals. The amphibians, reptiles, birds and mammals are **tetrapods**, meaning they are vertebrates with four limbs. Humans are tetrapods and mammals; as mammals, they have the following characteristics:

1. Possession of hair or fur somewhere on their bodies at some time in their lives
2. Are warm-blooded (endothermic)
3. Are born alive (viviparous)
4. Produce milk through maternal mammary glands
5. Have a more complex brain than other vertebrates.

More specifically, humans are placental mammals. This distinguishes them from the non-placental monotremes (egg-laying mammals such as the platypus and echidna) and marsupials (mammals with pouches, such as kangaroos, koalas and Tasmanian devils). The current fossil record suggests that monotremes first evolved around 150 Ma (Late Jurassic Period), marsupials appeared around 130 Ma and the placental mammals appeared about 110 Ma (both in the Cretaceous Period). Thus, the current paleontological evidence suggests that these extant taxa of mammals all arose in the Mesozoic Era, although the monotremes might date back slightly earlier when they evolved from their 'mammal-like reptile' (therapsid) ancestors. The therapsids that gave rise to the mammals were the **cynodonts**. The cynodonts led to the earliest mammals in the Late Triassic, apparently around 225 million years ago. And among the once-numerous non-mammalian therapsids, only the cynodonts themselves survived the Triassic extinction.

The earliest fossil 'mammals' were the **morganucodontids**, small shrew-like animals that lived in the Triassic, around 210 million years ago. Whether the morganuco-dontids can truly be called mammals is debatable. Unlike their reptilian (cynodont therapsid) ancestors or their genuine mammalian descendants, the morganucodontids had a double articulation joint the back of their jaws. Modern mammals have an articulation between the squamosal and dentary bones (called the temporal and mandible bones respectively in humans—thus the 'temporomandibular joint'). Modern reptiles and the extinct therapsids have a jaw joint between two different bones—the articular and quadrate bones. In the transitional morganucodontids, both of these jaw articulations were present. In modern mammals, including humans, the old articular and quadrate bones get pushed back further behind the jaw and become the middle ear ossicles called the malleus and incus respectively. The anatomical feature of a squamosal-dentary jaw joint is considered a key characteristic of mammals and is used in cladistics as a defining synapomorphy. (If you are a vertebrate but do not have this particular type of jaw joint, you do not qualify as an authentic mammal.)

Humans belong to the mammalian order Primates. Primate taxonomy has changed substantially over recent years. At one time it was standard to divide all primates into two major groups, the **prosimians** (lorises, lemurs, tarsiers and aye-ayes) and **anthropoids** or simians (monkeys, apes and humans). However, recent molecular cladistical analyses indicate that the prosimians are a polyphyletic group, since the tarsiers are more closely related to the Old-World monkeys than to lemurs and lorises.

Hence, the preferred modern taxonomy of primates defines the **lemuriforms** (lemurs and lorises) as the suborder **Strepsirrhini**, while the other major group (suborder **Haplorrhini**) includes the tarsiers along with the monkeys, apes and humans—that is, the tarsiers plus the anthropoids. The extinct lemur-like primates or Adapiformes are also members of the suborder Strepsirrhini.

Monkeys are classified in two superfamilies—the Old-World monkeys or **Cercopithecoidea** and the new world monkeys of superfamily **Ceboidea**. The remaining anthropoids constitute superfamily **Hominoidea**, the hominoids. Hominoids includes the tailless primates—apes and humans—and their extinct relatives. Hominoids differ anatomically from the other anthropoids (Old-World and New-World monkeys) by the absence of a tail and a larger brain as well as a slower rate of maturation. The latter, by necessity, is accompanied by more prolonged parental care. Hominoids also have a relatively longer lifespan.

The hominoids are divided into family Hylobatidae (the 'lesser apes'—gibbons and siamangs) and family Hominidae (the 'great apes'—orangutans, gorillas, bonobos and chimps, along with humans). But at this point, the topic is confusing thanks to multiple changes in classification over the years. At one time, the great apes (orangutans, gorillas, chimps and bonobos) were lumped together in family Pongidae while humans remained separate in **family Hominidae**. Now however, all are grouped into family Hominidae. But among these hominids, the group containing the orangutans (subfamily Ponginae) is separated from the other extant **subfamily, Homininae**. Thus, (according to this classification scheme) homininae includes gorillas (genus Gorilla), chimpanzees (genus Pan) and humans (both extant and extinct). For further granularity, the homininae may be broken down further into the Gorillini, which includes gorillas and the Hominini (chimps, humans and extinct relatives). As might be anticipated, there is a great deal of debate on human taxonomy and one alternative grouping is to exclude genus *Pan* (chimpanzees and bonobos) from Hominini and reserve that name (that is, the hominins) exclusively for the taxonomic tribe that includes only *Sahelanthropus, Ardipithecus, Australopithecus, Kenyanthropus* and *Homo*. Thus, according to this classification, the hominins would comprise the group that includes modern humans but branched off from the chimpanzees somewhere around 8.5 million years ago. Although paleoanthropology is one of the most contentious disciplines in all of science, it is worth recalling that humans and chimps share nearly 99% (best estimate is 98.7%) of their DNA sequences. Interestingly, humans share about 1.6% of their DNA sequences with bonobos that we do not share with chimps; conversely, about 1.6% of human DNA is shared with chimpanzees but not shared with bonobos. These initially surprising statistics should be tempered by the observation that humans also share about 98% of our DNA with pigs, 96.9% with orangutans, 80%

with cows and roughly 70% with chickens. In fact, human DNA and banana DNA are about 60% homologous!

The earliest fossil ape known at this time is *Proconsul*, which lived between 22 and 16 million years ago in the Miocene Epoch. Proconsul was a natural quadruped but had a **pronograde** posture, meaning that its back was horizontal (parallel to the ground) when it was walking on all fours. *Kenyapithecus* was another fossil ape who lived about 14 million years ago and might have been the first ape to venture beyond the African continent. *Dryopithecus* was another early ape who clearly lived outside Africa; it resided in Europe around 12 million years back. Also residing outside Africa, this time in India, was *Sivapithecus*. Early fossil ape specimens from the Siwalik Hills region were once called *Ramapithecus* but subsequent work revealed these specimens to be essentially identical to Sivapithecus (perhaps males and females of the same species in another example of sexual dimorphism). Thus, *Ramapithecus* is no longer a valid genus. *Sivapithecus* is now believed to be the ancestor of today's orangutan and not an ancestor to humans.

As demonstrated above, primate taxonomy is complicated and confusing but for our purposes, we shall use a simplified terminology in which 'hominoid' is the most inclusive grouping and includes all apes (both extinct and extant) along with humans (both extinct and the present modern human species, *Homo sapiens*); 'hominid' refers to the next most inclusive group (African great apes—chimps, bonobos and gorillas, along with humans and related extinct species but not the lesser apes (gibbons and siamangs) nor orangutans), and 'hominin' to humans and fossil relatives after the evolutionary split from the apes (which presently is estimated to have occurred about 8.5 Ma).

Using this definition of the hominin tribe, the earliest known member in the fossil record is *Sahelanthropus tchadensis*. *Sahelanthropus* was a biped who walked West Central Africa in the Miocene Epoch, around 7 Ma. The species name, *tchadensis*, stems from the country Chad. It should be mentioned in passing that although *Sahelanthropus* is presently considered the oldest confirmed fossil hominid, it has been argued that *Graecopithecus* was an extinct hominid that lived in southeast Europe during the late Miocene around 7.2 Ma, making it the first hominin. If true, it suggests that the human lineage arose in southern Europe rather than Africa, as traditionally presumed. This contested viewpoint is sometimes colloquially called the 'North Side Story'. (A long-standing maxim in physical anthropology is that the hominins originated in Africa between 4 and 7 million years ago and remained in Africa up until about 2 million years ago. Therefore, this 'Northside Story' is quite controversial and is hotly contested.)

Less controversial (although still not universally accepted) is the inclusion of *Orrorin tugenensis* in the homininae. *Orrorin* lived in the Late Miocene/Early Pliocene Epochs about 6.1 to 5.7 Ma in what is now the Tugen Hills of Kenya (thus its species name). Another early member of the hominin family was *Ardipithecus*. There are two known species of *Ardipithecus*—*Ardipithecus ramidus* and *Ardipithecus kadabba*. Specimens of *Ardipithecus* have been found in Ethiopia dating to between 5.8 and 4.4 million years ago; *Ardipithecus* ramidus was the species that dates most recently, at the 4.4 Ma mark. Ardipithecus was about the size of a chimpanzee and had a cranial capacity of about 300–350 c.c., which is like modern female chimpanzees and bonobos. (Adult male chimpanzee cranial capacity averages nearly 400 c.c.) Ardipithecus had relatively small and unimposing canines. Therefore, unlike chimps who show their large, intimidating canine teeth in a threat display or for overt aggressive intent, *Ardipithecus* was perhaps more human-like in its social organization and facial displays.

The inclusion of *Sahelanthropus, Orrorin* and *Ardipithecus* in the hominin family is largely based on a reduction in canine size and absence of the **C/P3 honing complex**. This is a dental arrangement that hones the upper (maxillary) canines (C) against the lower (mandibular) anterior premolars (P3) to keep them sharp. Although modern chimps and gorillas do possess such dental features, modern humans have obviously lost this. It is hypothesized that the loss of pronounced (and dangerous) canines is why displaying the teeth now has opposite meanings in chimps versus humans. A human smile (which generally indicates a friendly attitude) shows our small canines; in chimps, a toothy display indicates the potential (and perhaps plan) to attack and bite with bad intentions! There are other anatomical adaptations in the 'postcranial skeleton' (especially in the pelvis and lower limbs) of *Sahelanthropus, Orrorin* and *Ardipithecus* allowing for better bipedal locomotion. The pelvis of modern humans is characteristic of **constitutive bipedalism**, meaning that humans are ill-equipped to walk on all fours; we essentially are anatomically committed to bipedal locomotion. On the other hand, these early hominids seemed suited for **facultative bipedalism**, meaning that they could be comfortable on all fours but stand erect if they so desired. When on all fours, their spines would be roughly parallel to the ground. This contrasts with modern apes in which the back is slanted upwards because their arms are longer than their legs or humans in which the back would be slanted downward because the arms are much shorter than the legs. Also, the location of the **foramen magnum**, the opening in the base of the skull through which the spinal cord exists, is more anterior located in these early hominids compared to chimpanzees.

Following *Ardipithecus*, numerous **australopithecines** appeared on the African continent from between 4.4 to 1.4 Ma (during the Pliocene and Pleistocene Epochs). While inclusion of *Sahelanthropus, Orrorin* and *Ardipithecus* as genuine hominins is debated, there is less controversy about the inclusion of the australopithecines as hominins. Obligate bipedality is a defining feature of the hominins and this appears to be the case for both genera of hominins: *Australopithecus* and *Homo*. (Another possible hominin genus is *Kenyanthropus*.)

Among the most famous australopithecine fossils is 'Lucy', a remarkably complete and well-preserved specimen of *Australopithecus afarensis* from the Afar region of Ethiopia dating back 3.2 million years. There were many species of australopithecines during this span of time. A short list in general chronological order includes:

1. *Australopithecus anamensis*
2. *Au. afarensis*
3. *Australopithecus aethiopicus*
4. *Australopithecus bahrelghazali*

5. *Australopithecus africanus*
6. *Australopithecus gahri*
7. *Australopithecus sediba*.

For the most part, these australopithecines were small and had a constellation of human-like and ape-like anatomical features. They were bipedal and had small canines like humans but had relatively small brains (compared to humans). Incidentally, the small brain was a surprise to early anthropologists, many of whom predicted that the evolution of the human family tree would have started with the development of a larger brain, with the other anatomic features evolving subsequently. The australopithecine fossils have overthrown this hypothesis.

Not included in the above list is a recently discovered specimen attributed to another species—*Australopithecus deyiremeda*, which was found in the Afar region of Ethiopia and dates back to 3.5–3.3 Ma during the Pliocene. Since the only specimens are three partial jawbones, it is presently debated as to whether these constitute another species or if they might belong to the more well-established species, *Australopithecus afarensis* to which Lucy belongs. The earliest known *Australopithecus* species was *Au. anamensis*, found in northern Kenya near Lake Turkana; the specimens date back to between 4.2 and 3.9 Ma. The latest (most recent) known (gracile) *Australopithecus* fossil is an *Au. sediba* specimen that dates to around 1.8–2.0 Ma.

It was mentioned above that 'australopithecines' lived from 4.4 to 1.4 million years ago, but the latest specimen of *Australopithecus* was *Au. sediba*, only 1.8 Ma years ago. This is because there were two anatomically distinct groups of australopithecines roaming Africa over this timespan—the so-called 'gracile' species and the 'robust' species. Thus far, we have limited the discussion to the gracile species. But it was the robust group that survived until as recently as 1.4 ma.

The robust species are often given a separate genus—*Paranthropus*. But adding to the already confusing nomenclature, recently *Paranthropus* is again often being included in genus *Australopithecus*. For this text, we shall use the genus *Paranthropus* for the robust genera of australopithecines. The first of these powerfully jawed, robust australopithecines was *Paranthropus aethiopicus*, which lived from 2.7–2.3 Ma. More recent robust australopithecines include *Paranthropus robustus* and *Paranthropus boisei*; this last species left fossils as recently as 1.4 million years ago. These hominins had very large molars, a prominent sagittal crest (for attachment of large temporalis muscles) and strong masseter muscles, all of which indicate adaptations for chewing and grinding tough plant material. Despite their large jaws, their bodies were not very large—maybe 40 kg (88 lb) for *P. robustus* males and 50 kg (110 lb) for males of the larger species, *P. boisei*. The females of both were considerably lighter. In height, they were probably only around 4 feet tall—about the same height as the gracile species.

Another hominin member was *Kenyanthropus platyops*, which was found near Lake Turkana in Kenya and dates to around 3.2 Ma in the middle Pliocene Epoch. This putative genus is surrounded by controversy. On the one hand, it remains uncertain as to whether *Kenyanthropus* should be a truly separate genus from

Australopithecus or on the other extreme, whether the much later *Homo rudolfensis* might just be another species of *Kenyanthropus*.

28.30 Overlapping hominins and Gause's exclusion principle

Over several million years, a variety of hominins evolved and went extinct. On occasion, more than one hominin species inhabited the same environment simultaneously. This seems to be paradox since it appears to violate **Gause's exclusion principle**. This principle (also called **the ecological competitive exclusion principle** or **Gause's law**) states that no two species can occupy the same ecological niches at the same time. When resources are limited (and this is virtually always the case in natural situations) they will compete, and one species will be eliminated.

For the apparently overlapping hominins, the paradox is resolved by the observation that despite their superficial anatomical similarities, these similar hominins probably occupied different ecological niches in terms of their diets. For example, the **robust australopithecines** had very large, specialized molars and powerful jaws that were capable of cracking nuts (hence the nickname, '**nutcracker man**' for *P. boisei*). Isotopic analysis of the carbon ratios in the discovered bones and teeth is consistent with this proposed dietary difference.

Technically, the carbon-13:carbon-12 ratio or $\delta^{13}C$ is consistent with a robust australopithecine diet rich in so-called **C4 plants**—those plants exploiting the **C4 pathway of carbon fixation** in photosynthesis, also called the **Hatch–Slack pathway**. Most (~95%) modern plants are **C3 plants**, which solely use the **Calvin cycle** for production of sugars in the so-called dark phases of photosynthesis and do not have the Hatch–Slack pathway. The name C3 alludes to the fixation of carbon dioxide into a three-carbon sugar compound (3-phosphoglycerate) by the world's most abundant enzyme '**rubisco**' or **ribulose bisphosphate carboxylase** as the first step in the Calvin cycle. While this is obviously a successful pathway, it is inefficient at high oxygen concentrations because of a competing chemical reaction in which molecular oxygen rather than carbon dioxide is used by rubisco. This alternative reaction is called **photorespiration**. Thus, in conditions of high oxygen concentrations (which might happen when there is too much light), photorespiration can reduce the overall efficiency of photosynthesis in C3 plants. Additionally, C3 plants can lose a great deal of water through **transpiration** through their open stomates or stomata (leaf pores), especially in warm and bright conditions. Hence, C3 plants do not like too much light and heat.

In contrast, C4 plants conserve water by more tightly regulating the opening of their stomata and compartmentalizing the fixation of carbon dioxide into sugars into two distinct cell types that are anatomically separated. Their **mesophyll cells** take up atmospheric carbon dioxide and fix it into a four-carbon compound oxaloacetate rather than the three-carbon compound 3-phosphglycerate. This four-carbon compound is then converted into another four-carbon compound, malate, which is transported into the **bundle sheath cells**. There, the malate is decarboxylated (into pyruvate) and the carbon dioxide takes part in the conventional Calvin cycle for sugar synthesis. In this fashion, C4 plants are better adapted to drier and intensely sunny climates.

Modern primates—with the notable exception of humans—have a diet that consists mostly of C3 plants and the insects that eat these plants. In contrast to today's non-human primates, modern humans consume a good deal of farmed C4 plants (e.g. in the form of corn and rice) and we eat the meat of animals that largely graze on C4 plants such as grasses. These dietary habits are reflected in the carbon isotope ratios.

Biochemical processes prefer the lighter isotope when given a choice. Thus, C3 plants have less carbon-13 in their cells than what is naturally found in their environments. C4 plants also have less carbon-13 in their cells than in their environments; however this biological fractionation is not as pronounced. The net result is that C4 plants tend to have more carbon-13 in their cells than C3 plants do. Thus, the isotopic signatures of C3 and C4 plants differ. C3 plants, with only the Calvin cycle, have low $\delta^{13}C$ values (-24 to $-34‰$), while C4 plants, with the Hatch-Slack pathway, have relatively higher $\delta^{13}C$ values (-6 to $-19‰$). Incidentally, the so-called **crassulacean acid metabolism (CAM) plants** are also adapted to dry, sunny environments but use another variation on the theme of C4 plants. CAM plants keep their stomata completely closed during bright, hot daylight and only open them up for carbon dioxide intake (and oxygen release) at night. By keeping their stomata closed during the day, they minimize water loss through evaporation from the stomata (i.e. through transpiration). In any case, CAM plants like C4 plants also have relatively higher $\delta^{13}C$ values than C3 plants. Examples of CAM plants include cacti, jade plants and pineapples.

By analyzing the carbon isotopes in the collagen of teeth and bones of fossils through mass spectroscopy, one can determine the diet of extinct animals, including ancient hominins. Although not perfectly clear-cut, it does appear that thanks to their tooth and jaw adaptations, the robust australopithecines were able to readily switch their diets to tough grassy material when needed. Overall, the data indicate that *P. boisei* was probably eating 75% grass and grass products, which was more than any other hominin. Thus, although several different hominin species lived in the same place at the same time, they did not directly compete for food resources and there was no real violation of Gause's exclusion principle.

28.31 Genus *Homo*—the dawn of man

Among the early species belonging to genus *Homo* was **Homo habilis**. Early specimens were found at Olduvai Gorge in Tanzania and specimens have been found throughout Africa from 2.4 to 1.6 million years old. The oldest stone tools date back to about 2.6 million years ago and may have been made by *H. habilis*. Some of these stone tools (which were made by knocking chips or flakes off larger stones by striking them with harder rocks) were found coincidentally with *H. habilis*, thus the name, which means 'handy man'. In archeology, these earliest stone age tools were examples from the Lower Paleolithic or Old Stone Age. Specifically, these tools made near Olduvai Gorge are called the **Oldowan Industry**. Such tools are called **secondary tools** since one tool is used to create another, more sophisticated one. It has been proposed that secondary tool manufacturing is a defining behavioral feature of the genus *Homo*. In any case, these tools appear to be useful for butchering meat. Thus, Homo habilis probably consumed more meat than the contemporaneous, nut-eating robust australopithecines.

Homo habilis specimens had cranial capacities ranging from 500–900 c.c. This is considerably larger than the robust australopithecines (*P. robustus* and *P. boisei*) which had cranial capacities ranging from 410–530 c.c. This early species of genus *Homo* had a brain size averaging about 25% larger than that of any australopithecine species. Furthermore, the braincase suggests expansion in a region of the frontal lobe of the cerebrum known as **Broca's area**. Broca's area (**Brodmann areas** 44 and 45), is an important language center that is essential for verbal expression in modern man, ***Homo sapiens***. In addition to the larger and potentially restructured brain, *H. habilis* had a more vertical face and smaller teeth. These anatomical changes may all be intimately correlated and linked to the inactivation of a gene called *MYH*16 (myosin heavy chain gene 16) through a frameshift nonsense mutation that inactivated this gene in the **temporalis** muscle. This was a huge and powerful muscle in the robust australopithecines and still is in gorillas. But the loss of this gene in the temporalis shrank the muscle in humans and allowed greater encephalization (brain growth) in early *Homo*.

It has been proposed that *H. habilis* led to ***H. rudolfensis***, but the ~1.8 million year old fossil specimens of the latter are restricted to a single partial skull and some femur fragments. The skull specimen, called KNM-ER 1470, had a cranial capacity of about 750 c.c. This is consistent with the brain size of *H. habilis*. However, *H. rudolfensis* is generally distinguished from *H. habilis* by a larger overall body size. For this reason, it is also argued that this 'species' is not really a species of *Homo* at all but actually just represents larger male specimens of *H. habilis* (assuming that this species exhibited substantial sexual dimorphism). Another alternative perspective is that *H. rudolfensis* is not a true member of genus *Homo* but rather should be placed under *Australopithecus* or *Kenyanthropus*. The Uraha jawbone (UR 501) is another specimen tentatively assigned to *H. rudolfensis*. It is from the Uraha region in Malawi and dates to as far back as 2.4 Ma. If this is verified, it would make *H. rudolfensis* the earliest yet identified species of *Homo*. The youngest (i.e. most recent) potential *H. rudolfensis* specimen (KNM-ER 819) dates to 1.65–1.55 Ma.

28.32 *Homo erectus* and *Homo heidelbergensis*

The oldest known fossil specimen of ***Homo erectus*** is a 2.04 million year old partial skullcap (called DNH 134) found at the Drimolen Main Quarry in South Africa. *Homo erectus* ('upright man') had astonishing longevity as a species, persisting until as recently as 108 000 years ago in the form of a subspecies population in Java known as *Homo erectus soloensis*.

Homo erectus was well adapted to walking and was tall compared to the australopithecines and earlier species of Homo. Perhaps thanks to their upright posture and walking abilities, *Homo erectus* was the first human species to spread throughout Eurasia with a range spanning from the Iberian Peninsula to Java. It is presently speculated that through **insular dwarfism**, isolated populations of *Homo* *erectus* may have been ancestral to a dwarf species of humans known as ***Homo floresiensis*** (which was only about 3 feet 7 inches or 1.1 m tall) and left fossilized bones dating from 100 000 to about 50 000 years ago. Previous claims of this species persisting until as recently as 12 000 to 18 000 years ago have been disproven and attributed to stratigraphic anomalies. *Homo erectus* may have also led to another putative small species, ***Homo luzonensis*** (from the Late Pleistocene Philippines) as well.

Homo ergaster may have been a separate species or might have been a subspecies of *Homo erectus*. A remarkably well-preserved skeleton from the Lake Tukana region of Kenya is known as WT-15000, also called **Turkana boy** or **Nariokotome boy**. The fossil is about 1.6 million years old and belonged to a boy around 10 or 11 years old. As he was already 5 feet 3 inches tall (1.6 m), he probably would have reached about 5 feet 11 inches as an adult. The greater height and stride length would have enabled *Homo erectus* (or *Homo ergaster* if he ultimately proves to be assigned to that species) to walk or run further distances for hunting and foraging before becoming dehydrated.

Early Pleistocene fossils and stone tools recovered at Dmanisi, Georgia range in age from 1.85–1.77 million years old may have belonged to yet another species, ***Homo georgicus***. These 'Dmanisi people' might alternatively have been a subspecies of *Homo erectus* (*H. erectus georgicus*) or of *Homo ergaster* (*H. erectus ergaster georgicus*). Their relatively short stature and small brain size (545–775 c.c., averaging 631 c.c.) contrasts with later species of *Homo*, including most specimens of *Homo erectus* (which averaged around 900 c.c.). One Dmanisi specimen (Dmanisi skull 4) was, save one single remaining tooth, completely edentulous. The fact that this person survived and thrived (based on bone growth in the tooth sockets, which would not have been evident if the teeth were lost shortly before death) suggests a social structure in which the old were cared for by others in the group.

Several species of humans probably descended from *Homo erectus*, such as *Homo heidelbergensis* and *Homo antecessor*. It is presently believed that subsequently, the Neanderthals, Denisovans and modern humans may have evolved from *H. heidelbergensis*.

Homo heidelbergensis was perhaps a descendant of *Homo erectus* and lived from 700 000 to 200 000 years ago, primarily in Europe. It appears that this species might have also ventured into Asia and Africa. *H. heidelbergensis* was probably the first human species to inhabit colder climates. In accordance with **Allen's rule** (that animals residing in colder climates tend to have shorter appendages and stouter bodies than those living in warmer climates), their short, stocky bodies were likely an adaptation for conserving heat. They had big brains, with an average cranial capacity of around 1200 c.c.

It is possible that *H. heidelbergensis* dates back as far as 1.3 million years ago, but the very early fossils from Spain (Atapuerca) that have been attributed to *H. heidelbergensis* might better be grouped under a separate species, ***Homo antecessor***. Similarly, other early fossils and archeological evidence in the range of 800 000 to 1 million years ago from Ceprano, Italy might also belong to a different species, ***Homo cepranesis***. Hence, *Homo antecessor* was possibly another species of early humans living from 1.2 to 0.8 Ma (the Early Pleistocene). Initially, it was felt that this species led to *Homo heidelbergensis* and was the last common ancestor (LCA) of modern humans and Neanderthals but nowadays the view of many is that *Homo antecessor* was an offshoot of *Homo erectus* (or 'African Homo erectus', also known as *Homo ergaster*) that led to an evolutionary dead end, and was not an ancestor of *Homo heidelbergensis*, **Homo neanderthalensis** or modern man. Presently, the LCA of modern man and Neanderthals is believed to be *Homo heidelbergensis.*

Analysis of Neanderthal and modern human DNA indicates that the two lineages diverged from a common ancestor (probably *Homo heidelbergensis*) at some point between 350 000 and 400 000 years ago. The European branch led to the Neanderthals while the African branch led to modern humans. Along the base of the African branch that ultimately evolved into modern humans was an archaic group of humans sometimes referred to as **Homo rhodesiensis**. According to this interpretation, 'Rhodesian Man' was a species or perhaps a subspecies (known as **Homo sapiens arcaicus** or **Homo sapiens rhodesiensis**) that was the immediate precursor to modern man, **Homo sapiens sapiens**. But not all authorities concur with this perspective. But it is presently believed that *Homo sapiens* evolved in Africa from *H. heidelbergensis* directly or maybe indirectly through *H. rhodesiensis* during a time of drastic climate change around 300 000 years ago.

28.33 The evolution of human vocalization and language

Various milestones can be used to reconstruct human history. As mentioned above, the earliest stone tools date back to about 2.6 million years ago. The first documented evidence of human-made and controlled fire was around 1.5 million years ago. But the question of when spoken language first arose is more difficult since the advent of speech did not leave tangible traces. Nevertheless, we can infer a timeline based on anatomical changes. The larynx and hyoid bones of *Australopithecus* species were similar to those of chimpanzees, indicating that, while they could resonate and make hooting sounds, they were probably incapable of making the nuanced sounds of human language. But the hyoid bones of early *Homo* species showed anatomical advances, and the middle ear bones (the ossicles—malleus, incus and stapes) grew more refined, suggesting a synchronized improvement in both vocalization and auditory language comprehension. The middle ear ossicles of modern humans seem more fine-tuned for discrimination of middle-frequency sounds (as in discerning the shades of human speech) in comparison to the middle ear anatomy of chimps, which is better suited for hearing high-pitched shrieks.

28.34 FOXP2 and the Neanderthals

Another major advancement in language involved the **FOXP2 gene**. (The gene name *FOX* comes from 'forkhead box' in Drosophila fruit fly genetics; this and similar facts often do not help students remember the name nor are they very instructive in human genetics, thus the etymology of other oddly-named genes will often be omitted in this text.) Two mutations in *FOXP2* may have been critical in the development of language skills in modern humans. Specifically, the mutations are a threonine to asparagine substitution at amino acid position 303 (that is, a T303N point mutation) and an asparagine to serine substitution at position 325 (N325S). *FOXP2* is a **transcription factor** (that is, a protein that binds to DNA and stimulates gene expression). *FOXP2* is expressed in various organs including the heart, lungs and digestive system, but it is in the brain where *FOXP2* exerts its influence on language abilities. While these particular *FOXP2* mutations are not found in other mammals or primates, they are found in our Neanderthal cousins.

Whether both Neanderthals and modern humans independently acquired these advantageous mutations or inherited them from a common ancestor is unsettled. Deleterious mutations in this gene in modern people cause a severe language disability called **developmental verbal dyspraxia**. The condition can be passed on in an autosomal dominant pattern, although most cases are due to de novo mutations.

The Neanderthals (*Homo neanderthalensis*) primarily lived in Europe, from around 150 000 to 39 000 years ago. (Incidentally, the 'h' in Neanderthal is often silent and sometime the name is spelled without the 'h' at all—Neandertal.) Neanderthals had thicker, denser and slightly curved bones compared to ours. This may indicate adaptations to physical stresses during locomotion or other physical exertion. The bone differences were probably largely genetic but also might have been partly due to **Wolff's law**. Wolff's law states that a bone or bones will adapt to the stresses placed on them. A load placed on a particular bone will induce remodeling, and that bone will become stronger (more capable of resisting the load). The initial adaptive changes are seen in the architecture of the internal bone trabeculae. Following this, thickening of the external (cortical) portion of the bone may occur. Classic examples are seen in the bones of athletes, especially those involved in sports predominantly using one arm such as in pitching or tennis. Wolff's law works in reverse when bone stresses are diminished, as in the bone demineralization seen in bedridden patients and astronauts exposed to prolonged microgravity.

28.35 Our Neanderthal ancestry?

Although initial studies using only mitochondrial DNA (mtDNA, which is passed along maternally only) did not reveal any trace of Neanderthal ancestry, more comprehensive studies of nuclear DNA have confirmed the presence of Neanderthal genes in many modern people. This finding established the fact that along the way, when modern humans (*Homo sapiens*) encountered Neanderthals, there were occasional productive matings. While indigenous African people carry no (or nearly no) Neanderthal genes, non-Africans have on average, 2%–3% of their genomes derived from Neanderthal ancestry; and East Asians have around 20% more Neanderthal DNA than Europeans. But a fascinating fact about this is that it is not the same 2%–3% in all people—the specific Neanderthal DNA from one non-African person to another may be quite different. If all the Neanderthal DNA scattered about in the genomes of present-day people were gathered, it might amount to nearly 90% of the Neanderthal genome.

Neanderthals in Asia probably interbred with early modern humans somewhere between 50–60 000 years ago. This is more recently than the time of their overlap in Europe. Amorous encounters between modern humans and Neanderthals that left genetic traces in their descendants possibly occurred as far back as between 316 000 and 219 000 years ago, but more likely occurred around 100 000 years ago. Other possible encounters leaving traces in the genomes of living people occurred again 65 000 years ago. But in addition to these interbreeding episodes, the two human populations potentially encountered one another regularly for up to 20 000 years, starting in Asia about 60 000 years ago and ending about 39 000 years ago in Europe when the last Neanderthals went extinct. There is evidence of a climate shift that led to colder, drier conditions across Europe, which might have decreased food supplies for the Neanderthals. Regardless, by 39 000 years ago the (pure) Neanderthals became extinct.

The Neanderthal DNA segments found in the modern human genome are called **archaic single nucleotide polymorphisms (aSNPs)**. These aSNPs are DNA variations that are found in non-African populations (and in other archaic human species such as the Denisovans) but are not present in the African Yoruba population (which serves as a standard). The phenomenon wherein some Neanderthal DNA remains in our modern human genome thanks to past encounters is called archaic **DNA introgression**. Some regions of the modern human genome are largely devoid of Neanderthal genes ('Neanderthal deserts'), suggesting a selective pressure against retention of deleterious DNA derived from our Neanderthal ancestors.

The persistent Neanderthal DNA in non-Africans shows some interesting quirks including the absence of a Neanderthal-derived Y chromosome (which is only derived from males—patrilineal) and an absence of mtDNA which is strictly maternally inherited (matrilineal). Additionally, there is an underrepresentation of Neanderthal X chromosome DNA. Such observations might indicate diminished fertility or occasional sterility of some hybrids.

The role this Neanderthal legacy DNA plays in human health is presently under study in **genome-wide association studies.** Some studies have suggested a role in the development of health disorders such as type 2 diabetes, certain autoimmune diseases and cancer, among several others. Neanderthal gene variants associated with Graves' disease and rheumatoid arthritis appear to make carriers more susceptible to these autoimmune diseases. Different aSNPs derived from Neanderthals might increase susceptibility to addictions, allergies and severe depression. Another recent implication of Neanderthal DNA has been an increased susceptibility to serious respiratory complications in severe cases of COVID-19. Not all of these passed-along Neanderthal genes are harmful. Specifically, Neanderthal gene variants linked to prostate cancer on chromosome 2 (*rs12621278*) may be protective against prostate cancer.

Conversely, Neanderthals were apparently susceptible to some of the diseases we modern humans suffer today. For example, one Neanderthal rib from around 120 000 years ago had a type of benign tumor known as **fibrous dysplasia**. Since the Neanderthal skeleton was incomplete, it is impossible to say whether this was truly a case of monostotic fibrous dysplasia (meaning it was in one and only one bone) or if it was part of a polyostotic case. Polyostotic fibrous dysplasia may be seen in certain syndromes such as the sporadic McCune–Albright syndrome. Fibrous dysplasia is often asymptomatic and remains benign, but malignant transformation to sarcoma is possible. Such cancerous transformation has been reported in patients with a history of radiation therapy to that bone. It is worth recalling that the doses used in radiation therapy, albeit highly localized, are typically dozens of times higher than the radiation doses that astronauts would receive during a trip to Mars.

Neanderthal DNA is not the only non-*Homo sapiens* DNA found in modern humans. Many people possess some **Denisovan** DNA as well. The proportion of Denisovan DNA is highest in the Melanesian population (4%–6%), slightly lower in other Southeast Asian and Pacific Islander populations and very low or absent elsewhere. Whether any of the residual Neanderthal or Denisovan aSNPs make individuals more resistant or more susceptible to the effects of radiation that they might encounter as astronauts remains unknown. But ongoing debate focuses on how to select astronauts most appropriately for space missions that might expose them to higher amounts of radiation. Whether radiation susceptibility and resistance are immutable genetic features or whether radiation resistance might be induced (via the **Yonezawa effect** for example) is uncertain. The Yonezawa effect is seen when a small 'priming dose' of radiation causes cells to become better able to cope with subsequent doses of radiation, including high, potentially damaging doses. This appears to occur through the induction of more efficient repair of direct DNA damage and a reduction in mutation frequency after irradiation. The Yonezawa effect is a classic example of the radiobiological 'adaptive response' phenomenon.

28.36 The Denisovans

The Denisovans were yet another group of archaic humans, however they have not yet been officially assigned a separate species. This is partly due to the incomplete, fragmentary nature of the fossil specimens, which include only a few molars and bones/bone fragments. Nevertheless, some of these teeth and bones have yielded adequate DNA for in-depth analysis. In fact, the Denisovans are the first and only formal 'species' of archaic hominins to be revealed by DNA analysis alone (as opposed to classification based on fossils). One mandible was dated (through uranium series daughter analysis of rocky material attached to the mandible) to over 160 000 years ago. This specimen was found at an altitude of 3280 m or around 10 761 feet. This finding could be meaningful since modern Tibetans have more Denisovan DNA than their neighboring Han Chinese populations and they appear to be naturally well adapted to living at higher altitude than most people. Ethnic Sherpas have inherited a Denisovan gene variant in the *EPAS1* gene that allows them to thrive at higher altitudes than most other folks. DNA analysis indicates that the genomes of Tibetans, Melanesians and Australian Aboriginal peoples carry about 3%–5% Denisovan DNA. Genetic analysis of people living in Indonesia and Papua New Guinea suggests that *Homo sapiens* may have interbred with Denisovans as recently as 30 000 years ago, and possibly as recently as 15 000 years ago. The available, but limited, fossil evidence suggests that the Denisovans lived from around 300 000 to around 30 000 years ago.

If one examines these timelines (and if they are truly accurate), it appears that *Homo sapiens* overlapped chronologically with several other species of *Homo*, including *Homo heidelbergensis, Homo neanderthalensis, Homo erectus, Homo floresiensis* and the Denisovans. But unlike the prolonged peaceful coexistence seen (or presumed) between *H. habilis* and the robust australopithecines, Gause's exclusion principle seems to have eventually held up; today there is only one species of **Homo** remaining. Whether these other species of humans were simply out-competed, actively exterminated, subsumed into our own species by interbreeding or just went extinct without our help at all may never be fully known.

28.37 Cranial capacities of hominins

An average of five skulls of *Au. afarensis* (which lived from 3.9–2.9 Ma) showed that the cranial capacity of this species was about 445 c.c., which exceeds the estimated brain volume of Ardipithecus (∼350 c.c.) or chimpanzees (∼360–400 c.c.). *Au. africanus*, living from about 3.3–2.1 Ma, was a bit more recent than *Au. afarensis*, had a cranial capacity of between 420–510 c.c., averaging about 448 c.c. Some robust australopithecine (*A. robustus*) skull specimens indicate a cranial capacity of about 523 c.c., which exceeds the brain size of the gracile species. But it was with the advent of genus *Homo* that brain size began to noticeably increase.

Although brain size clearly correlates with intelligence, it is also well known that birds such as corvids (jays, crows and ravens) and grey parrots are quite intelligent even though they have small brains. As birds are descendants of (or surviving members of) the dinosaurs, it makes one wonder about dinosaur intellectual capacity. Curiously, brain size (and presumably intelligence) has increased not only in primates but also in other mammalian groups over the Cenozoic. Dolphins, for example, are very intelligent animals and examination of fossil specimens has shown that the braincase volume of extinct dolphin species expanded in two evolutionary phases, with the first phase occurring around 39 million years ago and a second burst around 15 million years ago. Modern dolphins have very large brains that also display a high degree of cerebral cortical convolution (meaning that the surfaces of their brains are not smooth but folded in on itself repeatedly to maximize surface area). From a gross anatomical perspective, modern dolphin brains exceed human brains in terms of both size and external complexity.

The largest brains are those of sperm whales, weighing in at about 8 kg (18 lb). Rivaling sperm whales are elephants with brain weights just over 5 kg (11 lb). The bottlenose dolphin is another contender with a brain mass of 1.5 to 1.7 kg (3.3–3.7 lb), In comparison, the human brain averages around 1.3 kg or 1300 g (2.9 lb). Of course, there is an obvious trend for brain sizes to vary in proportion to overall body size. The brain to body mass relationship is not simply linear, however. A commonly used metric is the **encephalization quotient**, which is the ratio of an animal's brain mass compared to *what would be expected* from its body mass. One equation commonly used for such estimates is **Jerison's formula**:

$$E = CS^{2/3}$$

Where E is the encephalization quotient, S is body mass and C is the cephalization factor (which fits best to 0.12).

Biology students might note that this is simply an adaptation of **Snell's equation of allometry** ($M_i = CM_b^{2/3}$ where M_i is the mass of the organ in question, M_b is the body mass and C is a scaling constant). The encephalization quotient for modern humans is between 7.4 and 7.8, which is the highest among all living mammals. Thus, although bottlenose dolphins, blue whales, African elephants and sperm whales have larger brains than man in absolute terms, modern humans have far higher encephalization quotients. The bottlenose dolphin is the nearest rival to humans in encephalization quotients, at about 5.3. It is noteworthy that the Neanderthals, who became extinct about 39 000 years ago, had larger brains than modern *Homo sapiens*. However, it is believed that Neanderthals probably weighed more than modern *Homo sapiens*, and therefore their value of E may have been equivalent to or slightly lower than that of modern *Homo sapiens*.

28.38 A close call for extinction of *Homo sapiens*?

Roughly around 150 000 years ago, during the **Marine Isotope Stage 6 (MIS-6) period** (which spanned between 195 000 and 123 000 years ago) there was a very significant genetic bottleneck in the *Homo sapiens* population. The cause of the population crisis was climatological. MIS-6 had unusual effects on the African continent, where *Homo sapiens* was geographically restricted to at the time. Much of the continent had become desert-like and was basically uninhabitable for early humans.

The **marine isotope stages** are determined by measuring the ratios of oxygen isotopes in calcite ($CaCO_3$) specimens from the sea floor. The oxygen-18:oxygen-16 ratio reflects the relative availability of these two oxygen isotopes for the formation of calcite in seawater. That, in turn is proxy for the climate. Cold climates lead to a great deal of snowfall and glacier formation over polar regions. Since snow has a relative abundance of the lighter oxygen-16 isotope (because the lighter H_2O molecules evaporate more readily than the heavier variety of water molecules), this enriches the oceans in oxygen-18–containing water. Thus, during glacial periods, the calcite formed by marine organisms will have slightly more oxygen-18 in it. This palaeoclimatological prediction was first proposed by Harold Urey, the discoverer of deuterium.

According to one hypothesis, a very small group of perhaps under 1000 people living on coastal regions of southern Africa made it through the crisis by surviving largely on seafood, mammals and available vegetable matter (notably the plentiful endemic fynbos and renosterveld plants). One such site was a cave called **Pinnacle Point 13B** near Mossel Bay, South Africa. Importantly for these hardy survivalists, this region of present-day South Africa was an Eden when it came to the available edible plants. This area harbored the world's greatest diversity of edible, digestible, calorie-rich and nutritious tubers, bulbs and corms (i.e. geophytes).

Since sea levels fall during glacial periods, the ancient refuges for these hardy humans are probably underwater today and thus erased or invisible to present-day archaeologists. But essentially everyone alive today is a descendent of the population who survived this genetic bottleneck. And if it is true, *Homo sapiens* was indeed an endangered species during this timeframe.

28.39 The Toba eruption and near extinction of humanity

But the MIS-6 catastrophe was not the only time *Homo sapiens* was on the endangered species list. Another severe genetic bottleneck in the human population may have occurred about 74 000 years ago. This time, the alleged proximate cause was the eruption of a supervolcano on Sumatra—the **youngest Toba eruption**. As a **supervolcano**, this eruption was a **category 8 explosion** on the **volcanic explosivity index (VEI)**. The VEI ranges from 0 (with erupted volumes under 1000 m^3) to VEI 8 (**supervolcanoes** with erupted volumes over 1000 km^3). The VEI scale is logarithmic except for the jump from VEI 1 to VEI 2, which is an increase by a factor of 100, rather than the increase by a factor of ten for all other steps.

In general, volcanic eruptions may be categorized as **explosive** or **effusive**, with effusive eruptions representing a simple outpouring of lava without any significant explosive component. Among explosive volcanoes however, the explosivity can be quantified by the VEI and/or described qualitatively by descriptors such as:

1. Hawaiian
2. Strombolian
3. Vulcanian
4. Pelean
5. Plinian
6. Ultra-Plinian
7. Supervolcanic.

These are listed in increasing volume of ejecta and decreasing frequency of eruptions. For instance, VEI 0 Hawaiian eruptions eject less than 1000 m^3 of material but may occur daily; VEI 6 eruptions such as Krakatoa and Mount Pinatubo eject >10 km^3 but erupt once every 50–100 years on average; and VEI 7 ultra-Plinean eruptions such as Tambora blast out over 100 km^3 of material but erupt only once every 500–1000 years or so. And supervolcanoes such as Toba and Yellowstone eject greater than 1000 km^3 of material but explode only once every 50 000 years or so. The Toba super-eruption blasted out between 2800 and 5300 km^3 rock equivalent volume of magma, generating an eruption column and cloud that reached an altitude of 30 to 40 km into the atmosphere. This was the largest known volcanic eruption in the last 2 million years. For the sake of comparison, the infamous eruption of Krakatoa in 1883 blasted out about 12 km^3 of magma.

At what is now Lake Toba, there was a series of at least four huge volcanic eruptions over the last 1.2 million years, with the latest one occurring around 74 000 years ago (thus the name youngest Toba eruption). According to what is called the **Toba catastrophe theory**, this supervolcano eruption caused a severe, planet-wide **volcanic winter** that lasted up to a decade and triggered a millennium-long climate change. According to the theory, the consequences to humanity were devastating. Another drastic genetic bottleneck may have resulted, and the worldwide human population might have fallen to as few as 1000–10 000 people. This implies that all the people in today's population are the descendants of a small group of survivors of this volcano-initiated genetic bottleneck. Recent evidence however casts doubt on the severity of the volcanic winter and its relationship to the genetic bottleneck; this challenges the entire the Toba catastrophe theory. The data supporting a major genetic bottleneck remain, but the putative cause by the Toba supervolcano eruption 74 000 years ago is now disputed.

A molecular comparison of the genomes between any two random people from distant different parts of the world will demonstrate that their DNA is far more similar to each other than that between any pair of chimpanzees or gorillas from different populations. This holds true even if the chimps or gorillas are from regions relatively nearby geographically. The relative lack of diversity in the *Homo sapiens* population as a whole is presumably because of the genetic bottlenecks over humanity's history. Overall, people who are native to the continent of Africa are the most genetically diverse. For instance, people outside Africa lack several genetic variants found only in Africans. And it has been demonstrated that the farther away from Africa that a group of native people lives, the less genetic and morphological diversity that population will possess. In terms of evolutionary genetics, this observation is explained by the idea that modern humans arose in Africa and migrating humans lost much of their genetic diversity in at least one dramatic episode of the bottleneck effect. The net result is that present day non-African people are all descendants of a small population of survivors who made it through a drastic population squeeze and subsequently migrated out of Africa. The precise causes of the human population bottlenecks are still undetermined, especially now that the Toba catastrophe theory is in doubt.

28.40 The Laschamps geomagnetic excursion

Earth's geomagnetic field is not always dipolar, and the magnetic north and south poles wander from time to time. Additionally, the field strength varies substantially over time. Finally, the entire field may occasionally collapse and re-form with a reversed orientation (meaning north becomes south and vice versa). These changes are called geomagnetic reversals if they persist for an arbitrarily defined 'long time' and are called geomagnetic excursions when they are shorter.

During geomagnetic field transitions (reversals and excursions), the morphology of the geomagnetic field grows highly complex and the geomagnetic field strength becomes significantly weaker. Under such circumstances, energetic charged particles from space are not as readily repelled and can more readily reach and damage Earth's atmosphere. More cosmic radiation may reach the biosphere as well.

The Laschamps event was a geomagnetic excursion that occurred between 42 200 and 41 500 years ago. The transition period from the normal field to the reversed field took approximately 250 years. The geomagnetic field then remained reversed for about 440 years.

During the transition, Earth's magnetic field strength decreased dramatically. The nadir may have been as low as 5% of the present field strength. Furthermore, during the 440 year interval when the field was fully reversed, the magnetic field strength was only about 25% of its current strength. This sharp reduction in geomagnetic field strength allowed more cosmic rays to interact with the atmosphere and reach the Earth's surface. And unlike the present arrangement (where the field is stronger at lower latitudes than at the poles), the data indicate that during the middle of the Laschamps excursion, this higher influx of solar and galactic energetic particles was virtually independent of latitude—all regions were susceptible. All this is documented by the greater quantities of the cosmogenic radioisotopes beryllium-10 and carbon-14 formed and deposited at this time. The increased radiation may have also caused a decrease in the stratospheric ozone layer.

A study that used a high-resolution continuous geomagnetic field computer model known as LSMOD.2 showed that the Earth's atmosphere was indeed exposed to more energetic particle radiation during the Laschamps excursion. The model confirmed that the energetic solar and galactic cosmic radiation dose rate and cosmogenic isotope production rate both increased substantially during this time.

The temporary loss of the geomagnetic shield may have affected the climate and possibly contributed to the extinction of the Neanderthals. This is a hotly contested concept currently and there is a lot of active ongoing research.

Further reading

[1] Karam P A and Leslie S A 1999 Calculations of background beta-gamma radiation dose through geologic time *Health Phys.* **77** 662–7
[2] Karam P A, Leslie S A and Anbar A 2001 The effects of changing atmospheric oxygen concentrations and background radiation levels on radiogenic DNA damage rates *Health Phys.* **81** 545–53
[3] Coogan L A and Cullen J 2009 Did natural fission reactors form as a consequence of the emergence of oxygenic photosynthesis during the Archean? *Geol. Soc. Am. Today* **19** 4–10
[4] Ossa F O *et al* 2022 Moderate levels of oxygenation during the late stage of Earth's Great Oxidation Event *Earth Planet. Sci. Lett.* **594** 117716
[5] Dodd M, Papineau D and Grenne T *et al* 2017 Evidence for early life in Earth's oldest hydrothermal vent precipitates *Nature* **543** 60–4
[6] Tschopp E, Mateus O and Benson R B J 2015 A specimen-level phylogenetic analysis and taxonomic revision of Diplodocidae (Dinosauria, Sauropoda) *PeerJ* **3** e857
[7] Southwood R 2003 *The Story of Life* (Oxford: Oxford University Press)
[8] Cabreira S F, Schultz C L, da Silva L R, Lora L H P, Pakulski C and do Rêgo R C B *et al* 2022 Diphyodont tooth replacement of Brasilodon—A Late Triassic eucynodont that challenges the time of origin of mammals *J. Anat.* **00** 1–17
[9] Kraft B *et al* 2022 Oxygen and nitrogen production by an ammonia-oxidizing archaeon *Science* **375** 97–100

[10] Suwa G *et al* 2021 Canine sexual dimorphism in *Ardipithecus ramidus* was nearly human-like *Proc. Natl Acad. Sci. USA* **118** e2116630118

[11] Fuss J, Spassov N, Begun D R and Böhme M 2017 Potential hominin affinities of *Graecopithecus* from the Late Miocene of Europe *PLoS One* **12** e0177127

[12] Benoit J and Thackeray F J 2017 A cladistic analysis of *Graecopithecus South Afr. J. Sci.* **113** 2

[13] Smith E I *et al* 2018 Humans thrived in South Africa through the Toba eruption about 74,000 years ago *Nature* **555** 511–5

[14] Black B A, Lamarque J F, Marsh D R, Schmidt A and Bardeen C G 2021 Global climate disruption and regional climate shelters after the Toba supereruption *Proc. Natl Acad. Sci. USA* **118** e2013046118

[15] Herries A I R *et al* 2020 Contemporaneity of *Australopithecus*, *Paranthropus*, and early *Homo erectus* in South Africa *Science* **368** eaaw7293

[16] Zeberg H and Pääbo S 2020 The major genetic risk factor for severe COVID-19 is inherited from Neanderthals *Nature* **587** 610–2

[17] Dannemann M 2021 The population-specific impact of neandertal introgression on human disease *Genome Biol. Evol.* **13** evaa250

[18] Taravella Oill A M, Buetow K H and Wilson M A 2022 The role of Neanderthal introgression in liver cancer *BMC Med. Genomics* **15** 255

[19] Den Ouden N, Reumer J W F and Van Den Hoek Ostende L W 2012 Did mammoth end up a lilliput? Temporal body size trends in Late Pleistocene Mammoths, Mammuthus primigenius (Blumenbach, 1799) inferred from dental data *Quat. Int.* **255** 53–8

[20] Michael S, Fong L, Mika K, Chigurupati S, Yon L, Mongan N P, Emes R D and Vincent J L 2016 TP53 copy number expansion is associated with the evolution of increased body size and an enhanced DNA damage response in elephants *eLife* **5** e11994

[21] Gao J, Korte M, Panovska S, Rong Z and Wei Y 2022 Effects of the Laschamps excursion on geomagnetic cutoff rigidities *Geochem. Geophys. Geosyst.* **23** e2021GC010261

[22] Fornalski KW, Adamowski Ł and Dobrzyński L *et al* 2022 The radiation adaptive response and priming dose influence: the quantification of the Raper–Yonezawa effect and its three-parameter model for postradiation DNA lesions and mutations *Radiat. Environ. Biophys.* **61** 221–39

IOP Publishing

Space Radiation
Astrophysical origins, radiobiological effects and implications for space travellers
James S Welsh

Chapter 29

Radiation and climate

As mentioned elsewhere, a hypothesis involving space radiation called the **Svensmark hypothesis** holds that increased incoming cosmic rays could contribute to cooling of the climate. This is allegedly because more cosmic rays would create more condensation nuclei in the atmosphere. These condensation nuclei facilitate the formation of clouds. In turn, the increased cloud cover diminishes incoming sunlight, causing a colder climate. Of course, the increased clouds can also lead to increased precipitation, including more snow. If sufficient snow falls and accumulates as glaciers, it will possibly contribute to a positive feedback loop wherein the high-albedo snow (along with the clouds) reflects incoming sunlight, further cooling the planet. In this manner global cooling might materialize. Conversely, a period with a paucity of cosmic rays could be associated with a diminution in cloud cover and a warmer climate. It should be emphasized that the Svensmark hypothesis is presently highly controversial and is actively debated among experts.

Generally speaking, when the magnetic field strength of either the Sun, the Earth or both is diminished, galactic comic ray influx might correspondingly increase. It is known that the Earth sporadically flips its geomagnetic polarity such that the north pole becomes a south pole and vice versa. These geomagnetic reversals occur on average (in recent times) about once every 200 000 years. Hence, according to the Svensmark hypothesis, during the transition periods when the magnetic field collapses while the north and south poles swap places, cosmic ray influx increases, and the climate should be cooler.

Since the last formal geomagnetic reversal was the **Brunhes–Matuyama reversal** of about 781 000 years ago, we are way past due. Formally recognized geomagnetic reversals are, by current convention, those that last at least 2000 years. Shorter-duration pole flips, such as the **Laschamps excursion** of around 41 000 years ago, may last only hundreds of years. By present definitions, such polarity reversals are called **geomagnetic excursions** rather than full reversals.

The transition period during which the polarity actively reverses was once believed to average about 7000 years but recent data indicate that the range of such transitions may be from well under 1000 years (maybe under a century) to as long as 28 000 years. During this transition phase, Earth's neat magnetic dipole collapses into a complicated multipolar arrangement, and the overall field strength decreases dramatically—by some estimates up to 90%. During these transition periods of diminished geomagnetic shielding (which again, may last centuries to many thousands of years), there would be a noticeable increase in cosmic rays striking the atmosphere and reaching the surface. Of course, the atmosphere will still filter out most primary galactic cosmic rays and the secondary cosmic rays (mostly muons). Nevertheless, the increase in incoming cosmic rays will mean more will reach Earth's surface.

It may be a mind-stimulating mental exercise to speculate (within reason) about what the climatological consequences might be if there were increased cosmic ray flux thanks to a geomagnetic reversal, especially if solar output happened to be particularly low. Conceivably (if the Svensmark hypothesis proves correct), there could be a tendency towards global cooling during geomagnetic reversals. This could be corroborated or discounted by careful correlations with the palaeoclimatological record.

As mentioned in another chapter, the radiobiological consequences of such increased radiation during a geomagnetic transition are debatable. It is extremely unlikely that such increased radiation was ever linked to any mass extinction events. However, it remains possible that such transition periods, when high-LET radiation was increased, might be associated with increased rates of biological evolution. Further research is needed to confirm or refute such hypotheses. Similarly, the idea that increased cosmic rays during the transition periods associated with geomagnetic reversals result in a cooler climate is debatable and further research is needed.

29.1 The Little Ice Age

The possibility of a climate and cosmic ray connection is supported (weakly) by one of the alleged mechanisms behind the climatological escapade called the '**Little Ice Age**' (which was between 1300–1850, or 1500–1850, depending on one's definition). Among the proposed causes of the Little Ice Age is **solar forcing** through prolonged **grand solar minima**. Basically, solar forcing implies a change in Earth's climate due to changes in the Sun. But that change is not attributed to a substantial change in gross solar output; it is associated with more subtle changes in the solar wind and the frequency of solar energetic particle events. Presently, the total solar irradiance or solar constant (G_{SC}) at the top of Earth's atmosphere is 1362 W m^{-2} during solar maximum and 1361 W m^{-2} during solar minimum. Averaged over the entire globe and over an entire year, this amounts to an average solar irradiance per square meter on the order of 340 W m^{-2}. But the small difference between insolation at solar maximum and solar minimum (1362 versus 1361) is much less than 1%. This is insufficient to cause a climatological change in and of itself, even if the solar minimum were prolonged for decades. However, the accompanying decrease in deflective solar magnetism may have played a role. Decreased heliomagnetism could have potentially affected Earth's climate by allowing more galactic cosmic rays to enter our Solar System and our atmosphere.

Sunspots (a surrogate for general solar activity including solar wind, solar prominences, solar flares, coronal mass ejections and overall solar magnetic field strength) were noticeably absent during much of the so-called Little Ice Age. Prolonged periods of decreased sunspot counts (and by association, decreased solar wind) are called **grand solar minima**. Three grand solar minima—the **Spörer Minimum** (1450–1540), the **Maunder Minimum** (1645–1715) and the **Dalton Minimum** (1790–1820) occurred during the Little Ice Age. These grand solar minima were documented by a marked diminution in sunspot counts in the post-telescope era (and via proxies in ice cores before the telescope was invented in 1609 by Hans Lippershey). Since solar minima correspond with a decrease in the deflective solar wind and heliomagnetic shielding, it is believed that such grand solar minima were accompanied by increased incoming galactic cosmic rays on Earth. Overall, the idea that the Little Ice Age was truly linked to a change in solar output (and thus increased cosmic ray influx) is a controversial subject.

29.2 Geographical and geological correlations with ice ages

The sea level was about 120 m or 400 feet lower during the last ice age, which ended 11 700 years ago. This allowed more animal migration but also permitted more erosion. Erosion increases during periods of lower sea levels because of the greater difference in height between the mountains (and other high regions on the continents) and the oceans into which water is flowing. Waters run faster and more furiously at these times. Additionally, the colder climates favor the evolution of stouter mammals with higher volume-to-area ratios. **Allen's rule** states that a cold climate leads to shorter extremities and stockier bodies. This is consistent with the square-cube law, which dictates that larger objects (including animal bodies) lose heat at a slower rate than smaller objects, thanks to the fact that volume goes by r^3, whereas surface area (through which heat is radiated away) goes as r^2.

Various mechanisms affect Earth's climate and work over vastly different timescales. Factors that work over tens to hundreds of millions of years to influence climate include:
1. plate tectonic motion
2. mountain weathering and
3. the carbonate–silicate cycle.

There are other factors (collectively lumped under the category of '**orbital forcing**') that operate over tens of thousands to hundreds of thousands of years to affect our climate such as:
1. Changes in the shape of Earth's orbit
2. Changes in Earth's rotational obliquity (axial tilt)
3. Precession of the equinox (that is, precession of Earth's rotational axis).

Collectively, these three factors that may alter our climate and induce ice ages are called the **Milankovitch cycles**.

Then there are other factors that can affect climate quickly, over just years to centuries, including:

1. Solar cycles ('solar forcing')
2. Ocean currents and circulation
3. Air quality (e.g. dust from volcanoes and greenhouse gases).

These various cycles may work in conjunction to accelerate climate change, or they may be out of synchrony and work against each other.

29.3 The present ice age

It may be that Earth has been prone to intermittent glaciation for the last 40 million years or so, thanks to the presence of substantial and stable ice caps over Antarctica. The development of these 'permanent' ice caps over the South Pole seem to have correlated with the formation of the Himalayan Mountain chain. (Mass mountain formation or **orogeny** promotes a cooling of the climate through extensive chemical **rock weathering**, which removes atmospheric carbon dioxide, a powerful greenhouse gas.) The presence of a stable ice sheet over the polar regions can serve as a catalyst, so to speak, for the positive feedback loops needed to precipitate an ice age.

Earth's most recent ice age spanned from about 70 000–10 000 years ago (although some assertions are that the ice age began even earlier—115 000–100 000 years ago). It is helpful to keep in mind that global glaciation has far-reaching ecological effects other than just the cold, glacier-covered regions. For example, more glaciers mean lower sea levels—and lower sea levels mean smaller coastal ecological zones. This is because the presently extensive, shallow, underwater continental shelf is not as available when sea level is very low. And the adjacent, but very steep, continental slope is a much harsher oceanic environment. Additionally, the larger amounts of continental land (thanks to lower sea levels) favor monsoons, with greater discrepancies between dry seasons and rainy seasons. Such conditions increase rock weathering. Rock weathering in turn removes atmospheric carbon dioxide and incorporates it into the carbonate–silicate cycle. Depending on the prevailing conditions at the time, increased rock weathering may facilitate geological homeostasis by removing excess global warming carbon dioxide or it may tip the scales too far in a positive feedback loop and initiate a new ice age.

About 11 700 years ago, the last ice age ended relatively abruptly and previously available land bridges between Asia and Australia, among many other places, became submerged. Thus, during our present interglacial period, sea levels are relatively high and the once available land bridges are gone. This time (about 11 700 years ago) of grand glacial retreat is called the Younger Dryas.

Over the last few hundred thousand years, several patterns have emerged. One is the background 'ice age' that we are presently in, which is punctuated by intermittent warm periods called **interglacials**. The glacial periods are generally longer than the interglacial warm phases. The glacial periods are colder but have highly variable climates. The Pleistocene Epoch (~2.6 Ma to 11 700 years ago) is sometimes called the **Great Ice Age** because of the multitude of glacial periods during this time.

All this is known through the careful examination of **climate proxies**, which are clues from the environment that serve as surrogate thermometers (which are discussed below). It should be emphasized that our knowledge is only as good as these proxies, which themselves are sometimes subject to criticism and controversy.

29.4 Climate proxies

Parts of the glaciers over Greenland and Antarctica are as old as 1 million years. While the snow was falling every winter over the poles and steadily accumulating in the polar glaciers, the rest of the world may not necessarily have been quite so cold. Study of the deuterium concentration in deep cores of this ancient glacial ice can be very informative about the climate during which the snow fell. An increase in deuterium concentration indicates a warmer climate at the time the ice was laid down in the polar glaciers.

This is because the vapor pressure of deuterium-containing water (D_2O and HDO, collectively called deuterated water) is lower than the vapor pressure of ordinary H_2O. (HDO is far more common than D_2O because of the natural abundances.) Incidentally, H_2O has a higher vapor pressure than water containing heavy oxygen (oxygen-18). That is, H_2O^{16} has a higher vapor pressure than H_2O^{18}. A higher vapor pressure implies that that chemical form is more likely to evaporate and get into clouds which eventually precipitate in the form of rain or snow. During warm periods, the deuterated water will be more likely to evaporate than during colder times. Hence, the clouds that will drop snow over the polar regions are more likely to contain higher concentrations of deuterium during warmer conditions than during cold spells. For this reason, a higher concentration of deuterium in ancient polar ice is indicative of warmer times.

Another proxy in paleoclimatology is found in the inorganic chemistry of fossilized calcareous creatures. For example, clam shells and the exoskeletons of other mollusks are made of calcium carbonate ($CaCO_3$). The carbonate comes from carbon dioxide, which is ordinarily in the form of carbon dioxide gas. Just as HDO has a lower vapor pressure than H_2O, CO^{18}_2 has a lower vapor pressure than the lighter CO^{16}_2. Similarly, H_2O^{16} has a higher vapor pressure than H_2O^{18}. Therefore, far more H_2O^{16} and CO_2^{16} evaporate than their heavier counterparts.

During warm periods, the evaporation rate increases. The result is the loss of lighter isotopes, leaving the heavier isotopes behind in the ocean. Hence, during warm spells the seas are relatively enriched in heavier isotopes. Corals, clams and coccolithophores must make use of whatever is available to them during the construction of their calcareous shells. Thus, during times of warmth, their calcium carbonate shells have more oxygen-18.

Note that in the first scenario (the deuterium in polar ice) we are measuring the heavy isotope that *had left the ocean* and was deposited as snow, whereas in the second scenario we are measuring the heavy isotope that *remained in the ocean* and was deposited in calcareous shells. This point can be confusing to students, but the problem is resolved by recalling where we are making our surrogate proxy measurements—out of the ocean versus within the ocean. Both surrogates seem scientifically robust, and results correlate well with each other. The data from the concentration of deuterium in polar ice cores and the oxygen-18 concentration in marine fossils yield similar results.

Thus, we have two overlapping and correlating climate proxies that go back nearly 1 million years. Of course, the data from marine shells alone can go back far further. Additionally, on the near side we have multiple other lines of short-term overlapping, correlative evidence that do not extend as far back as the polar ice cores, such as tree rings, lake sediments and radiocarbon data. On the far side, fossilized tropical coral reefs, which create calcium carbonate layers annually, can reveal clues about the temperature at the time of their formation. Information from ancient coral reefs may extend back hundreds of millions of years.

29.5 The Milankovitch cycles

Over tens to hundreds of thousands of years, subtle changes in Earth's orbit and axial tilt vary. Such variations are barely perceptible but nevertheless can have profound impacts on the global climate. These changes exhibit a repeating regularity called the **Milankovitch cycles**. Interestingly, the regularity and limited variability of these changes are conferred in large part by the Moon. The Earth–Moon system has a large combined angular momentum; this combined angular momentum is more difficult to perturb than just the angular momentum of the Earth alone would be. Without a large, stabilizing moon, Mars is more susceptible to the gravitational whims of Jupiter and shows far larger swings in its axial tilt. This affects its long-term climate far more dramatically than here on Earth.

1. **Eccentricity of Earth's orbit**

   Consistent with Kepler's Laws, Earth moves slower in its elliptical orbit when it is farther from the Sun. Therefore, if Earth's orbit grows more eccentric (because of Jupiter's gravity for instance), it will spend more time further from the Sun. The overall insolation will decrease, and Earth's climate will cool.

   Earth's eccentricity cycle (oscillation between more circular and eccentric orbits) exhibits a dominant period of roughly 100 000 years. Importantly, long-term climate proxies confirm a periodicity of about 100 000 years between major interglacial periods. Currently, Earth's orbit is approaching its least elliptical (i.e. most circular) and is very slowly decreasing. Hence, we are in a warm period as far as this part of the Milankovitch cycle goes.

2. **Axial tilt (obliquity)**

   Earth's axis of rotation is inclined to the orbital plane it takes around the Sun (the ecliptic). Presently, the axial tilt angle is about 23.5° From the normal to the **ecliptic**. However, over time the obliquity has varied between 22.1° and 24.5°. This variability repeats over a cycle of about 41 000 years. Generally speaking, the greater Earth's axial tilt angle, the more extreme our seasons are.

This increased obliquity amounts to each hemisphere receiving more solar radiation during its summer (when the hemisphere is tilted toward the Sun) and less during its winter (when the hemisphere is tilted away from the Sun). In other words, the greater the angle of inclination, the greater the difference in temperature between summers and winters.

Larger tilt angles cause more extreme variability between the seasons whereas smaller tilt angles favor less dramatic seasonal changes. At 23.5°, Earth's obliquity is presently about halfway between the two extremes. The angle of obliquity is very slowly decreasing in a cycle that spans about 41 000 years. Therefore, as far as this aspect of the Milankovitch cycles is concerned, we are headed for warmer times and less intense seasonal changes.

3. **Precession**

Finally, Earth's axial tilt wobbles or precesses like a spinning top over a period of around 26 000 years. (To be more precise, it is closer to 25 771.5 years.) The north pole is presently aimed at the north star, Polaris. Thanks to Earth's axial precession, in another 12 000 years or so, it will be directed at Vega. When that happens, the Northern Hemisphere will experience winter in July and the Southern Hemisphere will have its summer at that time. Although axial precession does not alter Earth's overall insolation, it nevertheless can have profound effects on climate. This is because of the uneven distribution of continents. There is much more continental mass in the Northern Hemisphere presently. As one example of how this works, continents are more likely to accumulate massive, heat-reflecting glaciers whereas dark blue oceans are heat absorbing. Therefore, an asymmetry in the distribution of continents can have an impact on climate.

Based on all three individual cycles, Milutin Milankovitch himself calculated that ice ages should occur approximately every 41 000 years. It has since been confirmed that major glacial periods did occur at 41 000 year intervals between 1 and 3 million years ago. But beginning about 800 000 years ago, the major glaciation cycle lengthened to about 100 000 years. This period corresponds best to Earth's elliptical (eccentricity) cycle.

29.6 The greenhouse effect

The Earth is much warmer than it would be in the absence of an atmosphere or if it had an atmosphere devoid of certain heat-retaining gases. This phenomenon, wherein the planet is warmed by the particular characteristics of its atmosphere, is called the greenhouse effect. (It should be noted that the greenhouse effect works through a different mechanism from how greenhouses retain heat and functions more akin to how blankets work.) There are many minor constituents of the atmosphere that serve as greenhouse gases including carbon dioxide, methane, ozone, chlorofluorocarbons, nitrous oxide (and other nitrogen oxides) and water vapor. Carbon dioxide, at 400 parts per million, is considered the most important of these. Water vapor is also a very powerful greenhouse gas, and its concentration can vary tremendously over very short intervals (hours to days). A molecule of water remains in the atmosphere for about 9 days on average. Then there is methane, a far more efficient greenhouse gas than carbon dioxide in reabsorbing infrared (about 26 times more effective on a molar basis) but it is far less abundant. Additionally, it is not as durable, being oxidized into carbon dioxide in our oxygen-rich atmosphere; methane has a half-life of around 9–12 years in our air. This translates into a **global warming potential** estimated to be between 30 and 82 for methane. Ozone in our stratosphere is crucial to life on Earth, as it screens out harmful ultraviolet radiation (especially UVC wavelengths). However, when present at lower altitudes, that is in our troposphere, ozone is not our ally; it does double duty as a pollutant and a greenhouse gas.

The Sun radiates most of its electromagnetic radiation in the form of visible light and our atmosphere is largely transparent to these wavelengths. Thus, much of the Sun's visible light passes through the atmosphere and reaches Earth's surface. Nevertheless, about 30% of the incident sunlight across the globe is reflected by clouds and other atmospheric components. And another 20% of what is absorbed is promptly re-emitted in all directions—including back into space. Overall, when the temperature is stable, the incoming radiation must equal the outgoing radiation. If this condition is not met, the planet will not be in thermal equilibrium and the temperature will change. The electromagnetic radiation that passes through the atmosphere and strikes the surface heats it by exciting atomic electrons and increasing the kinetic energy of atoms and molecules. The heated surface then radiates at longer wavelengths (infrared) consistent with a blackbody at about 15 °C.

Unlike the transparency to visible light, the tropospheric greenhouse gases may reflect, absorb or transmit the re-emitted infrared. Of the infrared that is absorbed by greenhouse gas molecules, some will be re-emitted outwardly and lost to space while some will be redirected back to Earth and warm the planet. This results in global warming, with a temperature that is substantially higher than what would be expected in the absence of an atmosphere (or an atmosphere devoid of greenhouse gases). For climate stability, the equilibrium equation (radiation energy in = radiation energy out) must be met, regardless of which wavelengths we are dealing with. The solar input is about 240 W m^{-2}. Recall that the output of a blackbody, as dictated by the Stefan–Boltzmann Law, is proportional to the fourth power of the temperature. For Earth, the equilibrium temperature should be around 0 °F or −18 °C. Thanks to the greenhouse effect, the actual temperature is a significant 60 °F (or 33 °C) warmer than it would be if we did not have these greenhouse gases. Thus, Earth's average global temperature is about 60 °F or 15 °C.

The concentration of carbon dioxide in our atmosphere has varied tremendously over Earth's history and such variations have correlated with global glaciations (Snowball Earth episodes) and Hothouse Earth periods. For instance, during parts of the Cretaceous Period, the Earth was unusually warm. About 100 million years ago there were no polar ice caps (which by some definitions means Earth was not in an ice age as it is now). The temperatures were high and sea levels were high. The estimated carbon dioxide concentration was over twice what we have presently (that is, it was over 800 parts per million). Earth's greenhouse gas content has thus fluctuated substantially over its long history, as has its climate. Examining the greenhouse effect on other planets may be instructive in this regard.

29.7 A tale of two planets—greenhouse effects on Mars and Venus

It is well known that Venus is scorching hot, and Mars is frigid. But this is not all due to their relative distances from the Sun. Venus, with its greenhouse gas-filled atmosphere that is about 110-fold as dense as ours, is far hotter than it would be without an atmosphere (or if its present atmosphere were replaced with ours). Mars' atmosphere is only about 1% as dense as Earth's. Regardless of chemical composition, it cannot produce much of a greenhouse effect. Thus, Mars is colder than it would be if it had no atmosphere. A calculation of Mars' temperature without any atmosphere is about −50 °C; its actual temperature averages about −60 °C (although there is great variability throughout the day and night). The actual temperature is slightly cooler than predicted because of reflection of sunlight.

Mars has a very weak and localized magnetic field, as opposed to a strong global dipole like Earth's. This is presumably because its liquid spinning core stalled around 4 billion years ago. Thus, beginning sometime about 4 Ga, the solar wind began scrapping away Mars' previously lush atmosphere. It is believed that Mars once had a warm climate with abundant liquid water on its surface. But as its atmosphere was steadily eroded, the greenhouse effect was lost and the temperature progressively fell. Today, Mars is cold and barren. Its original atmosphere is almost all gone. Apparently, all its liquid water evaporated away along with the rest of the atmosphere (although some H_2O may still be stashed away in the form of subterranean ice). When water vapor is exposed to space radiation such as the solar wind, it can break down into hydrogen atoms, electrons, radicals and ions that are too light for Mars' weak gravity to retain. But without a thick atmosphere enriched with greenhouse gases, Mars is now far less hospitable than it once was.

Venus has an even weaker global planetary magnetic field than Mars does. It is virtually non-existent. (This is probably related to its slow, retrograde rotation, which was likely caused by a colossal cosmic collision that literally flipped Venus upside down and had a dampening effect on its internal liquid metal magnetic dynamo.) Thus, one might expect Venus's atmosphere to have suffered the same fate as Mars (and maybe even sooner since it is closer to the Sun and subjected to even more intense solar wind).

However, extreme vulcanism, a greater planetary mass and importantly, the absence of plate tectonics have conspired to create a extraordinarily thick atmosphere made nearly entirely of carbon dioxide. This carbon dioxide was expelled from the tremendous volcanic activity over vast timespans and recent data suggest that some volcanoes on Venus are still active today. Given the greenhouse effects of carbon dioxide, Venus is much hotter than it would be without an atmosphere or if it had an atmosphere like ours. In fact, Venus has a hotter average temperature than Mercury despite Mercury being much closer to the Sun. Calculations suggest Venus would be about 55 °C or 131 °F without an atmosphere. But thanks to its runaway greenhouse effect, its measured temperature is nearly 500 °C or almost 900 °F.

This may have occurred because of evaporation of Venus's early oceans, which dumped huge amounts of water vapor into its atmosphere (and water vapor is a very potent greenhouse gas). This water vapor, accompanied by volcanic carbon dioxide, caused Venus to reach a tipping point—a point where positive feedback loops became unstoppable, and a runaway greenhouse effect became inevitable.

29.8 Plate tectonics as a global thermostat

Unlike Mars (which has no unequivocal volcanic activity) and Venus (whose active vulcanism was only recently confirmed) Earth has many active volcanoes. But that begs the question—why did Earth not follow the same pattern as Venus and become a permanent hothouse? The answer is plate tectonics. Plate tectonics serves as our global thermostat. Plate tectonics keeps our climate stable over spans of tens to hundreds of millions of years. It does so through the **carbonate–silicate cycle** (which is covered in greater detail below). Briefly, greenhouse gas carbon dioxide is removed from the atmosphere and dissolved in the lakes and oceans as bicarbonates. The bicarbonates are converted into biologically made and non-biologically made **carbonates** such as limestone in the oceans. The limestone is then drawn underground and buried in the mantle at deep sea **subduction zones**. It stays sequestered deep underground for perhaps hundreds of millions of years. Eventually the carbon dioxide may be returned to the atmosphere via volcanic activity. Such volcanic activity and release of greenhouse carbon dioxide is probably how Earth broke free from the grips of global glaciation in its 'Snowball Earth' episodes.

29.9 Earth as a privileged planet?

Incidentally, the ongoing volcanic and plate tectonic activity as well as our strong geomagnetic field are because Earth still has a hot, spinning, liquid iron outer core. Earth is unique among all the terrestrial planets in this regard. Both Mars and Venus have lost most or all of their planetary magnetism because their molten cores have frozen and ceased to function as magnetic dynamos. Thus, neither planet has much tectonic activity, and neither has any plate tectonics at all. ('Tectonics' is simply large-scale geological activity that can include volcanoes and earthquakes. Plate tectonics, on the other hand, is the lateral movement of large masses of crust across the surface of a planet.) As far as we know, only Earth has genuine plate tectonics.

The rotating molten iron in our outer core powers a magnetic dynamo, which generates our geomagnetic field—which in turn deflects incoming charged particles in the solar wind and galactic cosmic rays. In so doing, it also protects us from solar particle events and preserves our atmosphere. Exactly why Earth still has such a generous amount of internal heat is uncertain. One possibility is that Earth was blessed at birth with a greater endowment of primordial radioisotopes than its fellow terrestrial planets. These abundant radioisotopes have continued to decay and release heat deep in the interior over billions of years. This extra heat has kept Earth literally molten hot inside and maintained the liquid state of the outer core. This in turn allows the geomagnetic dynamo to continue functioning at full capacity. Earth is thus clearly distinguished from its neighboring terrestrial planets in several regards.

Complementing this generous supply of internal heat, plate tectonics allows a slow cooling of the interior, which may have been important in initiating plate tectonics. The slow (many millions of years) convection cycles in the mantle are the mechanical motors that move the continents in plate tectonics. But this convection also allows heat from deep inside to move to the surface and radiate away. Thus, mantle convection cools Earth's mantle. And the core (where the geodynamic resides) can only cool as fast as the insulating mantle does. If the core were too hot, it would not be as capable of working as a magnetic dynamo. It appears that the rate of cooling of the mantle and outer core was optimal for creating the liquid iron state needed for Earth's strong and steady geomagnetic field. And the relative abundance of radionuclides in Earth's interior serves a role in maintaining this radiation-shielding magnetic field and allowing the plate tectonics-driven planetary thermostat to continue.

29.10 Potential feedback loops

There are several potential positive feedback loops that could theoretically trigger a runaway greenhouse effect and set us up for a repeat of the **Permian Disaster**, or worse still, set us on a course copying Venus. Fortunately, there are also several negative feedback mechanisms as well, which serve as checks and balances and promote global climatic homeostasis. Below is an example of one possible positive feedback cycle:

1. Warm weather melts polar ice. If there is prolonged polar warming, the glaciers melt. This could trigger a domino effect because polar ice has a very high albedo and reflects incoming sunlight and heat. The product of glacial melting, dark blue ocean water, has very low albedo—it absorbs sunlight and solar heat.
2. Less sunlight reflection and more absorption of solar energy heats the surface and the atmosphere. This in turn, can lead to more polar ice melting and start a vicious cycle.
3. Compounding the problem is the fact that warm air can hold greater amounts of certain greenhouse gases. For example, the carrying capacity of warm air for the potent greenhouse gas, water vapor, is a strong function of temperature. For example, 1 kg of air at 0 °C can hold 3.5 g of water, air at 20 °C can carry 15 g kg^{-1}, air at 30 °C can hold 26 g kg^{-1} and air at 40 °C can hold nearly 50 g kg^{-1}. Hence, over a range of just 40 °C, the carrying capacity increases over an order of magnitude. In practice, one looks up the carrying capacity of air for water vapor over varying temperatures in tables. But because the 'carrying capacity for water in air at different temperatures' may be equivalently expressed as the 'vapor pressure of water as a function of temperature' one might also approximate the value using the **Antoine equation** for water:

$$\log_{10} p = A - B/(C + T)$$

where p is the vapor pressure, T is the temperature and A, B and C are constants depending on the specifics (water vs ethanol for instance).

Students are advised that these constants are not constant. For instance, the values of A, B and C are different above 100 °C compared to below it. In the range from 0 °C to 100 °C, commonly used values of A, B and C are 8, 1730 and 233 respectively.

The **dew point** or **saturation point** increases in a non-linear manner as a function of temperature. If air at the saturation point is cooled, it can lead to cloud formation, provided there are sufficient condensation nuclei on which the water droplets can form. As condensation occurs, gas is converted into liquid, which releases latent heat (the heat of vaporization). This may then heat the cloud and cause it to expand, or alternatively the cloud might evaporate because the heated air is no longer below the saturation point. Importantly, clouds have high albedo, and their net effect is a cooling of the underlying atmosphere and surface.

4. If the temperatures are sufficiently high and sustained, another reservoir of greenhouse gases might be tapped. Frozen **methane clathrates** in the oceans and permafrost could sublimate, leading to increased atmospheric methane (which again, is more potent than carbon dioxide as a greenhouse gas and can chemically transform into carbon dioxide). Dumping this bolus of methane into the atmosphere could greatly exacerbate the positive feedback loop.

It is believed that this type of uncontrolled positive feedback loop was set in motion during the end of the Permian Period about 250 million years ago. As discussed elsewhere, the **Permian Disaster** was the worst of the five major mass extinctions during the Phanerozoic Eon. It is believed that the proximate trigger of the vicious cycle was a profound and prolonged outgassing of carbon dioxide from the Siberian Traps. (Of interest, although Olympus Mons is oft cited as the largest volcano in the Solar System, the roughly 1 million cubic miles of lava ejected by the Siberian Traps make it a good rival.)

29.11 The carbonate–silicate cycle: Earth's negative feedback system

As mentioned, plate tectonics keeps our climate stable over a time frame of hundreds of millions of years. It does so through the **carbonate–silicate cycle**, also called the **inorganic carbon cycle**. The carbonate–silicate cycle is Earth's primary long-term negative feedback mechanism, which counterbalances the potential positive feedback loops discussed previously. It is an intricate and intertwined process that shuttles carbon back and forth between four reservoirs:

1. The atmosphere
2. The hydrosphere
3. The biosphere
4. The crust and mantle (i.e. the lithosphere, asthenosphere and mesosphere).

The carbonate–silicate cycle may be briefly summarized as a multi-phase process:

1. Carbon enters the atmosphere in the form of carbon dioxide from multiple sources such as animal respiration, volcanic activity and combustion of carbon-containing materials. Carbon dioxide is a potent greenhouse gas that warms the planet.
2. Atmospheric carbon dioxide dissolves in raindrops as **carbonic acid** and enters the hydrosphere in the form of bicarbonates. An equilibrium is reached between the atmosphere and the hydrosphere.
3. Chemical rock weathering caused by carbonic acid in rainwater slowly dissolves rocks, creating aqueous solutions containing various cations and anions. Because most rocks are composed of silicate minerals, among these anions in solution are silicate and bicarbonate ions. All these ions eventually get deposited into the oceans via streams and rivers.
4. In the oceans, **calcareous** marine organisms such as clams, corals and coccolithophores (a form of planktonic life with calcium carbonate shells or 'tests') lead to vast underwater reefs and ocean floor deposits ('calcareous ooze'). Non-biological mechanisms also make limestone and other carbonate rocks that accumulate on the ocean floors as well. In essence, silicate rocks and minerals have been converted into carbonate rocks and minerals that sit on the bottom of the sea.
5. Like a conveyor belt, plate tectonic activity moves the oceanic crust from spreading centers at mid-oceanic ridges laterally to deep-sea trenches.
6. At the trenches, plates collide and the denser oceanic crust undergoes subduction and is buried in the lithosphere and mesosphere (i.e. in the mantle).
7. Volcanic activity returns some of this buried carbon to the atmosphere through the outgassing of carbon dioxide. It may take many millions of years to complete the cycle.

The importance of this cycle cannot be overstated. Were it not for the carbonate–silicate cycle, Earth's atmosphere would chemically resemble Venus and Mars, both of which are over 90% carbon dioxide (about 96% and 95% respectively). Instead, our present atmosphere is about 400 parts per million carbon dioxide. Most of Earth's surface inorganic carbon is dissolved in the hydrosphere (oceans and lakes) as carbonic acid. Actually, since the pKa (~6.4) of the first ionization of carbonic acid is usually below the pH of most bodies of water, most of this carbonic acid is chemically in the form of bicarbonates. This hydrospheric reservoir holds an estimated 40 000 gigatons of carbon. The atmosphere, at its present concentration of 400 parts per million, contains an estimated 800 gigatons. The biosphere holds another 500 gigatons. These figures are absolutely dwarfed by the amount of carbon in 'deep Earth' (in the form of carbonate rocks and minerals). The lithosphere, asthenosphere and mesosphere (i.e. the crust and mantle) contain a whopping 1.85 billion gigatons of Earth's carbon. This means that in total, less than 1% (approximately 0.2%) of Earth's carbon is located in the superficial compartments;

the vast majority (99.8%) is sequestered in the crust and mantle. Of this 1.85 billion gigatons, ~1.5 billion gigatons are in the lower mantle, 315 million gigatons are in the lithosphere and about 30 million gigatons are in the upper mantle. The latter two subcompartments partake in the so-called '**slow carbon cycle**'.

The deep, subterranean component of the carbonate–silicate cycle is also sometimes called the 'slow carbon cycle' since it can take about 100–200 million years to form the carbonate rocks, move the carbon between the rocks in these deep-Earth compartments and release it again in the form of carbon dioxide from volcanoes. The three superficial compartments (atmosphere, hydrosphere and biosphere) participate in the so-called '**fast carbon cycle**' component of the carbonate–silicate cycle.

After carbonate rocks and minerals are made from dissolved bicarbonate by either non-biological or biological means (e.g. in the form of coral reefs, etc), these solid forms of carbon are buried via subduction at deep-ocean trenches and other subduction zones. The carbon is thereby taken out of the fast carbon cycle and entered into the slow carbon cycle, starting with subduction and ending with vulcanism. As mentioned, this multi-million year process removes greenhouse carbon dioxide from the atmosphere and sequesters it, thus limiting the amount of carbon dioxide available for the atmosphere, hydrosphere and biosphere. Hence, thanks to this complicated carbonate–silicate cycle, rather than the 95%–96% carbon dioxide atmosphere of Mars and Venus, our atmosphere is far less than 1% carbon dioxide (about 0.04%).

Although our atmospheric carbon dioxide does not vary very much thanks to the carbonate–silicate cycle, it does vary slightly, and such variations can have profound effects on the climate. As mentioned, there were times in Earth's geological past when carbon dioxide concentrations were far higher than today, and such times were associated with far warmer conditions than today. And depending on the particular arrangements of the continents during such times, they may have conspired to push Earth beyond the brink. The classic example is when the atmosphere was simultaneously inundated with greenhouse gases during the late Permian Period when the supercontinent of Pangea and its associated super-ocean, the Panthalassic, were in existence. Under such unusual circumstances, the oceanic conveyer belt stalled, resulting in worldwide stagnation (**euxina**). Although the **Keeling Curve** clearly shows a progressive increase in global atmospheric carbon dioxide and this corresponds to an increased global temperature over the last century or so, there is hope that the present geographical arrangement of the continents may allow continuation of the global oceanic conveyor far longer than what was possible with the single Panthalassic Ocean of the Permian Period.

Another interesting and instructive geological time was the interval between 180 and 150 Ma (the Late Triassic and Early Jurassic Periods). Over this time frame Earth's climate was progressively warming. It was a period of prolonged volcanic activity, specifically hot-spot vulcanism (similar to what created the Hawaiian island archipelago). This unmitigated volcanic activity expelled enough carbon dioxide to trigger a positive feedback loop of global warming. This led to the entire Cretaceous Period (145–65 Ma) being characterized by a relatively hot, tropical climate.

Conversely, several prolonged cold climatic snaps over the Phanerozoic Eon have been revealed by proxies. Many such cold spells were precipitated by continental collisions. Colliding continental plates produce new mountains. This orogeny is necessarily accompanied by chemical rock weathering. The carbonic acid-mediated chemical weathering converts silicate rocks into soil. (Rocks are mostly made of silicate minerals; weathering converts these rocks into clay minerals (soil) and bicarbonates.) But this rock weathering process pulls carbon dioxide out of the atmosphere as it generates bicarbonate anions that wash into the rivers and oceans. The bicarbonates are eventually converted into carbonate rocks such as limestone (made of **calcite** minerals (calcium carbonate)) and **dolostone** (made of **dolomite** minerals (calcium-magnesium carbonate)) either through non-biological mechanisms or biological means (such as creation of coral reefs). These carbonate rocks may sit in place for thousands of years or they may be subducted and buried for millions of years. Either way, large volumes of carbon dioxide gas are removed from the atmosphere, diminishing the global greenhouse gas pool. Hence, the overall impact of colliding continents and mountain building is a cooling of the climate. One notable example was the collision between the Indian and the Asian continental plates, which led to the formation of the Himalayan mountain chain. The growth of the Himalayas coincides with the present 'ice age' that began 40–50 million years ago. (By some liberal definitions, any time there is long-lasting ice over the polar regions, we are in an 'ice age'.) Thus, over the last 40 million years or so, we have been in an ice age, initiated by the formation of the Himalayas and punctuated by regular glacial and interglacial episodes driven by the Milankovitch cycles.

29.12 Snowball Earth episodes

The most pronounced periods of global cooling were exemplified by the 'Snowball Earth' episodes. During these times, Earth might have been covered by glaciers all the way to the equator (or fairly close to it), with the ocean surfaces frozen over. These Snowball Earth episodes lasted for surprisingly long lengths of time. The last global glaciation (called the **Cryogenian Period**) stretched from 720 to 635 million years ago, for a total timespan of 85 million years. Some data suggest that for about 65 million out of the total 85 million years, Earth's surface was completely or nearly completely covered by ice. The Cryogenian Period is broken down into substages. Thus, the **Sturtian glaciation** lasted about 57 million years and after a brief interglacial period this was followed by the **Marinoan glaciation** that lasted for another 15 million years. Exactly what triggered these extreme glacial periods is uncertain but weathering of the **Laurentian basaltic province** is a strong suspect. As with mountain formation and subsequent weathering, weathering of the Laurentian province (and other flood basalts of the time) extracted large amounts of carbon dioxide from the atmosphere, thereby reducing the greenhouse effect for a prolonged period. Whether Earth was a true snowball (covered entirely with ice) or a '**slushball**' (with some areas of cold but liquid water) is a subject of ongoing debate. The biological record suggests that somewhere, small patches of liquid water and thin transparent ice must have existed to allow photosynthetic microorganisms to survive and continue the chain of life.

29.13 Biology-provoked global climate change

As mentioned in another chapter, photosynthetic microorganisms themselves might have precipitated earlier Snowball Earth episodes through their consumption of carbon dioxide and, more importantly, through their reckless release of toxic molecular oxygen. The environmental effects of oxygen were unimaginable for ancient Earth. In addition to directly poisoning most fellow lifeforms in existence, the oxygen gas reacted with the then-abundant atmospheric methane. Oxygen and methane cannot coexist for very long and eventually the greenhouse methane gas was oxidized out of existence as oxygen levels continued mounting. The loss of this potent greenhouse gas probably contributed to the earlier Snowball Earth episodes.

And just as with runaway global warming, positive feedback loops may be set in motion when glaciers grow and reflect more sunlight. Thanks to their high albedo, a vicious cycle was initiated that caused the Earth to first cool and then grow colder still, eventually culminating in a prolonged global glaciation.

29.14 Breaking free from the grasp of glaciation

As mentioned, without a greenhouse effect, Earth's equilibrium temperature would be about 0 °F. Thus, this temperature might be considered as the baseline global temperature or the natural setpoint for the thermostat. So, after millions of years in this stable but frigid pattern (by human standards), how did Earth ever break free from the grips of global glaciation? Volcanoes.

This time the massive volcanic activity was not our enemy (as it did not initiate a Hothouse Earth like it did during the Permian Disaster). Instead, prolonged volcanic activity was our ally and rescued Earth, unfettering it from the grasp of an 85 million year long winter. The volcanic release of carbon dioxide heated the Earth up just enough to allow the release of fellow greenhouse gas, methane, from equatorial permafrost by melting of methane clathrates. As the equatorial regions became ice-free, dark dust (which has low albedo but was practically non-existent for millions of years) was blown from these low latitudes into other regions. This lowered the global albedo and sped up the deglaciation process. The last of the Snowball Earth episodes set the stage for the explosion of life that was to come during the Ediacaran and Cambrian Periods.

Further reading

[1] Herrick R R and Hensley S 2023 Surface changes observed on a Venusian volcano during the Magellan Mission *Science* **379** 1205–8
[2] Ueno Y, Hyodo M and Yang T *et al* 2019 Intensified East Asian winter monsoon during the last geomagnetic reversal transition *Sci. Rep.* **9** 9389
[3] Shaviv N J, Svensmark H and Veizer J 2023 The Phanerozoic climate *Ann. New York Acad. Sci.* **1519** 7–19

[4] Svensmark H, Svensmark J and Enghoff M B *et al* 2021 Atmospheric ionization and cloud radiative forcing *Sci. Rep.* **11** 19668
[5] Svensmark H 2023 A persistent influence of supernovae on biodiversity over the Phanerozoic *Ecol. Evol.* **13** e9898

IOP Publishing

Space Radiation
Astrophysical origins, radiobiological effects and implications for space travellers
James S Welsh

Chapter 30

Mass extinctions and radiation

The divisions in the paleontological record are based on changes in the fossil flora and fauna. For instance, the end of the Paleozoic Era and the beginning of the Mesozoic Era are demarcated in the sedimentary rock record by a dramatic change in fossils. Similarly, the end of the Mesozoic Era and the start of the Cenozoic Era are also signified by a drastic change in the fossil record. In the first case, the close of the Paleozoic Era coincides with the end of the Permian Period and the fossil record documents a major mass extinction. In fact, this was the most profound mass extinction in the history of life on Earth. In a similar vein, the close of the Cenozoic Era coincides with the end of the Cretaceous Period and this coincides with another mass extinction that led to the demise of the dinosaurs.

The true number of major mass extinctions over the Phanerozoic Eon (that is, over the last 543 million years or so) is debated and ranges from as few as five to over a dozen. This is largely due to different definitions of what constitutes a 'major' mass extinction event. One classification categorizes mass extinctions such that:

1. 20%–30% of known species go extinct during a 'minor' mass extinction,
2. Roughly 50% of species go extinct during an intermediate mass extinction and
3. >80% of species going extinct defines a major mass extinction.

To qualify as a mass extinction, the event must be global and extend across a range of ecologies—it cannot be geographically limited or ecologically restricted to one biome for instance. Furthermore, a mass extinction must be geologically sudden and span a short time interval (maybe a million years or less). Overall, there is consensus that there were five truly major mass extinctions (although there were many more minor and intermediate extinction events that provide fine structure to the paleontological record).

The '**Big Five**' major mass extinctions were, in chronological order:

1. The Ordovician extinction—440 Ma

2. The Devonian extinction—365 Ma
3. The Permian extinction—250 Ma
4. The Triassic extinction—210 Ma
5. The Cretaceous extinction—65 Ma.

Aside from the Devonian extinction event, all of these occurred at the end of the period in question. For instance, the Ordovician extinction is also called the end-Ordovician extinction or the Ordovician–Devonian extinction.

However, the Devonian extinction (which occurred around 379 Ma and is also known as the Kellwasser event or the Frasnian–Famennian extinction) did not coincide with the formal end of the Devonian. It was not the only extinction event in the Devonian; the end of the Devonian Period was accompanied by a second, slightly less extensive mass extinction called the Hangenberg event or the end-Devonian extinction. In fact, these various mass extinction events often literally define the end of a period and the start of another. (And in the case of the Permian and Cretaceous, they also define the end of entire geological eras.)

Until relatively recently, the etiology of extinction events was a mystery that seemed to go against the prevailing view of gradualism throughout geologic history. The other extreme view—catastrophism—had fallen out of favor over much of the 20th century. Nevertheless, the so-called Alvarez hypothesis on the extinction of the dinosaurs brought back catastrophism with a vengeance. As will be discussed in greater depth below, this hypothesis holds that a large asteroid struck Earth around 65 million years ago and triggered a host of after-effects that ultimately wiped out a huge number of species, including the non-avian dinosaurs. Thus far, the Cretaceous extinction was the only one that could trace its origin to an asteroid impact. Although the details are being worked out, presently a common denominator of many of these extinction events appears to be relatively sudden and dramatic climate change. However, the proximate precipitator of this climate change varies from case to case, with volcanic eruption coupled with facilitating continental plate location being a common theme. For instance, the 'mother of all extinctions', the Permian extinction event, was probably mainly due to massive and prolonged eruptions of the Siberian Traps, which belched out enormous volumes of greenhouse gas carbon dioxide over up to a million years. However, it is unclear about whether the effect on the biosphere would have been as great were it not for the permissive continental arrangement of all the continents agglomerated into the supercontinent Pangea at the time. Thus, the primary cause might have been enabled or accentuated by the prevailing continental arrangement.

It should be pointed out that although it was once fashionable to speculate about 'cosmic radiation' as a potential cause of major mass extinctions, there is little supporting evidence for such hypotheses. We shall explore the possibility of unusual circumstances such as a decrease in the geomagnetic field strength coinciding with an increase in incoming cosmic rays, but such ideas will remain mostly in the realm of speculation. However, there is strong and solid evidence supporting the concept of 'death from the skies' in the form of a bolide at the close of the Cretaceous. It is also possible that a gamma ray burst contributed to the end-Ordovician extinction around 440 million years ago. Here we shall review a few of the major mass extinctions and tie them into space radiation where applicable.

30.1 Do mass extinctions occur with a regular periodicity?

Upon scrutiny of mass extinctions, there is a (slight) hint of some regularity or periodicity. Raup and Sepkowski suggested a 26 million year pattern that did not seem to correlate with any particular repeating plate tectonic activity or climatological changes. They then explored the possibility of extraterrestrial causes. Among the proffered solutions was the idea of a regular perturbation of the Oort Cloud that could cause comets to come hurtling in periodically. Muller proposed small and dim companion star ('Nemesis') which could conceivably toss comets our way and depending on the orbital characteristics of this proposed star, the comets might arrive every 26 million years. This aspect of the hypothesis is not very far-fetched since a sizeable fraction of stars in our Milky Way are part of binary systems. Massive type O and B stars tend to be binaries most of the time (~80%) whereas sunlike stars (such as our G-type Sun) are binaries around half the time. But the majority of stars in the Milky Way are small, faint M-type red dwarfs (maybe 85% of all stars in the Milky Way). And these smaller stars are often single. Thus overall, maybe a third of stars in the Milky Way are part of a binary or higher system.

Anyway, red dwarf stars are the most abundant stars in the Milky Way but are typically less than 10% of the Sun's mass, so poor visibility of such a companion to our Sun would not be entirely surprising. Such a red dwarf companion star to the Sun could be hard to detect partly because of its duskiness but also because it would not be moving much with respect to the Sun since it is part of the same system. Nevertheless, thus far all searches for Nemesis have yielded nothing.

Along the same lines, an alternative to a companion star is the concept of a giant, undiscovered planet ('Tyche') with a mass of up to fourfold that of Jupiter. Problems with this proposal include the fact that the calculated location would be consistent with a period of 1 million years, which would not account for the observed 26 million year periodicity. And regardless of the orbital period, it might not be massive enough to really hurl a sufficient number of comets our way to produce regular mass extinctions. Additional challenges with these ideas include some skepticism about the 26 million year periodicity. Studies of craters on the Earth and Moon have not yet revealed any clear temporal pattern supporting a 26 million year repeating bombardment.

Another possible regularly repeating pattern of extinctions might trace back to the way the Solar System revolves around the Milky Way. The Solar System orbits the Galaxy once about every 225 million years. But in addition to revolving around the Galaxy as a whole, the Solar System bobs up and down as it orbits in a sinusoidal wave. The Sun and Solar System reach maximum amplitude (that is, furthest distance from the galactic plane) every 63.6 million years. Presently, the Solar System is near the Milky Way galactic plane but in another 31 million years it will be at a maximum distance from the galactic plane. Medvedev and Melott have suggested that as the Solar System ventures further from the galactic plane it is less well protected by the Milky Way's magnetic field—and thus more subjected to unmitigated cosmic rays.

Although the idea of a relentless bath of cosmic rays when Earth is farthest from the galactic plane every 31 million years sounds intimidating, we must keep in mind the fact that cosmic rays are charged particles that would still be deflected by both the Earth's and the Sun's magnetic fields. Additionally, an intact atmosphere would do a good job of shielding the biosphere from most of the secondary cosmic rays, despite the increased flux of primary cosmic rays. Nevertheless, in times of increased primary galactic (or extra-galactic in this case) cosmic ray flux, there will be an accompanying increase in secondary muons striking the surface. Additionally, increased incoming cosmic rays may create NO and NO_2 through conventional radiochemical processes when ionization and radical formation occur in atmospheric nitrogen and oxygen molecules. Furthermore, nitrogen oxides can also be generated by cosmic ray-triggered lightning discharges. Such nitrogen oxides can be mutagenic and destroy stratospheric ozone.

And there could be other, more subtle ways that large increases in cosmic ray flux might affect life. For instance, according to the controversial **Svensmark hypothesis**, increased cosmic rays could cause more atmospheric water vapor to condense into clouds; more clouds would block incoming sunlight and the climate cools. Additionally, more clouds might cause more storms, more storms in a colder climate means more snow and more snow leads to more reflection of incoming sunlight. This causes the climate to grow even colder. If the continental arrangement was conducive, all of this could set the stage for an ice age. Although there are a lot of 'ifs' there, the point is that increased cosmic rays could have indirect effects that could alter the biosphere without directly irradiating it.

Just what is deflecting cosmic rays and minimizing their concentrations along the galactic plane is uncertain. But recently a similar situation has been confirmed around the central portion of the Milky Way Galaxy. This inner region, called the central molecular zone (CMZ), is relatively devoid of cosmic rays in the GeV–TeV energy range based on gamma ray data collected by the Fermi Large Area Telescope.

According to current models, galactic cosmic rays are accelerated by shock waves in supernova remnants or by stellar winds emanating from massive stars. The accelerated charged particles should then diffusely propagate throughout the galactic magnetic field, occasionally experiencing re-acceleration, deceleration (energy loss) and spallation along their paths. This would account for the observed large-scale, quasi-steady-state 'cosmic ray sea'.

For some reason, the observed cosmic ray energy density in the CMZ is significantly lower than the cosmic ray sea component at large, suggesting the existence of some sort of presently uncharacterized barrier that inhibits the penetration of the particles from the cosmic ray sea into the CMZ. Just what the mechanism is remains unclear, but it could be analogous to how the Sun deflects low-energy incoming galactic cosmic rays through its own magnetic field and the solar wind. Hypothetically, a strong magnetic field might exist in the CMZ that deflects cosmic ray sea particles from entering. Additionally, there may be magnetized charged particle winds (generated perhaps by activity near the Milky Way's central supermassive black hole Sagittarius A*) which are analogous to our solar wind and also block low- and moderate-energy cosmic rays from entering the CMZ. If this model holds true, there could be an analogous, even larger-scale phenomenon in which the Milky Way as a whole deflects energetic extra-galactic cosmic rays and prevents them from entering. But when the Solar System and Earth bob up and down in its orbit around the Galaxy, we might stray a bit beyond the protective shielding of the Milky Way and become exposed to 'universal cosmic rays' originating from beyond our Galaxy. It is worth recalling that the most energetic cosmic rays detected appear to originate not from within the Milky Way but beyond it. For instance, a set of cosmic rays with energies exceeding 8×10^{18} electron volts, were recorded with the Pierre Auger Observatory. It appeared that these cosmic rays originated outside the Milky Way. Similarly, the Telescope Array Collaboration recently detected a cosmic ray with an astonishing 240 EeV. This particle also appears to have originated from outside the Milky Way galaxy (perhaps in the Local Void, which is an area of empty space just beyond the Local Group).

The Pierre Auger Observatory comprises 1600 Cherenkov particle detectors spread over 3000 km^2 in Argentina along with 27 fluorescence telescopes located in four separate sites in observatory region. The Cherenkov detectors pick up the myriad secondary cosmic ray (shower) particles created by primary cosmic ray collisions with atomic nuclear in our atmosphere, while the fluorescence detectors detect fluorescent light that is emitted when shower particles interact with nitrogen in the atmosphere. The results indicate that EeV primary cosmic rays are not 'galactic' cosmic rays in the sense that they fly in from beyond our Galaxy. Thus, when excursions of Earth above and below our galactic plane occur during our merry-go-round orbit, we may encounter more of the very highest-energy cosmic rays in existence.

30.2 The Cretaceous extinction

Historically, the extinction event at the end of the Cretaceous Period is the one that has been associated with the greatest curiosity and speculation and has had the greatest scientific scrutiny. This is quite understandable since the most massive land creatures of all time—the dinosaurs—abruptly disappeared from the fossil record at this time. The boundary between the Cretaceous Period and the following Paleogene Period is now called the K–Pg boundary. The 'K' stands for Cretaceous since 'C' was already used for the Cambrian Period. Up until recently, this dividing line in the sedimentary rock record was called the K–T boundary since the first period of the Cenozoic was formerly called the Tertiary Period. Nowadays, K–Pg is more customary. In any case, this dividing line is often grossly visible to the naked eye. Upon closer scrutiny, one can see that below the dividing line (that is before 65 million years ago), there were numerous and impressive giant reptiles in the form of flying pterosaurs, monstrous marine inhabitants such as mosasaurs and plesiosaurs and of course, on land, the dinosaurs. And although diversity might have been waning, overall the dinosaurs were not truly in decline at the end of the Cretaceous. This was the time when the terrifying Tyrannosaurus rex and Triceratops were roaming the North American continent and members of the largest dinosaurs of all time, the titanosaurs, were still in existence. But, rather dramatically, just above the K–Pg layer, there are no such impressive giants. They really did vanish rather suddenly (from a geological time perspective). In addition to the most obvious disappearance of the dinosaurs, ammonoids that flourished in the oceans for hundreds of millions of years suddenly vanished, along with many brachiopods, microscopic coccolithophores and foraminiferans (which are important planktonic components of the food chain). The disappearance of the latter meant not only a collapse of the marine food chain but also that chalk was no longer being deposited en masse on the seafloor. And since these microbes make their calcareous tests from atmospheric carbon dioxide, an important regulator of our climate was temporarily out of commission thanks to the K–Pg extinction. Carbon dioxide was more likely to remain in the atmosphere as a global warming greenhouse gas since the foraminifera and coccolithophores declined. Somehow, the microbial siliceous shell forming diatoms and radiolarians made the cut and survived, while their calcareous cousins did not. In any case, an important base of the marine food chain had been severely compromised.

And while some small mammals made it through, the fact is that some 35% of mammalian species went extinct. Close to 80% of North American plant species went extinct. Curiously, many of the botanical survivors were deciduous. Because they could 'go dormant' by shedding their leaves, their survival (at the expense of non-deciduous plants) meant a paucity of leaves for vegetarian animals to eat. The ecological chain reaction led to the demise of the dinosaurs, the marine reptiles and the flying pterosaur reptiles. Among the survivors were birds, crocodiles and alligators and turtles. It is known that the latter two groups are capable of surviving without food for extended periods and can scavenge. Additionally, turtles and alligators are capable of brumation or hibernation, prolonged periods of inactivity and fasting during inclement weather. Curiously, aside from some crocodilians and sea turtles, it looks like no animal species larger than around 50 lb or 23 kg survived the K–Pg event.

Insects, which largely made it through this extinction event, are also often capable of scavenging and can sometimes exhibit a durable dormancy stage. Some beetle larvae, for example, have been known to survive in dead wood for up to 40 years before emerging as adults. Just how and why the birds made it through remains unsettled. Among the marine microorganisms that survived were diatoms and dinoflagellates. One feature that both shared was the capacity of going into a dormant cyst stage and resting on the ocean floor until conditions improved.

Among the first land plants to rebound and recolonize the barren landscapes after the K–Pg extinction event were the ferns. This pattern (called the fern spike) was seen after other mass extinctions as well, including the Permian extinction.

Importantly, the Mesozoic was a time of good and plenty, which allowed many creatures to specialize. The enormous size of the herbivorous dinosaurs might be one example of such specialization. Such specialized animals are well suited for the specific circumstances of their environment, but they might not fare well when things suddenly change for the worse. If one were to contemplate the fates of koala bears (which feed almost exclusively on eucalyptus leaves) versus rats in the face of a global disaster, which would you bet on? Or giant pandas that feed largely on bamboo versus omnivorous pigs? Similarly, when the world changed 65 million years ago, small generalists who could hibernate during rough times probably had a distinct advantage over enormous beasts that had to eat massive quantities of specific foods to keep themselves going.

30.3 The Alvarez hypothesis

Luis Alvarez won the Nobel Prize in Physics for his contributions to particle physics (discovery of various resonance particles) through the use of a liquid hydrogen bubble chamber and he was also credited with the discovery of electron capture as a mode of beta radioactive decay. He was also involved in an effort to explore the innards of Egyptian pyramids through the use of cosmic ray imaging analysis. Alvarez's doctoral advisor was Arthur Compton, himself a Nobel Prize winner for his discovery that photons possess particle-like behavior (the Compton effect). Compton, with his then-graduate student Alvarez, conducted experiments that conclusively showed that cosmic rays were positively charged particles. Despite this remarkable career in physics, it was Luis Alvarez's paper in 1980 with his son Walter Alvarez that is probably most well known to the general public. Walter Alvarez was a geologist who, along with his famous physics father and chemists Frank Asaro and Helen Vaughn Michel, used the technique of neutron activation analysis to confirm that the layer between the Cretaceous and Paleogene periods contained a concentration of iridium hundreds of times higher than normal. As this K–Pg layer demarcated the demise of the dinosaurs, any unusual geochemical characteristics would be worthy of deeper examination.

It was known that extraterrestrial bodies such as meteorites contain more iridium than most earthly materials. Volcanic material also has a relative abundance of iridium but also has higher levels of chromium and nickel, which were not found in abundance within the K–Pg layer.

The discovery of disproportionate iridium in the K–Pg boundary layer was first noted in rocks of Gubbio, Italy and then also found in samples from Denmark and New Zealand. Subsequently, the excess iridium deposits were confirmed, essentially around the world. This so-called iridium anomaly was found to be in greatest concentration in areas that were closer to the Yucatán Peninsula of Mexico at the time (around 66 Ma) and concentrations fell off with distance from this area. All this was consistent with the idea of a massive bolide (e.g. an asteroid or comet) crashing into the Yucatán Peninsula. This was proposed as the ultimate cause of the Cretaceous extinction and is called the **Alvarez hypothesis**. Calculations based on the distribution of iridium around the globe, indicated that the asteroid would have been about 10 km in diameter and would have left a crater nearly 150 km wide. The consequences would have been the immediate incineration of anything in the vicinity of the impact, with debris being flung around the world along with tsunamis that would flood and destroy coastlines for half the globe. An accompanying cloud of dust would have been created that likely would block out sunlight for months afterwards causing the equivalent of a massive and prolonged 'nuclear winter' effect.

Twelve years after the proposal, the Chicxulub crater on the northern coast of Yucatán, Mexico was proffered as proof of the collision. Craters that are presumed to be created not by volcanism but rather an extraterrestrial impact from a meteor, comet or asteroid are sometimes called astroblemes. Dating through the argon-40/argon-39 method proved the crater to be approximately 65 million years old, highly consistent with the K–Pg extinction. Shocked quartz crystals (that are found at modern meteorite sites and atomic bomb blast craters) were found in the K–Pg boundary layer at many sites in North America but are relatively rare at equivalent sites in Europe, Asia and the Pacific; this finding favors an enormous impact blast in or near North America. Further evidence was found in the form of tektites (small glass beads and spherules created when molten silica is ejected into the atmosphere and cools off). Such tektites have been found in a geographically nearby area including Haiti, Colorado and New Mexico. Additionally, ejecta, rock blasted directly from the crater with chemical fingerprints of rock underlying the Yucatán has been found as far away as Canada. Finally, there is evidence of a tremendous tsunami near Mexico and Texas with land plant leaves being dragged far out to sea and buried in marine deposits. All of this is counted as evidence of a massive blast, possibly from an asteroid, in the northern Yucatán around 65 million years ago.

The Chicxulub crater itself provides a 'smoking gun'. The astrobleme is highly enriched in iridium making it a suitable suspect. The distribution of cenotes (circular sinkholes in the limestone often filled with fresh water scattered across the Yucatán Peninsula) are consistent with a massive impact. A semicircular concentration of cenotes in the northern edge of the Yucatán Peninsula seems to connect with the now-submerged portion of the Chicxulub crater in the Gulf of Mexico.

Initially the Chicxulub crater was thought to be about 110 miles or 177 km across, but subsequent studies showed that it is a 'complex crater' with a ring within a ring configuration. An additional ring was identified outside the initially discovered ring. The crater is now estimated to be an astounding 190 miles (300 km) in diameter. Present calculations indicate that the asteroid was about 6–9 miles (10–15 km) in diameter and traveling at a velocity of approximately 12.8 to 21.6 km s^{-1} (28 500 to 48 500 mph) when it struck. The amount of energy released by the impact is estimated to be roughly 100 million megatons of TNT (trinitrotoluene) or around 6.25 billion Hiroshima atomic bombs (which was around 16 kilotons).

Collectively, the evidence is quite strong. The consequences of such a crash would have been the obvious immediate effects of the blast and tsunami, along with global effects of the lingering dust cloud that would have probably blocked out sunlight for many weeks and maybe reduced its intensity for years. This could have significantly reduced primary production in the ecosystems via a slowdown in photosynthesis and diminished plant availability.

But while the acute effects of the asteroid impact were severe and predictable, it was the sudden shift in climate coupled with the late toxicity of the blast that was most devastating. The initial effect of the collision was probably a global cooling thanks to the nuclear winter effect of increased soot and cloud cover. Much of these effects could be attributed to the release of sulfur dioxide, which upon entry into the atmosphere becomes highly reflective and bounces incoming sunlight back into space. The rocks in the Yucatán were unusually endowed with sulfate rocks which upon impact, immediately vaporized into vast amounts of sulfur dioxide. Current estimates indicate about 325 gigatons of sulfur were instantaneously injected into the atmosphere, along with massive amounts of sun-blocking soot and dust. The amount of sulfur injected into the atmosphere by extraterrestrial impacts and volcanoes depends heavily on the landscape where the event occurs. For example, in recent times, two large volcanoes—Mount Saint Helens in 1980 and El Chichón in 1982—were comparable in size but only El Chichón led to slight global cooling via a transient volcanic winter. The probable explanation is that Mount Saint Helens ejected only 1 megaton of sulfur dioxide whereas El Chichón released 13 megatons of this high-albedo gas. El Chichón was located in a sulfur-rich limestone region thanks to abundant anhydrite (anhydrous calcium sulfate ($CaSO_4$)) while Mount Saint Helens was located in a region of low-sulfur, remelted igneous rock.

Potentially adding to the climate catastrophe was smoke and soot, released by rampant wildfires enabled by the dead, dry vegetation. The result was a quick cooling of Earth's average surface temperature by as much as 26 °C (47 °F). The resulting near- and sub-freezing temperatures persisted for at least 3 years following the blast. Possibly adding insult to injury was acid rain, thanks to the large amount of sulfur dioxide and carbon dioxide set free. This would have had a severe impact on organisms making shells from acid-labile calcium carbonate, and indeed the calcareous microorganisms such as foraminiferans and coccolithophores took a heavy hit. Acid rain would have had a negative impact on freshwater life as well, and the freshwater sharks became extinct at this time.

But afterwards, a long-lasting global warming probably occurred. Much of this sudden climatic lurching between sudden extreme cold and persistent heat might have been due to simple serendipity—had the bolide crashed into another location, the dinosaurs might still be around today. In addition to an extraordinary amount of sulfate rock, the Yucatán Peninsula is endowed with an inordinate amount of limestone. Limestone is a carbonate rock made largely of calcium carbonate. When all this calcium carbonate was instantly vaporized, there was the release of maybe between 425 and 1400 gigatons of carbon dioxide into the atmosphere. Besides the limestone-rich impact site, the asteroid itself was suspected to be a carbonaceous chondrite (meaning it was rich in carbon that upon disintegration would yield even more carbon dioxide). Thus, a large mass of carbon-containing material the size of Mount Everest was instantly converted into a colossal quantity of carbon dioxide upon vaporization.

Thus, once the acute effects of sun-blocking soot and sun-reflecting sulfur dioxide subsided after a few years, the long-term effects of greenhouse carbon dioxide were poised to take over. The net result of this bolus of carbon dioxide was a global heating (Hothouse Earth phenomenon) in which temperatures rose by 5 °C (9 °F) or perhaps even higher around the world, lasting for around 100 000 years. And it remains possible that this was a classic case of the asteroid striking the 'wrong place at the wrong time' (although there probably is never a good time for a massive asteroid impact). The geology of the Yucatán impact site is unusual—only approximately 13% of the Earth's surface could have yielded such a 'perfect storm' thanks to the abundance of sulfur and carbon at the blast site.

If the asteroid hit a place without as much underlying sulfate and carbonate, would the outcome have been very different? Did the composition of the asteroid (presumably a carbonaceous chondrite) make a difference by adding to the volume of greenhouse gas carbon dioxide? Presently, these questions are still unanswered.

But the asteroid impact might not have been the sole cause or even the primary cause of the extinction according to some experts. The Deccan Traps of Eastern India are the end product of unmitigated volcanism that persisted for up to a million years. These volcanoes were erupting around the same time and could have made a major contribution to the K–Pg extinction. The sheer volume of ejected lava was staggering. Presently, the Deccan Traps cover about 200 000 square miles or 500 000 km^2 but before erosion and burial of much of the lava over the past 65 million years, they might have covered four times this area. In terms of volume, the surviving remnant is over a mile thick in regions and amounts to about 1 million cubic kilometers (~240 thousand cubic miles), but the initial volcanic volume was far greater. Unlike explosive supervolcanoes with ultra-plinean style eruptions such as Toba and Yellowstone, the eruption style of the Deccan Traps was probably more Hawaiian, strombolian and vulcanian. However, what they lacked in explosivity, they more than made up for in duration and overall ejected volume. Using a conservative estimated eruptive duration of 400 000 years, the amount of erupted carbon dioxide might have been as much as 4.14×10^{17} moles of carbon dioxide. Incidentally, the hot spot that yielded the Deccan Traps 65 million years back is not dead yet. It presently resides under the island of Réunion in the Indian Ocean.

30.3.1 Could the two events be somehow related?

The sheer energy imparted by the asteroid impact certainly would have had major geological consequences. Among those consequences were mega-earthquakes reaching magnitudes of 12–13 on the Richter scale that might have persisted up to months afterwards. But it is natural to speculate that, as in a neurological coup-contrecoup intracranial injury (where impact on one side of the head can lead to brain injuries both directly under the impact site and on the opposite side due to a ricochet effect), a tremendous impact on one side of the planet might induce an effect on the opposite side of the globe. While the Yucatán Peninsula and western India are not opposite one another today, things were geographically different at the end of the Cretaceous. And the incoming asteroid probably did not come in at a 90° orthogonal angle but instead at a steep angle maybe closer to 60°. Could this have triggered the Traps? The chronology does not support a direct cause since the present data indicates that the Deccan Traps were erupting for an estimated 350 000 years before the collision. Interestingly however, prior to the impact, the eruption rate was rather modest. Then at the moment of the asteroid's arrival, the eruption rate drastically increased. Although the volcanoes were erupting 350 000 years before and 500 000 years after the impact, 70% of the lava came blasting out in the 50 000 years after the asteroid impact. Whether the asteroid or the Deccan Traps or a combination of the two was the ultimate cause of the extinction remains a topic of scientific debate.

30.4 The Permian extinction

Around 250 million years ago, the largest of all mass extinctions was documented in the sedimentary rock record. This is known as the Permian extinction, the end-Permian extinction event, the Permian–Triassic mass extinction and, more colloquially, the Permian Disaster, the 'mother of all extinctions' or the 'great dying'. Estimates of the loss of life range from 70%–80% of all land species and up to 96% of marine species. The extinction event was so dramatic and so pronounced in the rock record that the Permian–Triassic boundary marks not just the division between the two periods but also the boundary between the Paleozoic and Mesozoic Eras. This is also the case with the K–Pg extinction event which defines the Cretaceous/Paleogene Period boundary as well as the Mesozoic/Cenozoic Era boundary. This was the greatest loss of life in the history of the Phanerozoic Era, the 543 million years since abundant animal life was preserved in the fossil record.

The extinction wiped out some durable survivors such as the trilobites, who were around since the Cambrian Period and made it through the Ordovician and Devonian extinctions. The sea scorpions (eurypterid) vanished for good at this time as well. Echinoderms were hit very hard, with 98% of families going extinct, including the blastoids and many crinoids. Over 90% of brachiopods went extinct as well. Coral reefs and land forests both collapsed; this strange combination of disappearing ecosystems was not seen in any of the other extinctions. Up to the Permian extinction, rugose and tabulate corals were thriving and building massive reefs. After the Permian extinction, the world was devoid of active reefs and remained that way for nearly 15 million years. Only then did new types of corals emerge (scleractinian corals) and once again begin building reefs.

On land, the early Triassic fossil record contains a marked 'coal gap' which was caused by the disappearance of forests. In stark contrast to the Carboniferous and early Permian, when vast forests existed and were leaving behind immense deposits of coal, immediately after the end of the Permian Period, forests became scarce and coal formation came to a halt for quite a while. Seed ferns were among the dominant flora from the Devonian through the Permian but like so many others, they were gone forever after the Permian extinction. As one might predict, if plants became scarce, insects might have suffered as well. In fact, the Permian extinction was the only mass extinction that severely curtailed insect groups, with nearly one third of insect species checking out for good.

Curiously, one set of ecosystems that seemed spared were aspects of lacustrine (also called lentic) microbial ecologies. This alludes to the algae, archaea and bacteria surviving and thriving in lakes. And, as explained later, these thriving microbes might have exacerbated the overall situation though their waste products.

As with other extinction events, a large negative $\delta^{13}C$ was seen in the carbonate rocks laid down just after the Permian disaster. Since life likes the lighter isotope of carbon (carbon-12), a negative $\delta^{13}C$ excursion indicates the release of a large amount of biological carbon into the system. The lighter carbon-12 was no longer sequestered in biological organisms and was dumped into the pool of available carbon for inorganic carbonate production. This massive release of lighter carbon-12 resulted in a change on the $\delta^{13}C$ of about -6%.

The most favored explanation for the end-Permian extinction event was a very long-duration volcanic eruption. And although most volcanoes and earthquakes occur at tectonic plate boundaries, this particular one was different. Starting around 252 Ma, a 'flood basalt' type of volcano began pouring out lava in what is now known as an igneous province called the Siberian Traps. This type of phenomenon is also called trap magmatism. Instead of being associated with collisions between plates, the flood basalts creating the Siberian Traps were probably the result of a mantle plume, an eruption of magma stemming from as deep as the core–mantle interface 1800 miles deep.

In present-day Siberia, this igneous rock formation covers a vast expanse of land. The coverage is approximately 7 million square kilometers or 3 million square miles, which is nearly 60% of the Siberian craton. The total volume of rock in the present-day Siberian Traps is on the order of 1 million cubic miles or 4 million cubic kilometers. Thanks to weathering over the last 250 million years, the current volume and surface area are substantially smaller than they were originally, which gives one an appreciation of just how expansive this volcanic flood plain initially was.

As mentioned, the ultimate source of the Siberian Traps has been attributed to a mantle plume or hot spot deep in the mantle, perhaps as far down as the mantle–core interface. This mantle plume might have been present for an extremely long time and as the plates above it moved over time, possibly also gave rise to other lava outpouring such as the Viluy Traps in eastern Siberia (which has been implicated in the Devonian extinction) as well as the volcanoes of present-day Iceland. Another potential explanation was a still-unconfirmed asteroid impact in Antarctica. The proposed impact might have coincided with the birth of the Traps since it was geographically antipodal to them (meaning it was on the opposite side of the globe, and through Newton's Third Law, provoked the volcanic eruption).

It now appears that the end-Permian extinction was not a very abrupt event but rather was a prolonged extinction that could have come in two waves. The first wave started around 260 million years ago, some 8 million years before the second, larger and more devastating wave. This first wave was called the Guadalupian extinction and most severely affected marine brachiopods, corals and foraminiferans but many (74%–80%) of terrestrial species went extinct as well, at least in certain regions. The cause of this first extinction was volcanic activity in China, leaving behind what are now called the Emeshian Traps. Although enormous, the Emeshian Traps are dwarfed by the Siberian Traps which kicked into high gear 8 million years later. But the biospheric weakening caused by the Emeshian Traps during the Guadalupian extinction set the stage for the coup de grâce to follow. Although the Siberian Traps might have erupted for a million years, more recent analyses suggest that the volcanoes erupted most violently and intensely for a relatively brief period of only 50–80 000 years.

Through relentless output of greenhouse gases, a series of unfortunate events ensued eventually causing the Earth to become essentially uninhabitable. Among the key permissive factors was the continental configuration. During the Permian, all the continents were agglomerated into the single supercontinent of Pangea. The consequences of volcano-induced global warming were magnified by this arrangement. In today's world, an undersea conveyor current circulates ocean water across the planet. It travels from pole to pole, across the equatorial region and back again. During the Permian, the single-continent and single-ocean arrangement was susceptible to disruption of the global oceanic conveyor that set the stage for ecological disaster.

Today's deep, global ocean conveyor belt is motorized by a combination of temperature and salinity differences in ocean water at different locations. Cold and salty water is dense and sinks while warm and fresher water is less dense and moves to the surface. As the ocean conveyer moves near polar regions, it cools and sinks whereas when it travels over equatorial regions it will want to warm and rise. But simultaneously, as the conveyer currents move near the poles, additional fresh water is injected into the system by melting glaciers. The resultant thermohaline potentials can be complicated but drive the deep waters along a regular path in the deep ocean, while wind-driven currents direct the surface.

Although arbitrary, the ocean conveyor may be said to 'start' in the Norwegian Sea, where warm water from the Gulf Stream heats the atmosphere in the cold northern latitudes. This loss of heat to the atmosphere makes the water cooler and denser, causing it to sink. This warm Gulf Stream water simultaneously causes some glacial melting, which introduces some cold fresh water into the system. But as the warm water from the Gulf Stream moves north, it displaces the sinking cooler water, which then moves south. This cold bottom water flows south roughly along the middle of the Atlantic between North America and Europe and northern Africa, but then hugs the eastern coast of South America as it traverses the equator all the way down to Antarctica. It then takes an east turn and splits in two with one branch moving into the Indian Ocean where it takes a clockwise loop. The other branch continues east along the northern coast of Antarctica as a deep cold-water current until it reaches the mid-Pacific where it turns north. It continues northward and crosses the equator under the middle of the Pacific and eventually makes an eastward turn near the Bering Straits. From here it follows a clockwise pattern and travels as a superficial, warmwater current through Indonesia and north of Australia and eventually rejoining the branch that broke off in the Indian Ocean. From there, the oceanic conveyor continues as a warm-water current following the west coast of Africa in the South Atlantic but then crosses over, travelling westward, towards the Caribbean. From here it moves northeast as the Gulf Stream and reaches the starting/finish line in the Norwegian Sea.

Along the way, cold bottom water returns to the surface through density-driven mixing and wind-driven upwelling. In this fashion, the global oceanic conveyor belt continues to circulate water around the globe. And this constant motion ensures adequate oxygen levels from place to place and even at great depths.

This constant circulation and associated oxygenation cycle is critically important for undersea life. This oxygenation is obvious when one examines the ecology of deep lakes. In such lakes, the upper layers (epilimnion) are oxygen-rich and habitable by fish and other animals as well as sunlight-requiring plant life. However, below the thermocline lies the hypolimnion. The thermocline is a density barrier that keeps cold, hypoxic waters down below while warm, oxygenated waters remain above. In this fashion, the deep, hypoxic hypolimnion stagnates and becomes largely devoid of macroscopic life; lake life is largely confined to the superficial epilimnion. This statement of course is restricted to macroscopic life—certain microbes that thrive in anoxic environments. And as this anoxic microbial life proliferates, it might generate noxious gases that pose a serious problem for oxygen-dependent animal life.

In this fashion, modern deep lakes might serve as a microcosm of the Late Permian ocean. Thanks to the supercontinent cycle and Wilson cycle, super-continents such as Pangea, Pannotia, Rodinia, Columbia (Nuna), Kenorland and others amalgamated every 300–500 million years. The continental arrangement of Pangea with a single giant ocean (the Panthalassic) during the Permian Period made the ocean conveyor belt inherently more susceptible to major disruptions. In addition to potentially disrupting the critically important global conveyor belt, the single supercontinent geography allowed exposure of the continental shelves through a worldwide drop in sea level. Exposure of the continental shelves led to the direct killing of shallow water species such as corals.

And the massive amount of outgassed carbon dioxide from the Siberian Traps led to a global warming that tended to disrupt the conveyor belt system. The two phenomena conspired to magnify the negative consequences. The net result was a dramatic drop in oceanic oxygen—worldwide stagnant seas. Computer models that raised the Panthalassic Ocean surface temperatures by 10 °C (18 °F) suggest that the oceans lost about 80% of their oxygen content. About half the oceans' seafloor, especially at deeper depths, became completely anoxic. Like deep lakes today, the deep ocean at the end of the Permian became uninhabitable (by animals). But although fish and other marine animals could not survive, anaerobic microorgan-isms flourished and generated toxic gases, further making a bad situation worse.

The volcanoes not only spit out lava and carbon dioxide, but also dust and vast amounts of sulfur dioxide. The volcanic dust could have blocked incoming sunlight and disrupted photosynthesis. Atmospheric sulfur dioxide from the volcanoes interacts with moisture to generate sulfates, which have a high albedo. These sulfates would work to cool the planet by reflecting incoming sunlight. Thus, the dust and sulfur from the volcanoes would have diminished insolation and had consequences on primary production through photosynthesis.

Additionally, the atmospheric sulfur dioxide would mix with rainwater to create relatively strong solutions of sulfuric acid. And of course, carbon dioxide is acidic when dissolved in water as carbonic acid. The resultant acid rain might have had a pH as low as 2 (the equivalent of lemon juice). In the ocean, which is normally alkaline with a pH of about 8.1, presently, this acidification might have caused the pH to fall by 0.6 or 0.7 units. This catastrophic change killed off pH-sensitive calcareous organisms such as mollusks, sponges, corals and planktonic microbes with calcium carbonate exoskeletons. Zooxanthellae and zoochlorelae are algae that symbiotically grow within coral cells and generate oxygen and food via photosynthesis in exchange for a safe place to live. But under elevated temperatures, these symbionts abandon their hosts, leaving the corals vulnerable. The outcome was predictable as the reef systems collapsed.

On land, this acid rain could have killed trees and other plants. In addition to a direct effect on plants, prolonged exposure to acid rain could have had an indirect effect on terrestrial botanical life by affecting soil chemistry. The lowering of pH may have increased the availability of aluminum, which is generally toxic, while simultaneously decreasing the availability of essential calcium and magnesium in the soil for plants. This could have set up a domino effect in which the soils, normally firmly anchored by these plants and trees, were swept away by rains. The result would have been a denuded landscape that was less able to sustain plants and the animals that depended on them. Evidence supporting this hypothesis is seen at the end of the Permian by a change in river and stream patterns from a winding and meandering pattern (which is seen when tree roots and plants are abundant) to a more braided but straight series of fast-moving waters (which is characteristic of streams where tree roots are scarce). This loss of vegetation accelerated the long-term global warming process by disrupting the normal feedback loop in which plant life mops up excess carbon dioxide. Without the normal floral sink of carbon dioxide, atmospheric carbon dioxide accumulated even further in a positive feedback loop. The Siberian Traps also outgassed enormous amounts of chlorine and fluorine. Based on analyses of trapped gases in inclusion melts (pockets of air in the lava left behind), it has been estimated that the Siberian Traps spewed out as much as 7800 gigatons of sulfur, 8700 gigatons of chlorine and 13 700 gigatons of fluorine. And while these are upper limits, even lower amounts, if they reached the stratosphere, they could have wreaked havoc on the biosphere by ozone depletion.

30.5 A contribution from coal

While the volcanic dust and these other gases certainly did not help the environment, it was volcanic carbon dioxide that was probably the most significant problem thanks to the long-term greenhouse effect. The ensuing global warming is documented by a sharp decrease in oxygen isotope ratios (oxygen-18/oxygen-16) in fossils from the time. But the volcanoes had help when it came to carbon dumping. The Carboniferous and Permian Periods were times of immeasurable amounts of coal formation thanks to unbridled plant growth. And central Siberia was an area filled with major coal deposits near the end of the Permian. Thus, when the Siberian Traps erupted, they not only released carbon dioxide through volcanic outgassing but also set ablaze vast coal fields that poured gigatons of carbon dioxide into the atmosphere. And it is worth recalling that coal ash itself is quite toxic.

Whether the amount of carbon released directly by the volcanoes or the coal ash from the coal burning was greater remains uncertain. Nevertheless, the combined quantity might have set up yet another positive feedback loop that exacerbated the global warming even further—and melting of the methane clathrates.

30.6 Methane clathrate melting

This excessive heating set up additional positive feedback loops that conspired to make the Hothouse Earth even hotter. Most notable among these effects was a tipping point beyond which underwater methane clathrates began to melt. Clathrates are hydrated, solid, deep oceanic gas-rich deposits that exist primarily in offshore continental margins. Thanks to the depths in such seafloor regions, the water is cold even in the tropics. The cold temperatures and high pressures keep the clathrates (which are methane molecules trapped in a cage of water molecules) in a solid state. Modern-day clathrates contain much microbial-produced methane and sequester this methane on the ocean floor. By some estimates, there is about three times as much methane sequestered in the form of methane clathrates as there is in the form of fossil fuel sites.

As the ocean water temperature continued to rise at the end of the Permian, eventually it reached a point where the frozen clathrates melted and released their methane gas into the atmosphere. As mentioned elsewhere, methane is a far more potent greenhouse gas than carbon dioxide (maybe 25- to 34-fold more potent on a per unit weight basis). Thus, through carbon dioxide from the Siberian Traps, the warming atmosphere heated the oceans and the heated oceans (through released methane gas) then further heated the atmosphere. Additionally, methane consumes oxygen as it is oxidized into carbon dioxide. By some estimates, the total amount of carbon dioxide in the atmosphere thanks to the combined effects of the volcanoes, coal fires and clathrate melting might have been around 36 teratons, leading to an atmospheric carbon dioxide concentration ranging up to 10 000 parts per million (in contrast to today's current value of 400 parts per million).

As warm water is less capable of holding dissolved oxygen than cold water, this Hothouse Earth phenomenon, coupled with a drastic reduction in oxygen production thanks to the demise of plants, contributed further to a hypoxic worldwide environment.

30.7 The rise of methanogens

A basic question is, where do methane clathrates come from in the first place? The answer is, from microorganisms, specifically, microbes known as methanogens. Methanogens are archaea (not bacteria) that might have evolved somewhere during the Late Permian. An interesting hypothesis is that the Emeshian Traps and Siberian Traps volcanism triggered the growth of Methanosarcina, an Archean methanogenic microbe that then dumped large amounts of methane into Earth's atmosphere and contributed to methane clathrate creation. The proposed mechanism whereby volcanic eruptions stimulated the growth of Methanosarcina is a massive release of nickel into the environment (volcanoes spit out a lot of nickel and chromium). The link is that the methanosarcina require nickel in their enzymatic active sites. The enzyme is methyl cofactor M reductase, which contains cofactor F430; cofactor

F430 uses a coordinate nickel atom in its center. In any case, the theory holds that the newly plentiful environmental nickel encouraged the evolution of these methanogens. These methanogens in turn introduced methane into the environment and allowed the formation of methane clathrates. But these clathrates might not have had the chance to hang around very long as the temperatures rose and melted them, pouring their methane into the atmosphere and causing a worldwide greenhouse effect.

30.8 Poisonous fumes—hydrogen sulfide accumulation

The theory of oceanic stagnation is supported by evidence in the sedimentary rocks laid down on the ocean floor around this time. Early Triassic sediments tend to be black and full of pyrite (iron (II) disulfide or FeS_2). Such sediments are consistent with an anoxic environment. And these dark, sulfurous sediments contain very few animal fossils. But unlike metazoan animals who depend on oxygen (except for the loriciferans), anaerobic bacteria such as the sulfate-reducing bacteria thrive under such conditions. Therefore, warm, stagnant, anoxic and non-circulating oceans were the perfect environment for sulfate reducers.

Humans and other animals breathe in O_2 and reduce it to H_2O through the highly efficient process of aerobic respiration. Molecular oxygen is the final electron acceptor in the involved electron transport chain within mitochondria. In contrast, sulfate-reducing bacteria breath sulfate instead of oxygen and reduce it not to H_2O but rather into H_2S through the process of anaerobic respiration. In this form of anaerobic respiration, sulfate serves as the final electron acceptor instead of oxygen. The end product H_2S (hydrogen sulfide) is evident in the pyrite formed in abundance around the end of the Permian. But the important thing biologically is that hydrogen sulfide is extremely toxic to metazoan life. In a strange twist of evolutionary fate, the microbes who were banished from the surface by the generation of oxygen were now poised to take back the world through the generation of hydrogen sulfide.

The situation in which hypoxia is combined with an overabundance of hydrogen sulfide is known as euxinia. Thus, the oceans during the end of the Permian were euxinic. But as the ocean waters became saturated, hydrogen sulfide gas was emitted into the atmosphere. Atmospheric hydrogen sulfide poisoned plants and animals alike as well as possibly weakening the ozone layer. The bacterially produced hydrogen sulfide, along with the volcanic emissions of chorine, fluorine and sulfur might have all co-conspired to diminish the stratospheric ozone layer.

In modern **limnology** (lake science), a **meromictic lake** is a type of lake in which the shallow and deep layers (**epilimnion** and **hypolimnion** respectively) do not mix. Meromictic lakes contrast with **holomictic lakes** in which there is regular mixing between the epilimnion and hypolimnion at least once per year. Meromictic lakes might serve as a good model for the Late Permian Panthalassic Ocean. Sulfate-reducing bacteria thrive in the deep waters of meromictic lakes, where they produce hydrogen sulfide. This hydrogen sulfide is then available for other anaerobic bacteria —the green sulfur and purple sulfur bacteria—that use hydrogen sulfide as an electron donor in their photosynthetic biochemistry. Evidence of this was left behind in the form of chemical biomarkers such as okenane and chlorbactane in the rocks made around the Permian extinction. In addition to the biological consequences of euxinia, the atmosphere was probably grossly altered. Rather than a present day weather forecast of 'blue skies with a bit of haze' the probable daily forecast back then was 'green skies with a lot of haze'. With a fetid stench.

30.9 The gradual rebound

The biosphere was grossly altered for about 15 million years after the Permian–Triassic boundary. While the most obvious impact was the loss of animal and plant species, some organisms benefited from the chaos. Among those who flourished were the fungi. Not surprisingly, since fungi are saprophytic (meaning they live on and derive nourishment from dead organisms and decaying organic matter) they had plenty of food with little competition. Fungi might be described as 'disaster taxa' and 'opportunistic taxa' meaning that they survived the disaster and actually thrived thanks to being opportunists who could benefit from the new circumstances. Among the first plants to rebound were ferns. This was seen after other extinctions as well indicating that ferns were disaster taxa or 'Lazarus taxa' that rebounded back after practically disappearing. The therapsid genus *Lystrosaurus* has long been claimed to be a disaster taxon that made it through the extinction event and did well despite the setbacks. Recent research challenges this view, however.

30.10 The Ordovician extinction

Aside from the great diversification of life during the Cambrian (the Cambrian explosion), another pronounced episode of adaptive evolutionary radiation or divergent evolution occurred during the Ordovician period. The Middle Ordovician period was a time of major proliferation of new life forms. In fact, this surge of new life forms rivaled the Cambrian explosion and goes by the name, the Great Ordovician Biodiversification Event. An important example of such expansion was provided by the **graptolites** (a type of hemichordate) which underwent rapid diversification in the Ordovician. Graptolites adopted a planktonic lifestyle in the Ordovician and sank in vast numbers upon their death, sometimes almost completely covering the deep ocean floors. Upon burial and petrification, these organisms became of immense value for providing fine structure to the fossil record through **Smith's Second Law** (the **law of stratigraphy based on fossils**, which states that each stratum can be identified by its distinct set of fossils. For reference, **Smith's First Law** states that older strata are deeper than younger strata). Graptolites, through Smith's Second Law, allow Ordovician sedimentary rock layers to be accurately correlated even if they are not contiguous. Graptolites were the kings of correlation in stratigraphy during this time.

Also appearing during the Ordovician were burrowing bivalve mollusks, elongated crinoid echinoderms, the first coelenterate coral reefs and many more. Early agnathan fish called conodonts were among the first vertebrates. The conodonts first appeared in the Cambrian but diversified extensively during the Ordovician and reached their peak during this period. The last phylum of animals to evolve (or at least the last to appear in the fossil record) were the bryozoans or moss animals, who first left fossils in the Early Ordovician.

But all this would become disrupted around 444 million years ago when the first of the Big Five major mass extinctions occurred. Exactly what led to this extinction event remains unclear. But since one of the proposed explanations is a **gamma ray burst**, we shall include a discussion of this mass extinction in this chapter.

30.11 A changing climate near the end of the Ordovician

It is believed that dramatic climate change was at the root of the Ordovician extinction. The first substantial ice age of the Phanerozoic Eon took place in the Late Ordovician and it is possible that marine animal life was simply not yet equipped to handle such a challenge. But just why an ice age came about probably has a lot to do with plate tectonics and the permissive continental arrangement at the time.

Fragmented remnants of **Rodinia** (the second most recent supercontinent) which had been splitting up since about 750 million years ago were wandering about the planet. **Gondwanaland** comprised parts of Africa, South America, Antarctica, India and Australia while **Laurentia** comprised much of North America. Other continents included **Baltica**, which later formed much of northwest Europe, and **Avalonia**, which gave rise to today's southern Britain, Atlantic Canada and the East Coast of the United States. Over millions of years during the Ordovician Period, Gondwanaland moved towards—and then over—the South Pole. Unlike when flowing ocean water is over the poles, when a large landmass is over polar regions, snow readily accumulates and glaciers grow. Major glaciation occurred during the Late Ordovician. Eventually (like present-day Antarctica), part of Gondwanaland became completely icebound. Huge glaciers formed for the first time since complex metazoan life first evolved. Glaciation was new to this early burgeoning biosphere. None of the Ordovician creatures were preadapted to an ice age (as later species would be because their ancestors had gone through one). Thus, the casualties to the biosphere were inordinately high.

30.12 Proof of an Ordovician ice age

How do we know there was an ice age? Isotopic analysis of oxygen isotopes is a method of detecting a glacial period. The lightest oxygen isotope (oxygen-16) is the most common but there is a small but measurable proportion of oxygen-18. Water (H_2O) with the light oxygen-16 isotope is preferentially evaporated. Therefore, rain (and snow) is predominately made from the lighter oxygen-16. Normally all this gets washed back into the oceans. But during the Late Ordovician (or any) glacial period, lighter water (H_2O with the light oxygen-16 isotope) got locked up in glaciers. It became sequestered in polar ice and did not get back to the oceans.

Over time, whenever a large amount of snow accumulates on land in the form of glaciers, the oceans will develop a relative enrichment of the heavy isotope oxygen-18 in the remaining water. This isotopic signature was recorded in the chemistry of creatures that secrete calcareous shells (e.g. the Ordovician graptolites). Biogenic calcium carbonate ($CaCO_3$) is mostly oxygen (both on a molar basis and by weight) and this oxygen is ultimately derived from water in the ocean. Calcium carbonate shells from the Late Ordovician contain more of the heavy oxygen-18 isotope than normal, thereby confirming the presence of an ice age.

The isotopic data from calcareous shells indicate that this glacial period was short —probably under 1 million years in duration. The extinction appears to have occurred in two waves. Since the Ordovician extinction is also called the **Late Ordovician Mass Extinction Event (LOME)**, the first extinction **interval** is abbreviated **LOMEI-1** whereas the second interval is called **LOMEI-2**.

The extinction first began in the last **stage** of the Ordovician Period (the **Hirnantian stage**) with the initial extinction phase (LOMEI-1) occurring at the **Katian/Hirnantian stage boundary**. LOMEI-2 then occurred in the late Hirnantian stage of the Ordovician Period.

(The Hirnantian was the last stage of the Ordovician Period, spanning from 445.2 to 443.8 million years ago. It was preceded by the **Katian stage**, which began 453 million years ago and lasted until the beginning of the Hirnantian, 445.2 million years ago.) Overall, the extinction event was about 1.9 million years long, taking place approximately from 445 to 443 Ma.

30.13 The effects of the Ordovician extinction

By count of lost species, the Ordovician extinction was the second most devastating of the Big Five, behind only the Permian Disaster. All told, up to 85% of all species were gone by the time the extinction event was over. It is worth recalling however that life was restricted to water during the Ordovician; aside from the suggestion of some small plants based on spores, all life was in water back then.

The extinction was rather severe for reefs and reef faunas and anything adapted to warm water. For instance, 50 of the 70 known genera of tabulate and rugose corals suddenly vanished. The cooler water creatures started migrating toward the equator to avoid the freezing cold. But the life forms adapted to equatorial tropical climates had nowhere to go except for extinction.

30.14 The effect of the ice age on sea levels

When water becomes locked up in ice in the form of glaciers on the continents, not only does the ratio of oxygen isotopes in the ocean change, the ocean water is physically prevented from getting back into the oceans. Consequently, sea levels fall. Evidence of this is seen in the severe erosion of many Ordovician sediments. Rivers ran fast and furious in the Late Ordovician. This occurred as rivers tried to come to equilibrium with the new, but much lower sea levels.

30.15 The second wave

The second pulse of extinction (LOMEI-2) came at the end of the glacial period as temperatures and sea levels rose again. LOMEI-2 was not as severe as the first extinction pulse, LOMEI-1. It principally affected newly cold-adapted forms that struggled to rapidly switch back again to a warm environment.

30.16 Why was there a delay in the onset of the ice age?

Overall, the consensus is that glaciation was the most likely cause of the Ordovician extinction (at least the first wave). However, Gondwanaland had moved over the South Pole millions of years before the major glaciation and extinction occurred. So, the question arises—why did the glaciers not start forming back then? Something else (besides just the movement of Gondwanaland over the South Pole) must have changed in order to shift Earth to an ice age. One possible explanation is a simultaneous reduction in greenhouse gases. During the Ordovician Period, Baltica and Avalonia moved toward Laurentia. The Iapetus Ocean between them was shrinking. As the Iapetus Ocean closed and the continents approached one another, mountains started to rise (orogeny) thanks to tectonic activity.

Whenever new mountains form, they become subjected to intense rock weathering. This erosion and rock weathering occurred as the new mountain ranges formed in the Ordovician. The weathering of rocks uses up CO_2 in the atmosphere through silicate weathering and the carbon–silicate rock cycle.

Usually, the consumption of atmospheric CO_2 by rock weathering is balanced by the addition of new atmospheric carbon dioxide from volcanoes. But in the Late Ordovician, volcanic activity had slowed for some reason. Thus, there was a net drop in atmospheric CO_2. This, along with Gondwanaland being over the South Pole and encouraging glaciers, the world also experienced a drop in greenhouse gas levels. A diminished greenhouse effect, plus the particular location of the continents, was the key combination. In essence, the location of the continents over the south pole was a necessary, but not sufficient, condition for the initiation of an ice age. It was this *combination* that drove Earth towards a sudden ice age. Hence, there was a delay in the onset of the Late Ordovician ice age.

All of this discussion so far ignores the important role of **orbital forcing** in the climate. The **Milankovitch cycles** involve periodic changes in Earth's orbit around the Sun, its axial tilt and precession of the axis. The Earth's orbital eccentricity has an overall cycle of about 100 000 years, which is a combination of effects from components with 405 000 years, 95 000 years and 124 000 years. Earth's axial tilt or obliquity also varies periodically between 22.1° and 24.5° over a roughly 41 000 year cycle. Our present axial tilt is about 23.5° or more precisely, 23.44°. Finally, there is a periodic precession of the Earth's axis over a period of roughly 26 000 years (or maybe more precisely 25 770 years). Presently our northern axis points towards the North Star, Polaris. But this was not always the case. For instance, thanks to this precession of the equinoxes, in the year 3000 BCE, Thuban in the constellation Draco was the North Star. In the year 14 500 AD, Vega in the constellation Lyra will be the North Star. But the relevant issue here is that during the Late Ordovician, these parameters were slightly different. And in addition to the combined effects of continental location over the South Pole and decreased greenhouse gases mentioned above, there might have been important consequences of orbital forcing over and above these two other factors. Interested students can independently explore the newly emerging details of Milankovitch cycles during the Ordovician Period.

30.17 Another contributor to the cooling climate?

As discussed elsewhere, a hypothesis involving space radiation called the **Svensmark effect** or **Svensmark hypothesis** holds that increased incoming cosmic rays could contribute to cooling of the climate. The Svensmark hypothesis asserts that more cosmic rays would create more condensation nuclei in the atmosphere. These condensation nuclei facilitate the formation of clouds. In turn, over prolonged periods, the increased cloud cover reflects incoming sunlight, resulting in a colder climate. The increased cloud cover can also lead to increased snow. If sufficient snow falls and accumulates as glaciers, it will possibly contribute to a positive feedback loop wherein the high-albedo snow reflects incoming sunlight, further cooling the planet. In this manner a genuine global glaciation might materialize. Of course, obtaining proof of such increased cosmic ray influx in the Late Ordovician is difficult.

Generally speaking, when the magnetic field strength of either the Sun, the Earth or both is diminished, galactic cosmic ray influx will correspondingly increase. It is known that the Earth sporadically flips its geomagnetic polarity such that the North Pole becomes a South Pole and vice versa. These geomagnetic reversals occur on average (in recent times) about once every 200 000 years. During the transition period, the magnetic dipole collapses into a multipolar arrangement and the overall field strength decreases (perhaps as much as 90%). During these geomagnetic transition periods with diminished geomagnetic shielding (which may take centuries to thousands of years), there would be a noticeable increase in cosmic rays striking the atmosphere and reaching Earth's surface. Of course, the atmosphere will still filter out most primary galactic cosmic rays as well as the secondary muons but nevertheless, the increase in incoming cosmic rays will mean more radiation will reach the surface. Secondary muons are the most abundant energetic charged particles from space. At sea level they arrive with an average flux of around 1 muon per square centimeter per minute. The mean energy of cosmic ray muons at sea level is about 4 GeV. At these energies (and higher, up to a few hundred GeV), muons mainly interact with matter through interactions with atomic electrons (that is, through ionization). These muons lose energy at the fairly constant rate of about 2 MeV per g cm^{-2}.

The radiobiological consequences of such increased radiation are uncertain. Cosmic rays contribute only about 8% of all ionizing radiation exposure at this time (including medical exposures), so even doubling that amount would not drastically increase the overall annual dose to terrestrial organisms at sea level. In the United States today, the average annual radiation dose from cosmic rays is about 26–28 millirem (0.26–0.28 mSv) per year but ranges up to 70 mrem (0.7 mSv) for those residing at high altitudes (8000–9000 feet). In other countries that have regions with higher altitudes than the United States of America, especially at higher geographical latitudes, the annual dose from cosmic rays would be higher. In present-day society, a person living at high altitude, residing in a country nearer the poles, who flies about 100 000 miles per year via jet (which confers around 100 mrem (1.0 mSv) annually) could receive somewhere in the vicinity of 200 millirem

(2.0 mSv) of cosmic rays per year. To date, such individuals have not been found to have any health or genetic consequences of this increased cosmic ray radiation.

As discussed in greater detail in chapter 29, three grand solar minima—the **Spörer Minimum** (1450–1540), the **Maunder Minimum** (1645–1715) and the **Dalton Minimum** (1790–1820) occurred during the climatological anomaly called the Little Ice Age. These solar minima were documented by a marked diminution in sunspot counts (and via ice core proxies in the pre-telescope era). (Climate proxies are covered in greater depth in another section.) Since solar minima correspond to a decrease in deflective solar wind and heliomagnetic shielding, it is believed that such grand solar minima were accompanied by increased incoming galactic cosmic rays on Earth. Whether or not any such solar forcing was in effect during the Late Ordovician ice age is unknown.

30.18 A perfect storm?

As long as we are exercising our unbridled imagination, one can take the scenario to its logical extreme and wonder 'what if'…

1. What if the Earth was undergoing a geomagnetic reversal and the magnetic field strength was at a nadir?
2. What if the Sun was in a grand solar minimum so that the solar wind and heliospheric strength were also at a nadir?
3. What if the Earth, Sun and Solar System were at the maximum distance from the galactic plane at this time and hence the galactic magnetic deflection of extra-galactic cosmic rays was similarly at a nadir?
4. What if a nearby massive star then went supernova, flooding the skies with an abundance of cosmic rays?

What effects such a perfect storm might have on our biosphere and climate are the subject of speculation. If the climate can indeed be influenced by increased cosmic rays, such a quadruple-whammy scenario certainly could cause challenges. On the other hand, increased chronic radiation exposure from space, even under such extreme circumstances, is unlikely to lead to an extinction event through radiological or radiobiological mechanisms in the opinion of some. However, the fact remains that such an unusual situation would increase natural background radiation exposure and therefore one might speculate about the effects on evolution. Charles Darwin admitted that the ultimate source of the variability seen in species upon which natural selection operates remained a mystery to him. It would be interesting to see if a population of organisms chronically exposed to moderate levels of this type of high linear energy transfer radiation would be more capable of evolving, if the environmental conditions favored such change.

30.19 A gamma ray burst?

Although it is only a minority view at present, among the proposals for the Ordovician extinction was a **gamma ray burst (GRB)**. GRBs will be covered in greater detail in the following chapter. Although GRBs are as a rule extremely distant (on the order of billions of light years away), this postulated extinction-provoking GRB would have been relatively nearby in our Milky Way Galaxy, somewhere within 6000 light years of Earth. Such a nearby and intense GRB would have not caused an extinction through direct radiological effects or radiobiological consequences of the blast. But a 10 second blast, if sufficiently intense, would have promptly depleted the stratospheric ozone layer by up to 50% and our protective ozone layer would have remained inadequate for years to come. The radiochemical mechanism behind this is like the familiar radiochemical reactions behind radio-biology, except they took place in air rather than in aqueous solution. Just as in water, gamma rays in air will break covalent chemical bonds and create free radicals and other highly reactive chemical species. Since the atmosphere is presently 78% molecular nitrogen and 21% molecular oxygen, the most common chemically reactive products made would have been reactive nitrogen species (RNS) and reactive oxygen species (ROS). Gamma rays would have broken the triple bonds in molecular nitrogen (N_2) and created atomic nitrogen. The nitrogen atoms would be very reactive free radicals since they have unpaired valence electrons. These nitrogen atoms (N·) would have then reacted with molecular oxygen (O_2) to generate nitric oxide (NO). Nitric oxide attacks ozone (O_3) to yield nitrogen dioxide (NO_2) after the ozone-destroying reactions. Nitrogen dioxide can then react with more molecular oxygen to form additional ozone-eliminating NO, creating a chain reaction cycle. Some computer models have shown that around half the ozone layer would be destroyed in weeks and at 5 years later, over 10% would still be missing.

This would have exposed superficial marine life forms (recall that there were essentially no terrestrial life forms yet) to unmitigated ultraviolet radiation. Between 20%–60% of the planktonic biomass might have been killed by the intense and prolonged ultraviolet exposure. And during the Ordovician lakes and seas were probably less turbid than in today's world, allowing deeper penetration and increased biological effectiveness of the ultraviolet. Furthermore, it is probable that many of the larval forms of adult arthropods (such as benthic trilobites) were planktonic. Hence, coupled with the larger vulnerabilities imposed by the serious climate changes, entire generations of animals could have been wiped away, setting the stage for a genuine mass extinction.

Further reading

[1] Huang X, Yuan Q and Fan Y Z 2021 A GeV-TeV particle component and the barrier of cosmic-ray sea in the Central Molecular Zone *Nat. Commun.* **12** 6169

[2] Pierre Auger Collaboration 2017 Observation of a large-scale anisotropy in the arrival directions of cosmic rays above 8 × 1018 eV *Science* **357** 1266–70

[3] Raup D M and Sepkoski J J 1984 Periodicity of extinctions in the geological past *Proc. Natl Acad. Sci. USA* **81** 801–5

[4] Medvedev M V and Melott A L 2007 Do extragalactic cosmic rays induce cycles in fossil diversity? *Astrophys. J.* **664** 879–89

[5] Cui Y, Li M, van Soelen E E, Peterse F and Kürschner W M 2021 Massive and rapid predominantly volcanic CO2 emission during the end-Permian mass extinction *Proc. Natl Acad. Sci. USA* **118** e2014701118

[6] Modesto S P 2020 The disaster taxon *Lystrosaurus*: a paleontological myth *Frontiers Earth Sci.* **8** 610463

[7] Thomas B C *et al* 2005 Gamma-ray bursts and the Earth: exploration of atmospheric, biological, climatic and biogeochemical effects *Astrophys. J.* **634** 509–33

[8] Dai S and Luo J *et al* 2019 The Ordovician magnetostratigraphy and cyclostratigraphy: a review *Acta Geologica Sin.* **93** 94–7

[9] Sinnesael M 2022 Ordovician Cyclostratigraphy and Astrochronology *Geol. Soc. Spec. Publ.* **532**

[10] Artemieva N and Morgan JExpedition 364 Science Party 2017 Quantifying the release of climate-active gases by large meteorite impacts with a case study of Chicxulub *Geophys. Res. Lett.* **44**(20) 10.180–8

[11] Bermúdez H 2022 Presentation: The Chicxulub Mega-Earthquake: Evidence from Colombia, Mexico, and the United States *The Geological Society of America Connects 2022* 54 Geological Society of America Abstracts with Programs *(Denver, CO, 9-12 October 2022)* https://gsa.confex.com/gsa/2022AM/meetingapp.cgi/Paper/377578

[12] Thomas S, Tobin C, Bitz M and Archer D 2017 Modeling climatic effects of carbon dioxide emissions from Deccan Traps volcanic eruptions around the Cretaceous–Paleogene boundary *Palaeogeogr. Palaeoclimatol. Palaeoecol.* **478** 139–48

[13] Telescope Array Collaboration *et al* 2023 An extremely energetic cosmic ray observed by a surface detector array *Science* **382** 903–7

Chapter 31

Gamma ray bursts

31.1 History

The first gamma ray burst (GRB) was detected on 2 July 1967 by the Vela 3 and Vela 4 satellites as part of 'Project Vela' (which was designed to detect nuclear detonations to ensure compliance with the Nuclear Test Ban Treaty of 1963). Equipped with neutron, x-ray and gamma ray detectors, the Vela satellites were scanning for traces of nuclear emissions during the Cold War. The gamma ray signals detected however, were different from any then-known nuclear weapon signatures. It was soon realized that these gamma rays were not coming from terrestrial nuclear weapon detonations but rather were coming from outer space. By measuring the varying arrival times of the gamma rays by different satellites in different locations, investigators were able to approximately determine the sites of origin. These gamma rays were clearly not coming from Earth, or even the Sun. The paper, *Observations of Gamma-Ray Bursts of Cosmic Origin* by Klebesadel, Strong and Olson confirming that the source of these gamma ray bursts was beyond our Solar System, was published in 1973 in the Astrophysical Journal.

Upon closer subsequent analysis using the highly sensitive **Compton Gamma Ray Observatory's Burst and Transient Source Explorer (BATSE)** instrument, it was determined that these gamma rays were not even coming from our own galaxy. If the gamma ray sources were from within the Milky Way, they would be expected to have a distribution reflecting the bulk density of the galaxy—that is, they would be concentrated along the main disc. Pulsars, which are rapidly rotating neutron stars, are distributed largely along the plane of the Milky Way, where the massive stars that formed them are. But these bursts of gamma rays were not confined to the plane of the Milky Way; they were isotropic, meaning that they were emanating from sources seemingly randomly distributed. It was still possible that GRBs might be coming from the halo of our the Milky Way and giving the illusion of being randomly or isotopically distributed. However, if this were the case, we should see GRBs in the halo of other galaxies such as Andromeda. This concentration around Andromeda or any other nearby galaxy was not seen, however. Nevertheless,

Compton proved to be extremely valuable in the understanding of GRBs as it detected and mapped the (approximate) positions of around 2700 GRBs during its lifetime between 1991 and 2000.

The Compton Gamma Ray Observatory was one of NASA's 'Great Observatory' series, which included the Hubble Space Telescope, the Spitzer Space Telescope and the Chandra X-ray Observatory. (Joining these four is the recently launched James Webb Space Telescope, which detects mainly infrared radiation.) Compton was the second one of the four to launch. It was also the first space telescope to ever undergo an intentional, controlled de-orbiting after one of its gyroscopes failed. This failure meant it might become unstable and crash land into populated areas. At the time, it was the heaviest space telescope payload ever flown (17 000 kg or 37 000 lb) and its unpredictable crash could cause trouble. So, it was de-orbited and sent into the Pacific Ocean on 4 June 2000. During its reign, Compton resided in low Earth orbit (LEO) to avoid the radiation-filled Van Allen belts and was capable of detecting photons with energies between 20 keV and 30 GeV.

Based on the Vela and Compton data, it was concluded that these GRBs were 'cosmological' in distance, meaning from beyond our galaxy, and therefore must have been incredibly intense to be detectable here on Earth with our technology. The rate that astronomers were observing GRBs—about one per day—was another mind-boggling puzzle.

Since the GRBs were just that—short bursts rather than prolonged emissions—they were hard to track in other wavelengths. While the Compton BATSE could detect GRBs, it did not have fine spatial resolution and only provided roughly a 5° region that optical telescopes could examine. Given the vast distances of GRBs, this was not a very fruitful effort. Furthermore, by the time the telescopes would train their optics to the general area from which the gamma rays came, the gamma rays were often long gone, and a precise location could not be established. The fact that GRBs were so slick is testimony to how brief they are. As will be discussed below, GRBs come in two versions: short GRBs which are less than 2 s and long GRBs which are more than 2 s.

31.2 BeppoSAX

The Dutch–Italian low-Earth orbit satellite **BeppoSAX** was launched in 1996 and deactivated in 2000. This special satellite had a set of narrow-field x-ray detectors (capable of detecting photons with energies between 0.1 and 300 keV) as well as a wide-field gamma ray burst monitor with gamma ray detection capabilities from 100 to 600 keV. This wide-field camera had a very broad field of view of nearly all the sky and was ideal for the detection of GRBs. (Incidentally, BeppoSAX was named in honor of Italian physicist Giuseppe 'Beppo' Occhialini. SAX stands for 'Satellite for X-ray Astronomy' in Italian.) It should be mentioned that ground-based x-ray astronomy is impossible because our thick atmosphere effectively blocks out all incoming x-rays as well as nearly all gamma rays. In February 1996, the BeppoSAX wide-field camera detected a GRB and its x-ray camera caught a glimpse of fading x-rays from the same site. Then, based on these coordinates, the **William Herschel Telescope** (an optical/near-infrared reflector telescope on the Canary Islands, Spain built in 1987) saw a fading optical counterpart about 20 h afterwards. This particular GRB (called GRB 970228) was the first one to have an optical 'afterglow' detected.

Incidentally, GRBs are named for their date of discovery with the first two digits representing the year of discovery, the next two digits representing the month and the next two digits signifying the day and finally there might be a letter indicating the order of discovery on that particular day (with A being the first, B being the second and so forth). The fact that such a system is in place informs one of just how many GRBs there really are out there! In recent years with our advanced technologies, about one per day is detected.

Another important breakthrough came on 8 May 1997 (thus, the name GRB 970508). This time, the afterglow was observed only 4 h after the GRB and redshift measurements were made. These revealed that the GRB had a redshift of $z = 0.835$, corresponding to an astonishing 6 billion light years away. This confirmed that GRBs truly were 'cosmological' distances away, originating in extremely distant galaxies. In 1998, another GRB (GRB 980425) was observed to be coincident in location with a supernova. This was the first clue that GRBs might be connected in some way with the death of extremely massive stars.

Following BeppoSAX, the multi-institutional, international **High Energy Transient Explorer (HETE)-2** was designed to study GRBs in multiple energies, including gamma rays, x-rays and ultraviolet, using various instruments mounted on a single spacecraft. HETE-2 was launched on 9 October 2000 and deactivated about seven and a half years later in March 2008. This was gratifying given its anticipated duration of only 18 months and the fact that its predecessor, HETE-1, which was launched on 4 November 1996, failed in orbit because of inability to properly disconnect from its carrier vehicle. HETE-1 was unable to deploy its solar panels, lost power just a few days after launch and ultimately re-entered on 7 April 2022. HETE-2 discovered GRB 030329, which firmly established the connection between long GRBs and supernovae. Among other important contributions, HETE-2 confirmed the existence of a subclass of GRBs with lower energies, now called **x-ray flashes**. Like other long GRBs, x-ray flashes are believed to stem from hypernova explosions and last about 90–200 s. It is presently believed that x-ray flashes are nothing more than ordinary long-duration gamma ray bursts that happen to be at a greater angle from our line of sight.

31.3 Types of GRBs

Based on plots of the number of GRBs discovered versus time, a bimodal distribution was observed. The first peak in the curve corresponded to GRBs lasting under a second. The second peak appeared at a time corresponding to about half a minute.

We now understand that GRBs come in two basic forms, the short-duration and long-duration types. An arbitrary cutoff time of about 2–3 s historically divided the two types but it is now evident that there is some temporal overlap in the two types, which are believed to originate through very different mechanisms. Short-duration GRBs last for an average duration of about 300 ms; long-duration GRBs persist for an average duration of about 20–30 s and may then fade over the course of several minutes. As a generalization, short-duration GRBs tend to have higher-energy photons than long-duration GRBs. Hence, GRBs are sometimes categorized as 'short and hard' GRBs versus 'long and soft' GRBs (although the gamma rays in long-duration GRBs are not really very soft).

Although most GRBs fall into one of these two basic classes, like fingerprints and snowflakes, no two GRBs are identical. Upon scrutiny of their 'light curves' (a plot of intensity or luminosity on the ordinate (vertical axis or y-axis) with time on the abscissa (horizontal axis or x-axis)), subtle differences are seen between each and every GRB. As mentioned, the duration of gamma ray emission can vary from milliseconds to many minutes (even over 2 h in some very rare examples). There might be a single gamma ray pulse or there might be a series of sub pulses. Sometimes there might be a short, relatively faint burst of x-rays seconds to minutes before the actual GRB. This rare x-ray precursor is called a **GRB precursor**. There are many variations on the theme. Most GRB light curves have a 'spiky' appearance with multiple close peaks and valleys rather than a smooth continuous curve.

31.4 Long-duration GRBs

Thanks to their longer persistence (and easier visualization of their longer-wavelength afterglows) the long-duration GRBs are better characterized. Further contributing to their better understanding is the fact that the majority of GRBs (about 70%) are of the long-duration type.

It became apparent that long-duration GRBs were associated with distant galaxies, confirming their unbelievable distance. Further characterization showed that long-duration GRBs are found in galaxies with rapid star formation, especially in regions rich in very massive stars. A sizeable fraction of long-duration GRBs are clearly associated with a core-collapse supernova, which definitively links them with the death of very massive stars.

Long-duration GRBs have a slightly lower average energy than short-duration GRBs as they tend to emit fewer very-high-energy gamma ray photons.

31.5 Ultra-long GRBs

Some rare long-duration GRBs last longer than 10 000 s (2 h 46 min) and might constitute another class of GRBs called ultra-long GRBs. They are presently poorly understood and might be the products of ultra-massive blue supergiant collapse or newborn magnetars (which are a type of highly magnetized neutron star associated with soft gamma repeaters).

31.6 Kilonovae and short-duration GRBs

About a third of all GRBs are of the short-duration type. Their very brief gamma ray emissions (averaging about 200–300 ms) make them inherently more challenging to study since their afterglows are more difficult to track down and examine. NASA's Neil Gehrels Swift Observatory (previously called the Swift Gamma-Ray Burst Explorer), which was launched in 2004, was able to accumulate sufficient data to generate the model that short GRBs occurred when two ultra-dense objects such as neutron stars coalesced.

The most widely supported theory behind their formation is through a **kilonova** event. Kilonovae come from the **merger of two neutron stars** or the **merger of a neutron star with a black hole**. Confirmation of this mechanism came from GRB 170817A, which was detected 1.7 s after the detection of gravitational waves from GW170817. The set of gravitational waves (which are ripples in the 'fabric' of space-time) was the signal from the merger of two neutron stars in a kilonova event, thereby corroborating the link between short-duration GRBs and a kilonova (at least in the case of GRB170817A).

Parenthetically, gravitational waves have curious 'signatures' that stretch objects in one direction while compressing them in an orthogonal direction. Then they compress the object in the initial direction while stretching it in the other direction. To detect gravitational waves, the arrangement must be very large and capable of amplifying the natural resonant frequency of elongation and compression in two perpendicular directions. In this manner, large, L-shaped vacuum tubes were constructed through which laser beams are passed. The evacuated arms of the L-shaped structures are each 4 km long and can host five interferometers apiece. The laser beams can precisely measure the distance between mirrors mounted at the ends of the long tubes. When a gravitational wave passes through the setup, one arm of the L elongates while the other shrinks and the lasers can detect tiny changes that are amplified by the setup. **LIGO**, which stands for **Laser Interferometer Gravitational Wave Observatory**, consists of a pair of these large L's 3000 km apart, in Washington (Hanford) and Louisiana (Livingston) of the United States of America. A similar gravitational wave interferometer detector laboratory is called Virgo and is housed in Italy.

Short-duration GRBs tend to emit more very-high-energy gamma ray photons than long-duration GRBs do and therefore have a slightly higher average energy.

31.7 GRBs mark the births of black holes

It should be stated clearly that in either case, long-duration GRBs caused by the collapse of very massive stars ('collapsars') or short-duration GRBs caused by kilonovae, their end product is a black hole. Thus, it can be correctly stated that GRBs signify the births of black holes.

31.8 The fireball mechanism

One of the early obvious questions about GRBs was, how could they possibly be so luminous? What was the secret behind their incredible gamma ray intensity? Some of the brightest GRBs are detectable despite distances of billions of light years. NASA's Neil Gehrels Swift Observatory detected one GRB that was an incredible 12.8 billion light years away. And although most of the energy is emitted in the form of gamma rays, some GRBs also have extremely luminous afterglows in the x-ray and optical ranges. Given the extreme distance of these objects, their mechanism of action was a major mystery. For example, GBR 080319B had an optical afterglow that was comparable to a dim star that would have been visible with the naked eye or a pair of binoculars. Given its calculated distance of 7.5 billion light years, if this were a spherical object that exploded in a uniform fashion in all directions (that is, isotopically), it would have had to have nearly the energy equivalent of the Sun undergoing conversion of all its mass into energy via $E = mc^2$ instantaneously. No such mechanism was known at the time that GRBs were first being studied. Wild speculation about just what GRBs were went unrestrained for a while, with well over a hundred proposed ideas, some of them quite extreme. Among the suggested mechanisms were alien civilizations blowing themselves up in a nuclear Armageddon (which was very disturbing since they were being observed nearly once every single day!). Alternatively, it was suggested that GRBs were perhaps collisions between stars or planets made of matter with stars or planets made of antimatter. Another fascinating proposition was that GRBs were exploding miniature black holes. This **Hawking radiation** was predicted by Stephen Hawking based on quantum mechanical concepts and will be reviewed in another chapter. However, the observed pulsed, spiked or saw-toothed light curve patterns of GRBs did not match the smooth, exponentially rising pattern predicted for gamma ray emissions from explosively evaporating mini-black holes.

Further analysis led to the **fireball model**, which holds that instead of an isotropic explosion in all directions, the gamma rays and afterglow from a GRB are fired out in narrow jets going opposite directions. These jets are very streamlined and akin to lasers. If (and only if) the beam is pointed in our direction, would we see it with our instruments. Therefore, although still extremely energetic, the fact that we can see them from so far away makes more sense than if GRBs originated from any sort of uniform explosion.

Although the fireball model applies to both short-duration GRBs (which are associated with neutron star mergers) as well as long-duration GRBs (that are associated with collapsars), it may be easier to describe when focusing on a single collapsing giant star here. The beams originate in some sort of rotating internal engine within an ultra-massive dying star that has weak spots at the north and south poles. As this monstrous star collapses when it finally runs out of nuclear fuel, ultra-energetic charged particles then plow their way through these weak spots at the poles and are ejected at relativistic speeds along the axes. As these relativistic charged particles collide with one another, they create internal shock waves and generate gamma rays. Then as the particles travel further and crash into the interstellar medium, they generate the afterglow, which consists of x-ray, ultraviolet, optical radiation and other longer wavelength electromagnetic radiation. Thus, internal collisions or shocks between the relativistic particles make the gamma rays of a GRB, while external shocks between the energetic particles and the gas and dust of the interstellar medium make the afterglow radiation (figure 31.1).

This fireball process would occur in stars that are simply too massive to explode as conventional supernovae and end as neutron stars. Instead, in this collapsar model, the massive star dies in a **collapsar** (known as a **hypernova**) explosion, creating a black hole instead. Hypernovae are super-luminous supernovae that form when extremely massive stars (>30 $M_\odot$) collapse into black holes. During the

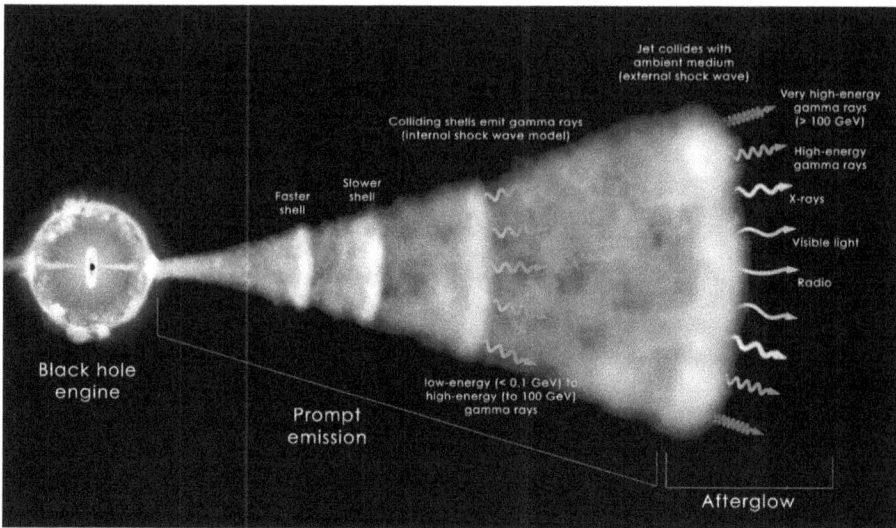

Figure 31.1. This illustration aims to show the internal mechanisms behind the most common type of GRB, the collapse of an extremely massive star into a black hole (the collapsar model). Collapsars are associated with long-duration GRBs—meaning they persist longer than 2 s. On the left, the core of a massive star has collapsed and formed a black hole. This 'engine' drives a jet of charged particles out through the poles of the collapsing star and into space at nearly the speed of light. The prompt emission, which typically lasts a minute or less, may arise from the jet's interaction with gas near the newborn black hole and from collisions between shells of fast-moving gas within the jet (internal shock waves). The afterglow emission occurs as the leading edge of the jet sweeps up its surroundings (creating an external shock wave) and emits radiation across the electromagnetic spectrum for some time—months to years, in the case of radio and visible light, and several hours at the highest gamma ray energies yet observed. These gamma ray photons may exceed 100 GeV in some cases. Short-duration GRBs last less than 2 seconds and are even richer in extremely energetic gamma photons. They appear to result from the merger of two neutron stars into a black hole, or the merger of a neutron star with a black hole to form a larger black hole. But this general fireball model applies to both short-duration and long-duration gamma ray bursts. Credit: NASA/Goddard Space Flight Center/ICRAR.

collapse phase, an accretion disc forms around the rotating black hole, while twin energetic jets of relativistic charged particles are fired out of the rotational axes. These charged particles then create the gamma ray beams that we observe through internal shocks along with the afterglows at longer wavelengths through external shocks with the interstellar medium. Hypernovae are also called super-luminous supernovae since they usually have a luminosity tenfold higher than most super-novae do. (But students should be aware that the term super-luminous supernova sometimes also includes very bright supernovae of other origins beside the collapsars, such as **circumstellar material model supernovae** and **pair-instability supernovae**.)

Super-luminous supernovae that produce GRBs tend to be of the type Ic variety but with broader than average spectral lines, consistent with extraordinary expansion velocities. This unusually high kinetic energy is why the term hypernova is used for such scenarios.

Type Ic supernovae exhibit no hydrogen in their spectra (and therefore are type I) but they also lack helium lines at 587.6 mn wavelengths distinguishing them from type Ib supernovae. In any case, type Ic supernovae are the result of the deaths and collapses of very massive stars that have lost their outer envelopes of hydrogen and helium along the way and are not the end products of accretion of material onto a white dwarf where it eventually detonates after exceeding the Chandrasekhar limit (which is what leads to a type Ia supernova). Type Ic supernovae are examples of stripped, core-collapse supernovae and as such, are the most likely to generate GRBs since there is less material (such as the hydrogen and helium still present on other collapsing stars) for the jets to plow through on their way out from the poles.

Among the well-characterized collapsars that have led to super-luminous super-novae and generated long-duration GRBs is SN 1998bw, which was associated with GRB 980425. Consistent with the model, SN 1998bw was a type Ic supernova but it produced a shockwave with over ten times as much energy than an ordinary supernova, thereby qualifying it as a **hypernova**. A more quantitative definition of a hypernova is a supernova with ejecta energy exceeding 10^{45} J. This means that the massive amount of ejected material will be travelling at up to 0.99c and the overall kinetic energy of the ejecta is an order of magnitude larger than that of an ordinary core-collapse supernova. SN 1998bw was not very ordinary, however. It was not just a typical type Ic supernova without hydrogen or helium, it seemed to be exception-ally stripped and maybe even part of its carbon–oxygen layer was lost before the blow-up. Furthermore, the associated GRB did not have an optical afterglow. While this made the supernova, which showed up around a week after the GRB, more obvious, the constellation of abnormalities left some to still question the mechanism. Fortunately, just a few years later in 2003, a very similar coincidence of a long-duration GRB and a stripped, core-collapse supernova occurred but this GRB had a regular afterglow. The absence of anomalies made a convincing connection between long-duration GRBs and collapsars and this has been confirmed several times since then.

Long-duration GRBs are found in galaxies that are forming vast numbers of extremely massive stars (e.g. population I stars). And since collapsars are the end products of extremely massive stars, the collapsar fireball model does fit. Collapsing massive stars create jets of ejecta along their axes. These jets punch their way through the outer shells of the stellar remnants and then generate beams of gamma rays and the afterglow. Since type Ib and Ic supernovae have lost their outer shells of hydrogen, they have less material for the jets to penetrate and therefore are the supernova types that are most likely to lead to GRBs.

31.9 Kilonovae: the mechanism behind short-duration GRM

Compared to long-duration GRBs, the brief period of gamma ray emissions of short-duration GRBs made study of their afterglows difficult. But over the decades, dozens of short-duration GRBs have had their afterglows detected and scrutinized. The data indicate that short-duration GRBs are found in areas with little to no star formation such as old, large, elliptical galaxies with little interstellar dust and gas.

Such regions are where metal-poor population II stars are found and perhaps even the very oldest stars in the Universe—the so-called **population III stars** (which should be nearly devoid of metals).

Most elliptical galaxies are composed of older, low-mass stars and there is generally little new star formation. This reality rules out an association between short-duration GRBs with dying massive young stars (as is the case with long-duration GRBs). Consistent with this observation is the lack of any link between short-duration GRBs and supernovae.

Instead, short-duration GRBs are probably created during the merger of two neutron stars or a black hole and a neutron star in binary systems. Mergers between the super-dense corpses of massive stars lead to kilonova explosions. Another less often used name for a kilonova is **macronova**. In addition to the gamma rays of the short-duration GRB they produce, kilonovae also are bright in gamma rays thanks to the radioactive decay of innumerable heavy radionuclides (such as thorium, uranium and plutonium) made though the **r-process** that are ejected during the merger. In fact, kilonovae could be the major source of r-process elements in the Universe. Here on Earth, nuclides with the most neutrons are generally the ones made through the r-process (including silver, gold and platinum).

31.10 The r-process of nucleosynthesis in kilonovae and core-collapse supernovae

As a short aside on the r-process (the rapid neutron capture process of nucleosynthesis), until just a few years ago it was believed that the r-process only occurred in supernovae (specifically in core-collapse supernovae which include types Ib, Ic and II). But it is now recognized that the r-process also occurs in kilonovae when neutrons stars coalesce into black holes. Big Bang nucleosynthesis made only helium, deuterium, traces of helium-3 and lithium. Beryllium and boron, along with some lithium and helium-3, are cosmogenic, meaning they are created by cosmic ray spallation of larger elements. The nuclides from carbon up to iron, cobalt and nickel are made in stars through stellar nucleosynthesis—the energy-releasing fusion reactions of lighter elements into heavier ones. But beyond iron-56, fusion no longer releases energy, it consumes it. Therefore, ordinary fusion cannot create heavier elements in the cores of stars. Two processes involving the addition of neutrons allow the creation of heavier elements: the **rapid process (r-process)** and the **slow neutron capture process (s-process)**. Since some red giants host nuclear reactions that generate and maintain a population of free neutrons, the s-process can occur in red giants (more specifically in the helium-burning shells of asymptotic giant branch stars). The s-process creates elements up to bismuth.

The s-process of nucleosynthesis is responsible for about half of the elements heavier than iron. The other half of the heavy elements trace their origins to the r-process. Basically, in the s-process neutrons are added to a seed nucleus creating a nuclide of one more atomic mass. In other words, A becomes $A + 1$. If this product is stable, additional neutrons can be captured creating heavier isotopes of the original seed nucleus. If the nuclide is unstable (that is, if it is a radionuclide), it will decay via beta emission and create a nuclide with a higher atomic number. In other words, Z becomes $Z + 1$. Because we are dealing with the s-process (slow) here, there is sufficient time for these radionuclides to actually decay before another neutron is added to the nucleus. This contrasts with the situation in the r-process where neutrons are so rapidly added that the radionuclide does not have time to decay via beta emission. The r-process involves extremely high neutron flux scenarios such as core-collapse supernovae and neutron star mergers. But these high-flux scenarios are also short-lived. Therefore, the neutron flux will eventually die down. And when it does, the heavy, neutron-rich radionuclides created by the r-process will finally get their chance to decay into stable nuclei. As expected, the stable heavy nuclei produced by this process wind up accumulating around certain **magic numbers**: 50, 82 and 126. In a plot of nuclide abundance on the ordinate versus atomic mass or atomic number on the abscissa, one sees peaks and valleys representing abundances and paucities of the nuclides in nature. Abundances hover around the magic numbers which are 2, 8, 20, 28, 50, 82 and 126. (Another peak of r-process elements centers around $A = 196$.) But the peaks are paired, meaning they typically have two prongs with a small valley in between. These twin peaks are due to the r-process and the s-process—the >r-process nuclei are to the left of the s-process nuclei. That is, the r-process nuclides are slightly lighter than the s-process nuclides in these pairs of peaks. The nuclide peaks for the r-process include Se, Xe, Te and Pt. Peaks of the s-process include Sr, Zr, Ba and Pb.

Curiously, the heaviest four isotopes of every naturally occurring heavy element can be made via the r-process, and the two heaviest natural isotopes of every heavy element are made exclusively through the r-process (and are therefore called **r-process only nuclides**). The relative proportion of heavy elements made through the two known mechanisms of the r-process—core collapse supernovae and kilo-novae—is under investigation.

31.11 Photon energies of GRBs

In less than a second, a GRB might emit as much energy as our Sun does over its entire 10 billion year lifetime. But beyond the incredible intensity (power output), GRBs emit some gamma rays with immense energies as well. One recent GRB (GRB 190114C) was the first GRB shown to eject photons in the TeV range. 190114C was initially detected by the Niels Gehrels Swift and Fermi Gamma-ray Space Telescopes. (The Fermi Gamma-ray Space Telescope was formerly called the Gamma-ray Large Area Space Telescope or GLAST. It was launched on 11 June 2008 and presently operates in LEO. Its main instruments are the Large Area Telescope and the Gamma-ray Burst Monitor.) GRB 190114C was a long-duration GRB so the coordinates determined by the Niels Gehrels Swift and Fermi Gamma-ray Space Telescopes could be relayed to various land-based telescopes.

The twin **Major Atmospheric Gamma Imaging Cherenkov (MAGIC)** land tele-scopes on La Palma in the Canary Islands, Spain registered secondary products from GRB 190114C consistent with an energy up to 1 TeV, making this GRB the producer of the highest-energy photons ever from a GRB. The **MAGIC Telescopes** are also called the **MAGIC Florian Goebel Telescopes**. MAGIC does not detect the gamma ray photons themselves but rather detects the Cherenkov radiation generated by particle showers induced by high gamma rays in the atmosphere.

GRB 190114C is the new record holder after besting the previous record-holder GRB 180720B in terms of maximum photon energy. GRB 180720B occurred in a galaxy about 6 billion light years away and was also first detected by NASA's orbiting Fermi and Neil Gehrels Swift Gamma-ray Telescopes. They immediately sent out an electronic alert to other telescopes to search for its afterglow in various wavelengths. One of these was **High-Energy Spectroscopic System (HESS)** in Namibia. Like MAGIC, HESS is an 'Imaging Atmospheric Cherenkov Telescope' for the investigation of cosmic gamma rays. HESS is actually an array of five telescopes. Like MAGIC, it does not detect gamma rays directly, but is sensitive to the Cherenkov light glow in the atmosphere when high-energy gamma photons strike atmospheric molecules and set energetic charged particles free. These charged particles will travel faster than the speed that light can travel in air. As such, they emit Cherenkov radiation, blue flashes akin to optical shock waves that are covered in greater depth in another chapter. HESS can detect this Cherenkov radiation and reconstruct the direction from where the original gamma rays came, as well as what their energies were. Incidentally, the acronym HESS honors Victor Hess, the discoverer of cosmic rays. HESS is able to detect gamma photons in the energy range of 30 GeV to 100 TeV. HESS detected gamma photons from GRB 180720B reaching 440 GeV. This record was shattered only a year later by GRB 190114C and its 1 TeV photons.

But maybe the most amazing and perplexing finding was that these ultra-high-energy photons did not stem from the burst itself but rather were part of the afterglow. The first burst of gamma rays caused by the internal shocks of charged particles in the ejected jet are responsible for photons in the 100 MeV to 100 GeV range but the ultra-high photons (UHPs), ranging up to 1 TeV, come from the external shocks between the ejected jet and the interstellar medium that produces the afterglow. Ironically, the afterglow contains the most energetic of all gamma photons from a GRB as well as the lowest-energy photons, which may be in the radio wave range.

It is presently believed that the low- and moderate-energy gamma photons from a long-duration GRB are generated via synchrotron radiation. This is the emission of electromagnetic radiation when charged particles are accelerating in a strong magnetic field. However, for the high-energy record-breaking radiation, synchrotron emission by ultra-high-energy protons is unlikely, thanks to the relatively low radiative efficiency. Another mechanism is required. To account for the observed similarity between the radiated power and temporal characteristics of the TeV gamma photons and the x-ray bands, the most plausible mechanism for creation of these ultra-high-energy photons is **inverse Compton upscattering**. In the inverse Compton effect, a photon gets slammed by a relativistic jet electron. The high-energy electrons transfer their energy to the photons, thereby boosting the photon's energies to extreme values. Through this mechanism, the photons gain energy, and wind up the gamma ray range. The high-energy electrons involved in this phenomenon are generated by the intense magnetic fields found near the newborn black hole. These magnetic fields accelerate electrons (and other charged particles) to incredible speeds and energies and shoot them out from the poles in the jets. If an already high-energy photon is struck by one of these ultra-energetic electrons, that photon might gain enough energy to reach the TeV range and break records. Although this is the most likely mechanism behind the UHPs found in the afterglows of certain GRBs, it is not yet proven.

And while inverse Compton is a plausible means of getting photons to reach levels in the TeV range, GRB 190114C is still not fully understood since present models suggest that the gamma ray intensity should have decreased more rapidly than what was actually recorded. Obviously, this is a very exciting time in GRB research.

Figure 31.1 is a diagram from NASA that illustrates the fireball mechanism of GRBs. The black hole engine might be a collapsar or a kilonova. The collapsar may or may not be associated with a supernova. The kilonova might be the merger of two neutron stars or the merger of a neutron star and a black hole, thereby creating an even larger black hole. Regardless of the specifics, first burst of gamma rays (called the **prompt emission** in this diagram) is caused by internal shock waves generated by faster charged particles in the ejected jets overtaking and crashing into slower charged particles in the jet. Further out, the jet of charged particles eventually collides with the interstellar medium and creates additional electromagnetic radiation—the afterglow. Whereas the prompt emission is all gamma rays, the afterglow spans the entire electromagnetic spectrum. Perplexingly, the very highest-energy gamma rays (>100 GeV) appear to originate from external shocks in the afterglow rather than from the internal shocks of the prompt emission.

31.12 The brightest of all time

Incredibly high energy is not the only amazing feature of some GRBs. The unbelievable power is similarly mind-bending. And a new record has recently been set. The most powerful GRB ever, GRB 221009A, was observed on 9 October 2022. Being the 'brightest of all time' it is nicknamed the **BOAT**. GRB 221009A was especially bright because its narrow jet of photons was pointing directly at us. But the amount of energy aimed our way was not small; it was the equivalent of taking the entire Sun and converting it instantly into energy via the Einstein relation. While most other GRBs yield far fewer than 1 million detected gamma ray photons per second, GRB 221009A's detection rate approached 7 million gamma rays per second based on reconstructed data from the Fermi Gamma-ray Space Telescope.

GRB 221009A triggered detectors on the Fermi Gamma-ray Space Telescope, the Neil Gehrels Swift Observatory and **Wind** spacecraft, among several others. (NASA's Wind was launched on 1 November 1994 and generally collects data on the solar wind. It orbits around Earth's first Lagrange point, which is approximately 930 000 miles away.)

GRB 221009A was estimated to be relatively close at 1.9 billion light years away. Afterglow radio wave data confirmed that GRB 221009A was 70 times brighter than

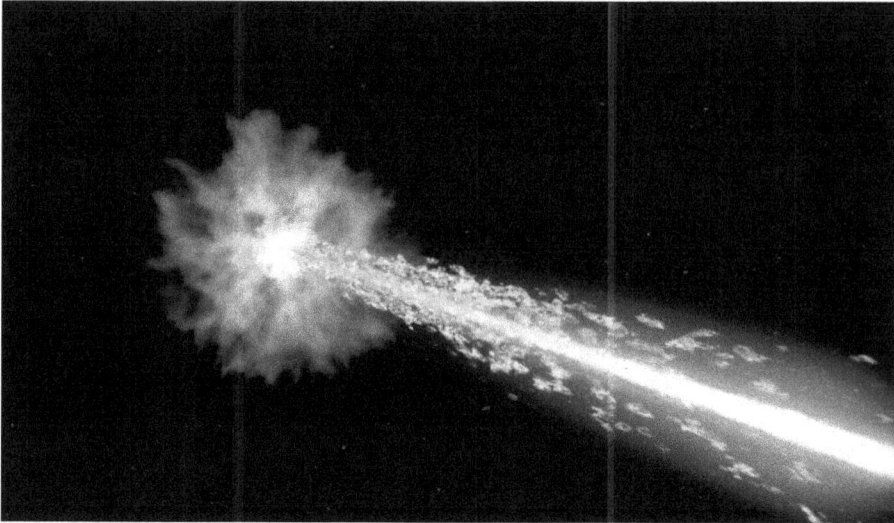

Figure 31.2. An illustration provided by NASA representing a GRB. GRBs, including the 'BOAT' or brightest of all time, GRB 221009A, represent the birth of a new black hole within the core of a collapsing star. Such 'collapsars' are believed to be associated with long-duration GRBs. The newly formed black hole powers jets of charged particles traveling close to the speed of light. These jets pierce through the remnants of the star near the poles and generate gamma rays when they stream into space and strike each other (internal shocks) and the interstellar medium (external shocks). Credit: NASA/Swift/Cruz deWilde.

any previously seen GRB. Its intensity was such that it was perhaps a 'once in 10 000 years' episode, meaning that it was probably the brightest GRB in the history of human civilization. GRB 221009A's gamma photon energy was not exceptionally high but the jet aiming that energy outward was quite narrow and directed straight at Earth, making it appear especially bright.

Despite lasting over 300 s and clearly falling into the long-duration GRB camp, no associated supernova has been spotted by the James Webb and Hubble Space Telescopes. Thus, it is possible that the star's core collapsed straight into a black hole rather than exploding. However, GRB 221009A was in a part of the sky that lies just above the plane of the Milky Way, so thick clouds of interstellar dust might have blocked incoming light from an associated supernova (figure 31.2).

31.13 Short-/long-duration 'hybrids'

The short-duration GRB 170817A was detected soon after the detection of gravitational waves (GW170817) coming from the same site. It appears that the gravitational waves were generated by the merger of two neutron stars in a binary system. This observation confirmed the link between short-duration GRBs and kilonovae.

However, as with many 'rules' in science, exceptions may materialize. An unusual observation of a long-duration GRB made by the Neil Gehrels Swift Observatory on 11 December 2021 was associated with a kilonova rather than a supernova. At a distance of 346 megaparsecs or 1.1285 light years, GRB 211211A lasted nearly

1 minute, making it a long-duration GRB on paper. Given the duration, other telescopes had time to point in that direction and examine the event in other wavelengths. But the afterglow in ultraviolet, visible and infrared was most consistent with a kilonova rather than the expected collapsar.

Thus, the assertion that long-duration GRBs stem from collapsars which lead to supernovae, while short-duration GRBs originate in neutron star mergers that lead to kilonovae may not hold in all cases. Some GRBs might belong to a hybrid category.

It is presently believed that the anomaly might have to do with one of the two neutron stars involved in the merger being either a white dwarf or an unusual type of neutron star known as a **magnetar**. Magnetars are neutron stars with super-strong magnetic fields that are associated with another astrophysical phenomenon that spits gamma rays into space—**soft gamma repeaters**. These will be covered in the next chapter.

Further reading

[1] Ioka K, Hotokezaka K and Piran T 2016 Are ultra-long gamma-ray bursts caused by blue supergiant collapsars, newborn magnetars, or white dwarf tidal disruption events? *Astrophys. J.* **833** 110

[2] Abbott B PLIGO Scientific Collaboration and Virgo Collaboration *et al* 2017 GW170817: observation of gravitational waves from a binary neutron star inspiral *Phys. Rev. Lett.* **119** 161101

[3] Tanvir N R, Levan A J, Fruchter A S, Hjorth J, Hounsell R A, Wiersema K and Tunnicliffe R L 2013 A 'kilonova' associated with the short-duration γ-ray burst GRB 130603B *Nature* **500** 547–9

[4] Troja E, Fryer C L and O'Connor B *et al* 2022 A nearby long gamma-ray burst from a merger of compact objects *Nature* **612** 228–31

[5] Pian E *et al* 2017 Spectroscopic identification of r-process nucleosynthesis in a double neutron-star merger *Nature* **551** 67–70

[6] Troja E, Fryer C L and O'Connor B *et al* 2022 A nearby long gamma-ray burst from a merger of compact objects *Nature* **612** 228–31

[7] MAGIC Collaboration 2019 Teraelectronvolt emission from the γ-ray burst GRB 190114C *Nature* **575** 455–8

IOP Publishing

Space Radiation
Astrophysical origins, radiobiological effects and implications for space travellers
James S Welsh

Chapter 32

X-rays and gamma rays in space

Our skies are literally filled with x-rays and gamma rays. In fact, there is a cosmic x-ray background radiation that permeates the skies in a manner analogous to the cosmic microwave background. Space x-rays and gamma rays come from a variety of mundane and exotic sources, ranging from reflected solar x-rays from the Moon to supermassive black hole-generated x-rays and gamma rays from blazars and quasars. When young, the visually stunning and radiant planetary nebulae are often bright in soft x-rays. They are illuminated by the hot white dwarf remnant in their centers, which shine in x-rays and ultraviolet, causing their magnificent clouds of gas to glow spectacularly (figure 32.1).

32.1 History of x-ray and gamma ray astronomy

X-ray astronomy got a relatively late start because x-ray photons are unable to penetrate our thick atmosphere. Therefore, the early work in this field was conducted by rockets and balloons. Some special balloons reach altitudes as high as 40 km above sea level. This means they ascend well into the upper stratosphere, nearly reaching the mesosphere, implying that they are above 99% of our atmosphere by mass. Despite this, x-ray photons <35 keV still cannot penetrate the gossamer exosphere, thermosphere and mesosphere and reach even the highest balloons. Thus, balloons have their limits, but so do rockets. The biggest drawbacks of rocket flights is their short duration (typically measured in minutes above the atmosphere) and their limited field of view. X-ray satellites and orbiting space telescopes with x-ray vision provided the needed technology for major advances in the field of x-ray astronomy. Riccardo Giacconi received the Nobel Prize in Physics in 2002 for ushering in the new field of x-ray astronomy, starting with the discovery of **Scorpius X-1** by his team in 1962. NASA's **Uhuru**, which was launched on 12 December 1970, was the first satellite specifically designed and launched for x-ray astronomy.

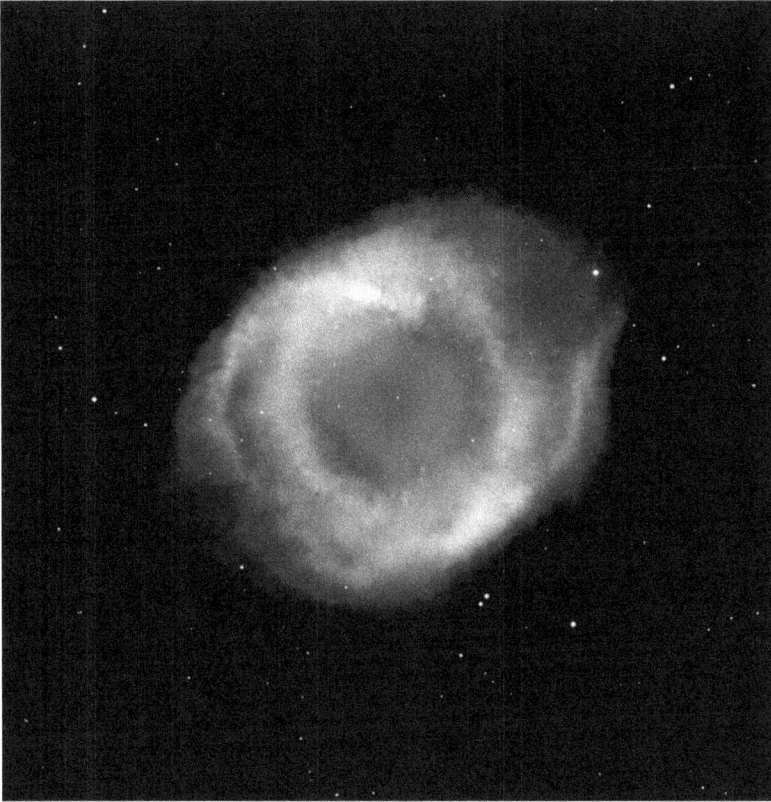

Figure 32.1. A photograph of the planetary nebula known as the Helix Nebula ('Eye of God') taken by NASA's Hubble Space Telescope. Planetary nebulae are the final products of stars like our Sun, which are not massive enough to end their lives in supernova explosions and leave behind neutron star or black hole corpses. The outer remnants of the former star have been blown into space and are fluorescing from the ultraviolet and x-rays emitted by the central white dwarf. Although they are no longer undergoing thermonuclear fusion some young white dwarfs can have surface temperatures well in excess of 100 000 K. According to Planck's Radiation Law and Wein's blackbody displacement law, objects with such temperatures emit ultraviolet and x-rays. These high-energy photons ionize the gases in the ejected stellar remnants leading to the spectacular fluorescent beauty that is seen in the visible range. The magnificent, brilliant colors are due to glowing oxygen (blue) and hydrogen and nitrogen (red). Credit: NASA, NOAO, ESA, the Hubble Helix Nebula Team, M Meixner (STScI), and T A Rector (NRAO).

32.2 Instruments specifically designed for x-ray and gamma ray astronomy

A short, partial list of the many space telescopes and satellites that have yielded valuable x-ray information is as follows:

Vela	Launched 1963
Uhuru	Launched in 1970
Einstein Observatory (HEAO-2)	Launched in 1978
Roentgen Satellite (ROSAT)	Launched in 1990
Compton Gamma Ray Observatory	Launched in 1991
Yohkoh satellite	Launched in 1991
Beppo SAX	Launched in 1996
Chandra X-ray Observatory	Launched in 1999
Neil Gehrels Swift Observatory	Launched in 2004
Fermi Gamma-ray Space Telescope	Launched in 2008
NuSTAR	Launched in 2012
NICER	Launched in 2017
eROSITA	Launched in 2019

32.3 The imaging atmospheric Cherenkov technique

Additionally, another means of detecting high-energy photons is through land-based Cherenkov telescopes. As discussed earlier, these telescopes do not directly detect the x-rays or gamma rays themselves but instead they detect the Cherenkov radiation generated when secondary charged particles (ejected from atmospheric atoms struck by the gamma rays) travel faster than the speed of light in our atmosphere. In May 2021 China's Large High Altitude Air Shower Observatory (LHAASO) reported the detection of a dozen remarkable, ultra-high-energy (UHE) gamma rays. These photons had energies exceeding 1 PeV, including one at 1.4 PeV. The latter is the highest-energy photon ever observed. The 'PeVatrons' responsible for such UHE gamma rays have not yet been firmly localized and identified; they remain a subject of great interest. Incidentally, the term 'very-high-energy gamma ray' (VHEGR) was historically applied for those exceptional photons with energies between 100 GeV and 100 TeV. As exotic astrophysical sources were found firing out photons in the PeV range, the upper limit of the definition had to be expanded—or joined by a new category, the UHEs. And as further studies continue, there is a possibility that the upper limit of space gamma rays might extend to near the EeV range.

Other instruments using this so-called **Imaging Atmospheric Cherenkov Technique (IACT)** or similar methods include Very Energetic Radiation Imaging Telescope Array System (VERITAS), High Energy Stereoscopic System (HESS), Major Atmospheric Gamma Imaging Cherenkov (MAGIC) Telescopes, the High-Altitude Water Cherenkov (HAWC) Observatory, the Pierre Auger Observatory, the Major Atmospheric Cherenkov Experiment Telescope (MACE) and the First G-APD Cherenkov Telescope (FACT). The Cherenkov Telescope Array (CTA) is a planned multinational project.

32.4 General sources of high-energy photons in space

Many bodies in our Universe emit, reflect or fluoresce x-rays or gamma rays. Some are nearby such as our **geocorona**, which is caused by collisions between energetic ions in the solar wind and our outermost atmosphere. A charge exchange phenomenon occurs in such collisions and x-rays are emitted when the exchanged electrons (from neutral atoms in our atmosphere to heavy charged particles in the solar wind such as carbon, nitrogen and oxygen ions) move from higher-energy orbitals and shells into lower-energy ones. Another nearby source of x-rays are solar corona x-rays reflected off the Moon.

The Moon also fluoresces in x-rays when highly energetic solar x-rays strike lunar atoms and excite them to higher-energy states. Upon de-excitation, these lunar atoms release fluorescent x-rays of slightly lower energies than the original solar photons. Fluorescent lunar x-rays can reveal the chemical composition of the Moon's surface. But far beyond our Solar System lie other quite unusual and poorly understood sources of high-energy photons. Among the numerous astrophysical x-ray and gamma ray sources are black holes in active galactic nuclei (AGNs; such as quasars, Seyfert galaxies and blazars), supernova remnants, main sequence stars (especially large, young, hot stars such as type O stars but even smaller, cooler ones like our Sun), white dwarfs and various types of binary stars systems. These binaries might contain a white dwarf (as in **cataclysmic variable stars** and **super soft x-ray sources**), or a neutron star or black hole (as in **x-ray binaries**, including **accretion-powered x-ray pulsars** and **x-ray bursters**). On a far larger and more distant scale, galaxy clusters are x-ray sources (figure 32.2).

As discussed in chapter 31, **gamma ray bursts** signify the birth of a black hole through either the collapse of a hypergiant star (the **collapsar model**) or through the merger of two neutron stars (or a neutron star with a black hole). The second scenario is called the **kilonova mechanism**. But either way, the underlying **fireball** mechanism gushes particles from the poles of the newly formed black hole, and these energetic charged particles then spawn the narrow but intense beams of gamma rays that might be seen by us (if we happen to be in the line of fire). Another source of gamma rays from space are **soft gamma repeaters**, which are caused by starquakes on magnetars. As the magnetars age (and they age quite quickly for astronomical entities), the soft gamma repeaters evolve into lower-energy versions called **anomalous x-ray pulsars**. These sources of x-rays and gamma rays are covered in chapter 33. In this chapter we shall discuss 'non-anomalous' x-ray pulsars and a handful of the many other sources of x-rays and gamma rays out there. Dozens of sources of ionizing radiation have been discovered thanks to x-ray and gamma ray astronomy. Some of the many additional unusual x-ray and gamma ray sources are discussed below.

32.5 The Fermi bubbles and eROSITA bubbles

As introduced in earlier chapters, our Milky Way Galaxy has two huge and odd radiation-emitting lobes bubbling out of its central bulge. The lobes are orthogonal to the galactic plane. They are directed up and down with respect to the galactic disc

Figure 32.2. A composite image that includes the **Rosette star formation region**, located about 5000 light years from Earth. X-ray data obtained from the **Chandra X-ray Observatory** are artificially colored red and outlined by the white lines. These x-rays represent the radiation from newly formed stars. Optical data from the Digitized Sky Survey and the Kitt Peak National Observatory (outside the red colored regions) reveal gas and dust, including pillars that remain behind after the intense radiation emanating from bright, massive stars has eroded the more diffuse gas. Credit: X-ray (NASA/CXC/SAO/J Wang *et al*), Optical (DSS and NOAO/AURA/NSF/KPNO 0.9-m/T Rector *et al*).

and spring forth from the central axis. These recently discovered blobs are filled with gamma radiation and are called the Fermi bubbles since they were first detected with the Large Area Telescope (LAT) on the Fermi Gamma-ray Space telescope. The LAT is the most sensitive and highest-resolution gamma ray detector thus far. These two bubbles of radiation extend 25 000 light-years 'north' and 25 000 light-years 'south' of the galactic center. As huge as these are, it seems surprising that they were only recently noticed. This is partly because of an obscuring haze of background gamma radiation caused by relativistic charged particles colliding with galactic interstellar gas. This background of relatively low-energy gamma radiation is largely confined to the pane of the Milky Way. The gamma ray emissions from the Fermi bubbles are far more energetic and localized than this diffuse gamma ray emission seen throughout the Galaxy. Through computer modelling and refined analysis of the data, the LAT is capable of compensating for this diffuse galactic gamma ray emission, thereby uncovering the previously camouflaged Fermi bubbles.

Surprisingly, the Fermi bubbles appear to have well-defined edges. The structure's overall configuration and gamma emission pattern hint it was formed as a result of a large, relatively recent and rapid release of energy. Although the mechanisms remain uncertain, one possible explanation for the Fermi bubbles is a jet of charged particle blasting away from the supermassive black hole at the center of the Milky Way. Powerful galactic magnetic fields may then further accelerate the charged particles in these jets. When the jets strike hydrogen and other atoms in galactic (and intergalactic) gas clouds, radiation is emitted (figure 32.3).

Astronomers have observed energetic particle jets streaming away from the central region of other galaxies as their central black holes consume prodigious amounts of matter. Such jets are powered by infalling matter as it spirals into the galaxy's supermassive central black hole. There is no trace of such jets coming from the region of our Milky Way's central black hole at present. However, such jets may have existed in the relatively recent past and the radiation-filled blobs we see today may be the aftermath.

Earlier hints of the Fermi bubbles might have been glimpsed through x-ray observations by the Roentgen Satellite, ROSAT. In addition to Fermi's LAT,

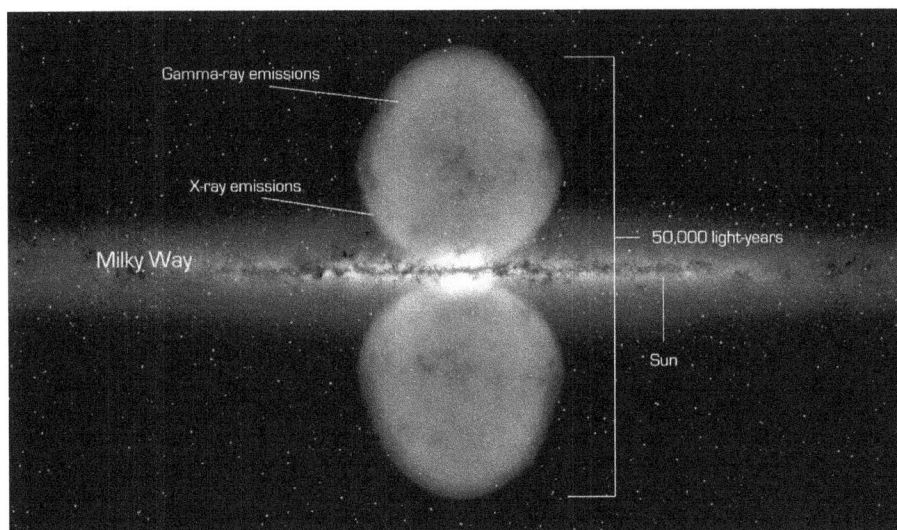

Figure 32.3. An illustration of the Milky Way Galaxy and the Fermi bubbles. Based on data from the Fermi Gamma-ray Space Telescope, enormous, previously unknown bubble-like structures arising from the galactic bulge and extending to 25 000 light years above and below the galactic plane have been identified. These bubbles are the source of x-rays (blue) and gamma rays (magenta) that are presumably made through inverse Compton scattering of extremely energetic electrons that have interacted with lower-energy photons. The origin of these relativistic electrons is presently unknown. Credit: NASA's Goddard Space Flight Center.

these radiation bubbles have been sensed by the eROSITA x-ray instrument. But what the two dedicated instruments have detected could be slightly different entities and thus they are sometimes given separate names—Fermi bubbles and **eROSITA bubbles**. Recent data are emerging that, while they overlap geometrically, the two are indeed different. The eROSITA bubbles might be approximated as two 'teardrops' or spheres of 45 000 light years diameter each, bulging to the north and south of the galactic plane and stemming from the center. Thus, they largely encase the 25 000 light year diameter Fermi bubbles. And in this double-bubble arrangement, the smaller Fermi bubbles contain higher-energy gamma rays while the larger eROSITA bubbles contain only x-rays. This difference probably reflects the relative energies of the charged particles contained by the two types of bubbles. The eROSITA bubbles contain hot plasma while the charged particles in the Fermi bubbles are cosmic rays.

Both sets of bubbles are probably the result of a prolonged outburst of charged particles from the region around the supermassive black hole at the center of the Milky Way, beginning 2.6 million years ago. This outburst would have inflated the bubbles and proceeded in stages. The first stage would have been the infall of vast amounts of debris into the Sagittarius A* supermassive black hole. During the inward swirl, some of that matter was channeled into enormous, energetic jets directed to the north and south of the galactic plane. The gravitational energy of the black hole served as a galactic particle accelerator and converted that energy into the kinetic energy of the jets. The charged particles were thus transformed into high-energy cosmic rays. These charged particles (now qualifying as cosmic rays thanks to their enormous energies) streamed into space and filled the bubbles. In so doing, they created a tremendous shock wave as they crashed into and pushed aside the interstellar gas in the way. This shock wave created by the collision led to superheated gas that glows in x-rays. This x-ray glow is what we see today as the eROSITA bubbles, which are shooting out of the sides of the Fermi bubbles.

Earth has its trapped torus of radiation in the form of its Van Allen belts. Other planets like Jupiter have even more radiation in their extensive surrounding radiation belts. In a (somewhat) analogous manner, the Milky Way Galaxy has its radiation-filled Fermi and eROSITA bubbles.

32.6 The cosmic and galactic x-ray background radiation

Just as the skies shine in microwaves thanks to the ubiquitous **cosmic microwave background**, there is an x-ray background (although the origin is completely different). The low-energy component of this x-ray background (0.3 keV) is the **galactic x-ray background**, whereas the higher-energy component (>0.3 keV) is called the **cosmic x-ray background**.

The **galactic x-ray background** is produced largely by energetic emissions from plasma in the Local Bubble of the Orion Arm of our Milky Way Galaxy. This is within 100 parsecs or 362 light years from our Solar System. The **cosmic x-ray background** originates from beyond the Galaxy and data from the Chandra X-ray Observatory reveal that a significant majority (roughly 80%) of this cosmic x-ray background radiation simply stems from unresolved extra-galactic x-ray sources. Most of these are likely AGNs.

Furthermore, the intergalactic space between galaxy clusters is filled with a very sparse but extremely hot gas. The average kinetic energy of the gas atoms and molecules corresponds to a temperature between 100 million and 1 billion K. Since x-ray (or gamma ray) emission can be expected from anything with a temperature above 1 million K, intergalactic space is itself full of x-ray photons.

32.6.1 Ultraluminous x-ray sources

Ultraluminous x-ray sources (ULXs) were first discovered by the Einstein Observatory (HEAO-2) X-ray Telescope (which was launched on 13 November 1978 and historically was the first space x-ray telescope). Subsequent studies were made by ROSAT (a joint NASA, German and British project), the Chandra X-ray Observatory and ESA's XMM-Newton (also known as the High Throughput X-ray Spectroscopy Mission). Characterization by these and other sources has demonstrated that a ULX is an astronomical source of x-rays that is less luminous than an AGN but is more consistently luminous than any known stellar process (assuming that the ULX source is isotopic rather than beamed like a gamma ray burst). Among galaxies that have ULXs, the average is one per galaxy, but some galaxies can have several. Generally, they tend to be more common in galaxies with active star formation. Thus far, no ULX has been confirmed in our Milky Way Galaxy.

Exactly what powers ULXs is presently unknown. ULXs are scientifically intriguing since they appear to exceed the **Eddington limit** for neutron stars and stellar-mass black holes. Recall that the Eddington limit is the maximum luminosity a star or other astrophysical entity can attain when that star is in a state of hydrodynamic equilibrium where there is balance between the outward-directed force of radiation and the inward-directed gravitational force. If a star or other body exceeds the Eddington limit, the radiation will blow a very strong stellar wind from its outer layers. Presently known astrophysical entities that exceed their Eddington limits include novae, supernovae and gamma ray bursts. Such cosmic creatures rapidly lose mass through this radiation-driven wind of particles. For accretion-powered sources (such as accreting neutron stars or **cataclysmic variables** which are accreting white dwarfs), the Eddington limit may reduce or even halt accretion through a balance of inflow and outflow. Theoretical 'super-Eddington emission' due to very fast accretion onto stellar-mass black holes is one possible model for ULXs. Another model is accretion onto intermediate mass black holes (black holes of hundreds to thousands of solar masses). Of course, a more mundane explanation such as beamed emission of x-rays from uncharacterized stellar mass objects remains possible. Some ULXs have been found to be nothing more than background quasars shining through a nearer galaxy.

32.7 X-ray binaries

In simple terms, x-ray binaries are just binary stars that shine in the x-ray part of the spectrum. Such binary systems consist of a larger **donor** that is dumping material onto a smaller **accretor**. The accretor might be a neutron star or black hole. With such massive but compact accretors, infalling matter from the donor can convert astonishing amounts of gravitational potential energy into x-rays. In extreme cases, the amount of potential energy converted can amount to a substantial percentage of the rest mass of the infalling object. Furthermore, as infalling gas is rapidly transferred from donor to accretor and furiously spirals inward, gravitational energy heats and accelerates the infalling matter to the point where pair production is possible. Huge numbers of electron–positron pairs are created and destroyed, leaving the distinctive 511 keV gamma photons. This phenomenon appears to be happening in the galactic center where our supermassive black hole (Sagittarius A*) is consuming and energizing vast amounts of infalling matter. But some **low-mass x-ray binaries (LMXBs)** also display similar, asymmetric positron-containing plasmas, suggesting a similar mechanism. In one model, over 10^{41} positrons per second can escape from a typical LMXB system. For some, this explanation was welcomed since it sidesteps the need for more exotic explanations (such as dark matter) for the astrophysical phenomenon.

X-ray binaries come in various forms but one simple classification system divides x-ray binaries into those with high-mass donors (**high-mass x-ray binaries (HMXBs)**) or low-mass donors (**low-mass x-ray binaries** or **LMXBs**).

LMXBs include soft x-ray transients, symbiotic x-ray binaries, super soft x-ray sources and **accreting millisecond X-ray pulsars (AMXPs)**. In these x-ray binaries, the accretor is a neutron star or black hole while the donor is the less massive member of the binary pair. This donor may be a main sequence star, a white dwarf or a red giant. Regardless of type, because the two members of the pair have grown closer, gravitational forces have led to the donor overfilling its Roche lobe, which then spills material onto its binary companion. Most LMXBs emit nearly all their electromagnetic radiation as x-rays with relatively little in the optical range (often less than 1%).

Scorpius X-1 is an LMXB system located about 9000 light years away. Aside from our Sun, Scorpius X-1 is the most intense source of x-rays (<20 keV) in the sky. The Scorpius X-1 system consists of a neutron star of approximately 1.4 $M_\odot$ and donor star of 0.42 $M_\odot$. Historically, Scorpius X-1 was the first extrasolar x-ray source ever discovered (12 June 1962). Approximately 200 LMXBs have been identified in our Milky Way Galaxy and many distant galaxies have been shown to harbor them as well. Some variable LMXBs are **x-ray pulsars** or **x-ray bursters**, which are discussed below.

HMXBs are strong x-ray sources in which the compact accretor is a neutron star or black hole while the donor is a high-mass companion such as a type O or B star, a blue supergiant, a Wolf–Rayet star or a red supergiant. Matter in the stellar wind of the giant donor spills onto the accretor and generates x-rays as it is heated and accelerated during the infall. In HMXBs, the massive donor companion may be extremely luminous and readily detectable. **Cygnus X-1** is one of the most intense sources of x-rays measured and is part of an HMXB system. Since the compact companion (i.e. Cygnus X-1) is estimated to have a mass of 21.2 $M_\odot$ but is too small and dim to be a normal star (or else it would be very visible), it is presumed to be a black hole (since this mass greatly exceeds the Chandrasekhar limit for a white dwarf or the Tolman–Oppenheimer–Volkoff limit for a neutron star). The black hole is in a close orbit (only 0.2 astronomical units) around a blue supergiant (HDE 226868) in this HMXB system. The powerful stellar wind from the blue supergiant provides the infalling matter for the accretion disc around the black hole. That stellar wind matter is accelerated by Cygnus X-1 and heated to millions of degrees and produces the x-rays of the HMXB. Other HMXBs include Vela X-1 and 4U 1700-37.

Although named for their x-ray emissions, some HMXBs also emit gamma rays and these gamma rays may reach incredible energies. For example, some of the radiation emitted by Cygnus X-3 has been measured in the PeV range. Cygnus X-3 is a rare example of an HMXB containing a Wolf–Rayet star as the donor and it may be an example of a **microquasar.** (Microquasars are discussed below.)

As with LMXBs, some variable HMXBs are x-ray pulsars. But unlike the situation with LMXBs, variable HMXBs are not x-ray bursters. Some HMXBs can evolve into a double neutron star binary system. In theory such a neutron star binary might then someday become a kilonova, creating a hard, short-duration gamma ray burst.

32.8 X-ray bursters

X-ray bursters are binary systems in which the donor is a main sequence star while the accretor is a neutron star. The periodic blasts of radiation from **type I x-ray bursters** are caused by continued accretion and heating of hydrogen or helium onto the surface of the neutron star until the threshold for thermonuclear detonation is reached. **Type II x-ray bursters** are the result of the conversion of gravitational potential energy into radiation. Type I x-ray bursters are analogous to novae. One of the main differences is that in a nova, the compact accretor is a white dwarf whereas in an x-ray burster, it is a neutron star. But in both cases, hydrogen accumulates on the surface, heats up and periodically undergoes thermonuclear detonation. Incidentally, since black holes do not have a real surface (infalling mass simply keeps falling until it disappears beneath the event horizon), they cannot serve as the compact member of a type I x-ray burster through this type of mechanism. Regarding nomenclature, the binaries that produce the x-rays are called x-ray bursters; the emitted pulses of radiation from such a system are called x-ray bursts.

The bursts of x-rays from most x-ray bursters repeat themselves with periods ranging from hours to days. The bursts themselves typically consist of an abrupt rise in x-ray output over 1 to 10 seconds followed by a decrease in intensity and spectral softening thereafter. For type I x-ray bursters (those due to thermonuclear fusion on the surface of the neutron star), the accreted material from the donor is compressed and heated to the point that the CNO cycle starts. As more and more mass is accumulated, the triple-alpha process can occur, leading to a helium flash. In very rare cases, carbon ignition may also take place, leading to 'super bursts'.

32.9 X-ray pulsars

The first x-ray pulsar to be discovered was Centaurus X-3, in 1971 with the Uhuru x-ray satellite. X-ray pulsars come in three varieties:

1. Accretion-powered pulsars
2. Rotation-powered pulsars
3. Anomalous x-ray pulsars.

32.9.1 Accretion-powered pulsars

The type of x-ray pulsars known as accretion-powered pulsars are a type of x-ray sources with periods ranging from under a second to several minutes. These x-ray pulsars are different from ordinary pulsars or anomalous x-ray pulsars in that accretion-powered x-ray pulsars are always part of binary systems. (Recall that anomalous x-ray pulsars are a stage in the evolution of magnetars where they start out as soft gamma repeaters but later develop into anomalous x-ray pulsars as they age.) In contrast to such single-star systems, in the accretion-powered binary x-ray pulsar systems, one of the members is a strongly magnetized neutron star. Such neutron stars have surface magnetic field strengths of about 10^8 Tesla, which are incredible but maybe not quite sufficient to qualify as magnetars. The donor in the accretion-powered pulsar binary is usually an ordinary main sequence star.

Gas is accreted onto the neutron star from the donor stellar companion. Although it is often alluded to as 'gas' in texts and papers, it is superheated and ionized as it spirals from donor to accretor and therefore is a plasma. As a plasma, it is subjected to the magnetic forces of the neutron star accretor and is thus funneled towards the neutron star's magnetic poles. Analogous to the situation on Earth that creates aurorae near the magnetic poles, the flowing ions create 'x-ray hot spots' near the magnetic poles of the neutron star. But the plasma producing these x-ray hot spots are far hotter than the Earth's auroral zones and highly energetic photons are emitted. In fact, the infalling gas (plasma) might attain speeds up to $0.5c$ before it crashes onto the neutron star. The conversion of gravitational potential energy into heat (which can reach millions of degrees) leads to the emission of thermal radiation in the x-ray range. Additionally, the sudden deceleration of the accreted ions as they crash onto the neutron star's surface combined with the circling of these ions around the neutron star as they spiral inward lead to bremsstrahlung and synchrotron radiation respectively. The intensity of the radiation emitted from these x-ray hotspots (which are relatively tiny in area at about 1 km^2) can be thousands of times more luminous than the Sun.

Just as with a plain pulsar, as the neutron star rotates, pulses of electromagnetic radiation are seen from our vantage point when the magnetic pole points our way. The difference is that in a pulsar, the electromagnetic radiation is largely radio waves whereas in an x-ray pulsar, the radiation is in the form of x-rays. As with pulsars, this mechanism requires that the magnetic and axial poles are offset from each other and that the magnetic pole must intermittently intersect our line of sight as it swings around and around. X-ray radiation is periodically detected as the x-ray hotspots move into and out of view during rotation of the neutron star. As mentioned, the periods range from under a second to a few minutes; the present records are 1.6 ms to over 10 min.

An interesting difference between x-ray pulsars and ordinary (radio) pulsars is that in the latter, the period is always increasing. In other words, regular pulsars are always slowing down. In contrast, x-ray pulsars may occasionally show an increase in their spin rates. This is because (accretion-powered) x-ray pulsars are always composed of a binary system and the neutron star member (which emits the x-rays) may gain or lose angular momentum through interactions with its binary donor partner. In contrast, radio pulsars are not generally part of a binary system that can transfer angular momentum from one partner to another. Regular radio pulsars are usually solitary, or when they are part of a binary system, their partner is often an inert white dwarf.

The **Neutron Star Interior Composition Explorer (NICER)**, which is aboard the International Space Station, discovered two stars that revolve around each other every 38 min. In other words, in this binary system, J17062, a year is only 38 min long. One of the stars is an AMXP and this incredibly rapid revolutionary rate is the present record holder for such a system. The NICER data show that J17062's stars are only about 186 000 miles (300 000 km) apart (which is closer than the Earth–Moon distance of ~250 000 miles). The center of mass (barycenter) of the binary system is a point around 1900 miles (3000 km) from the pulsar. Based on the binary pair's orbital period and separation distance, the second star is probably a hydrogen-poor white dwarf. This white dwarf donor is relatively small at only around 1.5% of the Sun's mass. The neutron star accretor, on the other hand, is much heavier, at around 1.4 $M_\odot$. Data from the **Rossi X-ray Timing Explorer (RXTE)** showed x-ray pulses coming from J17062 at 163 times a second. This translates into 9780 revolutions per minute, making this a very unusual and rapidly rotating millisecond x-ray pulsar.

32.9.2 Rotation-powered x-ray pulsars

It should be clearly stated that the x-ray pulsars discussed above are in the category of accretion-powered x-ray pulsars, which are part of binary systems. But there are some rare x-ray pulsars that are not members of binary pairs. In these cases, a young, solitary, highly magnetized, rapidly rotating neutron star will generate incredibly high electrical potentials that accelerate charged particles. The accelerated charged particles in turn produce beams of electromagnetic radiation that emanate from the magnetic poles. If the magnetic pole is geographically offset from the axial pole, the magnetic pole—and the radiation firing from it—will wobble about the spin axis in a cone. That cone may occasionally cross our line of sight and we may detect that radiation. In most cases, what we detect are radio waves in the form of a conventional pulsar. Roughly a thousand such pulsars have been catalogued. A small subset of these pulsars also emit electromagnetic radiation in other wavelengths ranging all the way up to x-rays and gamma rays. The pulsar in the Crab Nebula is one such pulsar. Born in 1054 CE, this is one of the youngest and most energetic known pulsars. It has been observed to pulse in wavelengths ranging from radio to x-ray and gamma ray wavelengths. Thus, the Crab Nebula pulsar is a rare example of a rotation-powered x-ray (and gamma ray) pulsar. Not only is the Crab Nebula pulsar a rotation-powered gamma emitter, but it is also an extraordinary gamma emitter—recent measurements of the energies of some of the gamma photons from this pulsar have reached over 1 PeV. The extreme energies of these gamma photons can be explained by a combination of synchrotron radiation and inverse Compton scattering from the relativistic electron–positron pulsar wind. A few dozen rotation-powered pulsars have been seen to pulse in x-rays and a half-dozen are known to pulse in gamma rays.

32.10 Anomalous x-ray pulsars

Starquakes on magnetars lead to magnetar giant flares (MGFs) on soft gamma repeaters. But as the gargantuan magnetic field repeatedly causes periodic restructuring of the neutron star's crust and release of gamma rays, it dissipates energy and loses field strength over time. Eventually, the energy of the photons emitted from these intermittent starquakes will no longer remain in the gamma ray range. But photons might still emanate from the weakening magnetar in the form of x-rays. In such cases, the x-ray pulsar is called an anomalous x-ray pulsar.

32.11 Brown dwarfs as sources of x-ray flares

Without a nuclear fusion source in their cores for internal heating, the interior of brown dwarf stars remains in a fully convective state. This convection can lead to tangled magnetic lines near a brown dwarf's surface. The complexity of these tangled lines of force is magnified by the often-rapid rotation rate of brown dwarfs. Snapping of these tangled magnetic field lines may be the explanation for an x-ray flare from **LP 944–20** observed by the Chandra X-ray Observatory. Turbulence of the highly magnetized material beneath the surface could have led to a release of energy into the atmosphere of the brown dwarf. This in turn may have created strong atmospheric electric currents and generated the x-ray flare from LP 944–20. The Chandra X-ray Observatory also detected x-rays from another brown dwarf (TWA 5B) that is in close proximity to its sunlike companions in a multiple-star system.

32.12 Gamma rays from a nova

If a white dwarf is created by the collapse of a star in a binary system and its partner later evolves into a red giant, that red giant may dump matter onto the smaller white dwarf companion. This is known as Roche lobe overflow. If so much mass accumulates on the white dwarf that it exceeds the Chandasekhar limit (which depends on the elemental composition of a white dwarf but theoretically is around 1.4 $M_\odot$ for a non-rotating idealized case), the white dwarf cannot withstand the inward pressure of gravity and the whole system collapses. The result is a type IA supernova.

As mentioned throughout the text, stars are locked in a fierce fight between the relentless inward crush of gravity and the outward push of thermonuclear-generated heat. Gravity gradually wins most of these wars as the thermonuclear fuel runs out, leaving behind stellar corpses such as white dwarfs and neutron stars. In the most decisive gravitational victories, gravity triumphantly and almost arrogantly compresses all the former star's matter into a single, dimensionless point— a black hole.

But type IA supernovae represent situations in which gravity is not the victor. In these types of supernova explosions, the threshold for carbon and oxygen fusion (or other elements depending on the particular scenario) on the accreting white dwarf is suddenly exceeded and thermonuclear runaway ensues.

Unlike in other supernova explosions where the crushed corpse of the former star is left behind as a neutron star or black hole, in type IA supernovae the aftermath at the scene of the explosion is... nothing. Gravity is resoundingly overcome and the formerly compressed matter in the white dwarf star is liberated and spread across the galaxy as the ashes of its former self, transformed into new, heavier elements.

Of course, not all binaries involving a white dwarf end in such dramatic supernova fashion. A classical nova is the result of matter accreting onto the surface of a white dwarf and causing a more modest thermonuclear detonation. Novae are now considered forms of **cataclysmic variables** or **cataclysmic binaries** since they consist of a white dwarf and a mass-transferring companion (the donor star). As infalling matter from the donor (mostly hydrogen) circles the white dwarf and creates an accretion disc, gravitational potential energy and friction heat up the infalling matter, which then emits ultraviolet and x-ray radiation. Then as the hydrogen continues to accumulate on the surface of the white dwarf, it might eventually exceed the threshold for fusion of hydrogen into helium and lead to intermittent nuclear explosions. This detonation provides the conditions needed to accelerate charged particles to great energies. And as in many other astrophysical phenomena, accelerated charged particles then generate high-energy photons. The MAGIC telescopes detected high-energy photons (in the 100 MeV to 10 GeV range) along with VHEGRs (in the 60–250 GeV range) from a 2021 outburst on RS Ophiuchi, a recurrent nova involving a white dwarf with a red giant companion. The data from MAGIC and the Fermi LAT are consistent with a model in which protons are accelerated to hundreds of GeV by the nova shock, which then make bubbles of increased cosmic ray density centered around the underlying cataclysmic binary star system.

32.12.1 Microquasars

Microquasars or radio-jet x-ray binaries are special x-ray binary systems which launch highly collimated relativistic jets blasting out in opposite directions from a central rotating source, perpendicular to an accretion disc. As with other x-ray binary systems, the compact accretor is a neutron star or a black hole. This compact accretor is not very massive at only a few $M_\odot$. The donor is a normal (main sequence) star but may be a very massive star of spectral type O or B. Such binaries are called **high-mass microquasars**. However, lower-mass donors (such as type A, F or G stars) are possible and such systems are called **low-mass microquasars**.

The accretion disc itself creates soft x-rays while the corona can make hard x-rays and gamma rays. The jets also generate high-energy photons reaching the gamma ray range. For instance, the gamma rays from the jets of Cygnus X-1 can be over 60 MeV.

Some microquasars include SS 433, 4U 0614+091, GRS 1915+105, Cygnus X-1, Cygnus X-3, LS I +61 303 and Scorpius X-1. SS 433 was the first discovered microquasar and is composed of a type A star donor and a black hole accretor circling one another every 13 days. The opposing jets of protons above and below the plane of the accretion disc have been tracked at a speed of 0.26c. The HAWC Observatory in Mexico recorded some gamma ray photons exceeding 25 TeV.

4U 0614+091 is technically a microquasar since it produces jets, but is also an LMXB system with a neutron star and a low-mass donor companion. It was the first microquasar identified with a neutron star rather than a black hole as the compact accretor.

On the other end of the spectrum, GRS 1915+105 is an x-ray binary star system with an ordinary star donor and a black hole accretor that also qualifies as a microquasar thanks to its jets. But the black hole in GRS 1915+105 is the heaviest known stellar-mass black hole in our Milky Way Galaxy at an estimated 10–18 $M_\odot$.

Cygnus X-1 is one of the strongest x-ray sources detected. The compact accretor in this HMXB is estimated to weigh about 21.2 $M_\odot$ and is too small to be any type of normal star and is too massive for a white dwarf or neutron star. Thus, Cygnus X-1 was the first suspected black hole. It orbits the donor, HDE 226868 which is a variable blue supergiant. The progenitor to the black hole was probably around 40 $M_\odot$ before it collapsed. It is likely that if it exploded as a core-collapse supernova, it would have blown away its companion star HDE 226868. Thus, it probably lost most of the missing mass in the form of a strong stellar wind. The matter falling into the black hole now creates an accretion disc reaching temperatures of million of degrees, thereby emitting the detected x-rays. As a microquasar, Cygnus X-1 also has the defining associated pair of radiation-generating relativistic jets spouting perpendicular to the accretion disc.

32.13 Quasars and Seyfert galaxies

A quasar (short for quasi-stellar object or quasi-stellar radio source) is an extremely luminous type of active galactic nucleus or AGN. Historically, these objects were first discovered in the 1950s and appeared star-like (thus the term, 'quasi-stellar') but demonstrated confusing spectral lines that were later interpreted as ordinary hydrogen lines with extraordinary redshifts. This indicated that although they appeared very small and star-like in appearance, they were incredibly distant, posing a puzzle as to their luminosity. Further study, especially through data from the Hubble Space Telescope has confirmed that quasars are not really stars but rather are galactic centers. We now understand that quasars are a subset of **AGNs**.

There are now well over a million known quasars. Some quasars are part of larger collections of quasars called large quasar groups. The closest quasar on record is Markarian 231 at a distance of around 600 million light years. Some might quibble over this claim since technically Markarian 231 is a Seyfert galaxy. Seyfert galaxies, like quasars, are luminous galaxies with AGNs but unlike quasars, their host galaxies are visible. Thus, the difference between Seyfert galaxies and quasars is just that the former are the active sites of galaxies that can be detected whereas quasars are the active nuclear regions of galaxies that cannot be discerned. This simply implies that Seyfert galaxies are much closer than quasars. But the nuclei in Seyfert galaxies emit roughly the same amount of visible light as the sum of the host galaxy's stars. Quasars in contrast have nuclei that emit over a hundred times as much visible light as the rest of the host galaxy's constituent stars.

Seyfert galaxies are classified as type I and type II Seyfert galaxies. In addition to their very bright luminosities in optical wavelengths, type I Seyfert galaxies are intense sources of ultraviolet and x-ray radiation. Type II Seyfert galaxies are bright in the infrared wavelengths.

Both quasars and Seyfert galaxies probably host exceptionally large and actively-eating supermassive black holes as their engines. Regarding terminology, both Seyfert galaxies and quasars are **active galaxies**, which are defined as galaxies hosting an AGN. Most active galaxies show large redshifts, indicating great distance and therefore great age. This implies that active galaxies were more common in the early Universe. Quasars were clearly more common in the distant past and went through a peak period around 10 billion years ago.

Currently, the farthest known quasar is QSO JO313–1806, which contains a supermassive black hole 1.6 billion times as massive as our Sun. (Recall that the Milky Way's central supermassive black hole, Sagittarius A* is only about 4.1 million $M_\odot$.) This quasar is well beyond 13 billion light years away, with a redshift of $z = 7.64$, indicating that the light we are seeing now originated at a time when the Universe was only 670 million years old.

Quasars are among the most powerful astrophysical entities in the Universe. Some quasars emit about a thousandfold more energy than our Milky Way Galaxy. The brightest known quasar in terms of apparent magnitude is 3C 273, which is around 2.4 billion light years from us. It has an absolute magnitude of −26.7 with a luminosity of around 4×10^{12} that of the Sun or roughly 100 times that of the entire Milky Way Galaxy. This means that if we replaced our galactic center with this quasar, a bright spot in Sagittarius would shine as bright in the night sky as the daytime Sun. Brighter still in absolute terms is the hyperluminous quasar APM 08279+5255 with an initial absolute magnitude estimate of −32.2. But further examination has confirmed that there is gravitational lensing that is magnifying this quasar by an order of magnitude. The enormous amount of radiation released along with the equivalent of a solar wind ('galactic wind'?) and of course the jets flooding from the core seem to have halted new star formation in these quasars. Thus, quasars, which seem to be a thing of the distant past in our Universe are quite different from modern galaxies. This evidence of evolution of the Universe is taken as support for the Big Bang theory and as a mark against the previous competing steady-state hypothesis (since the steady-state hypothesis holds that the Universe should be essentially static and homogeneous in all directions).

The electromagnetic radiation of quasars spans the spectrum with a peak in the ultraviolet and some quasars being bright sources of radio waves and gamma rays. Curiously, although the peak output is actually near the 121.6 nm Lyman-alpha line in the ultraviolet, the observed peak is in the near infrared thanks to the tremendous redshift. As with all active galaxies, quasars may be robust x-ray emitters. Some quasars (**radio-loud quasars**) may also emit even more energetic photons in the form of gamma rays. Such energetic photons probably arise through the inverse Compton effect, in which the radio wave-emitting electrons in the quasar jets transfer energy to photons and thereby boost them into the x-ray and gamma ray range. The x-ray-producing jets can be quite extensive. The Chandra X-ray Telescope showed that the jets from the quasar PKS 1127–145 stretch over a million light years from the quasar itself.

At the core of a quasar is a supermassive black hole. Infalling matter spirals around this central black hole and accelerates due to gravity. In addition to the conversion of gravitational potential energy into kinetic energy, friction further heats up this infalling matter. And the conversion of gravitational potential energy into energy is most impressive—depending on specific circumstances, up to 32% of the rest mass of an infalling object may be converted into energy. To put into context just how impressive this figure is, nuclear fission of uranium-235 converts about 0.1% of the mass into energy and nuclear fusion of hydrogen in the core of the Sun (the proton–proton chain) converts about 0.7% of the rest mass into energy. The precise amount of rest mass converted into radiation energy via gravitational potential energy before it enters the event horizon depends on the rotational rate of the black hole; the faster the rotation, the more energy released.

Another feature of quasars is their variable luminosity. The variation ranges from months to hours. This variability has been a great clue to their size. For example, if a quasar exhibits luminosity variability in weeks, the quasar must be no more than that many light weeks wide. In the extreme case of hours, such variability indicates that the entire object has to be only light hours across. For such small objects to have the luminosity they do, an energy source far more efficient than nuclear fission or nuclear fusion is needed. As mentioned above, black hole-powered conversion of gravitational energy into radiation is one such solution with efficiencies in the range of 6% to 32%. This appears to be the only possible process that can power quasars on their observed long-term basis. It seems likely that after a quasar has consumed all the fuel in its vicinity, it will grow quiescent and become an ordinary galaxy. But before that point, quasars can consume mass at a prodigious rate—the brightest quasars consume about 1000 Sun-sized stars per year (or the equivalent of 10 Earths eaten per second!). Students should remain aware of the fact that this is not the rate at which the central black hole gains mass given the amazing efficiency of the accretion disc converting mass into radiation energy. If 32% of the rest mass is converted into energy during the plunge, the black hole can acquire a gain in mass of only 68% of that infalling object's initial rest mass.

32.13.1 Blazars

Blazars are another intense but variable source of radiation from the very distant skies. The name may originate from the two subclasses of blazars: **BL Lacertae objects** and **optically violent variable quasars**. Blazars in turn may be a subcategory of **radio-loud quasars**.

The variability in brightness of blazars changes with periods ranging from minutes to years. The amplitude of the variability as well as the luminosity appear to be magnified by **relativistic jetting**.

The Fermi LAT has detected about 1000 gamma-emitting blazars so far bringing the total known to nearly 3000. High-resolution images indicate that most blazars lie in the centers of elliptical galaxies. The supermassive black holes at the cores of these gamma ray blazars can be a billion times as massive as our Sun, which begs the question of how such supermassive black holes could have formed in a mere 2 billion years, since some of the blazars are well over 12 billion light years away. Some blazars are very intense in gamma emissions and one (3C 454.3) outshines all other recorded gamma ray sources when it reaches it peak brightness during its cycles. Given its distance of over 7 billion light years, the output is phenomenal. And the energies of these gamma photons can be quite high. For instance, Markarian 501 and S5 0014+81 emit extraordinarily energetic gamma rays and are sometimes called TeV blazars because some of their photons reach the TeV energy range. Markarian 501 exhibits a high-energy photon spectrum with two peaks. The lower-energy peak is <1 keV and consists of soft x-rays. The other peak consists of hard gamma rays that can exceed 1 TeV. During flares and outbursts the peaks increase in both power and photon energy. The MAGIC imaging atmospheric Cherenkov telescopes have recorded high-energy flares from Markarian 501 lasting around 20 min. A curious and still unexplained observation is that the highest-energy photons (e.g. those over 1 TeV) were delayed by 4 min compared to the less energetic high-energy photons (e.g. those of 250 GeV). From a distance of 3.7 billion light years, another blazar (TXS 0506+056) has been shown by the IceCube Neutrino Observatory project to be a source of high-energy neutrinos as well as photons. During a gamma ray flare of this blazar, a 290 TeV muon neutrino was detected along with high-energy gamma rays. It is believed that this muon neutrino originated from the decay of a charged pion that was made by the interaction of an energetic blazar proton or nucleus (cosmic ray) with matter or a radiation field.

It is presently believed that blazars and quasars are really the same astrophysical entities viewed from slightly different angles. In fact, quasars, blazars and radio galaxies are similar phenomena based on the so-called unified scheme of radio-loud AGNs. All three types of AGNs produce relativistic jets of ionized matter generated by infalling matter in a whirling accretion disc surrounding a supermassive black hole at the center of a galaxy. As the matter in the surrounding accretion disc accelerates under the force of gravity from the central black hole it heats up. This heating is augmented by friction in furiously fast swirling matter. The energy in this accretion disc is what produces the electromagnetic radiation generally emanating from these AGNs. Nothing that actually enters the supermassive black hole can escape, not even photons. It is also the energy in the accretion disc that powers the twin relativistic jets of ions blasting out of the north and south poles of this rotating system. Again, the jets are created outside the black hole since nothing can escape from the black hole itself. These relativistic jets of plasma are then collimated by the strong magnetic field of the AGN and shoot out of the poles, perpendicular to the accretion disc. The difference between blazars, quasars and radio galaxies is the orientation of the beamed jets of ionized plasma flaring from the poles.

When the relativistic jet beam is directed straight at us, the AGN is a blazar. When it is not pointing directly at us but is still aiming in our general direction, we see a quasar. When the jets are not oriented in our direction (for example, if they are pointed perpendicular to us) we observe a radio galaxy. The brightness of blazars is boosted by a phenomenon called relativistic beaming or Doppler boosting; the head-on orientation of the relativistic jets amplifies the gamma ray output of blazars because of relativistic beaming. If a 0.999c jet were fired from an AGN at say, 5° from our line of sight, the observed brightness might be dozens of times higher than the actually emitted luminosity. But if that same beam were aimed right at us at 0°, the jet could look several hundred-fold more luminous.

Relativistic beaming also, through time dilation, increases the apparent frequency of bursts. For instance, if a blazar is firing jets at us once every minute from its frame of reference, we might observe a far higher frequency of jets from our frame of reference because of relativistic beaming.

Further reading

[1] Strohmayer T E *et al* 2018 NICER discovers the ultracompact orbit of the accreting millisecond pulsar IGR J17062-6143 *Astrophys. J. Lett.* **858** p.L13

[2] Weidenspointner G, Skinner G and Jean P *et al* 2008 An asymmetric distribution of positrons in the Galactic disk revealed by γ-rays *Nature* **451** 159–62

[3] Cao Z, Aharonian F A and An Q *et al* 2021 Ultrahigh-energy photons up to 1.4 petaelectronvolts from 12 γ-ray Galactic sources *Nature* **594** 33–6

[4] The LHAASO Collaboration *et al* 2021 Peta–electron volt gamma-ray emission from the Crab Nebula *Science* **373** 425–30

[5] Yang H Y K, Ruszkowski M and Zweibel E G 2022 Fermi and eROSITA bubbles as relics of the past activity of the Galaxy's central black hole *Nat. Astron.* **6** 584–91

[6] Su M, Slatyer T R and Finkbeiner D P 2010 Giant gamma-ray bubbles from Fermi-LAT: active galactic nucleus activity or bipolar galactic wind? *Astrophys. J.* **724** 1044

[7] Crocker R M, Macias O and Mackey D *et al* 2022 Gamma-ray emission from the Sagittarius dwarf spheroidal galaxy due to millisecond pulsars *Nat. Astron.* **6** 1317–24

[8] Acciari V A, Ansoldi S and Antonelli L A *et al* 2022 Proton acceleration in thermonuclear nova explosions revealed by gamma rays *Nat. Astron.* **6** 689–97

[9] Abeysekara A U, Albert A and Alfaro R *et al* 2018 Very-high-energy particle acceleration powered by the jets of the microquasar SS 433 *Nature* **562** 82–5

[10] Wang F, Yang J, Fan X, Hennawi J F, Barth A J and Banados E *et al* 2021 A luminous quasar at redshift 7.642 *Astrophys. J. Lett.* **907** abd8c6

[11] IceCube Collaboration *et al* 2018 Multimessenger observations of a flaring blazar coincident with high-energy neutrino IceCube-170922A *Science* **361** eaat1378

[12] IceCube Collaboration *et al* 2018 Neutrino emission from the direction of the blazar TXS 0506+056 prior to the IceCube-170922A alert *Science* **361** 147–51

IOP Publishing

Space Radiation
Astrophysical origins, radiobiological effects and implications for space travellers
James S Welsh

Chapter 33

Neutron stars, magnetars and soft gamma repeaters

33.1 Soft gamma repeaters

Aside from gamma ray bursts (GRBs), another source of gamma rays in deep space are **soft gamma repeaters (SGRs)**. Initially SGRs were believed to be another variety of GRB, but thanks to their unique properties and completely different means of making gamma rays, these high-energy photons producing entities are now given their own separate category.

It is now known that SGRs do not mark the birth of a new black hole but instead are a completely different astrophysical phenomenon with a completely different underlying mechanism of making their gamma rays. SGRs also differ from GRBs in that GRBs are one-time events whereas SGRs (as the name indicates!) repeat their gamma ray emissions periodically.

SGRs are believed to be associated with 'starquakes' on special types of neutron stars called **magnetars**. Starquakes on magnetars may occasionally be associated with giant x-ray flares as discussed below. Starquakes involve massive restructuring of the neutron star crust along with the intense magnetic fields of these highly magnetized neutron stars. Before going into details, we shall briefly review neutron stars.

33.2 Neutron star basics

After a star exhausts its supply of nuclear fuel and reaches the endpoint for fusion, it will no longer be able to continue its battle against the relentless inward pressure of gravity. For the Sun and solar-mass stars, the end will come after fusion of hydrogen and helium leads to carbon and oxygen. Temperatures of these stars will never reach the threshold for fusion of these elements into heavier ones, so this is the end of the line. The final corpse of these stars will be a white dwarf. White dwarfs cannot exceed the Chandrasekhar limit which represents the limit for electron degeneracy pressure. (Recall that electrons and other fermions obey the Pauli exclusion principle and cannot occupy the same quantum state in the same atom or nucleus. This resistance to being crowded together is what constitutes degeneracy pressure. All the lower-energy quantum states become fully occupied, leaving only the higher energy states available. These high-energy states correspond to more rapid motion of the particles (electrons or neutrons) involved, and more rapid motion corresponds to greater pressure. Degeneracy pressure can also be understood on a quantum level by the Heisenberg uncertainty principle, $\Delta x \cdot \Delta p > h/2\pi$. As Δx grows smaller and smaller with compression, Δp must correspondingly increase. Greater momentum means greater velocity and thus greater pressure.) But what if the mass of the imploding star's core exceeds the Chandrasekhar limit? Electron degeneracy pressure cannot overcome the strength of such gravity and the next stage is reached.

And supergiant stars (with total mass between 10 and 25 $M_\odot$) exert a lot of pressure! The endpoint for fusion is often stated to be iron-56 because it is near the peak of the nuclear binding energy curve. Students might recall that nickel-56 is technically higher on this curve, but it decays first into cobalt-56, which then decays into iron-56. In any case, once the core of a massive star starts making iron, the game is over since fusion of iron does not yield energy but rather consumes it. Unlike all previous exothermic fusion steps with lighter elements, fusion beyond iron is endothermic. Thus, in practically an instant, after mega-years of generating out-wardly directed heat and pressure through fusion of lighter elements, the core is suddenly incapable of continuing the fight against gravity. Gravity seizes the moment and pounces, crushing the core with all its might. The product will exceed the Chandrasekhar limit, so electron degeneracy pressure cannot stave off further collapse. The various elements produced in the core such as helium, carbon, oxygen, silicon and all the way up to iron will disintegrate as their protons are squeezed together with electrons to form neutrons. And as the protons and electrons merge under the force of gravity, the electron capture process will create neutrinos. This outward flood of neutrinos will contribute to the massive explosion known as a supernova when the imploding core suddenly stops and rebounds off the incom-pressible newly formed neutron star. Countless additional neutrinos are created by a process called thermal emission wherein energy at temperatures near 100 billion K is transformed, via pair production, into neutrino–antineutrino pairs. This makes an even larger deluge of neutrinos than the electron capture process does. In fact, the majority of the 10^{46} J released in a core-collapse supernova explosion is in the form of these neutrinos. But what is left behind is a neutron star, the remnant of the former supergiant star. Most neutron stars are in the range of 1.4 $M_\odot$ and around 10 km (~6 miles) in radius. At a density 10^{14} times that of water, neutron star stuff has a density comparable to that in atomic nuclei. A single spoonful of neutron star material would weigh in the vicinity of 3 billion tons. Neutron stars are supported by neutron degeneracy pressure against any further collapse caused by gravity.

Maybe even more amazing than their mass and densities is the magnetic field strength associated with some neutron stars. As will be discussed below, such neutron stars, called magnetars, have magnetic fields between 10^8 and 10^{15} times the strength of Earth's magnetic field.

Of course, just as white dwarfs and electron degeneracy pressure have their limits (the Chandrasekhar limit), neutron stars and neutron degeneracy pressure have their limits—the Tolman–Oppenheimer–Volkoff (TOV) limit. The present estimate for the TOV limit is in the range of 2.2–2.9 $M_\odot$. Beyond this limit, not even neutron degeneracy pressure and repulsive nuclear forces can stave off gravity and the neutron star will collapse into a black hole. The most massive neutron star yet detected is PSR J0952-0607, which is estimated to be 2.35 ± 0.17 $M_\odot$. This challenges the present estimated mass range for the TOV limit.

33.2.1 Pulsars

Neutron stars were first predicted by Fritz Zwicky and Walter Baade in 1933 but they remained theoretical until 1967 when Jocelyn Bell discovered the first 'pulsar' or pulsating radio star. We now know that all pulsars are neutron stars—but not all neutron stars are pulsars. Pulsars may or may not be visible in the optical wavelengths, but they invariably emit regular pulses of radio waves. The pulsar detected by Bell has a radio wave periodicity of 1.337 3011 s, demonstrating the typical high precision and consistency of most subsequently discovered pulsars. Other pulsars display different periodicities but regardless of the period, they are extremely self-consistent.

Initially thought to possibly communications from little green men, subsequent pulsars were found to be mostly confined to the plane of the Milky Way Galaxy, where the most massive stars reside. But this location, combined with the highly regular (and therefore not very information-rich) repeating pattern of the pulsars made it less likely that these were signals from intelligent aliens. Furthermore, the absence of a periodic Doppler shift (that would be expected from a planet or spaceship orbiting a central star) made it even less likely that these were deliberately released alien signals. Additional detailed analysis indicated that it was very unlikely that pulsars could be the products of oscillation (as in Cepheid variables or RR Lyrae stars) since the repetition rate was far too fast for such a mechanism for ordinary stars, or even for white dwarfs. Conversely, the intrinsic vibration rate was too short for what would have been expected with a then-still-hypothetical neutron star. Another idea that did not stand up to scrutiny was the concept of two rapidly orbiting stars in a very tight orbit. Ordinary stars or white dwarfs could not maintain such a tight orbit for long thanks to the overwhelming centrifugal forces associated with the repetition rates of pulsars. And while neutron stars could theoretically be in such a tight, fast orbit around their barycenters, they would emit gravitational waves and energy, thereby leading to a steady and perceptible slowdown in periodicity. Over the time course of human lifetimes (that is decades of observations), pulsars show remarkable constancy in their periodicity, which argues against the binary star mechanism.

Nevertheless pulsars do slow down over time. The average rate of pulsar slowdown is about 0.000 001 s yr^{-1} (a millionth of a second per year). One parameter used to gauge pulsar slowdown is its **characteristic age**. The formula for a radio pulsar's characteristic age is:

$$\tau = \frac{P}{2\dot{P}}$$

where τ it the characteristic age, P is the pulsar period and $\dot{P}$ is the time derivative of the period.

In general, the characteristic age approximates the real age. For instance, the pulsar in the Crab Nebula (the Crab Pulsar) was born in a supernova in 1054 CE and its characteristic age is 1240 years. However, there are some notable exceptions, and the discrepancy can be off by over 4000 years.

Overall however, it began to look like the best theory to explain pulsars was the rapidly rotating neutron star model.

33.3 Millisecond pulsars

So, what about the idea of a white dwarf or a (then-still-hypothetical) neutron star rapidly spinning on its axis? The periodicity of some pulsars (called millisecond pulsars) has been found to exceed a thousand hertz. Discovered in 2006, the millisecond pulsar called J1748−2446ad has a period of just 1.396 ms, which corresponds to a spin rate of 716 times s^{-1}. This is the current record holder among millisecond pulsar rates.

Given the existence of such millisecond pulsars, students are invited to calculate the tangential linear velocity of such objects (through $v = r\omega$) if they were the size of the Sun. (The answer exceeds the speed of light c, making such a mechanism impossible for a Sun-sized object.) For smaller objects, even if Einstein allowed a white dwarf to rotate at such rates, the centrifugal forces of a white dwarf with a period of 300 ms would tear it asunder. In contrast, the estimated high densities of then-undiscovered neutron stars would be capable of withstanding such swiftly spinning centrifugal forces. Thus, the only pulsar candidate still standing was the neutron star. In this manner, it was concluded that Jocelyn Bell's pulsar discovery was the confirmation of the long-sought-after neutron star.

At this time, the best model for pulsars is a rapidly rotating, highly magnetized neutron star in which the rotational and magnetic axes are slightly offset from one another. In such a system, the north and south magnetic poles would be tracing out a cone in space as the neutron star spins on its rotational axis as in the diagram below (figure 33.1).

Students might recall from basic electricity and magnetism that a rapidly rotating magnet will create a current. In a pulsar, the rapidly rotating magnetic fields generate electric fields and currents that accelerate electrons to relativistic velocities. And accelerating charged particles, such as electrons, radiate electromagnetic radiation. The electromagnetic radiation emanates from the precessing magnetic poles in beams that trace out a cone into space. When that cone intersects our line of sight, we can detect the pulsar. The phenomenon is akin to a lighthouse, in which we only see the light periodically—when the light happens to be directed our way—even though it is always on. Incidentally, young pulsars emit visible, ultraviolet and x-ray electromagnetic radiation but as they age, the higher energy radiation subsides, leaving in most cases only the radio signals. Each rotation of the neutron star about

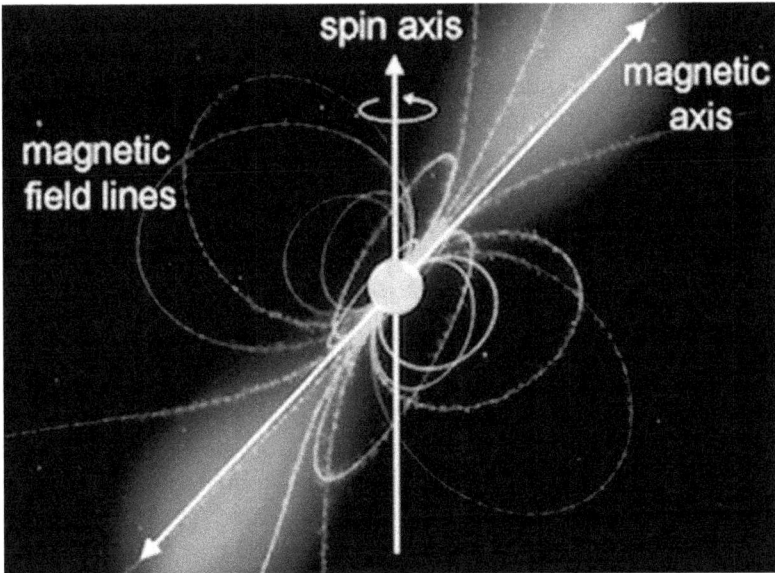

Figure 33.1. An artistic rendition provided by NASA depicting a gamma ray pulsar that lies within the supernova remnant, CTA 1 which is located about 4600 light-years away in the constellation Cepheus. It was discovered with NASA's Fermi Gamma-ray Space Telescope. In pulsars, a cone of radiation beams from the magnetic poles in a lighthouse-like fashion as the central neutron star rotates about its spin axis. Charged particles follow the pulsar's magnetic field lines (blue) and create the beams of energetic photons that emerge from the two poles. This unusual gamma-ray-only pulsar, which formed about 10 000 years ago, sweeps its beam in Earth's direction every 316.86 milliseconds. (Credit: NASA/Goddard Space Flight Center Conceptual Image Lab).

its spin axis leads to a sweeping cone of photons emanating from the slightly offset magnetic poles, and if we happen to be in the path of that cone of photons, we might periodically detect them. Given this mechanism behind pulsars, for every one we can see, there must be many we cannot see since their cone of light is not directed our way.

Furthermore, only on a neutron star, where the density is enormous and gravity is more than 100 billion times as great as on Earth, can matter withstand the magnetic forces of a magnetar. Yet even there the neutron star's crust can break apart if the strain is great enough.

33.4 Pulsars with planets?

In 1991 a millisecond pulsar was noted to have a regular irregularity in its pulse. The variability was traced to a Doppler effect that caused pulse rate to sometimes hasten while at other times slowing down. The explanation appears to be some sort of solar system wherein the pulsar and at least three orbiting planets are circling a common center of mass or barycenter. Thus, at times the neutron star is approaching Earth and at other times it is receding. Although the planets seem similar in size to real planets in our Solar System, they nevertheless must be very different. For one thing, they probably did not form concurrently with the original star that begat the pulsar. Any genuine planets in existence before the star went supernova would have been blasted to smithereens by the explosion. Rather than forming simultaneously with the pulsar progenitor in the initial acquisition disk, these planets most likely are 'Lazarus' planets that probably formed from the disk of debris surrounding the neutron star after the supernova. Most ironically given their scarcity (and the plethora of exoplanets discovered by the Kepler Mission), these extreme, exotic and truly strange planets were the first exoplanets to be discovered.

33.5 Do pulsars slow down? Do magnetars lose their strength?

There is a pulsar at the center of the Crab Nebula, which is the remnant of a supernova that was witnessed in 1054 AD. As with all pulsars, there are jets of particles and electromagnetic radiation firing out from its poles. But in addition to that, this pulsar produces a steady wind of radiation analogous to the solar wind. This pulsar wind energizes and ionizes gases in the Crab Nebula, causing it to glow. Consistent with the First Law of Thermodynamics, the energy lost by the pulsar in the form of pulsar wind and jets of particles must equal the energy gained by the Crab Nebula. The gradual loss of energy by the pulsar has slightly slowed down its pulsation rate. While pulsars seem remarkably constant in their periods, over time they do lose energy and decrease their rotation rates. This slowdown, coupled with energy loss and a diminution in magnetic field strength means that after a few million years a pulsar might grow too dim for detection.

Thus, while every pulsar is a neutron star, not every neutron star is an active pulsar. Some may have fizzled out over time. In others, the axis of rotation might be oriented in a manner that does not permit the electromagnetic radiation beams to intersect our line of sight.

The average pulsar period is approximately one spin per second. Some pulsars will rotate once every 10 s while others spin 10 times per second. However, some extreme pulsars can rotate hundreds of times per second, with periods measured in milliseconds, thus the name millisecond pulsars. Some of these frequencies fall into the hearing range for human ears. For instance, middle C is often taken as 262 Hz. The varying frequency of pulsars has inspired some musically inclined individuals to translate these frequencies into audible notes and combine them into tunes. While such sounds might technically be true 'music of the spheres' (celestial spheres, that is), thus far Bach's fugues and Beethoven's symphonies do not seem to be in any danger of being dethroned.

33.5.1 Magnetars

The intense magnetic field is the source of the pulsar's electromagnetic radiation, but what is the source of the high-intensity magnetic field? The most logical explanation is the collapse of the star and its associated original magnetic field. The shrinkage of the supergiant into a neutron star concentrates the magnetic field. And as the rotation rate increases during the collapse, additional magnetic effects occur.

Rotating neutron stars with super-strong magnetic fields are called **magnetars**. Just how strong are the magnetic fields of magnetars? Recalling that 1 T equals 10 000 G or 1 G equals 100 µT, our own geomagnetic field strength, which can move a compass needle, is roughly half a gauss or in the 30–65 µT range. Of note, the geomagnetic field strength varies on the surface from place to place, with values ranging from <0.3 G (<30 µT) in much of South America and parts of southern Africa, to >0.6 G (>60 µT) in polar regions such as northern Canada, Siberia or southern Australia. Furthermore, at the depth of Earth's liquid outer core about 1800 miles deep, the magnetic field strength is estimated to be about fiftyfold higher than at the surface: ≈25 G (2500 µT or 2.5 mT). Having said all that, the magnetic field strength of a decent magnetar is well over a trillion times as strong as Earth's.

For further reference, a small toy magnetic might have a strength of 100 G (or 10 000 µT or 10 mT) while a strong neodymium magnet might reach 2000 G or 0.2 T. A medical magnetic resonance imaging (MRI) unit that can dangerously fling ferromagnetic objects across a room, easily erase computer hard drives and wipe out credit cards might have a magnetic field strength of 3 T. In comparison, magnetars can reach up to 100 billion T.

As mentioned earlier, rapidly rotating pulsars possess magnetic field strengths of around a trillion gauss or 10^8 T. In fact, the magnetic energy density of a 10^8 T field is extraordinary and vastly exceeds the mass–energy density of any ordinary matter found on Earth. As strong as such fields are, the magnetic fields of magnetars are far stronger still. If one of these powerful magnetars was as far away as halfway to the Moon, it could wipe out all the credit cards on Earth. Magnetic fields of this magnitude can induce astoundingly weird phenomena such as polarizing the vacuum to the point that it becomes birefringent. Photons can then merge or they might split in two. Furthermore, virtual particle–antiparticle pairs are produced in the presence of such field strengths. Such super-strong magnetic fields can change electron energy levels in atoms and thereby alter the shapes of atoms and molecules. With this in mind, at a distance of 1000 km, the magnetic field would make all life impossible by disrupting biological chemistry (since the electron clouds of both small and macromolecules would be severely distorted). These magnetars, with magnetic field strengths up to 10^{15} G or 10^{11} T are the strongest known magnets in the Universe.

Like many other neutron stars, magnetars are the final products of the collapse of massive stars in the range of 10–25 $M_\odot$. And like most neutron stars, they are roughly 12 miles or 20 km in diameter. But magnetars tend to rotate slower than most neutron stars. At a rate of once every 2–12 s (with most in the once every 5–8 s range), they contrast with pulsars which can rotate once in milliseconds in the extreme but more typically at a rate of 1–10 times s^{-1}.

Furthermore, magnetars do not live as long as pulsars or most other astrophysical phenomena. One SGR (SGR 1806-20) has a spin-down rate of 0.26 s per century. This spin-down rate is far faster than that of pulsars. The relatively fast deterioration of rotation rate is attributed to magnetic braking induced by the Brobdingnagian magnetic field. Such a rapid spin-down rate means that SGRs have very short lifespans and their super-strong magnetic fields decay in only about 10 000 years.

Thereafter, the periodic bursts of x-rays and gamma rays characteristic of magnetars cease.

The relatively rapid reduction in rotational rate of magnetars is because, unlike in an ordinary pulsar, the pace of magnetar spin-down is directly driven by its magnetic field. This causes greater deceleration of the rotation rate compared to regular pulsars. Additionally, the magnetic field is strong enough to crack the crust. Fractures of the magnetar crust are starquakes, and starquakes are the source of giant flares of x-rays and gamma rays. These starquake-induced bursts of gamma rays may be very luminous (figure 33.2).

All magnetars are neutron stars but not all neutron stars are magnetars. Similarly, all pulsars are neutron stars but not all neutron stars are pulsars. But some rare neutron stars with just the right geographical arrangement (meaning the radio signals emanating from their poles intersects our line of sight) can be both magnetars and pulsars. The present count for such unusual dual entities is six (figure 33.3).

Finally, it should be noted that all SGRs are magnetars but not all magnetars are SGRs. Over time, magnetars slow down their rotation rates and decrease their magnetic field strength, and it is believed that eventually the SGRs will evolve into **anomalous x-ray pulsars**. The pulse periods, spin-down rates and quiescent x-ray emission properties of these two subtypes of magnetars (SGRs and anomalous x-ray pulsars) are similar, with the main difference being the level of activity observed and

Figure 33.2. A powerful burst of x-rays and gamma rays erupting from a magnetar (an intensely magnetic neutron star) during a starquake. Such phenomena are believed to be the mechanism behind the occasional giant flares of x-rays and gamma rays associated with SGRs. Credit: NASA's Goddard Space Flight Center/ Chris Smith (USRA).

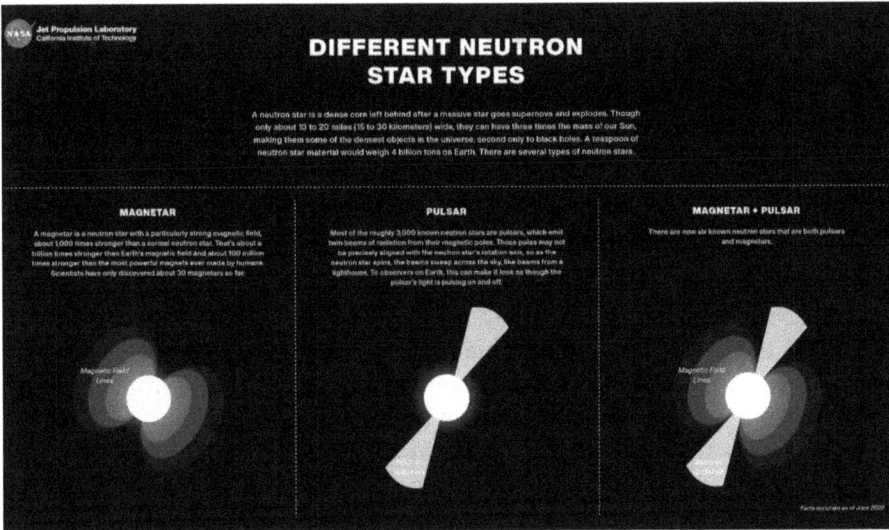

Figure 33.3. All pulsars are neutron stars and all magnetars are neutron stars. However, not all neutron stars are pulsars or magnetars. Nevertheless, some very unusual neutron stars happen to be both pulsars and magnetars. Magnetars are rapidly rotating, young neutron stars with incredibly powerful magnetic fields. Their magnetic fields, about a trillion times as strong as a refrigerator magnet, are strong enough to split x-ray photons and significantly distort the shapes of atoms. When their magnetic field lines intermittently stretch, grow tangled and eventually snap, enormous amounts of radiation are released in the form of gamma rays. This 'starquake' phenomenon is the proposed mechanism behind **SGRs**. Study of an event that occurred on 5 March 1979 in the N49 supernova remnant in the Large Magellanic Cloud suggests that **giant gamma ray flares** from magnetars may be responsible for a special type of phenomenon that is virtually indistinguishable from a **short-duration GRB**. Another giant gamma ray flare in the Sculptor Galaxy detected on 15 April 2020 further strengthened the idea that some strange events that appear just like short-duration GRBs are caused by plasma bursts associated with starquakes on magnetars. Credit: NASA/JPL-Caltech.

the energy of the emitted photons. Of the dozen or so known anomalous x-ray pulsars, the magnetic field strengths are still incredibly strong but slightly less than that of SGRs.

The precise mechanism behind the super strong magnetic fields of magnetars remain unclear. The most popular model is conservation of the neutron star's original magnetic field flux during compression from the progenitor supergiant star down into the tiny neutron star state. Simply stated, if the large star has a certain magnetic flux on its surface and that surface area shrinks down dramatically as it becomes a neutron star, conservation of flux means that the magnetic field strength correspondingly increases dramatically as well. This phenomenon is sometimes known as flux freezing. But the astronomical strengths of some magnetars pose some uncertainties with this simple mechanism.

33.6 Gamma and x-ray production from magnetars

On 5 March 1979, the first SGR was recorded, with gamma rays arising from a supernova remnant in the Large Magellanic Cloud. Initially thought to be a GRB, it was soon realized that it had several differences from typical GRBs. For one, its photons were less energetic than most GRBs and fell into the soft gamma and hard x-ray range. But more uncharacteristically, repeated bursts came from the same region, which was a feature never associated with a GRB.

This new entity and others like it were reclassified as soft gamma repeaters. Like short-duration GRBs, their blasts of gamma rays are very brief—about 0.1 s or 100 ms. But unlike any known GRB, in those SGRs that were also pulsars, repeating pulses of gamma rays might be directed our way and intersect our line of sight. Such bursts of gamma photons then head our way over a course of a few minutes with a period matching the rotation rate of the pulsar–magnetar (usually in the 5–8 s range).

Also, very uncharacteristic of genuine GRBs, blasts of high-energy photons from SGRs may recur over the years at irregular intervals. Of all astrophysical burst phenomena, SGRs are the brightest known bursts that may be recurrent. During the quiescent phase between bursts, SGRs might emit blackbody (thermal) x-rays that are enhanced by over an order of magnitude by the influence of the very strong magnetic field on the thermal structure of the neutron star envelope. As magnetars are very young neutron stars, they are extremely hot and therefore it is not surprising that they persistently emit x-rays as part of their blackbody emission spectrum. Most known neutron stars have surface temperatures in the vicinity of 600 000 K whereas magnetars that are SGRs have higher surface temperatures that range up to a million K. The energy of these SGR thermal x-ray photons is typically in the 2–10 keV range.

Of note, 'x-ray pulsars' are a different entity that are discussed separately in another chapter.

The fireball generated by the sudden release of energy as magnetic field lines break is momentarily confined by the magnetar's field but eventually flows forth from the poles. Then, just as in a pulsar if and only if the gamma ray emission is aligned with our line of sight do we get a glimpse of those gamma photons. The (occasionally multiple) bursts of soft gamma ray emission typically have a period of 5–8 s and last for a few minutes.

The young age (approximately 900 years) of hybrid pulsar/magnetars detected by the Rossi X-ray Timing Explorer supports the notion that neutron stars begin as magnetars and slowly dissipate their excess energy and evolve into pulsars. In May to July 2005, one such pulsar emitted five bursts of x-rays and gamma rays. Each blast lasted less than one second but released tens of thousands of times more energy than the Sun. This observation is consistent with the concept that young neutron stars start out as pulsars with magnetar qualities but over time they lose their intense magnetic properties and persist solely as pulsars (provided that we can see them at all via their lighthouse effects).

After they lose their strength, starquakes stop and so do the gamma ray emissions. But before fading altogether, magnetars/SGRs may evolve into anomalous x-ray pulsars. Thus, anomalous x-ray pulsars are probably an intermediate stage between magnetars and pulsars (figure 33.4).

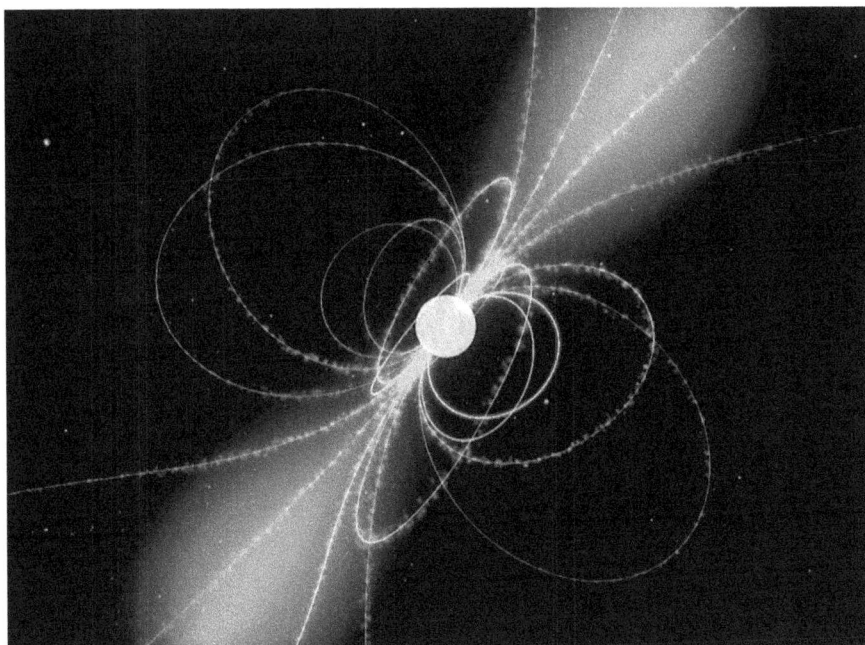

Figure 33.4. An artistic rendition provided by NASA depicting a gamma ray pulsar that lies within the supernova remnant CTA 1 which is located about 4600 light years away in the constellation Cepheus. It was discovered with NASA's Fermi Gamma-ray Space Telescope. In pulsars, a cone of radiation beams from the magnetic poles in a lighthouse-like fashion as the central neutron star rotates about its spin axis. Charged particles follow the pulsar's magnetic field lines (blue) and create the beams of energetic photons that emerge from the two poles. This unusual gamma-ray-only pulsar, which formed about 10 000 years ago, sweeps its beam in Earth's direction every 316.86 ms. Credit: NASA.

33.7 Giant x-ray flares associated with SGRs

Rearrangements of their ultra-intense magnetic fields are probably what causes the neutron star's crust to quake, not vice versa. But when a magnetar undergoes the equivalent of a seismic shift in its magnetic field, not only does the neutron star's crust crack and quake, tremendous amounts of energy are also released in the form of x-rays and gamma rays. In most instances, the photons are of only modest energies—x-rays somewhere around 10 keV (hence the name soft gamma repeaters). Less often, SGRs may generate **giant gamma ray flares**. It is estimated that a given SGR might do this once every 30–50 years. These relatively rare giant flares arise due to instabilities in the magnetic field on the magnetar that leads to occasional large-scale rearrangements. This is the presumed mechanism behind a giant flare in 1998 from SGR 1900+14.

The proposed mechanism of such scenarios remained theoretical for a while but the giant gamma ray flare on 27 December 2004 is now believed to be another major magnetar starquake with large-scale reorganization, confirming the proposed mechanism. It was the brightest gamma ray flare ever encountered from beyond our Solar System and surged out of Sagittarius for about a tenth of a second, making it initially indistinguishable from a short-duration gamma ray burst. During this 0.1 s, the giant flare unleashed energy at a rate of about 10^{40} W. Thus, in one tenth of a second 10^{40} J was released—as much as the Sun's output over 150 000 years. This giant flare arose from neutron star SGR 1806–20, about 42 000 light years away. The neutron star and known magnetar SGR 1806–20 was discovered in 1979 and still holds the record as most magnetic object in the Universe at 10^{11} T.

During ordinary bursts from an SGR, the gamma rays reach up to around 500 keV. If one employs the convention that gamma rays are those with the rest energy of an electron or above (which means 511 keV), these photons barely qualify as real gamma rays and if so, they are certainly 'soft'. But the gamma photon energies of giant flares are higher than that of ordinary SGR outbursts. Such flares routinely make photons with >1 MeV and can get up to 10 MeV (or even higher as mentioned in the next section). Although the giant flares still come from 'soft' gamma repeaters, these energies certainly qualify as 'hard' gamma rays by any definition.

33.8 Starquakes on magnetars

During a magnetar starquake, the neutron star's crust undergoes the equivalent of the seismic shift and cracks. The shift in star crust snaps tangled magnetic field lines (or mechanistically, probably vice versa). During such starquakes, enormous amounts of energy are instantly released upon the restructuring of the neutron star's magnetic field and its surface. The unleashed magnetic energy ejects electrons and positrons at relativistic speeds and also produces the sudden burst of soft gamma rays and hard x-rays typical of SGRs. The charged particles lead to gamma rays that barrel out of the north and south magnetic poles as the neutron star spins. This is like the mechanism of pulsars, but with far higher energy. The power might be in the range of 10^{30} to 10^{35} W, but lasting only about 1 second in most cases.

As mentioned, it is probably the reorganizing of the ultra-intense magnetic field that is causing the neutron starquake and cracking the crust, rather than vice versa. The enormous energy released during the magnetic field reconfiguration not only cracks the crust, but it also unleashes the vast amounts of energy seen in the x-rays and gamma rays of the SGR. In most cases, the energies of the individual gamma ray photons are only modest, and many of the emitted photons do not even qualify as gamma rays since their energies might hover around 10 keV and just be x-rays. But on occasion, giant gamma ray flares with high fluences of truly hard gamma rays are emitted.

The Large Area Telescope on the Fermi Gamma-ray Space Telescope recently detected seriously hard gamma photons from an SGR. These GeV gamma rays appeared after the initial detection of MeV photons in a magnetar giant flare on 15 April 2020. It appears that some GeV photons were generated by an ultra-relativistic outflow that first radiated a blast of prompt MeV-energy photons, and then deposited its energy far from the stellar magnetosphere. Subsequently, the outflow then struck interstellar gas and created shock waves that accelerated electrons to extraordinary energies. These high-energy electrons would have then emitted the GeV gamma rays through synchrotron radiation. This observation suggests the possibility that magnetars are indeed the engines behind many events that have been labelled as short-duration GRBs. One such short-duration GRB that could have in reality been a magnetar giant flare was GRB 200415A near the Sculptor Galaxy. Thus, magnetar giant flares might be virtually indistinguishable from some short duration gamma ray bursts, but magnetars have the possibility of repeating whereas true gamma ray bursts are a one-time event.

Fascinating from a scientific perspective but somewhat disconcerting from an anthropocentric perspective, the gamma rays from one giant gamma ray flare activated sensors on several orbiting satellites and partially ionized our upper atmosphere (as demonstrated by interference with radio transmissions). It was fortunate that the magnetar was 50 000 light years away, since if it was within 10 light years of our Solar System, the damage to our atmosphere would have been severe. The consequent damage to Earth's biosphere from such close-proximity gamma ray emissions (fortunately) remains untested.

33.9 Internal structure of neutron stars

Many magnetars repeat these radiation-releasing events, thus the name soft gamma repeaters (SGRs). The whole concept of magnetars and neutron starquakes begs the question of neutron star internal anatomy. The intense magnetism of SGRs hint of a solid surface with a liquid interior. Can sub-surface neutrons be packed together even more tightly than is seen on the surface of neutron stars? Are more massive neutron stars smaller and thus denser? If so, it implies density variability. If some are denser than others it implies variation in internal structure. Presently, the true internal structure of neutron stars remains uncertain, but theories include a core composition of hyperons, kaons and pions, with intermediate layers made of neutrons in a superfluid state. The outer 1 km (0.6 miles) is mostly solid neutrons (despite the temperatures that can reach 1 million K) and the most superficial layer might be composed of an extremely dense state of iron.

But following some logical reasoning, if white dwarfs can be squeezed down further into the denser state of neutron stars, one might ask if neutron stars can be squeezed down into a state of matter denser still? We know that if massive enough, neutron stars will collapse into black holes, but is there an intermediate state of sorts? Neutrons are not fundamental particles—they are composed of smaller up and down quarks (two down and one up, specifically). And we know that if atomic matter is squeezed tightly enough, the constituent electrons and nuclei are converted into a big ball of neutrons (a neutron star)... if neutrons are squeezed tightly enough, will they convert into a big ball of their constituent quarks? Because of quark confinement, individual quarks essentially do not exist; they are always found in hadrons as triplets (baryons) or pairs (mesons). If one tries to pull quarks apart from each other while they are in a baryon like a neutron, the attractive force increases sharply with distance. If more force is applied, the attraction between the quarks in question eventually grows so strong that new particles will materialize out of the vacuum. And these new particles would include new quarks that will quickly pair up with the separated quark to create new hadrons. So under familiar circumstances, breaking down neutrons into constituent quarks seems impossible. But neutron stars are clearly not familiar circumstances.

One proposed intermediate state of matter with a density between neutron stars and black holes is a strange quark star. Although under ordinary conditions strange quarks decay into lighter up and down quarks, under the extraordinary conditions found in neutron stars strange quarks might combine into triplets with up and down quarks to create what are called strangelets. Strange matter made of strangelets is expected to exhibit some (literally) strange behavior as discussed below.

33.10 Strange endings to some neutron stars?

According to a still-unproven hypothesis, strangelets that strike a neutron star might cause the neutrons to undergo a major phase change. But unlike familiar phase changes from solid to liquid or liquid to gas, in this phase change, ordinary matter might become strange matter.

Strangelets are fragments of strange matter, meaning they are made of not just up and down quarks (like all familiar baryonic matter) but also roughly equal numbers of strange quarks. Under ordinary circumstances, such as the temperatures and pressures found on Earth, strange particles are highly unstable. Particles such as the lambda family of baryons, with one up quark, one down quark and one strange quark (or another higher-generation quark) quickly lose their strangeness because their strange quarks rapidly decay into the lighter up and down quarks. The environment on neutron stars or other highly condensed matter states, however, might bestow a new degree of stability to strange matter. According to the strange matter hypothesis, the lowest energy state is that with roughly equal numbers of up, down and strange quarks—that is one composed of strangelets. The Pauli exclusion principle favors this three-quark setup, since with three rather than two available quark types, more arrangements leading to lowest-quantum-energy states can be created without having to move into higher energy states. Therefore, in theory, ordinary matter made of nuclei (which are made of triplets composed of only up and down quarks) should favor conversion into strange matter (which is made of triplets composed of equal numbers of up, down, and strange quarks). If strange matter were to come into contact with ordinary matter it might catalyze the conversion. Just a tad of strange matter could in theory set off a chain reaction in which all the ordinary matter is quickly converted into strange matter. It might be akin to the medical condition of spongiform encephalopathy, in which prions (which are ordinary proteins but with a different, lower-energy three-dimensional conformation) induce their normal conformation counterparts to adopt the prion protein conformation. This sort of chain reaction is seen in incurable neurological conditions such as Creutzfeldt–Jakob disease, mad cow disease, kuru and fatal familial insomnia.

Fortunately, under conditions found on Earth, although energetically favorable, this conversion from ordinary matter into strange matter has an essentially insurmountable activation energy and would probably not set off a world-ending chain reaction. Additionally, strangelets in their ground states are predicted to be positively charged and thus would be repelled by nuclei in ordinary matter.

But in the condensed matter of a neutron star, an uncontrolled chain reaction conversion seems more plausible. A neutron star may be considered to be a giant atomic nucleus but without a positive electric charge. Thus, it would not repel incoming positively charged strangelets through coulombic deflection like an atomic nucleus would. So, strangelets could come in undeterred. Upon striking a neutron star, a strangelet would first convert only a small area of the neutron star into strange matter. But through the unrestrained chain reaction process, that area could grow and gradually spread across the entire neutron star. In this manner, a neutron star would be converted into a **strange quark star**.

Further reading

[1] The Fermi-LAT Collaboration 2021 High-energy emission from a magnetar giant flare in the Sculptor galaxy *Nat. Astron.* **5** 385–91

[2] Sedaghat J, Zebarjad S M, Bordbar G H, Eslam Panah B and Moradi R 2022 Is the remnant of GW190425 a strange quark star? *Phys. Lett.* B **833** 137388

[3] Svinkin D, Frederiks D and Hurley K *et al* 2021 A bright γ-ray flare interpreted as a giant magnetar flare in NGC 253 *Nature* **589** 211–3

[4] Sokol J 2021 A star is torn *Science* **372** 120–3

IOP Publishing

Space Radiation
Astrophysical origins, radiobiological effects and implications for space travellers
James S Welsh

Chapter 34

Black holes and Hawking radiation

Gamma ray bursts, which shoot incredibly intense beams of ionizing radiation into space, arise during the birthing process of black holes. Ironically, the death of black holes might also spew gamma rays into outer space as well. But unlike the well-documented case of gamma ray bursts, which are now detected at the rate of about one every day, the hypothetical intense bursts of gamma rays have not yet been detected from the demise of any black hole. Thus, the so-called **Hawking radiation** which marks the explosive death of a black hole, remains purely theoretical.

Black holes are among the most fascinating topics in astronomy, if not all of science. Among the many strange and extreme concepts regarding black holes is the fact that gravity ultimately triumphs over matter and essentially crushes it out of existence. In a black hole, under the force of gravity a huge amount of material has been crushed to zero volume. This implies that the density becomes infinite. This theoretical point of zero volume and infinite density is called a **singularity**.

34.1 Black holes as rips in the fabric of spacetime

Space and time are represented by a four-dimensional construct called spacetime. Spacetime is warped by mass. In fact, according to general relativity, the effects of gravity can very accurately be explained by the degree of curvature of spacetime induced by the distribution of mass (or equivalently, energy through the Einstein relation, $E = mc^2$). The 'Einstein equation' equates the tensors for curvature of spacetime ($G_{\mu\nu}$) and the distribution of mass and energy ($T_{\mu\nu}$) through a set of constants:

$$G_{\mu\nu} = \frac{8\pi G}{c^4} \, T_{\mu\nu}.$$

Basically, the greater the amount of mass or energy in a given location, the greater the consequent curvature of spacetime. Without going into any details, one can conclude conceptually that the curvature in spacetime induced by a black hole with

doi:10.1088/978-0-7503-5444-8ch34

infinite density and zero volume must be extreme. When one is considering such infinities, a natural conclusion might be that spacetime is not only curved by a black hole but in fact 'torn' by it. The idea of such rips and tears in spacetime lead to the concept of **wormholes** where vastly distant parts of spacetime might be joined together through a conduit of sorts. Thus, black holes might (through the formation of wormholes) create shortcuts across spacetime and allow 'faster than light travel', not by exceeding c but rather through a severe distortion of spacetime itself—connections from one part to another. In other words, wormholes quite literally create shortcuts through spacetime. Obviously, blackholes, wormholes and related concepts are a favorite topic among science fiction fans and writers.

34.2 Properties of black holes

In contrast to popular myths and misrepresentations, black holes do not reach out, suck in and devour everything in their vicinity. But anything that does venture too close to them will be lost forever. This point of no return around a black hole is called the **event horizon**.

The event horizon is essentially the surface area around a black hole that corresponds to a radius at which the escape velocity approaches the speed of light. Since nothing can travel faster than c, nothing (including photons travelling at c) can escape a black hole if they fall within this radius. This radius is called the **Schwarzschild radius** and is given by the equation:

$$r_s = \frac{2GM}{c^2}.$$

From straightforward Newtonian mechanics, the escape velocity from a planet, star any or other astronomical body is given by:

$$v_e = \sqrt{\frac{GM}{r}}$$

where r is the distance to the center of mass.

If one sets v_e equal to c for a black hole, the equation becomes:

$$c = \sqrt{\frac{GM}{r}}.$$

Next, one can obtain the Schwarzschild radius of a black hole by squaring both sides, rearranging and solving for r:

$$r_s = \frac{2GM}{c^2}.$$

Somewhat amazingly and serendipitously, this solution is the same solution that one would obtain through a more rigorous treatment using general relativity and the Einstein equations. Karl Schwarzschild was the first to obtain an exact solution to this problem using the Einstein field equations. Thus, the radius of a black hole event horizon is named in his honor.

Starting with the basic equation for the Schwarzschild radius, $r_s = \frac{2GM}{c^2}$, this can be approximated as a scaling relation to our Sun's mass ($M_\odot$) through:

$$r_s = 3\frac{M}{Mo}\ km.$$

Thus, one can see that the Schwarzschild radius is directly proportional to the mass of a black hole. Hence, by $A = 4\pi r^2$, the surface area of a black hole's event horizon is also proportional to the mass. This seemingly obvious fact will become more relevant later in this chapter.

From the perspective of an outsider, anything that approaches too close to a black hole could fall in. And if an object actually falls within the Schwarzschild radius, there is no hope of turning back, since the amount of force required to reverse course approaches infinity. From the perspective of an insider, the escape velocity of a black hole exceeds the speed of light, meaning that there is no hope of escape. Even a high-energy photon aimed straight out will become gravitationally redshifted to the point of infinite wavelength (and thus zero energy by the Planck relation, $E = h\nu$). A wavefront of light will fail to escape a black hole but instead will appear to permanently remain just beneath the event horizon at the Schwarzschild radius. In other words, photons trying to escape would just hover under the event horizon at the Schwarzschild radius forever. Mindbogglingly, this freezing of photons implies that time has stopped. Indeed, this is just what would be observed from the outside as something falls into a black hole. From the perspective of an external onlooker, nothing actually ever crosses the event horizon. Instead, infalling matter becomes frozen in time, just at the point of crossing the event horizon. This is because, based on general relativity, time slows down in a strong gravitational field. Hence according to general relativity, a clock here on Earth moves slower than a clock in space a great distance away from our gravitational field. This is similar to the more familiar situation from special relativity in which clocks slow down when moving at a very fast pace (as on a spaceship). Incidentally, this difference in time thanks to general relativity must be accounted for in the GPS system since the satellites are so far from the full gravitational effects of Earth. In the extreme gravitational field of a black hole, time slows down immensely. And beneath the event horizon (within the Schwarzschild radius), mathematically, time stops.

The Schwarzschild radius defines the event horizon, and the event horizon defines the boundaries of the black hole. Note that this is not synonymous with the spacetime singularity at the center that is causing all of this.

From the perspective of infalling matter however the speed at which one approaches the black hole increases steadily under the steady pull of gravity. Eventually gravitational tidal forces will become so much greater at one's feet than at one's head that a human body would be stretched out. This elongation process is called 'spaghettification'. In reality, toes, feet and legs would be torn off one by one as the covalent bonds holding molecules together in these biological tissues are overwhelmed and broken mercilessly by gravity.

34.3 Black hole formation

Throughout the Universe there is a constant battle between gravity (which draws matter together) and the various forces that keep matter apart. In familiar daily situations we see that gravity draws us downward but the ground beneath our feet prevents us from sinking all the way to the center of the Earth. Electrostatic repulsion between the electron clouds on the outsides of atoms and molecules maintains this obvious state of balance. Such is the case for larger bodies such as the Moon and Earth, as well as all larger planets in this Solar System and others. And brown dwarf stars (those bodies that are not quite massive enough to initiate thermonuclear fusion of hydrogen into helium as a main sequence star, although deuterium fusion might be possible) also are largely supported by atomic electrostatic repulsion in their fight with gravity with a limited amount of internally generated heat also aiding electrostatic repulsion. (Note that some giant exoplanets put even our mighty Jupiter to shame and the dividing line between giant exoplanets and brown dwarfs can become blurred.) Such celestial bodies are in a state of equilibrium or a truce between the attractive force of gravity and the repulsive forces that fight it. As more mass is agglomerated in one spot by gravity, another repulsive force is called into action—the outward pressure generated by the heat of thermonuclear fusion in the core of main sequence stars. But as discussed earlier in the text, eventually this fuel supply will run out (especially quickly in very massive stars who live fast but die young). At that point, gravity will resume its inward crush and lead to various types of stellar corpses depending on the star's initial mass. Our Sun and stars like it will end as white dwarfs in which a final balance between gravity and electron degeneracy pressure is attained. White dwarfs are in the size range of a few thousand kilometers. For larger stars, electron degeneracy pressure may not be capable of staving off the force of gravity and the stellar remnant will be squeezed beyond the white dwarf stage and into a neutron star. In a neutron star, quantum electron degeneracy pressure is overcome and the electrons are pushed right into the protons of the white dwarf stage until they become neutrons (plus neutrinos). Neutron stars are thus the equivalent of white dwarfs except they are smaller, more massive, more dense and held up against gravity by neutron degeneracy pressure instead of electron degeneracy pressure. Despite usually having more mass than the Sun, neutron stars might have a radius of merely 10 km. But if the mass of the stellar remnant is great enough, even neutron degeneracy pressure cannot withstand the gravitational compression. Scientists debate if there is an intermediate step after neutron stars with even greater density such as strange quark stars but eventually, if the mass is great enough, gravity will squash the stellar remnant into a black hole with zero volume and infinite density—the spacetime singularity.

Unlike the situation with white dwarfs where the maximum mass is limited by the fairly well-understood Chandrasekhar limit, the exact limit for the maximum mass of a neutron star (or a strange quark star), the so-called Tolman–Oppenheimer–Volkov limit, is less certain since the internal structure and characteristics of matter are poorly understood at such incredible densities. The best estimate at this point is around 2–3 $M_\odot$ and it may be close to 2.2 $M_\odot$. It should be noted that this applies to stationary neutron stars and black holes; in the case of a rigid, uncharged, rapidly spinning neutron star, the weight limit might increase by up to 20%.

Just which stars lead to black holes is also unsettled. Superficially, one might assume that any very massive star might become a black hole, but it seems that very massive stars (e.g. those beyond 40 $M_\odot$) could lose too much mass through stellar winds and radiation pressure to end as black holes. The outer layers of these very massive stars essentially evaporate away, and the remaining star's core is disproportionately small and incapable of contracting into a black hole. Similarly, some supermassive stars might be in binary systems and transfer away a good deal of their mass to their companions, leaving a core too light to compress itself all the way down to a black hole. Thus, somewhat paradoxically the very largest stars (40–100 $M_\odot$) might end up as neutron stars instead of black holes. Incidentally, there seems to be an upper limit to the maximum size and mass of main sequence stars—but just what this limit really is remains debated. After around a mass of 120 $M_\odot$ of material has been accreted in a star, the surface temperature will be so hot that any new, infalling matter will be repelled as fast as it falls in. This theoretically limits the size of a newly formed star. As far as pre-existing stars go, another limit (the **Eddington limit**) dictates that the radiation pressure from a star reaching 150 $M_\odot$ would be so strong that it would exceed the gravitational attractive force holding matter to the star. Thus, new mass could not be added, and the star would have reached its theoretical limit. Having said this, there are several stars that have been documented to have exceeded these limits, indicating that there are other factors involved. (For instance, the above accretion limits and Eddington limits apply to metal-rich **population I stars** as opposed to metal-poor **population II stars**). Massive stars up to 20 $M_\odot$ might also become neutron stars rather than black holes. Thus, it is the group of main sequence stars in the 20–40 $M_\odot$ range that might be most likely to wind up as black holes. There is still much to be learned about black hole origins.

This is the description of how **stellar-mass black holes** arise. As will be discussed below, there are other categories of black holes based on mass, including miniature black holes, intermediate mass black holes and supermassive black holes.

34.4 Black holes as blackbodies

If one shines a light onto a black hole, nothing is reflected. There will be no trace of that incoming light if it crosses the event horizon. Similarly, if someone on the other side of a black hole were to shine a light at us, no light would be transmitted out of that black hole. A black hole thus absorbs all light shined on it, reflects nothing and transmits nothing. In essence, it fits the definition of the hypothetical ideal blackbody. Thus, one might be tempted to speculate about whether black holes might exhibit other features of blackbodies such as having a temperature and radiating heat. However, according to classical theory, black holes not only do not transmit or reflect any incoming radiation, they also do not emit any. This is in stark contrast to blackbodies, which are characterized by the thermal radiation they emit. But when black holes are considered from the quantum mechanical perspective, there possibly are situations in which black holes might radiate. And if they radiate, they lose energy; if they lose energy, they lose mass. This is the basis behind black hole evaporation, which will be discussed in further detail below in the context of Hawking radiation.

34.5 The no-hair theorem

When one attempts to describe another person, certain physical features might be helpful such as height and weight, approximate age, hair color and length or other things like a beard or mustache. But regardless of what has fallen into a black hole, its external features do not reflect any clues as to what is inside. All information about what has entered appears lost forever. This loss of information is perplexing and is a source of scientific debate. But from a simple perspective, a black hole has no hair—it has no distinguishing features.

The only characteristics of black holes are their:
(1) Mass
(2) Angular momentum
(3) Electrical charge.

34.6 Subtypes of black holes

Black holes can be categorized in two different ways. The first way is based on mass or size. The other means is based on angular momentum and electrical charge. Then of course there are permutations leading to subcategories within these basic groups.

When it comes to size or mass, black holes are classified in four groups:
1. Supermassive
2. Intermediate mass
3. Stellar mass
4. Miniature.

Supermassive black holes span a very large mass range between 10^5 and 10^{10} $M_\odot$. These entities are found at the centers of large galaxies such as our Milky Way.

Intermediate mass black holes may weigh between 10^2 and 10^5 $M_\odot$. The first suspected intermediate mass black hole was the strong x-ray source called HLX-1.

Stellar mass black holes have masses between about 4 and 100 $M_\odot$. These black holes are the products of core collapse of massive stars at the end of their lives.

Miniature black holes are also called **primordial black holes** since these (still purely hypothetical) bodies might have been formed through the gravitational collapse of regions of high density at the time of the Big Bang. Since they would have been born shortly after the Big Bang they have been around since the dawn of time and are given the name 'primordial'. Such miniature or primordial black holes are theorized to have masses comparable to or less than the mass of the Earth.

When it comes to spin or angular momentum, black holes may be classified as:
1. Stationary
2. Rotating.

Classification of black holes may also be based on the presence or absence of electrical charge. Thus, the four permutations involving spin and charge are:

1. Uncharged, stationary
2. Uncharged, rotating
3. Charged, stationary
4. Charged, rotating.

The first of these groups, the uncharged, stationary black hole is also called a **Schwarzschild black hole** (also known as a **static black hole**). Schwarzschild black holes are characterized solely by their mass. The absence of any angular momentum or spin makes such Schwarzschild black holes a bit unrealistic but of great theoretical value. **Kerr black holes** (uncharged but spinning) represent a more realistic scenario. However, rotating black holes are far more complicated. The surrounding spacetime is stretched and dragged around the spinning black hole, creating a weird whirlpool. And instead of a point, the singularity becomes an infinitely thin disc or ring; essentially a two-dimensional singularity. The event horizon also changes, with a twofold (inner and outer) aspect to it.

Charged black holes are probably rare and might not even exist in nature since any excess electrical charge would likely attract opposite charged particles to enter and neutralize things. Nevertheless, charged black holes can be of two subtypes. The charged, non-rotating (stationary) black holes are known as **Reissner–Nordstrom** black holes. The charged, rotating black holes are called **Kerr–Newman black holes**.

34.7 The black hole area theorem and the Second Law of Thermodynamics

As described above, a black hole's Schwarzschild radius is proportional to its mass and therefore, by $A = 4\pi r^2$, a black hole's event horizon or surface area is also proportional to its mass. And a basic postulate is that a black hole's surface area cannot decrease, it can only increase (or remain the same). This assertion, also called the **black hole area theorem**, was first ventured in 1971 by Stephen Hawking. Worded in this way, Hawking's black hole area theorem is remarkably similar to the Second Law of Thermodynamics, which can be stated analogously as: 'the entropy of a closed system cannot decrease, it can only increase (or remain the same)'.

The event horizon of a black hole is a unique and special sort of surface in that information (and everything else) only crosses it in one direction. But the 'arrow of time' in conventional thermodynamics (marking an increase in entropy) is also a one-way street. The parallels between the surface area of a black hole and entropy (as well as the statements of the black hole area theorem and the Second Law of Thermodynamics) are quite apparent.

Additionally, when a massive star collapses into a black hole, its entropy dramatically increases. For instance, the entropy of a black hole is far greater than the entropy of the Sun or any other equally massive object. Furthermore, when two black holes merge, the surface area of the new black hole is always greater than or equal to the sum of the previous two black holes; it is never smaller. Black hole entropy may be related to the amount of information, energy and matter that has vanished as it crossed the event horizon and become forever lost in the black hole.

Further reasoning led Jacob Beckenstein to the conclusion that the entropy of a black hole is mathematically proportional to the surface area of its event horizon. The formula for this black hole entropy (S) can be expressed as:

$$S = \frac{kc^3}{4\hbar G}A$$

where k is Boltzmann's constant and G is the gravitational constant.

By $A = 4\pi r^2$ and the expression for the Schwarzschild radius, this can also be written in terms of the black hole's mass as:

$$S = \frac{4\pi k G}{\hbar c}M^2.$$

The total entropy of a black hole is therefore proportional to the cumulative quantity of information that has crossed the surface of the black hole since its inception.

Despite having first articulated the black hole area theorem, Hawking initially did not accept this thermodynamics-based concept of black hole entropy. After all if a black hole has entropy and energy, like any classical blackbody, it must also have a non-zero temperature. And if a black hole has a temperature, like a blackbody or any other object, heat can flow from it to colder surroundings. But if heat energy is transferred from a black hole to something else that happens to be colder, by definition, it would be radiating and losing energy (and mass). Since this was patently impossible, Hawking asserted that black holes must have a temperature of exactly absolute zero. And thus, Hawking initially concluded that black holes cannot have entropy.

34.8 The Four Laws of Black Hole Mechanics

However, Hawking subsequently explored this idea in greater depth from the perspective of quantum mechanics. He eventually concluded that black holes could indeed have a temperature. This is now known as the **Hawking temperature**. Even though at first Beckenstein himself rejected these quantum mechanics-based concepts, they seemed to confirm his original hypothesis—black holes do indeed have entropy and temperatures. And as discussed below, a black hole's Hawking temperature is inversely proportional to its size. That is, the larger the event horizon, the lower the temperature. Conversely, the smaller the surface area, the hotter a black hole is and the more rapidly it loses energy and mass.

Based on the successful analogy between the black hole surface area theorem and the Second Law of Thermodynamics, four 'Laws of Black Hole Mechanics' or 'black hole dynamics' have been proclaimed that are manifestly analogous to the Four Laws of Thermodynamics:

1. A non-rotating black hole has equal surface gravity at every site on its event horizon. This is analogous to the Zeroth Law of Thermodynamics which

states that if two systems are in thermal equilibrium with a third system, they are in thermal equilibrium with each other. In simple terms, they are all the same temperature. Implicit in this law is an analogy between temperature and black hole surface gravity. Surface gravity (κ) is an important quantity in black hole dynamics, just as temperature is an important quantity in thermodynamics.

2. The change in energy of a black hole is the sum of three factors:

$$dE = \kappa/8\pi dA + \Omega dJ + \Phi dQ$$

where E is the energy, κ is the black hole surface gravity, A is the horizon area, Ω is the angular velocity, J is the angular momentum, Φ is the electrostatic potential and Q is the electric charge.

This is analogous to the First Law of Thermodynamics, which states that energy cannot be created nor destroyed but it can be relocated or converted from one form into another. Another way of saying the same thing is the change in internal energy (U) of a system equals the net energy transfer by heat (Q) into the system plus the net work (W) done *to or on* the system. Mathematically, it is often stated as:

$$\Delta U = Q + W$$

where delta U is the internal energy of the system, Q is the sum of all heat transferred into and out of the system and W is the net work done on the system.

Students are advised to be very careful about the positive and negative signs in this expression of the First Law, especially since conventions in texts and classes are changing!

3. The surface gravity or surface area of a black hole always increases or at best remains the same; it never decreases. Mathematically,

$$\frac{dA_s}{dt} > 0.$$

This is analogous to the Second Law of Thermodynamics which states that entropy in a closed system is always increasing or at best remains the same; it never decreases. Mathematically that is expressed as $\frac{dS}{dt} > 0$.

4. The Third Law of Black Hole Mechanics states that it is not possible for a black hole to have zero surface gravity. That is, $\kappa = 0$ cannot be achieved. An extreme black hole would have zero entropy, but such an extreme black hole is not possible. This is analogous to the Third Law of Thermodynamics which can most concisely be stated as 'a temperature of absolute zero cannot be achieved'. A more thorough statement is: as the temperature of a system approaches absolute zero, the absolute entropy of that system approaches a constant minimum value. Alternatively, it can be worded as 'it is not possible for any process to bring the entropy of a given system to zero in a finite number of operations'.

The **surface gravity** of a black hole, κ (which is the value of the gravitational field strength at the event horizon) is a critically important parameter. From the Zeroth Law, surface gravity in black hole dynamics is analogous to temperature in thermodynamics. From these Laws of Black Hole Dynamics, one can see why various phenomena make sense (such as the fact that when two black holes merge, their surface areas and surface gravities increase such that the new surface is greater than the sum of the parts (or at best is equal to it)). In thermodynamics, energy (E) and entropy (S) are related to each other through T, the temperature, as the proportionality constant:

$$\Delta E = T \Delta S.$$

In black holes, the mass is analogous to thermodynamic energy, black hole surface area (event horizon surface area, A_s) is analogous to entropy and the proportionality constant is surface gravity (κ) in analogy with temperature T in thermodynamics. Thus, the equation might be written as:

$$\Delta m = \kappa \Delta A_s.$$

As mentioned above these 'Laws of Black Hole Mechanics' are very analogous to the Laws of Thermodynamics. But importantly, through a detailed quantum description, Hawking subsequently discovered a deficiency in the Second Law—black holes might be able to decrease their surface areas after all. His analysis allows black holes to have a temperature and therefore be capable of losing mass and energy. If a black hole can lose mass, its surface area will decrease, which is a direct violation of the Second Law.

34.9 Black hole evaporation

According to classical theory, including general relativity, 'black holes are forever'. Nothing can, or ever will, escape and therefore they are permanent fixtures of the Universe. Hawking, however, delved deeper into the quantum mechanics of black holes and concluded that black holes might indeed be capable of 'evaporating', especially when small. This discovery led to many fascinating consequences.

According to Hawking's theory, the lifetime of a black hole is dependent on its event horizon surface area. This means it is also directly related to its mass, with smaller black holes evaporating far faster than larger ones. For example, a black hole of 1 $M_\odot$ takes 10^{67} years to evaporate (much longer than the current age of the Universe). Supermassive black holes such as those associated with galactic centers might take 1 googol (10^{100}) years to evaporate. On the other hand, a miniature black hole of only 10^{11} kg would evaporate within 3 billion years. For reference, a rough estimate for the mass of Mount Everest is about 8.10×10^{14} and the mass of Earth is nearly 6×10^{24}. Thus, small primordial black holes made during the Big Bang would have long since evaporated. On the other hand, mountain-sized primordial black holes would be blowing up right about now (since they go out with a blast of gamma rays).

34.10 Black hole temperature

According to the Hawking model, a black hole may be modeled as a blackbody with a **Hawking temperature** T_H of:

$$T_\mathrm{H} = \frac{\hbar c^3}{8\pi k G M}$$

where G the universal gravitational constant, k is Boltzmann's constant and M is the mass of the black hole.

Alternatively, this can be expressed in terms of solar masses $M_\odot$ as:

$$T_\mathrm{H} = 6 \times 10^{-8}\,(M_\odot/M)\ \text{Kelvins}.$$

It should be pointed out that these equations only hold for electrically neutral, non-rotating (i.e. Schwarzschild) black holes, where the only state variable is the mass.

As a black hole emits this blackbody (Hawking) radiation, the black hole would be expected to lose mass very slowly. This decrease in mass would in turn decrease the surface gravity and area of the black hole. This then increases the rate of radiation emission. By using the Stefan–Boltzmann Radiation Law $J = \sigma T^4$ where J is the rate of energy loss per area, and substituting the event horizon for the area, one can obtain:

$$\frac{dm}{dt} = \frac{\hbar c^4}{15,360\pi G^2}\frac{1}{M^2}.$$

Solving for dt and integrating yields the time (t_{ev}) it takes for a black hole to evaporate:

$$t_{ev} = \frac{5,120\pi G^2}{\hbar c^4}M^3 \approx 10^{-16}\,M^3.$$

In terms of solar masses, a black hole's evaporation time can be written as a scaling relation:

$$t_{ev} = 2.1 \times 10^{67}\,(M/M_\odot)^3\ \text{years}.$$

Thus, the rate of evaporation is inversely proportional to the black hole's mass and directly proportional to its temperature. And black holes are cold. For example, a black hole with 1 $M_\odot$ would be colder than the Universe at large ($\approx$2.73 K) and thus would absorb more from the cosmic microwave background radiation (CMB) than it would radiate. Thus, a 1 $M_\odot$ black hole would tend to grow larger rather than evaporate. On the other extreme, a miniature black hole would have a temperature higher than the CMB and thereby lose heat to the surrounding Universe.

As a black hole loses heat energy, it would shrink and its surface area would decrease. Recalling the relationship between surface area, surface gravity and entropy, this means that a shrinking black hole's entropy would decrease.

Furthermore, a small black hole would evaporate at an ever-increasing rate, and eventually it would explode in a flash of gamma rays. The radiation rate near the end of a black hole's life becomes very high at around 10^{23} J released in the last 100 ms.

Thus, when gamma ray bursts were first discovered, one of the proposed explanations was that we were discovering the 'Hawking radiation' of exploding black holes. As explained earlier, gamma ray bursts do not signify the violent deaths of black holes but instead mark their births through the collapse of a massive star or the fusion of two neutron stars.

34.11 Quantum mechanical basis of Hawking radiation

According to classical theory, nothing can escape from a black hole, including photons. Thus, black holes cannot emit radiation according to conventional principles. But based on quantum mechanics, the region right around a black hole (or anywhere else for that matter) might be capable of creating virtual particle pairs (matter–antimatter pairs such as electrons and positrons or pairs of photons of positive and negative energy). This concept stems from the non-zero energy value of the various fields (e.g. gravity, electromagnetism, etc) suffusing the Universe, in conjunction with the Heisenberg uncertainty principle. The latter states that one cannot determine, with infinite precision, the exact energy of a system and the time at which that measurement was made. This is expressed as:

$$\Delta E \Delta t \geqslant h/2\pi.$$

What this implies is that it is possible for virtual pairs of particles to emerge from the vacuum energy of space ('out of nothing') for a fleeting instant of time and then undergo mutual annihilation to vanish back into energy. The only requirement is that the product of the uncertainties in energy and time is restrained by $h/2\pi$. This phenomenon is called **quantum fluctuation**.

It is speculated that quantum fluctuations result in a sea of virtual particles that pop into and out of existence for a brief instance and provide the vast void of empty space with a non-zero vacuum energy. In other words, empty space is quite full. The question of whether this vacuum energy adequately explains the surprising expansion of the Universe remains an unanswered question. Presently, the calculations based on the vacuum energy for the accelerating expansion of our Universe are orders of magnitude greater than what is observed. Thus, the **dark energy** causing the accelerating expansion of the Universe remains uncharacterized.

The bizarre prediction of virtual particles associated with quantum fluctuations do have supporting evidence in the form of the **Casimir effect** and the **Lamb shift** of the hydrogen spectrum. The former is manifested as a sort of pressure pushing two very close parallel plates together in a vacuum. This pressure is due to a reduction in the number of virtual particles forming in the confined space between the two plates. It is often taken as experimental evidence of the validity of virtual particles in the vacuum.

According to a popular explanation of Hawking radiation, in the vicinity of the event horizon of a black hole, a member of the pair of virtual particles might fall into the black hole while the other member does not. In such a situation, a virtual electron might fall below the event horizon and vanish, whereas the virtual positron persists outside the event horizon and no longer has its partner to annihilate with.

In this case, the virtual positron has become a real positron. It then finds another electron to annihilate with and a pair of 511 keV gamma photons emerges. These photons appear to be coming from the black hole. One could imagine the same scenario in which the positron fell into the black hole, but the virtual electron did not. The virtual electron thus becomes a real electron and carries energy away from the black hole.

From our external perspective, the particle that falls into the black hole (whether matter or antimatter) enters the black hole with negative energy. The escaped particle (whether matter or antimatter) carries positive energy. In this fashion, a black hole loses energy through this Hawking radiation. And if it is losing energy, it is losing mass. Another way of viewing this is that through quantum tunneling, a particle has somehow managed to penetrate the seemingly impenetrable barrier of the event horizon.

The end product of this process, in which particles escape and accelerate past each other (thereby emitting photons if the particles were charged) is a thermal or blackbody distribution of radiation. Additionally, particle–antiparticle pairs annihilate and generate more photons. Hawking radiation is indeed akin to blackbody radiation that would be seen from a black hole with a finite temperature. Hence, black holes do have a temperature and they do evaporate at a mathematically predictable pace.

A black hole will gradually lose mass through this process. As it loses mass, its surface area will decrease and its temperature will increase. But as its temperature increases, it will radiate faster and therefore lose mass at an even quicker pace. This vicious cycle will accelerate until the black hole's mass approaches zero and the radiation rate approaches infinity. The result is that the black hole ultimately explodes in a bright flash of gamma rays.

But this popular explanation for the lay public, which largely stems from Stephen Hawking himself, leaves much to be desired. For instance, a still poorly understood aspect of Hawking radiation centers on quantum entanglement. The virtual particles which become separated, with one falling into the black hole and the other escaping to create Hawking radiation, should be permanently entangled. Just what the implications of this quantum entanglement are in the context of one member of the virtual pair being engulfed below the event horizon while the other member escapes as a real particle remains a mystery. A full understanding of Hawking radiation, especially at the very end of a black hole's existence when the rate of radiation and the energy are highest, awaits a more comprehensive theory of quantum gravity. Nevertheless, the popular description of Hawking radiation and black hole evaporation is incomplete if not outright misleading. Therefore an alternative explanation is provided below.

34.12 Three depictions of Hawking radiation from the quantum perspective

The quantum mechanics of black hole evaporation and Hawking radiation is quite complex and confusing. Part of the confusion lies with Stephen Hawking's own somewhat misleading explanation in some of his popular writings. Although his layman's explanation lent itself nicely to diagrams and videos, the details differed somewhat from his more in-depth treatment. We shall go through three different ways of describing the Hawking radiation here.

In the vicinity of the event horizon of a black hole, a member of the pair of virtual particles might fall into the black hole while the other member does not. For instance, in such a situation, a virtual electron might fall below the event horizon and vanish, whereas the virtual positron persists outside the event horizon and no longer has its partner to annihilate with. In that case, the virtual positron becomes a real positron. It then finds another electron to annihilate with and a pair of 511 keV gamma photons emerges. These photons appear to be coming from the black hole. One could imagine the same scenario in which the positron fell into the black hole but the virtual electron did not. In that case, the virtual electron becomes a real electron and carries energy away from the black hole.

But the astute reader will ask the question, 'If the black hole emits an electron (or positron) but simultaneously gains a positron (or electron), how it the net sum not zero?'

One proposed solution to the dilemma holds that from our external perspective, the particle that falls into the black hole (whether matter or antimatter) enters the black hole with *negative energy*. The escaped particle (whether matter or antimatter) carries *positive energy*. In this fashion, although an electron or positron falls into the black hole while its virtual partner remains outside the event horizon, the black hole still loses mass and energy when a particle with negative energy enters. But a real particle with positive energy has left, and hence the black hole loses energy through this Hawking radiation. And if it is losing energy, it is losing mass.

Another way of viewing black hole evaporation is that it is an example of quantum tunneling. Through the tunnel effect, a particle has somehow managed to penetrate the seemingly impenetrable barrier of the event horizon. Thanks to the wave–particle duality, despite the energy barrier imposed by the event horizon, a particle's wavefunction has a non-zero probability of being outside the event horizon. In essence, it has found itself on the other side of the barrier and once outside, makes a run for it! But one might wonder how the number of those tunneling out is not equal to those entering for a net sum of zero.

There is a serious deficiency with the explanation involving virtual particle–antiparticle pairs popping up through quantum fluctuations and one member falling in while the other escapes. For one, it suggests that Hawking radiation would consist of particles and antiparticles. In fact, it is composed of electromagnetic radiation rather than a mix of particles and antiparticles. Second, according to the popular explanation, the Hawking photons are emitted from the surface of the event horizon, but in reality, they would stem from a relatively large region outside the event horizon. And in contrast to the expectation that such photons would all be highly energetic (to escape the fierce gravity just outside the black hole), the emitted photons exhibit a wide range of energies—a range represented by a Planck curve for blackbody radiation. Finally, the construct of positive energy photons and particles emerging from the black hole and reducing its energy and mass while negative energy versions fall in and also reduce the energy and mass of the black hole may seem unfamiliar and nonsensical. Because it is.

34.13 An alternative explanation of Hawking radiation

According to the theory of relativity, different observers in different reference frames may see reality quite differently. For example, observers in relative motion to each other (that is, accelerating relative to each other) will perceive things differently. Likewise, observers in regions where the curvature of spacetime is very different thanks to differences in gravity will see discrepancies. From a distant vantage point, the zero-point energy of the vacuum far from a black hole (and any other gravitational source) will be recorded as a certain value. Likewise, the zero-point energy in spacetime recorded by an observer right at the event horizon may be the same exact value. But if these two observers could somehow compare their readings, it would become apparent that there is a disagreement. And that disagreement is related to the degree of gravity-induced curvature of spacetime around the two observers.

In Hawking's original paper, he performed the calculations comparing the differences between an observer at infinity versus an observer near the event horizon and found that a form of radiation will emerge not only from the event horizon but from the entire volume of gravitationally curved space around the black hole (amounting to a volume covering as far out as 20 Schwarzschild radii). The calculations also inform us of the expected spectrum of emitted quanta—a perfect blackbody depicted by a Planck curve. Furthermore, the analysis suggests that Hawking radiation should also come from other celestial bodies that severely warp spacetime such as white dwarfs and neutron stars.

So, it is the gravitationally warped spacetime that emits the Hawking radiation. And the more severely distorted spacetime is, the more intense the emitted radiation. Thus, small black holes with highly curved event horizons emit Hawking radiation faster than large ones with lower degrees of spacetime curvature. In this fashion, there is no need for negative and positive energy. And there is no confusing discrepancy between some virtual particle pair members falling into the black hole yet not adding to the black hole's mass budget while those falling to the outside of the event horizon somehow subtract from the black hole.

In this description, it is the vigorously 'bent' spacetime around the black hole that is the source of the radiation. The energy for this radiation comes from the black hole—and the emitted energy causes the curvature of spacetime in the vicinity to slowly but steadily 'straighten out' over time. And the result is a slowly but steadily shrinking black hole.

34.14 Unruh radiation

This portrayal of Hawking radiation—that it arises from the curvature of spacetime —leads one to immediately wonder about the relationship between Hawking radiation and **Unruh radiation**. Unruh radiation stems from the act of acceleration. And since Einstein showed that gravity and acceleration are equivalent, one might speculate that Hawking and Unruh radiation are really the same thing viewed from different perspectives.

From the perspective of a stationary observer, empty space is truly empty. But from the perspective of an accelerating observer, that same empty space will be filled with blackbody radiation. And the temperature of the space with that blackbody radiation is proportional to the acceleration. The photons (and particle–antiparticle pairs if energetic enough) seen by an accelerating observer are now called Unruh radiation for the discoverer of the phenomenon (the Unruh effect), William Unruh.

The equation for Unruh radiation energy for a uniformly accelerating observer can be expressed as:

$$E = kT = \frac{\hbar a}{2\pi c}$$

where a is the constant acceleration of the observer

34.15 Hawking radiation as blackbody radiation

The smaller a black hole is, the faster the rate of radiation emission. The faster the radiation rate, the higher the temperature. The higher the temperature, the higher the frequency of the emitted radiation of a blackbody (by $E = h\nu$). This is odd. Normally, the more a body radiates, the colder it gets. The opposite holds true for black holes.

The end product of this process, in which particles escape and accelerate past each other (thereby emitting photons along the way if the particles are charged) is a thermal or blackbody distribution of radiation. Additionally, some particle–antiparticle pairs will annihilate and generate more photons. Hawking radiation is indeed akin to blackbody radiation that would be seen from a black hole with a finite temperature. Hence, black holes do have a temperature and they do evaporate at a mathematically predictable pace as described in the sections above.

Although Hawking radiation is still conjectural, the calculations show that a black hole with the mass of the Sun would have a temperature of only one ten-millionth of 1 K. Such a black hole would be far colder than the CMB temperature of the Universe (about 2.73 K) and therefore would be absorbing rather than radiating energy. In contrast, a proton-sized miniature black hole would have a temperature of 120 billion K. Such a hot primordial black hole would be releasing energy at an astounding rate of 6 GW.

Thus, an object the size of a proton would be cranking out the energy equivalent of six nuclear power plants. And at this temperature, this miniature black hole would be capable of creating electron–positron pairs, neutrinos as well as massless particles such as photons and maybe gravitons (if they indeed exist). This diffuse spectrum of gamma rays should be detectable by dedicated gamma ray space telescopes and satellites. The photon energy of the gamma rays routinely emanating from a radiating primordial black hole could be on the order of 100 MeV.

34.16 The end of a black hole

A black hole will gradually lose mass through the evaporation process. As it loses mass, its surface area will decrease and its temperature will increase. But as its temperature increases, it will radiate faster and therefore lose mass at an even quicker pace. This vicious cycle will accelerate until the black hole's mass and surface area approach zero and the radiation rate approaches infinity. The result is that the black hole ultimately explodes in a bright flash of gamma rays. The energy of these gamma rays, right at the end of the big blast, would be on the order of just under 1 GeV.

34.17 Experimental modelling of Hawking radiation?

Black holes and their event horizons might be represented experimentally by a fast-flowing fluid in which one end (the narrower one) is moving much faster than the other end (the wider end). In such differentially flowing fluids, a 'horizon' or boundary exists where above this boundary sound waves can escape but below which sound cannot (since the speed of the fluid on that side of the boundary is supersonic). Such experimental setups might provide clues about how black holes and their event horizons (which use light rather than sound) behave.

Further reading

[1] Bekenstein J D 1973 Black holes and entropy *Phys. Rev.* D **7** 2333–46

[2] Hawking S 1974 Black hole explosions? *Nature* **248** 30–1

[3] Unruh W G 1981 Experimental black-hole evaporation? *Phys. Rev. Lett.* **46** 1351–3

[4] Steinhauer J 2016 Observation of quantum Hawking radiation and its entanglement in an analogue black hole *Nat. Phys.* **12** 959–65

[5] Cowen R 2014 Hawking radiation mimicked in the lab *Nature* https://doi.org/10.1038/nature.2014.16131

[6] Isi M *et al* 2021 Testing the black-hole area law with GW150914 *Phys. Rev. Lett.* **127** 011103

[7] Ethan S 2022 Ask Ethan: how does Hawking radiation lead to black hole evaporation? Big Think. https://bigthink.com/starts-with-a-bang/hawking-radiation-black-hole-evaporation/

Chapter 35

The Big Bang and radiation—the beginning and the end

35.1 Olbers' paradox

One of the vexing riddles of early 20th century astronomy was, 'why is the nighttime sky dark?' While the dark of night is obvious to everyone, this simple fact posed a mystery to astronomers who believed that the Universe was infinite and static. In principle, between any two given stars in the night sky, there should be another, and between that other star and any other, there should be yet another, and so on. Therefore, one might expect the night sky to be fully illuminated. The actual observation of a dark nighttime sky punctuated by a smattering of dim stars is known as **Olbers' paradox**. The ultimate solution to Olbers' paradox lies in an expanding, rather than static, Universe. This is because, even if there are stars and more stars in the gaps between bright evident stars, the incessant expansion causes the light from these receding stars and galaxies in the gaps to grow dimmer and less energetic due to red-shifting. Thanks to the Doppler effect, light from stars and galaxies that are moving away from us are shifted towards the red, less energetic end of the spectrum. According to the Planck–Einstein relation, $E = h\nu$, the energy of such photons is lowered. For extremely distant stars and galaxies, the energy is red-shifted out of the visible range.

35.2 The expanding Universe and Hubble's law

It was Vesto Slipher who was the first to document that the Universe was expanding, based on observations that the vast majority of galaxies are receding from us—and each other. Edwin Hubble then showed that the further away a galaxy is, the faster its rate of recession. This is now known as Hubble's Law (or the **Hubble–Lemaître Law**).

In 1931, Abbé Georges Lemaître examined in earnest the **Friedmann equation** (which suggested an expanding Universe and is covered later in this chapter) and drew the logical conclusion that as one went back in time, the Universe must have been smaller, denser, and hotter. He surmised that at one point there was a 'day without yesterday', to use his resplendent phrase. At some point in the very distant past, the Universe must have existed in a highly contracted state that he called **ylem** or the **primeval atom**. From this condensed state, the Universe suddenly exploded. Thus, Lemaître is considered the father of the **Big Bang theory**.

As mentioned, Slipher and Hubble independently used redshift data to discover that the Universe was expanding. This was determined using **standard candles**. Just as a candle is dimmer the further away it is, so too are stars and galaxies. And just as one can determine the distance of a candle of known luminosity by measuring its apparent brightness, one can determine the distance of a star or galaxy if the absolute luminosity is known.

Luminosity (L) is measured in watts. At a distance, d, the luminosity of a source is spread across a sphere with surface area $A = 4\pi d^2$. To observers of that sphere at distance d, the **observed brightness** (b) is related to absolute luminosity by:

$$b = L/A = L/4\pi d^2.$$

In this fashion, by measuring the observed brightness of objects with known absolute luminosity L, we can calculate distance. And there are several astronomical bodies that tend to have set absolute luminosities and can thus function as standard candles.

35.3 Standard candles in cosmology

Several astronomically valuable standard candles exist, including:
1. Cepheid variable stars
2. Type Ia supernovae
3. Planetary nebula luminosity function
4. Tip of the red giant branch
5. The Tully–Fisher method.

The last of these methods, the Tully–Fisher relation, applies to spiral galaxies. Basically, this standard candle method capitalizes on the observation that brighter spiral galaxies rotate faster than less luminous ones. Through Doppler measurements, one can estimate the rotation rate of a given galaxy. Thus, the measured rotation rate can be mathematically correlated with an absolute luminosity. The measured brightness (apparent luminosity or apparent magnitude) can then be compared to the absolute luminosity (absolute magnitude) to provide a distance. The relationship between the intrinsic luminosity (L) of a spiral galaxy and the circular velocity (v) of that galaxy generally fits. The equation may be written as:

$$L \propto v^4/(M/L)^2 I \quad \textbf{Tully–Fisher relation}$$

where v is the circular velocity of the spiral galaxy, L is the intrinsic luminosity, I is the intensity (or luminosity per unit area, $I = L/4\pi r^2$), and (M/L) is the mass:light ratio.

It is worth stating that although the velocity exponent was written here as 4, in more rigorous treatments this is replaced by α, which does not have a unique value;

it is affected by the experimental details of the photometric and spectroscopic measurements made. The Tully–Fisher method works over relatively long distances, reaching around 300 million light years.

Cepheid variable stars (named after the prototype Delta Cepheus) are unusual stars that vary in luminosity over time in a regular manner. In 1912, Henrietta Swan Leavitt determined the period-luminosity relationship of cepheid variables. Simply stated, those with longer periods were more luminous. Thus, by measuring the period of a cepheid variable, one could estimate the absolute luminosity. Then by comparing the measured brightness against that absolute luminosity, one could calculate the distance to it. Of course, the Cepheid variable method only works if the galaxy in question has obvious Cepheid variables. This approach has been used to calculate distances of galaxies as far as 65 million light years away.

The Type Ia supernova method takes advantage of the fact that when a white dwarf undergoes thermonuclear detonation due to accumulating too much mass from its companion star (probably a red giant), the ensuing supernova explosion is nearly always of a certain absolute luminosity. These standard candles have an amazing reach of nearly 10 billion light years! Incidentally, over the last few decades, careful analyses of these Type Ia supernovae have shown that the Universe is not only expanding, it is expanding at an accelerating rate. From this, cosmologists have proposed the existence of expansion-inducing **dark energy** which might constitute up to 75% of the mass-energy in our Universe.

Returning to the 20th Century, thanks to standard candles, one could calculate the distances to stars and galaxies. Hubble was thus able to show that nearly all observed galaxies were regressing. Not only were they moving away from us, the rate at which they were separating was linearly proportional to the distance away:

$$\mathbf{v = Hd} \quad \textbf{Hubble–Lemaître Law}$$

where v is the velocity of recession, d is the distance from Earth, and H is the Hubble's constant.

It was in this manner that astronomers were able to observationally confirm that the Universe was expanding.

35.4 General relativity and Hubble's law

Albert Einstein finalized his theory of general relativity in 1915. This theory of gravity relates the curvature of space-time to the density of matter and energy. As succinctly summarized by John Wheeler, space-time tells matter how to move while matter tells space-time how to curve. The original **Einstein equation** of general relativity is:

$$G_{\mu\nu} = 8pG/c^4 \times T_{\mu\nu}$$

where the tensor $\mu\nu$ subscripts denote the four dimensions of space-time, thereby yielding a total of 16 equations.

$G_{\mu\nu}$ is a tensor that represents the curvature of space-time while $T_{\mu\nu}$ is another tensor that represents the distribution of mass and energy. Given that our Universe is so homogeneous and isotropic, the sixteen equations can be reduced to just two thanks to symmetry. At the time that Einstein first deduced his equations, the prevailing wisdom was that the Universe was 'static'—it was neither expanding nor contracting. However, the Einstein equations were not compatible with a static situation; they predicted a Universe that would collapse down upon itself. To address this discrepancy, Einstein introduced a simple solution—a driver of expansion (or a resistor of condensation) he called the **cosmological constant**. The cosmological constant would mathematically ensure compatibility with a static universe. It is this cosmological constant that Einstein later famously described as his biggest blunder. Ironically, this 'blunder' might not have been such a disastrous mistake after all, since the recent discovery of an accelerating rate of expansion, which implicates the existence of so-called **dark energy**, is mathematically compatible with a cosmological constant.

Among the first to formally suggest an alternative solution was Alexander Friedmann. In 1922 he used the Einstein equations to examine a Universe that was not just static but actively expanding. As will be described below, Friedmann developed two equations that obviated the need for a cosmological constant. If the Universe was either expanding or contracting, the Friedmann equations could quantitatively depict the evolution of the Universe over time.

35.5 The Friedmann equation

Given the observationally confirmed expansion of the Universe, it was natural to speculate on the logical consequences of this expansion. As Lemaître initially proposed, if the Universe is currently enlarging, it must have started in a smaller, highly contracted state. Taken to its logical extreme, this contracted state at the dawn of time must have been extremely small, dense, and hot. Using Hubble's Law and Einstein's equations, Friedmann was able to derive an equation of state that provides the size and temperature of the Universe as a function of time. In fact, Friedmann was the first to solve the Einstein equations for the Universe as a whole. The **Friedmann equation** may be written as:

$$(dS/dt)^2 - (8\pi G\rho/3) \times S^2 = -k(c^2/R_0)^2$$

where R_0 is the present curvature radius of the Universe, and S is its size or scale factor.

The **scale factor $S(t)$** is a mathematical means of following the size of the Universe over time. One can assign a value of $S(t) = 0$ at the origin (i.e. at the Big Bang) and a value of $S(t) = 1$ for today. The Friedmann equation accurately describes all epochs of the Universe—from the present into the future, as well as from the Big Bang to the present day.

35.6 Evidence supporting the Big Bang theory

In the past, rival proposals such as the **Steady State hypothesis** carried as much weight as the Big Bang hypothesis (which incidentally was given its name in a derisory fashion by Fred Hoyle during a radio interview). Hoyle, as was discussed earlier, was one of the principal masterminds behind the current understanding of stellar nucleosynthesis. The Steady State hypothesis conceded that the Universe was indeed expanding, but to maintain stasis, new matter would mysteriously crop up in the voids formed as galaxies receded from one another. Today, the Big Bang theory has significant supporting evidence and is widely accepted, whereas alternative models, lacking supportive evidence, remain as merely speculative hypotheses. Nevertheless, what happened before the Big Bang, what initiated the rapid expansion, what the mechanism behind inflation is, and many more questions remain unanswered. The data supporting the Big Bang can be listed as follows:

1. The light from galaxies essentially everywhere exhibits a redshift indicating that they are receding from one another. The further away galaxies are from each other, the faster the rate of recession (Hubble's Law). This is consistent with an expanding Universe. An expanding Universe implies that, at one time, it must have existed in a condensed form that has been enlarging since its inception.
2. If there was a Big Bang, there should be remnant radiation pervading the entire Universe from the initial explosive episode. This proposed omnipresent residual radiation has been detected in the form of the cosmic microwave background (CMB) radiation. This thermal (blackbody) radiation fits the expectations remarkably well.
3. Even in the very oldest stars, helium is found throughout the Universe at approximately 25% abundance (by mass). This amount of helium is not possible through stellar nucleosynthesis alone but is consistent with its formation through so-called Big Bang nucleosynthesis. The observed 25% abundance of helium across the Universe fits predictions quite well.
4. The appearance and characteristics of very old (that is, very distant) galaxies differs substantially from younger galaxies. This is consistent with an evolving Universe. The concept of an evolving Universe is more compatible with the Big Bang Theory than the static Steady State hypothesis.
5. The recent detection of extremely distant (i.e. old) metal-free gas clouds is consistent with their formation directly from material created in the Big Bang (rather than from material created by stellar nucleosynthesis, which would be contaminated with elements such as carbon, oxygen, and silicon). These gas clouds could thus be **primordial**, that is, made directly from the material created during Big Bang nucleosynthesis.

Based on simple physical principles of ideal gases, expansion leads to cooling whereas compression leads to heating. Thus, as one goes back in time, one also goes up in temperature. And since temperature is a measure of kinetic energy, the emerging field of high energy physics (HEP) in the 20th century was a window into the very first moments of the Big Bang. Although it was not fully appreciated at the time, the accelerators that were developed to study subatomic particles were achieving energies and temperatures corresponding to early moments after the Big Bang. A useful conversion is:

$$1 \text{ eV} = 1.6022 \times 10^{-19} \text{ J} = 11604 \text{ °K}$$

Thus, 1 eV is roughly 10^4 kelvin or 1 °K $\approx 10^{-4}$ eV. From this, one can see that a modest particle energy of 1 MeV corresponds to a temperature of approximately 10^{10} K or 10 billion degrees. Particles accelerated to energies of 1 GeV correspond to a temperature of around 10^{13} K or 10 trillion degrees, and those with energies around 1 TeV (as in the Large Hadron Collider or LHC at CERN) correspond to a temperature of 10^{16} K. Students may enjoy using $E = 3/2\ kT$ for the kinetic energy of a particle and relating that to kinetic energy ($E = \frac{1}{2}\ mv^2$) to calculate the 'temperature' of a 200 MeV proton used in clinical radiotherapy.

Applying the Friedmann equation and the principles outlined above allows us to go back in time and get a glimpse of the younger Universe. The Friedmann equation tells us the relative size of the Universe during earlier epochs, and that condensed size, in turn, can be translated into temperatures:

$$t_{\exp} = \left[\frac{2.7K}{T}\right]^{3/2} \sqrt{\frac{3c^2}{8\pi g\rho_0}}$$

where t is time and T is temperature.

In this manner we can use the Friedmann equation to tell us that the temperature of the Universe was 3 billion degrees kelvin at 1 min and 10 billion K at 1 s. Additional information helps calibrate the time-temperature curve. For instance, the cosmic microwave background or CMB radiation tells us that the Universe was ~3000 K at 380 000 years after the Big Bang. This is because, prior to that time, the temperature was too hot for atoms to exist and the plasma of free electrons and protons was opaque to photons. At a temperature of 3000 K 'recombination' occurred and the free electrons and protons then formed atoms, which were opaque to photons. From that moment (and temperature) on, the photons roaming the Universe were set free and are seen today as the CMB radiation, or just CMB. Additionally, the relative abundance of helium (the hydrogen:helium ratio) tells us that the Universe was about 100 million K at 3 min. This will be explained in greater detail below. From these data points we can reconstruct the early history of the Universe, and it tells an amazing tale.

35.7 The Planck epoch

The Planck epoch spans the interval from $t = 0$ to the **Planck time**, which is 10^{-43} s. The Planck time is defined as the time it takes for a photon (moving at c) to travel a distance equal to the **Planck length** L_P (which is 1.62×10^{-35} m, which comes from $L_P = (hG/2\pi c^3)^{1/2}$). This can be written:

$$tP \equiv \sqrt{\frac{\hbar G}{c^5}} = 5.39124 \times 10^{-44}\ \text{s} \approx 10^{-43}\ \text{s}.$$

At 10^{-43} s the temperature calculates to 10^{32} kelvin (the so-called Planck temperature) and the Universe was a point. At this time and at such energies, gravity was believed to take on a quantum nature and was governed by a still-unelucidated quantum theory of gravity. Extremely little is known or can be about the Universe at this time. Presently, our most powerful particle accelerators can reach energies around 14 TeV, which correlates with a temperature of roughly 10^{17} K or a time around 10^{-14} s. Thus, in addition to having no adequate theory to explain the Universe at the Planck time or before, there is a huge gap between our present experimental experience and this extremely poorly understood Plank time.

Nevertheless, based on what we do know through the Standard Model, there are no other particles that would have existed before 10^{-14} s. If this proves true, that would imply that there were no new particles across an enormous range of roughly 10^{10} in temperature, 10^{20} in time, and 10^{40} in density. This hypothetical void is called **the particle desert**.

However, there are some theories which incorporate 'physics beyond the Standard Model' that postulate the existence of additional, yet unidentified, particles. One such theory is **supersymmetry**.

35.8 Supersymmetry

In supersymmetry, each fermion and each boson has a partner of the opposite sort called a **supersymmetric partner** or **super-particle partner**. Thus, each fundamental fermion would have a corresponding supersymmetric boson and that boson's name would start with an 's'. For instance, there would be **squarks** and **selectrons** according to supersymmetry. Likewise, each boson would have a supersymmetric fermion partner and those partner's names would end in '-ino'. Thus, there would be **photinos**, **Zinos**, and **Winos** in supersymmetry. As with all particles, heavy super-symmetric particles would decay into lighter ones, and so all supersymmetric particles would have eventually decayed into the lightest—but stable—super particle. If correct, these lightest remaining super-particles should remain in existence and be potentially detectable someday. Although to date, no super-symmetric particles have been discovered, the concept remains popular since these hypothetical stable super-particles could be candidates for the **dark matter** that makes up approximately 85% of the matter in our Universe (with the remaining 15% of matter being the familiar **baryonic matter** or **atomic matter** that we, and everything else we know of, is made of). In addition to the still-unidentified dark matter, the Universe as a whole consists of roughly 68% **dark energy**, which powers its presently accelerating expansion. In other words, the Universe is 68% dark energy, 27% dark matter, and only about 5% familiar baryonic matter. Regardless of whether supersymmetry proves correct, the search remains active for dark matter candidates and the mystery of dark energy remains. It is thus clear that there must be physics beyond the Standard Model that simply remains unknown at present.

35.9 Grand unified theories and theories of everything

Although we have no machines capable of accelerating particles to energies coming anywhere close to temperatures providing proof, there are some candidate **theories of everything** or **TOEs** that postulate that all forces of nature (including gravity) can be governed by a single all-encompassing unified theory. This TOE would mathematically describe the Universe in terms of a single force rather than the present four forces of nature which manifest at our current temperatures. The TOE would operate at the unimaginable temperatures which existed at a tiny fraction of the Universe's first second of existence. Following this brief instant, gravity was the first force to 'freeze out' as a separate entity. At temperatures around 10^{28} K or 10^{16} GeV, which correspond to 10^{-36} s, the strong force was still united with the electroweak force creating a '**grand unified force**' that operated according to a a **grand unified theory** or **GUT**. Although not yet experimentally confirmed, the theory is supported by the observation that the strong force grows weaker at increasing temperatures (allowing **quark deconfinement,** as will be discussed below). Furthermore, it seems like the coupling constants of the strong, electromagnetic, and weak forces become almost equal at energies near 10^{16} GeV. Such GUTs might dictate different laws of nature at this energy and temperature range, with the three forces all following these laws as one—the grand unified force law. Present versions of GUTs seem to adequately explain the W:Z boson mass ratio, and why the electron and proton have equal electric charges, but they also call for instability of the proton; to this date, all experiments have failed to detect any instability of the proton. Currently, data indicates that if it does decay at all, it must have a half-life beyond 1.67×10^{34} years.

35.10 Freeze-outs and broken symmetry

In earlier chapters, when discussing Noether's Theorem, it was mentioned that the ubiquitous conservation laws that abound in physics imply the existence of under-lying **symmetries** or **invariances**. At the ultra-high energies that we have been talking about here, the various laws of nature seem to converge and exhibit greater symmetry than is evident at the normal temperatures of today. Colloquially speaking, it appears that the laws of nature demonstrated greater symmetry at the high temperatures of the early Universe and nature was more 'tolerant'—it tended to ignore variables that presently make a difference. In other words, at the ultra-high temperatures during the first fractions of a second, there was greater invariance and more symmetry than we see today. But as time marched forward and temperatures fell, such perfect (or nearly perfect) symmetry was broken, and fewer things were 'ignored' by nature. The breaking of symmetry that allowed the gravitational and strong forces to freeze out and depart from the perfect symmetry governed by TOEs and GUTs are but two examples of symmetry-breaking as the temperature cooled.

Our present understanding is that as the temperature of the Universe cooled through adiabatic expansion, first the gravitational force 'froze out' (maybe—some people believe that gravity may have never really been part of the gang), and next the strong force froze out leaving the unified electroweak force. As we know from today's world, the electromagnetic and weak forces have also separated out, and this shall be discussed below. But before that, another phenomenon occurred that had profound influence on our Universe—for reasons unknown, the Universe enormously expanded many orders of magnitude through a transient phenomenon called inflation.

35.11 Inflation

The cosmic inflation theory was independently developed by Alan Guth of the United States and Russian scientists Andrei Lind and Alexei Starobinsky. Inflation was conceived to explain that the Universe must have suddenly and inexplicably expanded exponentially just a tiny fraction after its formation.

According to the theory, the Universe briefly expanded by an uncertain but phenomenally high factor—possibly on the order of $\sim 10^{26}$ or about e^{60}. This staggering expansion is believed to have occurred during the time interval from 10^{-36} to 10^{-32} s. (This estimate is constantly being revised, but at the time of this writing in 2024, this was considered current.) Invoking inflation solves many vexing problems in cosmology which will not be covered here (such as the **flatness problem**, the **magnetic monopole problem**, the **smoothness problem**, and the **horizon problem**), but there is no presently comprehensible mechanism or explanation for the faster-than-light phenomenon.

At the tremendous temperatures and pressures prevailing at the dawn of time, an 'anti-gravity' repulsive force might have been in effect. General relativity allows for a repulsive force, which could have caused the Universe at large to expand at this incalculable rate. Recall that special relativity stipulates that nothing in the Universe can travel faster than or convey information faster than c, but it doesn't preclude the Universe itself from growing faster than c.

35.12 Electroweak symmetry breaking

The Standard Model calls for a symmetry in which the carriers of the weak force should all have the same (zero) mass to allow the unification of the electromagnetic and weak nuclear forces into a single electroweak force. Indeed, as will be described below, in the hot early Universe above 10^{15} K, the W and Z weak gauge bosons were massless members of the pervading 'quantum soup'. But as the temperatures fell and these particles acquired mass through the **Higgs mechanism**, they could no longer be freely formed from energy via $E = mc^2$ in the quantum soup. Hence, at a certain time and temperature, the previously unified electroweak interaction differentiated into the separate components of the electromagnetic force (mediated by the photon) and the weak interaction (mediated by the newly-massive W and Z bosons). This symmetry breaking might have occurred at around 10 picoseconds (10^{-11} s), which means that it overlapped the so-called Quark Epoch.

35.13 Quark Epoch

At about 10^{-12} s the **Quark Epoch or Quark Era** began. Although extremely short, it spanned several orders of magnitude, from about 1 picosecond to 1 μs. The corresponding temperature was about 10^{15} K at 10^{-12} s (*'one petakelvin at one picosecond'*). Although the four forces had already frozen out and separated, the ambient temperatures and kinetic energies were still too high for quarks to condense into hadrons. Nevertheless, at these high temperatures (e.g. $> 10^{14}$ K), the strong force was no longer very strong, and quarks were 'unglued' from one another.

Unlike now, where **quark confinement** in hadrons is the rule, at such temperatures, **deconfinement** was the standard at that time and temperature. Quarks (and gluons, the strong force mediators) were freely swimming about unbound in an ultra-hot **'quark soup'** or, more technically, a **quark-gluon plasma**. It could be said that during the Quark Epoch it was still 'too hot for hadrons' and the bound state was not yet energetically favored. Protons and neutrons could not exist under these circumstances. In this primordial particle and antiparticle soup, all the basic building blocks of the Universe abounded in roughly equal numbers.

Quark-gluon plasma has been experimentally produced through ultra-high-speed collisions of heavy atomic nuclei at facilities such as the Relativistic Heavy Ion Collider at Brookhaven National Laboratory. As mentioned before, the strong force is not quite so strong at astronomically high temperatures. At the close of the Quark Epoch, near one (to maybe ten) microseconds post-Big Bang, the temperature fell sufficiently for the strong force to gain its full strength and live up to its name. Quark confinement thus began, and from that moment on, quarks only lived in little packets called hadrons. Thus, the close of the Quark Epoch ushered in the Hadron Epoch.

35.14 Hadron Epoch

The time interval roughly from about 1 μs to 1 s was the **Hadron Epoch** or **Hadron Era**. (Note that although the temperatures and kinetic energies corresponding to events such as quark confinement are reasonably well understood based on high energy physics experiments, the times they correlate with after the Big Bang are approximations that are subject to revisions as more data emerges and theories are refined.) As time marched on and the Universe continued to expand and cool during the Hadron Epoch, the temperature fell to around ten billion degrees (10^{10} K) by about 1 s. At this point, the quark soup of the previous epoch had sufficiently cooled down to the point where quarks and gluons were finally (and forever) bound into hadrons such as protons and neutrons. In other words, quark confinement had commenced. Another way of expressing this is by stating that the ambient temperature had fallen to below the **Hagedorn temperature**.

From that point forward, quarks were essentially forever found only in bound states. The transition from free quarks and gluons in a hot, quark gluon plasma into a quantum particle gas of bound quarks is called **hadronization**. The days of free quarks roaming the Universe in a quark-gluon plasma were now over for good. They were given a lifetime jail sentence of confinement in hadrons (and given that protons may be eternal, that is a long sentence!). With the exception of exotic astrophysical phenomena and some high-energy physics laboratories, almost all quarks have been continuously combined with other quarks in color-neutral baryons (or paired with antiquarks in colorless mesons) held together by colored gluons. Of course, one place where quarks and gluons may also be set free from their hadron jailhouses is in deep space where galactic cosmic rays with far more energy than anything made in our physics labs crash into matter and other particles.

35.15 Matter–antimatter asymmetry during the Hadron Era

It was at the close of the Hadron Epoch that the very slight preponderance of matter over antimatter (i.e. the **baryonic matter–antimatter asymmetry** or, more simply, the **baryon asymmetry**) began with the elimination of all anti-hadrons. It is estimated that the mutual annihilation of protons with antiprotons (and all other baryon–antibaryon pairs in existence at the time) left only one-in-a-billion ordinary protons behind; all antiprotons were eliminated in the mass massacre. Yet these proton survivors have gone on to constitute *all* of the hadronic matter of today's known Universe.

At the very high temperatures during the Hadron Epoch (a few trillion kelvins), photon energy was sufficient for constant pair production though interactions with the vacuum where virtual particles resided. Given a sufficient injection of energy, these virtual particles could become real particles. Thus, protons and antiprotons would collide and annihilate to release gamma photons, and those gamma photons could then, through pair production, lead to new protons and antiprotons in a perpetual cycle.

This constant creation of real particle–antiparticle pairs of hadrons was the norm during the Hadron Epoch. But as the temperature fell to around 10 billion kelvins near the 1 s mark (or maybe slightly sooner at the 100-µs mark), it became progressively more difficult to create hadron–antihadron pairs from the vacuum. Nevertheless, mutual annihilation could of course continue unimpeded. The natural expectation would be for all hadrons and antihadrons to have annihilated one another, leaving only gamma radiation as a trace of their former existence. Fortunately for us, being made of baryonic matter, this did not occur; for every billion antiprotons there was a billion and one protons. We are the products of those relic 'leftover' protons.

One might be tempted to speculate that maybe there really was perfect symmetry and what we see today with our telescopes are some stars and galaxies made of matter and others, identical in appearance, made of antimatter. If that were the case, however, we would expect to see high-intensities of gamma rays emanating from the edges of galaxies where ongoing mutual annihilation was occurring. Such photons would have very characteristic energies—938 MeV gamma rays from proton–antiproton annihilations and 511 keV gamma rays from electron–positron annihilations. Despite searches, these photons are not among the confirmed sources of cosmic radiation. Given this absence, we must conclude that the Universe is solely made of matter, and that all the antihadrons were slaughtered at the end of the Hadron Epoch. The leftovers of that slaughter include the one-in-a-billion protons that make up every atom in every molecule in every cell of our bodies. So, thank goodness for that asymmetry.

35.16 The Lepton Epoch

The **Lepton Epoch** (also called the **Lepton Era** or the **Age of Leptons**) began with the carnage that eliminated all antimatter hadrons, along with most matter hadrons (except one-in-a-billion protons). This was the annihilation event signifying the end of the Hadron Epoch and the dawn of the Age of Leptons. The temperature at this time was near 10 billion degrees and is believed to have occurred somewhere between 100 microseconds and 1 s after the Big Bang. From that moment on, there were no free hadrons and antihadrons flying about, but there were still electrons and antielectrons galore. All the other leptons (and their antilepton partners) were in abundance as well, in approximately equal amounts.

Thanks to the disappearance of most of the more massive hadrons/antihadrons at the end of the Hadron Epoch, leptons became the dominant form of mass-energy in the Universe during the Lepton Epoch. Unlike before, leptons now comprised most of the matter in the Universe. Although protons weigh about 1836 times as much as electrons (938 vs 0.511 Mev/c^2), the mass murder (mutual annihilation) of all the Universe's antiprotons and 99.999 9999% of all protons allowed the lightweight contingency to gain the upper hand.

As during the Quark Era (when free quarks and antiquarks abounded), and in the Hadron Era (when all sorts of mesons and baryons and their antimatter counterparts abounded), during the Lepton Era all sorts of leptons and antileptons swarmed the Universe. Electrons, muons, and tau leptons were flying about and colliding with (and annihilating) their antimatter counterparts. In addition to these charged leptons, neutral neutrinos and antineutrinos of the three different flavors also flourished. Of course, all these leptons were constantly in flux, momentarily appearing from the vacuum through pair production and swiftly disappearing through mutual annihilation.

The Lepton Epoch lasted from roughly around 1 s to about 10 s, when the temperature fell to around 2.4 billion degrees K (not the 11.9 billion K that one might calculate from the 1.022 MeV pair production threshold, as explained below). Below this temperature, electrons and positrons could no longer be readily made from the vacuum through pair production; nevertheless, mutual annihilation of electrons and positrons could proceed unimpeded. Thus, the Lepton Era ended (like the immediately preceding Hadron Epoch) with the mass slaughter of nearly all leptons through the mutual annihilation of leptons and antileptons, save one-in-a-billion electrons. And for every surviving electron, there must have been about a billion pairs of gamma photons made through electron–positron mutual annihilation. The resulting gamma photons led to the **Radiation Epoch**, which followed the Age of Leptons and persisted for hundreds of thousands of years.

35.17 Neutrino decoupling, primordial Big Bang neutrinos, and the cosmic neutrino background

Up to the close of the Lepton Epoch, leptons such as electrons, muons, and tauons were constantly encountering their antimatter counterparts and undergoing mutual annihilation. The ambient temperatures were also sufficiently high for incessant pair production of new lepton–antilepton pairs from the vacuum. In addition to the charged leptons, the various flavors of neutrinos and their antineutrino counterparts were also doing the same thing. The three flavors of neutrinos and antineutrinos (electron neutrinos, muon neutrinos, and tauon neutrinos and their counterparts) were being absorbed, annihilated, and recreated regularly. But unlike today where neutrinos hardly interact with anything, in the extremely energetic and dense early Universe, neutrinos certainly did interact with everything possible. One might describe the early Universe as 'foggy', even to neutrinos. They couldn't get very far before encountering something to interact with. Thus, much like gamma photons formed in the core of the Sun, these leptons had very short pathlengths before being consumed or annihilated and reformed.

But it was also during the Lepton Epoch that things changed forever for these **primordial neutrinos**. At some point during the Lepton Epoch, the temperature fell to around 2.3 $\times10^{10}$ K ($\approx$ 2 MeV), and the primordial neutrinos 'decoupled' from matter. **Neutrino decoupling** meant the 'fog' had lifted. Below that certain temperature, neutrinos ceased interacting with matter and were 'set free', so to speak. From that point onward, these **Big Bang neutrinos** were forever free to roam about the Universe in their characteristically antisocial manner. Recall that a beam of neutrinos might be attenuated by only 50% by a block of lead 1 light year thick. These neutrinos and antineutrinos thereby escaped the fate of the hadrons and other leptons that underwent mass mutual annihilation.

Due to expansion of the Universe, the Big Bang neutrinos have been red-shifted and diminished in energy over time. Their calculated temperature today is 1.95 K ($\approx$ 168 µeV), which is $[4/11]^{1/3}$ or 71.4% of the CMB radiation temperature (for reasons partially described in the next section). The primordial neutrinos continue to fly about and have only recently been experimentally detected in 2015 (very indirectly via an imprint on the CMB). These omnipresent primordial neutrinos are said to constitute a **cosmic neutrino background radiation** (**CNB** or **CνB**) akin to the more famous and meticulously measured CMB radiation or CMB.

In addition to neutrino decoupling, another major event in the Lepton Epoch, which defines its end, was the mutual annihilation of all antimatter leptons. That is, at approximately 10 s after the Big Bang, essentially all positron–electron pairs were annihilated, leaving only an infinitesimal excess of electrons (about one in a billion). Incidentally, the excess of electrons exactly matches the excess of protons and provides the electrical neutrality of today's Universe. This minuscule excess of electrons later paired up with the leftover protons (from the mutual annihilation of all baryon–antibaryon pairs at the end of the Hadron Epoch) to produce all the atomic matter in the known Universe.

35.18 Energy density of the early Universe

Most students are familiar with the concept that tiny bits of matter can yield enormous amounts of energy in nuclear weapons via the Einstein relation, $E = mc^2$. For example, only 2.33 kg of hydrogen actually undergo conversion into energy in a 50-megaton hydrogen bomb, and less than one gram (perhaps 700 mg) of uranium-235 was actually converted into energy in the 15 kiloton Little Boy fission device. Given that 1 ton of TNT = 4184 MJ, this truly is a tiny amount of mass being converted into an extraordinary quantity of energy. This gives one perspective on the amount of energy released during the early Universe, not through the fractional losses of mass seen in fusion or fission, but the complete conversions of mass into energy through mutual annihilation during the Lepton and Hadron Epochs.

At 1 s post-Big Bang, just at the close of the Hadron Era, the temperature is said to be somewhere roughly above ten billion degrees. Under these conditions, the density of the matter in the Universe (essentially all protons) was like the density of liquid water—about 1 g ml^{-1}. In contrast, the density of the *radiation* itself at this time (predominately gamma and x-rays) was about one *ton* per milliliter. One cubic centimeter of any region within the 1 s old Universe contained the mass-energy equivalent of one ton of matter. Considering the tiny quantities of mass (milligrams to kilograms) converted into energy in nuclear bombs, converting this amount of mass (tons) into energy via $E = mc^2$ yields a very large amount of energy indeed!

Near the close of the Lepton Epoch at around 10 s, the Universe was considerably cooler and less dense thanks to expansion, but still consisted of some amazing stuff—each milliliter contained about 100 mg of protons, 2 kg of electrons, 2 kg of positrons, and 2 kg of photons (mass-energy equivalent, that is, since photons are massless). But at the end of the Lepton Epoch, the temperature fell below the threshold for electron–positron pair formation. Electrons and positrons could no longer be created out of the vacuum as the temperature fell further. Then in an instant, literally all the positrons and all except one-in-a-billion electrons underwent mutual annihilation into gamma radiation. One might suspect that this colossal discharge of energy must have accelerated the Universe's expansion or done something similarly dramatic, but in fact it did not. Given the state of the Universe at that time (and the fact that it wasn't expanding into free space but rather the whole of space-time itself was expanding), the net result was a modest 40% increase in temperature—a mere blip on the curve. Nevertheless, the gamma photons set free during this final phase of carnage between the primordial electrons and positrons has left a permanent mark. It is this leftover radiation that we presently detect now as the **cosmic background radiation** or CBM. (Of course, most of this radiation started out as monoenergetic 511 keV photons, but over the following 380 000 years, through interactions with the free electrons and protons sailing through the Universe, this radiation broadened into a full spectrum of blackbody radiation.)

Since the primordial neutrinos had already decoupled (at about 1 s) and left the scene before this mass slaughter of the remaining leptons, they did not benefit from this additional boost of energy. Thus, their energies are slightly lower than the energies of the photons released later. For that reason, the primordial Big Bang photons (which constitute the CMB), display a blackbody spectrum corresponding to a temperature of 2.73 K, but this is slightly warmer than the (still not directly detected) cosmic neutrino background of primordial Big Bang neutrinos, which are estimated to have a temperature of 1.95 K ($\approx$ 168 μeV). This figure is $[4/11]^{1/3}$ or 71.4% of the photon's energy.

35.19 How do we calculate the temperatures of the early Universe?

It is commonly stated that temperature at the start of the Lepton Epoch was slightly above 10 billion K and fell to about 1 billion degrees (10^9 K) at the close. However, for all of the various flavors of leptons to be in thermal equilibrium with their surroundings and freely undergoing mutual annihilation and pair production, we know that the instantaneous temperature in the vicinity of such pair production must have been much higher than this nominal figure. For example, if a tau lepton has a mass of roughly 1776 MeV/c^2, then for production of tau–antitau pairs, the ambient temperature must exceed the threshold of twice this figure or roughly 3552 MeV/c^2. Recalling from earlier chapters that **1 eV = 11 600 degrees K**, one can estimate that:

(3552 MeV) $\times$ (10^6 eV/MeV) $\times$ (11 600 Kelvin/eV) = 4.1 $\times$ 10^{13}

or about 41 trillion Kelvin.

Similarly, to create mundane electron–positron pairs, one might imagine the threshold energy to be twice the rest mass of 0.511 MeV or 1.022 MeV, which corresponds to:

(1.022 MeV) × (10^6 eV/MeV) × (11 600 Kelvin/eV) = 11 855 200 000
or nearly 12 billion K.

This 12 billion degree calculation is at odds with what was described above. Thus, there is something missing from these simple calculations. The paradox is resolved by recalling that temperature reflects a spectrum of photon energies rather than a single figure. Given that we are talking about converting photon energy into pairs of particles, we should recall that blackbody (thermal) electromagnetic radiation exhibits a spectrum of energies, and although this spectrum has a peak, or most probable photon energy (given in wavelength) that is provided by **Wien's displacement law**, $\lambda = b/T$, the more proper way to perform these calculations is to take the entire broad spectrum of wavelengths into consideration. We can combine Wein's law ($T = b/\lambda$ or $T = 0.0029/\lambda$) with the Planck–Einstein relation ($E = h\nu$), along with the fact that $c = \nu\lambda$. Rearranging and substituting (and recalling the units!), one can show:

$$E = h\nu = hc/\lambda = hcT/b$$

or

$$T = Eb/hc$$

Or, putting in conversion units:

$T = (E$ in eV$) \times (0.0029$ K $-$ m$) \times (1.6 \times 10^{19}$ J/eV$)/(6.626 \times 10^{-34}$ J $-$ s$)$
$\times (3 \times 10^8$ m/s$)$.

Plugging in a value of 938 MeV/c^2 for the rest mass of a proton, and realizing that for pair production we need twice this value (thus we need a factor of 2), we can calculate the temperature:

$$T = 2(E) \times (b)/(h) \times (c)$$

$T = 2(938$ MeV$) \times (10^6$ eV/MeV$) \times (0.0029$ K $-$ m$) \times (1.6 \times 10^{-19}$ J/eV$)/$
$(6.626 \times 10^{-34}$ J $-$ s$) \times \left(3 \times 10^8$ m s$^{-1}\right) = 4.4 \times 10^{12}$ **Kelvin**.

In other words, for proton–antiproton pairs to form from the freely available radiation, the temperature must be about 4 trillion degrees outside. A pretty hot day.

If we do the same calculation for electron–positron pairs, recalling that they each weigh 511 keV or 511 000 eV:

$T = 2(511\ 000$ eV$) \times (0.0029$ K $-$ m$) \times (1.6 \times 10 - 19$ Je/V$^{-1})/$
$(6.626 \times 10^{-34}$ J $-$ s$) \times (3 \times 10^8$ m s$^{-1}) = 2.4 \times 10^9$ K.

In other words, *to create electron–positron pairs from the ambient radiation, the temperature must be a bit above 2 billion degrees.* Although this is still extremely hot, it is a far cry from the nearly 12 billion degree naïve calculation initially made.

It is from these types of calculations, based on amounts of energy required to generate pairs, that one can estimate what the temperature was at these early stages of the Universe. From such calculations, the rough figures quoted above and in the table are derived (table 35.1).

Table 35.1.

Epoch	Temperature	Time after Big Bang (approximately)	Energy	Events
Electroweak Epoch	10^{20}–10^{33} K	10^{-32} s to 10^{-12} s	10^{16}–10^{29} eV	Weak and electromagnetic forces freeze out from a grand unified electroweak-nuclear force, leaving gravity, the strong force, and an electroweak force in existence. With further temperature reduction, electroweak symmetry breaking leads to all four present forces being manifest. Fundamental particles acquire their present masses thanks to the Higgs mechanism. (Prior to this time (i.e. above this temperature) all were massless.)
Quark Epoch	10^{16}–10^{20} K	1 picosecond to 1 μs (10^{-12} to 10^{-6} s)	10^{12}–10^{16} eV	All four forces exist but the temperature is still too high to allow quarks to bind together as hadrons. The Universe was filled with quark–gluon plasma, containing photons, quarks, gluons, leptons and their antiparticles. The Quark Epoch ends with quark confinement.
Hadron Epoch	10^{12} K (1 trillion) to 10^{16} K	Approximately 1 μs to 100 μs (or maybe 1 s)	10^{8}–10^{12} eV	Begins with quark confinement ('hadronization' or the formation of bound states of quarks). Birth of baryons and mesons. Hadrons dominate the mass of matter in the Universe. Ends with annihilation of all antiprotons.

Lepton Epoch	10 billion to 1 billion K	100 μs (or maybe 1 s) to 10 s	86 keV — 1.022 MeV	A rough rule: '10 billion K at 1 s, 1 billion K at 10 s' Neutrino decoupling occurred at about 1 s. Neutrinos no longer regularly interacted with matter from then on but were set free as the cosmic neutrino background. Electrons, positrons, and other leptons dominate the mass of matter in the Universe until the mass mutual annihilation at the close of the Lepton Epoch.
Radiation Epoch (Photon Epoch)	≈ 4 billion K to ≈ 3000 K (more precisely, about 3481 K).	10 s–380 000 years	1.022 MeV to 0.3 eV	Begins with the mass destruction of all antielectrons (positrons) through mutual annihilation with all electrons except one in a billion. Some say the Photon Epoch ends with recombination (at 380 000 years) but others argue that the Photon Epoch ends when radiation's dominance over matter ended (in terms of mass-energy density) at approximately 60 000 years post Big Bang.

35.20 The Radiation (Photon) Epoch

At somewhere around the 10 s mark, the **Photon Epoch** or **Radiation Epoch** began. Much happened during this period, including what is termed **Big Bang nucleosynthesis**, but basically, this time interval is defined by *the dominance of radiation over matter as the primary form of mass-energy in the Universe*. Interestingly, matter does not diminish in rest-mass energy over time and expansion; it remains constant via $E = mc^2$. In contrast, radiation loses energy over time because of the red-shifting imposed by expansion of the Universe. The photons decrease their energy by the Planck–Einstein relation, $E = h\nu$. Thus, it is natural to expect that as time rolled on and expansion continued, that there would come a time when radiation's dominance over matter would end.

This Universal radiation energy loss can be quantified through the scaling factor, S, which reflects the linear dimensions of the Universe over time. Volume is proportional to S^3; temperature is inversely related to S: $T = 1/S$. Thus, temperature decreases as the cube of the linear dimensions.

Recall that S assumes a value of 0 at the time of the Big Bang and 1 for today. So, going backwards in time, as the Universe shrank, we know that the temperature increased. But since the amount of mass in matter is constant, matter density increased in the familiar relationship between size and mass (density = mass/volume) as $1/S^3$. But radiation energy is not constant over time or S factor—by the Planck–Einstein relation, photon energy decreases with increasing wavelength, and wavelengths increase with S and time. Thus, going backwards in time, radiation energy density increased faster than matter density as $1/S^4$.

The actual number ratio of photons to protons today is about 1.6 billion to 1. The number of protons was set at the end of the Hadron Era when the mass annihilation of all the antiprotons left a tiny excess of matter protons (of one in a billion). More photons were added at the end of the Lepton Epoch when all the antielectrons were annihilated, leaving a tiny excess of matter electrons behind (the exact same one in a billion as the protons). This photon:proton number ratio does not change much over time or the size of the Universe. However, the energy density ratio of photons to protons does change dramatically over time thanks to the $1/S^4$ versus $1/S^3$ dependence on scale factor (for photon energy density versus mass energy density, respectively). For example, at about 400 000 years after the Big Bang, the linear dimensions were about 1000 times less than today, and the volume was about 1 billion (1000^3) times less than today's Universe. Since today's proton density is roughly 1 proton per 4 m^3, and since the Universe was one billion times smaller at the 400 000 year mark, the proton density was one billion-fold higher, amounting to approximately 250 protons cm^{-3}. The temperature at this time was around 3000 K, corresponding to a sky that would look a bright orange-red. But although the photon:proton number ratio was the same as today, the photon energy density was far higher than it is now because photon energy density goes as $1/S^4$, and since S was 1/1000 at 400 000 years, the photon mass-energy density was higher (by 10^{12}).

Going back even further, to around 10 s after the Big Bang, the proton (i.e. matter) density was about the same as it is today in Earth's atmosphere: about 10^{12} cm^{-3}. On the other hand, the radiation energy density at that time was tremendous—over 200 times the density of lead.

At a time around **57 000 years after the Big Bang** the mass-energy density of photons was the same value as the mass-energy density of protons. Some experts use this time to define the end of the Radiation Epoch, whereas others say the close of the Radiation Epoch occurred around 380 000 years when the temperature fell to a mere 3000 K and **recombination** occurred.

35.21 Recombination and the CMB

Recombination is very analogous to neutrino decoupling and can in fact be called **the photon decoupling event**. Just as with neutrinos before decoupling, prior to the 380 000 year mark the early Universe, consisting of a plasma of electrons, protons, and light nuclei, was 'foggy' to photons. In other words, during the Radiation Epoch, no photons could get very far before being absorbed, vanishing, and then reforming in this all-permeating plasma. Unlike today's Universe, which is largely transparent to photons, the hot, dense early Universe was opaque.

But upon commencement of recombination, *for the first time*, electrons could join protons to form the bound state known as hydrogen atoms. ('Recombination' is thus a poor name for the phenomenon since it implies something is happening *again*, when in fact it happened for the first time.)

Anyway, at temperatures below 3000 K (that is, after 380 000 years) ionization would no longer be the rule, and electrons could remain bound to nuclei from that time and temperature forward. But because atoms are relatively transparent to light compared to free electrons and nuclei, this also meant that photons would become unobstructed and could roam about the Universe joining the primordial Big Bang neutrinos. These **primordial photons** (also known as 'relic radiation') are now permeating the cosmos as the **CMB radiation**. Although we commonly say that recombination occurred at 3000 K, that is the temperature threshold at which essentially *all* hydrogen and helium atoms are completely un-ionized. In practice, it might be better to describe the temperature at which roughly 50% of the atoms were ionized. That temperature corresponds closer to 4000 K. Based on Wien's Law ($\lambda = b/T$) where b is Wein's displacement constant (2.9×10^{-3} m K^{-1}) and T is temperature, one could do the calculation and come up with 724 nm, which corresponds to far red light in the optical part of the electromagnetic spectrum. Biology students might recall that this wavelength is at the limit for human vision. Also, it is instructive to recall that this wavelength is not monoenergetic; it is just the peak in a spectrum.

Another way of working this out is via one form of the **Saha equation** specific for hydrogen:

$$\frac{n_p n_e}{n_H} = \left(\frac{m_e kT}{2\pi\hbar^2}\right)^{3/2} e^{-E_1/kT}$$

where n_e is the number density (the number per unit volume) of free electrons, n_p is the number density of free protons, n_H is the number density of neutral hydrogen atoms, m_e is the mass of the electron, k is Boltzmann's constant, T is the temperature, $\hbar$ is the reduced Planck constant (h/2π), and E_1 is the ionization potential of hydrogen (13.6 eV).

Students may define x_e as the fraction of free electrons such that $x_e = n_e/(n_e + n_p)$ and rewrite the Saha equation as:

$$\frac{x^2}{1-x} = 1/n\left(\frac{m_e kT}{2\pi\hbar^2}\right)^{3/2} e^{-13.6/kT}.$$

If we define recombination as an ionization fraction of 10% ($x_e = 0.1$), the temperature at recombination, T_R, is 0.3 eV or 3481 K. But one can also define recombination at different degrees of ionization fraction. For instance, if one plugs in 60% ionization into the above equation, the temperature calculates to 4000 K.

3481 K corresponds to 0.3 eV, or via Wien's Law, to a thermal photon spectrum with a peak wavelength of 832.5 mn which is near the limit of the red end of the visible spectrum (or it may be in the near-infrared, depending on one's definition). Lower temperatures such as the oft-quoted value of 3000 K correspond to longer wavelengths, but one must remember that the peak value calculated via Wein's Law is part of a spectrum and that spectrum will include photons in the visible portion.

So, then why is our night sky not a uniform faint red color? Recall that over time the Universe has continued to expand, and that expansion has led to red-shifting of all electromagnetic radiation. The reddish optical photons set free during recombination have been red-shifted down in energy to the microwave portion of the spectrum. Just as Olbers' paradox was resolved by red-shifting of visible starlight in an expanding Universe, the red-shifting of the reddish optical glow of recombination has also been stretched in wavelength and diminished in energy to below the visible range by an expanding Universe.

One final point worth mentioning about recombination is that it did not all occur instantly. Electrons and protons *began* to combine into hydrogen atoms only when the temperature finally fell sufficiently, but not all did so simultaneously. As seen from the Saha equation, as the temperature falls, only a certain fraction at any specified temperature will remain ionized, and a complementary fraction will have combined into neutral hydrogen. The same concept of course holds true for deuterium, helium, and other atoms. The estimated total time span of cosmological recombination was probably on the order of tens of thousands of years.

35.22 Big Bang theory history

In the way of history, following Lemaître's 1931 suggestion of 'a day without yesterday', there was a fairly long interval before any more serious speculation took place about the Big Bang. But, between 1948 and 1953, George Gamow, Ralph Alpher, and Robert Herman wrote several key papers addressing **Big Bang nucleosynthesis**—the creation of elements in the furnace of the first few minutes. With the exception of a minute amount of lithium, however, they were unable to account for the origin of the elements beyond helium (i.e. the 'metals') through this **primordial nucleosynthesis**. Nevertheless, they did predict that the original blast should have left a remnant, which should today be evident as a pervasive microwave radiation that permeates all of the Universe in all directions. Unfortunately, neither Lemaître nor Gamow, Alpher, and Herman had the tools to detect this omnipresent microwave background, and for several decades the concept was all but forgotten. In 1964, Arno Penzias and Robert Wilson, working at the Bell Labs in New Jersey with a new microwave telescope, could not account for about 55% of their detected microwaves. Even after recleaning the receiver repeatedly to remove the suspicious 'white dielectric material' (pigeon poop), the microwave background noise remained. Around the same time, Yakov Zeldovich in Russia and Jim Peebles and Robert Dicke of Princeton had independently 'rediscovered' Gamow's idea of a residual cosmic microwave signal. In fact, they began building a special microwave detector for this purpose. Unfortunately for them, Penzias and Wilson had already serendipitously discovered this CMB radiation and beaten everyone to the punch. Penzias and Wilson were awarded the 1978 Nobel Prize in Physics for their discovery of the CNB, which firmly established the Big Bang model as a solid and evidence-based theory of cosmology.

In a spectacular display of theory matching experiment, the photons that were set free during recombination were predicted to have red-shifted down in energy to the microwave energy range, and in 1990, NASA's **Cosmic Background Explorer** (or **COBE**) satellite had analyzed the CMB and created a spectrum. This observed spectrum essentially *exactly* overlapped the predicted blackbody spectrum. If the photons set free during recombination were 3000 degrees at that time (380 000 years after the Big Bang), red-shifting should have converted these optical photons into microwaves with a thermal spectrum corresponding to 2.73 degrees K today. The COBE satellite data yielded a microwave spectrum corresponding to a temperature of 2.725 ± 0.002 degrees K. A more perfect match couldn't be asked for. This data provides additional robust evidence for the Big Bang theory. It strongly supports the idea that the early Universe consisted of a smooth, hot, glowing gas (plasma) that eventually cooled and finally let its photons fly freely in the form of the omnipresent CMB.

A final fine point regarding the CMB is that, although what we detect now are microwave photons that started out as reddish photons when they were set free by recombination starting around 380 000 years after the Big Bang, the ultimate origin of these photons was far earlier. They trace their roots to the end of the Lepton Epoch, around 10 s post Big Bang. Recall that the Lepton Epoch ended in the mass destruction of all remaining positrons and other antileptons. This massacre also eliminated just about every remaining electron, except for the lucky one-in-a-billion that has survived to this day. The energy of mutual annihilation of electrons and positrons was released in the form of pairs of 511 keV gamma photons that shot off in opposite directions. But, because of the density and temperature of the early Universe, these gamma photons were constantly being absorbed and recreated, instead of 'decoupling' and flying free and unimpeded. The young Universe was opaque rather than transparent. It wasn't until the temperature of the Universe fell sufficiently for recombination that the photons could finally decouple and constitute the CBM. But by then, their initial gamma energies had degraded through red-shifting from cosmic expansion into red and infrared photons corresponding to 3000–4000 degrees K. And of course, that energy degradation has continued to this day until they finally constituted the 2.73 K CMB.

Studying the CBM is quite literally looking back in time. Given the present age of the Universe at around 13.7 billion years, examining the microwave glow that originated 380 000 years after the Big Bang is the equivalent of looking at a human zygote during the first half-day following fertilization. We are practically looking at the moment of creation. The radiation that we now see in the form of the CBM emanates from the so-called **surface of last scattering**, a hypothetical spherical surface from which the photon recombination event occurred. We would never hope to see any electromagnetic radiation dating from earlier in time than these micro-wave photons. The surface of last scattering is sometimes called the **cosmic photo-sphere** in an analogy to the visible 'surface' of our Sun. The present energy density of the CMB is 0.260 eV cm^{-3}, which corresponds to roughly 411 photons cm^{-3}.

35.23 Big Bang nucleosynthesis

At one time it was proposed that all the heavy elements were synthesized in the Big Bang. However, calculations showed that although the Universe was extremely hot and dense at the start, the very rapid expansion and cooling would have precluded any significant heavy element nucleosynthesis. It is now known that essentially all elements heavier than boron were created in stellar interiors via **stellar nucleosynthesis** and that Big Bang nucleosynthesis was mostly restricted to the creation of helium, deuterium, and traces of lithium. (Of note and relevance to this text, it is believed that much of today's lithium, beryllium, and boron—the so-called LiBeB elements—were created via collisions of cosmic rays with heavier nuclei. Thus, these elements are spallation products that were created in interstellar space by cosmic rays and are called **cosmogenic** nuclei.)

Although helium is produced in stars like our Sun, there is far too much of it in the Universe to account for through stellar nucleosynthesis alone. Thus, one of the most perplexing problems to early cosmologists was the helium abundance conundrum. Based on stellar nucleosynthesis, the Universe as a whole has far more helium than it should. The solution to this riddle lies in comparisons of relatively young stars and extremely old stars. Both show roughly the same amount of helium— approximately 25% by mass. New stars contain far more heavy elements than very old stars, which contain practically none. Nevertheless, all stars display roughly 25% helium abundance no matter what their age. This suggests that the helium was primordial—there right from the beginning—and therefore created in the Big Bang.

If we say the helium comes 'from the Big Bang', to be legitimate the theory should quantitatively account for the relative abundances of helium and hydrogen. Does it?

Unlike in the cores of stars, free neutrons were in abundance during the early moments of the Big Bang. These neutrons were born via quark confinement during the Hadron Epoch. The temperatures had fallen from the Quark Epoch (when all subatomic particles were in existence in roughly equal numbers and were freely swimming about in the quark soup) to the point where quarks 'recombined' into hadrons like protons, neutrons, pions, and others. The availability of free neutrons made Big Bang nucleosynthesis very different from stellar nucleosynthesis.

Based on the scale factor, at time $t = 1$ s, the temperature of the Universe was 10 billion degrees, and at that time, there were 2 neutrons for every 10 protons. This ratio deviates from 1:1 because the neutron is slightly heavier than the proton. Heavier particles can spontaneously decay into lighter ones, but the opposite reaction will require energy. In this manner, thanks to the slight mass imbalance, the conversion of a neutron into a proton releases energy (is exergonic), whereas converting a proton into a neutron is endergonic

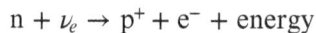

$$n + \nu_e \rightarrow p^+ + e^- + \text{energy}$$

$$p + \bar{\nu}_e + \text{energy} \rightarrow e^+ + n.$$

Under the extreme conditions of the Hadron Epoch, when neutrinos and antineutrinos abounded (and still interacted with matter), neutrons could readily convert into protons and vice versa. Before 1 s, there was sufficient energy to allow an equal number of protons and neutrons to exist as they freely interconverted. But, as the temperature and available energy decreased, by the 1 s mark (or maybe by 100 μs according to some) the ratio began to skew in favor of more protons. Protons could still be made from neutrons, but new neutron production had first slowed and then come to a halt. Left alone, the decreasing temperatures would have led to a complete absence of neutrons very quickly as they converted into protons through the exergonic neutrino capture reaction above but failed to re-form through the endergonic one.

Continuation of neutron-consuming reaction without the converse neutron-forming reaction would be expected to lead to a depletion of free neutrons down to zero in about 10 s. Recall, however, that at a temperature of about 10 billion degrees, which was at approximately 1 s, the neutrinos decoupled. They fled the scene and ceased to interact with matter of any sort as they went on their way as primordial Big Bang neutrinos. So, instead of the neutron:proton ratio falling to 0 by 10 s, thanks to neutrino decoupling, the ratio of neutrons:protons became fixed at that moment in time. That ratio at that freeze out temperature was probably 1:6

Thus, the small mass imbalance between protons and neutrons, and the neutrino decoupling event at around the 1 s (10 billion K) mark led to a slight preponderance of protons over neutrons, but this ratio would not remain fixed. Another complicating factor is that free neutrons are unstable—they undergo beta decay (into a proton, electron, and electron antineutrino) with a half-life of about 14 min and 39 s (or maybe 14 min and 47 s depending on which method of determination is used—the so-called bottle vs beam controversy). Also, recall the half-life is the time to fall to one-half the original value whereas the mean lifetime is the time to fall to 1/e of the original value; the half-life is 1.44 times longer than the mean lifetime. In any case, starting with 1 neutron for every 6 protons at $t = 1$ s, neutrons began to radioactively decay. By the time Big Bang nucleosynthesis was in full swing, the ratio had fallen to 1:7.

Unlike in stellar nucleosynthesis, where no free neutrons are available, during Big Bang nucleosynthesis, protons and neutrons could combine directly into deuterium. That deuterium could then fuse into helium.

A lot was going on at this time—the primordial neutrinos had decoupled, free neutrons were decaying away, and deuterium was being synthesized from the fusion of protons and neutrons. However, deuterium is relatively fragile. At these high temperatures, it was simultaneously undergoing photonuclear disintegration back into protons and neutrons. This last point—the fragility of deuterium—is another reason why Big Bang nucleosynthesis couldn't proceed right from the start. The fusion reactions that lead to helium had to wait for the temperature to fall enough for deuterium to become stable. This delay is called the **deuterium bottleneck**.

Unlike stellar nucleosynthesis, which begins with the fusion of two protons into deuterium via a very slow weak interaction in the proton–proton chain, Big Bang nucleosynthesis could begin with the fusion of a proton and a neutron:

$$p + n \rightarrow {}^2d + \gamma.$$

This reaction is far more kinetically favorable compared to the weak force-mediated fusion of two protons in stars. Although it is a fast reaction that is not hindered by the weak interaction, the product, deuterium, is delicate; it was readily destroyed through photodissociation by the swarms of energetic gamma photons at the 10-billion-degree temperatures. Only when the temperatures fell to between 900 million K (9×10^8 K) and 3 billion K (3×10^9 K), did the number of high-energy gamma photons plummet sufficiently to overcome the deuterium bottleneck. At that point, deuterium could accumulate sufficiently to participate in further fusion reactions. This probably occurred at around the 10 s mark. (Although much of this is uncertain and subject to revision, an adage in place at the time of this writing in 2024 is 'ten billion kelvin at one second and one billion kelvin at 10 seconds.')

Hence, after the 10 s mark, the abundance of deuterium nuclei began to dramatically increase, and other reactions could take place. However, as the Universe continued to expand and rapidly cool, eventually it would become too cool for fusion of deuterium into helium and the reactions would cease altogether. Thus, Big Bang nucleosynthesis is believed to have taken place in the interval from roughly 10 s to 1000 s, since this corresponds to the temperature range when it was cool enough for deuterium to survive but still hot and dense enough for fusion to occur at a significant rate. But only by the 1 min mark, when deuterium was truly abundant, did things really step into high gear.

However, during this same period, the number of free neutrons was slowly decreasing through beta decay, and the number of protons was increasing through the same neutron-to-proton conversion process. So, during this period of nucleosynthesis, the proton:neutron ratio had changed from the initial 6:1 to 7:1 through neutron decay.

Fusion reactions then proceeded at full speed for a few minutes, but then, by the half-hour mark, the temperature and density of the early Universe had fallen to where all such reactions had ceased; it was all over. By 1000 s (16.67 min) the temperature had already fallen to around 400 million K and, statistically, particle energies began dropping well below Coulomb barrier energies (even accounting for the quantum tunnel effect). Primordial nucleosynthesis was coming to a close. The temperature drop, coupled with the decreasing pressure thanks to expansion meant that it was probably essentially done just before the 20 min mark, and all over by a half-hour.

35.24 The significance of the hydrogen:helium ratio

It has been said that the present relative abundance of helium (25% by mass) tells us that the Universe was about 100 million degrees K at 3 min. Calculating the quantitative yield of Big Bang nucleosynthesis can confirm the veracity of these figures.

After the deuterium bottleneck was overcome, protons and neutrons began fusing in profusion. As the temperature fell to 9×10^8 K, essentially all free neutrons combined with protons to make deuterium with a starting proton:neutron ratio of 7:1. Students should carefully note that a '7:1 ratio' means one in every *eight* nucleons was a neutron —not one in every *seven*! Starting with this 7:1 ratio, out of every 16 nucleons, 4 (specifically, 2 neutrons and 2 protons) will combine into one helium-4 nucleus, leaving behind 12 protons (hydrogen nuclei). This translates into 1 helium for every 12 hydrogens, which could be said to represent about 8% helium abundance by atoms or moles. Nevertheless, given the ultimate mass ratios, the cosmology literature often alludes to a 25% helium abundance *by mass*. The actually measured amount of helium in the Universe corresponds remarkably well to this prediction. This correlation is considered to constitute additional strong evidence in favor of the Big Bang model.

Finally, the number of neutrino flavors provides another constraint on the Big Bang. The Standard Model provides for three flavors of neutrinos—the electron neutrino, the muon neutrino, and the tau neutrino (along with their antineutrino counterparts). If there were more than three flavors of neutrinos, decoupling would have occurred at a different time and temperature. This would have altered the proton:neutron ratio; it would not have been 1:7 at the moment of decoupling. Consequently, the observed abundance of helium would reflect that different ratio. Specifically, if there were more flavors of neutrinos, the Universe would have cooled quicker, and the deuterium bottleneck could have been overcome sooner. The net effect would have been a lesser impact of neutron decay on primordial helium synthesis; more neutrons would have meant more helium. Therefore, the more neutrino flavors, the more helium there would have been in the Universe. The observed 25% abundance seems to fit best with *three and only three* flavors of neutrinos. The observation of only three flavors of neutrinos is thus consistent with the Big Bang nucleosynthesis model.

It might be noted that in just a few minutes, a quarter of all the existing nucleons had undergone a prodigious blast of thermonuclear fusion into helium. One might have expected this to have an enormous impact on the expanding early Universe. In fact, it did not. Just as the preceding mass mutual annihilation reactions of all the Universe's positrons with nearly all its electrons at the end of the Lepton Epoch (and the mass mutual annihilation reactions of all the Universe's antiprotons with nearly all its protons at the end of the Hadron Epoch), this enormous amount of fusion did not cause a tremendous acceleration of the Universe's expansion or immensely elevate its temperature. Given the background temperatures at these time points, these events were mere blips on the curve.

35.25 Why didn't the Big Bang create heavier elements? The beryllium barrier

As mentioned, the oldest stars seen still contain roughly 25% helium. This is far more than can be accounted for through stellar nucleosynthesis. Thus, one can conclude that the helium must have been there from the start and this helium was made in the Big Bang. Younger stars also contain 25% helium but also have more metals. Thanks to the immense heat and pressure, stellar nucleosynthesis can create these metals as described in previous chapters. But the Big Bang also had tremendous heat and pressure (far higher in fact than any stellar interior). So, why couldn't nucleosynthesis of carbon, oxygen, and beyond occur in the Big Bang? For starters, there is one thing available in the cores of stars that wasn't available in the Big Bang—time.

In the core of stars, the proton–proton chain begins with a weak interaction: the fusion of two protons into deuterium plus a positron and an electron neutrino. This reaction, mediated by the weak force is very slow and has a very small probability. It is estimated that a typical proton in the Sun's core might take over a billion years to undergo this reaction. Given the Sun's ample number of protons and its estimated lifespan of about ten billion years, this is not a problem. The Big Bang on the other hand was accompanied by rapid expansion and cooling. The incredible heat and pressures in the early phases gave way to cooler and more rarified conditions only minutes into the expansion. The window of time was extremely brief.

But there was another hurdle to surmount—the **beryllium barrier**. Although helium-4 is extremely stable with a magic number (2) of both protons and neutrons, there are no stable nuclides with $A = 5$. Thus, there is no stable ^{5}He, ^{5}Li, or ^{5}Be. All such nuclei rapidly decay. The same situation exists for all nuclides with $A = 8$. Unstable beryllium-8 promptly decays into two more stable alpha particles (helium nuclei). Be-8 is so unstable that it decays in 10^{-17} s and is considered an 'unbound resonance' rather than a real radionuclide. Thus, to synthesize carbon, nitrogen, oxygen, etc, one must build on deuterium and add protons, neutrons, other deuterons, or alpha particles to bypass the unstable $A = 5$ bottleneck. In so doing, the processes must also yield $A = 6$ or $A = 7$, but not $A = 8$. Unfortunately for Big Bang nucleosynthesis, such options simply were not possible. Aside from helium, the end of the line included just traces of deuterium, helium-3, and an exceedingly small amount of lithium.

Although the hot, dense and long-lasting cores of stars allow the triple-alpha reaction to sidestep the beryllium barrier, during the early Big Bang the environmental conditions precluded such a bypass. The beryllium barrier was insurmountable to primordial nucleosynthesis. Again, the observed abundances of elements in the Universe are quite compatible with the Big Bang model.

35.26 From dust to dust, from radiation to radiation

It should be noted that the Radiation Era previously described (the period from the end of the Lepton Epoch to the time of recombination), is not the only way to define this time interval. As mentioned in section 35.20, some cosmologists define the Radiation Epoch as the time period when radiation was the dominant form of mass-energy in the Universe; that is, the interval between the end of the Lepton Epoch and the time when the mass–energy density of matter and radiation were roughly at around 57 000 years after the Big Bang. There are other chronological classifications as well. In one arrangement, time after the Big Bang is divided into two major chronological divisions: the **Radiation Era** and the **Matter Era**. Using this scheme, the Radiation Era came first and spanned from the Big Bang to around 57 000 years after the Big Bang (when the mass-energy density of photons fell to the point where it equaled the mass-energy density of protons). The Matter Era thus began when matter first started to dominate over radiation as the predominant form of mass-energy in the Universe. In this scheme, each Era is composed of several shorter epochs. The Radiation Era is separated into eight epochs: **Planck Epoch**, **Grand Unified Epoch**, **Inflationary Epoch**, **Electroweak Epoch**, **Quark Epoch**, **Hadron Epoch**, **Lepton Epoch**, and the **Nuclear Epoch**. Big Bang nucleosynthesis (which was discussed earlier in this chapter) occurred during the Nuclear Epoch in this system. The subsequent Matter Era then consists of three epochs: the **Atomic**, **Galactic**, and **Stellar Epochs**. The Stellar Epoch includes the present day. Not surprisingly, there are other classification systems beyond this one as well.

Another chronological classification calls the time *interval between recombination and the formation of the first stars*, the cosmological **Dark Ages**. Although there was a transient red-orange glow all around thanks to the release of the relic radiation after recombination, there was no other light for hundreds of millions of years. The Dark Ages ended between 200–400 million years after the Big Bang as dense regions of gas collapsed under their own gravitational attraction and formed the first stars, producing the first starlight. These earliest stars were big and bright. They made metals in their cores and sprayed them about the early Universe relatively quickly upon exploding as supernovae (probably with lifetimes of just a few million years). Not only did these brilliant stars and their explosive endings put an end to the Dark Ages, but they also 'overshot' the mark and re-ionized their surroundings once again. Thanks to their size, temperature, and luminosity, they were bright in x-rays and ultraviolet radiation, which ionized the hydrogen that was previously neutral since recombination. The early galaxies that hosted these first, ultra-bright stars are now extremely distant and emit much ultraviolet in the form of Lyman-alpha radiation. Even today, aside from some neutral galactic clouds, most of the hydrogen in the Universe remains in a low-density ionized state. Hence, following the Dark Ages came the **Epoch of Reionization** using this chronological account. By somewhere around 500 million years after the Big Bang, essentially all the free hydrogen and helium had once again become ionized.

Yet another chronology was originally outlined by Fred Adams and Gregory Laughlin. This system employs the concept of *cosmological decades,* which describe the Universe's age in terms of powers of ten. For instance, at decade 9 the Universe would be 10^9 or one billion years old. In this manner, the **Primordial Era** occurred up until the Universe was roughly eight decades or 10^8 (one hundred million) years old. During this period, there were no stars. When stars could finally form, around 200–400 million years after the Big Bang, the **Stelliferous Era** began. We are still in this star-forming stage of our Universe, and we shall remain here until around decade 14. By that time, the hydrogen-burning stars will have all burned out and there will no longer be sufficient hydrogen to regenerate new stars. Without starlight, the Universe will then grow colder and darker. It will then enter the **Degenerate Era**. This timespan will be dominated by condensed degenerate matter such as white dwarfs, black dwarfs (that is, burnt out white dwarfs), neutron stars, and some brown dwarfs (smaller 'stars' which never had enough mass to make it to the main sequence). Along with these degenerate forms of matter, dark matter, dark energy, and some stray planets might litter the landscape. Dark matter annihilation may release radiation, and this could become the dominant source of energy in the Universe during the Degenerate Era. The Degenerate Era is expected to span from 10^{14} to 10^{40} years or decades 14–40. After the Degenerate Era will come the **Black Hole Era**. This might span from decades 40 through 100 (10^{40} to 10^{100} years) and will be dominated by black holes thanks to the eventual decay of protons that ended the Degenerate Era. (One must keep in mind that although theory calls for proton instability, there is presently no indication that the proton truly does decay.) Nevertheless, theorists predict that by 10^{40} years, all the Universe's protons will decay, and the Universe will transition from the Degenerate Era to the Black Hole Era. But just as protons will eventually decay, so too shall black holes. Through Hawking radiation (which, as discussed in chapter 34, is still unproven), black holes will eventually decay and evaporate. In their last seconds, the rate of evaporation will escalate dramatically, and black holes will explode into gamma rays. Predictions hold that stellar mass black holes will decay by about 10^{65} years (decade 65), and supermassive black holes by about 10^{100} years (decade 100). After about a googol years, the Black Hole Era will end as the last black holes evaporate and explode, releasing gamma photons.

Then the Universe will enter the final phase—the **Dark Era** (not to be confused with the Dark Ages described earlier). The Dark Era will begin at 10^{100} (one googol) years or at cosmological decade 100. Just like near the beginning of time, this will again be a time of radiation—the Universe will consist of a sea of photons, neutrinos, electrons, and positrons. The positrons will have come from the decay of protons, and they can hook up with stray electrons to form positronium (an 'atom' consisting of an electron and positron orbiting one another). Eventually, through mutual annihilation, these positrons and electrons will turn into photons, and by cosmological decade 110, the Universe will be a huge but cold expanse of photons and neutrinos. Hence our Universe begins with—and ends with—radiation.

Further reading

[1] Fumagalli M, O'Meara J, Prochaska M and Xavier J 2011 Detection of pristine gas two billion years after the Big Bang *Science* **334** 1245–9
[2] Guth A and Steinhardt P 1984 The inflationary universe *Sci. Am.* **250** 116
[3] Follin B, Knox L, Millea M and Pan Z 2015 First detection of the acoustic oscillation phase shift expected from the cosmic neutrino background *Phys. Rev. Lett.* **115** 091301

www.ingramcontent.com/pod-product-compliance
Lightning Source LLC
Chambersburg PA
CBHW082114210326
41599CB00031B/5769